Notes on Numerical Fluid Mechanics and Multidisciplinary Design

Volume 150

Notes on Numerical Fluid Mechanics and Multidisciplinary Design publishes state-of-art methods (including high performance methods) for numerical fluid mechanics, numerical simulation and multidisciplinary design optimization. The series includes proceedings of specialized conferences and workshops, as well as relevant project reports and monographs.

More information about this series at http://www.springer.com/series/4629

Geert Degrande · Geert Lombaert ·
David Anderson · Paul de Vos ·
Pierre-Etienne Gautier · Masanobu Iida ·
James Tuman Nelson · Jens C. O. Nielsen ·
David J. Thompson · Thorsten Tielkes ·
David A. Towers
Editors

Noise and Vibration Mitigation for Rail Transportation Systems

Proceedings of the 13th International Workshop on Railway Noise, 16–20 September 2019, Ghent, Belgium

Springer

Editors
Geert Degrande
Department of Civil Engineering
KU Leuven
Leuven, Belgium

Geert Lombaert
Department of Civil Engineering
KU Leuven
Leuven, Belgium

David Anderson
Acoustic Studio
Stanmore, NSW, Australia

Paul de Vos
SATIS
Weesp, The Netherlands

Pierre-Etienne Gautier
SNCF Réseau
La Plaine Saint Denis, France

Masanobu Iida
Research & Development Promotion Division
Railway Technical Research Institute
Tokyo, Japan

James Tuman Nelson
Wilson, Ihrig & Associates
Emeryville, CA, USA

Jens C. O. Nielsen
Department of Mechanics and Maritime Sciences
Chalmers University of Technology
Gothenburg, Sweden

David J. Thompson
Institute of Sound and Vibration Research
University of Southampton
Southampton, UK

Thorsten Tielkes
DB Systemtechnik GmbH
München, Germany

David A. Towers
Cross-Spectrum Acoustics Inc
Burlington, MA, USA

ISSN 1612-2909 ISSN 1860-0824 (electronic)
Notes on Numerical Fluid Mechanics and Multidisciplinary Design
ISBN 978-3-030-70288-5 ISBN 978-3-030-70289-2 (eBook)
https://doi.org/10.1007/978-3-030-70289-2

This Springer imprint is published by the registered company Springer Nature Switzerland AG
The registered company address is: Gewerbestrasse 11, 6330 Cham, Switzerland

Preface

This volume contains the peer-reviewed contributions to the 13th International Workshop on Railway Noise (IWRN13), which took place in Ghent, Belgium, during 16–20 September 2019. The workshop was hosted by the Structural Mechanics Section of the Department of Civil Engineering of KU Leuven in collaboration with the Conference and Events Office of KU Leuven.

IWRN13 was also made possible by the support of Pandrol (Gold sponsor), Infrabel and Strukton Rail (Silver sponsors) and D2S International, GeoSIG and TUC Rail (Bronze sponsors).

The workshop was attended by 159 delegates and eight accompanying persons from 23 countries on four continents: Germany (20), Belgium (19), France (19), UK (17), China (16), The Netherlands (10), Sweden (8), Australia (6), Spain (6), Austria (5), Japan (5), Switzerland (5), USA (5), Czech Republic (3), Singapore (3), Denmark (2), Hong Kong (2), Italy (2), Norway (2), Canada (1), Hungary (1), South Korea (1) and Romania (1).

In comparison with other modes of transportation, rail transport is safe and environmentally friendly and is generally described as the most sustainable mode for regional and international transport. However, it is also recognised that the environmental impact of railway noise and vibration needs to be further reduced.

Since the first IWRN in 1976, held in Derby (United Kingdom) with 35 delegates, the workshop series has been established as a regular event (held every three years) that brings together the leading researchers and engineers in all fields related to railway noise and vibration. The IWRN workshops have contributed significantly to the understanding and solution of many problems in railway noise and vibration, building a scientific foundation for reducing environmental impact by air-borne, ground-borne and structure-borne noise and vibration.

Following the tradition from previous workshops, the scientific programme of IWRN13 was held as a single session event over three and a half days. The programme contained two keynote lectures, 55 oral presentations and 37 poster presentations. The poster sessions commenced with three-minute rapid-fire oral presentations to introduce each poster, and these sessions were very well attended.

This volume contains the peer-reviewed papers from 78 of these presentations, including two keynote papers on wheel-rail impact loads, noise and vibration and on interior noise in railway vehicles. IWRN13 covered nine different themes: high-speed rail and aerodynamic noise; interior noise; policy, regulation and perception; predictions, measurements and modelling; rail roughness, corrugation and grinding; squeal noise; structure-borne noise and ground-borne vibration; wheel and rail noise and bridge noise and vibration.

In parallel with the scientific programme, eight companies gathered at IWRN13 to display their technology and services in the area of railway noise and vibration: Pandrol, GeoSIG, Enmo, Gerb, Sekisui, Sigicom, 4Silence and ACSOFT.

There is no formal organisation behind the IWRN but rather an informal, committed international committee. It supports the chairman during the preparation process with the experience and expertise of its members. Assistance is given to formulate the scientific programme by reviewing the submitted abstracts, to act as session chairmen and to act as peer reviewers and editors of the IWRN proceedings published in this volume.

The international committee is grateful to Kristien Van Crombrugge, Ann Zwarts, Kurt Scherpereel, Cédric Van hoorickx and Pieter Reumers of the local committee for their great commitment and care in organising the workshop.

The editors of this volume are grateful to Professor Wolfgang Schröder as the general editor of the "Notes on Numerical Fluid Mechanics and Multidisciplinary Design" and also to the staff of the Springer Verlag (in particular, Dr. Leontina Di Cecco) for the opportunity to publish the proceedings of the IWRN13 workshop in this series. Note that previous workshop proceedings has also been published in this series (IWRN9 in volume 99, IWRN10 in volume 118, IWRN11 in volume 126 and IWRN12 in volume 139).

We hope that this volume will be used as a "state-of-the-art" reference by scientists and engineers involved in solving noise and vibration problems related to railway traffic.

August 2020

Geert Degrande
Geert Lombaert
David Anderson
Paul de Vos
Pierre-Etienne Gautier
Masanobu Iida
James Tuman Nelson
Jens C. O. Nielsen
David J. Thompson
Thorsten Tielkes
David A. Towers

Contents

Keynote Lectures

Wheel–Rail Impact Loads, Noise and Vibration: A Review of Excitation Mechanisms, Prediction Methods and Mitigation Measures 3
Jens C. O. Nielsen, Astrid Pieringer, David J. Thompson, and Peter T. Torstensson

Industrial Methodologies for the Prediction of Interior Noise Inside Railway Vehicles: Airborne and Structure Borne Transmission 41
Pascal Bouvet and Martin Rissmann

High Speed Rail and Aerodynamic Noise

Evaluation on Aerodynamic Noise of High Speed Trains with Different Streamlined Heads by LES/FW-H/APE Method 57
Chaowei Li, Yu Chen, Suming Xie, Xiaofeng Li, Yigang Wang, and Yang Gao

Comparisons of Aerodynamic Noise Results Between Computations and Experiments for a High-Speed Train Pantograph 66
Xiaowan Liu, Jin Zhang, David J. Thompson, Giacomo Squicciarini, Zhiwei Hu, Martin Toward, and Daniel Lurcock

Measurement and Reduction of the Aerodynamic Bogie Noise Generated by High-Speed Trains in Terms of Wind Tunnel Testing . . . 73
Yoichi Sawamura, Toki Uda, Toshiki Kitagawa, Hiroshi Yokoyama, and Akiyoshi Iida

Development of New Low-Noise Pantograph for High-Speed Trains . . . 81
Mitsuru Saito, Fumio Mizushima, Yusuke Wakabayashi, Takeshi Kurita, Shinji Nakajima, and Toru Hirasawa

The Influence of Track Parameters on the Sound Radiation from Slab Tracks 90
Jannik S. Theyssen, Astrid Pieringer, and Wolfgang Kropp

Pass-By Noise Assessment of High Speed Units by Means of Acoustic Measurements in a Perimeter Close to the Train 98
Gennaro Sica, Jaume Solé, and Pierre Huguenet

Interior Noise

Using a 2.5D BE Model to Determine the Sound Pressure on the External Train Surface 109
Hui Li, David J. Thompson, Giacomo Squicciarini, Xiaowan Liu, Martin Rissmann, Francisco D. Denia, and Juan Giner-Navarro

Acoustic Design of Rolling Stock for Comfortable Telephone Conversations 118
A. Bistagnino and Joan Sapena

Vehicle Modeling for High Frequency Vibration 126
Qi Wang, Xinbiao Xiao, Jian Han, and Yue Wu

Numerical Analysis on the Radiation Efficiency of an Extruded Panel for the Railway Vehicle Using the Waveguide Finite Element and Boundary Element Method 134
Hyungjun Kim, Jungsoo Ryue, David J. Thompson, and Angela D. Müller

Speech Intelligibility - Effects of Railway Tunnels at High Speeds 142
Christian H. Kasess, Thomas Maly, Holger Waubke, Michael Ostermann, and Günter Dinhobl

A Soundscape Approach to Assess and Predict Passenger Satisfaction 150
Paul H. de Vos, Tjeerd Andringa, and Mark van Hagen

Industrial Engineering Framework for Railway Interior Noise Predictions 158
Torsten Kohrs, Karl-Richard Kirchner, Haike Brick, Dennis Fast, and Ainara Guiral

Structure-Borne Noise Contributions on Interior Noise in Terms of Car Design and Vehicle Type 167
Alex Sievi, Haike Brick, and Philipp Rüst

Shift2Rail Research Project DESTINATE Interior Railway Noise Prediction Based on OTPA 176
Nathan Isert and Otto Martner

Virtual Test Method of Structure-Borne Sound for a Metro Bogie 186
Gang Xie, Martin Rissmann, Pascal Bouvet, Xiaowan Liu, David J. Thompson, Luis Baeza, Juan Moreno, Julián Martín Jarillo, Francisco D. Denia, Juan Giner-Navarro, and Ines Lopez Arteaga

Policy, Regulation and Perception

Railway Noise Mitigation Framework in Europe: Combining Policies with the Concerns of the Railways 197
Jakob Oertli

Short and Very Short Term Indicators to Characterize Train Pass-By Noise 206
Franck Poisson and Anne Guerrero

Cost Effectiveness of Noise and Vibration Mitigation Measures Using Life-Cycle Cost Analysis 214
Xiaohan Phrain Gu, Ziyan Ma, Anbin Wang, Longhua Ju, Xiaodong Chai, and Xiaogang Gao

Predictions, Measurements, Monitoring and Modelling

Application to Real Cases of a Methodology to Evaluate the Uncertainty of Train Exterior Noise Predictions 225
Eduardo Latorre Iglesias, A. L. Gomes Neves, A. Bistagnino, and Joan Sapena

On Measurement of Railway Noise: Usability of Acoustic Camera 234
Günter Dinhobl, Martin Petz, Sebastian Kümmritz, and Helmut Hutterer

Acoustic Monitoring of Rail Faults in the German Railway Network 242
Astrid Pieringer, Matthias Stangl, Jörg Rothhämel, and Thorsten Tielkes

Nonlinear Model of an Embedded Rail System for the Simulation of Train-Track Dynamic Interaction and the Analysis of Vibration Transmission 251
Andrea Collina, Roberto Corradi, Egidio Di Gialleonardo, and Qianqian Li

Wheel-Rail Contact Analysis with Emphasis on Wear (Measurements/Simulation) 259
Kris Decroos, Jonathan Ceulemans, Bert Stallaert, and Tom Vanhonacker

Evaluation of BBI Performance Indicator in a Full-Scale Test Bench 267
Hamid Masoumi, Bram Veelhaver, Manthos Papadopoulos, Fülöp Augusztinovicz, and Patrick Carels

Rail Roughness, Corrugation and Grinding

Tonal Noises and High-Frequency Oscillations of Rails Caused by Grinding Procedures 277
Jörg Rothhämel, F. Kendl, B. Lütke, S. Lange, T. Kempinger, and B. Asmussen

Extracting Information from Axle-Box Acceleration: On the Derivation of Rail Roughness Spectra in the Presence of Wheel Roughness 286
Tobias D. Carrigan and James P. Talbot

Numerical and Experimental Analysis of Transfer Functions for On-Board Indirect Measurements of Rail Acoustic Roughness 295
Anna Rita Tufano, Olivier Chiello, Marie-Agnès Pallas, Baldrik Faure, Claire Chaufour, Emanuel Reynaud, and Nicolas Vincent

Analysis of the Effect of Running Speed and Bogie Attitude on Rail Corrugation Growth in Sharp Curves 303
Andrea Collina, Roberto Corradi, Egidio Di Gialleonardo, and Qianqian Li

Rail Grinding and Track Vibration Resonances 312
Shankar Rajaram, James Tuman Nelson, and Hugh Saurenman

Measurement of Longitudinal Irregularities on Rails Using an Axlebox Accelerometer System 320
Stuart L. Grassie

Squeal Noise

Investigation of Railway Curve Squeal Using Roller Rig and Running Tests 331
Takeshi Sueki, Tsugutoshi Kawaguchi, Hiroyuki Kanemoto, Masahito Kuzuta, Tatsuya Inoue, and Toshiki Kitagawa

A Full Finite Element Model for the Simulation of Friction-Induced Vibrations of Wheel/Rail Systems: Application to Curve Squeal Noise 339
Van Vuong Lai, Olivier Chiello, Jean-François Brunel, and Philippe Dufrénoy

Structure-Borne Noise and Ground-Borne Vibration

Importance of a Detailed Vibratory Characterization of a Railway Line for the Propagation of Vibrations in an Eco-Neighborhood 351
Guillaume Coquel, Catherine Guigou-Carter, and Philippe Jean

An Efficient Three-Dimensional Track/Tunnel/Soil Interaction Analysis Method for Prediction of Vibration and Noise in a Building 360
Kazuhisa Abe, Koya Yamada, Sota Yamada, Masaru Furuta, Michiko Suehara, Sho Yoshitake, and Kazuhiro Koro

Prediction of Structure-Borne Sound in Buildings Consisting of Various Building Elements, Generated by Underground Rail Traffic . . . 368
Fülöp Augusztinovicz, A. B. Nagy, P. Fiala, Hamid Masoumi, and Patrick Carels

Ground-Borne Noise and Vibration Propagation Measurements and Prediction Validations from an Australian Railway Tunnel Project 377
Peter Karantonis, Conrad Weber, and Hayden Puckeridge

A Simplified Model for Calculating the Insertion Gain of Track Support Systems Using the Finite Difference Method 385
Rupert Thornely-Taylor, Oliver Bewes, and Gennaro Sica

Evaluation of the Vibration Power Transmitted to Ground Due to Rolling Stock on Straight Tracks 394
Michel Villot, Catherine Guigou-Carter, Philippe Jean, and Roger Müller

Vibration Excitation at Turnouts, Mechanism, Measurements and Mitigation Measures 403
Roger Müller, Yves Brechbühl, Stefan Lutzenberger, Samir Said, Lutz Auersch, Catherine Guigou-Carter, Michel Villot, and Roland Müller

Predicted and Measured Amplitude-Speed Relations of Railway Ground Vibrations at Four German Sites with Different Test Trains 411
Lutz Auersch

Prediction of Ground-Borne Vibration Generated at Railway Crossings Using a Hybrid Model 420
Evangelos Ntotsios, Dimitrios Kostovasilis, David J. Thompson, Giacomo Squicciarini, and Yann Bezin

FE Modelling as a Design Tool for Mitigation Measures for Railway Vibrations 429
Karin Norén-Cosgriff, Birgitte M. Dahl, Joonsang Park, and Amir M. Kaynia

Towards Hybrid Models for the Prediction of Railway Induced Vibration: Numerical Verification of Two Methodologies 437
Brice Nelain, Nicolas Vincent, and Emanuel Reynaud

A Methodology for the Assessment of Underground Railway-Induced Vibrations Based on Radiated Energy Flow Computed by Means of a 2.5D FEM-BEM Approach 445
Dhananjay Ghangale, Robert Arcos, Arnau Clot, and Jordi Romeu

Assessment of Dynamic Vibration Absorbers Efficiency as a Countermeasure for Ground-Borne Vibrations Induced by Train Traffic in Double-Deck Tunnels Using an Energy Flow Criterion 454
Behshad Noori, Robert Arcos, Arnau Clot, and Jordi Romeu

Increase in Force Density Levels of Light Rail Vehicles 462
Shankar Rajaram, James Tuman Nelson, and Hugh Saurenman

Predicting Vibration in University of Washington Research Facilities up to 500 m from a Light Rail Tunnel 470
Hugh Saurenman, Roberto Della Neve Longo, and James Tuman Nelson

Evaluation of the In-Situ Performance of Base Isolated Buildings 478
Raphaël Cettour-Janet, Benjamin Trévisan, and Michel Villot

Ground and Building Vibration Estimation for Health Impact Research 487
Mikael Ögren, Alf Ekblad, Peter Johansson, Arnold Koopman, and Kerstin Persson Waye

1 dB per Floor? How Does Noise and Vibration Propagate in High-Rise Buildings Near Railway Lines? 496
Dave Anderson, Sav Shimada, and David Hanson

Design of Railway-Induced Ground-Borne Vibration Abatement Solutions to be Applied in Railway Tunnels by Means of a Hybrid Modelling Approach 504
Robert Arcos, Dhananjay Ghangale, Behshad Noori, Hassan Liravi, Arnau Clot, and Jordi Romeu

Response of Periodic Railway Bridges Accounting for Dynamic Soil-Structure Interaction 512
Pieter Reumers, Kirsty Kuo, Geert Lombaert, and Geert Degrande

Identification of a Randomly-Fluctuating Continuous Model of the Ballasted Track Based on Measurements at the Pass-By of High-Speed Trains 521
Patryk Dec, Régis Cottereau, and Baldrik Faure

Numerical Prediction and Experimental Validation of Railway Induced Vibration in a Multi-storey Office Building 529
Manthos Papadopoulos, Kirsty Kuo, Matthias Germonpré, Ramses Verachtert, Jie Zhang, Kristof Maes, Geert Lombaert, and Geert Degrande

Rail Roughness Evolution on a Curved Track and Its Impact on Induced Structure Borne Vibration 538
Vincent Jurdic, Rob Lever, Adrian Passmore, and Mark Scotter

Vibration Emission of Underground Rail Systems: Experimental Assessment, Extrapolation and Parametric Study 546
Eric Augis, Pierre Ropars, and Alice Ly

Reduction of Ground Vibration Transmission by Means of Double Wall Barriers 555
Cédric Van hoorickx, Mattias Schevenels, and Geert Lombaert

Investigation of Vibration Mitigation by Concrete Trough with Integrated Under Ballast Mats for Surface-Railways 563
Rüdiger Garburg, Christian Frank, and Michael Mistler

Ground-Borne Vibration from Manchester Metrolink 571
James Block, Chris Jones, Steve Cawser, Conor Tickner, and Paul Shields

Wheel and Rail Noise

CRoNoS Railway Noise Prediction Tool: Description and Validation Based on Field Tests 581
Ainara Guiral, Maddi Olazagoitia, and Egoitz Iturritxa

Scale and Numerical Modelling of a Metro Rail Viaduct with Sound Absorption to Reduce In-Car and Wayside Noise 589
Conrad Weber, Terrence Wong, and Benjamin Panarodvong

A Comparison of Rolling Noise from Different Tram Tracks 597
Wenjing Sun, David J. Thompson, Martin Toward, Marcus Wiseman, and Evangelos Ntotsios

Numerical Study on Acoustic Characteristics of Embedded Track Used in a Metro 605
Jian Han, Gang Yang, Xinbiao Xiao, Xiaozhen Sheng, and Xuesong Jin

Modelling Wheel/Rail Rolling Noise for a High-Speed Train Running on a Slab Track 613
Xiaozhen Sheng, Gong Cheng, and David J. Thompson

Estimation of Track Decay Rates and Noise Based on Laboratory Measurements on a Baseplate Fastening System 621
Boniface S. Hima, David J. Thompson, Giacomo Squicciarini, Evangelos Ntotsios, and David Herron

Track Decay Rate Analysis and Rail Damper Noise Reduction for Slab Tracks 629
Briony Croft, Aaron Miller, and Christoph Gramowski

Rail Damper Composed of Tuned Mass Damper and Constrained Layer Adopted in Non-symmetric Rail 638
Sam Lo, Wilson Ho, and Cindy Cheung

Entering the Real Operation Phase: Design, Construction and Benefit Verification of Freight Wheel Noise Absorber 646
Christoph Gramowski and Thomas Gerlach

Half Installation of Rail Dampers 654
Wilson Ho, Cindy Cheung, Morgan Cheng, and Lyn Lin

Vehicle and Track Noise Separation Methodology Based on Advanced Transfer Path Analysis Technique 662
A. Malkoun, Egoitz Iturritxa, E. Cierco, Ainara Guiral, Joan Sapena, and F. X. Magrans

Rail Vibration and Rolling Noise Reduction Using Tuned Rail Damper for Vulcanized Bonded Baseplate 671
Ziyan Ma, Anbin Wang, Xiaogang Gao, Xiaohan Gu, Longhua Ju, and Zhiqiang Wang

Effect of Ground Conditions and Microphone Position on Railway Noise Measurement Results 680
David J. Thompson, Xianying Zhang, and Giacomo Squicciarini

Bridge Noise and Vibration

Comparison of Vibration and Noise Characteristics of Urban Rail Transit Bridges with Box-Girder and U-Shaped Sections 691
Qi Li, Hong-yao Huang, and David J. Thompson

A New Guideline to Reduce Noise Radiated by Railway Bridges in Germany 699
D. Stiebel, C. Gerbig, B. Schewe, M. Neudeck, Christian Frank, P. Tecklenburg, and B. Asmussen

Noise Radiation from Concrete Box Girders Generated by a High-Speed Train Running Along a Track-Bridge System as an Infinitely Long Periodic Structure 707
Yuanpeng He, Jian Han, Gong Cheng, Xiaozhen Sheng, and Qingsong Feng

Author Index 717

Keynote Lectures

Wheel–Rail Impact Loads, Noise and Vibration: A Review of Excitation Mechanisms, Prediction Methods and Mitigation Measures

Jens C. O. Nielsen[1(✉)], Astrid Pieringer[2], David J. Thompson[3], and Peter T. Torstensson[4]

[1] Department of Mechanics and Maritime Sciences, Division of Dynamics/CHARMEC, Chalmers University of Technology, 412 96 Gothenburg, Sweden
jens.nielsen@chalmers.se

[2] Department of Architecture and Civil Engineering, Division of Applied Acoustics/CHARMEC, Chalmers University of Technology, 412 96 Gothenburg, Sweden

[3] Institute of Sound and Vibration Research, University of Southampton, Southampton SO17 1BJ, UK

[4] Swedish National Road and Transport Research Institute, Regnbågsgatan 1, 417 55 Gothenburg, Sweden

Abstract. Railway noise and ground-borne vibration induced by wheel–rail impact loads are generated by discrete wheel/rail surface irregularities or local deviations in the nominal wheel–rail contact geometry. On the running surface of a rail, a discrete irregularity can be inherent to the railway design, for example at crossings or insulated joints. On the wheel or rail, the irregularity could also be the result of surface damage due to rolling contact fatigue cracking or a consequence of wheel sliding without rolling. This review describes the mechanisms of wheel–rail impact generated by wheel flats, rail joints and crossings. These can be a source of locally increased noise and vibration levels and increased annoyance, as well as of damage to vehicle and track components. The wheel–rail excitation at such irregularities, as indicated by the vertical wheel centre trajectory, leads to an abrupt change of momentum, potentially causing a momentary loss of wheel–rail contact followed by an impact on the rail. The resulting loading is a transient and often periodically repeated event exciting vibration in a wide frequency range with most of the energy concentrated below about 1 kHz. For the numerical prediction of high-magnitude transient loading and situations potentially leading to loss of contact, a non-linear wheel–rail contact model is required, implying that the simulation of contact force is carried out in the time domain. To avoid the need for large, computationally expensive models, a hybrid approach has been developed in which the time history of the contact force is transformed into an equivalent roughness spectrum; this is used as input to frequency-domain models for the prediction of noise and vibration. Since the excitation mechanism is similar to that for rolling noise, the same types of measures to mitigate wheel and track vibration can be applied. However, the main priority should be to control the irregularity by design and regular maintenance.

G. Degrande et al. (Eds.): *Noise and Vibration Mitigation for Rail Transportation Systems*, NNFM 150, pp. 3–40, 2021.
https://doi.org/10.1007/978-3-030-70289-2_1

Keywords: Wheel–rail impact noise and vibration · Excitation mechanisms · Mitigation

1 Introduction

In comparison with most other modes of transportation, railway traffic is safer and more environmentally friendly, and is generally described as the most sustainable mode for regional and international transport. However, with higher train speeds and axle loads, increased traffic density and sometimes a growing maintenance debt due to increasing rates of wheel and rail profile degradation, the issues of railway noise and vibration are becoming increasingly important. This results in a growing number of complaints from residents near existing lines as well as resistance to the building of new railways.

The many different forms of railway noise include rolling noise induced by the acoustic roughness (unevenness) on the rolling surface of wheels and rails, aerodynamic noise generated by air flow and turbulence around the train at high speeds, squeal noise due to frictional instability in the tangential wheel–rail contact in sharp radius curves, impact noise at discrete wheel/rail surface irregularities, brake noise, engine noise, etc. [1]. The understanding and modelling of the mechanisms of these noise sources have improved significantly over recent decades and solutions for mitigation are available and widespread. The state-of-the-art on rolling noise has been presented for example in refs. [2, 3], on aerodynamic noise in refs. [4, 5], and on squeal noise in refs. [6, 7]. An early contribution to the prediction of wheel–rail impact noise was presented by Vér et al. [8] and later extended by Remington [6]. Ground-borne vibration generated by railway traffic was surveyed in refs. [9, 10]. Several of these reviews were initiated and first presented within the International Workshop on Railway Noise (IWRN) community. Following this tradition, this paper is a review on noise and vibration generated by wheel–rail impact.

In this review paper, the term *impact* refers to a situation with a transient vertical wheel–rail contact loading resulting in a maximum of the contact force that is high relative to the static wheel load but not necessarily leading to a momentary loss of contact.

The impact between wheel and rail is generated due to a severe wheel/rail surface irregularity or a local deviation in the nominal wheel–rail contact geometry. In extreme cases this can also lead to a momentary loss of wheel–rail contact. A discrete irregularity on the running surface of a rail can be inherent to the railway design. This is the case for crossings, required to allow for intersecting tracks, and for insulated rail joints used to divide the track electrically for train detection as part of the signalling system. Discrete irregularities can also be the result of surface damage due to rolling contact fatigue of wheel and rail leading to crack propagation and subsequent breaking out of pieces of wheel/rail material (shelling, squats, etc.), or a consequence of wheel sliding without rolling (wheel flats).

Wheel–rail impact loads due to discrete surface irregularities lead to the generation of noise and ground-borne vibration. Furthermore, if the wheels and rails are not maintained on time, these irregularities may lead to damage to the track or vehicle due to repeated impact loading and plastic deformation, leading to a further growth of the irregularities. In this paper, wheel flats, rail joints and crossings will be used as three examples of irregularities generating wheel–rail impact loads, noise and vibration. The characteristics of each of these irregularities are briefly described in Sect. 2. Examples of field measurements quantifying the consequences of wheel–rail impact are presented in Sect. 3. Impact noise can be considered as an extreme form of rolling noise, in which the prescribed relative wheel–rail vertical displacement excitation is generated by a discrete irregularity. On this note, prescribed wheel trajectories based on simple formulae and used as input to models of dynamic wheel–rail interaction are surveyed in Sect. 4. Models of wheel–rail contact and dynamic wheel–rail interaction are treated in Sects. 5 and 6. Impact loads, noise and vibration are discussed in Sects. 7 and 8, while mitigation measures are summarised in Sect. 9.

2 Discrete Wheel/Rail Surface Irregularities

2.1 Wheel Flats

A discrete wheel tread defect is a local deviation from the nominal wheel radius along a short section of the wheel circumference. This deviation introduces a radial irregularity that, for each wheel revolution, may generate an impact load in the wheel–rail contact. One common discrete tread defect, the wheel flat, is developed due to sliding (without rolling) of the wheel along the rail, see Fig. 1(a). The reason for the sliding may be that the brakes are poorly adjusted, frozen or defective, or that the braking force is too high in relation to the available wheel/rail friction [11]. Contaminations on the rail surface, such as leaves, grease, frost and snow will reduce the friction coefficient and may aggravate the problem. As a consequence of the sliding, part of the wheel tread is worn off and locally the wheel temperature is raised significantly due to the dissipated friction energy. When the wheel starts rolling again, this is followed by a rapid cooling due to conduction into the large steel volume surrounding the flat. This may lead to the formation of martensite and residual stresses [11–13].

The initial flat with sharp edges will soon be transformed into a longer flat with rounded edges because of wear and plastic deformation of the wheel material at subsequent impacts with the rail, whereas the depth at the centre of the flat can be unchanged [14]. However, if martensite is formed, cracks may initiate and propagate in the brittle material due to the rolling contact loading and the repeated impacts. Due to the tensile residual stresses in the surrounding material, cracks may grow to considerable depths and pieces of the wheel tread may detach aggravating the irregularity.

Surface or subsurface initiated rolling contact fatigue resulting in cracking and the breaking out of clusters of pieces of tread material is another type of discrete wheel tread defect that may result in higher impact loads than generated by a wheel flat [15].

2.2 Rail Joints

Rail joints are used to allow for the expansion of rails in hot weather conditions (expansion joints) or for signalling purposes in continuously welded track (insulated joints).

One example of an insulated rail joint is shown in Fig. 1(b). To ensure the insulation, a nylon material is used in the gap between the two rails. In this example, the joint has been assembled using two 0.9 m long fish-plates (joint bars), one on each side of the rail, also separate from the rail by nylon insulators. The mass of the two fish-plates with six bolts is about 50 kg [16].

To allow for variations in temperature, the gap width between two rails in a modern track is 0–20 mm. Besides a small vertical misalignment between the two rails, there may be a vertical dip in the rail generating a dip angle at the joint. Welding of rails in the field may result in a similar cusp-like discontinuity [17]. It has been suggested that the dip can be approximated by a symmetric quadratic function on each side of the joint, see Sect. 4.2. Severe deformation of the two rail ends due to repeated impact loading and/or detached material is a safety issue as the insulating function may be lost.

In addition to the geometric irregularity, the joint leads to a local variation in dynamic track stiffness due to the two free rail ends and the weight and stiffness added by the fish-plate assembly. An example of a model for simulation of dynamic interaction between a vehicle and a track with a joint is shown in Fig. 2. The variation in beam properties $p(x)$ of the coupled rail/fish-plate structure is schematically sketched in the figure, where for example the rail bending stiffness changes from its nominal value to a higher value due to the coupling with the fish-plates. Over the gap, the rail bending stiffness, then only supplied by the fish-plates, drops to a lower value.

Examples of measured and calculated track receptances are presented in ref. [16]. For frequencies above around 300 Hz, the fish-plate assembly leads to an increased dynamic stiffness (lower receptance) compared to nominal track. The track resonance at around 250 Hz, where there is a relative motion of the rail and sleepers, is shifted to a lower frequency because of the mass of the assembly. Also, the receptance at the pinned-pinned resonance at around 1000 Hz is influenced by the joint.

2.3 Crossings

A fixed crossing, see Fig. 1(c), allows for trains to pass over two intersecting tracks. Since two different wheel paths intersect at one point, there is a flangeway on either side of the crossing nose to allow the wheel flanges to pass through the crossing in either the through route or the diverging route. The rails are therefore split into a crossing nose and two wing rails, see also Fig. 3. In the facing move (travelling direction from switch to crossing), the wheels roll on the closure rails towards the crossing. In the trailing move the traffic travels in the opposite direction.

For nominal wheel and rail profiles, according to design, the transfer of a wheel from the wing rail to the crossing nose (or vice versa) in the crossing panel should be relatively smooth. However, wheel profiles with different stages of profile wear will make the transition between wing rail and crossing nose at different positions along the crossing [18]. Wheels should not make contact near the tip of the crossing nose where it is too thin to carry the wheel load, see Fig. 1(c). For the same reason, situations where the field side of the wheel tread is making contact with the gauge corner of the wing rail should be avoided. These two constraints determine the length of the transition zone.

The conicity of the wheel in combination with the variation in rail geometry along the crossing panel, where the stock rail is curved laterally into the wing rail and the crossing

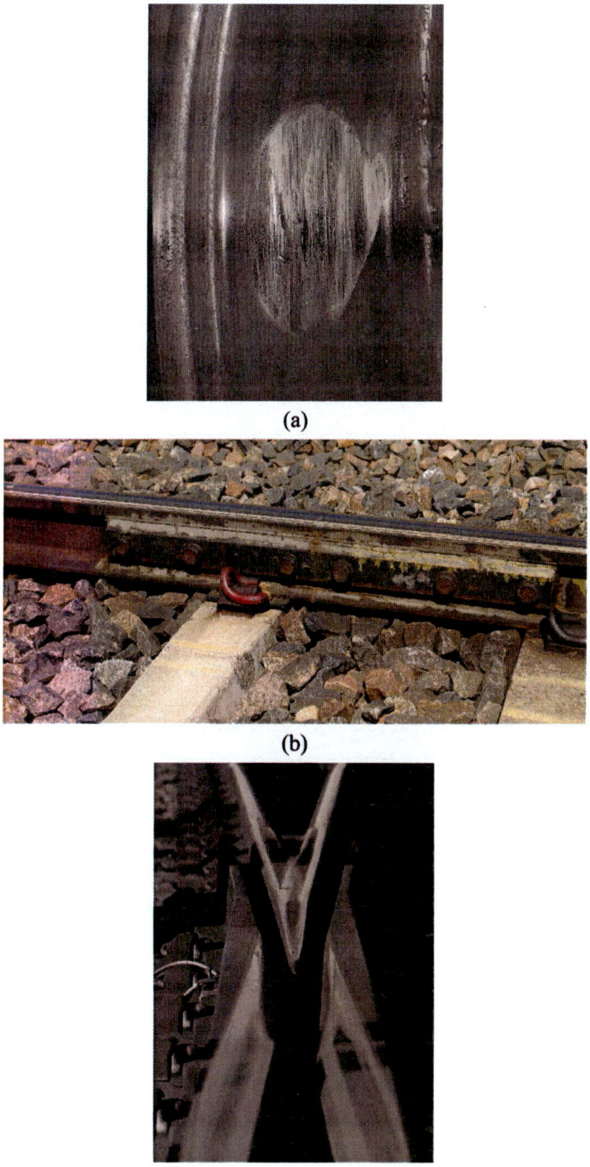

(a)

(b)

(c)

Fig. 1. (a) Wheel flat. Photo by Robert Fröhling, Transnet, South Africa. (b) Insulated rail joint with fish-plate assembly, from [16]. (c) Fixed crossing with crossing nose and two wing rails. Photo by Björn Pålsson, Chalmers University of Technology, Sweden.

nose has a vertical inclination, results in a wheel–rail excitation that is characterised by a dip angle in the vertical wheel centre trajectory, see Sect. 4.3. Wheels with different stages of profile wear result in different trajectories with different dip angles in the

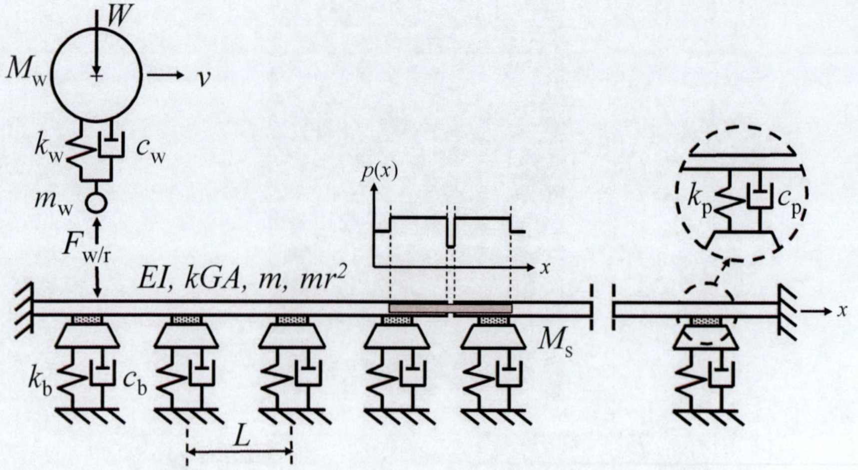

Fig. 2. Sketch of model for simulation of vertical dynamic interaction between a wheelset and a track with a fish-plated joint.

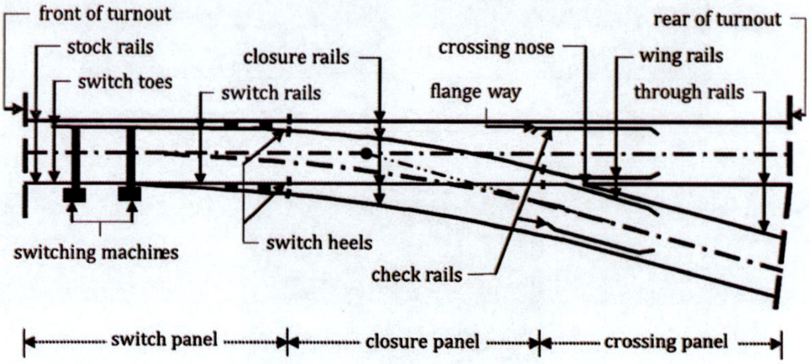

Fig. 3. Sketch of a switch & crossing (S&C) with notations. From [20].

same crossing. Wheel profiles deviating from the design conditions will generate severe impact loads. In particular, this is the case for hollow worn wheel profiles [19, 20].

In crossings, the variation in length of sleepers and the complex arrangement with a combination of several rails lead to a variation in track stiffness and mass [21]. Moreover, the symmetry of plain line track is lost leading to a significant variation and difference in track vertical stiffness at the inner and outer rails, see Fig. 4.

3 Field Measurements

To illustrate the influence of local deviations in the nominal wheel–rail contact geometry on impact loading, a few results from various tests carried out in the field are summarised in this section. Field tests of dynamic wheel–rail interaction involving a train with different types of wheel irregularities (out-of-roundness) have been reported in several studies, see for example refs. [22–26].

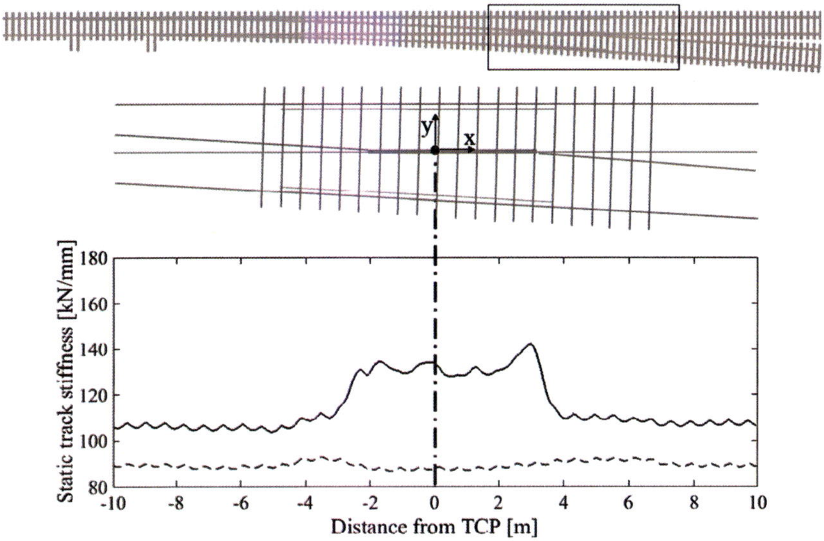

Fig. 4. Calculated static track stiffness at rail level along the through route in a crossing panel: — crossing rail, - - - outer rail. TCP indicates theoretical crossing point, i.e. the end of the crossing nose. From [21].

Results from one test are illustrated in Figs. 5 and 6 [23, 24]. Artificial wheel flats of length 40 mm and depth 0.35 mm were ground on the two wheels of one wheelset. The standard tangent track consisted of 60 kg/m rails, 10 mm studded rubber rail pads (here referred to as rail pad type A) and concrete monobloc sleepers on ballast. The standard rail pad with a dynamic stiffness of 120 kN/mm was replaced by a synthetic polymer-based pad (type B) along a track length of about 25 m. Laboratory measurements indicated that pad B was some ten times stiffer than pad A. The two test sites A and B were 70 m apart. An instrumented wheelset was used to measure the vertical wheel–rail contact force via strain gauges on the wheel discs [27]. The measured signals were low-pass filtered with cut-off frequency 1 kHz. An example of measured time history of the contact force is illustrated in Fig. 5 [23]. It is observed that the wheel flat results in a transient and periodic loading due to an impact for each revolution of the wheel. Initially, there is a partial unloading as the wheel flat enters the wheel–rail contact. During this phase, the wheel moves downwards (and the rail upwards) to compensate for the missing wheel material. Since the wheel and rail cannot completely compensate for the irregularity due to their inertia, there is a reduction in the contact force. After passing the centre of the flat, the wheel continues downwards because of its higher inertia. This results in a peak in the contact force, which is followed by a damped transient response.

The maximum contact force deviation from the static wheel load versus train speed is plotted in Fig. 6 for two different axle loads. Depending on the combination of rail pad stiffness and axle load, there is a local maximum at a speed between 20 and 70 km/h. Similar local maxima in wheel–rail contact force were reported in refs. [22, 26]. The local maximum is referred to as the P2 resonance [26], which can be described as the resonance frequency at which the unsprung mass of the vehicle and the equivalent track mass are vibrating in phase on the equivalent stiffness of the track. This corresponds to the fundamental frequency at which the combined receptance of the coupled vehicle–track system has a local minimum. The unsprung mass of the vehicle includes the mass of the wheelset with axle boxes and parts of the primary suspension, and if present the brake discs and parts of the mechanical drive system. In Fig. 6, the local maximum (at both test sites) is shifted to a higher vehicle speed when the axle load is increased. This appears to be explained by the higher stiffness of the rail support with increasing axle load.

The influence on the wheel–rail impact load in a crossing of train speed, route and traffic direction has been investigated in a field test, see ref. [28]. Also here, the contact force was measured using an instrumented wheelset. For all train routes, an increase in impact load with increasing train speed was observed. For a given speed, the magnitude of the impact load was considerably higher for the diverging route compared with the through route. Based on the results from another field test with vehicle speeds in the range 40–70 km/h (and different types of rail pads than in Fig. 6) [29], it was concluded that the magnitude of the impact load in the crossing was reduced when implementing softer rail pads.

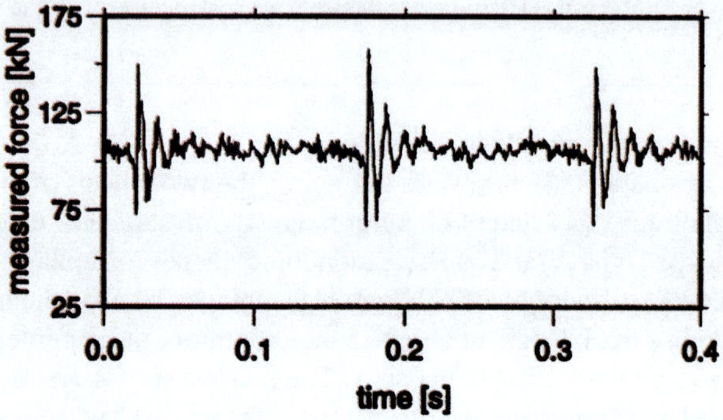

Fig. 5. Time history of wheel–rail contact force measured by an instrumented wheelset with a 40 mm rounded wheel flat. Vehicle speed 70 km/h and axle load 22 tonnes. From [23].

Pass-by noise and rail vibration generated by passenger trains in a fixed crossing (in the trailing move) have been measured [30, 31]. Noise was recorded using a microphone positioned at 7.5 m from the track centre and 1.2 m above the top of the rail. Vertical rail vibration was measured using an accelerometer attached to the underside of the wing rail at the position at which the wheel–rail contact transfers to the wing rail. Fig. 7 shows an example of measured rail vibration during a train passage, where the influence of the passing wheels can be observed as periodic maxima. However, these impacts were found to be less evident in the sound pressure signal. The spectrograms of the measured rail velocity and sound pressure are shown in Fig. 8. In the vibration data, the spectrogram is mainly characterised by horizontal bands. These correspond to each wheel impact and cause the rail to vibrate in a wide frequency range, with most of the energy concentrated between 20 and 1000 Hz. In contrast, the sound pressure spectrogram is mainly characterised by vertical bands. These are caused by the wheel/rail acoustic

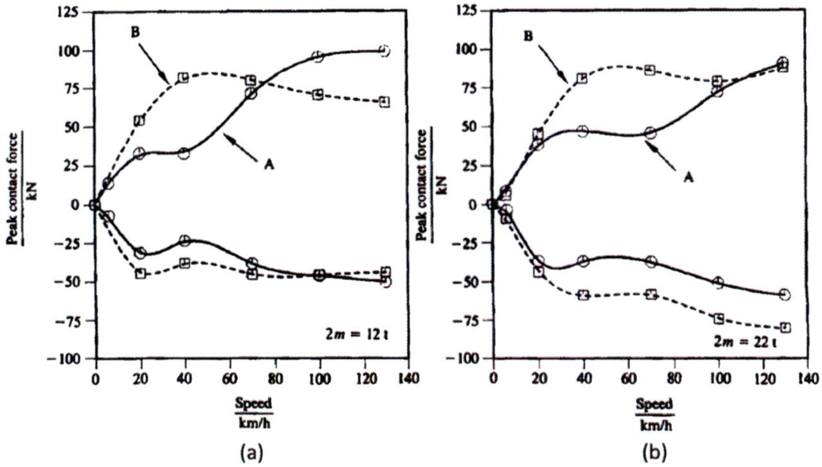

Fig. 6. Measured peak (maximum and minimum) contact force deviations from static wheel load versus train speed due to a 40 mm long wheel flat at test sites A and B for axle load (a) 12 tonnes and (b) 22 tonnes. From [24].

roughness and correspond to the rolling noise generated on the approach to and departure from the crossing. The acoustic roughness was measured as part of these tests and found to be much higher on the wing rail than outside the crossing, probably because grinding trains do not access the crossing. It was not possible to separate the rolling noise and impact noise and, although clearly audible, it was concluded that the contribution of the impact noise to the sound pressure spectra was small [31].

Results from a field test aiming to quantify the contribution of impact noise generated by different types of wheel tread defects on the pass-by noise of freight trains are presented in ref. [32]. The noisiest wheel with a 2 mm deep local crushing defect on the running surface led to an increased pass-by noise level (L_{Aeq} over the vehicle length) of 9 dB(A), measured at 7.5 m from a track with low acoustic roughness for train speeds 80–120 km/h. Wheel flats of depth 0.8 mm and 1.35 mm led to increases in noise level of 3 and 7 dB(A) respectively.

In ref. [33], a laboratory experiment was carried out in an anechoic chamber to examine the subjective reaction of 25 subjects to train noise containing components due to wheel–rail impact. Compared with railway noise without impact loading and at the same equivalent sound energy level, it was concluded that their recorded case of wheel flat noise increased the perceived loudness by 3 dB while the rail joint increased the loudness by 5 dB.

4 Quasi-Static Wheel Centre Vertical Trajectory – Point Contact Model

As discussed above, when the wheel runs over a discrete wheel/rail surface irregularity, this may lead to unloading of the wheel–rail contact and even a momentary loss of contact followed by the generation of an impact load. In a ‘point contact’ model, the

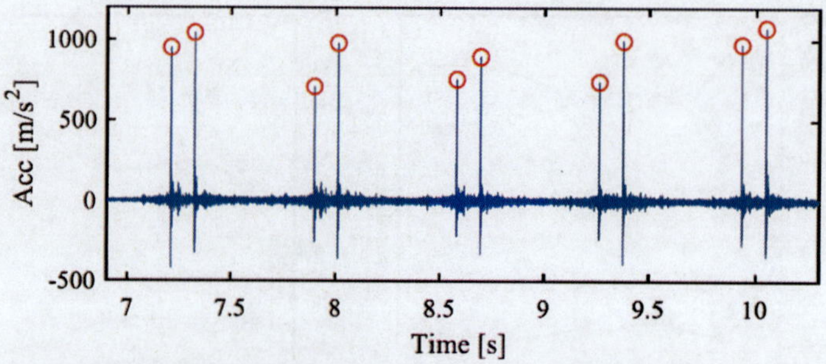

Fig. 7. Measured vertical acceleration of the wing rail. From [31].

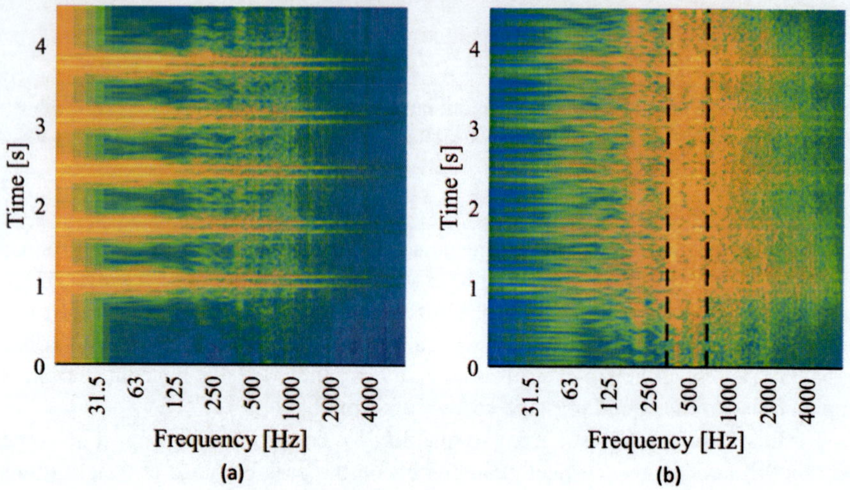

Fig. 8. Measured (a) rail vibration and (b) sound pressure spectrograms due to impact on the wing rail. From [31].

excitation of the wheel–rail system can be described by a relative displacement input between the wheel and rail that is equal to the quasi-static wheel centre trajectory, i.e., the path followed by the wheel centre if it ran very slowly over the surface irregularity [34]. In this section, expressions are presented for the quasi-static wheel centre trajectory, based on a simplified description of the geometry of each irregularity and assuming point contact between a rigid wheel and a rigid track. These formulae represent the prescribed relative wheel–rail vertical displacement excitation that occurs between a flexible wheel and a flexible track in the same way as the acoustic roughness, processed to account for the curvature of the wheel as described in EN 15610 [35], forms the input for rolling noise calculations [34, 36]. The actual motion of the wheel is the dynamic response to this excitation.

Although the simplified formulae given here are instructive in establishing trends of impact force, noise and vibration and their dependence on defect geometry, the actual geometry of the surface irregularity can vary considerably and will have a significant influence on the magnitude of the impact load [37], see also Sect. 6.2. When using a three-dimensional contact model, the actual geometry of the surface irregularity serves as input to the calculation of the dynamic vehicle–track interaction [37]. In this case, the quasi-static wheel centre trajectory is not needed.

In this section, the contact filter effect on the input to the prescribed relative wheel–rail vertical displacement excitation is neglected, whereas it is automatically included if a three-dimensional elastic contact model is used. The influence of the contact model on the calculated impact load for new and rounded wheel flats will be discussed in Sect. 6.2.

4.1 Wheel Flat

In most models of wheel–rail interaction due to wheel flats, the two-dimensional shape of the flat is described by a simple analytical function. A new wheel flat with sharp edges can be described as a chord of the wheel circumference, where the length l_0 and depth d are approximately related by

$$l_0 \approx \sqrt{8R_\mathrm{w}d} \tag{1}$$

Here R_w is the wheel radius and it is assumed that $d \ll R_\mathrm{w}$. The vertical wheel profile deviation x_nf, which is the difference between the rolling surface of the undamaged wheel and the surface of the wheel featuring the new flat, is approximately given as, [37],

$$x_\mathrm{nf} \approx d - \frac{z^2}{2R_\mathrm{w}}, \quad -\frac{l_0}{2} \le z \le \frac{l_0}{2} \tag{2}$$

where z is the circumferential distance from the centre of the flat. As discussed in Sect. 2.1, the initial flat is soon transformed into a longer flat due to wear and plastic deformation of the flat edges. Based on ref. [22], the vertical wheel profile deviation x_rf for a rounded flat can be approximated as

$$x_\mathrm{rf} = \frac{d}{2}\left(1 + \cos\left(\frac{2\pi z}{l}\right)\right), \quad -\frac{l}{2} \le z \le \frac{l}{2} \tag{3}$$

Here it is assumed that the depth of the new and rounded flats is the same but $l > l_0$, cf. Ref. [14]. The true two- and three-dimensional shapes of a wheel flat can be expected to be more complicated than indicated by Eqs. (2) and (3). In ref. [38], it is assumed that the three-dimensional shape of a new flat corresponds to the shape of the rail head on which it was formed. In this case, the parameter lines of the vertical wheel profile deviation in the rolling direction are of the type given in Eq. (2), while parameter lines in the transverse direction are circular arcs with rail head radius R_r.

Wheel flats introduce a relative displacement input to the wheel–rail system in a similar way as acoustic roughness. For a new flat, the wheel pivots around the two corners, and the wheel trajectory differs from the shape of the flat due to the curvature of the wheel. Assuming the rounded edges of the flat can be described by a quadratic function with smooth transitions and the lateral contact positions on the wheel and rail remain constant, the wheel centre vertical trajectory x_w can be written as, cf. Ref. [34],

$$x_\mathrm{w} \approx \begin{cases} 4d\left((2z + l/2l)\right)^2, & -l/2 \le z \le 0 \\ 4d\left((l - 2z/2l)\right)^2, & 0 \le z \le l/2 \end{cases} \tag{4}$$

Note that Eq. (4) can be used also for new flats by using the relation $d = l^2/8R_\mathrm{w}$, cf. Equation (1). Based on the wheel flat geometries described by Eqs. (2) and (3), examples of wheel centre trajectory for new and rounded wheel flats are given in Fig. 9. It is observed that the relative displacement excitation based on the vertical wheel centre trajectory differs from the geometric shape of the wheel flat. For the new flat in Fig. 9(a), it is observed that the slope of the wheel centre trajectory is discontinuous at the centre of the flat. This sharp discontinuity may have an influence on the magnitude of the impact load and a smoothing to account for the finite size of the contact (application of a contact filter) may be necessary, see Sect. 6.2.

For measured three-dimensional flats, a numerical procedure can be employed to determine the wheel centre trajectory, similar with that used in ref. [39] for roughness. The three-dimensional geometry of the flat and the lateral position of the running band relative to the centre of the flat influence the magnitude of the generated impact load. For example, in ref. [40] it was shown that the minimum circumferential curvature of the wheel tread along the wheel flat has a larger influence on the magnitude of the impact force than the flat depth.

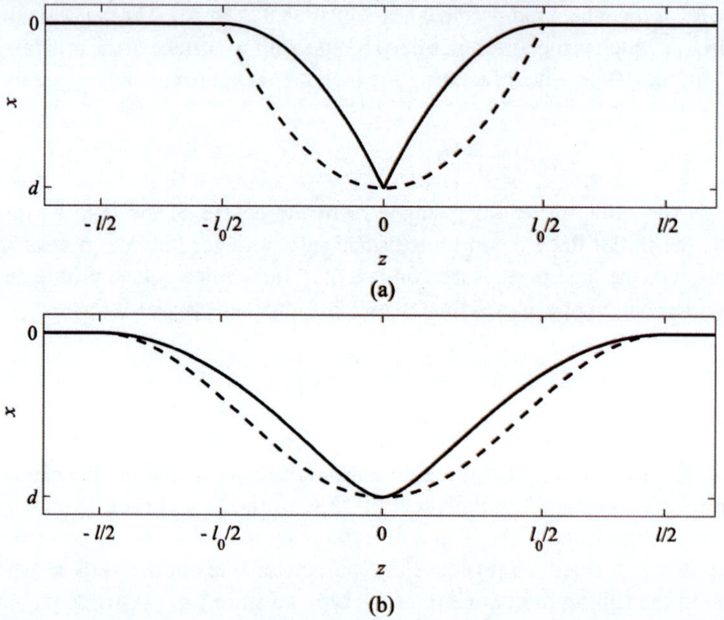

Fig. 9. Examples of relative displacement input: --- wheel profile deviation; – wheel centre trajectory: (a) new wheel flat with depth d and length l_0, (b) rounded wheel flat with depth d and length $l = 1.76l_0$. From [37].

4.2 Rail Joint

The corresponding equations for the relative vertical displacement excitation due to a rail joint were derived in ref. [41]. Again, this input is determined by the trajectory of a rigid wheel over the rail joint on a rigid track. Three different situations may occur depending on whether the wheel is in tangential contact with the rail on either side of the joint when making the transition to the other rail. Depending on the type of contact at the transition, the three situations can be described as: (I) tangential contact on both sides of the gap, (II) tangential contact on one side and non-tangential contact on the other, and (III) non-tangential contact on both sides of the gap. The first two situations are illustrated in Fig. 10. Thus for situation II, for example, if the wheel is not in tangential contact with the rail when making first contact with the rail on the other side of the gap, it will pivot about the contact point on that rail edge until it regains tangential contact.

As discussed in Sect. 2.2, on each side of the joint, there is often a vertical dip in the rail. The geometry of the rail joint is then determined by the gap width w and height difference h between the two rail ends and by the geometry of the dip. Considering these input data, the geometry of each of the two rails ($i = 1, 2$) can be described by a set of coordinates $(z_{\mathrm{r}i}, x_{\mathrm{r}i})$. To determine the relative displacement excitation input, the position of the wheel when making simultaneous contact with both rails and the type of contact situation (I to III) need to be identified. This can be accomplished by applying a search algorithm around the unloaded joint, where the longitudinal position of the wheel on the left rail is shifted forwards in small steps until there is a first overlap between the wheel and the right rail. Based on the set of coordinates ($z_{\mathrm{r}i}$, $x_{\mathrm{r}i}$) and assuming continuous tangential contact on both rails (situation I in Fig. 10), the wheel centre trajectory (z_{w}, x_{w}) on either side of the joint is calculated as, [41],

$$\theta_i \approx \tan\theta_i = \frac{\mathrm{d}x_{\mathrm{r}i}}{\mathrm{d}z_{\mathrm{r}i}} \tag{5}$$

$$z_{\mathrm{w}} = z_{\mathrm{r}i} + R_{\mathrm{w}}\sin\theta_i \tag{6}$$

$$x_{\mathrm{w}} = x_{\mathrm{r}i} + R_{\mathrm{w}}(1 - \cos\theta_i) \tag{7}$$

However, if $\theta \geq \theta_2$ at the transition to the other rail (situation II in Fig. 10), the wheel will pivot around the rail end until it regains tangential contact. During this transition stage, the wheel centre trajectory is calculated as, [41],

$$z_{\mathrm{w}} = z_{\mathrm{r}} - R_{\mathrm{w}}\sin\theta \tag{8}$$

$$x_{\mathrm{w}} = x_{\mathrm{r}} + R_{\mathrm{w}}(1 - \cos\theta) \tag{9}$$

As in Sect. 4.1, it is assumed that the lateral contact positions on the wheel and rail profiles remain constant. Examples of wheel centre trajectory due to two different rail joints are shown in Fig. 11. Equations for the wheel centre trajectory in situation III are given in ref. [41], but it is stated that this is a situation that rarely occurs in practice. Furthermore, for gap widths $w \leq 20$ mm, it was concluded that the wheel centre trajectory is more influenced by the height difference than by the gap width.

4.3 Crossing

In front of the crossing nose (when observed in the facing move), the stock rail is curved laterally into the wing rail, see Fig. 3. In a facing move, this means that the contact

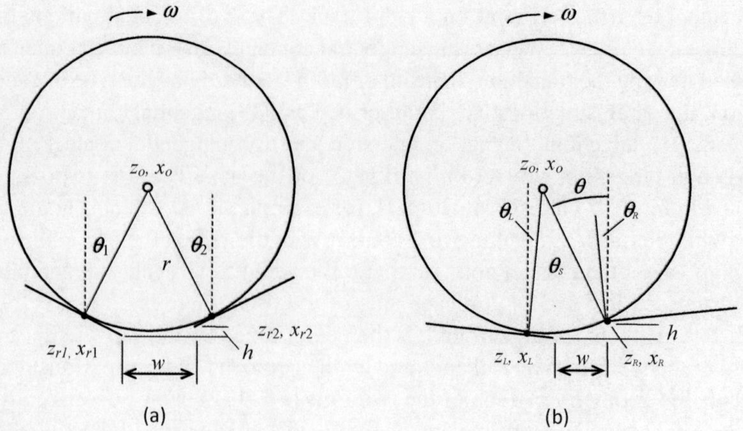

Fig. 10. Rolling contact geometry of a wheel over dipped rails at a joint: (a) situation I – tangential contact on both rails, (b) situation II – non-tangential contact on the right rail. From [41].

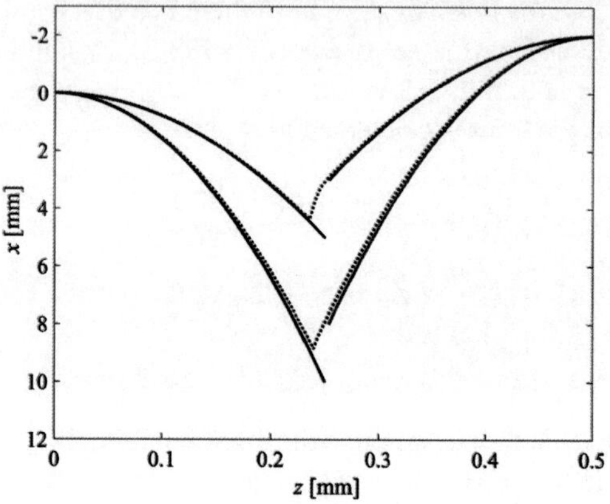

Fig. 11. Examples of relative displacement input for a wheel rolling over a dipped step-up rail joint with gap width $w = 7$ mm and vertical misalignment $h = 2$ mm: – dipped rail shape, --- wheel centre trajectory. Upper curves are for a 5 mm dip at the joint, bottom curves are for a 10 mm dip at the joint. From [41].

patch on the wheel tread moves outwards (away from the crossing nose) as long as the contact remains on the wing rail. Due to the decreasing wheel rolling radius, the wheel centre moves downwards until the wheel tread makes contact with the vertically inclined surface of the crossing nose. Due to this inclined surface, the wheel centre moves upwards again until the level of the stock rail is reached again (on the through rail). In contrast with Sects. 4.1 and 4.2, where it was assumed that the lateral contact position on the wheel profile remained constant, there is a significant variation in lateral contact positions on wheel and rail as the wheel passes over the crossing. The wheel centre trajectory is therefore influenced by the given crossing geometry as well as the specific wheel profile.

For a given combination of wheel profile and crossing geometry, the vertical trajectory of the wheel centre can be determined based on a kinematic (quasi-static) pre-study using a software for vehicle multibody dynamics. Examples are given in Fig. 12. It is observed that the wheel centre trajectory displays a kink, referred to as the dip angle, which is the reason for the change of vertical momentum and the impact load generated at the crossing. A linear model for deriving the wheel centre trajectory and dip angle in crossings was derived by Pålsson [42]. In this model, the wheel profile is taken as conical and it is assumed that the variation in contact point position on the wing rail and crossing nose along the crossing panel can be described using linear functions.

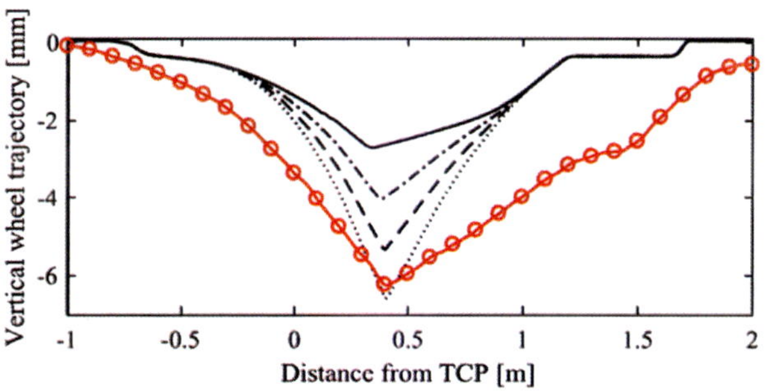

Fig. 12. Quasi-static vertical wheel trajectories during pass-by of a nominal S1002 wheel profile on a measured crossing geometry (—o—), and on generic crossings with dip angle 6 mrad (——), 12 mrad (–·–·–), 18 mrad (– – –) and 24 mrad (··········). Zero relative lateral wheel–rail displacement is assumed. TCP indicates theoretical crossing point, i.e. the end of the crossing nose. From [31].

5 Wheel–Rail Contact

To simulate dynamic vehicle–track interaction in the vertical direction when there is a large variation in the wheel–rail contact force relative to the static wheel load, a non-linear wheel–rail contact model is required and the interaction must be solved in the time domain. In particular, this is the case where a momentary loss of wheel–rail contact may occur. The most common model accounting for such situations is the Hertzian contact model [43], see e.g. the simulation models presented in refs. [22, 44, 45]. A comprehensive description of Hertzian contact theory is given in refs. [46, 47].

The Hertzian theory for normal contact between two solid bodies relies on several assumptions [47, 48]: (I) The two bodies are linear elastic solids. (II) The surfaces of the two bodies are non-conforming, i.e. contact is first made at a point (or along a line). (III) The dimensions of the contact area are small relative to the dimensions of the two bodies and their radii of curvature. For the calculation of stresses and deformations, this implies that the bodies can be considered as semi-infinite elastic solids with a plane surface (elastic half-space). (IV) The surfaces are perfectly smooth. (V) Hertzian surfaces, i.e. the surfaces can be described by quadratic functions near the contact. (VI) There is no friction in the contact area and the only transmitted load is normal to the contact.

Based on these assumptions, the contact area is an ellipse with semi-axes a and b, while the normal pressure distribution is ellipsoidal with maximum pressure

$$p_{\max} = \frac{3}{2}\frac{F_{\mathrm{w/r}}}{\pi ab} \tag{10}$$

Here $F_{\mathrm{w/r}}$ is the wheel–rail contact force in the direction normal to the contact, see Fig. 2. Formulae for determining a and b are given in refs. [46, 47]. The Hertzian spring model used in most time-domain models is a single non-linear spring (point contact model). The force–displacement characteristic of this spring is expressed as

$$F_{\mathrm{w/r}} = C_{\mathrm{H}}\langle\delta\rangle^{3/2} \tag{11}$$

where δ is the approach distance of two distant points on wheel and rail, while the factor C_{H} is a function of material parameters and the principal relative radii of curvature [46]. The Macaulay brackets are defined as $\langle\bullet\rangle = 0.5(\bullet + |\bullet|)$. Thus, $F_{\mathrm{w/r}} = 0$ if $\delta < 0$ (loss of contact).

In frequency-domain models, the contact model must be linear. The spring characteristic should then be linearised around the approach δ_0 corresponding to the static wheel load F_0. The stiffness of the linear Hertzian spring $k_{\mathrm{H,lin}}$ is obtained from the tangential gradient in the point (δ_0, F_0) as

$$k_{\mathrm{H,lin}} = \left.\frac{\mathrm{d}F_{\mathrm{w/r}}}{\mathrm{d}\delta}\right|_{\delta_0} = \frac{3}{2}C_{\mathrm{H}}\delta_0^{1/2} = \frac{3}{2}\frac{F_0}{\delta_0} \tag{12}$$

For cases with excitation by acoustic roughness (rolling noise) and without loss of contact, the difference between the non-linear and linear Hertzian contact models was investigated in ref. [49]. It was concluded that the difference between the two models is negligible if the root mean square (rms) of the roughness is less than about 0.35 times the static contact deflection, but rapidly becomes significant above this value. Moreover, it was concluded that the effects of non-linearities on the wheel–rail interaction are not noticeable at low frequencies up to 100 Hz because in this frequency region the dynamic stiffness of wheel and rail is much lower than the contact stiffness and thus the contact can be regarded as effectively rigid. Note that when using a point contact model, it is assumed that contact always occurs directly below the wheel centre. This is generally not true for impact excitation due to wheel flats and rail joints, but the significance of this effect on the contact force, noise and vibration is unknown.

The assumptions of Hertzian contact are generally not met for the cases of impact excitation discussed in this paper. Hertzian theory relies on constant radii of curvature of the (undeformed) bodies in the contact area, but at discrete wheel/rail surface irregularities these radii may vary significantly in the longitudinal and lateral directions. For example, for a wheel flat, the radii of wheel curvature change quickly and the wheel surface cannot be described by quadratic functions. Due to the variation in the radii of curvature, the contact stiffness varies along the flat. Baeza et al. [38] compared impact loads calculated with Hertzian and non-Hertzian models and found that the Hertzian model overestimated the peak force. In crossings, during the transition of the wheel from the wing rail to the crossing nose, there is a continuous variation in lateral contact position on the rail and the changes in curvature within the contact area lead to the formation of non-elliptical contact patches and pressure distributions deviating from the Hertzian distribution [19].

As discussed above, the half-space assumption is valid if the dimensions of the contact area are small in comparison with the characteristic dimensions of the contacting bodies. The half-space assumption is reasonable for wheel–rail tread contact but it is violated for flange contact, contact near the gauge corner of the rail, and for contact between a wheel and a rail joint. Standard Hertzian theory is limited to one single contact patch, but two- or multiple-point contact situations are common in S&C.

To solve the three-dimensional contact problem for arbitrary non-Hertzian geometries, the equations of elasticity for the two continua need to be solved. This is generally only possible numerically, e.g. by using the finite element method [50, 51].

For an elastic half-space, explicit influence functions describing the surface displacement field due a point load on the surface are known. This is the reason why the half-space assumption considerably simplifies the solution of the three-dimensional contact problem. The influence functions for an elastic half-space were derived by Cerruti [52] and Boussinesq [53]. Several non-Hertzian contact models are of a special type of boundary element methodology based on the Boussinesq–Cerruti expressions for the elastic half-space, of which Kalker's model is a well-known example [54]. It uses a variational method based on the principle of maximum complementary energy and applies an effective active-set method to solve the contact problem. The potential contact area is then discretised into rectangular elements in which the surface traction is assumed to be constant. Non-Hertzian contact models of this type have been used in the simulation models for dynamic vehicle–track interaction presented in refs. [37, 38].

In recent years, two trends have been observed in the development of contact algorithms for non-Hertzian geometries. One trend is advanced finite element formulations, see Sect. 6.3, while the other is approximate and fast solution methods. The finite element method is not limited to half-spaces and has the capability of including arbitrary contact geometries, non-linear material properties and plastic deformation. A disadvantage can be extensive computational time. Examples of fast and approximate methods have been developed in refs. [55–58]. One such model is based on a bedding of non-interacting elastic springs, the Winkler bedding, which is fast because the coupling between different points in the continuum is neglected. Examples of the application of Winkler beddings in the wheel–rail contact are the DPRS (Distributed Point Reacting Springs) models proposed by Remington and Webb [39] for the three-dimensional case and by Ford and Thompson [17] for the two-dimensional problem.

The contact models discussed above are quasi-static, which implies that the local dynamics due to the inertia in the contact area is not accounted for. Given the small region of influence, it may be assumed this effect is negligible.

6 Dynamic Vehicle–Track Interaction

6.1 Vehicle and Track Models

Knothe and Grassie [59] surveyed modelling of railway track and dynamic vehicle–track interaction in the frequency range 20–5000 Hz. A classification of track models was suggested based on the number of (up to three) layers in the model describing the rails, sleepers and ballast/subgrade and whether the rails are assumed to be continuously or discretely supported. It was concluded that improved models were needed to describe more accurately the dynamic and non-linear characteristics of rail pads and ballast/subgrade. For example, the dynamic stiffness of rail pads and ballast/subgrade is known to vary with frequency and with the magnitude of the applied load, but this is ignored in most models (some exceptions are described in Sect. 6.2). In a vehicle–track interaction model, the dynamic stiffness of railway track components should be used as input rather than the corresponding static stiffness [49]. Surveys of vehicle–track interaction models are presented also in refs. [60, 61].

The rail pads, in particular, have a strong influence on the high-frequency track dynamics. The stiffness and damping of studded rubber rail pads have been measured by Thompson et al. [62]. The rail pad was preloaded by a static force before a cyclic loading was applied. For preloads above 40 kN, it was shown that the dynamic stiffness increased significantly with increasing preload. A range of other rail pads were measured in ref. [63].

The dynamics of rail vehicles are discussed in many text books, see for example ref. [64], and several commercial software packages based on multi-body dynamics modelling of vehicle–track interaction in three dimensions are available. However, for the prediction of impact loads and railway noise, it is generally sufficient to model only the wheelset (unsprung mass) since the primary suspension isolates the vehicle from the wheelset in the frequency range of interest above 20 Hz. A finite element model is required to capture the eigenmodes of the wheelset. An alternative, simple vehicle model of one wheelset is shown in Fig. 2, see refs. [34, 65]. It contains two degrees of freedom (dofs), where the large mass M_W is the unsprung mass. The input data for the spring k_W, damper c_W and small mass m_W are selected to better approximate the dynamics of the finite element model. Most resonances and antiresonances of the wheelset cannot be captured by the simple model, but the average of the receptance at frequencies above 1 kHz is similar [65]. Thus, the 2-dof model in Fig. 2 is a better alternative than a 1-dof model since the latter leads to a very high dynamic stiffness at high frequencies.

Dynamic vehicle–track interaction can be simulated either in the frequency domain or in the time domain. The paper by Grassie et al. [66] is an early contribution to the frequency-domain modelling of high-frequency track dynamics (including the rail modelled as a Timoshenko beam on discrete supports) and vehicle–track interaction. However, for the prediction of loss of wheel–rail contact and impact loads, the simulation must be carried out in the time domain using a non-linear contact model. Nevertheless, for the prediction of impact noise in a post-processing step, frequency-domain models of wheelset and track dynamics have been applied by Wu and Thompson using a hybrid approach and the concept of equivalent roughness excitation, see Sect. 6.4.

6.2 Vehicle–Track Interaction with Impact Loading

An early contribution to the modelling and understanding of dynamic vehicle–track interaction and wheel–rail impact loading due to wheel flats and dipped rail joints was made by Jenkins et al. [26]. The terminology using the so-called P1 and P2 resonance frequencies to explain the dynamic response of the vehicle–track system was introduced.

In the time domain, the P1 resonance leads to an initial sharp peak in the contact force after less than 1 ms, which is followed by a broader peak corresponding to the P2 resonance after 5–10 ms. Here the P1 resonance is the dominating frequency of the transient response directly after the impact, which is determined by the resonance of the unsprung wheelset mass and the effective track mass on the contact spring stiffness, typically around 700–1000 Hz. As discussed in Sect. 3, the P2 resonance can be described as the resonance frequency where the unsprung mass of the vehicle and the equivalent track mass are vibrating in phase on the equivalent stiffness of the track, occurring typically around 50–100 Hz. In ref. [26], it was concluded that the impact load is closely proportional to the product of dip angle and train speed.

Newton and Clark compared different track models for the calculation of impact load due to wheel flats [22]. It was shown that a continuously supported rail on a single-layer support model is not sufficient as the rail pads are neglected and the sleeper mass is not correctly distributed along the rail. Further, to capture the high-frequency dynamics after impact, the rail should be modelled by Timoshenko beam theory, accounting for the influence of shear deformation, rather than by the more simplified Euler-Bernoulli beam theory. Tunna [67] extended the model developed in ref. [22] by implementing a wheel–rail contact model that, while the wheel centre is over the wheel flat irregularity, scans around the wheel circumference to detect the position of maximum wheel–rail overlap. Based on this overlap, the force is calculated and applied at the wheel centre position. An early model for the simulation of impact loads in railway crossings was developed by Andersson and Dahlberg [68].

The calculated influence of train speed and depth of rounded wheel flats on the maximum impact load is illustrated in Fig. 13 [37], cf. the experimental results in Fig. 6. A three-dimensional non-Hertzian contact model based on Kalker's variational method was applied. The dynamic vehicle–track interaction was solved with a moving impulse response function (Green's function) approach [48, 69, 70]. As expected, with increasing flat depth, the magnitude of the impact load increases and loss of wheel–rail contact occurs at a lower train speed. The speed dependence is essentially similar for different depths, but the local maximum and minimum of each curve are more pronounced for larger depths and are also shifted to higher speeds. It is observed that the smaller the flat depth, the earlier the curve flattens out at higher train speeds.

The influence of the contact model on calculated impact loads due to new and rounded wheel flats was investigated by Pieringer et al. [37], see Fig. 14. The three-dimensional non-Hertzian contact model based on Kalker's variational method, the two-dimensional non-Hertzian contact model consisting of a Winkler bedding of independent springs (DPRS), and a single non-linear Hertzian contact spring (point contact model) were compared. The relative displacement excitation used as input to the Hertzian model was either the wheel profile deviation, see Eqs. (2) and (3), or the pre-calculated wheel centre trajectory in Eq. (4). Both the two-dimensional model and the Hertzian spring model with the wheel centre trajectory as input were found to generate results rather similar to the three-dimensional model. However, the Hertzian model with the wheel profile deviation as input resulted in large errors. It was concluded that the correct representation of the longitudinal geometry of the wheel flat has a larger influence on the impact load than the choice of contact model. For the new flat in Figs. 14(b,d,f), it is observed that the Hertzian

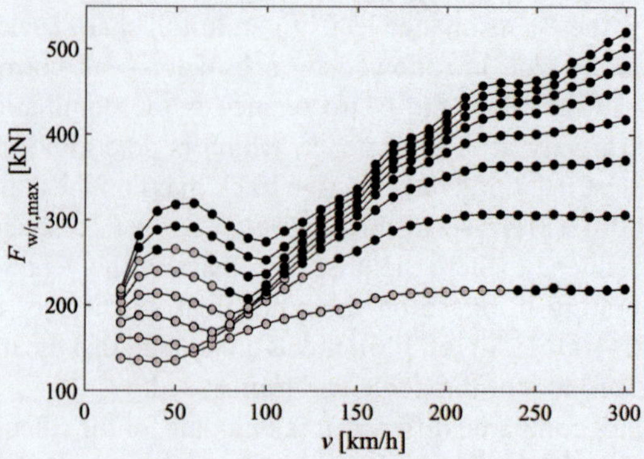

Fig. 13. Magnitude (including static wheel load 118 kN) of impact load due to a rounded wheel flat as a function of train speed and flat depth. The lines (from lower to upper) correspond to the depths [0.25, 0.50, 0.75, 1.00, 1.25, 1.50, 1.75, 2.00] mm. The maximum of the impact loads from eight simulations with different initial angular positions of the wheel is shown. Black circles indicate that loss of contact occurs for at least one of these eight simulations. Three-dimensional non-Hertzian contact model based on Kalker's variational method. From [37].

spring model with the wheel centre trajectory as input consistently leads to higher impact load magnitudes than the two- and three-dimensional models. It is possible that this is a consequence of ignoring the contact filter effect when applying Eq. (4) as input to the point contact model.

Wu and Thompson [71] studied the influence of non-linear rail pad and ballast/subgrade stiffness on the dynamic vehicle–track interaction. Non-linear load–deflection laws for pads and ballast/subgrade were considered as well as potential loss of contact between rail and pad or between sleeper and ballast. It was shown that these non-linear characteristics only affect a few rail supports on either side of the wheel load. Yet, it was observed that both impact loads and track vibration levels were noticeably higher for the non-linear track model compared with the linear model, and it was concluded that a linear track model is not sufficient for the prediction of wheel–rail impact. The calculated vibration levels were found to be higher with the non-linear model since the deformation of the track foundation is large at the early stage of loading due to the lower stiffness under a low load magnitude. The non-linear track model is mainly required in the case of severe wheel–rail impact loading as this may also lead to loss of contact between rail and sleeper and/or between sleeper and ballast. However, for wheel/rail acoustic roughness excitation, a linear track model is sufficient since the magnitudes of the wheel–rail interaction forces are moderate.

A finite element model of a railway track including the rail represented by a Timoshenko beam on discrete supports, and where loss of contact between rail and sleeper or between sleeper and ballast is considered by updating the system matrices in each time step of the simulation, is presented in refs. [25, 45]. Another vehicle–track interaction model accounting for general state-dependent track properties is described in ref. [72].

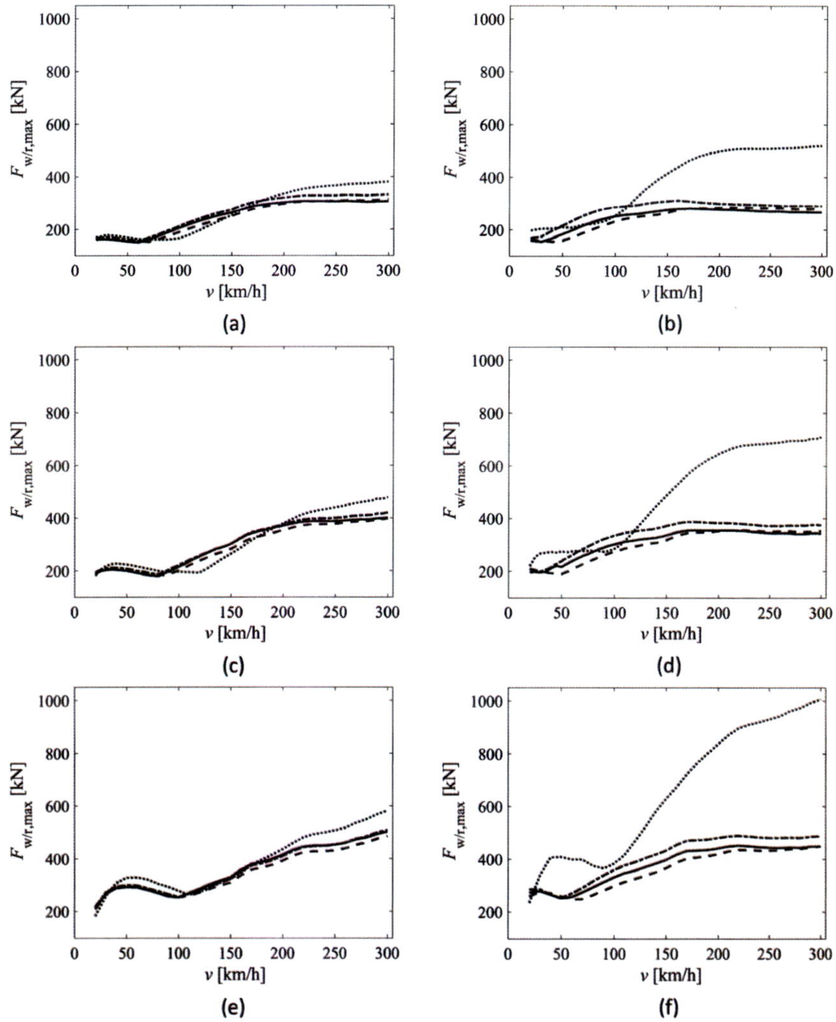

Fig. 14. Magnitude (including static wheel load 118 kN) of impact load due to (a) a rounded 0.5 mm deep wheel flat; (b) a newly formed 0.5 mm deep wheel flat; (c) a rounded 0.9 mm deep wheel flat; (d) a newly formed 0.9 mm deep wheel flat; (e) a rounded 1.75 mm deep wheel flat; and (f) a newly formed 1.75 mm deep wheel flat. Calculations with different contact models: —— 3D; – – – 2D; – · – Hertz (wheel centre trajectory); · · · Hertz (profile deviation). From [37].

6.3 Explicit Finite Element Models

To simulate wheel–rail impact at discrete rail surface irregularities, Zhao et al. [50] developed an extensive finite element model of a vehicle–track system. Wheelset, rail (including the irregularity) and sleepers were meshed using three-dimensional brick elements and the transient wheel–rail rolling contact was solved in the time domain using a surface-to-surface contact algorithm. In this case, the contact model accounts for the three-dimensional contact filter effect and the variation in contact stiffness along and across the defect (as in Kalker's variational method). It also accounts for the inertia within the contact.

Simulations with three-dimensional finite element models containing many dofs are time consuming and therefore the length of the implemented track model needs to be limited. Unless non-reflecting boundary conditions are applied in the model, spurious dynamic effects will appear where rail vibrations transmitted from the wheel–rail contact are reflected at the boundaries and returned to the contact. In Yang et al. [51], the three-dimensional finite element model was extended to account for non-linear material properties of wheel and rail. The model was applied to simulate high-frequency (up to 10 kHz) impact vibration and noise induced at an insulated rail joint. The length of the track model was 10 m with free boundaries for the rail ends at the joint and non-reflecting boundary conditions at the far ends of the rail.

6.4 Hybrid Approach Using Equivalent Roughness Excitation

When the train speed is much lower than the lowest speed of flexural wave propagation in the rail, and when the effects of discrete rail support or other variations along the track can be neglected, a so-called *moving irregularity model* can be applied [34, 59, 61]. In this case, the influence of vehicle motion along the track is neglected and the vehicle–track interaction can be solved in the frequency domain. The wheel and rail are excited by a roughness spectrum containing irregularities of different wavelengths and amplitudes; i.e. the relative wheel–rail displacement is prescribed. This is the concept used in the railway noise software TWINS [73]. On the other hand, as discussed above, a time-domain model is required to predict the wheel–rail impact loads. However, if the model is to cover the frequency range up to say 5 kHz in sufficient detail, and including a substantial length of the rail, very large computation times would be required. This calls for a hybrid approach where a time-domain model is first applied to calculate the wheel–rail contact force. The time history of the impact load is then transformed to the frequency domain and converted into the form of an equivalent roughness spectrum that can be used as input for the prediction of noise in a conventional frequency-domain model [34, 41].

The hybrid approach actually requires three models. The first is the time-domain model, including a non-linear contact model, that is used to calculate the wheel–rail contact force. Only the fluctuation of the contact force $F_{\mathrm{w/r}}(t)$ around the static wheel load F_0 results in impact noise. The dynamic component $F_{\mathrm{d}}(t)$ is obtained as

$$F_{\mathrm{d}}(t) = F_{\mathrm{w/r}}(t) - F_0 \tag{13}$$

The narrow-band spectrum of this force, $F_{\mathrm{d}}(\omega)$, can then be determined. The second model is a frequency domain model in which the wheel and track must have identical dynamic properties to those in the time domain model. The contact force per unit roughness can be calculated as

$$H_{\mathrm{rF}}(\omega) = \frac{\mathrm{i}\omega}{Y_{\mathrm{R}}(\omega) + Y_{\mathrm{W}}(\omega) + Y_{\mathrm{C}}(\omega)} \tag{14}$$

where the denominator includes the sum of rail, wheel and wheel–rail contact mobilities at circular frequency ω. As this is a linear model, the contact mobility corresponds to a linearised contact spring, see Eq. (12).

The equivalent roughness spectrum $r(\omega)$ is then obtained as

$$r(\omega) = \frac{F_{\mathrm{d}}(\omega)}{H_{\mathrm{rF}}(\omega)} \tag{15}$$

The equivalent roughness spectrum is the roughness input that would produce the same contact force spectrum $F_{\mathrm{d}}(\omega)$ if the contact spring was linear and there was no loss of wheel–rail contact [1]. This equivalent roughness spectrum is then applied in the third model in the frequency domain. In this model there is no requirement for the wheel and rail dynamics to correspond to those in the first two models. It has been shown in ref. [34] that the equivalent roughness can be calculated, for example, using a simple mass-spring model for the wheel and then used to calculate the response of a modal wheel model.

This is the approach that has been applied in a few studies of impact noise, see Sect. 7. Note that, for wheel flats, the magnitude of the impact load is influenced by the position at which the flat hits the rail within the sleeper bay and therefore the model should be evaluated at different positions within the sleeper bay [37]. It is important that roughness and not contact force is used to transfer the impact excitation from the time-domain model to the frequency-domain model [1]. For a wheel model containing lightly damped resonances, the force spectrum and contact force per unit roughness must be defined with a very fine frequency resolution around the wheel resonances.

When applying the hybrid approach, it is essential that the dynamics of the vehicle and track models at the wheel–rail contact in the first two models, in the frequency and time domains, are similar. To speed up the calculation of impact load in the time domain, Wu and Thompson [34, 41] replaced the full track model including rail and sleepers with an equivalent low-order state-space system representing the track response at the excitation point. The coefficients of the state-space system were determined by an optimisation problem, where the difference in frequency response of the two models was minimised. In Yang and Thompson [74], the hybrid approach was also adopted but without reducing the track model to a low-order state-space system. The vehicle–track interaction was solved in the time domain using a *moving mass model* and implementing a non-Hertzian and three-dimensional contact model to account for the variation in contact stiffness over the surface irregularity. Compared with Hertzian theory, the accuracy of the predicted impact vibration and noise at high frequencies was improved when using the detailed contact model. Sueki et al. [75] applied the equivalent roughness excitation methodology to predict impact noise generated at a rail joint.

7 Simulation of Wheel–Rail Impact Loads and Noise

In refs. [31, 34, 41], simulations of wheel–rail impact loads and noise have been performed for the three types of wheel/rail surface irregularities described in Sect. 2. The hybrid approach described in Sect. 6.4 was applied combining the simulation of non-linear dynamic vehicle–track interaction in the time domain with the prediction of noise using a linear frequency-domain model [73]. The excitation was transferred to the noise prediction model using equivalent roughness spectra, see Eq. (15). For the wheel flat and rail joint, the wheel trajectories described in Sect. 4 were used as input, while the three-dimensional contact model according to Kalker was applied for the crossing. The influence of the lateral shift of wheel–rail contact position on the generated noise was neglected in refs. [31, 34, 41]. This is a limitation of the hybrid approach.

The results from refs. [31, 34, 41] are summarised below and compared with the simple formulae derived by Vér et al. [8]. Note that Remington [6] extended the work by Vér et al. and introduced an equivalent roughness spectrum approach, but this was based directly on the geometry of the irregularity rather than on the calculated contact force spectrum.

7.1 Wheel Flat

In ref. [34], the discrete Fourier transform of the wheel–rail contact force was calculated based on the calculated time history corresponding to one revolution of the wheel, see also ref. [1]. Provided there is no loss of wheel–rail contact, it was concluded that the equivalent roughness can be derived directly from the wheel flat geometry if modified to account for the curvature of the wheel, i.e. the wheel centre trajectory. In this case, both the simulation of wheel–rail interaction and noise can be carried out in the frequency domain without involving two different models. However, if there is loss of contact, part of the surface irregularity where no contact is made is arbitrary and will not contribute to the excitation. In this case, the wheel–rail contact force must be solved in the time domain to determine the equivalent roughness spectrum. Further, it was shown that noise radiated by the high-frequency modes of the wheel can be accurately accounted for in the frequency-domain model using the equivalent roughness as input even if a simplified (rigid) wheel model is used in the time-domain calculation of contact force.

Equivalent roughness spectra calculated for a new 2 mm deep wheel flat at different train speeds were used as input to TWINS [73]. Since these spectra were found to drop quickly at high frequencies, there was no need to apply a contact filter in the calculation of the overall noise level. The predicted overall sound power spectra radiated by the vibration of one wheel and the associated track are shown in Fig. 15(a); both wheel and track models were linear. It is observed that noise at frequencies above 200–400 Hz increases with increasing vehicle speed, which is due to the shortening of the impact pulse and the fact that the time between pulses is reduced. For comparison, corresponding rolling noise spectra generated by a wheel with cast-iron tread brakes on good quality track (low-roughness rail) are shown in Fig. 15(b). For the given wheel/track combination and train speed interval, it was concluded that the noise generated by a flat with the significant depth of 2 mm exceeded the rolling noise at all speeds although the rolling noise level increased more rapidly with increasing speed.

The influence of train speed on the overall sound power level is illustrated in Fig. 16. Impact noise levels due to different sizes of flats are compared with the rolling noise generated by a wheel with cast-iron tread brakes on a low-roughness rail. The noise level from a new flat is greater than that from a rounded flat with the same depth. At train speeds leading to loss of wheel–rail contact, the impact noise generated by wheel flats increases at a rate of approximately 20log$v$. At lower speeds with maintained contact, the noise increases at about 30log$v$, which is the same rate as for rolling noise. It is observed that the impact noise increases with increasing train speed also when there is a momentary loss of contact. This is in disagreement with Vér et al. [8], where a constant noise level was predicted above the so-called critical speed, corresponding to the vehicle speed at which loss of wheel–rail contact starts to occur. In ref. [34], it was also concluded that the impact noise level increases with increasing axle load. For a doubling of the wheel load (50 kN to 100 kN), the noise level due to a 2 mm rounded flat increased by about 3 dB.

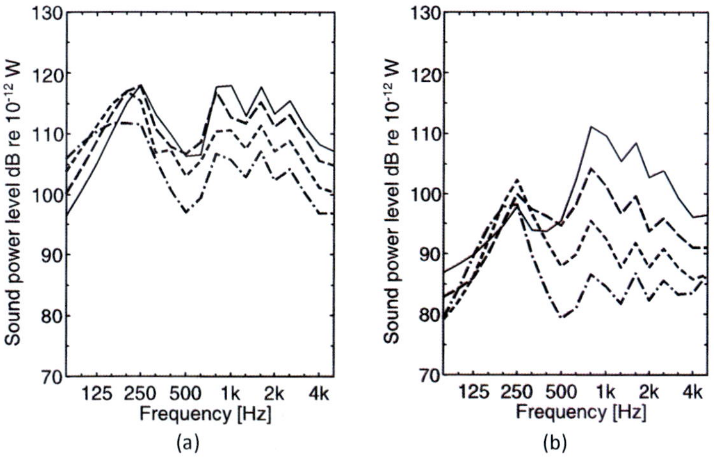

Fig. 15. Overall sound power level due to radiation by wheel and track (average over one wheel revolution). (a) From a 2 mm deep new flat. (b) From the roughness on a typical cast iron tread-braked wheel.: -·- 30 km/h, · · · 50 km/h, --- 80 km/h, – 120 km/h. From [34].

7.2 Rail Joint

A corresponding investigation of impact noise generated at dipped rail joints was presented in ref. [41], see also ref. [1]. In the time-domain simulation of dynamic wheel–rail interaction, the track model was modified to include a pair of semi-infinite Timoshenko beams supported by a continuous spring-mass-spring foundation. To account for the reduced bending stiffness at the joint, the two rails were pin-jointed with zero rail bending moment at the joint. Both wheel and track models were linear. An alternative track model from ref. [16] including the mass and bending stiffness of the fish-plates is illustrated in Fig. 2.

In ref. [41], it was concluded that the impact load at a dipped rail joint increases with increasing vehicle speed. For a 5 mm dipped rail (described by a quadratic function over 0.5 m on each side of the joint, see Fig. 11), impact loads generated at step-up joints increased with increasing height difference, whereas the influence of height difference was not very significant for step-down joints. For a 10 mm dipped rail, the impact was dominated by the larger dip angle, while the influence of height difference was smaller

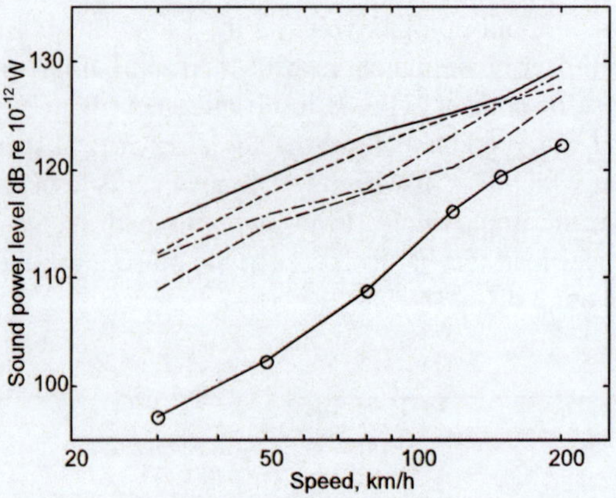

Fig. 16. A-weighted sound power level radiated by one wheel and the associated track vibration (average over one wheel revolution): – –1 mm rounded flat, --- 2 mm rounded flat, --- 1 mm new flat, – 2 mm new flat, o–o rolling noise due to roughness (cast iron tread brakes and low-roughness rail). From [34].

for both the step-up and step-down joints. As discussed in Sect. 4.2, the influence of the gap width on the wheel trajectory can be neglected compared with the height difference and dip angle.

To evaluate the influence of the impact on the short-term perception of noise, equivalent roughness spectra and sound power levels were calculated based on the time history of the impact load during 0.125 s [41]. Equivalent roughness spectra corresponding to different combinations of train speed and joint geometry were used as input to TWINS [73]. Since these spectra were found to drop quickly at high frequencies, there was again no need to apply a contact filter in the noise calculation in determining the overall noise level. It was concluded that the noise radiation increased at all frequencies with increasing train speed. The influence of train speed on the overall A-weighted sound power level generated by different joints is illustrated in Fig. 17. For the 5 mm dipped rail with a step-up joint, the predicted noise increased approximately at a rate of 20log$v$, while there was a lower rate of increase for the step-down joint. The noise radiation increased by up to 8 dB as the step-up height difference increased from 0 to 2 mm. According to the formulae derived by Vér et al. [8], noise generated at step-up and step-down joints increases at a rate of 20log$v$. However, for step-down joints, a constant noise level was predicted above the so-called critical speed.

In Fig. 17, the sound power induced at a rail joint during an averaging time of 0.125 s is also compared with the rolling noise generated by the wheel/rail roughness due to cast-iron tread brakes on a low-roughness rail. By adjusting the averaging time to the speed-dependent travelling time between two joints, it was concluded in ref. [41] that the average noise due to repeated impacts increases at a rate of 30log$v$ and that the equivalent noise from dipped rail joints is 0–10 dB higher than the rolling noise.

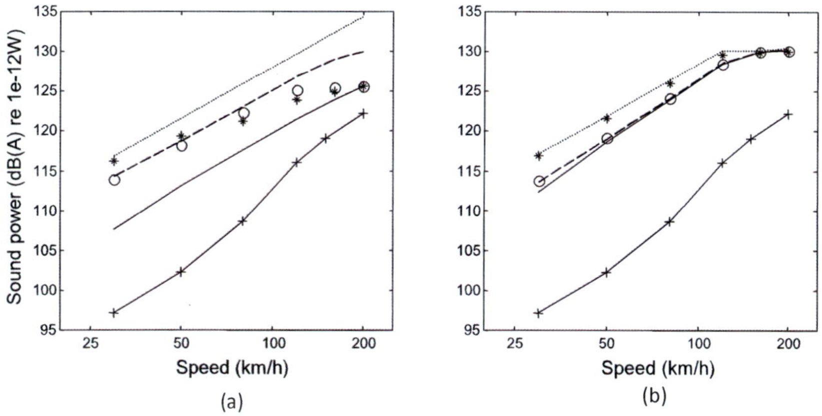

Fig. 17. A-weighted sound power level radiated by one wheel and the associated track vibration during 0.125 s due to a wheel passing over different rail joints with 7 mm gap width. (a) 5 mm dip. (b) 10 mm dip. · · · 2 mm step-up, --- 1 mm step-up, – no height difference, ∗ 2 mm step-down, ○ 1 mm step-down, + – + rolling noise due to roughness (cast iron tread brakes and low-roughness rail). From [41].

7.3 Crossing

Impact loads and noise generated at crossings with different dip angles and at different train speeds were investigated in refs. [30, 31]. The time-domain model included a moving impulse response function (Green's function) approach for solving the dynamic vehicle–track interaction [48, 69, 70, 76], while the three-dimensional contact model was based on Kalker's variational method. The continuously varying rail profile along the crossing was considered by interpolating the three-dimensional contact geometry between rail profiles measured at a sampling distance of 10 mm.

A linear finite element model of the track, consisting of a discretely supported rail described by Rayleigh-Timoshenko beam elements with spatially-varying beam properties, was developed for the through route of the crossing. At each position along the crossing, the crossing rail and the two wing rails were combined into one equivalent rail cross-section. For the tuning of the track model versus rail receptances measured in the field, the cross-sectional area and moment of inertia over the crossing were estimated to vary by up to between two and four times the corresponding values for a nominal 60 kg/m rail. The length and mass of the rigid sleepers were assumed to vary linearly, while the bedding modulus (stiffness per unit area) of the sleeper support was assumed to be constant. Two linear wheel models were compared: a rigid model and a flexible model based on modal superposition [30, 31]. The static wheel load was 100 kN. In ref. [30], it was concluded that the influence of the axle load on the generated impact noise levels is low when there is maintained wheel–rail contact over the crossing.

The calculated ratio of the maximum dynamic wheel–rail contact force to the static wheel load is illustrated in Fig. 18 [31]. It is observed that the impact load increases with increasing train speed and dip angle, and contact forces in the trailing move exceed those in the facing move. Impact loads are significantly lower when wheel structural

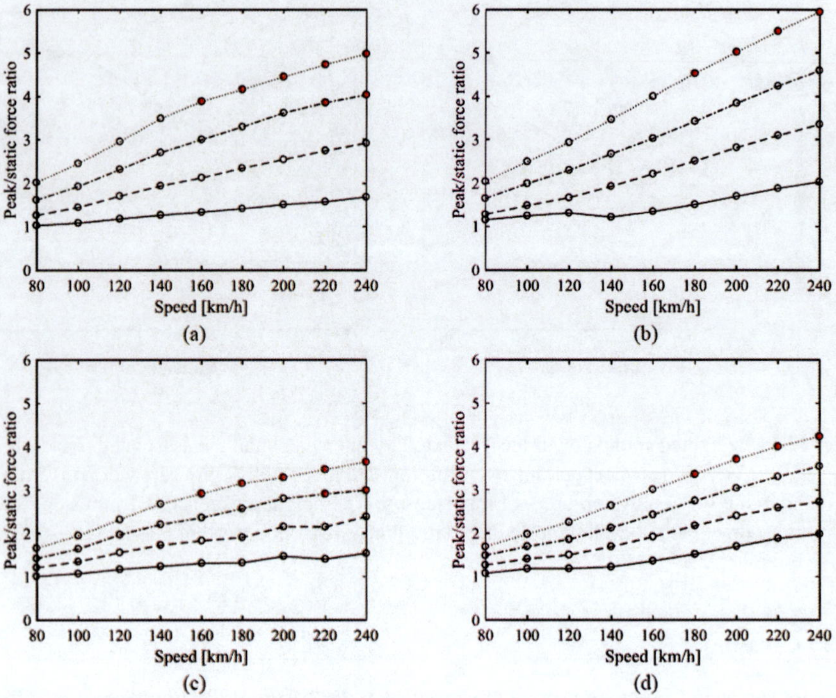

Fig. 18. Ratio of maximum impact load over static wheel load in a crossing with dip angles (from bottom to top) of 6 mrad (—o—), 12 mrad (- o -), 18 mrad (-·o·-) and 24 mrad (···o···). Cases with momentary loss of contact are indicated by filled circles. (a) Rigid wheel, facing move, (b) Rigid wheel, trailing move, (c) Flexible wheel, facing move, (d) Flexible wheel, trailing move. From [31].

flexibility is accounted for. For dip angles 18 and 24 mrad, Fig. 18 shows that loss of contact occurs at high vehicle speeds. In the facing move, it is observed that the speed dependence of the impact load changes when loss of contact occurs.

For each combination of dip angle and train speed, the calculated equivalent roughness spectrum was used as input to TWINS [73]. For simplicity the noise radiation model was based on a nominal rail. To evaluate the influence of the impact on the short-term perception of noise, equivalent roughness spectra and sound power were calculated based on one wheel and the time history of the impact load during 0.125 s. This corresponds to a travelling distance of about 3 m at the lowest train speed and 8 m at the highest. For comparison, rolling noise was also evaluated by using roughness measured on the test site described in ref. [31]. For the rolling noise calculation, the spectra were evaluated over a fixed length of 20 m and included the presence of two axles in each 20 m long vehicle (the vehicles were articulated).

The character of the impact and rolling noise spectra shown in Fig. 19 are different. All the impact noise spectra in Fig. 19(a) include a strong contribution below 500 Hz, while the rolling noise spectra in Fig. 19(b) are dominated by the higher frequency bands. With increasing speed, the rolling noise increases more at the higher frequencies,

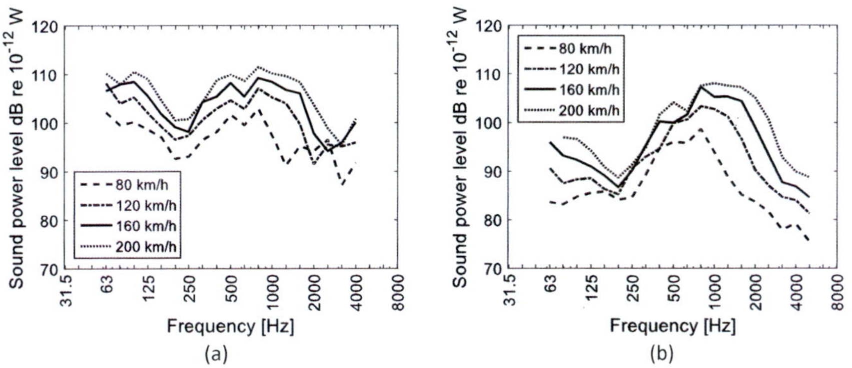

Fig. 19. Overall sound power level for impact noise and rolling noise at different train speeds. (a) Impact noise at 80–200 km/h for 18 mrad dip angle; (b) Rolling noise at 80–200 km/h.

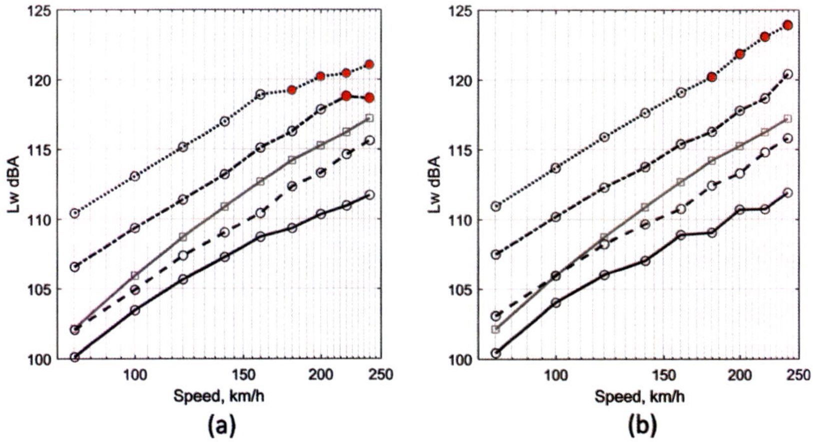

Fig. 20. A-weighted sound power level radiated by one wheel and the associated track vibration at different dip angles and train speeds. Dip angle 6 mrad (—o—), 12 mrad (- o -), 18 mrad (-·o·-), 24 mrad (···o···) and (—□—) rolling noise. Rolling noise was calculated for a time window corresponding to 20 m, while the impact noise was calculated for a time window corresponding to 0.125 s. Cases with momentary loss of contact are indicated by filled circles. (a) Facing move, (b) trailing move. The horizontal axis is given on a logarithmic scale.

whereas for the impact noise spectra there is an increase across the entire frequency range.

Fig. 20 shows the trend of the calculated A-weighted sound power level versus train speed and compares rolling noise with impact noise at different dip angles [31]. As expected, the impact noise increases with increasing train speed and dip angle. When there is no loss of wheel–rail contact, the noise levels in the trailing move are slightly higher than those in the facing move, but the difference is on average less than 1 dB. For combinations of dip angle and speed resulting in loss of contact, the difference is more substantial. The facing move shows the type of results attributed in Vér et al. [8]

to a step-down type of impact, where for speeds leading to loss of contact the peak sound pressure was found not to increase with increasing speed. For the trailing move, the results appear more similar to a step-up type of rail joint with the impact noise increasing with increasing speed independent of whether there is maintained contact or a momentary loss of contact. As a consequence, for the highest speed and dip angle, the noise levels in the trailing move were found to be up to 3 dB higher than for the facing move. For cases without loss of contact, the speed dependence of the impact noise is approximately 30log$v$, which is higher than the 20log$v$ found in previous research on impact at rail joints [8, 41]. It was concluded that the influence of the dip angle on the impact noise is considerable. For example, an increase in dip angle from 6 mrad to 24 mrad was found to correspond to an increase in radiated impact noise of about 11 dB(A) for the cases without loss of contact [31]. For comparison, by numerically evaluating the difference in trajectory slope on the wing rail and crossing nose close to the transition point, the measured dip angle in Fig. 12 is approximately 12 mrad.

8 Simulation of Ground-Borne Vibration Due to Impact Loading

Railway vehicle induced ground-borne vibration perceived as mechanical vibration of the human body has a frequency content in the interval 1–80 Hz, while higher frequency vibration between 16 and 250 Hz is radiated as noise inside buildings [1, 9]. The vibration is generated by the static and dynamic components of the vertical wheel–rail contact forces between the train and the track. The quasi-static excitation is determined by the static component of the moving wheel loads, axle distances and train speed, while the dynamic excitation is induced by wheel and rail unevenness (such as low-order wheel polygonalisation and deviations in longitudinal level), irregularities in track support stiffness and impacts due to discrete wheel/rail surface irregularities. For train speeds well below the wave speeds in the soil, the track (near-field) response is mainly determined by the contribution due to the quasi-static loading, whereas the free-field response is dominated by the dynamic excitation [9, 77–79]. In this section, the focus is on the influence of wheel–rail impact excitation on free-field ground-borne vibration.

Field measurements have shown that the time history of ground-borne vibration generated by a freight train has a more irregular character than for passenger trains [9]. This indicates that various forms of wheel out-of-roundness, such as wheel flats, are more prevalent on freight cars than on passenger vehicles and are significant for the induced vibration levels. Measures to reduce the number and severity of discrete wheel surface irregularities are therefore important to mitigate vibration levels [80]. Amplified vibration levels have also been measured at railway crossings [81]. However, since there can be a significant difference in ground properties from one site to another, it is difficult to make general conclusions about the contribution of impact excitation to vibration levels in a similar way as for air-borne impact noise.

Most computationally efficient models for the prediction of ground-borne vibration in layered soils are linear, assume continuous wheel–rail contact and are carried out in the frequency-wavenumber domain, see e.g. refs. [77–79]. On the other hand, time-domain vehicle–track interaction models do not include a half-space model of the soil and cannot be used for the prediction of ground vibration. Fully three-dimensional time-domain models including the ground, see e.g. ref. [82], are very computationally expensive. This suggests a hybrid approach for the calculation of vibration due to impact excitation in a similar way as for the prediction of impact noise. As discussed in Sect. 6.4, it is essential that the wheel and rail mobilities at the wheel–rail contacts used to determine the equivalent roughness should be directly equivalent to the dynamic models used in the time domain model. While this is not a problem for the wheel receptance if the same vehicle model is used in both domains, it may be a challenge for the rail receptance due to the significant differences between the two track models in the descriptions of ballast and ground. In particular, the low-frequency rail receptance below 50 Hz is difficult to model correctly in the time-domain model because the receptance at low frequencies is dominated by the (layered) soil conditions. In ref. [83], the input data to a five-parameter discretised model of the ground in the time domain were determined by using a genetic algorithm to minimise the difference in rail receptances calculated for the two track models. The unknown five parameters represented the input data for a repetitive set of soil masses connected by springs and viscous dampers in translation and shear. In this case, the physical meaning of the tuned input data for the time-domain model was not important as long as the rail receptance at the wheel–rail contact was in good agreement with the frequency-wavenumber domain model.

A hybrid model for the prediction of ground-borne vibration due to discrete wheel/rail surface irregularities was presented in ref. [83]. The 2-dof vehicle model in the time domain represented one wheelset, see Fig. 2, while the layered ground model in the frequency-wavenumber domain was based on measured ground properties. Since the focus was on ground-borne vibration, the equivalent roughness approach described in Sect. 6.4 necessary to predict wheel vibration was not applied although this would have been a more consistent methodology. Instead, the two models were integrated based on the Fourier transform of the wheel–rail contact force calculated in the time domain.

For new flats and the given set of measured track and soil conditions, the influence of wheel flat length and train speed on the maximum wheel–rail contact force and free-field velocity at 8 m is illustrated in Fig. 21. A local maximum in contact force is observed at a train speed of around 30 km/h, cf. Fig. 6. As expected, the free-field velocity increases with increasing wheel flat size, but the influence of train speed is not as evident. Higher train speeds resulted in loss of wheel–rail contact and severe impact loads, but the frequency content of these loads was shifted to higher frequencies which were less significant for the ground vibration because of lower magnitudes in the free-field transfer mobility. Furthermore, it was found that the magnitudes of the ground vibration generated by the studied wheel flats were considerably higher than those induced by a common level of irregularity in vertical track geometry (longitudinal level). The influence of dipped rail joints on the ground-borne vibration was investigated in ref. [84].

Kouroussis et al. [85] also used a two-stage approach to study the ground vibration induced by rail joints and wheel flats. Here the second stage model was a full 3D FE model of the ground.

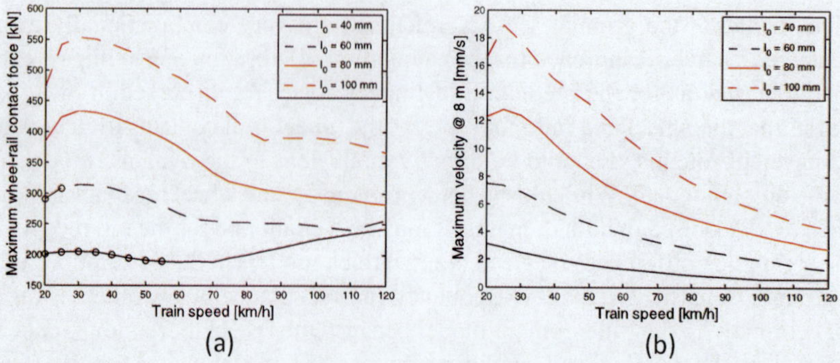

Fig. 21. Influence of train speed and length l_0 ($l_0 = \sqrt{8R_\mathrm{w}d}$) of a new wheel flat with depth d on (a) maximum wheel–rail impact load, and (b) peak particle velocity at 8 m from the track. Vehicle model including one wheelset with axle load 22 tonnes. Circles indicate continuous wheel–rail contact. From [83].

9 Mitigation Measures

As discussed above, the excitation mechanism of noise (and ground-borne vibration) due to wheel–rail impact is a relative vertical wheel–rail displacement, similar to that for rolling noise. Thus, impact noise is also generated by a relative vertical displacement excitation in the wheel–rail contact leading to vibration of wheel and rail. This means the same types of mitigation measures as for rolling noise can be applied [1]. This includes using wheel shape optimisation and damping treatments to reduce noise from the wheels, and increasing the track decay rate by implementing rail pads with a higher stiffness or tuned rail damping treatments to reduce noise from the rails. However, note that stiffer rail pads lead to higher wheel–rail impact loads and higher levels of sleeper vibration.

Still, the main priority should be to eliminate or at least minimise the discrete irregularities in the wheel–rail contact geometry. This has the benefit of reducing noise and vibration as well as vehicle and track damage. Wheel impact load detectors are used to identify wheels generating magnified wheel–rail contact forces to schedule preventive maintenance action in the case of rolling contact fatigue damage before the wheel out-of-roundness grows to unacceptable levels, and to take out of service wheels with flats for corrective maintenance. Improved brake system design and wheel material quality, and wheel slide protection, can be used to mitigate the generation of wheel flats and other discrete wheel tread irregularities. For the track, continuously welded rails, condition monitoring of track geometry at rail joints and crossings, and tight tolerances of welds to avoid dipped joints are recommended.

For crossings, impact loads and noise can be mitigated by designing and maintaining a crossing dip angle that is as small as possible. The optimisation of crossing geometry has been investigated in refs. [18, 86–88]. However, since the crossing will be subjected to wheels with different stages of profile wear, a certain dip angle is required for the crossing geometry to be compatible with a range of wheel profiles. In particular, hollow worn wheel profiles typically demonstrate a poor compatibility with crossing geometries and induce significant wheel–rail impact loads and rail damage [19, 30]. Thus, better maintenance of wheel profiles is recommended. Impact loads at crossings can also be reduced by using softer rail pads [28, 29].

The unsprung mass is the key vehicle-related parameter influencing the magnitude of impact loads and ground-borne vibration. Means for reducing the unsprung mass are discussed in refs. [26, 80] including alternative designs for the wheelset and the suspension of the mechanical drive system.

10 Concluding Remarks

The excitation mechanisms of railway noise and ground-borne vibration induced by wheel–rail impact loads have been reviewed. Impact loads are generated by discrete wheel/rail surface irregularities, such as wheel flats, and by local deviations in the nominal wheel–rail contact geometry, for example at rail joints and crossings. The resulting vertical loading is a transient and often periodically repeated event exciting vibration in a wide frequency range with most of the energy concentrated below about 1 kHz.

Models for the simulation of wheel–rail impact loads, noise and vibration have been surveyed. In situations with a momentary loss of wheel–rail contact, it is clear that the prediction of transient wheel–rail interaction must be carried out in the time domain using a non-linear contact model. In fact, it has been shown [49] that non-linear effects in the wheel–rail contact become noticeable already when the rms acoustic roughness exceeds 0.35 times the static contact deflection. In several applications, especially for S&C, it is necessary to solve the three-dimensional wheel–rail contact problem for non-Hertzian geometries and to account for the variation in contact stiffness across the irregularity [37]. This can be accomplished by applying a contact model based on boundary element methodology, such as Kalker's variational method [54]. However, if the contact geometry cannot be simplified to an elastic half-space model or if the influence of non-linear wheel/rail material properties is important, a three-dimensional finite element model with a high mesh density is essential. Moreover, a track model considering non-linear properties of rail pads and ballast/subgrade is required to account for the large dynamic range of the wheel–rail contact force and forces distributed in the track structure.

A hybrid simulation approach has been introduced in which a time-domain model is applied to calculate the transient wheel–rail contact force. The time history of the force is transformed to the frequency domain and converted into the form of an equivalent roughness spectrum used as input in a linear frequency-domain model for the prediction of noise [31, 34, 41]. A next step to improve the understanding of wheel–rail impact noise is to develop a time-domain model for the prediction of transient noise, where the non-linear vehicle, track and wheel–rail contact models are extended to be accurate for frequencies up to 5 kHz.

There is a need to develop criteria to assess the influence of transient impact noise on human perception. For example, in a field measurement of clearly audible impact noise at a railway crossing [31], it was concluded that the contribution of the impact noise to the sound pressure spectrum was small. On the other hand, in a laboratory experiment to examine subjective human reaction to train noise containing components due to wheel–rail impact, it was found that the contribution of impact noise due to wheel flats and rail joints increased the perceived loudness by 3–5 dB compared to noise without impact loading at the same equivalent sound energy level [33].

Since impact noise is also generated by a relative vertical displacement excitation in the wheel–rail contact, the same types of mitigation measures as for rolling noise can be applied. However, to minimise noise and ground-borne vibration, the main priority should be to eliminate or at least minimise the discrete irregularities in the wheel–rail contact geometry.

Acknowledgements. This work has been performed within the Centre of Excellence CHARMEC, see www.charmec.chalmers.se. Parts of the study have been funded within the European Union's Horizon 2020 research and innovation programme in the project In2Track2 under grant agreement No 826255. The contributions by co-workers Björn Pålsson, Giacomo Squicciarini and Tianxing Wu in studies leading up to this review paper are gratefully acknowledged.

References

1. Thompson, D.: Railway Noise and Vibration: Mechanisms, Modelling and Means of Control, 1st edn. Elsevier, Amsterdam (2009)
2. Thompson, D.J., Jones, C.J.C.: A review of the modelling of wheel/rail noise generation. J. Sound Vib. **231**(3), 519–536 (2000)
3. Thompson, D.J., Gautier, P.-E.: Review of research into wheel/rail rolling noise reduction. Proc. Inst. Mech. Eng. J. Rail Rapid Transit **220**, 385–408 (2006)
4. Talotte, C.: Aerodynamic noise: a critical survey. J. Sound Vib. **231**(3), 549–562 (2000)
5. Thompson, D.J., Latorre Iglesias, E., Liu, X., Zhu, J., Hu, Z.: Recent developments in the prediction and control of aerodynamic noise from high-speed trains. Int. J. Rail Transp. **3**(3), 119–150 (2015)
6. Remington, P.J.: Wheel/rail squeal and impact noise: What do we know? What don't we know? Where do we go from here? J. Sound Vib. **116**(2), 339–353 (1985)
7. Thompson, D.J., Squicciarini, G., Ding, B., Baeza, L.: A state-of-the-art review of curve squeal noise: phenomena, mechanisms, modelling and mitigation. In: Anderson, D., et al. (eds.), Proceedings of the 12th International Workshop on Railway Noise, September 2016, Terrigal, Australia. Notes on Numerical Fluid Mechanics and Multidisciplinary Design, vol. 139, pp. 3–41. Springer, Heidelberg (2018)
8. Vér, I.L., Ventres, C.S., Myles, M.M.: Wheel/rail noise – part III: impact noise generation by wheel and rail discontinuities. J. Sound Vib. **46**(3), 395–417 (1976)
9. Lombaert, G., Degrande, G., François, S., Thompson, D.J.: Ground-borne vibration due to railway traffic: a review of excitation mechanisms, prediction methods and mitigation measures. In: Nielsen, J.C.O., et al. (eds.), Proceedings of the 11th International Workshop on Railway Noise, September 2013, Uddevalla, Sweden. Notes on Numerical Fluid Mechanics and Multidisciplinary Design, vol. 126, pp. 253–287, Springer, Heidelberg (2015)
10. Kouroussis, G., Connolly, D.P., Verlinden, O.: Railway-induced ground vibrations – a review of vehicle effects. Int. J. Rail Transp. **2**(2), 69–110 (2014)
11. Jergéus, J.: Railway wheel flats – Martensite formation, residual stresses and crack propagation. PhD thesis, Department of Solid Mechanics, Chalmers University of Technology, Gothenburg, Sweden (1998)
12. Nielsen, J.C.O., Johansson, A.: Out-of-round railway wheels – a literature survey. Proc. Inst. Mech. Eng. J. Rail Rapid Transit **214**, 79–91 (2000)
13. Nielsen, J.C.O.: Out-of-round railway wheels. In: Lewis, R., Olofsson, U. (eds.) Wheel-Rail Interface Handbook. Woodhead Publishing Ltd., Oxford (2009)

14. Snyder, T., Stone, D.H., Kristan, J.: Wheel flat and out-of-round formation and growth. In: Proceedings 2003 IEEE/ASME Joint Rail Conference, Chicago, IL, USA, pp. 143–148 (2003)
15. Johansson, A., Nielsen, J.C.O.: Out-of-round railway wheels – wheel–rail contact forces and track response derived from field tests and numerical simulations. Proc. Inst. Mech. Eng. J. Rail Rapid Transit **217**, 135–146 (2003)
16. Kabo, E., Nielsen, J.C.O., Ekberg, A.: Prediction of dynamic train–track interaction and subsequent material deterioration in the presence of insulated rail joints. Veh. Syst. Dyn. **44**(Suppl), 718–729 (2006)
17. Ford, R.A.J., Thompson, D.J.: Simplified contact filters in wheel/rail noise prediction. J. Sound Vib. **293**, 807–818 (2006)
18. Pålsson, B.A.: Optimisation of railway crossing geometry considering a representative set of wheel profiles. Veh. Syst. Dyn. **53**, 274–301 (2015)
19. Li, X., Torstensson, P.T., Nielsen, J.C.O.: Simulation of vertical dynamic vehicle–track interaction in a railway crossing using Green's functions. J. Sound Vib. **410**, 318–329 (2017)
20. Pålsson, B.: Optimisation of railway switches and crossings. PhD thesis, Department of Applied Mechanics, Chalmers University of Technology, Gothenburg, Sweden (2014)
21. Li, X., Nielsen, J.C.O., Torstensson, P.T.: Simulation of wheel–rail impact load and sleeper–ballast contact pressure in railway crossings using a Green's function approach. J. Sound Vib. **463**, 114949 (2019)
22. Newton, S.G., Clark, R.A.: An investigation into the dynamic effects on track of wheelflats on railway vehicles. J. Mech. Eng. Sci. **21**(4), 287–297 (1979)
23. Fermér, M., Nielsen, J.C.O.: Wheel/rail contact forces for flexible versus solid wheels due to tread irregularities. Veh. Syst. Dyn. **23**(S1), 142–157 (1994)
24. Fermér, M., Nielsen, J.C.O.: Vertical interaction between train and track with soft and stiff railpads – full-scale experiments and theory. Proc. Inst. Mech. Eng. J. Rail Rapid Transit **209**, 39–47 (1995)
25. Dukkipati, R.V., Dong, R.: Impact loads due to wheel flats and shells. Veh. Syst. Dyn. **31**, 1–22 (1999)
26. Jenkins, H.H., Stephenson, J.E., Clayton, G.A., Morland, G.W., Lyon, D.: The effect of track and vehicle parameters on wheel/rail vertical dynamic forces. Rail. Eng. J. **3**(1), 2–16 (1974)
27. Gullers, P., Andersson, L., Lundén, R.: High-frequency vertical wheel–rail contact forces – field measurements and influence of track irregularities. Wear **265**, 1472–1478 (2008)
28. Kassa, E., Nielsen, J.C.O.: Dynamic interaction between train and railway turnout: full-scale field test and validation of simulation models. Veh. Syst. Dyn. **46**(Suppl), 521–534 (2008)
29. Pålsson, B.A., Nielsen, J.C.O.: Dynamic vehicle–track interaction in switches and crossings and the influence of rail pad stiffness – field measurements and validation of a simulation model. Veh. Syst. Dyn. **53**(6), 734–755 (2015)
30. Torstensson, P.T., Squicciarini, G., Krüger, M., Nielsen, J.C.O., Thompson, D.J.: Hybrid model for prediction of impact noise generated at railway crossings. In: Anderson, D., et al. (eds.) Proceedings of the 12th International Workshop on Railway Noise, September 2016, Terrigal, Australia. Notes on Numerical Fluid Mechanics and Multidisciplinary Design, vol. 139, pp. 759–769. Springer, Heidelberg (2018)
31. Torstensson, P.T., Squicciarini, G., Krüger, M., Pålsson, B.A., Nielsen, J.C.O., Thompson, D.J.: Wheel–rail impact loads and noise generated at railway crossings – influence of vehicle speed and crossing dip angle. J. Sound Vib. **456**, 119–136 (2019). https://doi.org/10.1016/j.jsv.2019.04.034
32. Cordier, J.F., Letourneaux, F., Fortmann, J.C., Müller, R.: Acoustic impact of defects on railway wheel running surfaces. In: World Congress on Railway Research, Montreal, Canada, June (2006)
33. Kaku, J., Yamashita, M.: Impact noise from railroads. J. Sound Vib. **120**(2), 333–337 (1988)

34. Wu, T.X., Thompson, D.J.: A hybrid model for the noise generation due to railway wheel flats. J. Sound Vib. **251**(1), 115–139 (2002)
35. EN 15610. Railway applications – Noise emission – Rail and wheel roughness measurement related to rolling noise generation (2019)
36. Thompson, D.J.: On the relationship between wheel and rail surface roughness and rolling noise. J. Sound Vib. **193**(1), 149–160 (1996)
37. Pieringer, A., Kropp, W., Nielsen, J.C.O.: The influence of contact modelling on simulated wheel/rail interaction due to wheel flats. Wear **314**, 273–281 (2014)
38. Baeza, L., Roda, A., Carballeira, J., Giner, E.: Railway train–track dynamics for wheelflats with improved contact models. Non-linear Dyn. **45**(3), 385–397 (2006)
39. Remington, P.J., Webb, J.: Estimation of wheel/rail interaction forces in the contact area due to roughness. J. Sound Vib. **193**, 83–102 (1996)
40. Steenbergen, M.J.M.M.: The role of the contact geometry in wheel–rail impact due to wheel flats. Veh. Syst. Dyn. **45**(12), 1097–1116 (2007)
41. Wu, T.X., Thompson, D.J.: On the impact noise generation due to a wheel passing over rail joints. J. Sound Vib. **267**(3), 485–496 (2003)
42. Pålsson, B.A.: A linear wheel–crossing interaction model. Proc. Inst. Mech. Eng. J. Rail Rapid Transit **232**(10), 2431–2443 (2018)
43. Hertz, H.: Über die Berührung fester elastischer Körper. J. für reine und angewandte Mathematik **92**, 156–171 (1882)
44. Nielsen, J.C.O., Igeland, A.: Vertical dynamic interaction between train and track – influence of wheel and track imperfections. J. Sound Vib. **187**, 825–839 (1995)
45. Dong, R.G., Sankar, S.R., Dukkipati, V.: A finite element model of railway track and its application to the wheel flat problem. Proc. Inst. Mech. Eng. J. Rail Rapid Transit **208**, 61–72 (1994)
46. Lundberg, G., Sjövall, H.: Stress and deformation in elastic contacts. Publikation no 4, The Institution of Theory of Elasticity and Strength of Materials, Chalmers University of Technology, Gothenburg, Sweden (1958)
47. Johnson, K.L.: Contact Mechanics. Cambridge University Press, Cambridge (1985)
48. Pieringer, A.: Time-domain modelling of high-frequency wheel/rail interaction. PhD thesis, Department of Civil and Environmental Engineering, Chalmers University of Technology, Gothenburg, Sweden (2011)
49. Wu, T.X., Thompson, D.J.: Theoretical investigation of wheel/rail non-linear interaction due to roughness excitation. Veh. Syst. Dyn. **34**, 261–282 (2000)
50. Zhao, X., Li, Z., Liu, J.: Wheel–rail impact and the dynamic forces at discrete supports of rails in the presence of singular rail surface defects. Proc. Inst. Mech. Eng. J. Rail Rapid Transit **226**, 124–139 (2012)
51. Yang, Z., Boogaard, A., Chen, R., Dollevoet, R., Li, Z.: Numerical and experimental study of wheel–rail impact vibration and noise generated at an insulated rail joint. Int. J. Impact Eng **113**, 29–39 (2018)
52. Cerruti, V.: Mem. fis. mat. Accademia dei Lincei, Rome (1882)
53. Boussinesq, J.: Application des Potentiels à l'Étude de l'Équilibre et du Mouvement des Solides Élastiques. Gauthier-Villars, Paris (1885)
54. Kalker, J.J.: Three-Dimensional Elastic Bodies in Rolling Contact. Kluwer Academic Publishers, Dordrecht (1990)
55. Pascal, J.P., Sauvage, G.: The available methods to calculate the wheel/rail forces in non-Hertzian contact patches and rail damaging. Veh. Syst. Dyn. **22**, 263–275 (1993)
56. Ayasse, J.-B., Chollet, H.: Determination of the wheel rail contact patch in semi-Hertzian conditions. Veh. Syst. Dyn. **43**(3), 161–172 (2005)
57. Piotrowski, J., Kik, W.: A simplified model of wheel/rail contact mechanics for non-Hertzian problems and its application in rail vehicle dynamics. Veh. Syst. Dyn. **46**(1–2), 27–48 (2008)

58. Sichani, M.S., Enblom, R., Berg, M.: A novel method to model wheel–rail normal contact in vehicle dynamics simulation. Veh. Syst. Dyn. **52**(12), 1752–1764 (2014)
59. Knothe, K.L., Grassie, S.L.: Modelling of railway track and vehicle/track interaction at high frequencies. Veh. Syst. Dyn. **22**, 209–262 (1993)
60. Popp, K., Kruse, H., Kaiser, I.: Vehicle–track dynamics in the mid-frequency range. Veh. Syst. Dyn. **31**(5–6), 423–464 (1999)
61. Nielsen, J.C.O., Lundén, R., Johansson, A., Vernersson, T.: Train–track interaction and mechanisms of irregular wear on wheel and rail surfaces. Veh. Syst. Dyn. **40**(1–3), 3–54 (2003)
62. Thompson, D.J., van Vliet, W.J., Verheij, J.W.: Development of the indirect method for measuring the high frequency dynamic stiffness of resilient elements. J. Sound Vib. **213**(1), 169–188 (1998)
63. Thompson, D.J., Verheij, J.W.: The dynamic behaviour of rail fasteners at high frequencies. Appl. Acoust. **52**(1), 1–17 (1997)
64. Iwnicki, S. (ed.): Handbook of Railway Vehicle Dynamics. CRC Press (2006)
65. Nielsen, J.C.O.: High-frequency vertical wheel–rail contact forces – validation of a prediction model by field testing. Wear **265**, 1465–1471 (2008)
66. Grassie, S.L., Gregory, R.W., Harrison, D., Johnson, K.L.: The dynamic response of railway track to high-frequency vertical excitation. J. Mech. Eng. Sci. **24**, 77–90 (1982)
67. Tunna, J.M.: Wheel/rail forces due to wheel irregularities. In: 9th International Wheelset Congress, Montreal, Canada, September (1988)
68. Andersson, C., Dahlberg, T.: Wheel/rail impacts at a railway turnout crossing. Proc. Inst. Mech. Eng. J. Rail Rapid Transit **212**, 123–134 (1998)
69. Heckl, M.: Proposal for a railway simulation program. In: A Workshop on Rolling Noise Generation. Institut für Technische Akustik, Technische Universität Berlin, pp. 128–148 (1989)
70. Nordborg, A.: Wheel/rail noise generation due to nonlinear effects and parametric excitation. J. Acoust. Soc. Am **111**(4), 1772–1781 (2002)
71. Wu, T.X., Thompson, D.J.: The effects of track non-linearity on wheel/rail impact. Proc. Inst. Mech. Eng. J. Rail Rapid Transit **218**, 1–5 (2004)
72. Nielsen, J.C.O., Oscarsson, J.: Simulation of dynamic train–track interaction with state-dependent track properties. J. Sound Vib. **275**, 515–532 (2004)
73. Thompson, D.J., Hemsworth, B., Vincent, N.: Experimental validation of the TWINS prediction program for rolling noise. Part I: description of the model and method. J. Sound Vib. **193**, 123–135 (1996)
74. Yang, J., Thompson, D.J.: Time-domain prediction of impact noise from wheel flats based on measured profiles. J. Sound Vib. **333**, 3981–3995 (2014)
75. Sueki, T., Kitagawa, T., Kawaguchi, T.: Evaluation of acoustic and vibratory characteristics of impact noise due to rail joints. QR RTRI **58**, 119–125 (2017)
76. Andersson, R., Torstensson, P.T., Kabo, E., Larsson, F.: An efficient approach to the analysis of rail surface irregularities accounting for dynamic train–track interaction and inelastic deformations. Veh. Syst. Dyn. **53**(11), 1667–1685 (2015)
77. Sheng, X., Jones, C.J.C., Thompson, D.J.: A comparison of a theoretical model for quasi-statically and dynamically induced environmental vibration from trains with measurements. J. Sound Vib. **267**(3), 621–635 (2005)
78. Auersch, L.: Ground vibration due to railway traffic – the calculation of the effects of moving static loads and their experimental verification. J. Sound Vib. **293**, 599–610 (2006)
79. Lombaert, G., Degrande, G., Kogut, J., François, S.: The experimental validation of a numerical model for the prediction of railway induced vibrations. J. Sound Vib. **297**(3–5), 512–535 (2006)

80. Nielsen, J.C.O., et al.: Reducing train-induced ground-borne vibration by vehicle design and maintenance. Int. J. Rail Transp. **3**(1), 17–39 (2015)
81. Müller, R., Nielsen, J.C.O., Nélain, B., Zemp, A.: Ground-borne vibration mitigation measures for turnouts: State-of-the-art and field tests. In: Nielsen, J.C.O., et al. (eds.) Proceedings of the 11th International Workshop on Railway Noise, September 2013, Uddevalla, Sweden. Notes on Numerical Fluid Mechanics and Multidisciplinary Design, vol. 126, pp. 547–554, Springer, Heidelberg (2015)
82. Galvin, P., Romero, A., Dominguez, J.: Fully three-dimensional analysis of high-speed train-track-soil-structure dynamic interaction. J. Sound Vib. **329**(24), 5147–5163 (2010)
83. Nielsen, J.C.O., Lombaert, G., François, S.: A hybrid model for prediction of ground-borne vibration due to discrete wheel/rail irregularities. J. Sound Vib. **345**, 103–120 (2015)
84. Nielsen, J.C.O.: Classification of track conditions with respect to vibration emission – Part 1: Classification of track irregularities and numerical simulations. RIVAS(SCP0-GA-2010–265754), Deliverable 2.1, p. 74 (2012)
85. Kouroussis, G., Connolly, D.P., Alexandrou, G., Vogiatzis, K.: Railway ground vibrations induced by wheel and rail singular defects. Veh. Syst. Dyn. **53**(10), 1500–1519 (2015)
86. Nicklisch, D., Kassa, E., Nielsen, J.C.O., Ekh, M., Iwnicki, S.: Geometry and stiffness optimization for switches and crossings, and simulation of material degradation. Proc. Inst. Mech. Eng. J. Rail Rapid Transit **224**, 279–292 (2010)
87. Wan, C., Markine, V.L., Shevtsov, I.Y.: Improvement of vehicle–turnout interaction by optimising the shape of crossing nose. Veh. Syst. Dyn. **52**, 1517–1540 (2014)
88. Wan, C., Markine, V., Dollevoet, R.: Robust optimisation of railway crossing geometry. Veh. Syst. Dyn. **54**, 617–634 (2016)

Industrial Methodologies for the Prediction of Interior Noise Inside Railway Vehicles: Airborne and Structure Borne Transmission

Pascal Bouvet(✉) and Martin Rissmann

Vibratec, Chemin Du Petit Bois, BP 36, 69131 Ecully Cedex, France
pascal.bouvet@vibratec.fr

Abstract. This paper presents computational methods that are used by rolling stock manufacturers to predict noise inside vehicles. For airborne transmission, which dominates in the medium-high frequency range, a four-step procedure is applied: source description by their emitted sound power level, propagation of noise to the train's exterior surface, panel transmission loss and acoustic response of the interior cavity. Reasonable agreement between computations and measurements is usually obtained, and the method makes it possible to rank the different source contributions and airborne transmission paths. Structure borne noise dominates in low frequencies. Finite Element models are used to improve car body design (dynamic stiffness at input points and carbody vibroacoustic transfers), but they do not cover the whole problem since the modelling of excitation from the bogie is not included. Recent research allowing the computation of blocked forces at car body input points and starting with wheel/rail interaction is briefly presented. Concerning source modelling, a focus is made on traction noise, including electromagnetic excitations in electric motors and mechanical excitations due to the meshing process inside gearboxes. Efficient computational methods and validation examples are presented. The coupling of these methods with optimization methods has great potential for improvement of motor noise and vibration design.

Keywords: Acoustic comfort · Airborne and structure borne transmission · Acoustic sources · Gearbox noise · Traction motor noise

1 Introduction

The comfort inside trains is one of the most important issues for railway operators, and studies have shown the importance of the acoustic aspect for passengers. The interior noise levels of new rolling stock are specified by railway operators from the tender stage, for various train operating conditions (standstill, acceleration and braking, running at constant speeds).

For interior noise prediction during the different vehicle development phases, rolling stock manufacturers need efficient and accurate modelling tools, taking into account the different sources, transmission paths, environmental (free field or tunnel circulation) and

G. Degrande et al. (Eds.): *Noise and Vibration Mitigation for Rail Transportation Systems*, NNFM 150, pp. 41–54, 2021.
https://doi.org/10.1007/978-3-030-70289-2_2

operating conditions. A description of computational methods for airborne and structure borne noise transmission in use in the railway industry is presented in Sect. 2 below.

Section 3 is dedicated to source modelling, with a focus on recent developments to model traction motor noise due to electromagnetic excitation and gearbox noise due to the meshing process.

2 Modelling Methods for Interior Noise Prediction

2.1 Relative Contributions from Airborne and Structure Borne Transmission Paths

Usually, at locations close to bogies, the contribution of airborne noise transmission is dominant at high frequencies whereas structure borne noise transmission dominates at lower frequencies (typically below 200–300 Hz).

Figure 1 gives an example of structure borne contribution to the total interior noise. In this case, the structure borne contribution was assessed using Experimental Transfer Path Analysis (TPA) and in-situ blocked force measurement at the bogie attachment point to the vehicle body [1]. Several examples of TPA analysis on different types of railway vehicles have shown a predominance of structure borne transmission in the low frequency range [2, 3].

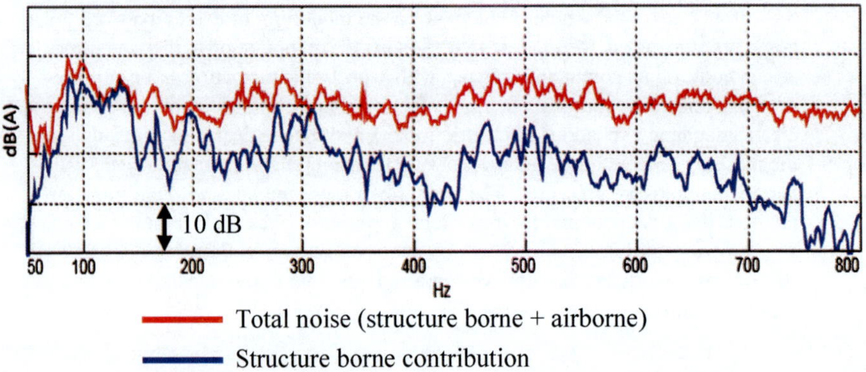

Fig. 1. Sound pressure levels inside an urban railway vehicle

2.2 Airborne Noise Model

For more than 15 years, rolling stock manufacturers have implemented and regularly improved computational methods to estimate the airborne contribution inside vehicles. A global overview and critical analysis of methods that can be used for interior noise prediction is given in ref. [4, 5].

For the prediction of interior noise, different sources (rolling noise, traction noise, equipment noise and aeroacoustic excitation), transmission paths and operating conditions must be taken into account. Airborne noise models are built from the tendering phase to roughly assess the acoustic solutions to be used in order to meet interior noise requirements. When possible, this first model is based on input data from previous or similar vehicles. In a second step, the model is used to establish acoustic target settings for each component (panel transmission loss (TL), source sound power, etc.). It is then regularly updated throughout the project when new acoustic data are made available from suppliers or from acoustic tests on components. The methodology applied to calculate the air-borne noise component is composed of four steps as shown in Fig. 2:

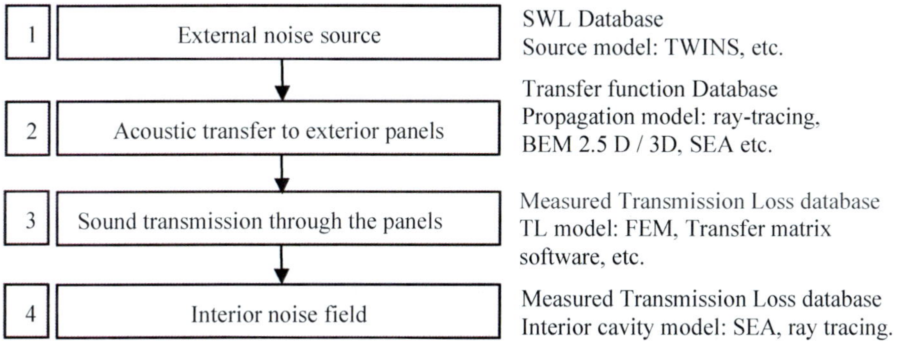

Fig. 2. Railway vehicle. airborne noise computation method

Step 1 - Noise source description of: Globally, external sources cover

- Mechanical sources (wheel/rail noise, traction noise, auxiliary noise, etc.);
- Interior sources (HVAC, ventilation, etc.)
- For high speeds: aerodynamic sources (such as turbulent flow generated around the bogie and inter-coach spacing), Turbulent Boundary Layer (TBL) around the coach.

Depending on source types, the source sound power levels can either be computed (TWINS calculations for wheel/rail noise, Computational Fluid Dynamics (CFD) computations for aerodynamic sources or TBL, Finite Element Method (FEM) computations for gearbox and traction motor noise) or extrapolated from measurement databases.

An extension of the validation of TWINS for cases like resilient wheels or track radiation in the low frequency range is still required [5]. For fans, acoustic source data can be affected by their installation on the vehicle, because of flow disturbances at the inlet that may be different between laboratory tests and vehicle conditions.

Step 2, Noise transmission to coach panels: The sound transmission from sources to the coach's external envelope is described by means of (Lp_i - Lw) transfer functions, Lp_i being the parietal sound pressure at any point i of the coach envelope and Lw being the source sound power. These transfer functions must be determined for various environmental conditions (free field and tunnel).

A global overview of methods that can be used for in ref. [6]. Several computational tools such as 3D Boundary Element Method (BEM) or ray tracing are sometimes used to compute these transfer functions [7]: this remains quite challenging with regard to the necessity to cover the whole 100 Hz – 5 000 Hz frequency range, the whole geometry of the vehicle and very different acoustic environments (reverberation in tunnels, free field diffraction effects, etc.), while keeping reasonable accuracy and computational time. Because of these difficulties, experimental databases of transfer functions measured with artificial sources at standstill or in rolling conditions on similar coach geometries are still used, with the main drawback that in the case of a new rolling stock geometry they can hardly be transposed from one vehicle to another.

Consequently, this is a field where research efforts are still needed: more recently a 2.5D boundary element method was used to predict the exterior sound pressure spectrum on the train sides when the train was in running operation. Reasonable agreement was found with measurements [8].

Steps 3 and 4, Transmission through panels and interior noise field: panel Transmission Loss (TL) is usually taken from a database, either from laboratory measurements or from in-situ measurements on the vehicle. Simple computational tools such as transfer matrix methods are often used to estimate tendencies, but with limited accuracy regarding the complexity of the car body's structure and trim panels. FE models are still too heavy in terms of time to set up models and compute acoustic TL. TL computation of panels under TBL excitation is not covered at this moment.

For acoustic propagation inside cavities, Statistical Energy Analysis (SEA) (Fig. 3 and ref [4, 9]) or ray-tracing methods [10] are classically used with a good level of accuracy and short computational time. Here, the main parameter to control is sound absorption inside the vehicle, which poses no specific difficulty.

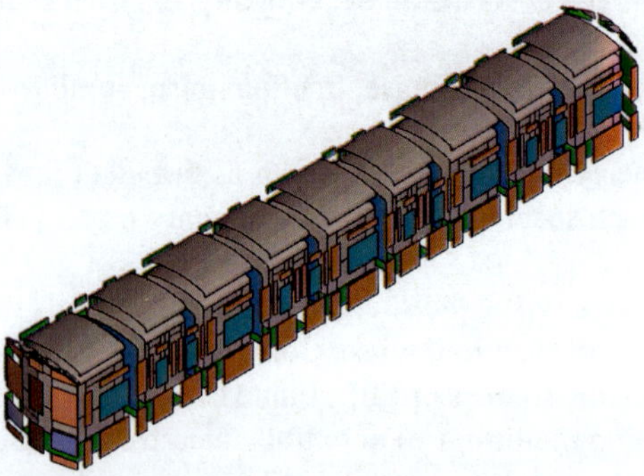

Fig. 3. Example of an SEA model for interior airborne noise prediction

The global accuracy of such models is quite good, with the capability to predict global interior noise levels with a maximum difference of 1 to 2 dB depending on the input parameters' quality. Post-processing of model results leads to a ranking of source and panel contributions to interior noise (Fig. 4). Even more important, the model is robust and fast for parametric studies: the effect of low noise solutions simulated with the model is usually in line with that obtained after implementation on the vehicle.

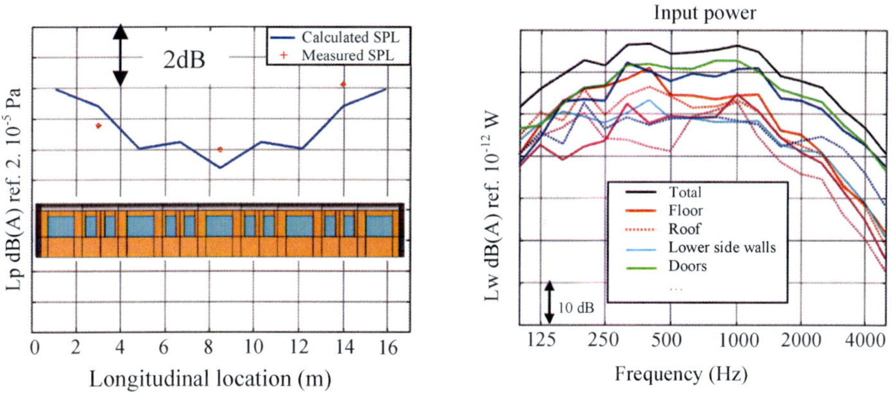

Fig. 4. Interior airborne noise prediction for Metro rolling stock. ***Left:*** comparison between measurements and computations ***Right:*** Panel contribution analysis

2.3 Structure Borne Noise Model

Structure borne noise is transmitted from the bogie to the car body through all the mechanical links, mainly dampers and traction bars. Secondary suspensions are usually a transmission path of lesser importance.

Structural Finite Element Models (FEM) are used by rolling stock manufacturers to control and optimize car body design. The quantities that are computed and compared with targets are the dynamic stiffness at input points (the real part of the mobility is sometime used), and the car body's vibroacoustic transfer functions P/F. Most of the time, simple models are used to take the acoustic cavity into account (hybrid FEM-SEA or analytical formula). Computations can then be done up to 1 kHz without any major difficulty, with a very good level of confidence for dynamic stiffness and an acceptable level of confidence for vibro-acoustic transfers. An example for a Light Rail Vehicle is given in Fig. 5. Full vibroacoustic models including a Finite Element model of the cavity can also be used to improve accuracy in the low frequency range (below 300 Hz) by taking into account the cavity's modal behavior, but this is rarely done in railway projects at this moment.

This type of computation is very useful to ensure a "healthy" car body Noise-Vibration-Harshness (NVH) design, with low vibroacoustic transfer functions. In a second step, these transfer functions must be combined with input dynamic forces at connecting points to predict the interior noise levels. The blocked force approach is more and more used, these blocked forces being usually obtained from in situ tests [1, 2] on vehicle or laboratory tests [12, 13] on components.

A virtual test method for structure borne noise generated from railway running gear is in progress to compute blocked forces transmitted to the car body starting with wheel/rail contact forces. A complete description of a method under assessment is given in [13]; only a brief summary is presented here. The Finite Element model includes the wheelsets, axle boxes and bogie (Fig. 6). Traction bars and dampers are represented by equivalent beam elements. All suspension elements and rubber bushings are modeled with frequency variable stiffness elements, the dynamic stiffness of these elements being

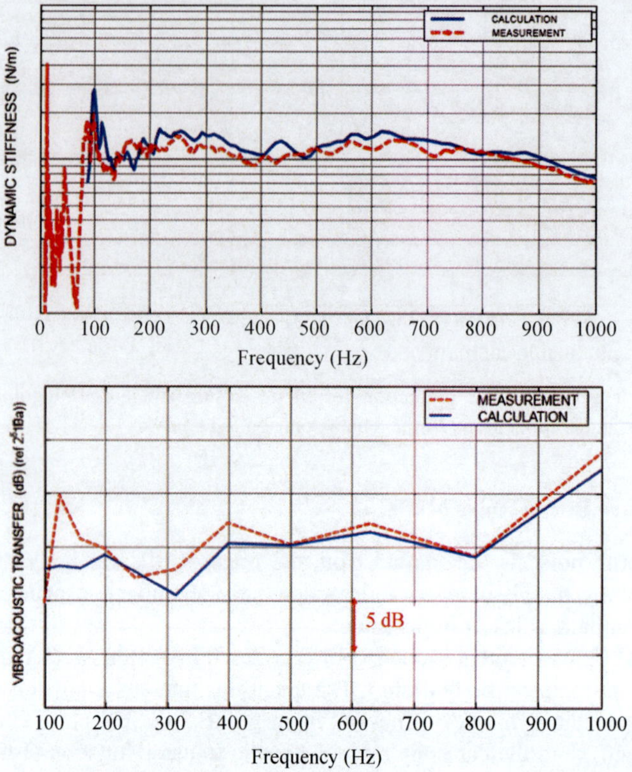

Fig. 5. Car body Finite Element Analysis computation results for structure borne noise control ***Upper:*** Input point dynamic stiffness ***Lower:*** P/F transfer function estimated with an analytical formula

previously measured in laboratory conditions. Wheel/rail contact forces are computed, taking into account the interaction with the track, wheel/rail roughness and contact filtering effect. They are then introduced into the FE model to compute blocked forces at the car body attachment points. In a final step, these blocked forces are combined with measured car body vibroacoustic transfers P/F to estimate structure borne interior noise.

3 Traction Noise

Different physical phenomena are at the origin of traction noise: electromagnetic excitation for electric motors, mechanical excitation due to the meshing process between pairs of teeth in a gearbox and aeroacoustics sources due to cooling unit fans. This section focuses on electric motor noise generated by electromagnetic excitation and gearbox noise generated by the meshing process.

3.1 Electric Motor Noise

Modelling Noise Due to Electromagnetic Excitation: The main excitation phenomenon is the Maxwell magnetic pressure acting in the air gap, on the stator and

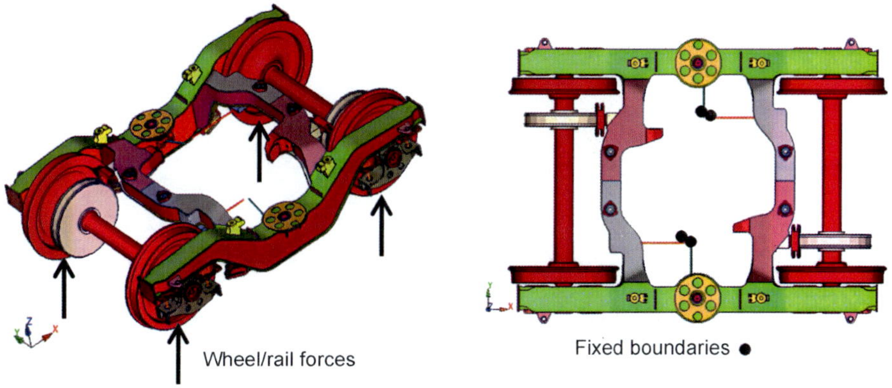

Fig. 6. The FE model of a complete bogie with dampers and traction bars, from ref. [13]

on the rotor. Many parameters influence the content of the excitation, such as the motor topology (induction or synchronous machine, number of slots and poles) or the motor drive (current shape including Pulse-With Modulation (PWM) effects).

Noise computation is a multiphysical problem for which it is necessary to combine electromagnetic, dynamic and acoustic Finite Element calculations (Fig. 7).

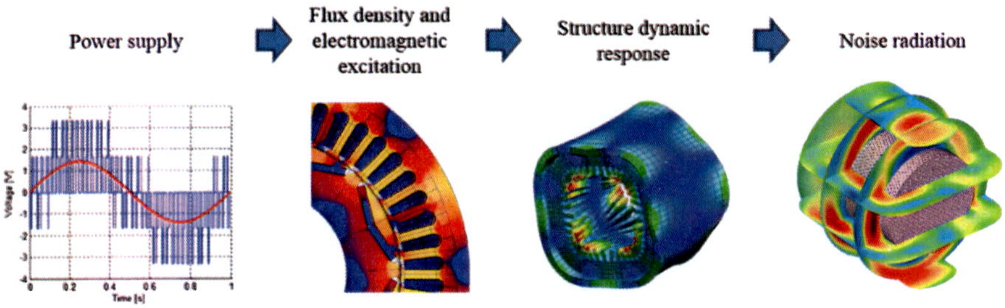

Fig. 7. Noise generation process for electric motor

This type of calculation has been presented by several authors e.g. [14], and has recently been validated against measurements in industrial applications [15]. A typical example of comparison between measurement and computation is presented Fig. 8: in this case the model gives representative order of magnitude of vibration levels, and is able to capture the dominant physical phenomenon (here the excitation of a stator mode by the electromagnetic excitation at a speed of about 4500 rpm).

Noise Minimization Strategies. Several low noise strategies are possible and can also be combined:

- Optimization of the structural design and/or of the PWM strategy, in order to avoid resonance phenomena, characterized by a spatial and frequency coupling between the electromagnetic excitation and the modal behavior of the stator [16].
- Optimization of the electromagnetic design in order to minimize Maxwell pressure inside the airgap of the machine. An example is given below.

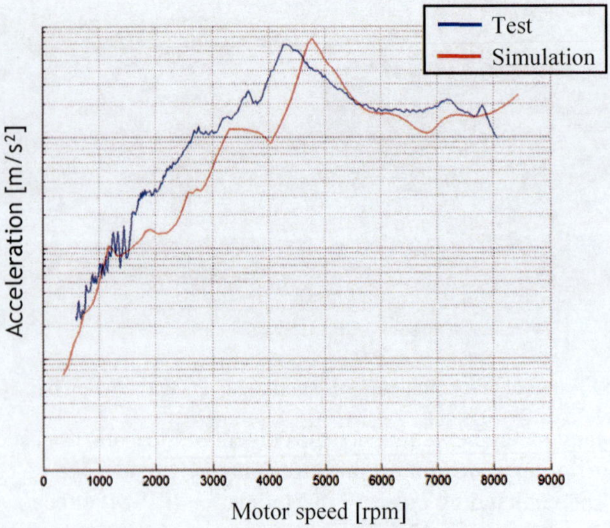

Fig. 8. Comparison of measured/computed normal mean squared acceleration of stator housing for the dominant engine order excitation during a run-up

Electromagnetic Design Optimization. Electric motor design optimization currently aims at minimizing a cost function corresponding to the acoustic power level of the machine, while respecting constraint functions so as not to deteriorate the motor's overall performance. In ref [17], the computational method presented in Fig. 7 is driven with an optimization software and different case studies are presented, including an Interior Permanent Magnet Synchronous Machine (IPMSM, Fig. 9).

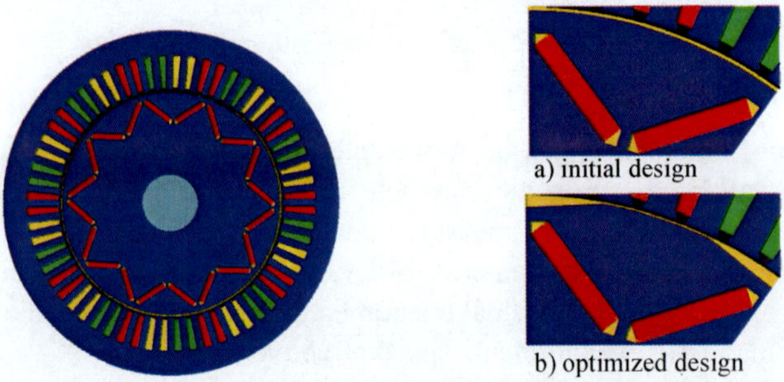

Fig. 9. Geometry of the initial and optimized Interior Permanent Magnet Synchronous Machine (IPMSM), from ref. [17]

The results after optimization show a very significant 14 dB decrease of the maximum Sound Power Level during an engine run-up (Fig. 10). Some inequality constraints are also defined, so that the mean torque produced by the motor is not reduced and torque ripple is not increased in comparison with the initial design. This reduction is achieved by small changes in the shape of the rotor poles depicted in Fig. 9 that do not affect the motor's price or weight.

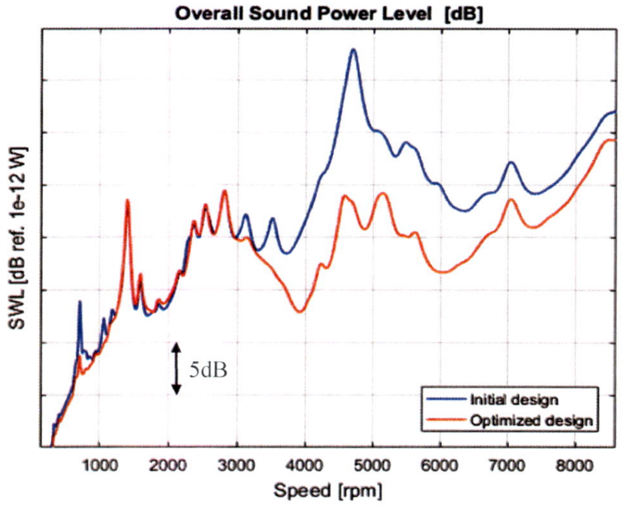

Fig. 10. Comparison of initial and optimized Interior Permanent Magnet Synchronous Machine (IPMSM) design SWL, from ref. [17]

3.2 Gearbox Noise

Gearbox Whining Noise Modelling: the gear meshing process is usually the main excitation source in gearboxes. Indeed, it is commonly assumed that Static Transmission Error (STE) and gear mesh stiffness fluctuations are responsible for radiated gearbox noise. They generate dynamic mesh forces which are transmitted to the housing through wheel bodies, shafts and bearings. Housing vibration is directly related to the noise radiated from the gearbox (whining noise).

The computational scheme is based on a 2-step procedure (Fig. 11):

- Step 1: computation of the excitation (transmission errors and mesh stiffness fluctuations), the input parameters being the teeth macro and microgeometries. Note that in certain situations, the static deflection of the shafts and housing can strongly influence STE computations and has to be taken into account. This can be done by means of static deflection computations including the complete model (gears, shafts and housing) to determine the equivalent helix deviation function of input torque: a unit torque is applied and therefore permit to determine the static deflection and the equivalent helix error as illustrated in Fig. 12.
- Step 2: computation of the vibration response of the gearbox, based on a Finite Element model of the gearbox and a modal approach. For gearbox whining, specific frequency algorithms (spectral iterative method) can be used to solve the dynamic equations with an iterative procedure [18]. These algorithms are much faster than time domain algorithms proposed in multibody software.

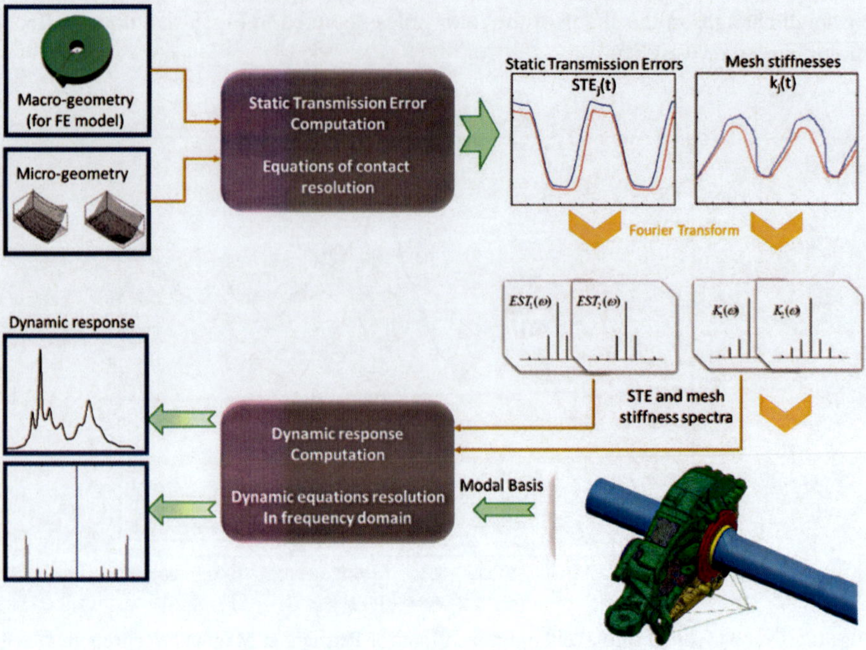

Fig. 11. Overview of the computational scheme

Fig. 12. Equivalent helix deviation due to static deflection generated by the transmitted torque

Ref [19, 20] respectively give examples of validation on an automotive and a railway application: a reasonable agreement between computations and measurements is achieved.

Figure 13 shows a comparison between measured and computed housing vibrations for a railway gearbox, under high torque condition. In this case, two STE computations are done, with and without the effect of static deflection: better vibration predictions are usually obtained when the effect of static deflection on STE is taken into account. Despite the uncertainty sources, a correct order of magnitude is provided for almost all measurement points on the housing.

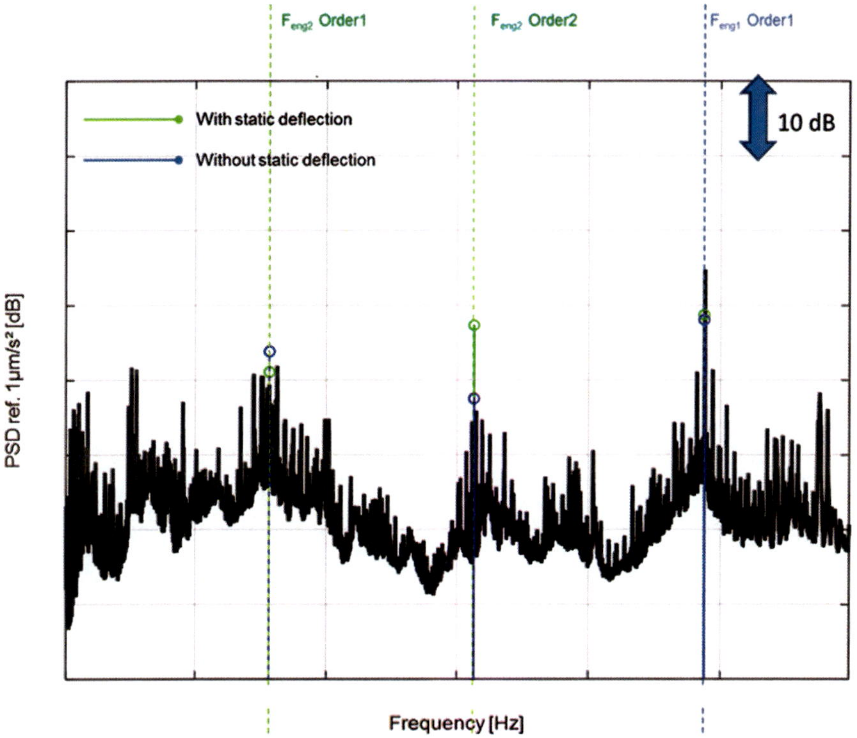

Fig. 13. Railway gearbox. Power spectral density of the housing normal mean squared acceleration (from ref. [12])

Noise Minimization: Gearbox noise can be minimized in different ways:

- Optimization of the housing's structural design, to avoid the excitation of local modes by meshing frequencies. This type of optimization is rather classical in NVH engineering and is not detailed here.
- Tooth microgeometry optimization in order to minimize STE. An example is given below.

Tooth microgeometry optimization in simple mesh gear systems for a given torque has been studied by many authors. However, for railway and automotive applications, this may not be sufficient as the torque transmitted by the gearbox is not constant. To overcome this limitation, an optimization strategy has been developed in order to minimize the STEs of multi-mesh gearboxes for a wide operating torque range [21].

An example of peak-to-peak STE values as a function of the torque is given in Fig. 14, before/after optimization. For the reference case, the STE was already minimized for medium and low torques (200–600 Nm). After optimization, different tooth microgeometries are proposed: for low and medium torque, they reach the same level of performance as the reference microgeometry, but they bring significant improvement for higher torque values (above 1000 Nm).

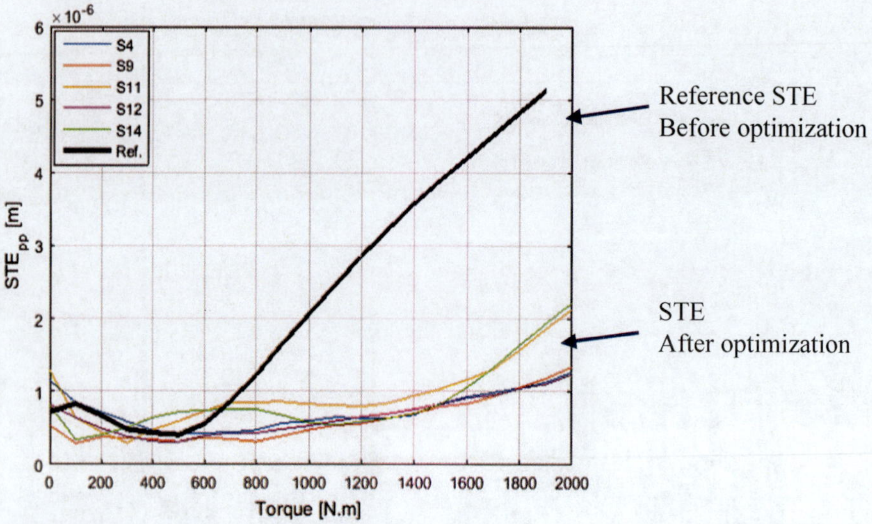

Fig. 14. Static transmission error computation (peak to peak values) function of torque before/after tooth microgeometry optimization.

A robustness analysis can then be carried out in order to select, among the different possible optimized microgeometries, the more robust with regard to manufacturing tolerances. This will ensure that noise reduction will still be obtained even if the selected optimized microgeometry will not be perfectly manufactured.

This robustness study is done with a Monte-Carlo method, i.e. 10000 STEs are computed, chosen randomly optimized teeth microgeometry parameters limited by the manufacturing tolerance intervals. This allows the establishment of the STE probability function of each possible optimized solution and the computation of statistical variables such as mean value and standard deviation of STE.

Figure 15 shows an example of STE probability density functions for the reference microgeometry and three possible optimized solutions. This illustrates how the final optimized solution is selected. The solution S1 has the smallest mean and maximum values. Although the mean value for the optimized solution S2 is also low, it is less interesting than the solution S1 because its STE maximum value is higher: the risk of deterioration of the noise because of manufacturing tolerances is higher. From a practical point of view, the mean value is the first criterion to consider. If the mean values are of the same order of magnitude (up to a difference of 15%), the solution with the smallest maximum value of the STE probability density function should be retained.

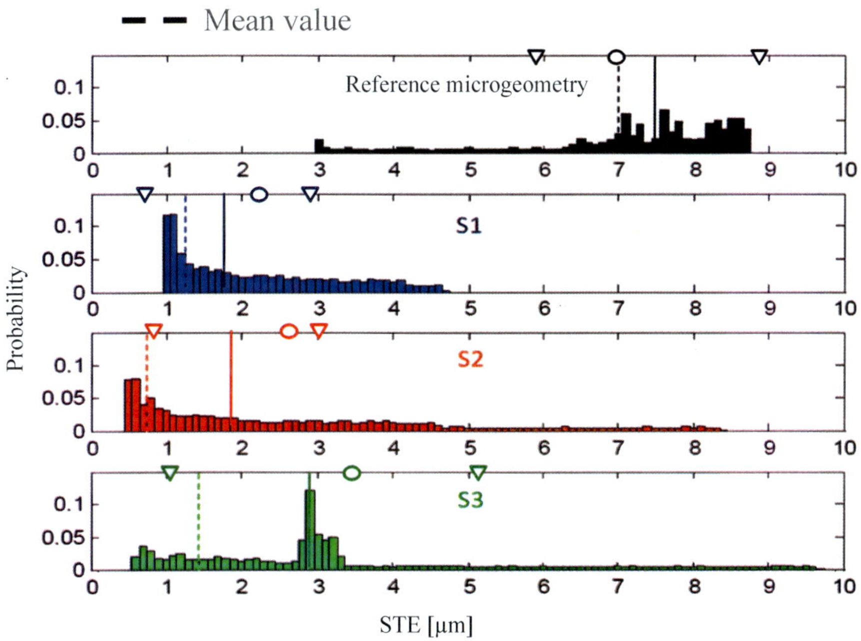

Fig. 15. STEs Probability density functions for the reference microgeometry and three optimized solutions S1, S2, S3 (from ref. [13])

4 Conclusions

Industrial methods used by rolling stock manufacturers to predict interior noise levels inside vehicles have been presented. For airborne transmission, global and pragmatic modelling strategies have been developed over the past 15 years, mixing experimental databases and acoustic computations with reasonable levels of confidence for most situations, while keeping low computational times compatible with railway projects. Some (non-exhaustive) fields of improvement have been mentioned in this paper.

Modelling structure borne noise transmission remains a challenge and a field of applied research for railway companies. Up to now, most of the work consists in controlling the car body design (dynamic stiffness and vibroacoutic transfer) independently from structural excitation coming from the bogie. Research on this topic is in progress, and validation and optimization of computational procedures are required.

Concerning traction motor noise, efficient computational methods are now available, with a sufficient level of accuracy to be used for industrial applications. The coupling of these methods with optimization software has a great potential for motor NVH design.

Acknowledgement. For structure borne models, financial support from the EU (grant no. 777564) and collaboration with CDH and ISVR are gratefully acknowledged.

For gearbox noise, the French National Research Agency has also supported research works through the joint laboratory LADAGE (ANR-14-Lab6-003) issued from the collaboration between

LTDS-Ecole Centrale de Lyon and VIBRATEC. The support from ALSTOM is also acknowledged (ref. [20]).

References

1. Moorhouse, A.: In situ measurement of the blocked force of structure-borne sound sources. J. Sound Vibr. **325**(4), 679–685 (2009)
2. Schleinzer, G., Kern, T.: Transfer path analysis on a siemens combino-plus tram in Almada – seixal (Lisbon). In: Proceedings of IWRN 11 (2013)
3. Poisson, F., Loizeau, T., Bouvet, S., Vincent, N., Transfer path analysis within a TGV Duplex coach. ICSV16, Krakov, 5–9 July (2009)
4. Report: Deliverable 7.1, Review of state of the art for industrial interior noise prediction, EU Shift2Rail project FINE 1 (2017)
5. Sapena, J., Latorre, E., Recorbet, S., Bistagnino, A., Vallespín, A.: Railway rolling stock - Validation is not a must, it is a support to reduce risk. In: proceedings of NOVEM (2018)
6. Kohrs, T., et al.: Sound propagation and distribution around typical train carbody structures. In: Proceedings of EURONOISE (2018)
7. Bistagnino, A., et al.: Computation of parietal pressures of rolling stock vehicles. In: Proceedings of IWRN 11 (2013)
8. Li, H., Thompson, D., Squicciarini, G., Liu, X., Rissmann, M., Denia, F., Giner-Navarro, J.: Using a 2.5D BE model to determine the sound pressure on the external train surface. In: Proceedings of IWRN13 (2019)
9. Poisson, F., et al.: Transfer path analysis within the TGV duplex driver cab, platforms, upper and lower rooms from 150 to 360 km/h. In: Proceedings of WCRR (2011)
10. Gabet, P., et al.: Prediction of acoustic comfort of a trainset. In: Proceedings of SIA Conferences (2008)
11. Sanei, N., et al.: Structure borne noise characterization of an air generation and treatment unit (AGTU) for a train by using blocked forces method. In: Proceedings of Internoise (2019)
12. Elliott, A., et al.: Structure borne noise characterization of an air generation and treatment unit (AGTU) for a train using the mobility method and sub-structuring. In: Proceedings of Internoise (2019)
13. Xie, G., et al.: Virtual test method of structure borne sound for a metro bogie. In: Proceedings of IWRN 13 (2019)
14. Marinescu, M., Marinescu, N.: Numerical computation of torques in permanent magnet motors by maxwell stresses and energy method. IEEE Trans. Mag. **24** (1), 463–466 (1988)
15. Humbert, L.: Electromagnetic and structural coupled simulation to investigate NVH behavior of an electrical automotive powertrain. In: Proceedings of ISNVH Conf. Graz (2012)
16. Dupont, J.-B., et al.: Noise radiated by electric motors: simulation process and overview of the optimization approaches. In: Proceedings of ATZ Conference (2017)
17. Dupont, J.-B., Jeannerot, M.: Multi-objective optimization of electrical machines including NVH, performance and efficiency considerations. In: Proceeding of SIA Powertrain Conference (2019)
18. Perret-Liaudet, J.: An original method for computing the response of a parametrically excited forced system. J. Sound Vibr. **196**, 165–177 (1996)
19. Carbonelli, A.: Modeling of gearbox whining noise. In: Proceeding of, ISNVH Conference (2014)
20. Carbonelli, A., Guerder, J.-Y.: Computing and minimizing the whining noise – application to a bogie gearbox. In: Proceedings of VDI Congress (2017)
21. Carbonelli, A., et al.: Robust optimization of a truck timing gear cascade: numerical and experimental results. In: 21ème Congrès Français de Mécanique (2013)

High Speed Rail and Aerodynamic Noise

Evaluation on Aerodynamic Noise of High Speed Trains with Different Streamlined Heads by LES/FW-H/APE Method

Chaowei Li[1], Yu Chen[2], Suming Xie[1](✉), Xiaofeng Li[1], Yigang Wang[2], and Yang Gao[3]

[1] Dalian Jiaotong University, Dalian 116028, Liaoning, China
ck_lcw@126.com, sumingxie@163.com
[2] Tongji University, Shanghai Automotive Wind Tunnel Center, No. 4800, Cao'an Highway, Shanghai, People's Republic of China
[3] CRRC, Changchun Railway Passenger Vehicle Co., Ltd., Changchun 130062, People's Republic of China

Abstract. Aerodynamic noise of the high speed train is essential to exterior and interior noise. Five 1:8 scaled high speed train numerical models were established. The large eddy simulation was used to obtain the body turbulent fluctuation pressure. Based on the FW-H and the APE equation, far-field noise and near-field noise were obtained respectively. The difference between the simulation results of the total sound pressure level in the far field and the wind tunnel test was less than 2.0 dB(A). Their spectrum change trends were the same, and the amplitude differences were relatively small, indicating the reliability of FW-H equation. The bogie cavity is the dominant noise source, and the OSPL of turbulent pressure level is 15 –25 dB(A) higher than that of acoustic pressure inside the bogie section. The head shape varies the flow and acoustic field of the bogie section, leading to the OSPL ranges from 77.1 dB(A) to 78.6 dB(A) in the far field.

Keywords: High speed train · Aerodynamic noise · LES · FW-H · APE method

1 Introduction

Aerodynamic noise problems accompanied by the speed-up of train system are, at present, receiving a considerable attention that should be urgently resolved [1]. The head car of high speed train (HST) is a predominant aerodynamic noise source, resulted from both field test and aero-acoustic wind tunnel test [2].

In the numerical method, the turbulent fluctuating pressure in the body area of high-speed trains are usually obtained by large eddy simulation, while the far-field noise is obtained by FW-H equation based on acoustic analogy. Zhu *et al.* [3] used Delayed Detached Eddy Simulation (DDES) and Ffowcs Williams-Hawkings (FW-H) equation to solve the unsteady flow field and far-field noise of a 1:5 simplified bogie section. It is found that a highly turbulent flow is generated within the bogie cavity and the ground

G. Degrande et al. (Eds.): *Noise and Vibration Mitigation for Rail Transportation Systems*, NNFM 150, pp. 57–65, 2021.
https://doi.org/10.1007/978-3-030-70289-2_3

increases the noise levels. The influence of head car shape on aerodynamic noise of HSTs is not clear.

Meanwhile, the acoustic pressure is caused by the pressure fluctuation and travels at the speed of sound. The energy distribution in wavenumber domain is different between turbulent and acoustic pressure, the latter has higher receptivity to transmit through the train body. Ewert *et al.* [4] deduced Acoustic Perturbation Equation (APE) to simulate the flow-induced sound field, and then it was applied to predict the sound generated by a cylinder in a cross flow. The turbulent fluctuating pressure and near field acoustic pressure are both excitations for interior noise, it is necessary to obtain near field acoustic pressure by APE.

In present study, aerodynamic noise generated by HST models with five different streamlined head was studied by Large Eddy Simulation (LES)/FW-H/APE method. The effects of head car shape to far field noise, turbulent pressure fluctuation level and acoustic pressure level in the near field were studied.

2 Methodology

The differences between real case and model case on aerodynamic noise are: Firstly, the Reynolds number effect; Secondly, the moving ground effect. There are scaling law of aerodynamic noise for the simple geometry, like the circular cylinder in free space. For the complex geometry like the HST, Lauterbach *et al.* [5] found that noise from the first bogie section can be characterized by cavity mode excitation, and it reveal only a weak Reynolds number dependence. The moving ground effect is still unclear according to published papers, because moving ground will generate additional background noise. As a result, the scaled train model and stationary ground condition were studied in this paper.

Numerical simulations are carried out using computational fluid dynamics and computational aeroacoustics method. HSTs operate at Mach number 0.2–0.3. Thus, the compressibility effects could be neglected [6]. The simulation was performed using the commercial software STAR-CCM+.

2.1 Large Eddy Simulation

To obtain unsteady flow field, Large Eddy Simulation is adapted. It is an inherently transient technique in which the large scales of the turbulence are resolved everywhere in the flow domain, and the small-scale motions are modeled.

Turbulent viscosity was modeled by WALE (Wall-Adapting Local-Eddy Viscosity) Subgrid Scale model. It uses a novel form of velocity gradient tensor in its formulation. This model is less sensitive to the value of model coefficient than the Smagorinsky model. Besides, it does not require any form of near-wall-damping, which automatically gives accurate scaling at walls. It is one of the most widely used turbulence models in recent years.

2.2 FW-H Equations

The near flied unsteady flow computation provides acoustic sources which are fed to FW-H equation for far field noise prediction. The solution of is based on Farassat's Formulation 1A [7]. For permeable source surfaces, it could account for the monopole, dipole and quadrupole noise sources within the permeable surfaces region. The total acoustic pressure $p^{'}(\overline{x}, t)$, resulting from the sum of the thickness noise $p_{T}^{'}(\overline{x}, t)$ and the loading noise $p_{L}^{'}(\overline{x}, t)$, is defined as:

$$p_T^{'}(\vec{x}, t) = \frac{1}{4\pi}\left\{\int\limits_{f=0} [\frac{\rho_0(\dot{U}_n + U_{\dot{n}})}{r(1-M_r)^2}]_{ret} dS + \int\limits_{f=0} [\frac{\rho_0 U_n(r\dot{M}_r + c(M_r - M^2))}{r^2(1-M_r)^3}]_{ret} dS\right\} \quad (1)$$

$$\begin{aligned} p_L^{'}(\vec{x}, t) = & \frac{1}{4\pi}\frac{1}{c}\int\limits_{f=0} [\frac{\dot{L}_r}{r(1-M_r)^2}]_{ret} dS + \frac{1}{4\pi}\int\limits_{f=0} [\frac{L_r - L_M}{r^2(1-M_r)^2}]_{ret} dS \\ & + \frac{1}{4\pi}\frac{1}{c}\int\limits_{f=0} [\frac{L_r(r\dot{M}_r + c(M_r - M^2))}{r^2(1-M_r)^3}]_{ret} dS \end{aligned} \quad (2)$$

where $L_r = \vec{L} \cdot \vec{r} = L_i \cdot r_i$, $U_n = \vec{U} \cdot \vec{n} = U_i \cdot n_i$, $\vec{r}$ and $\vec{n}$ are the direction of sound radiation and normal vector of the permable source surface respectively.

2.3 Acoustic Perturbation Equations

APE, which solves the flow-induced noise on the basis of unsteady flow simulation, was selected for the acoustic pressure calculation. The equations were firstly proposed in the terms of perturbation pressure $p^{'}$ and irrotational perturbation velocity u^{a}, and then acoustic pressure p^a was given as [4]

$$\begin{aligned} & \frac{1}{c^2}\frac{\partial^2 p^a}{\partial t^2} + \frac{2\overline{U}}{c^2} \cdot \nabla \frac{\partial p^a}{\partial t} + \frac{(\overline{U} \cdot \nabla)}{c^2}(\nabla \cdot \overline{U} p^a) - \nabla^2\left(p^a + \tau \frac{\partial p^a}{\partial t}\right) \\ & = -\left(\frac{\overline{p}}{c^2}\frac{\partial^2 P'}{\partial t^2} + \frac{2(\overline{U} \cdot \nabla)}{c^2}\overline{\rho}\frac{\partial P'}{\partial t} + \frac{(\overline{U} \cdot \nabla)}{c^2}(\nabla \cdot \overline{U} P')\right) \end{aligned} \quad (3)$$

where $P^{'} = P - P_{mean}$ represents the flow field pressure fluctuation, τ is the damping term, p^a is acoustic pressure. Therefore, from the LES derived pressure fluctuation, APE can be calculated to solve the term of p^{a}.

2.4 Mesh Setup

Five head cars and bogies CAD model were shown in Fig. 1. The power bogies were installed on middle car. The trailer bogies were installed on the head and tail car. The surface mesh sizes of train body and bogies range from 2 mm–4 mm.

Inside the noise zone, certain requirements have to meet for correct calculation of LES and APE. The Point per Wavelength (PPW) was defined as:

$$PPW = \frac{\lambda}{\Delta x} \tag{4}$$

where λ is the wavelength of interest and Δx is the mesh size. For the frequency of 3500 Hz and mesh size of 3 mm, the corresponding PPW = 32. PPW > 30 can be regarded as a satisfied standard. Trimmer mesh, which is mainly hexahedral cell, was used to fill the volume of computational region. The cell number is about 44 million. The details were shown in Table 1. In this study, the grid number of different heads range from 44 million to 46 million, which is the numerical results are comparable.

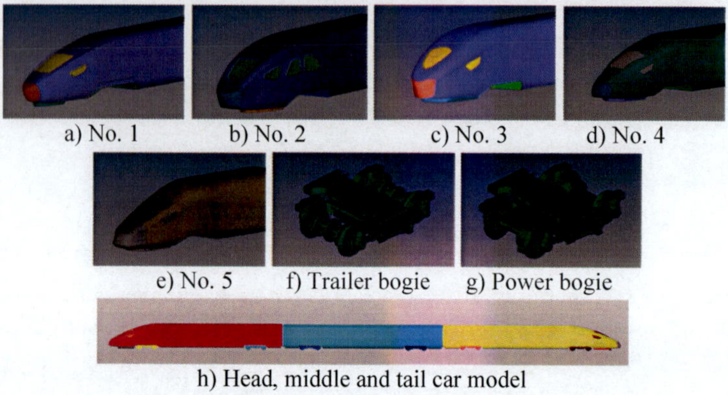

Fig. 1. Five different streamlined head car, bogies and train model

Fig. 2. Noise zone and noise source weighting

Table 1. Details of mesh setup

Cell size (mm)				First prism layer height (mm)			Number of prism layer		
Noise zone	Head region	Near train	Boundary	Body	Bogies	Ground	Body	Bogies	Ground
3	6	12	48	0.015	0.015	0.35	12	12	2

2.5 Calculation Setup

The 1:8 scaled models was 10.2 m in length, 0.44 m in width and 0.5 m in height. The model was placed in a block-like computational region, with a height of 2.5 m, length of 18.7 m and width of 5.5 m, as shown in Fig. 3. The incoming velocity is 250 km/h and the turbulence intensity is 0.2%. The pressure outlet is 0 Pa. The side and top boundaries was defined as inviscid wall. The train, bogies and ground were defined as viscid wall boundary condition, which are constant to the wind tunnel test. In calculation of APE, the inlet, outlet and inviscid wall were all defined as acoustically non-reflected boundary. The non-linear aerodynamic sources around the train were enclosed by the permeable source surface.

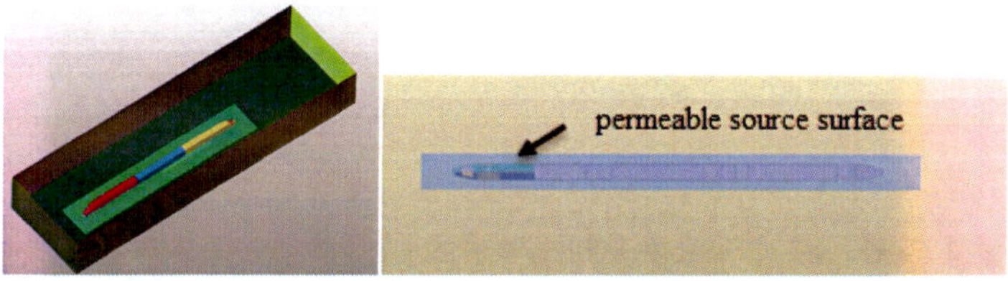

Fig. 3. Computational region and permeable source surface

The time-step size must be small enough to maintaining the proper value of acoustic CFL, defined as

$$Acoustic\ CFL = \frac{c_0 \Delta t}{\Delta x} \tag{5}$$

where c_0 is sound speed. The target of acoustic CFL is under the value of 6. Therefore, the size of time-step is set as 0.00005 s. This size corresponds to the acoustic CFL of 5.67 for the case, while ensure 10 steps inside a fluctuation cycle up to 5000 Hz. For the CFD solver, the temporal discretization scheme is second order. Normally, the spatial discretization for LES is bounding central differencing. As for the APE solver, the temporal and spatial discretization schemes are both second order.

The noise weighted function defined as Fig. 2. Inside the noise zone the value is 1. While outside the noise zone the value is 0, which means no sound would be produced there. In the transition band, with a thickness of 15 mm, a spatial Hanning window was applied to enable the smooth variation between 0 to 1. The noise damping function, which contributes to the term τ in Eq. 3, is the opposite of the noise weighted function. It has the value of 0 inside the noise zone and the value of 1 outside the noise zone.

Firstly, the RANS model SST k-ω was run to initialize the unsteady flow. Then LES started for a relatively large time-step of 0.0005 s, with the duration time of 0.5 s, which is roughly two times the flow-through time of the whole region or 60 times the flow-through time of the first bogie section. Hence, the time-step switched to 0.00005 s and ran for 1000 steps. According to APE, averaged pressure should be solved before APE calculation started. Therefore, averaged pressure has been recording 750 steps prior APE activated. The data sampling began after APE had run for 250 steps. During 2500 steps data sampling, flow field pressure obtained from LES and acoustic pressure obtained from APE were recorded.

2.6 Aeroacoustic Wind Tunnel Test

The experiment was carried out in the SAWTC aeroacoustic wind tunnel at Tongji University, where the background noise level is 72.4 dB(A) at the nozzle speed of 250 km/h. The wind tunnel is a 3/4 open section wind tunnel with a nozzle size of 6.0 m × 4.25 m, the turbulence intensity is 0.2%. The nozzle speed in the experiment is the same as the inlet velocity of simulation.

The wind test model is the same with No. 4. The train model is made of stiff wood and the exterior surfaces were processed Computerized Numerical Control (CNC). The bogies were made of ABS Engineering Plastics. Figure 4 shows a general view of the train model in aeroacoustic wind tunnel. The test model is the same with No. 4. The model was attached above a reflective ground. It was mounted on simplified rails, and the head car nose was 1 m away from the nozzle. The ground clearance between the train body and the ground 53 mm, which is 1/8 of a real HST ground clearance also. This relative movement can be simulated in wind tunnel tests using a moving belt under the train, but this has not been used here as it would generate additional background noise.

Three B&K free field microphones are located 7.5 m away from the center line of the train, and 0.8 m in height. The distances between F1, F2 and F3 are 3 m. The sampling frequency and sampling time of microphones are 48 kHz and 10 s respectively. Figure 5 shows the spectrum of the background noise and train model.

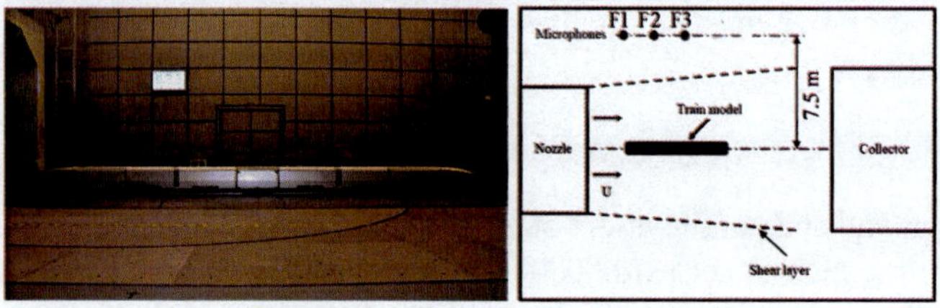

Fig. 4. Wind tunnel test scheme

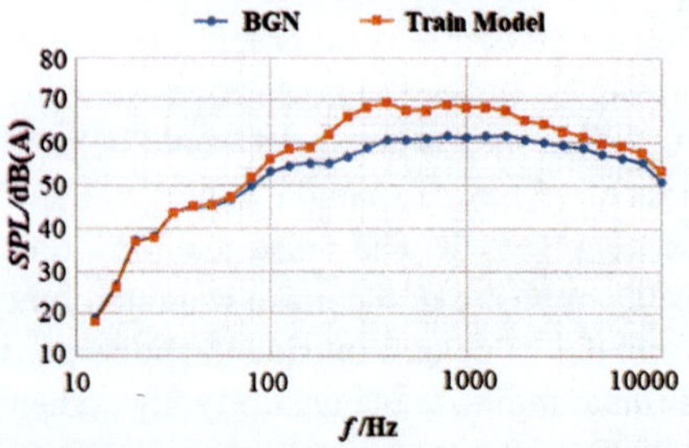

Fig. 5. Background noise and train model noise

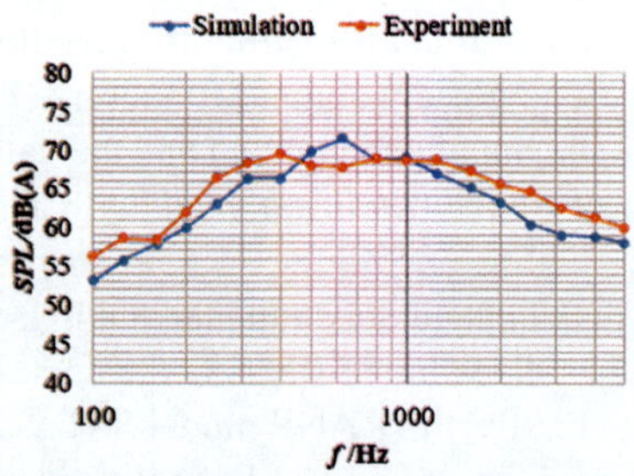

Fig. 6. 1/3 Octave band spectrum of F2

3 Results and Discussion

3.1 Validation of Simulation and Experiment

Due to a high background noise level in the low frequency range generated from the nozzle itself, simulation results are only considered above 100 Hz. The background noise of the anechoic chamber was also shown. Its noise level was by several orders of magnitude lower than those generated from the HST model in the frequency range between 100 Hz and 3500 Hz.

Figure 6 shows the spectrum of F2 obtained by wind tunnel test and numerical simulation. The numerical simulation results are consistent with the wind tunnel test results, and the amplitude is not very different. The amplitude of the SPL is 60 dB (A) in the lower and higher frequency bands. The amplitude of the SPL is 65–70 dB (A) in the middle frequency band (300 Hz–2000 Hz). There is no peak in the whole frequency band, indicating that the far-field noise is broadband noise.

The total SPL of each microphone are shown in Table 2. The differences between each microphone are within 2 dB(A). The results of simulation are consistent with those of wind tunnel test, which shows that the prediction of far-field noise based on FW-H equation is basically correct.

Table 2. Total sound pressure level of test points dB(A)

Items	F1	F2	F3
Simulation	78.5	78.4	80.6
Test	78.1	78.7	78.6
Δ(Simulation-Test)	0.4	−0.3	2.0

3.2 Influence of Head Shape on Pressure and Acoustic in the Near Field

The simplified model does not reflect every single feature of a real train. For aerodynamic noise studies, any component like air-condition inlets, wiper of head car, and small antennas are important noise sources. However, the simplified train and bogie models reflects the most important noise sources.

Compared with the background noise, the HST aerodynamic noise energy was mainly between 100 Hz to 3500 Hz. The Overall SPL in A-weighting was obtained for both turbulent pressure (TP) and acoustic pressure (AP) from 100 Hz to 3500 Hz. The contours of OSPL were shown in Fig. 7 and 8.

The distribution of TP level and AP level are similar for five heads. Both of them are much higher on the underbody structures. The OSPL of TP level is 15–25 dB(A) higher than that of AP inside the bogie section. The TP is strong in the rear edge of bogie cavity, cowcatcher and wheels, while it is weak in the top roof of bogie cavity. The AP is strong in the rear edge and top roof of bogie cavity. The vortex interaction between shear layer and rear edge of bogie cavity are not only the major source of TP, but also the major one of AP.

3.3 Influence of Head Shape on Far Field Noise

Table 3 shows the SPL of F2 in the far field. The OSPL ranges from 77.1 dB(A) to 78.6 dB(A), which means the different styling of head car affects the SPL in the far field.

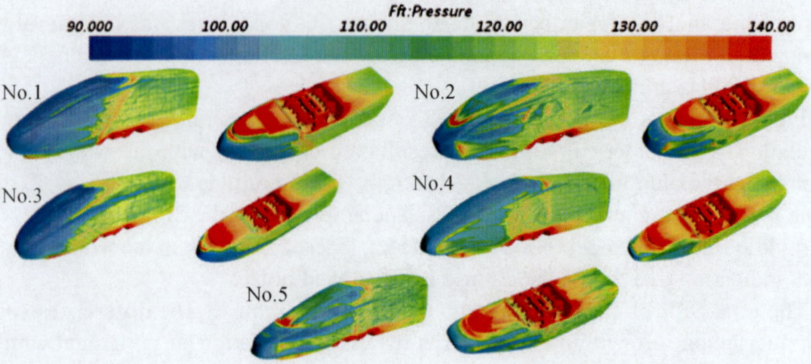

Fig. 7. Turbulent pressure fluctuation level obtained by LES dB(A)

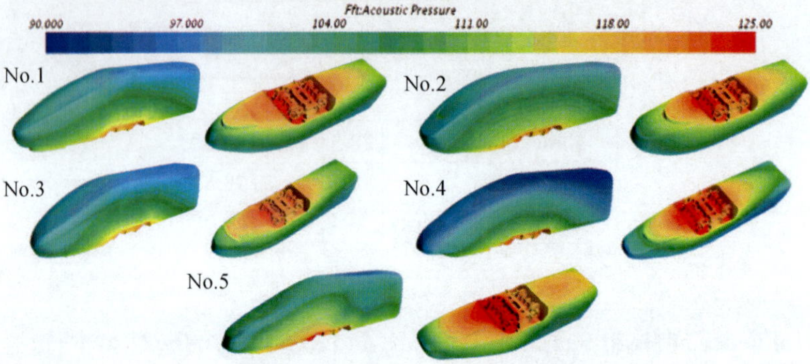

Fig. 8. Acoustic pressure level obtained by APE dB(A)

The turbulent and acoustic pressure levels on the upper head are much lower than that of underbody structure. That is to say, the head shape varies the flow and acoustic field of the first bogie section.

Table 3. Sound pressure level in the far field dB(A)

SPL	No. 1	No. 2	No. 3	No. 4	No. 5
F2	77.9	77.1	77.5	78.4	78.6

Figure 9 gives the spectrum in the far field. The trends of five streamlined head are similar. The No. 5 HST emits higher noise level in the frequency range 500 Hz–1000 Hz. The No. 2 and No. 3 HSTs have lower noise level in the frequency range 200 Hz–1250 Hz.

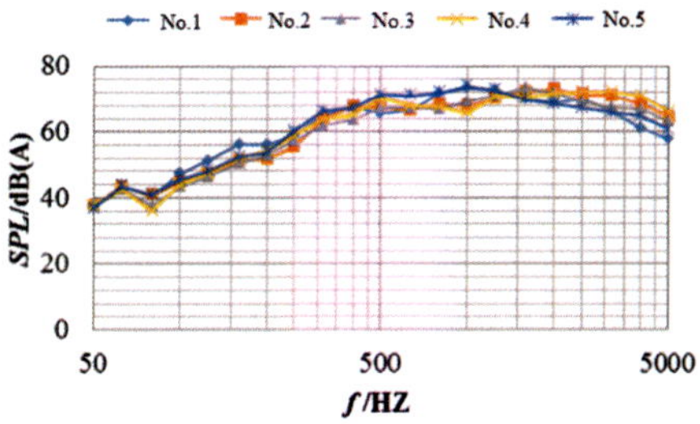

Fig. 9. The spectrum of F2

4 Conclusion

Hybrid LES/FW-H/APE methods were performed to investigate the aerodynamic noise of HSTs with five different streamlined heads. The No.4 HST with the same configuration was researched experimentally in the aeroacoustic wind tunnel at SAWTC. The SPL in the far field obtained by LES/FW-H was validated by the experimental measurement and it provided good agreement. The acoustic pressure gives the insight that the acoustic pressure is strong in the rear edge and top roof of bogie cavity, and different from the turbulent pressure level distribution.

The OSPL of turbulent pressure level is 15–25 dB(A) higher than that of acoustic pressure for every same place inside the bogie section. The different shapes of head car affect the underbody flow field, leading to the differences of SPL in the far field, turbulent and acoustic pressure level in the near field.

References

1. Thompson, D.: Railway Noise and Vibration: Mechanisms, Modeling and Means of Control. Elsevier, Amsterdam (2009)
2. Thompson, D., Iglesias, E., Liu, X., et al.: Recent developments in the prediction and control of aerodynamic noise from high-speed trains. Int. J. Rail Transp. **3**(3), 119–150 (2015)
3. Zhu, J., Hu, Z., Thompson, D.: The effect of a moving ground on the flow and aerodynamic noise behaviour of a simplified high-speed train bogie. Int. J. Rail Transp. **5**(2), 110–125 (2017)
4. Ewert, R., Wolfgang, S.: Acoustic perturbation equations based on flow decomposition via source filtering. J. Comput. Phys. **188**(2), 365–398 (2003)
5. Lauterbach, A., Ehrenfried, K., Loose, S., et al.: Microphone array wind tunnel measurements of Reynolds number effects in high-speed train aeroacoustics. Int. J. Aeroacoust. **11**(3 & 4), 411–446 (2012)
6. Ask, J., Davidson, L.: An acoustic analogy applied to the laminar upstream flow over an open 2D cavity. Comptes Rendus Mécanique **333**(9), 660–665 (2005)
7. Brenter, K., Farassat, F.: An analytical comparison of the acoustic analogy and Kirchhoff formulation for moving surfaces. AIAA J. **36**(8), 1379–1386 (1998)

Comparisons of Aerodynamic Noise Results Between Computations and Experiments for a High-Speed Train Pantograph

Xiaowan Liu(✉), Jin Zhang, David J. Thompson, Giacomo Squicciarini, Zhiwei Hu, Martin Toward, and Daniel Lurcock

Institute of Sound and Vibration Research,
University of Southampton, Southampton SO17 1BJ, UK
Xiaowan.Liu@soton.ac.uk

Abstract. Aerodynamic noise induced by interaction with the air flow becomes particularly important as train speeds increase. For modern electric high-speed trains, one of the main aerodynamic noise sources is the pantograph which normally sits in a cavity on the train roof. A component-based model, based on empirical constants obtained from available literature, is applied in this paper to predict pantograph noise in realistic conditions. The predictions are compared with field measurements obtained by using a microphone array. Comparisons are made for two types of train fitted with different pantographs, running at different speeds. Good agreement of the far-field noise spectrum with the measurements are obtained for both pantographs. The individual contributions from the panhead and knee-joint are analysed. The dominant source is found at the panhead whereas the knee-joint is of less importance. In addition, through the model, a detailed component spectral analysis is presented for the first pantograph.

Keywords: Train pantograph · Aerodynamic noise · Component-based model · Microphone array measurement

1 Introduction

Aerodynamic noise induced by interaction with the air flow becomes particularly important as train speeds increase. For modern electric high-speed trains, one of the main aerodynamic noise sources is the pantograph which normally sits in a cavity on the train roof. Lölgen [1] measured noise spectra for a DSA350 pantograph in a wind tunnel, and found that the spectra were characterized by the Aeolian tones associated with the vortex shedding mechanism. Latorre Iglesias et al. [2] conducted detailed measurements to study the vortex shedding noise characteristics from cylinders with different cross-sections corresponding to different components of the pantograph.

Numerical techniques have also been widely applied for the prediction of aerodynamic noise from pantographs. Liu [3] investigated numerically the effects of yaw angle

G. Degrande et al. (Eds.): *Noise and Vibration Mitigation for Rail Transportation Systems*, NNFM 150, pp. 66–72, 2021.
https://doi.org/10.1007/978-3-030-70289-2_4

and Reynolds number on the noise characteristics of yawed circular cylinders approximating the pantograph arms. This used a hybrid approach combining Computational Fluid Dynamics with an acoustic analogy method. However, the numerical methods require large computing resources and cannot be used routinely for a whole pantograph. Latorre Iglesias et al. [4] therefore developed a detailed semi-empirical component-based model in which empirical constants obtained from existing noise measurements on components were used to model the noise from the whole pantograph. This model can also be extended by the use of computational results [3].

Field measurements using microphone arrays can also be conducted to identify the contribution of the different noise sources, including the pantographs, on moving trains. This paper presents field measurement results for two types of train fitted with different pantographs. Measurements of the far-field noise spectra are compared with predictions using the component-based model. The dominant noise sources are identified based on the spectral analysis from the prediction model.

2 Microphone Array Measurement

Measurements were made using the microphone array shown in Fig. 1(a) located 10 m from the centre of the track. The design of this microphone array includes 90 microphones (see [5]) mounted on a carbon fibre frame, positioned in the arrangement shown in Fig. 1(b). Two types of train with different pantograph types were recorded (here named pantograph 1 and pantograph 2). Results for pantograph 1 are available at 273 km/h and for pantograph 2 at 177 km/h.

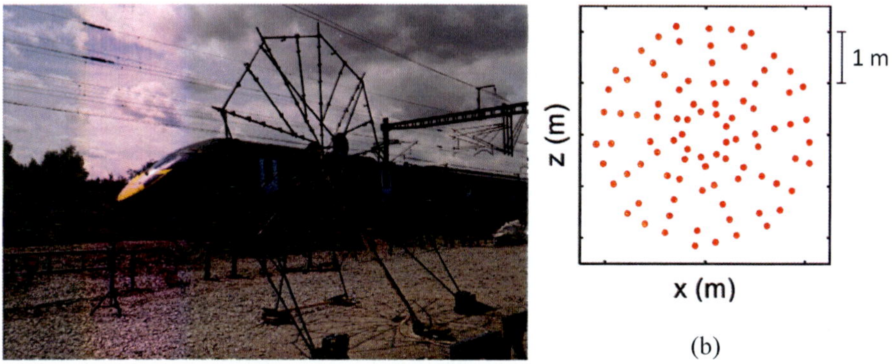

(a) (b)

Fig. 1. Microphone array setup, (a) field measurements; (b) microphone positions

The beamforming algorithm implemented for this study is based on the assumption that all the noise sources on the train can be represented by uncorrelated monopoles [6]. A virtual scanning grid is defined over the surface of the moving train identifying all the potential sources and each point on the grid is 'tracked' over a certain length. This is set to be 3.2 m in this study. Appropriate delays, based on the relative distance between the potential locations of the sources on the moving train and the array microphones, are applied to the measured signals to provide estimates of the source contributions from those locations. The results can be shown as a colour map (beamforming output) that can be superimposed on an image of the train. The regions with colours of higher intensity (main lobes) identify sources with higher noise contributions. In applying delays to the measured signals the Doppler effect is removed and the beamforming outputs are therefore shown in a frame moving with the train. The contribution from each source can be quantified by integrating the beamforming output over the area enclosing the sources of interest. Although the dominant aeroacoustic sources have a dipole nature, due to the limited length of the analysis window the difference in directivity this introduces can be neglected in the source reconstruction.

The quantification approach adopted in this work is based on the Source Power Integration (SPI) method [7]. It can be divided into the following steps:

- A frequency-dependent compensation factor is first obtained numerically by simulating the pass-by of a monopole over the length of a single car. This will be used to relate the beamforming output to the SPL contribution at a reference microphone, set at the centre of the array.
- The various sources on the map are identified and enclosed in a certain predefined area centred at the peak of the main lobe of that source.
- The acoustic map is integrated over these areas to give a frequency-dependent result for each of the sources identified on the map.
- The compensation factor is applied to the integrated beamforming result to give the spectrum of each source. The effect of the compensation factor is to rescale the noise spectra estimated from the beamforming output so that they represent noise spectra at the array centre averaged over the length of a single car.

More details on the quantification algorithm are given in [5].

3 Component-Based Model

If it is assumed that the noise from each component is uncorrelated, the far-field noise spectrum from the whole pantograph can be determined from the incoherent sum of the noise from each individual strut as follows [4],

$$\overline{p^2(f)} = \frac{\rho_0^2 U_i^6}{16 c_0^2 R_i^2} \sum_i \eta_i S_i F_i(f) \frac{D_{i\mathrm{rad}}(\phi)}{(1 - M\cos(\theta))^4}$$

where ρ_0 is the air density, c_0 the speed of sound, R the distance between the noise source and the observer, and U the incoming flow speed. $F_i(f)$ is a normalized spectrum at frequency f which needs to be determined appropriately for each component, i. η_i is the amplitude factor, given by $\eta = St^2 C^2{}_{L\mathrm{rms}} l_c LD$ for a noise source dominated by the lift dipole, where St ($= fD/U$ where D is the cylinder diameter) is the normalized frequency, also called Strouhal number, $C_{L\mathrm{rms}}$ the rms fluctuating lift coefficient, l_c the spanwise correlation length, normalized by D. $D_{\mathrm{rad}}(\phi) = \cos^2(\phi)$ is the directivity function and ϕ

is the angle between the lift force vector of each component and the position vector of the observer with respect to the component position. The factor $(1 - M\cos(\theta))^4$ accounts for the convective amplification for a dipole source, where θ is the angle between the flow direction and the observer position. To allow comparisons with the results obtained from the quantification procedure (Sect. 2), in which the pantograph is considered as a source moving over the length of a single car, the noise in the prediction is calculated for the pantograph at different locations within this length, and the average level is determined.

4 Results and Discussion

4.1 Pantograph 1

The noise map based on the frequency range 400–4000 Hz for a train fitted with the first pantograph passing at 273 km/h is shown in Fig. 2. The pantograph was operating with the knee-joint downstream. The region with the highest level appears at the panhead. The knee-joint region is also found to be important, although the level is about 6 dB lower than that at the panhead.

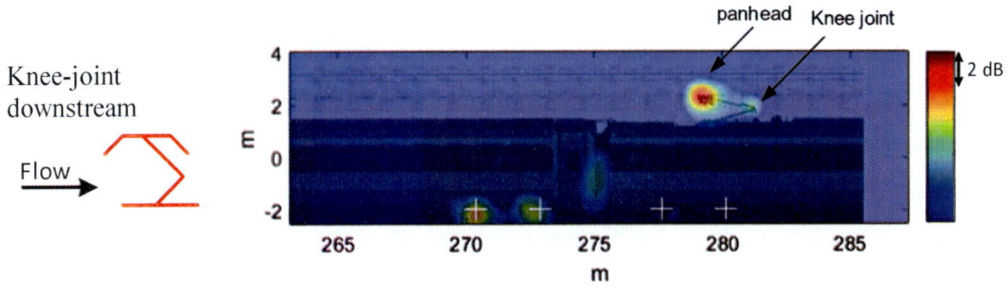

Fig. 2. Noise map over 400–000 Hz for pantograph 1 at 273 km/h with knee-joint downstream. The pantograph orientation is shown on the left.

The measured and predicted noise spectra are compared in Fig. 3(a). Good agreement is found between the prediction and the measurement above 800 Hz with a maximum difference of 7 dB at 3150 Hz where a peak occurs in the prediction but not in the measurement. At low frequencies, especially between 200 and 315 Hz, the predicted level is up to 10 dB higher than the measurement. The peak in the prediction at 250 Hz is caused by the upstream contact strip (see Fig. 3(b)) even though this radiates mainly in the vertical direction. Note that the predicted level was obtained at a specific location, 2.8 m below the panhead, corresponding to the microphone array centre, while in the measurements, a microphone array containing different microphone heights in the range −2 to 2 m relative to the array centre was used. This inconsistency in receiver location may lead to some discrepancies between the prediction and the measurement.

Figures 3(b)–(d) indicate the components contributing most to the overall noise of the pantograph. In Fig. 3(b) the noise from the central segment of the upstream contact strip can be seen to generate, not only a significant peak at 250 Hz, but also large broadband noise over the whole frequency range. Figures 3(c) and (d) show that the component responsible for the peak at 800 Hz is part of the horn, while the peak at 2500 Hz is due to the straps.

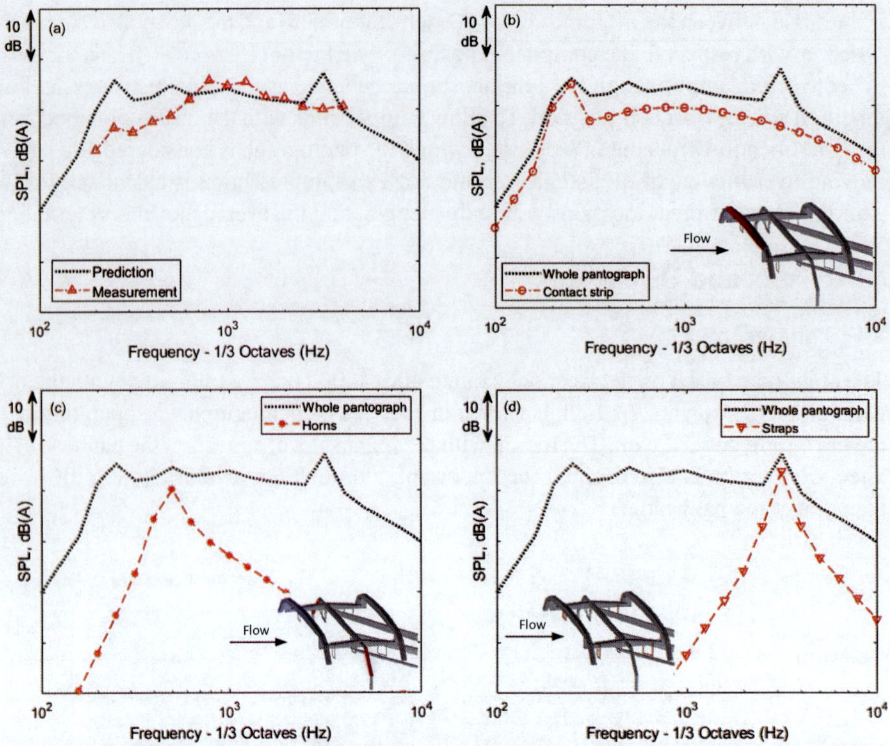

Fig. 3. (a) Comparisons between predicted and measured A-weighted noise spectra for the panhead of pantograph 1. (b), (c), (d) A-weighted noise spectra of different panhead components.

4.2 Pantograph 2

The noise map for the second pantograph at 177 km/h with knee-joint downstream is shown in Fig. 4. Available measurement data was obtained for frequencies from 200 to 4000 Hz, but the contributions from different pantograph components are difficult to separate below 400 Hz. Therefore, the measured results are presented for 400–4000 Hz. Similar to Fig. 2, the dominant sources for this pantograph are again found at the panhead and the levels at the knee-joint region are much lower.

The measured spectra for the dominant components, the panhead and the knee-joint, and the whole pantograph are presented in Fig. 5. The spectrum of the whole pantograph was obtained by taking the incoherent sum of the spectra of the panhead and the knee-joint. The noise spectrum is dominated by the contribution from the panhead in the region from 630 to 4000 Hz, where the levels from the knee-joint are at least 7 dB less than those of the panhead.

Figure 6 shows comparisons between the predictions and measurements for the noise spectrum generated by the panhead. The predicted noise spectrum for the panhead shows good agreement with the measurement over the frequency range 250–4000 Hz with the differences generally less than 5 dB.

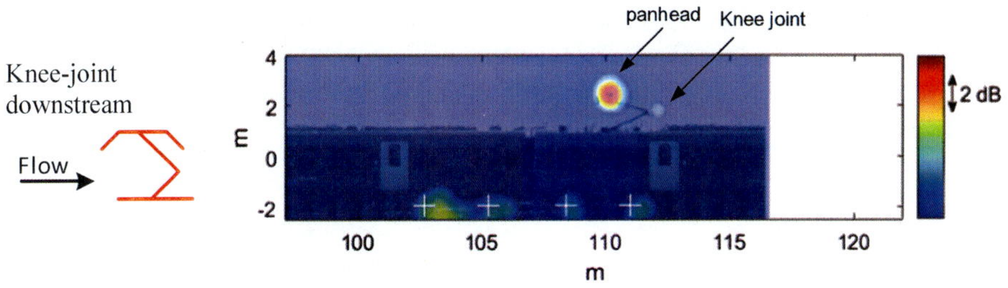

Fig. 4. Noise maps over 400–4000 Hz for pantograph 2 at 177 km/h with knee-joint downstream. The pantograph orientation is shown on the left.

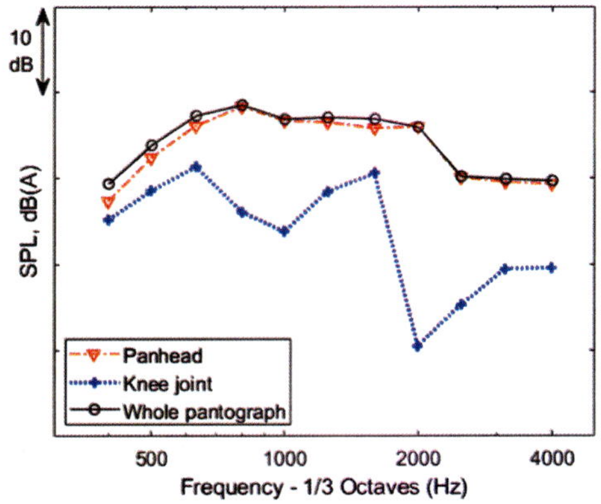

Fig. 5. Measured noise spectra of two source regions, the panhead and the knee-joint for pantograph 2 at 177 km/h.

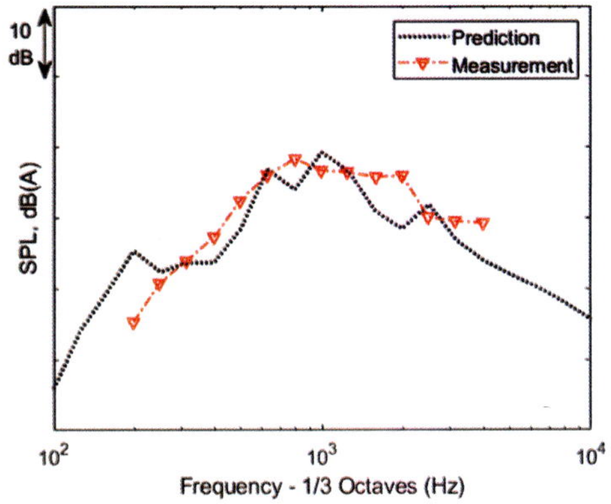

Fig. 6. Comparisons between predicted and measured A-weighted noise spectra for the panhead of pantograph 2 at 177 km/h.

The prediction for the knee-joint is not available in the current model. The geometry in the knee region is complex, which makes it more likely to cause strong flow interactions; therefore the corresponding noise generation cannot be simply considered as vortex shedding noise.

5 Conclusions

A component-based model has been applied to predict aerodynamic noise from two high-speed train pantographs. The predicted noise results have been compared with the measurements obtained with microphone array techniques. It is observed from the measurements that the panhead is the most significant source component for both pantographs whereas the knee-joint is less important.

For pantograph 1, the agreement between the prediction and the measurement is commendable, although some deviations are observed between 200 and 315 Hz, where higher noise levels are predicted due to the contribution of the contact strip. The predictions for pantograph 2 give good agreement over the frequency range 250–4000 Hz with spectral differences generally less than 5 dB. It should be noted that the predicted results were obtained at a specific height, 2.8 m below the panhead, corresponding to the microphone array centre, while in the measurements, a microphone array containing different microphone heights in the range –2 to 2 m relative to the array centre was used. This inconsistency may lead to some discrepancies between the predicted and measured noise results.

References

1. Lölgen,T.: Wind tunnel noise measurements on full-scale pantograph models. In: Joint ASA/EAA meeting, Berlin, Germany (1999)
2. Latorre Iglesias, E., Thompson, D.J., Smith, M.G.: Experimental study of the aerodynamic noise radiated by cylinders with different cross-sections and yaw angles. J. Sound Vib. **361**, 108–129 (2016)
3. Liu, X.: Aerodynamic noise from components of a train pantograph and its reduction, Ph.D. thesis, University of Southampton (2017)
4. Latorre Iglesias, E., Thompson, D.J., Smith, M.G.: Component-based model to predict aerodynamic noise from high-speed train pantographs. J. Sound Vib. **394**, 280–305 (2017)
5. J. Zhang, Implementations of microphone arrays for railway noise identification, Ph.D. thesis, University of Southampton (2018).
6. Dittrich, M.G., Janssens, M.H.A.: Improved measurement methods for railway rolling noise. J. Sound Vib. **231**(3), 595–609 (2000)
7. Brooks, T., Humphreys Jr, W.: Effect of directional array size on the measurement of airframe noise components. In: 5th AIAA/CEAS Aeroacoustics Conference and Exhibit (1999).

Measurement and Reduction of the Aerodynamic Bogie Noise Generated by High-Speed Trains in Terms of Wind Tunnel Testing

Yoichi Sawamura[1](✉), Toki Uda[1], Toshiki Kitagawa[1], Hiroshi Yokoyama[2], and Akiyoshi Iida[2]

[1] Railway Technical Research Institute, Environmental Engineering Division, Noise Analysis, 2-8-38 Hikari-cho, Kokubunji, Tokyo, Japan
sawamura.yoichi.26@rtri.or.jp

[2] Department of Mechanical Engineering, Toyohashi University of Technology, 1-1 Hibarigaoka, Tempaku, Toyohashi, Aichi, Japan

Abstract. The aerodynamic noise generated from bogie sections is one of the main sources of noise when high-speed trains run above 300 km/h. A new wind tunnel test method is proposed herein to determine the sound source distribution in detail and clarify the generation mechanism of the aerodynamic bogie noise in railways. Our wind tunnel test conducted for the latest Shinkansen vehicle indicated that the aerodynamic bogie noise at around 250 Hz to 315 Hz is generated by the cavity structure of the bogie section. Moreover, the noise source at around 400 Hz to 500 Hz is located at the traction motors and a gear unit. Three countermeasures were used herein to control the flow at the upstream of the bogie (i.e., two-dimensional deflector, three-dimensional (3D) deflector, and bottom lowering). Consequently, the aerodynamic bogie noise is properly reduced by the 3D deflector and bottom lowering.

Keywords: Aerodynamic noise · Wind tunnel test · High-speed trains · Sound source localization

1 Introduction

The noise generated by high-speed trains is an important environmental issue. The primary noise sources of high-speed trains are pantograph aerodynamic noise, lower-part noise (aerodynamic and rolling noise), upper-part aerodynamic noise, and bridge noise. For the lower-part noise, the aerodynamic noise generated from bogie sections become dominant above 300 km/h. The past study [1, 2] suggested that the aerodynamic bogie noise was dominant in the frequency range of 125–500 Hz. An attempt was made to control this noise using a wind tunnel test, in which some countermeasures were applied around the relevant section [3]. In studying the aerodynamic bogie noise, which is the

G. Degrande et al. (Eds.): *Noise and Vibration Mitigation for Rail Transportation Systems*, NNFM 150, pp. 73–80, 2021.
https://doi.org/10.1007/978-3-030-70289-2_5

main source of the lower-part noise, the noise generation mechanism must be clarified by means of wind tunnel experiments and numerical simulations [4]. The present study will perform the following procedures: (1) develop a new measurement method to estimate the aerodynamic noise using a wind tunnel test, (2) properly understand the relevant mechanisms, and (3) examine the effect of countermeasures on the aerodynamic noise.

2 Methodology

2.1 Wind Tunnel Test

A wind tunnel test was performed at the Railway Technical Research Institute's large-scale low-noise wind tunnel in Japan to simulate the aerodynamic bogie noise generated by a high-speed train. This wind tunnel has the ability to achieve a maximum wind velocity of 400 km/h. The background noise level is 75.6 dBA at 300 km/h. The nozzle exit is 3.0 m in width and 2.5 m in height. The test section is 7.0 m long; hence, longer railway vehicles can be tested. Figures 1 and 2 show an illustration of the wind tunnel test.

In the test, a one-seventh-scale train model with a bogie model was used, and a two-dimensional (2D) microphone array was set under the bogie. Figure 3 shows a schematic figure around the bogie. R.L. denotes the maximum height of rails in the figure. The flow velocity was 325 km/h. The model was fixed by two supports set in the front and rear of the train model. The ground is stationary. In other words, there is no speed difference between train body and the ground. This is different from the real case where train runs over ground. The flow between the underbody and ground is approximated by a Couette flow in real case and a Poiseuille flow in test case. However, this is acceptable. This is because in this tests, the main purpose is to understand the generation mechanism of the aerodynamic bogie noise, and it is the most important to simulate the airflow very close to the train model. If the ground is moved, the profile of the whole profile under the body will be well-simulated, but the airflow very close to the model is almost the same regardless of whether the moving ground is set or not. The horizontal profile of the flow speed at the bogie is controlled as possible to those obtained by the field tests [5]. An acoustic transmission plate, which lets the sound propagate while blocking the airflow, was set on the stage floor between the bogie and the microphone array. The average pore size of the porous plate was 150 μEm. The porosity rate was 89%. A 30 mm-thick porous plate was used to reduce the effect caused by the air flow, although the transmission loss increased with the porous plate thickness.

This measurement method is reasonable to understand aerodynamic bogie noise, nevertheless this method obtains the aerodynamic bogie noise radiating vertical direction and not lateral direction. If the observation point is located at the position lateral to the bogie, it is necessary to make a source model including the ground effect.

2.2 Transmission Loss of the Porous Plate

The acoustic properties of the porous plate were investigated before performing the wind tunnel test. A speaker was set downward at the bogie section, and the white noise emitted from the speaker was measured with an omnidirectional microphone at the center of the microphone array. Figure 4 shows the difference in the sound pressure level between the presence and the absence of the porous plate. A larger vertical axis means a larger transmission loss caused by the porous plate. The transmission loss was approximately a range from 10 to 20 dB in the scaled frequency range of 250 to 500 Hz, where the aerodynamic bogie noise becomes dominant.

2.3 Measurement of Aerodynamic Bogie Noise Using a 2D Microphone Array

The microphone array had a radius of 1 m and consisted of 66 microphones, which were radially arranged. The distance between the microphone array and the ground plate was 370 mm. The source plane was assumed to be the bottom of the vehicle, which was 60 mm above the ground plate. The total distance between the microphone array and the source plane was set at 430 mm. The sound wave convection across the wind must be considered, and the sound source must be properly modified in the wind tunnel tests. However, in the wind tunnel setup shown in Fig. 1, the wind convection will be small enough to be ignored because the bogie is close to the ground plate. In the measurement method, the actual vertical positioning of the aerodynamic noise source could be shifted because the source plane was assumed to be the bottom of the vehicle. However, this assumption is reasonable because high-speed flow exists only near the bottom of the vehicle.

The aerodynamic bogie noise distribution in the range of 800 Hz to 20k Hz (114 Hz to 2.9k Hz in a full-scale model) was obtained with a 16 Hz frequency resolution (2.3 Hz in a full-scale model). The signals of the microphone array were processed using one of the deconvolution algorithms implemented in the non-negative least-square method [6], which resulted in a sound source map at 5 mm intervals.

The measurement method can determine in detail the bogie sound source. A-weighted SPL was calculated as follows using a 2D microphone array to evaluate the aerodynamic bogie noise: (1) the SPL distribution on each one-third octave band was calculated; (2) the integral region was defined in the whole SPL distribution region (Fig. 5); (3) the integrated value over the integral region was calculated on each one-third octave band; (4) the value of the measurement frequency represented by one-third octave band was divided by 7 to convert the value obtained in (3) into a full-scale model; and (5) A-weighting was applied to the SPL obtained in (4). This evaluation method is reasonable for the aerodynamic bogie noise [7].

2.4 Countermeasures for the Aerodynamic Bogie Noise

The aerodynamic bogie noise is generated from the bogie components; hence, it is better to control the airflow around the bogie section. Accordingly, the three following countermeasures are proposed to control the airflow (Fig. 6): (1) a 2D deflector device, (2) a three-dimensional (3D) deflector device, and (3) bottom lowering. The 2D deflector device expected to deflect the airflow downward and decrease the flow velocity into the bogie section. 3D deflector device is expected to deflect the airflow not only downward but also lateral direction, and decrease the flow velocity into bogie section. These countermeasures are intended to reduce the inlet flow speed at the bogie cavity, while the flow speed increase near the ground. Therefore, another consideration is required for ballast section. Also issues of collection of snow and ice is the future study when deflector devices are applied to actual trains.

Bottom lowering is expected to narrow the flow path and decrease the flow velocity into the bogie.

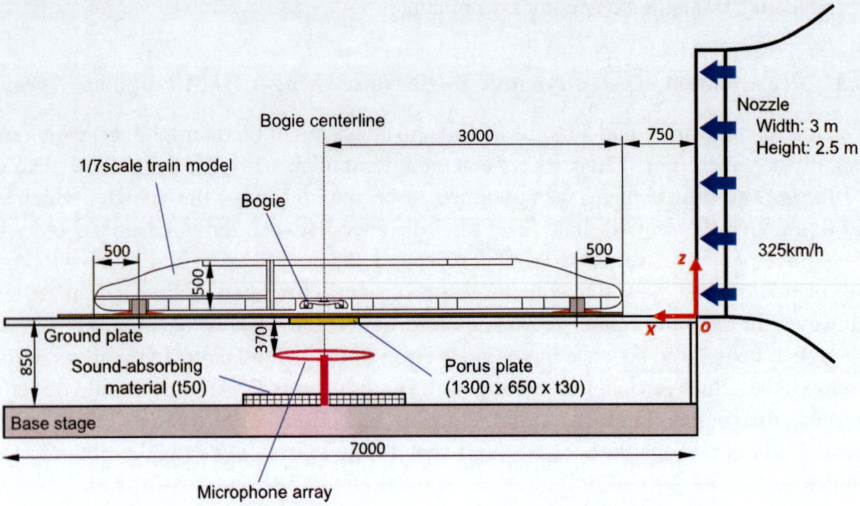

Fig. 1. Schematic side view of the wind tunnel setup

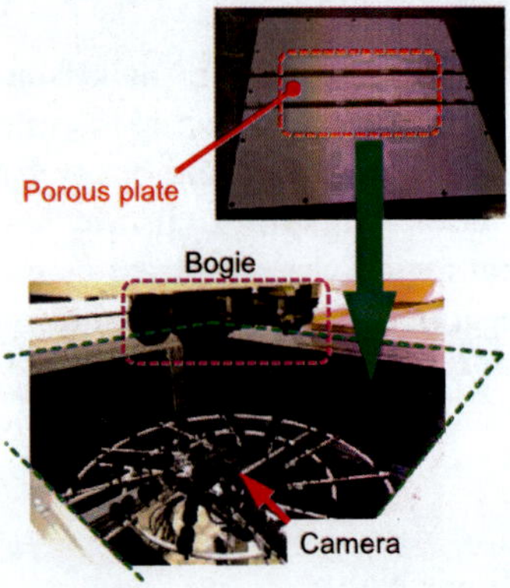

Fig. 2. Positional relation between the bogie and the microphone array

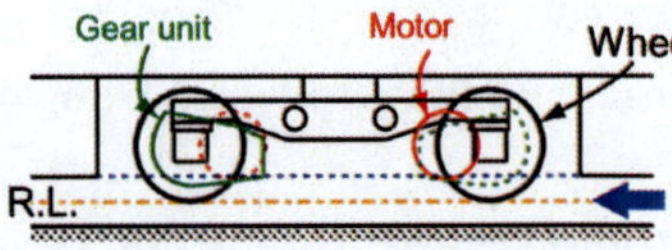

Fig. 3. Schematic side view of the bogie section

3 Results

3.1 Distribution of the Aerodynamic Bogie Noise

A wind tunnel experiment was performed to investigate in detail the sources of the aerodynamic bogie noise using the measurement method described in Sect. 2.1. Figure 7

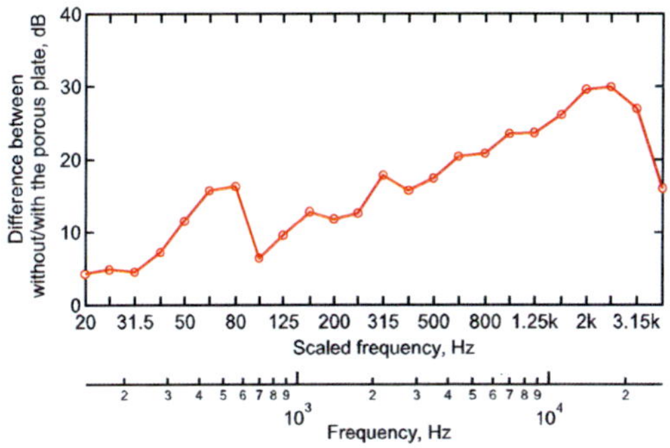

Fig. 4. Difference in the sound pressure levels without/with the porous plate

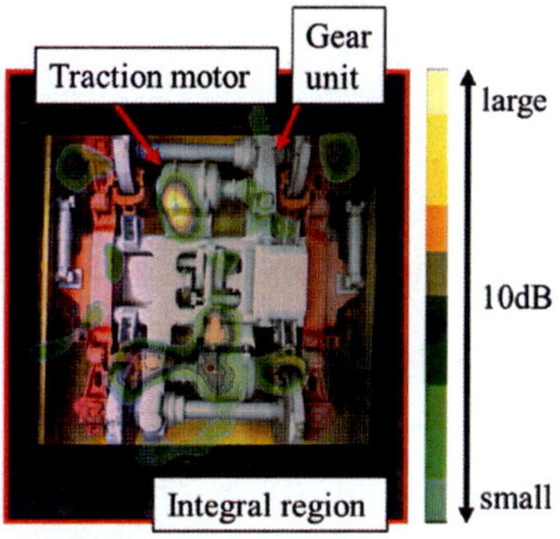

Fig. 5. Example of the source distribution of the aerodynamic bogie noise

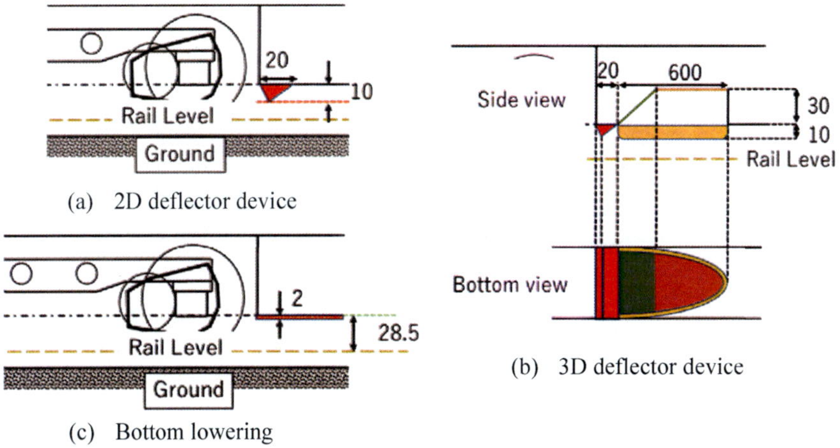

Fig. 6. Devices for the aerodynamic bogie noise reduction (units: mm)

shows the sound sources. The porous plate was not transparent; hence, a picture of the bogie without the porous plate was taken separately, and the calculated sound source distribution was superimposed on the picture. The sound source distribution was different at each frequency band. The main sources in the frequency range of 100 Hz to 125 Hz in Fig. 7(a) were located around the traction motors and axles, and that located downstream was greater. The source located most downstream in the frequency range of 250 Hz to 315 Hz in Fig. 7(c) became greater, suggesting that the cavity structure of the bogie section emitted an aerodynamic sound. The vortex sound is generated at the downstream edge of the cavity, and fluid-resonant oscillation with a wavelength corresponding to one-third the cavity length (4m in actual scale) are observed. This trend can also be seen under the condition without the bogie. Finally, at 400 Hz to 500 Hz in Fig. 7(d), prominent sound sources were observed around the traction motors and the gear unit located downstream because a part of these traction motors and gear units was located below the vehicle bottom line (Fig. 3) and exposed to a high-speed air flow. Therefore, one of the design concepts to reduce the aerodynamic bogie noise is to make the air flow speed lower in front of the bogie parts. This concept will be implemented by deflecting the air flow in front of each bogie section. The position of the traction motors and the gear units must also be modified such that they can be positioned more vertically upward.

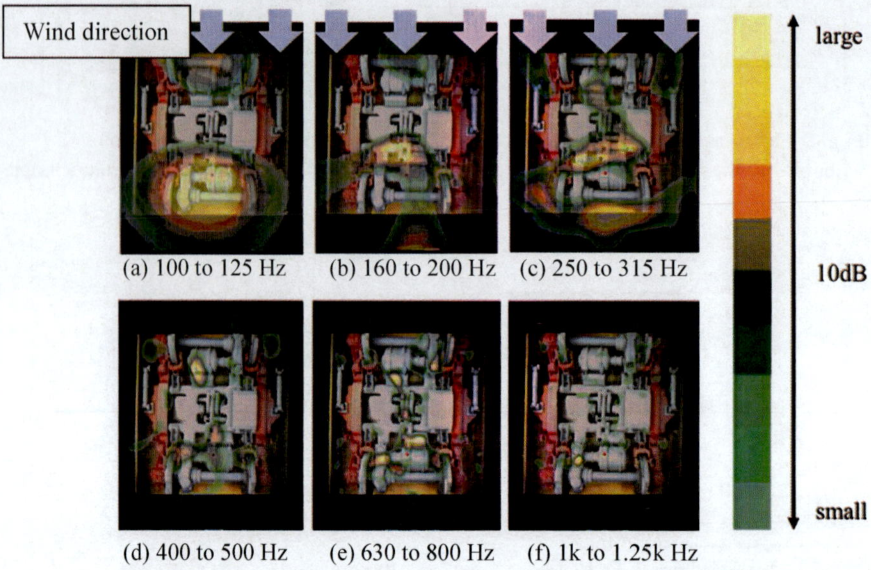

Fig. 7. Noise source distribution of the aerodynamic bogie noise

3.2 Noise Reduction of the Aerodynamic Bogie Noise by the Countermeasures

The wind tunnel tests were performed for the standard bogie condition and the three countermeasures introduced in Sect. 2.4 to evaluate the effect of the countermeasures. Figure 8 shows the frequency spectra in the three conditions. For the 2D deflector device, the SPL in the frequency range below 1000 Hz was smaller by up to 3 dB. For the 3D deflector device and bottom lowering, the SPL in the frequency range of 200–500 Hz was smaller by up to 6 dB than that in the standard bogie condition. The difference of the SPLs in this frequency range between the standard and 2D deflector device conditions was small.

Figure 9 shows the source distributions of the aerodynamic bogie noise in the frequency range of 400–500 Hz. For the 2D deflector device, the SPL near the traction motor on the upstream side was smaller than that in the standard condition (Fig. 9(a)). However, on the downstream side, the SPL remained greater near the wheel because the device only controlled the airflow on the upstream side. Therefore, a device must have an effect that controls the flow speed on the entire bogie to effectively reduce the aerodynamic noise. These results show that the measurement method can be effectively used to estimate the effects caused by the countermeasures as well as understand the necessary mechanisms for noise reduction. Figures 9(b) and (c) show that when using the measurement method, the SPLs near the traction motor on both the upstream and downstream sides were smaller because of the two countermeasures. This result is attributed to countermeasures (2) and (3) considerably reducing the airflow speed into the bogie section.

When these countermeasures are applied to the actual train, the noise reduction could be smaller than that of the wind tunnel test. This is because the turbulence intensity of inlet flow at the bogie cavity in the wind tunnel test could become weaker than that of the actual train case. Thus, frequencies of the aerodynamic bogie noise observed in actual trains could be broader than that of the wind tunnel test.

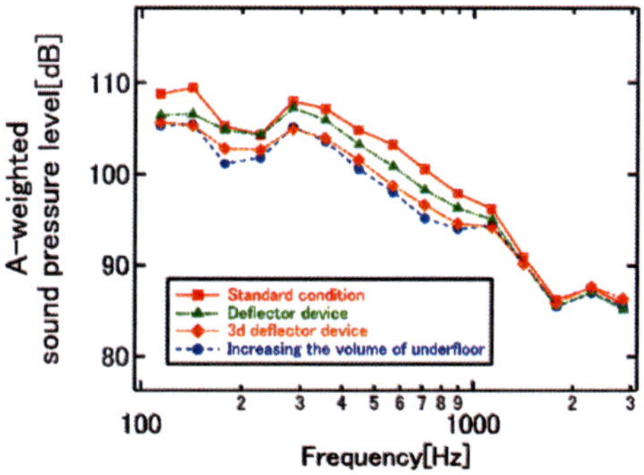

Fig. 8. Frequency spectra for the three countermeasures (standard condition is for reference)

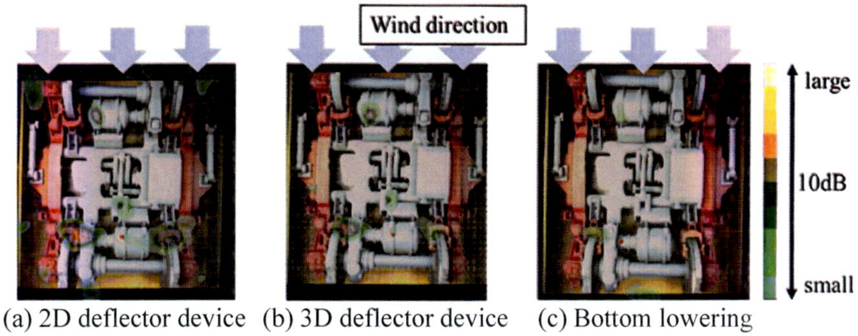

Fig. 9. Noise source distribution of the aerodynamic bogie noise (400 Hz to 500 Hz)

4 Conclusion

This study proposed a new wind tunnel test method to determine the sound source distribution in detail and clarify the generation mechanism of the aerodynamic bogie noise in railways. The wind tunnel test conducted for the latest Shinkansen vehicle indicated that the aerodynamic bogie noise at around 250 Hz to 315 Hz was generated by the cavity structure of the bogie section, and the noise source around 400 Hz to 500 Hz was located at the traction motors and a gear unit. The latter result implied that the basic concept of the aerodynamic bogie noise reduction deflected the air flow in front of each bogie part or modified the position of the traction motors and the gear units to make it more vertically upward.

References

1. Yamazaki, N., Kitagawa, T., Uda, T., Nagakura, K., Iwasaki, M., Wakabayashi, Y.: Evaluation method for aerodynamic noise generated from the lower part of cars in consideration of the characteristics of under-floor flows on Shinkansen trains. Quart. Rep. RTRI **57**(1), 61–68 (2016)
2. Mellet, C., Létourneaux, F., Poisson, F., Talotte, C.: High speed train noise emission: Latest investigation of the aerodynamic/rolling noise contribution. J. Sound Vibration, **293,** 535–546 (2006)
3. Yamazaki, N., Uda, T., Kitagawa, T.: Influence of bogie components on aerodynamic bogie noise generated from Shinkansen train. In: The Proceedings of the Symposium on Environmental Engineering (2016)
4. Zhu, J.Y., Hu, Z.W., Thompson, D.J.: Analysis of aerodynamic and aeroacoustic behavior of a simplified high-speed train bogie. Notes on Numerical Fluid Mechanics and Multidisciplinary Design, vol. 126, , pp. 489–496 (2015)
5. Uda, T., Wakabayashi, Y., Kurita, T., Iwasaki, M., Yamazaki, N., Ido, A.: Velocity profile under Shinkansen vehicle on slab track. Trans. JSME **81**(830) (2015). (in Japanese)
6. Ehrenfried, K., Koop, L.: Comparison of iterative deconvolution algorithms for the mapping of acoustic sources. AIAA J. **45**(7), 1584–1595 (2007)
7. Yamazaki, N., Nakayama, M., Nishiura, T.: A method to evaluate the aeroacoustic bogie noise of Shinkansen high-speed trains by considering the acoustic field. Trans. Japan Soc. Mech. Eng. **85**(869) (2019)

Development of New Low-Noise Pantograph for High-Speed Trains

Mitsuru Saito[1](✉), Fumio Mizushima[1], Yusuke Wakabayashi[1], Takeshi Kurita[1], Shinji Nakajima[2], and Toru Hirasawa[2]

[1] Advanced Railway System Development Center, R&D Center of JR East Group, East Japan Railway Company, 2-479, Nisshin-cho, Saitama City, Saitama, Japan
saitou-mitsuru@jreast.co.jp

[2] Design Division of Transportation System Works, Transportation Business Unit, Toyo Electric Mfg. Co., Ltd, 3-8, Fukuura, Kanazawa-ku, Yokohama, Kanagawa, Japan

Abstract. When increasing the speed of Shinkansen trains, it becomes more important to reduce noise generated from cars. In the past, East Japan Railway Company developed a low-noise pantograph (PS208 type) to equip on the Series E5 Shinkansen trains that was able to reduce aerodynamic noise emitted around pantographs. However, with plans for greater increases in speed, pantograph-based noise continues to be an issue, as it has been one of the major noise sources even on the Series E5. In this study, we first investigated the noise source of the existing PS208 and found that noise is mostly generated from the four following areas: base of the lower arm, hinge cover between the upper and lower arms, pantograph head, and pantograph head support. Next, as countermeasures for these noise sources, we conducted wind tunnel tests and numerical simulations to test for improved shapes. As a result, we developed a new pantograph (PS9040), with 1.5 dB less aerodynamic noise than the PS208.

Keywords: High-speed rail · Shinkansen · Pantograph · Aerodynamic noise · Wind tunnel test

1 Introduction

In Japan, Shinkansen noise standards require that wayside noise at 25 m away from the track to be 75 dB(A) or less. Therefore, noise levels at higher speeds can be no greater than those of existing commercial trains running at 320 km/h. For this reason, further noise reduction is one of the most important issues encountered when increasing speed. In particular, aerodynamic noise from pantographs is a significant issue to be addressed as it greatly affects the overall noise level.

In the past, the East Japan Railway Company developed a low-noise pantograph (PS208 type (Fig. 1)) for Series E5 Shinkansen trains that succeeded in reducing aerodynamic noise around pantographs. However, with plans for greater increases in speed, pantograph-based noise continues to be an issue, as it has been a major noise source

G. Degrande et al. (Eds.): *Noise and Vibration Mitigation for Rail Transportation Systems*, NNFM 150, pp. 81–89, 2021.
https://doi.org/10.1007/978-3-030-70289-2_6

even on the Series E5 (Fig. 2). In this study, we first investigated noise sources of the existing PS208 using microphone array, and then developed the countermeasures to be explained here.

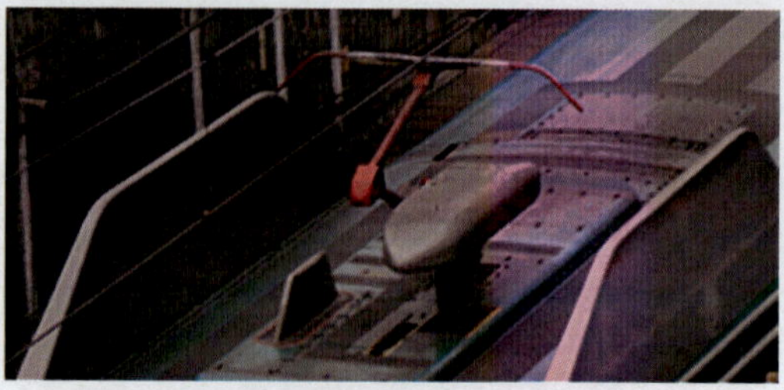

Fig. 1. Type PS208 pantograph

Fig. 2. Noise source distribution of series E5 Shinkansen using two-dimensional microphone array

2 Development of a Low-Noise Pantograph

2.1 Investigation of Noise Sources of the PS208

The noise sources of the existing PS208 were investigated in the Railway Technical Research Institute's large-scale low-noise wind tunnel (background noise level 75.6 dB at 300 km/h) using a two-dimensional microphone array (Fig. 3). Figure 4 shows the results of the investigation. We found that major noise sources are mostly generated from the four following areas: pantograph head, pantograph head support, base of the lower arm, and hinge cover between the upper and lower arms. Countermeasures for noise sources investigated by wind tunnel tests and numerical simulations are explained in the following section.

2.2 Countermeasures for Pantograph Noise

Pantograph Head Support. In our previous study, we conducted numerical simulation for the upper part of the PS208 [1]. We found that strong pressure fluctuation that generates aerodynamic noise is caused by interference between flow around the pantograph head and flow around the pantograph head support (Fig. 5); and that it is possible to reduce interference by changing the positional relationship of the pantograph head and head support.

Fig. 3. Wind tunnel tests using microphone array

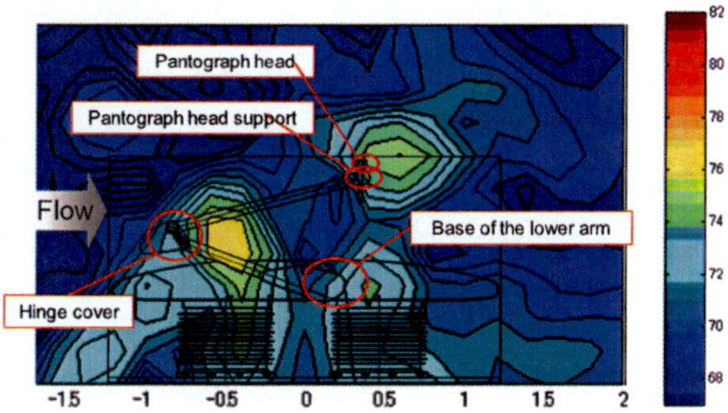

Fig. 4. Noise source distribution of the PS208 (1 kHz band)

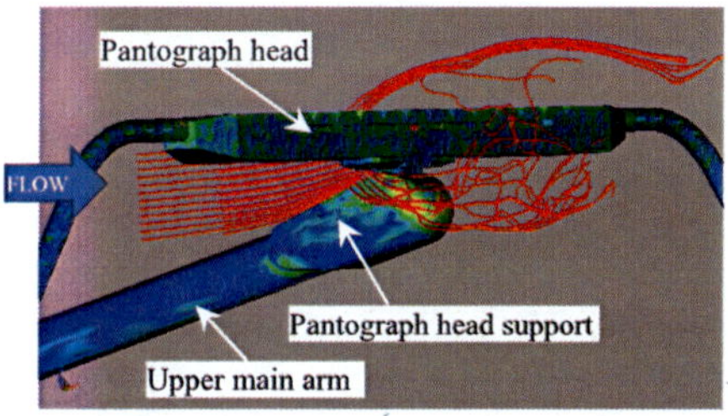

Fig. 5. Pressure fluctuation at the upper part of the PS208 pantograph at 1 kHz with stream line (Red line)

From the above findings, we proposed an improved shape in which the pantograph head is embedded into the pantograph head support (Fig. 6). Pressure fluctuation was also observed in the boundary area between the pantograph head support and upper main arm as shown Fig. 7, so we smoothened the boundary area.

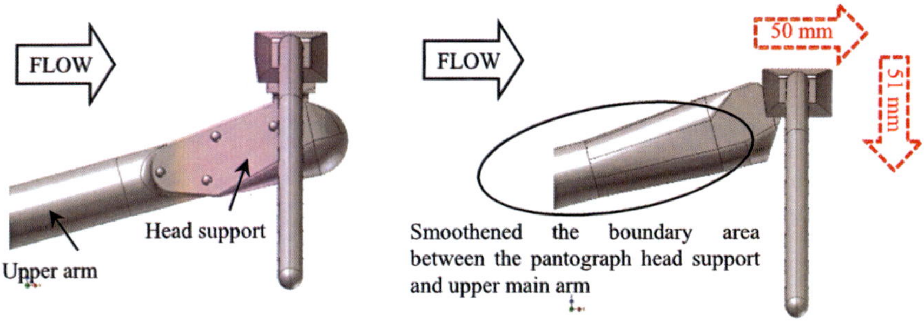

(a) Standard shape (PS208) (b) Improved shape

Fig. 6. The analysis model of the "improved shape", shifted the pantograph head 50 mm downstream (right) and lowered it by 51 mm at the tip of the upper arm

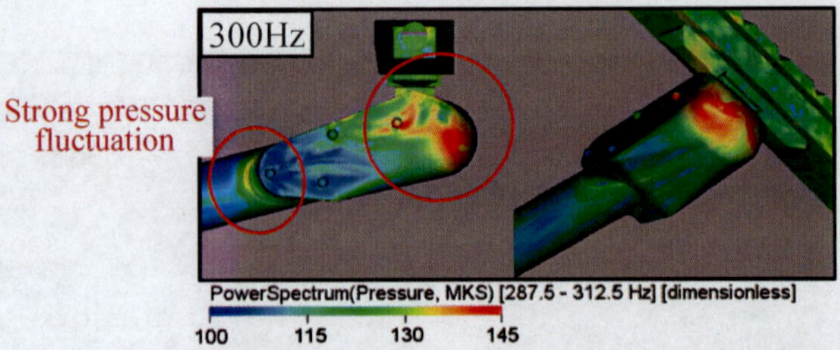

Fig. 7. Pressure fluctuation at 300 Hz on the surface of "standard shape" in side and bottom view

Figure 8 shows the sound pressure levels (SPLs) of the upper part of the pantograph obtained by numerical simulation. We confirmed that the improved shape reduced noise from the pantograph head support over a wide frequency range from 100 to 800 Hz, and reduced the overall noise from the upper part of pantograph by 1.6 dB.

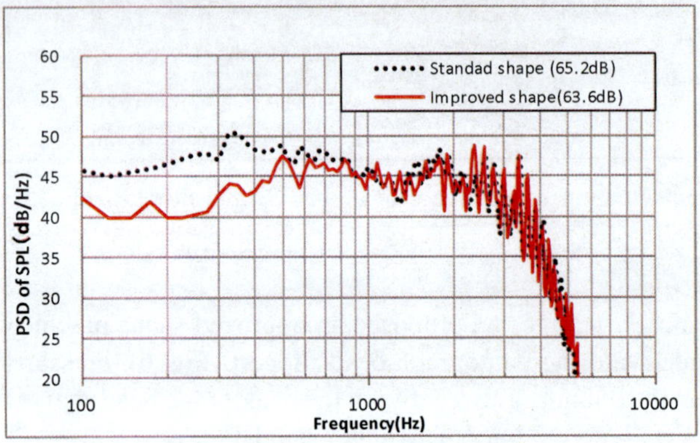

Fig. 8. SPLs for the standard and improved shapes

Pantograph Head. A multi-segment slider is used in the PS208 pantograph head as shown in Fig. 9 to improve the contact loss ratio of pantographs [2].

We conducted a numerical simulation for the PS208 pantograph head using the Lattice Boltzmann Method (LBM) via PowerFLOW© [3]. Simulation settings are shown in Table 1. Mesh size was determined in regards to both its location in the analysis space and the distance from surface of the pantograph head. Basically, the closer the mesh is to the surface of the pantograph head, the finer the mesh to be used. The finest level (0.2 mm) is applied to gaps between segments on the multi-segment slider of pantograph head. Figure 10 shows results of pressure fluctuation at both 500 Hz and 1 kHz. We found that strong pressure fluctuation on the pantograph head is generated on the upstream side of the pantograph contact strip. To reduce the pressure fluctuation, the contact strip was chamfered as shown in Fig. 11 and the effect was investigated by 1/3.15-scale wind tunnel as shown in Fig. 12 Comparison of the aerodynamic noise spectra of standard

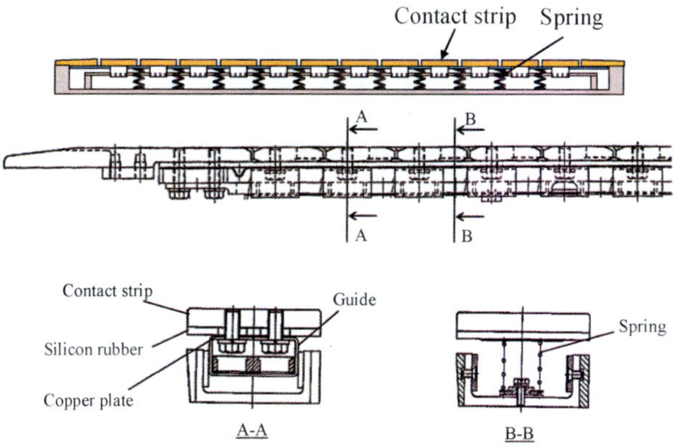

Fig. 9. Detail of multi-segment slider structure (PS208)

(PS208) and improved shapes in wind tunnel test is shown in Fig. 13. The sound pressure levels of the improved shape, chamfered at an angle of 30 degrees was reduced over a wide frequency range and was reduced by 1.6 dB in overall noise.

Table 1. Simulation Settings.

Mean velocity	320 km/h (88.9 m/s)
Fluid	air
Satndard pressure	atmospheric pressure
Analysis time	0.2 sec
Scale	1: 1
Target frequency range	200-2000 Hz

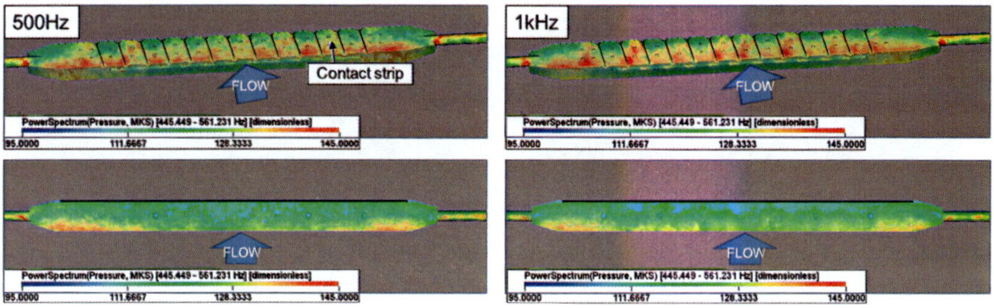

Fig. 10. Pressure fluctuation at 500 Hz and 1 k Hz on the surface of "standard shape" (PS208) in top view.

Base of the Lower Arm. We conducted numerical simulations for the PS208 pantograph as shown in Fig. 14. The analysis revealed that the strong pressure fluctuation around the base of the lower arm is caused by separating flow from the main axle cover hitting on the windshield cover. In an attempt to reduce the pressure fluctuation, we imple-mented a 1/10 scale wind tunnel test (Fig. 15).

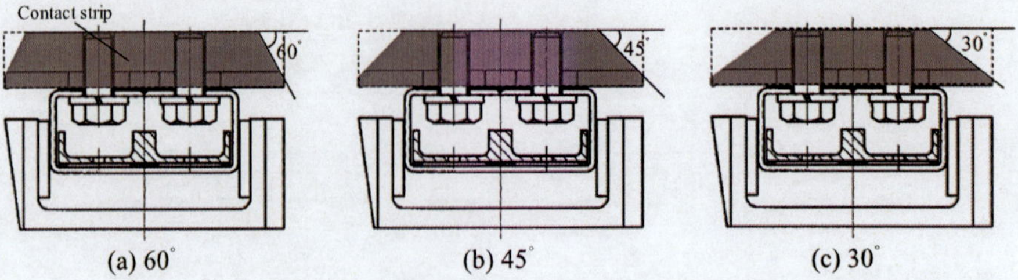

Fig. 11. Chamfered contact strips in the improved shape (cross section)

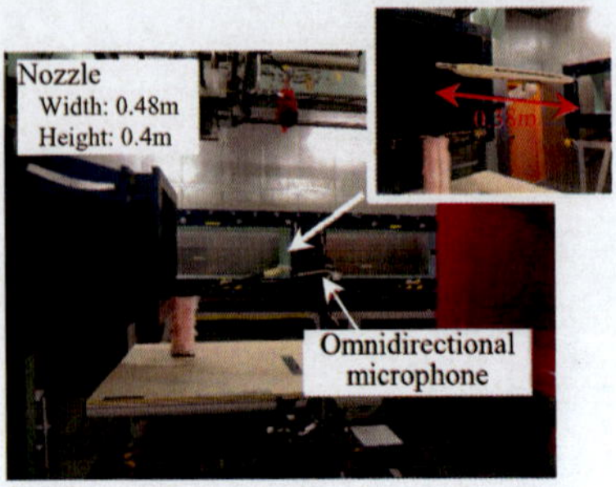

Fig. 12. 1/3.15-scale pantograph head model

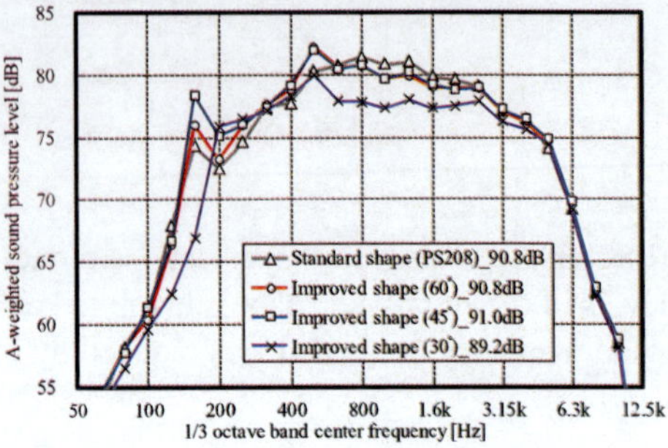

Fig. 13. Results of wind tunnel tests at wind speed of 80m/s

By trial and error, we found that the improved shape, where the supporting structure of the lower arms was changed as shown in Fig. 16, is most successful at reducing noise.

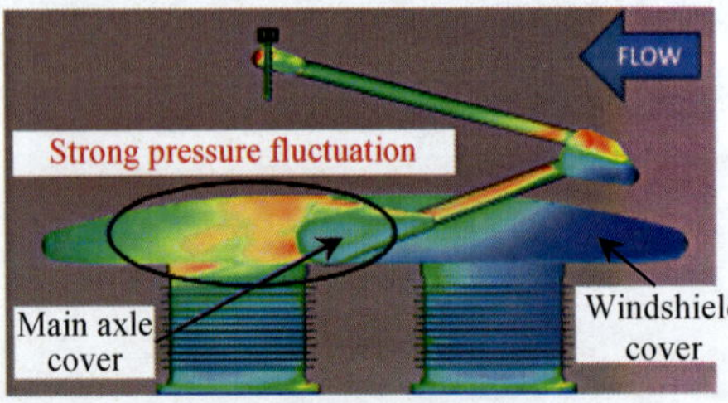

Fig. 14. Pressure fluctuation on the PS208 pantograph at 100 Hz to 2 kHz

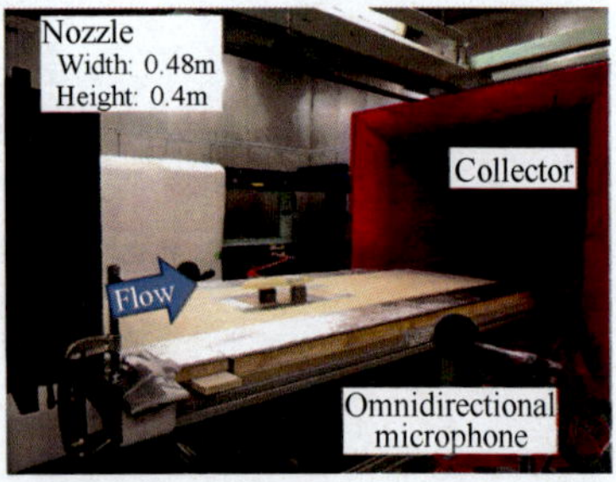

Fig. 15. 1/10-scale wind tunnel test

In addition, the two insulators were relocated and embedded into the pantograph's base frame as shown Fig. 17. The aerodynamic noise spectra of standard shape (PS208) and this improved shape are shown in Fig. 18. Sound pressure levels of the improved shape were reduced over a wide frequency range and overall noise with the improved shape was reduced by 1.8 dB.

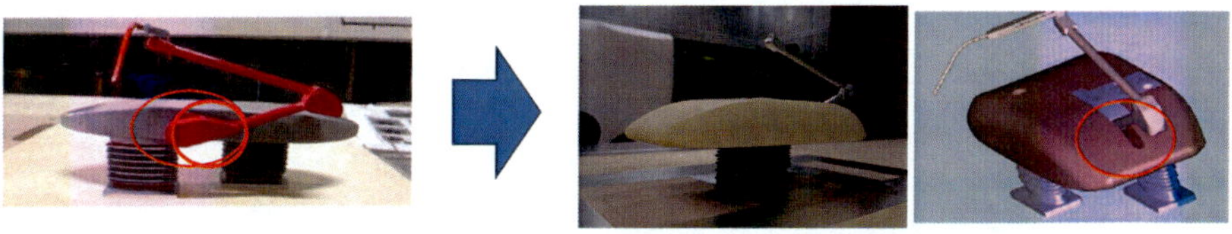

(a) Standard shape (PS208) (b) Improved shape

Fig. 16. Improved shape changed the supporting structure of the lower arm

Hinge Cover. The hinge cover was made more compact and thinner to reduce the projected area exposed to wind, as shown in Fig. 19.

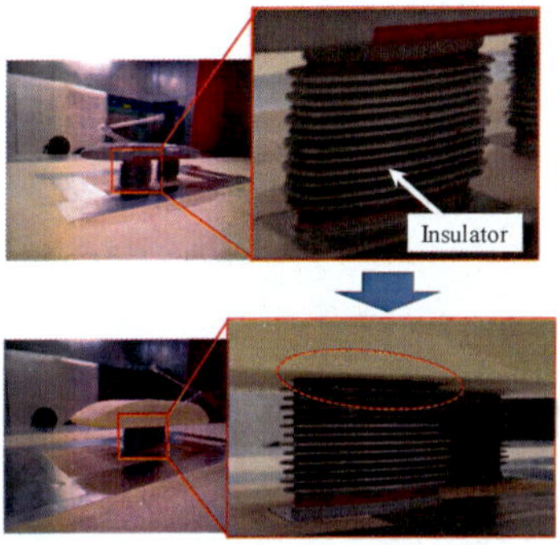

Fig. 17. Insulators embedded into the pantograph's base frame

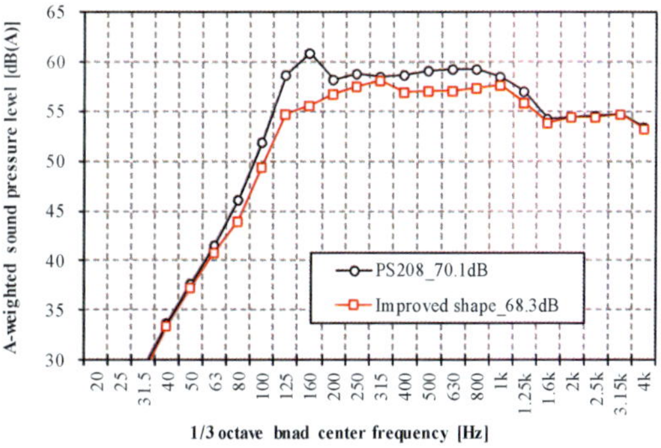

Fig. 18. Results of wind tunnel tests at wind speed of 80 m/s (without pantograph head)

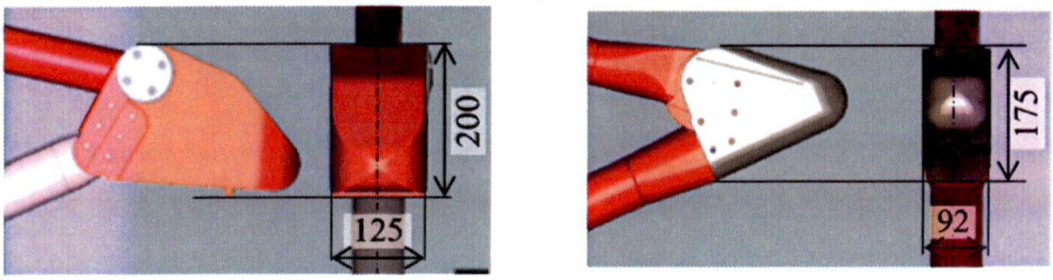

(a) PS208 (b) Improved shape

Fig. 19. Shape of hinge cover

2.3 Development of the New Low-Noise Pantograph

In order to examine the noise reduction performance by the above improvements, we have developed a new low-noise pantograph "PS9040" (Fig. 20). Noise reduction performance is verified by wind tunnel tests. The arrangement of the non-directional microphones is shown in Fig. 21. Measurement results and a comparison of aerodynamic noise spectrum are shown in Table 2 and Fig. 22, respectively. Overall, the PS9040 generated less noise than the PS208 in all microphones.

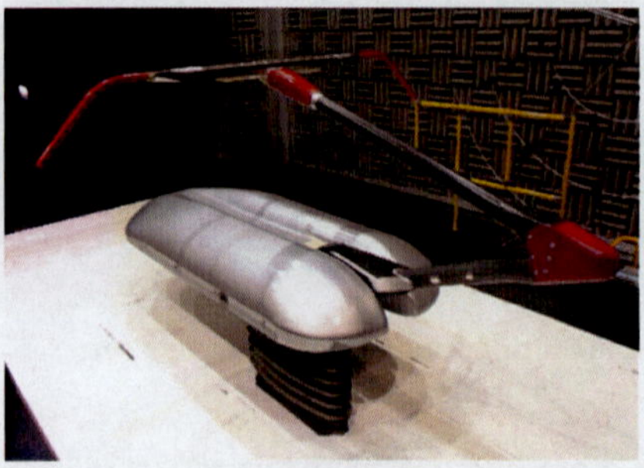

Fig. 20. PS9040 type pantograph

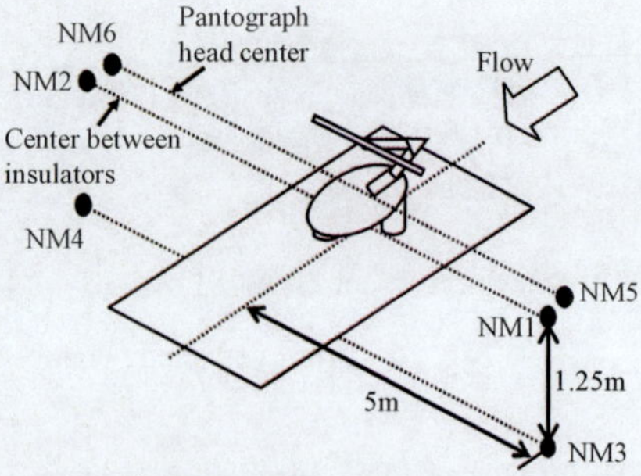

Fig. 21. Experiment equipment

Table. 2. Noise measurement results (overall differences in overall level from the PS208) Wind speed: 100 m/s

	Microphones (Unit: dB)					
	NM1	NM2	NM3	NM4	NM5	NM6
Knuckle forward	−2.0	−1.5	−1.9	**−1.5**	−2.2	−1.9
Knuckle backward	−2.9	−3.8	−2.2	**−2.8**	−2.5	−3.5

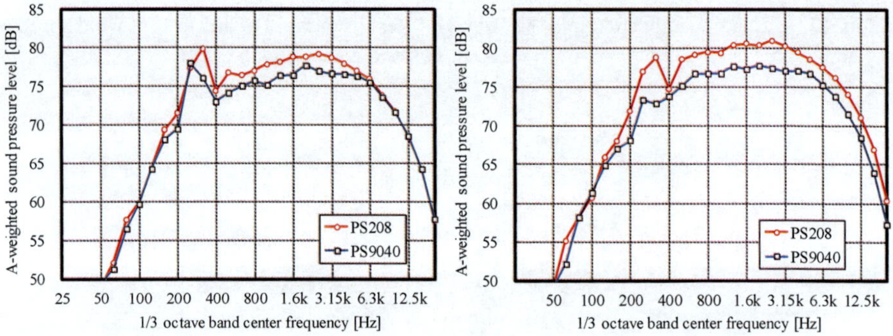

(a) knuckle forward (b) knuckle backward

Fig. 22. Comparison of the noise spectrum between PS208 and PS9040 at wind speed of 100 m/s (NM4)

3 Conclusion

In this paper, we investigated aerodynamic noise sources of the existing PS208 in use and examined improved shapes that would further reduce such noise. As a result, we developed a new pantograph (PS9040) that produces less aerodynamic noise than the PS208. Now we have just started running tests for a new high-speed Shinkansen test train "ALFA-X" [4] equipped with the PS9040. Other countermeasures for noise from pantographs and lower part of cars --, both major sources contributing to overall noise --, have also been adopted in the ALFA-X. The PS9040 and other newly proposed countermeasures in running trains will undergo a multifaceted evaluation in the near future to verify their effectiveness.

References

1. Shiraishi, N., Wakabayashi, Y., Kurita, T.: Numerical simulation for pantograph aerodynamic noise reduction. In: 7th International Symposium on Speed-up and Sustainable Technology for Railway and Maglev Systems, (STECH15) Proceedings CD-ROM (2015)
2. Kurita, T., Hara, M., Wakabayashi, Y., Mizushima, F., Satoh, H., Shikama, T.: Reduction of Pantograph Noise of High-speed Trains. J. Mecha. Syst. Transp. Logist. **3**(1), 63–74 (2010)
3. Chen, H., Kandasamy, S., Orszag, S., Shock, R., Succi, S., Yakhot, V.: Extended Boltzmann Kinetic Equation for Turbulent Flows. Science **301**, 633–636 (2003)
4. Edo, Y., Kawakami, O.: New Development in Next-Generation Shinkansen Research –New E956 Shinkansen Prototype–, Japanese Railway Engineering, No. 201 (2018), vol. 58, no. 3 (2018)

The Influence of Track Parameters on the Sound Radiation from Slab Tracks

Jannik S. Theyssen(✉), Astrid Pieringer, and Wolfgang Kropp

Applied Acoustics, Chalmers University of Technology, 412 96 Gothenburg, Sweden
jannik.theyssen@chalmers.se

Abstract. The influence of track parameters on the sound radiation has so far mainly been studied for ballasted tracks. The increasing usage of slab tracks in new railway lines worldwide makes a review of these parameters relevant. In this paper, the structural vibrations of rail and slab are evaluated based on two waveguide finite element models that are coupled in a finite number of positions. This discretely supported rail is compared to a rail on a continuous support. The sound radiation from rail and slab is evaluated based on a wavenumber boundary element method. The slab and rail contributions are evaluated separately. It is found that comparable to ballasted tracks, the rail pad stiffness has a large influence on the radiated sound power. For a continuously supported rail, the total sound power is reduced and the rail pad stiffness is less influential.

Keywords: Slab track · Infinite waveguide · Boundary elements · Discrete coupling · Sound power · Rail pad stiffness

1 Introduction

Slab tracks are a common type of track in high-speed railway lines. With the increasing length of high-speed lines in the world, more residents close to these lines are affected by noise emissions related to these tracks. Multiple references describe increased noise levels of slab tracks compared to ballasted tracks. A simulation of both track types reported in [1] showed an increase of 1.5 to 3 dB(A) for the slab track. Poisson [2] describes an observed increase of 3 dB(A) at 25 m distance from the track based on a number of measurements of different trains at different speeds. Thompson [3] describes the strong effect of the rail pad stiffness on the radiated noise for ballasted tracks. An increased pad stiffness leads to a decreased noise radiation. This decrease is described to be due to the stronger coupling of the rail to the sleepers, leading to a higher track decay rate. Since slab tracks use a lower pad stiffness by design, a higher sound radiation due to the weaker coupling is expected. However, slab tracks give the opportunity for supporting the rail continuously, increasing the coupling. Slab tracks furthermore have a larger mass than a system with sleepers, decreasing the receptance under the rail pad. These effects are investigated in the following.

G. Degrande et al. (Eds.): *Noise and Vibration Mitigation for Rail Transportation Systems*, NNFM 150, pp. 90–97, 2021.
https://doi.org/10.1007/978-3-030-70289-2_7

Different methods of calculating the radiated sound power from slab tracks have been proposed. In [1], the radiation from the rail is calculated using equivalent sources. The slab radiation is calculated for a continuously supported rail using a Finite Element model. The sound radiation from the rails and slab is treated separately in [4] as well. Here, the rail radiation is evaluated in a 2.5D boundary element model. The slab is modelled as a baffled plate. In the following, an integrated method of calculating the combined receptance and radiated sound power from rail and slab is introduced based on waveguide finite elements for the structural vibrations and a wavenumber boundary element model for the radiation.

2 Methodology

In this section, the waveguide finite element (WFE) method is described. Then the discrete coupling of multiple WFE-models is introduced. Finally, the coupling to the acoustic domain using the wavenumber boundary element (WBE) method is summarized.

2.1 Waveguide Finite Element Method

The following is a summary of the method, a more extensive derivation can for example be found in [5]. In this work, this method is applied to the geometry of a rail as well as the superstructure of a slab track.

A structure that is sufficiently large in one dimension with a constant cross-section can be approximated as a waveguide. The WFE method uses the assumption of travelling waves along the waveguide to reduce the finite element problem to the cross-section A of the structure. A lies in the y, z-plane and is discretized using conventional 2-dimensional, nine-node, iso-parametric quadrilateral elements with quadratic polynomials as shape functions.

A stationary motion at circular frequency ω is assumed with the time dependency $e^{j\omega t}$. A matrix equation for the relationship between the element nodal displacements u_i and the nodal forces f_i is derived for each element by applying Hamilton's principle. Assembling these matrices leads to the expression

$$\left[\mathbf{K}_2\frac{\delta^2}{\delta x^2} + \mathbf{K}_1\frac{\delta}{\delta x} + \mathbf{K}_0 - \omega^2\mathbf{M}\right]\hat{\mathbf{U}} = \hat{\mathbf{F}} \tag{1}$$

with the stiffness matrices $\mathbf{K}_i$ and the mass matrix $\mathbf{M}$. Using the Fourier Transform, the displacements $\hat{\mathbf{U}}$ and forces $\hat{\mathbf{F}}$ in spatial domain are expressed in wavenumber domain

$$\left[\mathbf{K}_2(-j\kappa)^2 + \mathbf{K}_1(-j\kappa) + \mathbf{K}_0 - \omega^2\mathbf{M}\right]\tilde{\mathbf{U}} = \tilde{\mathbf{F}} \tag{2}$$

Considering $\tilde{\mathbf{F}} = \overrightarrow{0}$ this represents a linear eigenvalue problem when specifying a wavenumber κ or a quadratic eigenvalue problem when prescribing a frequency ω. The solution of the quadratic eigenvalue problem produces n complex conjugate pairs of wavenumbers κ_n corresponding to propagating, decaying waves. The right and left eigenvectors $\mathbf{U}_{nR}$ and $\mathbf{U}_{nL}$ are superposed to calculate the total response for a force input at one node and $x = 0$ for each frequency,

$$\hat{\mathbf{U}}_{\mathbf{0}} = \sum_n A_n \mathrm{e}^{-j\kappa_n x} \quad \text{for } x \geq 0 \tag{3}$$

with

$$A_n = j\frac{\tilde{\mathbf{U}}_{nL}\tilde{\mathbf{F}}_0}{\tilde{\mathbf{U}}_{nL}\mathbf{D}(\kappa_n)\tilde{\mathbf{U}}_{nR}} \tag{4}$$

and

$$\mathbf{D}(\kappa_n) = -2\kappa_n\mathbf{K}_2 - j\mathbf{K}_1. \tag{5}$$

Equation 3 is expressed in wavenumber domain,

$$\tilde{\mathbf{U}}_0 = \sum_n A_n\tilde{\mathbf{U}}_{nR}\left(\frac{-1}{Im(\kappa_n) - j(\kappa + Re(\kappa_n))} + \frac{-1}{Im(\kappa_n) + j(\kappa - Re(\kappa_n))}\right) \tag{6}$$

with Re/Im denoting the real and imaginary part of the complex wavenumber.

2.2 Discrete Support

Two WFE models, the rail and the superstructure, can be discretely coupled at a finite number of locations x along the waveguide. The displacement at any point on the coupled rail u_i^r is, similar as described in [4], a superposition of the free WFE response due to the excitation force $\hat{F}_e$ and the response to the reaction forces $\hat{F}_c$ in the coupling points,

$$u_i^r = \alpha_{ie}^r\hat{F}_e - \sum_c \alpha_{ic}^r\hat{F}_c \tag{7}$$

where the transfer functions α_{i*} describe the free rail response at location x_i for an excitation at x_*. Likewise the displacement at any point of the track t is the result of the superposition of the reaction forces.

$$u_i^t = \sum_c \alpha_{ic}^t\hat{F}_c \tag{8}$$

The transfer functions α_{ic}^r and α_{ic}^t are evaluated for each WFE model, which in case of the track includes essential boundary conditions. At each of the n_c coupling locations along the waveguide, the structures are coupled in m degrees of freedom of the FE-mesh. The coupling condition is described by a receptance α^p. Here, this is achieved using linear springs. The vertical and lateral direction are considered uncoupled.

$$u_i^r - u_i^t = \alpha_i^p\hat{F}_i \tag{9}$$

Equation 9 links Eq. 7 with Eq. 8. The $n_c \cdot m$ transfer functions α^* for each system are assembled to form a system of equations

$$\left(\boldsymbol{\alpha}^p + \boldsymbol{\alpha}^r + \boldsymbol{\alpha}^t\right)\hat{\boldsymbol{F}} = \boldsymbol{\alpha}_e^r\hat{F}_e \tag{10}$$

which is solved for the vector of reaction forces $\hat{\boldsymbol{F}}$. These are then reintroduced in Eq. 7 and Eq. 8 to calculate the receptance in spatial domain at any point on either of the structures. The reaction forces at the discrete coupling locations are expressed as a wavenumber spectrum at the origin x_0.

$$\tilde{F}_{m,0}(k) = \sum_{i=1}^{n_c} \hat{F}_{m,i} e^{jkx_i} \tag{11}$$

This force spectrum is introduced as an excitation in Eq. 6 to calculate the receptance on each structure. For the rail, the excitation force is either included in the force spectrum, or Eq. 6 is used to calculate the free response as well as the response due to the reaction forces and their displacements are superposed.

2.3 Wavenumber Boundary Element Method

The wavenumber boundary element method is described in [5]. The wavenumber in air $k = \omega/c$ is divided into the in-plane component α and the component along the waveguide matching the wavenumber in the structure κ by

$$\alpha = \sqrt{k^2 - \kappa^2}. \tag{12}$$

The in-plane boundary element problem is solved for each κ based on the 2D Helmholtz equation [6]. The normal velocity $\tilde{v}_{ni}$ serves as the Neumann boundary condition

$$\tilde{v}_{ni} = j\omega\tilde{\mathbf{u}}_i\mathbf{n} \tag{13}$$

where the nodal displacement vector $\tilde{\mathbf{u}}_i$ is the direct result of Eq. 6 and $\mathbf{n}$ is the outward normal direction.

In the case of the discrete support, multiple radiating boundaries are included. Their geometries and nodal displacements are combined in one BE model. For the continuous support, the air gap between the rail and the slab does not exist and only one boundary needs to be considered. Ground reflection is included in the Green's function formulation [7]. An acoustically rigid ground is placed right below the model. The CHIEF method is used to prevent internal resonances. Having determined the pressure at the boundary, the radiated sound power P can be calculated by integrating the intensity over the surface. In the wavenumber domain this is a double integral over the wavenumber spectrum and the perimeter. Contributions from individual bodies to the total sound power can be calculated by integrating over their surfaces individually.

$$P = \frac{1}{2} Re \left(\int_{-k}^{k} \int_{\Gamma} \tilde{p}^*(\kappa)\tilde{v}_n(\kappa) \mathrm{d}\Gamma \mathrm{d}\kappa \right) \tag{14}$$

2.4 Model and Parameter Description

The cross-section of the geometry is shown in Fig. 1. For the discrete support, the slab and the rail are separate WFE-models which are connected via vertical and lateral springs in three points over the width: in the centre and on either side of the rail foot. In longitudinal direction the supports are evenly spaced with the support distance $d_s = 0.65\,\text{m}$.

The rail geometry is a standard UIC60 rail. The geometry of the track is a simple concrete slab that is resting on a soft ground with $100\,\text{MN/m}^3$ foundation stiffness. The width of the concrete slab and ground is 2.4 m, with the rail positioned half the standard track gauge from the centre. The material properties for these structures are given in Table 1. For the continuous support, the Young's modulus of the rail pad is adjusted to the average pad stiffness of the discretely supported rail

$$E_p = \frac{k_p h}{d_s w \lambda} \text{ with } \lambda = \frac{1-\nu}{(1+\nu)(1-2\nu)} \tag{15}$$

with dynamic rail pad stiffness k_p, width of the rail foot w, the height of the rail pad h and the rail seat spacing d_s, to achieve a similar input receptance in vertical direction at the top of the rail. The factor λ compensates for the cross-contraction of the material. The stiffness is assumed linear with respect to displacement and constant over frequency.

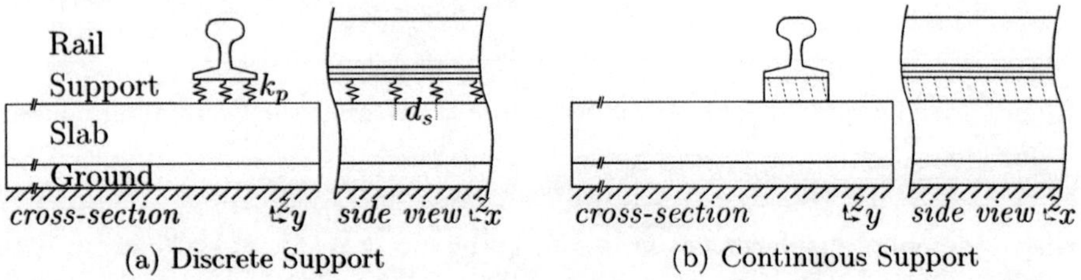

Fig. 1. Model setup for the discretely and continuously supported rail. (a) The discrete support case. The lateral springs are not shown here. (b) The continuous support.

Table 1. Material properties of the components in the WFE model: Young's modulus E, density ρ, Poisson ratio ν and complex damping coefficient η. The ground density is chosen low to avoid internal resonances.

	E (GPa)	ρ (kg/m^3)	ν (-)	η (-)
Rail	210	7850	0.3	0.01
Rail pad	E_p	1000	0.3	0.25
Slab	36	2400	0.2	0.1
Ground	0.0115	1	0.2	0.2

The researched parameters are the stiffness of the rail pad, the thickness of the slab and the type of support. The vertical rail pad stiffness k_s takes the values 10, 20, 40, 80, 160 or 320 kN/mm. The lateral stiffness is set to 10%

of these values. The slab thickness is 10 cm or 20 cm. It should be noted that some of these parameters can be considered unrealistic. The goal is solely to investigate their effect.

3 Results

In the following, the rail is excited using a unit force at the top of the rail, 3 mm off-centre towards the inside of the track. For the discrete support, the excitation is above a rail pad. The vertical and lateral direction are treated separately. The input and transfer receptance for both the continuous and the discretely supported rail are compared in Fig. 2. A good agreement is found below 1 kHz. The effect of the discrete support is visible for example at the pinned-pinned frequency 950 Hz in vertical direction.

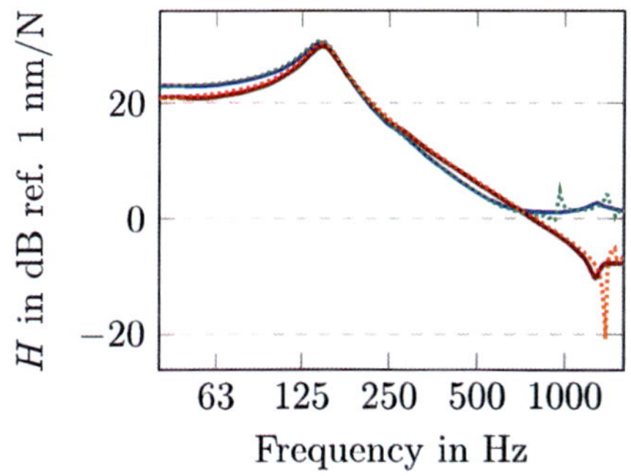

(a) Vertical Receptance

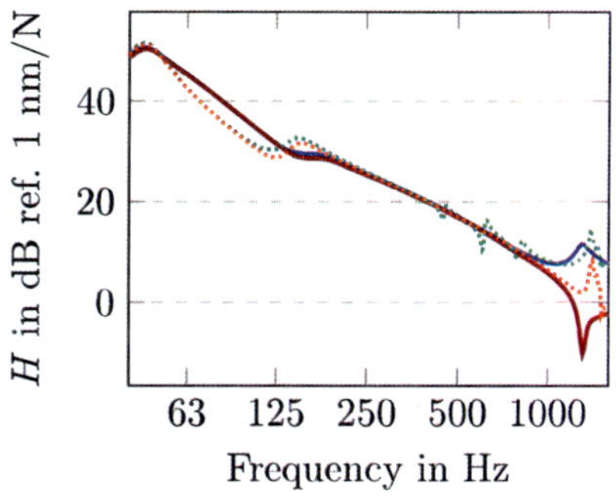

(b) Lateral Receptance

Fig. 2. Receptance H of the continuous support (input —, transfer —) compared to discrete support (input ....., transfer.....). The transfer receptance describes the receptance to the top of the rail at mid-span, or the equivalent distance for the continuous support.

The radiated sound power for the different rail pad stiffnesses and both types of support is shown in Fig. 3. For both supports and both excitation directions it is found that the radiated sound power is in general lower for stiffer rail pads. This is in line with the prediction made in [3] for ballasted tracks. The structural resonance at which the slab and the rail vibrate vertically out of phase produces a peak in the spectrum. It is the largest peak in the spectrum for the continuously supported rail. The slope of the curves in 3(a) is about 20 dB/octave. Figure 3(b) shows the same behaviour before reaching the structural resonance. Note that the rail pad stiffness does not seem to affect the radiated sound power much after this resonance occurs. For lateral excitation, the difference between the rail pad stiffnesses is less apparent for both supports. The main difference is around the first lateral bending resonance of the rail, below which a stiffer pad is beneficial.

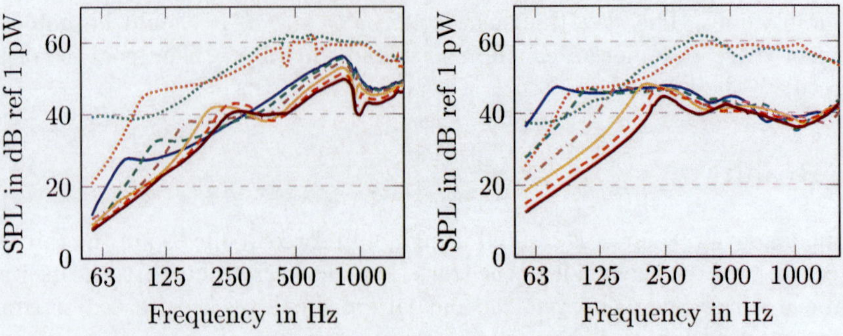

(a) Discrete Support (b) Continuous Support

Fig. 3. Radiation from a 20 cm slab at vertical excitation on the rail for different rail pad stiffnesses k_s: 10 kN/mm (—), 20 kN/mm (- - -), 40 kN/mm (- · -), 80 kN/mm (—), 160 kN/mm (- - -) and 320 kN/mm (—). Lateral excitation included for 20 kN/mm (.....) and 160 kN/mm (.....).

The contribution of the slab and the rail to the total sound power is evaluated separately and presented in Fig. 4 for the discretely supported rail with vertical excitation. In general, the sound power spectrum is dominated by the radiation from the rail. However, a stronger coupling between the rail and the slab leads to a larger contribution of the slab, especially below the structural resonance. It was found that the contribution of the slab is negligible when exciting the rail in lateral direction.

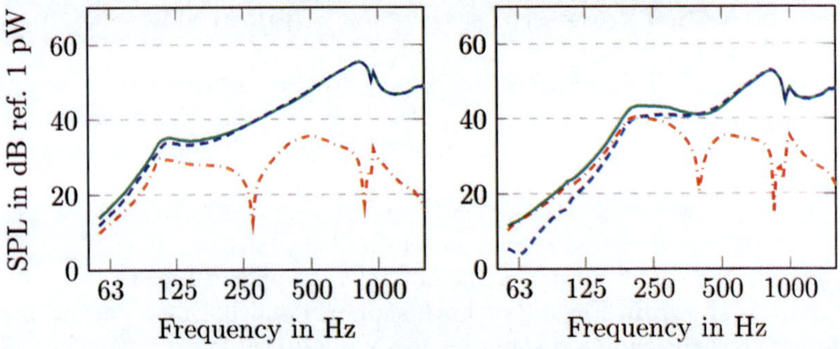

(a) $k_p = 20$ kN/mm (b) $k_p = 80$ kN/mm

Fig. 4. Contribution of rail (- - -) and slab (- · -) to the total radiated sound power (—) in case of a discretely supported rail on a 20 cm thick slab.

Different slab thicknesses are compared for the discretely supported rail with vertical excitation. Figure 5 shows that the slab has minor influence on the radiated noise for soft pads. With increasing stiffness of the pad, the modal behaviour of the thinner slab becomes more influential.

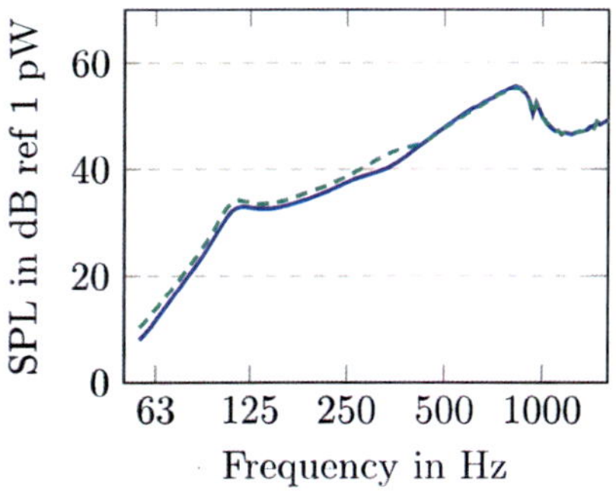

(a) $k_p = 20$ kN/mm

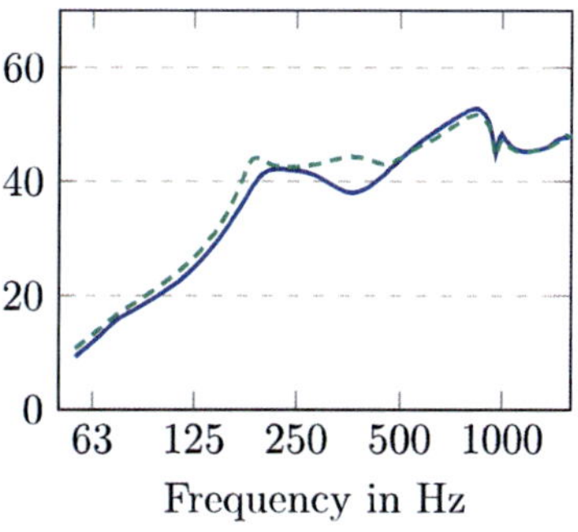

(b) $k_p = 80$ kN/mm

Fig. 5. Comparing the radiated sound power of a discretely supported, vertically excited rail on a 10 cm (- - -) and 20 cm slab (—).

4 Conclusion

A numerical method of calculating the structural response of slab tracks based on discretely coupled waveguides and a wavenumber BE method has been developed and implemented. The sound radiation from the combined rail and track vibration is modelled for a set of parameters. The results show a decrease in radiated sound power for higher rail pad stiffnesses. The rail is the main noise source. Its stronger coupling to the slab reduces the vibration levels on the rail. It leads to a relevant contribution of the slab to the total radiated noise for frequencies below the vertical structural resonance. Using a continuous support significantly decreases the total radiated sound power for higher frequencies.

Acknowledgements. This work is part of the research activities in the Chalmers Competence Centre in Railway Mechanics *CHARMEC*. Especially the support from Trafikverket (the Swedish Transport Administration) is acknowledged.

References

1. Van Lier, S.: Vibro-acoustic modelling of slab track with embedded rails. J. Sound Vib. **231**(3), 805–817 (2000)
2. Poisson, F.: Railway noise generated by high-speed trains. In: Nielsen, J.C.O., et al. (eds.) Noise and Vibration Mitigation for Rail Transportation Systems, pp. 457–480. Springer, Heidelberg (2015)
3. Thompson, D.J.: Railway Noise and Vibration. ISBN 9780080451473 (2009)
4. Zhang, X., et al.: A model of a discretely supported railway track based on a 2.5D finite element approach. J. Sound Vib. **438**, 153–174 (2018)
5. Nilsson, C.M., Jones, C.J.C., Thompson, D.J., Ryue, J.: A waveguide finite element and boundary element approach to calculating the sound radiated by railway and tram rails. J. Sound Vib. **321**(3–5), 813–836 (2009)
6. Wu, T.W.: Boundary Element Acoustics. ISBN 9781853125706 (2000)
7. Brick, H., Ochmann, M.: A half-space BEM for the simulation of sound propagation above an impedance plane. J. Acoust. Soc. Am. 123(5), 3418 (2008)

Pass-By Noise Assessment of High Speed Units by Means of Acoustic Measurements in a Perimeter Close to the Train

Gennaro Sica[1(✉)], Jaume Solé[2], and Pierre Huguenet[2]

[1] High Speed Two (HS2) Ltd, 1 Eversholt Street, Euston, London NW1 2DN, UK
gennaro.sica@hs2.uk.org
[2] SENER Ingeniería y Sistemas, C/. Creu Casas i Sicart 86-88,
08290 Cerdanyola del Vallès, Spain

Abstract. Characterization of rolling stock from the perspective of exterior noise emission is typically carried out by following the technical procedures detailed in the international standard ISO 3095. Whilst this standard can be considered adequate for conventional rolling stock, the advent of very high speed railway transport has introduced a number of challenges in terms of the characterization of noise emission. The most relevant one is the multiplication of noise sources, with the substantial increase in aerodynamic noise emissions from particular zones in the train including: pantograph, pantograph recess, inter-coach gaps, front of train, connection cables, low-level turbulences, high level noise, etc. This paper proposes an improved methodology for acoustic characterization that although more difficult than the procedures described in the standard ISO 3095, is not as difficult as acoustical analysis based on microphone arrays.

Keywords: High speed railway noise · Rolling stock · Measurements

1 Introduction

1.1 Formulation of the Problem

In comparison with the other mode of transports, railway is considered the most environmental friendly. One of the few concerns still associated with the development of a previously existing and/or an entirely new scheme is the social criticisms related to the fears of ambient noise impact.

To minimize this impact and comply with regional noise policies, noise requirements for new rolling stock fleets are set by promoter of new scheme or by existing infrastructure company in order to go beyond what is set within the European Technical Specification for Interoperability (TSI). One example of this approach can be found in the development of the new high speed railway line High Speed 2 (HS2). The HS2

G. Degrande et al. (Eds.): *Noise and Vibration Mitigation for Rail Transportation Systems*, NNFM 150, pp. 98–105, 2021.
https://doi.org/10.1007/978-3-030-70289-2_8

rolling stock procurement has specified pass-by noise to be as low as possible, establishing an incentivisation regime. This is seeking to achieve the noise predictions in the HS2 Hybrid Bill based on performance of both train and track, which is 3 dB lower than 2008 TSI at 360 km/h [9].

Very high speed railway transport has introduced a number of challenges in terms of characterization of noise emission. The most relevant one is the multiplication of noise sources, with the substantial increase in aerodynamic noise emissions from particular zones including: pantograph, pantograph recess, inter-coach gaps, front of train, connection cables, low-level turbulences, overall roof unevenness, etc.

Evaluation of rolling stock from the perspective of exterior noise emission is typically carried out by measurements according to the procedures detailed in the international standard ISO 3095 [13]. This standard adequately measures the overall noise produced by a trainset but does not allow to evaluate separately the noise coming from the different sources. The absence of standard methodologies for evaluating noise sources at very high speed makes very challenging to setup specific requirement and/or acceptance criteria on aerodynamic noise sources which can be verified with field test.

This paper propose an improved methodology for acoustic characterization that although more difficult than the procedures described in the standard ISO 3095, is not as difficult as acoustical approaches already presented in literature.

1.2 State of the Art

In order to design the proposed measurement methodology, several test approaches already presented in literature have been analyzed and summarized in Table 1 below, with some pros and cons in terms of being used for the evaluation of noise sources at very high speed.

Table 1. Comparison of the different test approaches

Test Approach	Advantages	Disadvantages
Close measurement onboard the train	Concept is easy to understand Quite objective and not prone to raise objections Repeatable in different places	Few precedents [10, 12] Difficult to get permissions
Close array measurements	Some precedents available [1, 4, 8] Provides comparatively more information than others Quite repeatable in different places	Requires complex processing Not so intuitively understandable Difficult to get permissions

(*continued*)

Table 1. (*continued*)

Test Approach	Advantages	Disadvantages
Far field measurements behind a noise barrier	Concept is easy to understand No permission is actually necessary (although it is recommended)	No precedents Difficult to exactly repeat the test in another place Easy to understand but potentially objectionable because of specific barrier characteristics and secondary noise captured (diffracted, transmitted)
Far field measurements with directional microphones	Concept is relatively easy to understand No permission is actually necessary (although it is recommended) Comparatively easier to repeat the test (effect of barrier particularities is considerably rejected)	Few precedents [15, 16] Some detailed calculation required to achieve accurate results (although approximate results are quite straightforward to achieve)

2 Methodology

2.1 Aims

Based on the analysis presented in Sect. 1.2, the aim of the proposed method are the following listed below:

1. Concept easy to understand;
2. Assure repeatability in different places;
3. In order to deliver point 2, the measure should be close to the source despite the difficulty to get permission; and
4. Results to be produced without complex processing.

2.2 The Requirements

In order to achieve the aims listed above and in addition to the requirements described in the international standard ISO 3095 [13], this proposed methodology has the following requirements:

- The placement of a number of microphones in a perimeter close to the train. Taking into account safety issues, an existing catenary portal has been proposed for the installation of 6 microphones, as shown in Fig. 1 (MIC1 to MIC6). Compared to other possible pre-existing structures (bridges, buildings, etc.) a catenary portal has the advantage of having a small acoustic cross section. This helps in minimizing acoustic reflections that could contaminate the measurements.

- The placement of accelerometers on the rail (see ACC1 and ACC2 in Fig. 1) in order to discriminate rolling noise emissions from the microphone at short distance from the lower part of the train (MIC6 in Fig. 1).
- The complementary use of a low-profile aerodynamic microphone installed in the sleeper (See MIC7 in Fig. 1).

Another additional requirement to the methodology is:

- The placement of different microphones to measure noise propagation in the far field up to 300m from the track center, in a free field test site having acoustic conditions comparable to semi-anechoic. Both near and far field measurement sensors are fully time synchronized.

3 Results

3.1 Test Site and Setup

A suitable test site with a catenary portal frame and free field propagation conditions was found on the Spanish high speed network managed by ADIF, close to the small town of Las Inviernas (Guadalajara, Spain). Commercial traffic at speeds between 250 km/h and 300 km/h was measured over 4 days, covering 4 different types of high speed rolling stock (Alstom TGV, Siemens Velaro E, Talgo 350 and CAF Alvia/ATPRD). Outside of commercial hours, a single dedicated Siemens Velaro E was studied at increasing speeds from 250 km/h to 350 km/h, both with the pantograph raised and lowered. Figure 1 shows the scheme and view of the sensors installed in the catenary portal and its surroundings, up to 300 m, as described in Sect. 2.2.

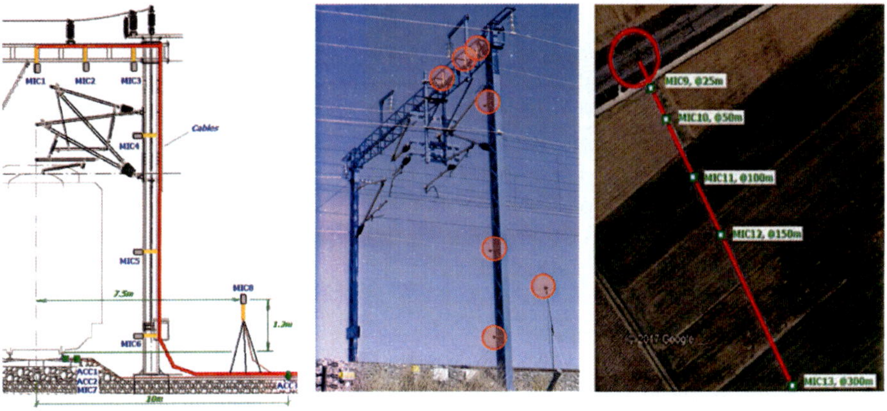

Fig. 1. Scheme and view of the main elements of the test setup.

3.2 Track Assessment (Roughness and Decay Rate)

In order to assess the potential influence of track characteristics on rolling noise, complementary measurement of rail roughness (Fig. 2) and track decay rate (Fig. 3) were carried out. Values obtained have been similar to the specified in the noise standard ISO 3095 [13] as reference; track has been therefore considered correct for tests.

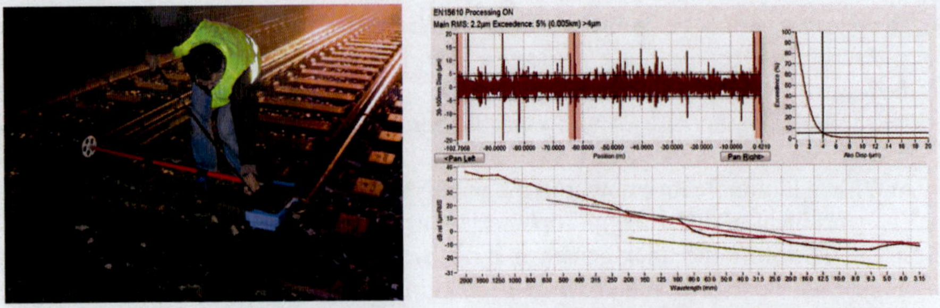

Fig. 2. Rail roughness measurements (left) and results obtained in the test site (right).

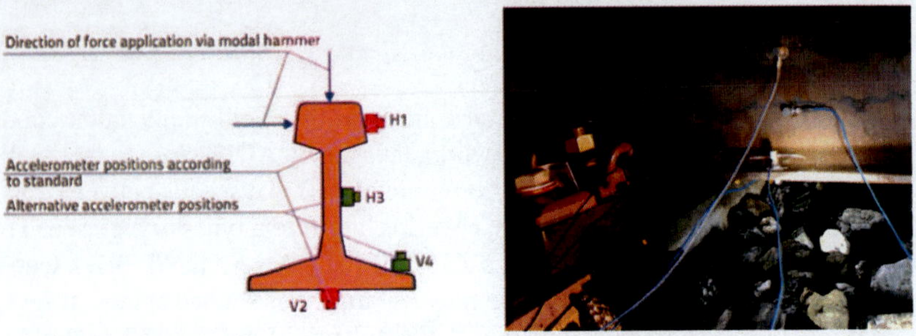

Fig. 3. Measurements of track decay rate in Las Inviernas according to standard [14].

3.3 Pass-By Measurements

Measurements at pass-by were carried out at different distances according to setup shown in Fig. 1. Reference measurement points at 7.5 m and 25 m as specified in European TSI were included. It was observed that for all speeds and rolling stock, noise levels were well below European TSI limits (see Fig. 4). From the results obtained, extrapolation at speeds up to 350 km/h using a logarithm law $50 \cdot log(V/V_{ref})$ has been shown to be reasonable. This extrapolation law (with $V_{ref} = 250$ km/h) is suggested by European TSI for speeds ranging from 250 km/h to 320 km/h; results obtained in the test show this law can be reliably extended to 350 km/h.

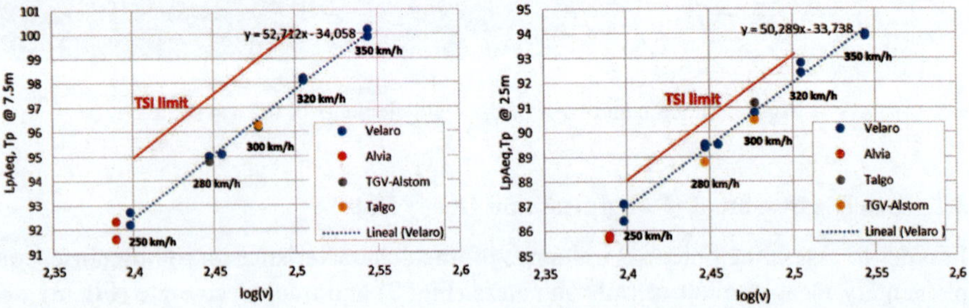

Fig. 4. Pass-by noise results at different speeds compared to TSI, as observed at 7.5 m of distance from track center and 1.2 m height (left) and 25 m from track center and 4m height (right).

3.4 Measurements in the Catenary Portal

Preliminary results for the measurements in the catenary portal showed a much cleaner definition of the high level noise sources, compared to pass-by measurements. This was observed in all microphones placed above 3m higher than the rolling plane, but was particularly clear in the case of the microphone installed just above the train (MIC1 in Fig. 1). Fig. 3 (left) shows the increase in noise source discrimination in with respect to usual measurement points at 7.5 m and 25 m. In the right of Fig. 3, distinct noise signatures and pantograph peak identification for different rolling stock are shown. Figure 4 (left) shows that the increase in discrimination allows a quite clear identification of high level noise sources. Accurate quantification of pantograph emission was achieved with the pantograph drop-off tests, as shown in Fig. 4 (right).

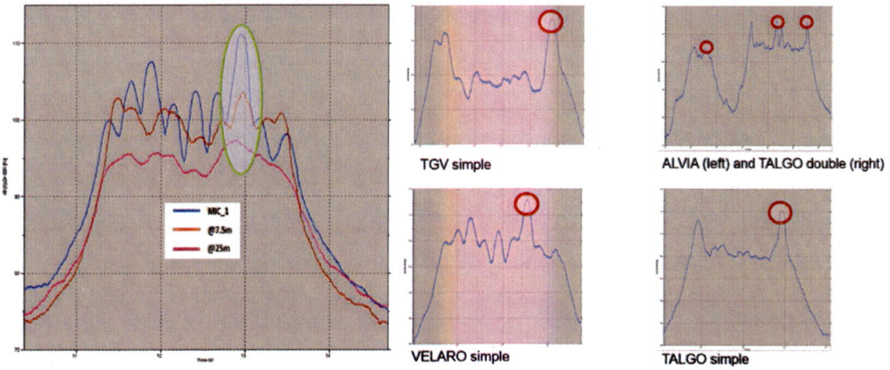

Fig. 5. Left: increase in noise source discrimination in MIC1, compared to usual measurement points at 7.5 m and 25 m (ISO 3095/Technical Specification for Interoperability in EU). Right: distinct noise signatures and pantograph peak identification for different rolling stock.

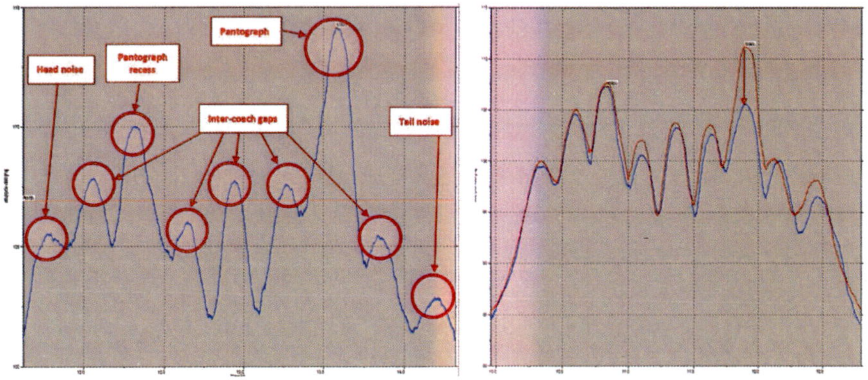

Fig. 6. Left, identification of high level noise sources in a Siemens Velaro E from noise lecture in catenary portal above the train. Right, effect of pantograph drop-off in a Siemens Velaro E running at 350 km/h.

4 Estimation of Pantograph Noise Directivity

From the readings in microphones MIC1 to MIC4 it has been possible to estimate approximate polar patters of noise emission directivity for the pantograph, in the plane perpendicular to the track (see Fig. 5). While being just approximations, they can be used to compare the behavior of different pantograph models at different speeds, and also can be used to estimate differences of noise emission in the directions that are of relevance for each acoustic study. Reference [11] shows details of the methodology followed to estimate these polar patterns.

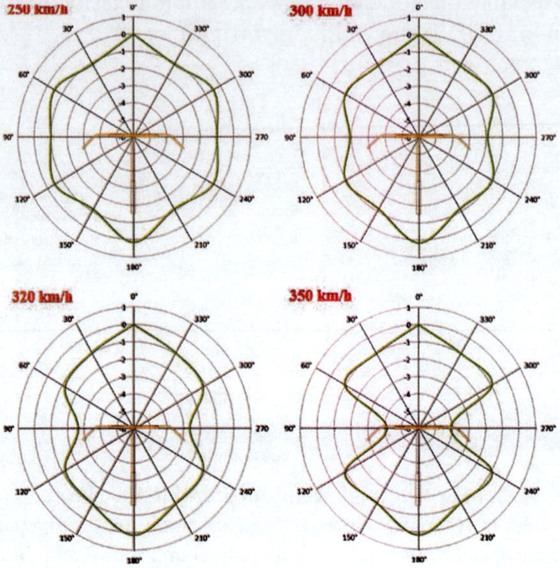

Fig. 7. Evolution of pantograph directivity with speed (Velaro E). Approximate curves obtained from readings in the microphones installed in the portal (see [11]).

5 Conclusions and Further Work

A new experimental methodology for characterization of high speed rolling stock has been presented. This methodology aims at achieving a more detailed acoustical characterization of high speed trains, without resorting to the use of sophisticated technologies with complicated post-processing (e.g. acoustic holography or beamforming). Preliminary results show promise, being possible to achieve (with a relatively simple setup) a level of noise source discrimination that had only been reached previously by using far more complex approaches. It is considered that with some further development, a consistent and repeatable characterization methodology can be developed from this experience.

Further results are expected from all the data gathered during the test. Particularly, it is expected to combine readings from the accelerometers installed in the rail together with readings in the aerodynamic microphone installed in the sleeper and also readings in the microphone installed in the lower part of the catenary portal (MIC6), in order to discriminate between rolling noise and aerodynamic noise in the lowest parts of the train. Also, the high number of spatial positions where noise recordings have been taken at each pass-by is expected to be used in order to improve acoustical models, in terms of transmission and directivity of the different noise sources involved.

References

1. Noh, H.M., Choi, S., Hong, S., Kim, S.W.: Investigation of noise sources in high-speed trains. J. Rail Rapid Transit **228**(3), 307–322 (2014)
2. Martens, A., Wedemann, J., Meunier, N., Leclere, A.: High speed train noise – Sound source localization at fast passing trains. Forum Acusticum Sevilla (2002)
3. Ross, J., Hanson, C.H., Meister, L.: Application of the federal railroad administration noise model to new high speed trainsets. Noise-Con 2011, Portland (2011)
4. He, B., Xiao, X., Zhou, Q., Li, Z., Jin, X.: Investigation into external noise of a high-speed train at different speeds. J. Zhejiang Univ. (Appl. Phys. Eng.) **15**(12), 1019–1033 (2014)
5. Gautier, P.E., Poisson, F., Letourneaux, F.: High Speed Trains external noise: a review of measurements and source models for the TGV case up to 360km/h" 8th World Congress on Railway Research, Seoul (2008)
6. Gautier, P.E, Poisson, F., Letourneaux, F.: High Speed trains external noise: recent results in the TGV case. In: 19th International Congress on Acoustics (2007)
7. Ivanov, N.I., Boiko, I.S., Shashurin, A.E.: The problem of high-speed railway noise prediction and reduction. Transportation Geotechnics and Geoecology, St. Petersburg, Russia (2017)
8. Yamada, H., Wakabayashi, Y., Kurita, T., Horiuchi, M.: Noise Evaluation of Shinkansen High-speed Test Train (2008)
9. European Union Agency for Railways: Technical Specification for Interoperability (2018)
10. Zongguang, C., Jianmin, G., Xinwen, Y., Yinxiao, L., Di, W.: Evaluation of pantograph noise of high speed trains. In: The 21st International Congress on Sound and Vibration-ICSV21, Beijing (2014)
11. Solé, J., Huguenet, P., Sica, G.: Evolution of pantograph noise directivity at increasing speeds. Internoise 2019, Madrid (2019)
12. Genescà, M., Solé, J., Romeu, J., Alarcón, G.: Pantograph noise measurements in Madrid-Sevilla high speed train. Internoise 2004, Prague (2004)
13. International Standards Organisation: ISO 3095 - Measurement of noise emitted by railbound vehicles (2013).
14. European Norm EN 15461: Characterization of the dynamic properties of track selections for pass by noise measurements (2011)
15. Hirota, T., Zenda, Y., Hosaka, S.: Measurement and analysis of vehicle noise on the yamanashi maglev test line. RTRI Quarterly Reports, Vol. 41 Issue Num. 2 (2000)
16. Loose, S., Lauterbach, A., Algermissen, S., Heckmann, A.: DLR Activities in Acoustics" Rail Technology Review Special – Next Generation Train, ISSN 1869–7801 (2011)

Interior Noise

Using a 2.5D BE Model to Determine the Sound Pressure on the External Train Surface

Hui Li[1](✉), David J. Thompson[1], Giacomo Squicciarini[1], Xiaowan Liu[1], Martin Rissmann[2], Francisco D. Denia[3], and Juan Giner-Navarro[3]

[1] Institute of Sound and Vibration Research, University of Southampton, Southampton SO17 1BJ, UK
Hui.Li@soton.ac.uk
[2] Vibratec, Chemin du Petit Bois, BP 36, Ecully Cedex 69131, France
[3] Centro de Investigación en Ingeniería Mecánica, Universitat Politècnica de València, Camino de Vera s/n, 46022 Valencia, Spain

Abstract. In this paper, a wavenumber-domain boundary element (2.5D BE) approach is adopted to predict the transmission of noise from the wheels, the rails and the sleepers to the train external surfaces. In the 2.5D models, only the cross-section of the vehicle is created by using boundary elements, while the third direction is taken into account in terms of wavenumbers. After the sound pressure on the train cross-section is obtained, an inverse Fourier transform is applied to obtain the spatial distribution of the sound on the train surfaces. To validate this approach, the 2.5D boundary element method was used to predict the sound distribution on the train surfaces due to a point source below the vehicle, and due to the vibration of the track. The prediction of the sound distribution from the 2.5D method shows the sound pressure levels on the train floor are 20 dB higher than the pressure on the sides, and the pressure on the train roof caused by the sources below the vehicle is negligible. The 2.5D boundary element method was also used to predict the sound pressure spectrum on the train sides when the train was in running operation. Reasonable agreement was found with measurements.

Keywords: 2.5D waveguide · Boundary element model · Train external surfaces · Rolling noise

1 Introduction

It is important to give attention to improve the interior acoustic environment in trains but this is a complex task. An important reason is that there are many noise sources which contribute to the interior acoustic environment. Moreover, the main noise sources are located outside the train, including rolling noise, aerodynamic noise, traction noise, fan noise from equipment as well as HVAC and so on [1]. In most cases, noise is transmitted to the interior by both airborne and structure-borne paths.

Among the main railway noise sources, the noise from the wheel and rail plays an important role for the interior noise [1]. To model wheel/rail noise and its propagation

G. Degrande et al. (Eds.): *Noise and Vibration Mitigation for Rail Transportation Systems*, NNFM 150, pp. 109–117, 2021.
https://doi.org/10.1007/978-3-030-70289-2_9

around the train body is an important step for predicting the interior noise; this can then be combined with the transmission loss of the floor, walls, doors and windows to determine the airborne sound transmitted to the interior. In this paper, a wavenumber domain (2.5D) boundary element model is used to calculate the sound transmission from the wheel and rail to the outside of the train surfaces.

2 Waveguide Boundary Element Model

The derivation of the waveguide boundary element method starts from a three-dimensional (3-D) problem. The boundary integral for a 3-D acoustic problem is [2]

$$p(x',y',z') = -\int_S (\mathrm{i}\rho\omega v(x,y,z)\psi(x,y,z|x',y',z') + p(x,y,z)\frac{\partial\psi(x,y,z|x',y',z')}{\partial n})\mathrm{d}S \quad (1)$$

where S is the surface of the vibrating boundaries, $p(x',y',z')$ is the sound pressure at a receiver **P**, ρ is the density of air, $v(x,y,z)$ and $p(x,y,z)$ are the normal velocity and pressure on the vibrating surface and $\psi(x,y,z|x',y',z')$ is the Green's function. If the geometry of the problem can be considered uniform and of infinite length in one direction, e.g. x, Eq. (1) can be conveniently solved in the 2-D domain for a range of wavenumbers in the x direction. To achieve this, the Fourier transform pair for the sound pressure $p(x, y, z)$ with respect to x can be expressed as

$$\tilde{p}(k_x,y,z) = \int_{-\infty}^{\infty} p(x,y,z)\mathrm{e}^{\mathrm{i}k_x x}\mathrm{d}x \quad (2)$$

$$p(x,y,z) = \frac{1}{2\pi}\int_{-\infty}^{\infty} \tilde{p}(k_x,y,z)\mathrm{e}^{-\mathrm{i}k_x x}\mathrm{d}k_x \quad (3)$$

where k_x is the wavenumber in the x direction. For a time-harmonic pressure, Eq. (3) becomes:

$$p(x,y,z,t) = \int_{-\infty}^{\infty}\left(\frac{1}{2\pi}\tilde{p}(k_x,y,z)\mathrm{e}^{-\mathrm{i}k_x x}\mathrm{d}k_x\right)\mathrm{e}^{\mathrm{i}\omega t} \quad (4)$$

Equation (4) indicates that the sound pressure at the receiver **P** has the form of a harmonic wave with angular frequency ω, propagating in the x direction with wavenumber k_x. The distribution of sound pressure can therefore be obtained from Eq. (4) once $\tilde{p}(k_x, y, z)$ is determined. This can be achieved through a 2.5D formulation of Eq. (1) where pressure, velocity and the Green's function are expressed as functions of k_x, y and z and the boundary integrals are solved over the perimeter of the boundary region Γ. The wavenumber domain integral equation therefore becomes:

$$\tilde{p}(k_x,y',z') = -\int_\Gamma (\mathrm{i}\rho\omega\,\tilde{v}(k_x,y,z)\,\tilde{\psi}\left(k_x,y,z|y',z'\right) + \tilde{p}(k_x,y,z)\frac{\partial\,\tilde{\psi}(k_x,y,z|y',z')}{\partial n})\mathrm{d}\Gamma \quad (5)$$

$v(k_x, y, z)$ in Eq. (5) is the normal velocity in the wavenumber domain, which is calculated by applying the Fourier transform to the velocity in the spatial domain.

$$\tilde{v}(k_x, y, z) = \int_{-\infty}^{\infty} v(x, y, z)e^{ik_x x}dx \tag{6}$$

The wavenumber in the x direction, k_x, is not dependent on y, z. The Green's function $\tilde{\psi}(k_x, y, z|y', z')$ takes the same form as the 2D fundamental solution [2]:

$$\tilde{\psi}(k_x, y, z|y', z') = -i\frac{1}{4}H_0^{(2)}(\sqrt{k_0^2 - k_x^2}r) \tag{7}$$

with $H_0^{(2)}(x)$ being the Hankel function of the second kind, and k_0 the wavenumber in air. If only a half space with a rigid ground is considered, the presence of the ground can be taken into account by means of an image source located symmetrically beneath the ground. The fundamental solution has two components and takes the form

$$G = -\frac{i}{4}H_0^{(2)}(\sqrt{k_0^2 - k_x^2}r) - \frac{i}{4}H_0^{(2)}(\sqrt{k_0^2 - k_x^2}r') \tag{8}$$

where r is the distance from the actual source to the receiver, and r' is the distance from the image source to the receiver. The phase difference between the two sources is therefore included automatically in this equation.

Partially absorbing boundaries are modelled through their surface normal impedance, which can be obtained by means of analytical or empirical models. The Delany-Bazley impedance model [3] is adopted in the current work to model the ballast

$$z_n' = 1 + 9.08(\frac{1000f}{\sigma_e})^{-0.75} - 11.9i(\frac{1000f}{\sigma_e})^{-0.73} \tag{9}$$

where σ_e is the flow resistivity, 50 kPa • s/m^2, and f is the frequency.

3 Waveguide Boundary Element Model for Rolling Noise

3.1 Wheel Noise

The sound radiation from a train wheel can be divided into two components: axial and radial [1]. The sound directivity of the radial component can be represented approximately as a monopole while that of the axial component approximates to a horizontal dipole. In the 2.5D model, the radial component is modelled by implementing a monopole-like source and the axial component is modelled by giving a lateral oscillation to the same fundamental source (equivalent to a dipole). In the boundary element calculation, the velocities of the sources are set to unity and the final results are re-scaled by the sound power levels calculated in TWINS.

In 3D, monopole sources can be represented with pulsating spheres, but this is not possible in a 2.5D waveguide approach. An equivalent result can be found by using a circle in the y-z plane. The boundaries of the circle are given unit normal velocity and a spatial window is applied in the x direction (see Fig. 1), with the shape of the spatial window set through its wavenumber spectrum. The circle should be small so that the non-vibrating part does not interfere with the sound field. The distribution of the velocity is chosen to be constant in the x direction over a length corresponding to twice the radius of the circle a, such that:

$$v_n(x) = \begin{cases} 1 \; x \in [-a, a] \\ 0 \; x \notin [-a, a] \end{cases} \tag{10}$$

In a similar way, a dipole source is created by applying the velocity pattern of an oscillating sphere to a circle in 2.5D. The normal velocity in the y-z plane is represented in Fig. 1(c), and it can be expressed as

$$v_n(y, z) = v\cos\theta \tag{11}$$

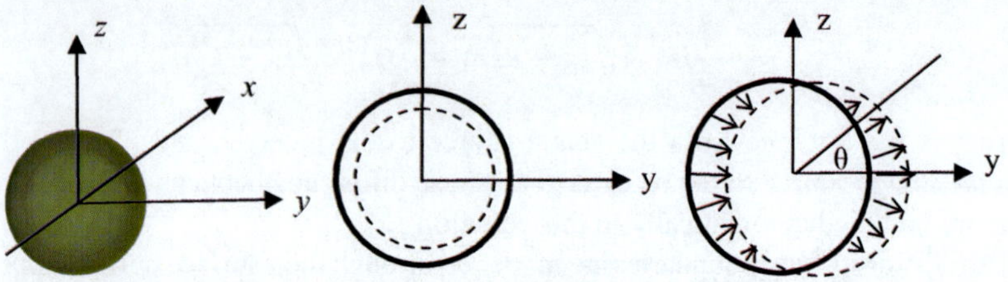

Fig. 1. Sphere and its cross-section in y-z plane.

The velocity distribution in the x-z plane is the same as the velocity distribution in y-z plane, just replacing y by x. The velocity distribution in the x direction can be expressed as Eq. (11) but only valid in the region [-a, a] in the x direction. For the convenience of applying the Fourier transform to the velocity in the x direction, the velocity distribution in the x-z plane is expressed as a function of x, as

$$v_n(x) = v\cos\left(\frac{2\pi}{4a}x\right) \tag{12}$$

The corresponding velocity in the wavenumber domain is obtained by applying a Fourier transform,

$$\tilde{v}_n(k_x) = \int_{-\infty}^{\infty} v_n(x)\mathrm{e}^{\mathrm{i}k_x x}\mathrm{d}x \tag{13}$$

3.2 Rail and Sleeper Noise

The noise radiation from a rail has both vertical and lateral components, which are dealt with separately in this work. The mobility $Y(x)$ of the rail at a distance x from the excitation point is derived from a Timoshenko beam model on a continuous two-layer spring-mass-spring foundation [4], expressed as

$$Y(x) = u_1 e^{-\mathrm{i}k_r|x|} - \mathrm{i}u_2 e^{-\beta|x|} \tag{14}$$

where k_r is the wavenumber corresponding to the propagating wave in the rail, β corresponds to the near field wave and u_1 and u_2 are the wave amplitudes. The vertical and lateral mobility of the rail are obtained in the same way. Applying a Fourier transform to the mobility in the spatial domain, the velocity in the wavenumber domain due to a unit force can be expressed as

$$\tilde{Y}(k_x, y, z) = u_1\left(\frac{1}{\mathrm{i}(k_r + k_x)} + \frac{1}{\mathrm{i}(k_r - k_x)}\right) - \mathrm{i}u_2\left(\frac{1}{\beta + \mathrm{i}k_x} + \frac{1}{\beta - \mathrm{i}k_x}\right) \tag{15}$$

where k_x is the wavenumber in the x direction introduced by the Fourier transform. k_x exists from -∞ to ∞.

The vibration of the sleepers is also needed to calculate their contribution to the noise in the 2.5D model. It can be derived from the rail mobility and the ratio of the sleeper displacement to that of the rail [1]

$$r = \frac{s_p}{s_p + s_b - \omega^2 m_s} \tag{16}$$

where s_p is the complex pad stiffness in vertical direction including damping, s_b is the complex ballast stiffness and m_s is the sleeper mass.

3.3 Set up for the 2.5D Model

Based on the analysis in Sects. 3.1 and 3.2, the treatments of the wheels, rails and sleepers are different, so they are dealt with in separate 2.5D models. They are shown in Figs. 2, 3 and 4.

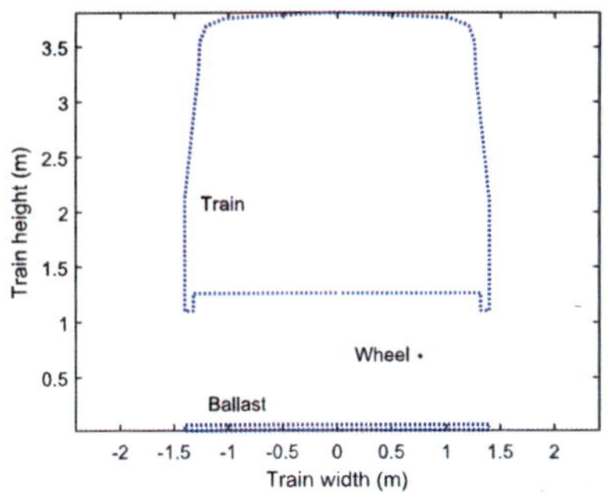

Fig. 2. 2.5D model for the wheel.

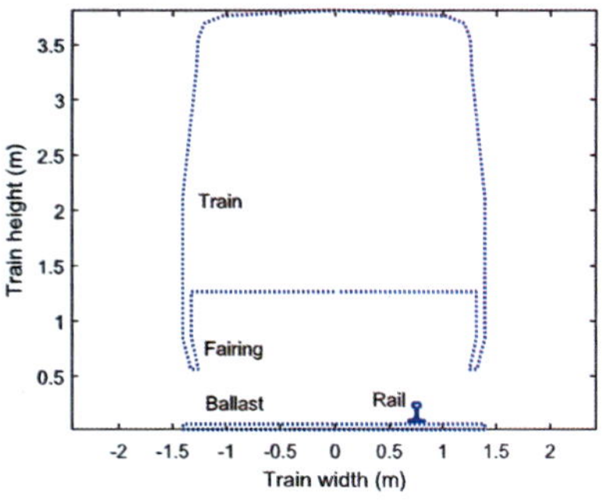

Fig. 3. 2.5D model for the rail.

Figure 2 shows the cross-section of the 2.5D model used for investigation of the wheel contribution. The ballast was modelled with boundary elements with an acoustic

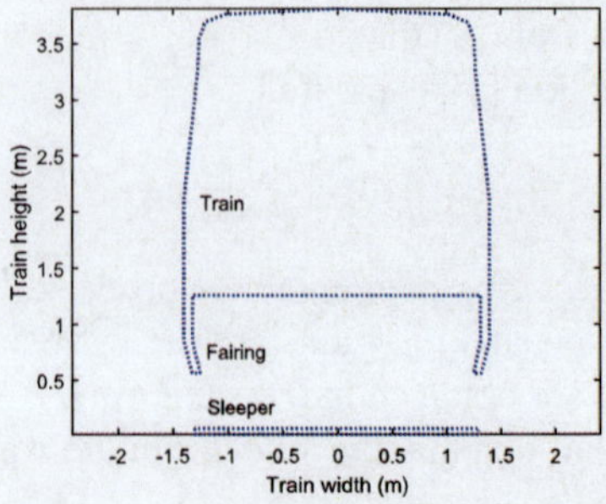

Fig. 4. 2.5D model for the sleepers.

impedance boundary condition while the car body is set to be rigid. The cross-sections adopted to calculate rail and sleeper contributions are shown in Figs. 3 and 4. The velocity of the sleepers was calculated as in Sect. 3.2 and assigned to the top surface of the sleepers. Allowance was made for the fact that they only cover part of the track surface due to their discrete nature. It can be noted that the train cross-section in Fig. 2 does not include the side fairings (as they are not present in the bogie region) while the fairings are present in Figs. 3 and 4.

The 2.5D models can give the sound pressure on the train cross-section in the wavenumber domain. After applying the inverse Fourier transform, the sound distribution in the spatial domain is obtained. However, the source strength in the 2.5D models for the wheel, the rail and the sleepers is either based on a unit velocity or a unit force. The predictions can give the decay pattern but cannot give the correct amplitude on the train external surfaces. To do so, the sound powers of the wheel (radial and axial), the rail (vertical and lateral) and the sleepers obtained from the TWINS model [5] are used to adjust the corresponding 2.5D model results.

3.4 Sound Pressure on Train Outside Surfaces

The total sound pressure on the train surfaces is composed of the contributions from the wheel, the rail, and the sleepers. These are combined as

$$\overline{p_{tot}^2} = \overline{p_{wheel}^2} + \overline{p_{rail}^2} + \overline{p_{sleeper}^2} \tag{17}$$

$\overline{p_{wheel}^2}, \overline{p_{rail}^2}, \overline{p_{sleeper}^2}$ are the mean-square sound pressures on the train external surfaces due to the wheel, the rail and the sleeper, obtained from the 2.5D models. Overall A-weighted sound pressure levels on the train external surfaces are shown in Fig. 5. The sound pressure distribution on the train outside surfaces roughly shows that the highest sound power is incident on the train floor area because sound can reach the floor directly and also due to the presence of the fairings. Noise can diffract around the edge of the fairing to reach the side surfaces of train, and it decays with increasing height. Noise on the train roof due to the wheel and track is much less significant.

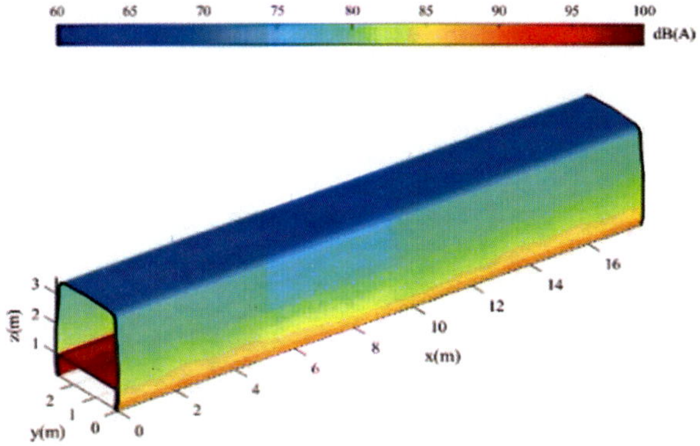

Fig. 5. Overall sound pressure levels on the train outside surfaces.

4 Experimental Validation

Field measurements were performed to measure the sound pressure distribution on the external surfaces of a train running at 50 km/h on a test track. Four positions (points 1006, 1009, 1010, 1011) were measured on the train side surface, see Fig. 6.

The numerical models are those given in Figs. 2, 3 and 4. The TWINS predictions were also verified using pass-by measurements of radiated noise and rail vibration. Sound pressure spectra on the train surfaces are compared with the measurements in Fig. 7. Their overall levels are compared in Table 1. The comparisons show that the 2.5D model is able to predict satisfactorily the sound pressure on the train external surfaces in a running operation.

Fig. 6. Field measurement positions on the train outside surface.

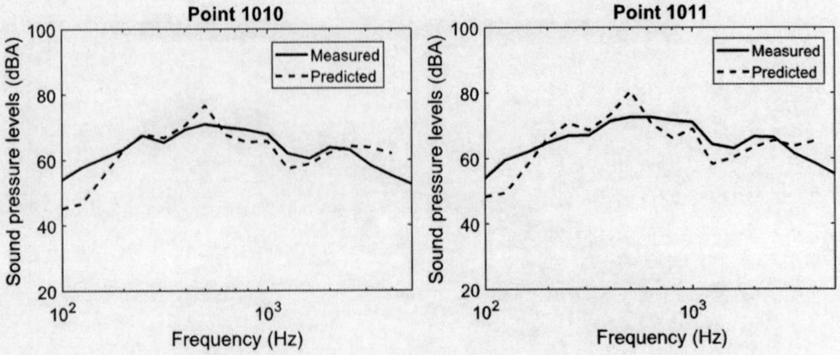

Fig. 7. Comparisons between the 2.5D predictions and the measurements.

Table 1. Overall sound pressure levels on the train sides (dB(A)).

	P 1006	P 1009	P 1010	P 1011
Prediction	83.2	83.1	79.7	82.9
Measurement	80.6	81.1	78.3	80.4
Difference	2.6	2.0	1.4	2.5

5 Conclusions

A 2.5D waveguide boundary element approach was developed to calculate the contributions from the wheel, rail and sleepers to the sound pressure at the external surfaces of a train. The wheel was modelled by using an approximation of a monopole and a dipole in the 2.5D models. The rail and the sleepers were modelled by their cross-section in the 2D domain allowing for their wavenumbers in the third direction. An inverse Fourier transform was used to convert the sound pressure from the wavenumber domain back to spatial domain. The TWINS output was used to adjust the 2.5D model predictions. The overall sound pressure on the train external surfaces was obtained by adding the various components incoherently. Measurements of sound pressure levels of a train in running operation were used to validate the model. The comparisons show that the predicted sound pressure levels agree satisfactorily with the measured ones. The 2.5D waveguide boundary element model can therefore be used to predict the sound pressure on the external surfaces of a train, which is an essential input quantity to evaluate interior noise.

Acknowledgements. The work presented in this paper has received funding from the Shift2Rail Joint Undertaking under the European Union's Horizon 2020 research and innovation programme under grant agreement no. 777564. The contents of this publication only reflect the authors' view and the Joint Undertaking is not responsible for any use that may be made of the information contained in the paper. The authors are grateful to Metro de Madrid for assistance with the field tests.

References

1. Thompson, D.:. Railway noise and vibration: mechanisms, modelling and means of control. Elsevier (2008)
2. Wu, T.W., ed. Boundary element acoustics: fundamentals and computer codes. WIT Press/Computational Mechanics (2000)
3. Delany, M., Bazley, E.: Acoustical properties of fibrous absorbent materials. Appl. Acoust. **3**(2), 105–116 (1970)
4. Kitagawa, T., Thompson, D.J.: The horizontal directivity of noise radiated by a rail and implications for the use of microphone arrays. J. Sound Vib. **329**(2), 202–220 (2010)
5. Thompson, D.J., Hemsworth, B., Vincent, N.: Experimental validation of the TWINS prediction program for rolling noise, part 1: Description of the model and method. J. Sound Vib. **193**(1), 123–135 (1996)

Acoustic Design of Rolling Stock for Comfortable Telephone Conversations

A. Bistagnino(✉) and Joan Sapena

Alstom, 48, Rue Albert Dhalenne, 93400 Saint-Ouen, France
andrea.bistagnino@alstomgroup.com

Abstract. The first objective of the acoustical design of rolling stock vehicles is the compromise between passenger comfort and the technical and economical means required to achieve it. Typically, global A-weighted sound pressure levels are used to evaluate acoustical such performance, but acoustical comfort is a complex phenomenon and it is difficult to find a single measure that fits all situations. This paper presents a novel approach to the study of one such special case: comfort experienced by passengers engaged in telephone conversations in a train. Noise inside the vehicle will affect the comfort of the telephone conversation, but we found a better measure of comfort than global dB(A) levels. Its definition, based on 1/3 octave band spectra, can be used since the beginning of the design of a new rolling stock to drive the acoustical design of passenger areas, including vestibules.

Keywords: Acoustic comfort · Rolling stock noise · Psychoacoustics

1 Introduction

Noise inside rolling stock vehicles is today one of the main drivers during the design of new trains. Indeed, comfort inside trains is related to interior noise, so operators willing to offer a comfortable trip to their customers must ask train manufacturers to develop low-noise solutions. A good balance between comfort and the associated costs must then be looked for. Usually, the study of acoustic discomfort is done by relating psychoacoustic parameters and jury tests; example are previous studies of discomfort in passenger areas [1, 2]. But an interesting case of discomfort is the one of noise generated by the passengers themselves: if there are noisy passengers onboard, how can a train manufacturer impact such noise sources to avoid the annoyance of other passengers? Today it is very difficult to answer to such questions, and it requires a close collaboration between the manufacturer and the operator. But there is one case where something can be done: that of telephone conversations.

Everyone will agree that listening to someone else's telephone conversations is not the most comfortable condition to be subjected to, especially when confined inside a rolling stock vehicle. That is why many operators ask passengers to do their calls in vestibules; in this way, other passengers are not disturbed and the comfort of people in

G. Degrande et al. (Eds.): *Noise and Vibration Mitigation for Rail Transportation Systems*, NNFM 150, pp. 118–125, 2021.
https://doi.org/10.1007/978-3-030-70289-2_10

passenger areas does not need to be addressed. The problem is that vestibules are, in general noisy. High background noise levels mean low comfort for passengers that are trying to have phone calls in these areas. The solution is simply to lower background noise, but vestibules are, from the acoustical point of view, a complicated area to treat; in general noise levels are higher than in compartments due to the closeness of the main nose sources (wheels, motor and gearbox) and the weak noise paths (doors and gangway). Reducing global Sound Pressure Levels can be very expensive or even technically unfeasible.

One way to solve this problem is to ask ourselves: do all types of background noise make telephone conversations uncomfortable? Are we able to determine the minimum requirements for background noise to make it acceptable for such use? And is it doable with a reasonable effort? The study presented here tries to answer to all these questions. A methodology based on a mix of online recordings, noise synthesis, laboratory measurements and jury tests was developed to study this issue. As a result, it was possible to define an objective, quantitative way to evaluate the comfort of telephone conversations based only on background noise spectra. This is in turn a clear target that can be used to achieve a good acoustic design with the right effort. Several interesting topics had also to be touched to achieve this result: the impact of telephone types and telephone technologies, the relative importance of listening and speaking effort, the Lombard effect and its consequences.

2 Methodology

The methodology proposed in this paper consists in the assessment of the comfort during telephone conversations by means of jury tests, when exposed to background noise representative of what can be typically found in train vestibules. When dealing with psychoacoustic studies, a major problem is to ensure the reproducibility and stability of the test conditions. In the case of rolling stock vehicles, it means that jury tests cannot be done directly on the train running. Another challenge is that the objective is to study the impact of several background noise types, including if possible some that cannot be found on today's trains. The following approach was thus defined:

1. Measurement of background noise samples on an existing Very High-Speed Train
2. Reproduction of the measured background noise in a laboratory
3. Generation of synthetic noise samples based on the measured ones
4. First round of jury tests to assess the most important effects
5. Second round of jury tests to quantify and evaluate results

It could have been tempting to approach this subject on the basis of STI or related methods (as described in [3]), but this would have restrained the study to the correlation of intelligibility and comfort. The results of this study show that discomfort exists even in the case of perfect intelligibility as it is more related to the talking effort than to the listening effort.

This approach has the strong positive point of being very flexible, because it is possible to generate several realistic synthetic noise samples to test. It requires careful measurements on the train, because the three-dimensional noise field must be correctly reproduced in the laboratory. Once this was achieved, everything was in place to treat the main question of the study: what really affects the comfort of telephone conversations in trains?

3 Experimental Setup

The first step of this study was the measurement of typical background noise spectra in a high-speed train. The train of choice was an Alstom Italo train of the AGV class, running on the Italian railway at a maximum speed of 300 km/h. These measurements were done in several positions inside a vestibule and the compartment, in various conditions (train speed, tunnel or free-field, etc. See Fig. 1, left). Measurements were done with an 8-channel microphone setup to record the acoustic field around an artificial head. Such measurements make it possible to reproduce in laboratory the exact background noise field that was found on the train, following the principles of [4].

Fig. 1. Measurement setup in the.Italo AGV train (left) and laboratory setup (right).

The laboratory itself is a semi-anechoic room with eight speakers, whose gain is adjusted so that the sound field measured in the center of the laboratory exactly matches the one measured on the train (see Fig. 1, right). In the center of the laboratory, people can be subjected to the sound field they would experience in the train's vestibule and have telephone conversations with someone sitting in the adjacent room. The jury members would then rate a score on the comfort of the conversation, as it will be detailed in Sect. 4.

Once the acoustic field measured on the train is reproduced in the laboratory, the obvious question is: how would the results change if the sound field were a bit different than the measured one? The decision was taken to investigate primarily the effect of the spectral shape of background noise on the comfort of the conversation: a sufficient number of samples was thus needed to explore the phase space of possible background noise spectra and assess the related comfort. This was obtained by creating synthetic noise samples with a wide (but reasonable) range of spectra.

The creation of synthetic noise samples followed this process:

1. Analysis of samples recorded on the train and choice of the reference ones (maximum speed in free-field and in tunnel).

2. Based on the reference sample (measured with the 8-microphones and artificial head setup), creation of two new samples from two other past measurements in vestibules of other Alstom High Speed Trains. This step is necessary to have a wider range of baseline spectra. These old measurements were taken with a single microphone, so the reference sample (all 8 channels) was equalized to match with the spectra of the old measurements.
3. Equalization of reference sample to obtain synthetic ones with the following characteristics: ± 3 dB overall, ± 6 dB overall, ± 3 dB on frequencies ≤ 250 Hz, ± 3 dB on frequencies ≥ 315 Hz, -6 dB on frequencies ≥ 315 Hz, -3 dB and -6 dB on frequencies ≥ 2000 Hz.
4. Other samples were created, notably on the impact of pure tones, rattling, etc. but these will not be treated in this paper.

A total of 38 samples was obtained in this way, all of them useable for jury testing in the laboratory.

An additional topic that was considered was the impact of the telephone type and telephone technology – at least at the time of starting of the study, in 2015. Five different telephones that at the time had a widespread use were tested. Apart from the number and quality of the microphone(s), of the speaker and in general of the build, there is one distinctive feature to be taken into account: two out of five feature a technology called NELE (Near-End Listening Enhancement), which consists of increasing the gain of the phone's speaker when ambient noise high at high frequencies. This increases the listening performance when in a noisy environment. The five telephones used for the study are: Apple Iphone 6, LG G2, HTC One, Samsung S4 mini (featuring NELE) and Nokia Lumia 735 (featuring NELE). As it will be shown in the rest of this paper, the telephone type and technologies like NELE have a very limited impact on the results, that are thus still valid today in 2019.

4 Jury Tests

The evaluation of the psychoacoustic impact of background noise is done by means of jury tests realized in the laboratory setup described in the previous section. As discussed in the previous section, about 38 noise samples were prepared, to be verified on 5 telephones. Due to the different constraints imposed by the number of samples, the environment and the testing procedure, it has been decided to follow the ITU-T P.831 [5] standard for jury testing. This standard is the reference of jury testing for telephone performances and it offers a lean but yet reliable approach for such tests. The jury was composed of seven "expert listeners". These persons were chosen to have normal hearing, already with experience in acoustic jury testing, unaware of the objective of the test. This is the suggested approach for such applications for the ITU-T P.831 standard; it made it feasible to test all samples in the same session, thus increasing the reliability of results.

Jury test components were asked to rate, on a scale from 1 to 5, the comfort of the telephone conversation. A rating of 1 corresponds to a communication "impossible without any feasible effort", a rating of 5 to a conversation with "complete relaxation possible, no effort required". Half-point answers were accepted. Results were then processed to obtain a Mean Opinion Score (MOS). For each sample, two questions were asked: "How do you rate the Talking Effort in the conversation?" and "How do you rate the Listening Effort to understand the person from the far end?".

Preliminary tests were done with all five telephones, but results were so comparable that only two telephones were kept for the following, one with NELE (Nokia) and one without (Iphone). For this first test, it was natural to compare results in the vestibule and in the compartment, where noise is lower and where we would like that passengers don't have telephone conversations. Results are shown in Fig. 2. The findings were:

- NELE appears to slightly improve only Listening Comfort, as expected.
- Comfort in compartments is significantly better than in non-treated vestibules
- Higher background noise relates to lower MOS and thus lower comfort
- Talking Comfort scores are significantly lower than Listening Comfort ones.

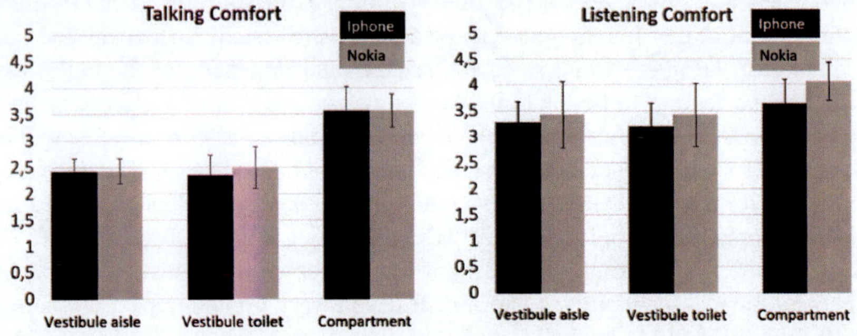

Fig. 2. Mean Opinion Score for Talking Comfort (left) and Listening Comfort (right) in three different positions, with two different telephones. Confidence intervals at 95% are also given.

The most important result is that all listeners consistently gave higher scores for Listening Comfort than for Talking Comfort. In the compartment, where background noise is low, the difference is not very high, but in vestibules the average difference is about 0.9 points, higher than standard deviation of results (about 0.5). We conclude from this result that discomfort during telephone conversations has a stronger link to the Talking Comfort than to the Listening comfort; to really improve comfort, it is important to focus on the Talking Comfort. For the sake of completeness, it is pointed out that the main reason for lower Talking Comfort when subjected to background noise should be looked for in the Lombard effect, which describes the fact that people naturally increase their vocal effort when subjected to background noise [6].

5 Results

As described in Sect. 4, a second round of jury tests was done on the full set of samples. As for the first round, only 7 "expert" listeners were used as jury. Only the most meaningful results will be presented here. Figure 3 shows Mean Opinion Scores for a few samples. As a first guess, it is possible to evaluate the correlation of typical quantities like loudness or global dB(A) level of background noise with MOS results; it is shown in Fig. 4. As expected, higher sound pressure levels or loudness lead to lower scores. Loudness correlates better than dB(A) but, especially for high MOS, correlation is not very good. As already stated in Sect. 1, the main hypothesis of this study is that background noise spectra are more meaningful to look at than global levels or similar global quantities.

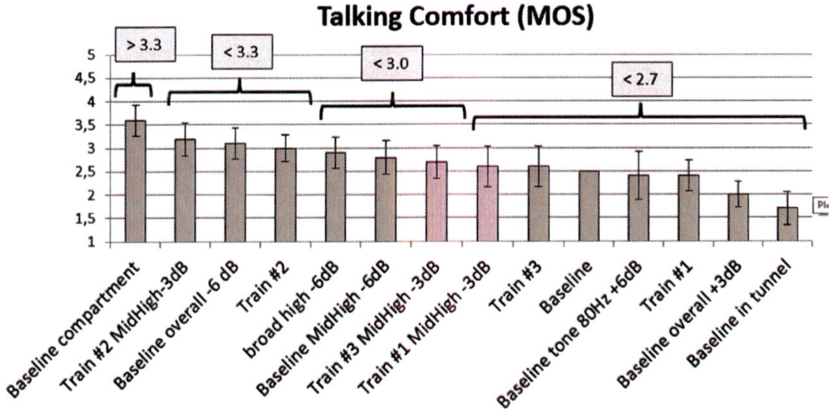

Fig. 3. Mean Opinion Scores for Talking Comfort for a set of meaningful samples with 95% confidence interval.

Going back to Fig. 3 and focusing on the samples with the highest scores (as we are interested in the design of comfortable vestibules), we can define at least 4 classes of background noise types, depending on their MOS: MOS $\geq$ 3.3, MOS $\geq$ 3, MOS $\geq$ 2.7, MOS < 2.7.

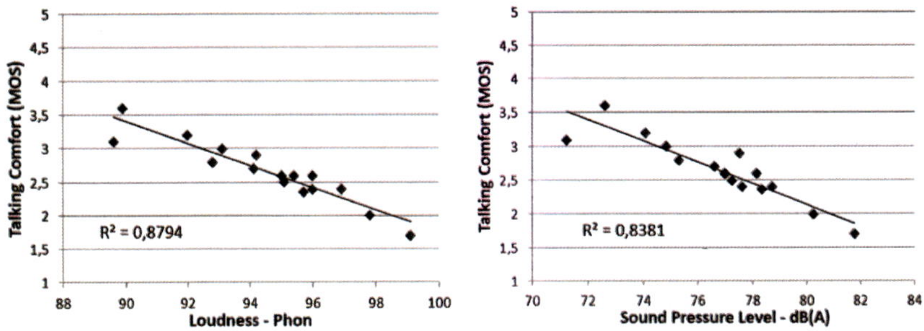

Fig. 4. Correlation of Talking Comfort Mean Opinion Scores with background noise Loudness and global dB(A) levels.

Looking at 1/3 octave band spectra of samples in these classes, we found that spectra in each class seem to have similar characteristics. More precisely, we could define for each of the above-mentioned classes a target spectrum in 1/3 octave bands. A background noise is evaluated to have a give MOS class if its spectrum lies entirely (for each band) below the target spectrum.

All of the spectra of the study, compared to the three target spectra, are thus correctly associated to their MOS class (to be precise, there is no target spectrum for MOS < 2.7, as it corresponds to not fulfilling the requirement for the MOS ≥ 2.7 class). Fig. 5 shows the three defined target spectra with an example spectrum measured on an existing train. Incidentally, notice that current trains in free-field already give Mean Opinion Scores that can be considered acceptable.

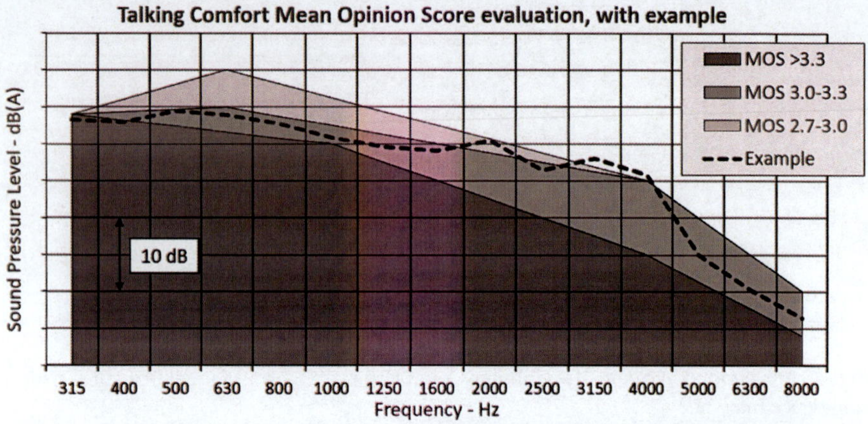

Fig. 5. Target spectra for the three classes of Talking Comfort Mean Opinion Score.

6 Acoustic Design of Railway Vehicles

The results discussed in the previous section are the foundation for a correct acoustic design of train vestibules. Today, car manufacturers have the tools to predict interior noise spectra in 1/3 octave band with reasonable precision. Typically, this is used to ensure the fulfillment of customer requirements often expressed as global noise levels in dB(A). While global dB(A) levels are often driven by energy in the low-middle frequencies (see for example Fig. 5), the target spectra proposed in the previous section are fairly demanding on the high frequencies. That happens because that is an important portion of the spectrum for human communication. This in turn means that, at least in some cases, to improve comfort it is better to focus on high frequencies, even if the global dB(A) level is not affected.

Let us make an example to show how results of Sect. 5 can be exploited. Figure 5 shows a spectrum measured in the vestibule of a real train at maximum speed. Given the definition of Sect. 5, this train is expected to receive a MOS lower than 2.7. Effectively, it was evaluated to be at 2.4. This scoring is rather low, but its analysis based on Fig. 5 shows that it would be sufficient to gain a few decibels in the bands 2000 Hz, 3150 Hz and 4000 Hz to reach a MOS between 3.0 and 3.3, much better. This gain would have a negligible impact on the global dB(A) level. Calculations or advanced testing (not realized on this particular train) will reveal how this could be achieved; maybe the contribution through the doors is such that increasing their Transmission Loss would be enough to achieve this target. This would be impossible without the analysis proposed in this paper.

This approach can be used to drive and validate the acoustic design of train vestibules. Already in tender phase, background noise spectra in vestibules can be calculated and compared to the target spectra shown in this paper. Usually, performances in low-middle frequencies turned out to be already good enough. Globally, this ensures good comfort for the final passengers, while focusing on high added value improvements.

7 Conclusion

The work carried out shows how background noise affects comfort of telephone conversations in train areas, mainly vestibules. This result makes it possible to do a real acoustical design of areas where telephone conversations shall take place: the metric being a target spectrum in 1/3 octave bands, it is even possible to do these studies in the early design phases of a new train and influence the train design to reach the desired comfort. This study is, to the best of our knowledge, the first of its kind for the rolling stock industry. Results of the study could be used to generate objective interior noise targets for vestibule areas without the need of very expensive, low noise solutions that in some cases are not needed to have comfortable phone conversations. While numerical values of target spectra shown in Fig. 5 cannot be shared due to Intellectual Property limitations, this approach could be used in similar contexts to try to focus on the well-being of passenger rather than purely achieving a global dB(A) level.

References

1. Boullet, I., Rabau, G., Meunier, S., Poisson, F.: Psychoacoustical indicators of acoustical comfort inside trains for background noise and emergent signals. In: 19th International Congress on Acoustics 2007, PACS 43, Madrid (2007)
2. Parizet, E., Hamzaoui, N., Jacquemoud, J.: Noise assessment in a high-speed train. Appl. Acoust. **63**, 1109–1124 (2002)
3. IEC 60268–16:2001: Sound system equipment – Part 16: Objective rating of speech intelligibility by speech transmission index
4. ETSI TS 103 224: Speech and multimedia Transmission Quality (STQ); A sound field reproduction method for terminal testing including a background noise database
5. ITU-T P.831:1998: Series P: Telephone transmission quality, telephone installations, local line networks. Methods for objective and subjective assessment of quality. Subjective performance evaluation of network echo cancellers
6. Lane, H., Tranel, B.: The Lombard sign and the role of hearing in speech. J. Speech Lang. Hearing Res. **14**(4), 677–709 (1971)

Vehicle Modeling for High Frequency Vibration

Qi Wang, Xinbiao Xiao(✉), Jian Han, and Yue Wu

State Key Laboratory of Traction Power, Southwest Jiaotong University, Chengdu 610031, China
602662339@qq.com, xinbiaoxiao@163.com

Abstract. Numerous experiments have shown that the wheelset-track noise can be transferred by the vibration through the primary suspensions to the bogie, which is called "structural-borne sound". The high frequency vibration transfer characteristics of primary suspensions, such as helical springs, arm bush and damper, is important. In this paper, a vehicle-track coupling dynamic model was developed considering high frequency vibration transferred by the primary suspensions. Based on the 3D finite element method, the modes of the bogie were simulated and a model of flexible bogie was established using mode superposition method. The 3D FE method was also used to develop the helical springs' simulation model. Based on the tests' result, the model of arm bush was established using transfer matrix method. Laboratory and field tests were carried out to validate the developed model. Short wavelength irregularities were adopted to excite the system to analyse the different influence on the structural-borne sound transfer characteristics. The results show that the vibration transmission rate is larger in 600–1000 Hz, which agrees with the test results. The model lays a foundation for analysing the sound transmission of structures.

Keywords: High frequency vibration · Primary suspensions · Dynamic stiffness model

1 Introduction

The wheelset-track noise is a significant noise source for the high-speed train. It can be transferred by the vibration through the primary suspensions to the bogie, which is called "structural-borne sound". The high frequency vibration transfer characteristics of primary suspensions, such as the helical springs, the arm bush and the damper, is important. Figure 1 shows the measured vibration of the axlebox and the bogie.

It can be seen in Fig. 1 that in the frequency range of 400–1000 Hz, the vibration attenuation of primary suspensions is less than that at other frequencies. The reason of this phenomenon may be related to the flexible vibration of the bogie and the dynamic stiffness of the primary suspensions. Therefore, the vibration transfer characteristics of primary suspensions and the bogie needs to be analysed.

The excitation of vibration is generated by wheel-rail contact when the vehicle is running. Considering the vibration frequency is larger than 20 Hz, the track should be considered in the model such as the rail, sleeper and track plate. The vehicle-track

G. Degrande et al. (Eds.): *Noise and Vibration Mitigation for Rail Transportation Systems*, NNFM 150, pp. 126–133, 2021.
https://doi.org/10.1007/978-3-030-70289-2_11

coupling dynamic model is a time domain which is a possible method to solve this problem. However, the application of vehicle-track coupling dynamic vibration research is still mainly focused on the low (less than 20 Hz) and medium (20–100 Hz) frequency ranges of the dynamic problem, and only in recent years, it is gradually extended to the high frequency (100–1000 Hz) region. Based on vehicle-track coupling dynamics model, Shuoqiao Zhong [1] considered the shaft bending of the flexible wheel on the model. The model can be used to analyse the wheel vibration at high frequency on the influence of wheel/rail contact behaviour. Liang Ling [2] established a rigid and flexible coupling dynamic model of high-speed train, which considered the flexible deformation within 50 Hz of the vehicle body and 1000 Hz of the bogie. However, the model does not consider the high frequency vibration transfer characteristics of the suspension elements.

For the transfer characteristics of the suspension, existing studies pay more attention to the transfer characteristics of a single part, either the spring or the rubber of the arm bush. To study the transfer of structural borne noise into the bogie, it is essential to build a vehicle-track coupling dynamic model based on a coupling model of primary suspensions and flexible bogie.

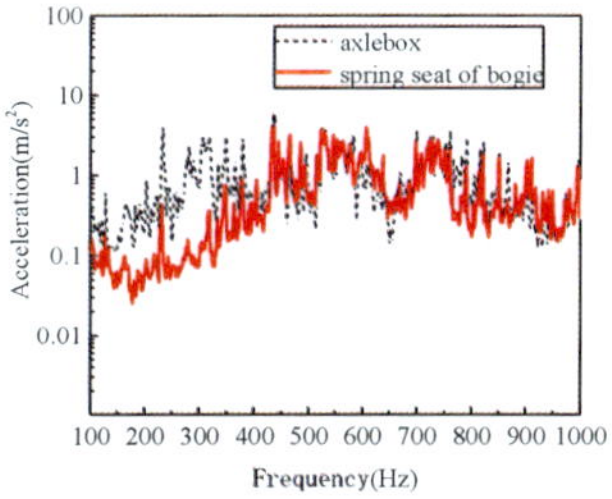

Fig. 1. Vertical vibration of high-speed train.

Fig. 2. Finite element model of bogie

Compared with vibration transfer at low frequency, the development of the new model has to do:

(1) the transfer simulation model of the suspensions' dynamic stiffness, especially of the arm at high frequency;
(2) the coupling model of the primary suspensions and the flexible bogie.

2 Model Establishment

2.1 Flexible Bogie

When high frequency vibration transfer characteristics is considered in the simulation model, it is necessary to consider the flexible vibration of the bogie. The vehicle model of multi-rigid continues to promote the vehicle model of rigid and flexible coupling. Considering the influence of the flexible deformation of the bogie, the hybrid coordinate method is used to describe the flexible deformation of the bogie. Basic modeling principle of the hybrid coordinate method is as following: a wide range of floating coordinate system of the rigid body coordinates and the node coordinates of flexible body (modal coordinates) is used to establish the dynamic model of the first in view of the structure of floating coordinate system. The deformation of the bogie is considered to be a wide range of floating coordinate system of rigid body motion and relative to the coordinate system of the flexible deformation of the stack. In the concrete modeling process, the floating coordinate system of the component is solidified, the frame is discretized by the finite element method according to the free-free boundary condition, as shown in Fig. 2, and the flexible vibration response of bogie is solved by the mode superposition method.

In order to verify the validity of the bogie model, it is necessary to conduct modal tests. Due to the large volume of the bogie, a 3D laser vibrometer is adopted to conduct the modal test of the bogie. During the test, one end of the bogie was lifted by an elastic rope and the other end was supported with elastic rubber to simulate free boundary. The bottom of the exciter was supported by elastic foam pad. A schematic diagram of frame layout is shown in Fig. 3. A number of positioning columns are placed next to the frame, which are kept still during the test. Fixed points are marked on the top of the positioning column for positioning and calibration of test coordinates after the laser emitter changes its position during the test. The modal frequency considered is less than 1000 Hz. There are 91 simulated modal frequencies and 23 test modal frequencies in Fig. 4.

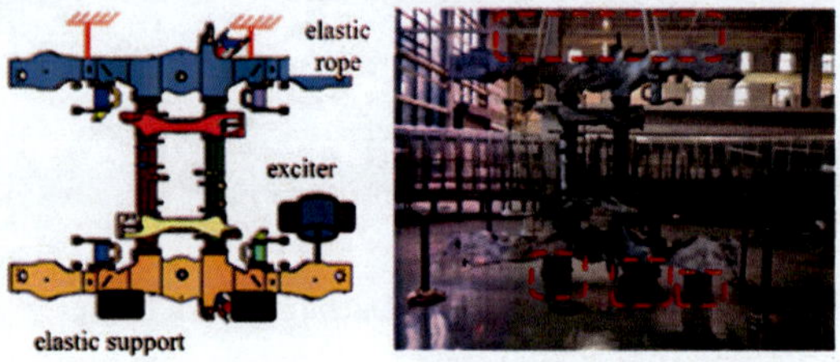

Fig. 3. Schematic diagram of frame layout

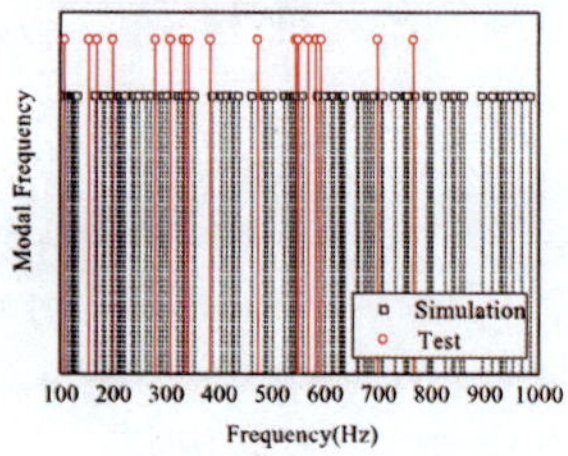

Fig. 4. Modal frequency of bogie

Figure 4 shows the modal frequency of bogie and Fig. 5 shows the modal shape of bogie.

The results in Fig. 4 and Fig. 5 show that due to the influence of environmental light and other factors and other reasons in the test process, the number of identified modes is less than the finite element calculation results. The finite element calculation results of the low frequency bogie mode are consistent with the test results, and errors in the predicted modal frequency is small. The finite element calculation results of the middle and high frequency bogie mode are consistent with the test results, and the modal frequency error is larger than the low frequency, but within the acceptable range. At about 585 Hz, the modal shape is the nod of the side beam, the reverse bending of the beam and the sharp torsion above the beam. The mode shape is identical, and the error of the modal frequency calculation result is 0.04%. So, the accuracy of the finite element model is verified by experimental modal analysis.

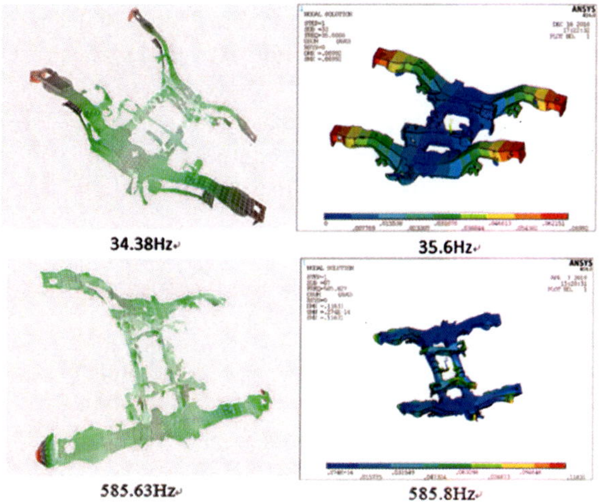

Fig. 5. Comparison of test modal and calculated modal modes

2.2 Flexible Spring

The same method was used to develop the helical springs' simulation model. Figure 6 shows the vertical FRF of the helical spring.

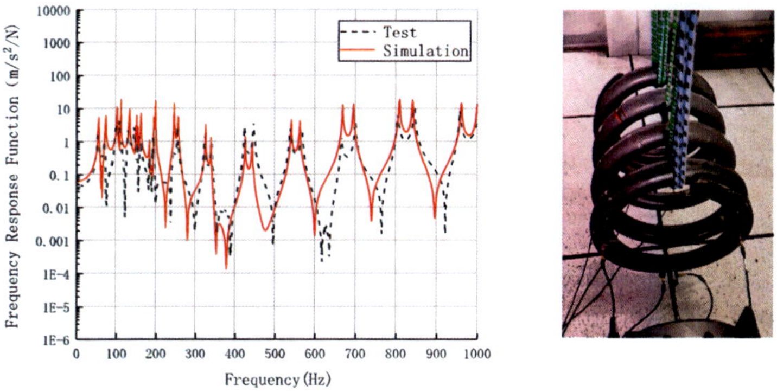

Fig. 6. FRF of helical spring

It can be seen that the measured modal frequency and the simulated modal frequency correspond to each other well in the low frequency band. Within 300 Hz, steel spring has many modes, which is sensitive to vibration isolation. In the frequency band of 400–1000 Hz, steel spring has fewer modes, but the large FRF value indicates poor vibration isolation.

2.3 Flexible Arm

In the previous dynamic modelling process, the arm is usually regarded as a simple spring/damping element, ignoring the influence of the modal characteristics of the arm on the transmission at high frequency. This simplified model will lead to large attenuation of high-frequency vibration. This section mainly introduces how to establish the dynamic stiffness model of arm bush [3].

A dynamic stiffness model is established as shown in Fig. 7. The model includes a linear element and a dynamic stiffness element. Each dynamic stiffness element corresponds to a complex modal frequency. The method for obtaining parameters of dynamic stiffness element is introduced below.

Suppose the relationship between force and displacement is shown in Eq. (1)

$$F = KX \tag{1}$$

Where F is the external force applied, X is the displacement response, and K is the dynamic stiffness.

In the Laplace space, Eq. (2) is fitted, as shown below

$$\frac{X(s)}{F(s)} = \frac{1}{K(s)} = \frac{a_0 s^m + a_1 s^{m-1} + a_2 s^{m-2} \cdots + a_m s^0}{b_0 s^n + b_1 s^{n-1} + b_2 s^{n-2} \cdots + a_n s^0} \quad m < n \tag{2}$$

Where a and b are parameters of numerator and denominator respectively, m and n are the highest order in the numerator and the highest order in the denominator. By inversed Laplace transform of Eq. (2), a differential equation of higher order results. Due to the high order, the solution of this equation in the time domain is relatively complicated. Therefore, in order to simplify the operation, partial fraction expansion of Eq. (2) is carried out to reduce its order to second-order equation:

$$\begin{aligned} \frac{X(s)}{F(s)} &= \frac{a_0 s^m + a_1 s^{m-1} + a_2 s^{m-2} \cdots + a_m s^0}{b_0 s^n + b_1 s^{n-1} + b_2 s^{n-2} \cdots + a_n s^0} = \frac{r_1}{s - p_1} + \frac{r_2}{s - p_2} + \cdots + \frac{r_n}{s - p_n} \\ &= \frac{1}{c_0 s - k_0} + \sum_i \frac{A_i s + B_i}{s^2 + \zeta_i s + \chi_i} \end{aligned} \tag{3}$$

Therefore, the relationship between force and displacement is

$$\frac{X(s)}{F(s)} = \frac{1}{c_0 s - k_0} + \sum_i \frac{A_i s + B_i}{s^2 + \zeta_i s + \chi_i} \tag{4}$$

Convert the equation to the time domain,

$$k_0 \varepsilon_0 + c_0 \dot{\varepsilon}_0 = F \tag{5}$$

$$\ddot{\varepsilon}_i + \zeta_i \dot{\varepsilon}_i + \chi_i \varepsilon_i = A_i \dot{F} + B_i F \tag{6}$$

$$\varepsilon = \varepsilon_0 + \sum_i \varepsilon_i \tag{7}$$

The symbols in Eqs. (3) to (7) are described as follows:

F —The force transmitted by the suspension element,

ε —Total deformation of suspension element,

ε_0 —The amount of deformation of a linear element,
k_0 —Stiffness parameters of linear elements,
c_0 —Damping parameters of linear elements,
ε_i —The deformation of the ith dynamic stiffness element,
χ —The stiffness parameter of the ith dynamic stiffness element,
ζ_i —Damping parameter of the ith dynamic stiffness element,
$A_i\,B_i$ —Force parameter of the ith dynamic stiffness element,
k_0, c_0, χ_i, ζ_i, A_i, B_i need to be calculated by the dynamic transfer characteristics of the suspension.

According to the theory of four-end parameter method, the relationship between the force and displacement at the origin and span point of the suspension member is shown in Eq. (8) [4].

$$\begin{bmatrix} F_1 \\ F_2 \end{bmatrix} = \begin{bmatrix} K_{11} & K_{12} \\ K_{21} & K_{22} \end{bmatrix} \begin{bmatrix} X_1 \\ X_2 \end{bmatrix} \tag{8}$$

Where F_1, F_2 represents the forces at the upper and lower interfaces, x_1, x_2 represents the displacement of the upper and lower interfaces. K_{11}, K_{22} represents the dynamic stiffness of the origin, K_{12}, K_{21} represents the cross-point stiffness, where $K_{11} = K_{22}$, $K_{12} = K_{21}$.

By combining Eq. (4) with Eq. (8), the dynamic force transmitted by the suspension member can be obtained.

As shown in Fig. 8, a certain type of arm bush was tested in the laboratory, and the axial and radial impedance of the arm bush were tested respectively, and the point impedance and the transfer impedance were obtained.

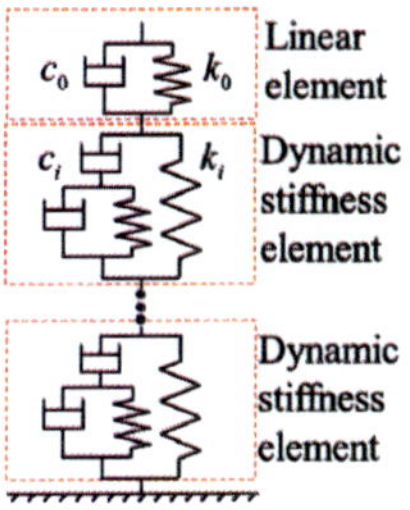

Fig. 7. Dynamic stiffness model

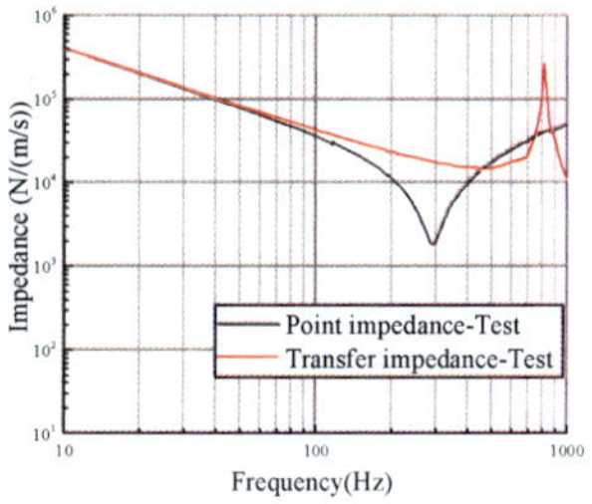

Fig. 8. The axial impedance of arm bush

It can be seen that the axial transfer impedance of arm bush has a peak value at around 810 Hz and the axial transfer impedance has a valley value at around 290 Hz, which means the arm bush has different transfer characteristics at different frequency. After the impedance is obtained, the parameters of the dynamic stiffness model were obtained by fitting the impedance curves.

As to the model of the arm bush, the spring model and the 'Berg model' are widely used. The former considers the arm bush as a stiffness and damping, while the latter considers the nonlinearity of the rubber (the frequency is generally below 100 Hz). From the FRF test, it can be seen that the FRFs of the arm bush above 100 Hz is not a constant, but a peak and a valley value. Therefore, it is necessary to adopt the more complex model in this paper.

The other parts, such as wheelsets and carbody are regarded as rigid bodies. The rail coupler is simulated by linear spring-damping element. The rail is regarded as continuous elastic discrete point support on a finite length Timoshenko beam. The sleeper and track plate are regarded as continuous elastic discrete point support rigid mass block. The wheel-rail interaction model is based on vehicle-track coupling dynamics. The parameters are the same as the model in paper [2].

3 Results of Simulation and Conclusion

To simulate the vibration transmissibility from axlebox to bogie, the vehicle is moving in a straight line with the speed of 250 km/h. Measured short wave irregularities were adopted to excite the system. Figure 9 shows the test and simulated vibration transmissibility.

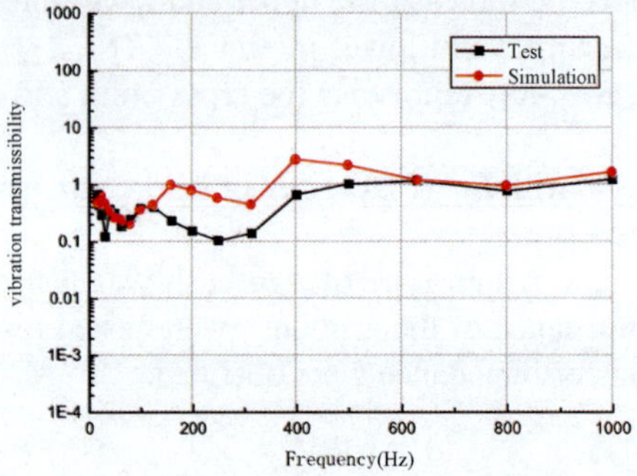

Fig. 9. Vibration transmissibility

It can be seen that the simulated vibration transmissibility curve is consistent with the measured curve at 600–1000 Hz, which means that the model is effective. At 200–400 Hz the simulated curve value is large than the test, but the curve trend are is similar. The reason may be as follows:

(1) When the model parameters are obtained from the arm bush experimental results, the validity around 100–1000 Hz is mainly concerned, and the matching effect less than 100 Hz and above 1000 Hz is less concerned. The dynamic characteristic at these frequency may be different between mdoel and reality.
(2) The damping at 200–400 Hz used in simulation model is different from the real damping, which lead the simulation is larger than the test.
A vehicle-track coupling dynamic model is developed considering high frequency vibration transferred by the primary suspensions. The developed model was validated by laboratory and field tests. Although the vibration transmissibility does not match well at all frequency, the work shows the spring and the arm bush's flexible deformation should be taken into account about the "structural-borne sound" of the vehicle. And the vehicle-track model can be used to simulate the issue.

Acknowledgement. This study was supported by the National Key R&D Program of China (2016YFE0205200).

References

1. Zhong, S.Q., Xiong, J.Y., Xiao, X.B., et al.: Effect of the first two wheelset bending modes on wheel-rail contact behavior. J. Zhejiang Univ. Sci. Appl. Phys. Eng. **15**(12), 984–1001 (2014)
2. Liang, L.: 3D rigid-flexible coupling dynamics of high-speed train/track system. Southwest Jiaotong University (2015)
3. Gerstberger, U., Knothe, K., Wu, Y.: Combined Modelling of Discretely Supported Track Models and Subgrade Models - Vertical and Lateral Dynamics. System Dynamics and Long-Term Behaviour of Railway Vehicles, Track and Subgrade. Springer, Berlin Heidelberg (2003)
4. O'Hara, G.J.: Mechanical impedance and mobility concepts. J. Acoust. Soc. Am. **41**(5), 1180–1184 (1966)

Numerical Analysis on the Radiation Efficiency of an Extruded Panel for the Railway Vehicle Using the Waveguide Finite Element and Boundary Element Method

Hyungjun Kim[1], Jungsoo Ryue[1(✉)], David J. Thompson[2], and Angela D. Müller[2]

[1] University of Ulsan, 93, Daehak-ro, Nam-gu, Ulsan 44610, Republic of Korea
jsryue@ulsan.ac.kr

[2] University of Southampton, Southampton SO17 1BJ, UK

Abstract. Complex-shaped stiffened panels are widely used nowadays in many structures such as railway vehicles, ships and offshore structures, in order to make them lighter and stronger. Therefore, it is essential to investigate vibration and radiation of the complex shaped panels to predict the vibro-acoustic performance of the large structures. In this paper, a coupled waveguide finite element and boundary element method, which can be applied where the cross-sectional properties of the panels are constant in one direction, is applied to predict the radiation efficiency of the extruded panel. The dependence of the radiation efficiency on two different types of excitation positions is investigated; a point force in the middle of the strip or on an interior stiffener. Finally, numerical results are validated through the comparison with experiments carried out with a 1.5 m length extruded panel.

Keywords: Extruded panel · Waveguide finite element and boundary element method · Radiation efficiency

1 Introduction

Complex-shaped stiffened panels are widely used nowadays in many structures such as railway vehicles, ships and offshore structures, in order to make them lighter and stronger. To predict the acoustic performance of these structures, it is essential to understand the vibro-acoustic behaviour of such complex-shaped panels. As an example of such panels, an extruded aluminium panel used in the floor of a railway vehicle, which has a uniform cross-section along its length, is considered in this study.

In this paper, the radiation efficiency of an extruded panel from a railway vehicle are investigated using the waveguide finite element and boundary element (WFE/BE) method. The WFE/BE approach has previously been used to predict the sound radiated from rails [1, 2] and stiffened double plates [3, 4]. In this work, the dependence of the radiation efficiency on two different types of excitation positions is investigated; a point force in the middle of the strip or on an interior stiffener. To give insight into the vibration

G. Degrande et al. (Eds.): *Noise and Vibration Mitigation for Rail Transportation Systems*, NNFM 150, pp. 134–141, 2021.
https://doi.org/10.1007/978-3-030-70289-2_12

and sound radiation behaviours, results will be presented in the wavenumber domain, from which the contributions of different types of waves can be identified. Finally, the numerical results are validated through comparison with experiments carried out with a finite length extruded panel.

2 The Waveguide Finite Element and Boundary Element Method

In this section, the waveguide finite element and boundary element (WFE/BE) method is described briefly. This method is known to be particularly useful for waveguide structures which have a uniform cross-section along one direction (x direction in this paper), such as rails, plates, pipes, ducts, etc. Details of the derivation of the governing equations are well-reported in the literature [4, 5].

The radiation efficiency (σ) of the waveguide structure can be determined by

$$\sigma = \frac{\Pi_{\text{rad}}}{\rho_0 c_0 \Gamma \left\langle \overline{v^2} \right\rangle_{inf}}, \tag{1}$$

where c_0 is the sound speed in the fluid, Γ is the perimeter of the cross-section in contact with the fluid and $\left\langle \overline{v^2} \right\rangle_{inf}$ represents the integral of the mean-squared velocity over the length and averaged over the width. This differs from the usual definition of σ due to the infinite extent of the panel in the waveguide direction. The radiated power, Π_{rad}, and average mean-squared velocity, $\left\langle \overline{v^2} \right\rangle_{inf}$, of the panel in Eq. (1) can be obtained from

$$\Pi_{\text{rad}} = \frac{1}{4\pi} \text{Re}\left\{ \int_{-k}^{k} \int_{\Gamma} \tilde{p}(\kappa, y) \tilde{v}_n(\kappa, y) dy d\kappa \right\}, \tag{2}$$

$$\left\langle \overline{v^2} \right\rangle_{inf} = \frac{1}{2\pi \Gamma} \int_{-\infty}^{\infty} \int_{\Gamma} \tilde{v}_n(\kappa, y) dy d\kappa, \tag{3}$$

where $k = \frac{\omega}{c_0}$ is the acoustic wavenumber and the superscript $*$ denotes the complex conjugate. In Eq. (2), the integration with respect to κ is limited to $\pm k$ because the panel does not radiate sound into the far field for $\kappa > k$. Since the sound pressure, $\tilde{p}(\kappa, y)$, and velocity, $\tilde{v}_n(\kappa, y)$, can be found from Eq. (6) in ref. [4], Eqs. (2) and (3) can be evaluated from the WFE/BE solutions without transforming the results from the wavenumber domain into the spatial domain.

3 Modelling the Panel and Fluid

In this study, the extruded panel shown in Fig. 1(a), which is a part of typical floor panel of a railway vehicle, is examined by WFE/BE analysis. The panel is made of aluminium but the top plate is covered with a heavy rubber mat. The cross-sectional model of the panel structure is illustrated in Fig. 1(b); this is composed of two-noded plate elements with a cubic shape function. The rubber mat attached on the top plate is included in the model by increasing the mass and damping of the top plate. The dimensions and properties of the panel are listed in Table 1. To consider the stiffening effects of the joints where the interior stiffeners are connected to top or bottom plate, the WFEs close to the joints are modelled with a thickness that is 20% larger than the remaining plates.

The frequency-dependent damping loss factors of the top and bottom plates of the panel measured by Zhang et al. [3] are used in this paper. The two damping loss factors for the top and bottom plates are evaluated by means of the reciprocal average of the values measured in each bay [4]. The interior stiffeners of the panel are assigned the same damping as the bottom plate.

In the WFE modelling for the radiation efficiency, the panel is set to have free-free boundary conditions at both ends for compatibility with the measurements described in Sect. 5 below. For simplicity, the WBEs enclose the panel in a rectangular shape, as shown in Fig. 1(c). Since the height of the panel (about 0.07 m) is relatively small compared with the width of the panel (1 m), it is expected that most of the sound will be radiated from the top and bottom plates. Hence, the WBEs on both sides are assigned to have zero velocity.

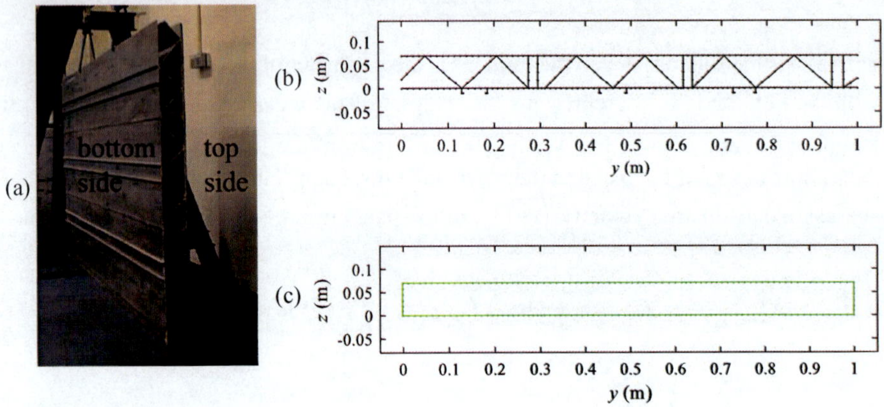

Fig. 1. (a) A specimen of an extruded floor panel for a railway vehicle. (b) Cross-sectional WFE model of the extruded panel and (c) WBE model for external air. (Dots represent nodes in each model.)

Table 1. Properties and dimensions of the extruded panel.

Material	Model	Thickness (mm)	Density (kg/m^3)	Poisson's ratio	Young's modulus (GPa)
Aluminium	Top plate	2.8	2700	0.3	70
	Bottom plate	2.7			
	Vertical stiffeners	2.3			
	Oblique stiffeners	2.6			
Rubber	Rubber mat	4.0	1500	–	–

4 Numerical Results for the Extruded Panel

4.1 Dispersion Relation

It can be seen from Fig. 2 that there are several waves at low frequencies. These correspond to global motions of the extruded panel, i.e. the vertical bending, torsion, lateral bending and longitudinal waves of the whole section, in descending order of the wavenumber at 100 Hz. The wave which cut-on around 130 Hz is flapping mode wave of left-hand side of the bottom plate. However, at frequencies above about 400 Hz and at high wavenumbers, many local modes occur in the plate strips.

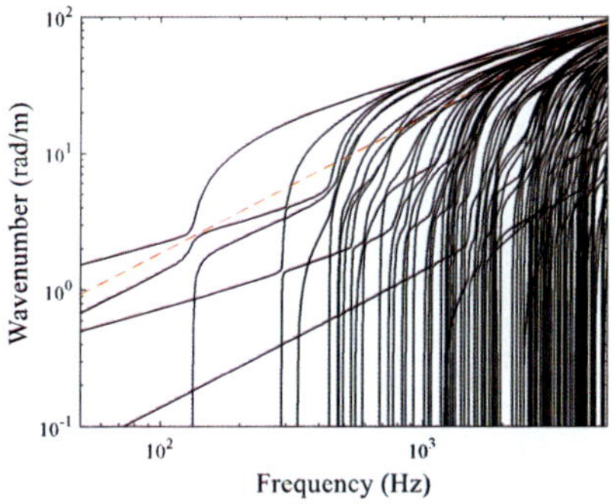

Fig. 2. Dispersion diagrams of the extruded panel. (Acoustic wavenumber is added with a thin dashed line.)

4.2 Radiation Efficiency of the Extruded Panel

In the prediction of the radiation efficiency, a unit amplitude point force is applied at two different types of location on the bottom plate; one is in the middle of a strip between two interior stiffeners and the other is where an interior stiffener is connected to the plate. These positions correspond to P1 (or P2) and P3 on the strip and P4 and P5 on the stiffener as specified in Sect. 5 below.

The average mean-squared velocity and the sound power radiated from both the top and bottom plates are illustrated in Figs. 3, 4 and 5, as image plots against the frequency and wavenumber. The results in Fig. 3 are obtained for a point force applied in the middle of a strip at P1 or P2 and the results in Fig. 4 are for P3, whereas the results in Fig. 5 correspond to a position on a stiffening rib at P4. The acoustic wavenumber is also plotted with a dashed line.

For the excitation in the middle of the strip, as shown in Fig. 3 and Fig. 4, a small number of global wave modes generate the vibration at frequencies below about 400 Hz and many waves having higher order cross-sectional deformations contribute above about 400 Hz for both the strip excitation at P1 (or P2) and P3. However, in terms of the radiated sound power, for the excitation at P1 (or P2), the waves which have wavenumbers larger than that of the acoustic wave do not contribute at all, see Fig. 3(b). On the other hand, for the excitation at P3, the waves which generate large vibration in Fig. 4(a) radiate

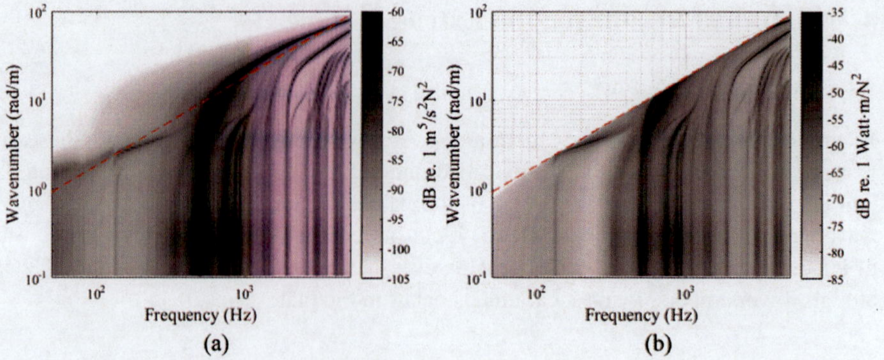

Fig. 3. Image plots of (a) the average mean-squared velocity and (b) radiated sound power of the panel for excitation in the middle of a strip at P1 or P2.

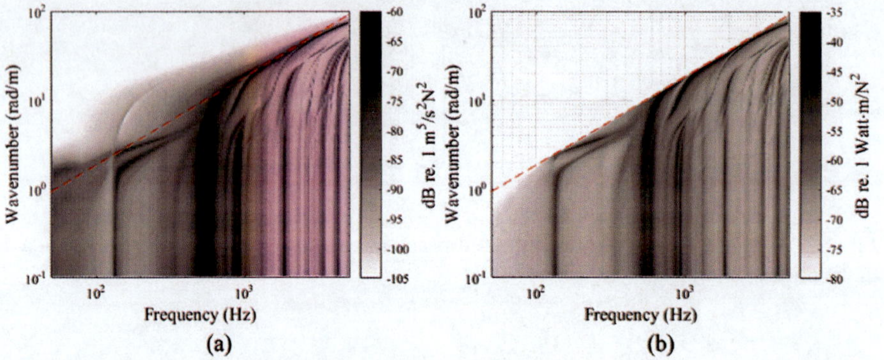

Fig. 4. Image plots of (a) the average mean-squared velocity and (b) radiated sound power of the panel for excitation in the middle of a strip at P3.

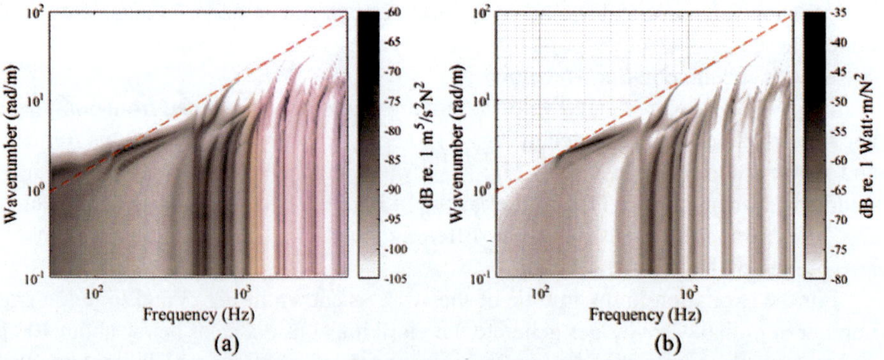

Fig. 5. Image plots of (a) the average mean-squared velocity and (b) radiated sound power of the panel for excitation on a stiffener at P4.

sound loudly, as shown in Fig. 4(b) because those waves have wavenumbers lower than that of the acoustic wave.

On the other hand, for the excitation on the stiffener at P4, it can be seen from Fig. 5 that relatively weak vibration and radiation occur compared with the strip excitation. In addition, Fig. 5 shows that the responses are mostly limited to low wavenumbers. This implies that the interior stiffeners govern the global rigidity of the extruded panel so that the panel deforms mainly globally by the stiffener excitation. Another numerical results for the stiffener excitation at P5 are similar with those for the stiffener excitation at P4.

The numerical results for the radiation efficiency are calculated from the averaged mean-squared velocity and radiated sound power. The results will be compared with the measured ones in Sect. 5.

5 Comparison of Numerical Results with Experiments

In this section, the results obtained from the WFE/BE approach are compared with those obtained experimentally on a finite panel. The extruded panel used in the experiment is shown in Fig. 1(a) and has a length of 1.5 m along the extrusion. Regarding the actual length of the train coaches and the assumption of waveguide, the length 1.5 m would not be long enough but chosen for lab experiments.

5.1 Experimental Set-Up of Radiation Efficiency

To measure the radiation efficiency, the panel was hung freely with elastic ropes in a reverberation chamber [6] so that that the coupling effects between the panel and chamber would be excluded. A mechanical excitation was applied with a shaker. This shaker was attached to the panel via a slender rod stinger connected to the force transducer. This was necessary to allow the panel to be excited in only one direction so that no rotation was introduced. The measurements were made of the surface-averaged panel vibration using a scanning laser vibrometer. The spatially-averaged sound pressure in the reverberation chamber was measured, from which the sound power could be determined.

In these measurements, two types of excitation point were chosen on the bottom plate as described in Sect. 4.2. One is in the middle of a bay (or a strip), and the other is

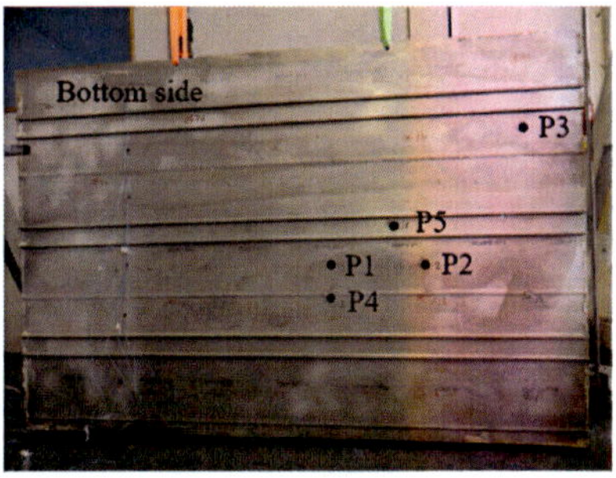

Fig. 6. The locations of five excitation points applied on the bottom plate.

Table 2. The coordinates of excitation points represented in Fig. 6. (The origin is at the top left-hand corner of the plate.)

Excitation point	x(m)	y(m)
P1	0.78	0.528
P2	1.10	0.528
P3	1.32	0.207
P4	0.78	0.635
P5	1.05	0.460

where an interior stiffener is connected. The specific locations of each excitation point are indicated in Fig. 6, and their coordinates are listed in Table 2.

The measured radiation efficiencies are compared with the predicted ones in Fig. 7. For the strip excitation, the measured radiation efficiencies are compared with the predicted ones in Fig. 7(a). The measured result by P3 show different level and tendency at high frequencies above 800 Hz compared with measured results by P1 and P2 which show similar level and tendency each other. The numerical results generally look similar with the measured ones and the discrepancy between P1 (or P2) and P3 is also observed in the numerical results above 800 Hz. The different trends in the radiation efficiency by the strip excitation at P1 (or P2) and P3 is caused by the location of the external stiffener in the bay. Regardless of this difference, it can be said that the WFE/BE approach is applicable in predicting the radiation efficiency of the finite length panel for the strip excitation.

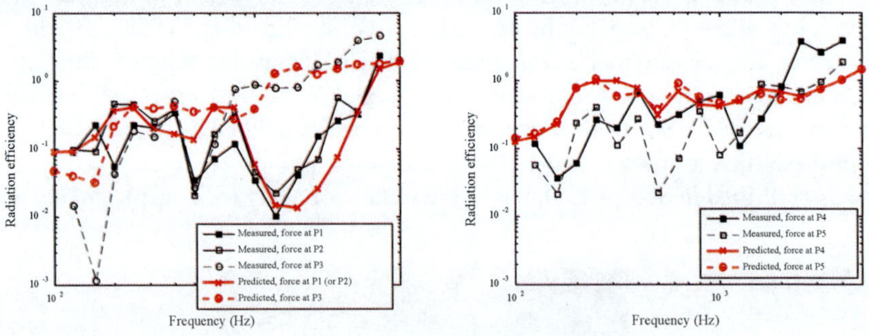

(a) (b)

Fig. 7. Comparison of radiation efficiencies obtained from numerical simulation and experiments for (a) strip excitations at P1, P2 and P3 and (b) stiffener excitations at P4 and P5.

On the other hand, for the stiffener excitation, the predicted results in Fig. 7(b) provide bigger discrepancy than the strip excitation. This may be attributed to the fact that the stiffener excitation generates mainly low wavenumber waves as shown in Fig. 5. Hence, the responses predicted by the WFE/BE approach for the stiffener excitation are likely to be discrepant from those measured from the short finite length panel. Therefore, the WFE/BE method becomes less suitable for the prediction of radiation efficiency from this finite length panel in case of the stiffener excitation.

6 Conclusions

The vibro-acoustic behaviour of a complex shaped extruded panel has been investigated by using the waveguide finite element and boundary element (WFE/BE) method.

Two different types of excitation position were considered; on a strip or on an internal stiffener. It was found that the strip excitation generates localised waves strongly while the stiffener excitation makes global wave modes which are limited to low wavenumbers.

The predicted results were validated from experiments using a 1.5 m length extruded panel. It was observed from this comparison that the predicted radiation efficiency for the strip excitation gives better agreement to the measured ones than that for the stiffener excitation. This is because the waves generated by the stiffener excitation have low wavenumbers. Hence, the waveguide assumption in WFE/BE analysis does not suitable to the 1.5 m length panel for the stiffener excitation.

References

1. Ryue, J., Jang, S., Thompson, D.J.: A wavenumber domain numerical analysis of rail noise including the surface impedance of the ground. J. Sound Vib. **432**, 173–191 (2018)
2. Nilsson, C.M., Jones, C.J.C., Thompson, D.J., Ryue, J.: A waveguide finite element and boundary element approach to calculating the sound radiated by railway and tram rails. J. Sound Vib. **321**, 813–836 (2009)
3. Zhang, Y., Thompson, D.J., Squicciarini, G., Ryue, J., Xiao, X., Wen, Z.: Sound transmission loss properties of truss core extruded panels. Appl. Acoust. **131**, 134–153 (2018)
4. Kim, H., Ryue, J., Thompson, D.J., Müller, A.D.: Application of a wavenumber domain numerical method to the prediction of the radiation efficiency and sound transmission of complex extruded panels. J. Sound Vib. **449**, 98–120 (2019)
5. Nilsson, C.M.: Waveguide finite elements applied on a car tyre. Ph.D. thesis, KTH (2004).
6. Müller, A.D.: Acoustical investigation of an extruded aluminium railway vehicle floor panel. M.Sc. thesis, University of Southampton (2004)

Speech Intelligibility - Effects of Railway Tunnels at High Speeds

Christian H. Kasess[1(✉)], Thomas Maly[2], Holger Waubke[1], Michael Ostermann[2], and Günter Dinhobl[3]

[1] Acoustics Research Institute, Austrian Academy of Sciences, Vienna, Austria
christian.kasess@oeaw.ac.at
[2] Institute of Transportation, Vienna University of Technology, Vienna, Austria
[3] Austrian Federal Railways, ÖBB-Infrastruktur AG, Vienna, Austria
https://www.kfs.oeaw.ac.at/

Abstract. While a train is passing through a tunnel, the interior noise level is elevated compared to a track in open country due to reflections from usually fully reflecting walls and track surfaces. The aim of this study was to investigate how speech communication is influenced inside a tunnel at speeds typical for Austria of up to 230 km/h using the speech transmission index for public address systems (STIPA). Although for the covered speed range the speech intelligibility is considerably decreased inside a tunnel in absolute terms the STIPA values still indicates a good intelligibility when passengers are sitting in adjacent seats. However, when the distance between speaker and listener is increased, at higher speeds complex information may become difficult to comprehend, depending on the speech level.

Keywords: Tunnel · Speech communication · Speech transmission index

1 Introduction

Railway tunnels are an important part of the infrastructure, especially in mountainous countries such as Austria. Unfortunately, when a train passes through a tunnel the interior noise level is elevated compared to a track in open country [1] due to reflections from usually fully reflecting tunnel walls and track surfaces. The interior noise of a train is, however, an important constituent of passenger comfort. Not only can an elevated noise level be a nuisance in itself resulting in increased annoyance [1], the interior noise may also affect the communication between passengers.

There are two main ways to assess speech intelligibility (for an overview see e.g. [2]): tests with human listeners who judge the intelligibility or indices to quantify the transmission channel from sender (speaker) to receiver (listener). The former is very time consuming, in particular when many different settings are to be evaluated. For the latter the two most widely used indices are the

G. Degrande et al. (Eds.): *Noise and Vibration Mitigation for Rail Transportation Systems*, NNFM 150, pp. 142–149, 2021.
https://doi.org/10.1007/978-3-030-70289-2_13

speech transmission index or STI [3] and the speech intelligibility index or SII [4]. While the SII allows more elaborate adjustments of the settings [2], the former has been shown to have a higher sensitivity under similar settings [5]. For railway vehicle interior noise the relation between speech level, background level, and speech intelligibility was previously investigated [6]. Although the main research question concerned the speech privacy, i.e. the effect of speech from nearby seats, the first experiment showed that a higher speech-to-noise ratio resulted in a higher intelligibility as derived from an actual intelligibility test and that the intelligibility correlated with the STI. Also important, the STI, or more precisely the STIPA (STI for public address systems which is a reduced and thus less time consuming version of the STI) has already been used in technical specifications for interoperability e.g. TSI for persons with reduced mobility (TSI-PRM [7]). Thus, the STIPA was chosen for the evaluation of the effect of train noise on verbal communication.

The aim of this study was to investigate to which degree the speech communication is altered inside a tunnel at speeds up to 230 km/h which are typical for Austria using the STIPA as a measure. The assessment in this study is done in terms of the train speed, the seating arrangements, as well as the speech level.

2 Methodology

2.1 Interior Noise Measurements

Measurements of the interior noise during operation were performed with three microphones (G.R.A.S 40AE, preamplifier 26CA; 24-channel ICP-module DIC24, Head Acoustics) as well as a head-and-torso-simulator (HATS, HMS IV, Head Acoustics). The placement of the microphones and the HATS is illustrated in Fig. 1 (upper panel). The positions of the HATS and the microphones related to the seat were chosen similar to the natural position of an average-sized person.

Recordings of the train noise were done while the train was running at different speeds (primarily 160 km/h, 180 km/h, 200 km/h, and 230 km/h) inside the tunnels and outside. For each train speed the interior noise was recorded for at least 30 s. Due to the positioning of the tunnels along the track and the limited length of the tunnels not all speed steps were achievable in every type of tunnel. The coach in which measurements were performed was without passengers with the air conditioning activated to obtain recordings of the interior noise close to reality. In case of a passing train or other disturbing noise during the recording procedure, the recording time was extended accordingly to get usable samples.

HATS recordings only took place in one position close to M1. In order to get an estimate of the noise in M2 and M3 for the HATS, third-octave band corrections were determined from the differences of the band levels of the microphone recordings of M2 and M3 compared to M1. For each band, train speed and infrastructure the median across all segments was determined. These corrections were applied to the third-octave bands of the corresponding HATS recordings. Due to a problem with the microphone at M1, these corrections were only possible for speeds up to 200 km/h. For the same reason, the 160 km/h measurement for

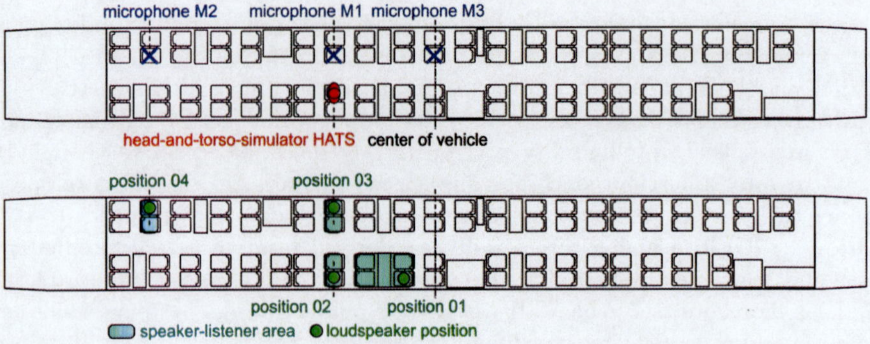

Fig. 1. Measurement setup. The upper panel shows the placement of the HATS and the microphones for interior noise measurements in a train during operation. The lower panel shows the placement of speaker and listener (loudspeaker and microphone/HATS, respectively) for STIPA measurements in a shut down train of the same type.

the double-track tunnel at M1 is also missing. Furthermore, it is important to note that this approach cannot capture any differences in directionality of the sound field which might affect the HATS recording.

2.2 STIPA Procedure

The measurement procedure as well as the signal generation and STI calculations are on the basis of the IEC 60268-16 [3]. Compared to the full STI for the STIPA instead of measuring all 14 modulation frequencies for each of the 7 octave bands (125 Hz to 8 kHz), the different modulations are distributed across the frequency bands (2 modulations per octave band) such that all modulations and frequency bands are measured simultaneously using a modulated noise carrier. From these, the values for the 14 modulation transfer functions can be determined which are then combined to a single index as described in [3]. STIPA signals were generated using the ArtemiS Software (Head Acoustics).

Before the measurements, all potential noise sources within the passenger car were identified and switched off. STIPA signals were recorded in quiet in different positions within the passenger car for different arrangements of speaker and listener (Fig. 1, lower panel), i.e. loudspeaker and microphone/HATS. The distances ranged from less than 50 cm (positions 02, 03, 04) to simulate passengers sitting next to each other to about 140 cm (position 01) which is a representative distance for passengers sitting diagonally opposite in a 4-people bay.

The STIPA signals were processed to control for the frequency response of the amplifier/loudspeaker setup using the minimum-phase impulse response [8] which was determined in a soundproof chamber (A-weighted background level 17 dB) at 0.5 m distance. First, the STIPA in quiet was calculated using different A-weighted STIPA signal levels as determined in 1 m distance in the soundproof chamber after correction ranging from 54 up to 72 dB. In the opposite seating arrangement (140 cm distance), these levels were measured to be about 6 dB

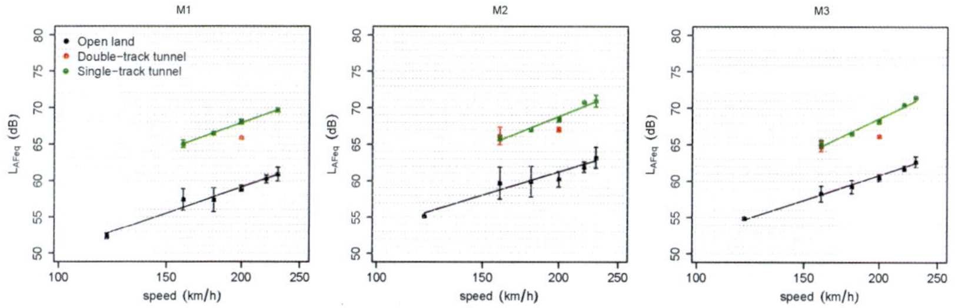

Fig. 2. L_{AFeq} and train speed. The panels show the L_{AFeq} as a function of train speed for the different recording positions derived from 10 s long sound segments.

higher due to reflections of window and ceiling. Different speech levels were generated by an amplitude scaling of the corrected STIPA signal instead of actually repeating the measurements at different levels. Thus, the signal-to-noise ratio (SNR) for the STIPA signals in quiet was the same for all levels.

Using the so-determined clean STIPA signals, the SNRs under real conditions were derived by comparing the octave bands of the scaled STIPA and the train noise band levels. Implementing the procedure prescribed in [3], the corrected STIPA-values under noisy conditions, i.e. when the train was in motion, were derived. Each audio recording during operation was divided into segments of 10 s and the SNR and STIPA were calculated for each segment separately. From the median across segments per recording the mean and standard deviation were calculated. For the purpose of comparison, the STIPA was also determined for manually segmented signals (only one segment per recording session) where signals with audible components unrelated to the typical interior noise were excluded. Furthermore, to validate the SNR procedure STIPA signals and audio signals of the interior noise were superimposed and the STIPA values under noisy conditions were determined directly from these noisy signals.

3 Results

Figure 2 illustrates the L_{AFeq} for different situations and speeds for the microphones in positions M1, M2, and M3. What is clear from all graphics is that the noise level in the tunnels (red and green symbols and lines) was significantly higher at the same speed than on an open track. In the double-track tunnel (red symbol) the L_{AFeq} was slightly lower, in particular at a speed of 200 km/h than in the single-track tunnel (green). Sound pressure levels towards the end of the car (M2) and in the middle (M3) are slightly higher at high speeds than in microphone position M1. For the HATS, the dependency between sound pressure level and train speed are similar.

Figure 3 illustrates the third-octave band spectra for the different situations and positions in the train including the background level present during the STIPA measurements. The main change of level happens between 100 Hz and

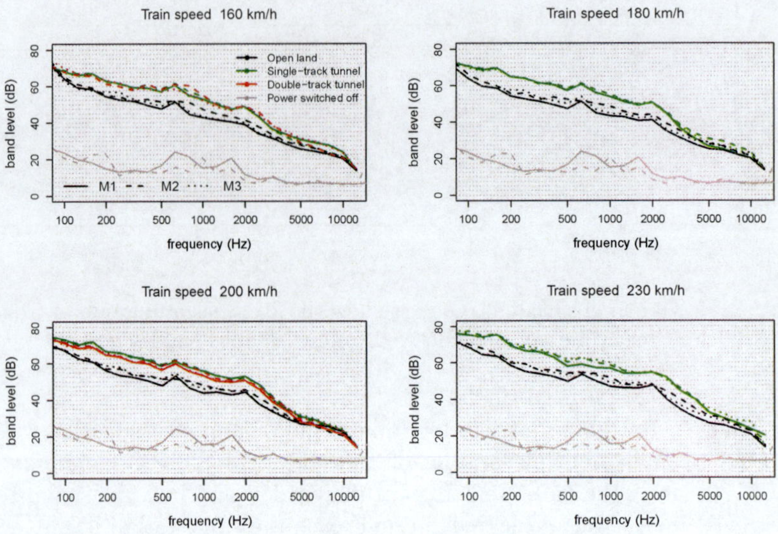

Fig. 3. Third-octave band spectra of interior noise. Shown are the spectra for different infrastructures (line color) and recording positions (line type) determined from the microphone recordings at various speeds. The gray lines show the measurements in quiet in between the STIPA measurements.

5 kHz which is consistent across different train speeds. Concerning intelligibility without any train noise, the STIPA in quiet was higher than 0.8 and often above 0.9 which implies a very good acoustic environment for speech communication. With the train in operation, the increase in the noise level caused a significant decrease in intelligibility depending on the track as well as the speed of the train (Fig. 4). The upper row shows again the data for the conventional microphone measurements. Comparing the opposite vs. the adjacent seating arrangement, it is clear that the STIPA is considerably higher for the adjacent seating. In this setting, for higher speech levels the difference between open track and tunnel becomes more and more negligible as the STIPA approaches high levels. For the HATS the difference between the types of seating is more pronounced. This is most likely due to the further decreased distance between the speaker and the listener by roughly half the head size. When comparing the A-weighted levels of the STIPA signal measured in the train (L_{AFSTIPA}) the difference for the microphone between the different positions is roughly 4 dB whereas for the HATS it is between 6 and 7 dB for the better ear, i.e. the ear with the higher STIPA.

For both measurement modalities and similar to the case of the A-weighted level, the double-track tunnel was slightly better in terms of speech communication. The manual segmentation as well as the superimposed signals yielded comparable results. It is also important to remember that positions M2 and M3 for the HATS were actually not measured but just derived from a spectral modification of the HATS measurement of M1 using the spectral differences between the microphone measurements at different positions.

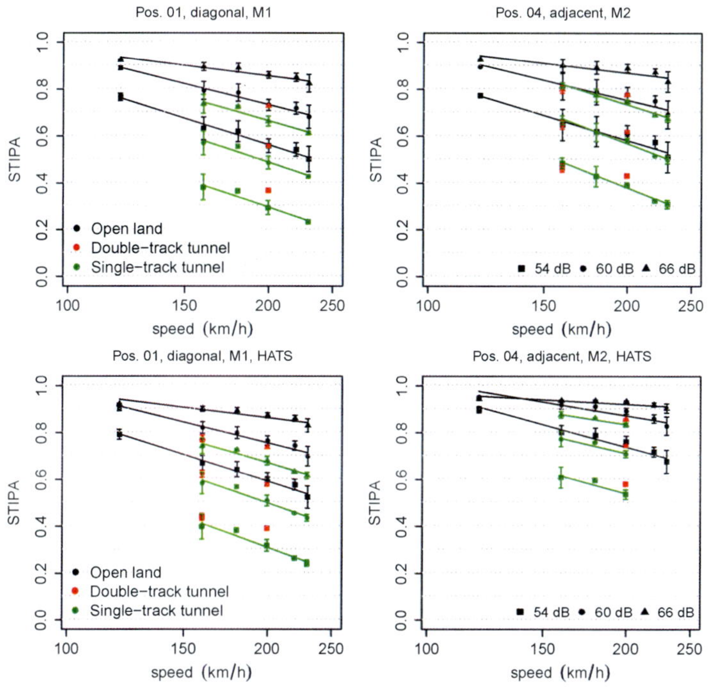

Fig. 4. STIPA and train speed. The upper panels show the STIPA values as a function of train speed and speech level as determined for the microphone in diagonal opposite seating (upper left) and adjacent seating (upper right). The lower panels show the same situations for the HATS measurements. Different symbols denote different free-field A-weighted speech levels, i.e. in 1 m in the soundproof chamber.

Figure 5 illustrates for the microphone data how the STI evolves with the speech level, i.e. the L_{AFSTIPA}. Here, the measured A-weighted level of the STIPA signal is used. For the opposite seating this is about 6 dB higher than in the soundproof chamber in 1 m distance. Comparing the opposite to the adjacent position it is easy to see, that the speech level is considerably higher in the adjacent situation, most likely due to the reduced distance. This in turn leads to a higher SNR and a higher STI as shown before. To reach e.g. an STI of around 0.6 at the receiver, the L_{AFSTIPA} has to be around 2–3 dB higher than the L_{AFeq} (c.f. Figs. 5 and 2). In [6] this value was around 0 dB for and STI of 0.6. However, they used actual speech which in contrast to the STIPA contains pauses and it is unclear whether those were considered. If not, the STI-standard [3] suggests as an estimate a 3 dB correction for speech which would account for the difference.

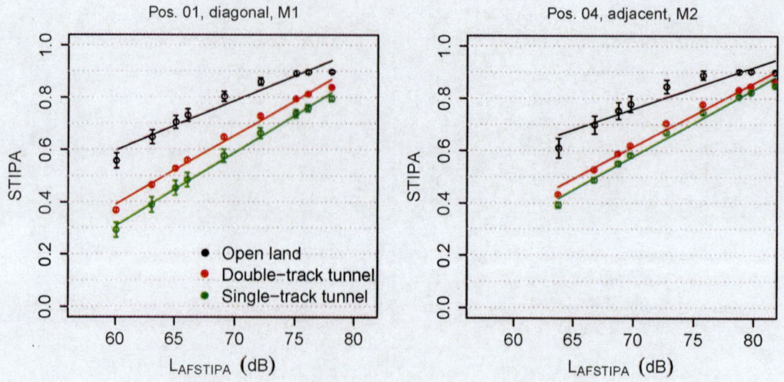

Fig. 5. STIPA and speech level. Shown are the STIPA values at 200 km/h as determined for the microphone in diagonal opposite seating (left) and adjacent seating (right) as a function of measured speech levels.

For the STI in terms of the noise level (Fig. 6) it becomes clear, that the main influence in the different settings is the A-weighted level as the data of different situations essentially seem to lie on the same line. This may be a consequence of the spectra mostly being only shifted upwards in the main frequency region.

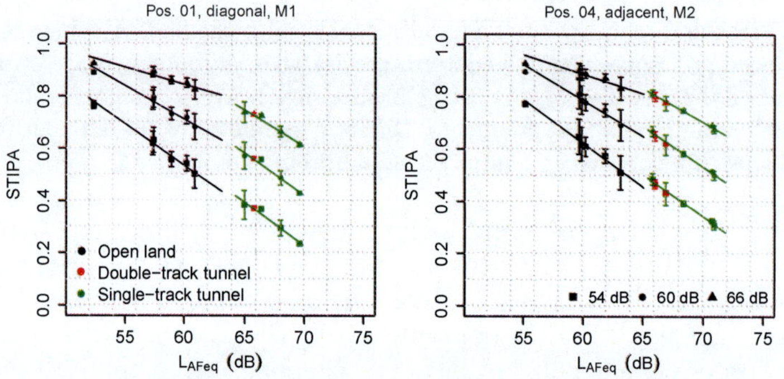

Fig. 6. STIPA and noise level. Shown are the STIPA values as determined for the microphone in diagonal opposite seating (left) and adjacent seating (right) as a function of measured noise levels across all speeds. Different symbols denote different free-field A-weighted speech levels, i.e. in 1 m in the soundproof chamber.

4 Conclusion

As was to be expected, at the covered speed range the speech intelligibility is decreased inside a tunnel which is caused by an increased interior noise level. The decrease is larger than the reported just-noticeable-difference for STI scores of 0.03 [9]. In absolute terms the STIPA still indicates a good intelligibility when passengers are sitting in adjacent seats. The data indicates also that although the procedure for HATS and microphone based STIPA measurements are essentially identical, STIPA values may differ, in this case most likely due to the further reduced distance between sender and receiver for the HATS. Thus, the recording equipment is important to consider when comparing the results to other data. As a consequence of the increased distance for opposite seating STIPA values for normal speaking drop significantly below 0.5 which means, that complex information will become difficult to understand. However, the STIPA clearly improves for higher speech levels. Different speech levels were simulated to accommodate the potential increase of vocal effort due to the increased background noise which is one of the consequences of the Lombard effect (see e.g. [10]). Although the standard [3] proposes an increase of 10 dB it does not provide clear guidelines for the dependence of the background noise which may be an important point for future versions of the standard.

References

1. Park, B., Jeon, J.Y., Choi, S., Park, J.: Short-term noise annoyance assessment in passenger compartments of high-speed trains under sudden variation. Appl. Acoust. **97**, 46–53 (2015)
2. Letowski, R., Scharine, A.: Correlational Analysis of Speech Intelligibility Tests and Metrics for Speech Transmission, Standard (2017)
3. International Electrotechnical Commission, IEC 60268-16:2011: Sound system equipment – Part 16: Objective rating of speech intelligibility by speech transmission index, Standard (2011)
4. Acoustical Society of America, ANSI/ASA S3.5-1997 (R2017): American national standard - Methods for the calculation of the speech intelligibility index, Standard (2017)
5. Zhu, P., Kang, J.: Experimental comparison between STI and SII metrics for the objective rating of speech intelligibility. In: Proceedings ICSV 21, Beijing, pp. CD–Rom (2014)
6. Jeon, J.Y., Hong, J.Y., Jang, H.S., Kim, J.H.: Speech privacy and annoyance considerations in the acoustic environment of passenger cars of high-speed trains. J. Acoust. Soc. Am. **138**(6), 3976–3984 (2015)
7. E. Commission, 1300/2014: Regulation on the technical specifications for interoperability relating to accessibility of the Union's rail system for persons with disabilities and persons with reduced mobility, Official Journal of the European Union L 356/110
8. Oppenheim, A.V., Schafer, R.W.: Discrete-Time Signal Processing, 3rd edn. Prentice Hall Press, Upper Saddle River (2009)
9. Bradley, J.S., Reich, R., Norcross, S.G.: A just noticeable difference in C50 for speech. Appl. Acoust. **58**(2), 99–108 (1999)
10. Zollinger, S., Brumm, H.: The Lombard effect. Curr. Biol. **21**(16), R614–R615 (2011)

A Soundscape Approach to Assess and Predict Passenger Satisfaction

Paul H. de Vos[1](✉), Tjeerd Andringa[2], and Mark van Hagen[3]

[1] Satis, Verlengd Buitenveer 9, 1381 NB Weesp, The Netherlands
info@satis.nl
[2] Sound Appraisal, Travertijnstraat 12, 9743 SZ Groningen, The Netherlands
[3] Nederlandse Spoorwegen, Laan van Puntenburg 100, 3511 ER Utrecht, The Netherlands

Abstract. Interior noise in trains is a significant contributor to passengers' comfort and satisfaction. In the design of new rolling stock, the A-weighted average sound level has generally been applied as the standard indicator. However, surveys show that the passengers' satisfaction varies for different types of rolling stock, even though the sound level is comparable. In order to set consistent design targets, the challenge is to relate emotional and subjective parameters such as comfort and satisfaction to technical and objective parameters such as for example sound level and reverberation time. This is relevant both for prediction and assessment of passenger comfort.

This paper presents an experiment, based on the standardized appraisal of soundscapes. A soundscape can be understood as "the acoustic space a person finds himself in" or "a collection of sounds that forms a particular environment". In the experiment the subjective appraisal of soundscapes inside trains is assessed, with subsequent analysis of the physical parameters involved in the same event, trying to relate these to one another. This approach shows promising results that could be used, after further development, to predict and improve passenger satisfaction. Further research and validation is required, but the experiment presented here may be the first step to a new system of indicators, assessment methods and design recommendations.

Keywords: Acoustic comfort · Soundscape · Interior noise in trains

1 Introduction

The Destinate project [1], a contribution to the European Union's Shift2Rail program [2], was finished in October 2018 after two years of work. In the project three universities (Berlin, Poznan and Manchester), one manufacturer (Stadler Rail, Valencia), two consultants (Müller-BBM and Satis) and two research institutes (Empa Switzerland and NLR Netherlands) worked together. The project focused on the assessment of cost and benefits of rail noise management and control. Benefits were considered to represent not only the reduction of socio-economic cost due to health effects of noise, but also, and more particular, socio-economic revenues thanks to an improved attractiveness of

G. Degrande et al. (Eds.): *Noise and Vibration Mitigation for Rail Transportation Systems*, NNFM 150, pp. 150–157, 2021.
https://doi.org/10.1007/978-3-030-70289-2_14

travelling by rail and an improved reputation of rail infrastructure managers as good neighbors.

2 Passenger Comfort

2.1 The Relevance of Perceived Noise and Acoustic Comfort

Sometimes, the relevance to passengers of acoustic comfort inside the train is questioned. There are however strong indications of its relevance. This was noticed in the early 21st century, for example in France, [3], particularly related to the design of TGV rolling stock.

The relevance was confirmed by passenger survey results (not published, from Netherlands Railways), showing significant differences in the comfort rating between different types of rolling stock in identical services. In other words: travelling is perceived as more pleasant in one rolling stock type than in another. Acoustic comfort appears to be an important factor. This underlines the hypothesis of a hierarchy in traveler experience, where dissatisfiers (such as poor punctuality) are at the base and satisfiers (such as comfort) are at the top of the hierarchy (Fig. 1). This implies, that negative emotions ("my train is delayed") can be overruled or compensated by positive emotions ("it feels good to be on this train").

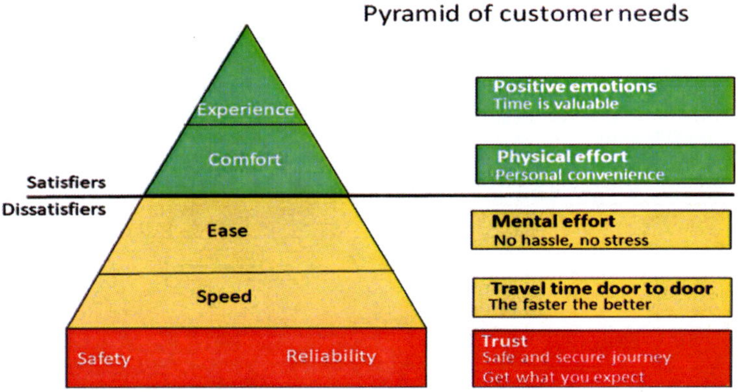

Fig. 1. The "pyramid of customer needs" introduced by Mark van Hagen of Netherlands Railways (NS) [4]

2.2 Interior Acoustics as a Design Parameter

In spite of the high level of standardization in rolling stock design, there are very few standards relating to interior noise in trains. There are no general requirements stating a particular quality, not even minimum requirements. Very few standards for prediction and assessment methods can be found. A single leaflet, published by UIC, is intended to ensure the technical compatibility of high speed trains [5]. It specifies maximum A-weighted sound levels inside the passenger compartments of the train at maximum and average speed. Similar dB(A) requirements are used in current specifications for new vehicles, although it is clear that acoustic comfort requires much more than just a low average sound level. In fact, a low level might even be counterproductive. Therefor there is a need for a more sophisticated and preferably common approach.

3 Soundscape Appraisal

3.1 Soundscapes Inside Trains

The application of soundscapes has developed over the last decades, particularly for the urban environment [6]. According to [7] a soundscape is defined as "an acoustic environment as perceived or experienced and/or understood by a person or people, in context". A soundscape is a concept or experience that can be described in words, in pictures, as a recording, or in other ways. It is always context (i.e. time and location) dependent.

The concept of "pleasantness" was introduced as a quality to rate different soundscapes, for example in [8]. In [9] the relevant sources and their impact on the soundscape were investigated. This approach considers the soundscape as an intermediate step between emotional parameters (such as pleasantness) and physical parameters (such as sound level, reverberation, speech intelligibility, frequency content, etc.). Once the relation between the emotional and physical parameters has been established, there is in principle a way to "engineer" a pleasant soundscape by tuning the physical parameters (Fig. 2).

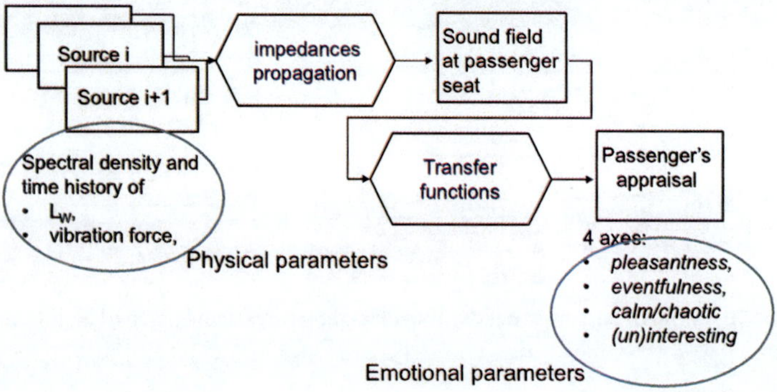

Fig. 2. The soundscape represents the relation between physical and emotional parameters.

A typical interior soundscape in a train is a dynamic mix of many sounds of different origin: rolling noise and traction noise from the train, ventilation noise, exterior noise from the train itself and other sources, warning signals for doors closing, loudspeaker messages, and last but not least the other passengers talking and playing their personal audio systems. It is virtually impossible to appropriately describe this mixture with a single indicator. The soundscape concept is expected to be more suitable.

3.2 Rating of Soundscapes

Roughly based on the Reversal theory of Michael Apter [10], which distinguishes two axles, i.e. the hedonic tone (between pleasant and unpleasant) and the arousal (between high and low arousal), a description of people's rating of soundscapes is presented in the following Fig. 3, which distributes the rating over two main axes: pleasant/unpleasant and eventful/uneventful. The two axes separate four areas: chaotic, lively, boring and calm.

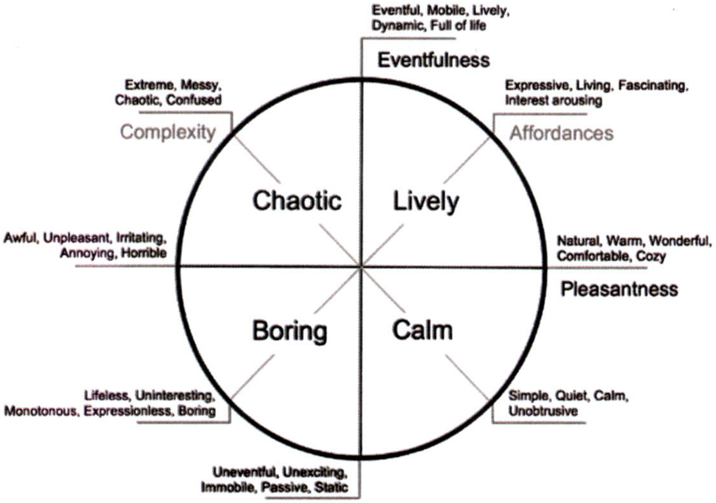

Fig. 3. Characterization of different aspects of a soundscape. Note the four quarters Chaotic, Lively, Boring and Calm. Author T. Andringa 2013

4 The Passengers' Satisfaction Survey

In Destinate, a pilot survey was carried out between 10 and 26 July 2018, in a range of 14 different NS train types. Roughly 75% were interregional trains, 25% intercity trains. In the survey, a mobile phone or tablet app called MoSart, developed by the Dutch company SoundAppraisal, was used to collect personal ratings of comfort and satisfaction on the train. During the automated interview, the app recorded ambient sound. These recordings were sent online to the SoundAppraisal lab for further analysis. In total approximately 180 passengers were interviewed.

Passengers were requested to characterize the sound in the train (as either Chaotic, Pleasant, Lively or Eventful), to classify the perceived sounds, to rate the train (busy, cozy, quiet), to rate their overall feeling (pleasant, bored, relaxed) and to rate the whole journey.

4.1 Results

It was found that rolling stock types differed significantly both in the pleasantness (from −0.16 to + 0.62 out of + or −1.0) and overall sound (from 6.0 to 8.0 out of 0 to 10). The further analysis of the results was concentrated on establishing correlation coefficients, both between physical and emotional parameters and between two different parameters of the same category. If the correlation coefficient between a physical and an emotional parameter was found to be low, it would be impossible to establish the transfer function between the two. If the correlation coefficient between two parameters of the same category would be strong, one of these parameters could be dropped. Strong correlation coefficients were found for example between "chaotic" and "not silent", as well as between "eventful" and "not silent".

There were remarkably high correlation coefficients of more than 0.6 between the sound quality and the overall rating of the train journey. This confirms the assumption that acoustic comfort is a relevant factor for the overall rating of a journey. The sound pressure level as such does hardly correlate (C = 0.2) with the rating of the sound. Neither is there a significant correlation between the position on the axis lively-boring and the overall sound rating. This is fully consistent with the reversal theory, claiming that when people are aroused, the level of stimuli should be reduced rather than the number of stimuli. Figure 4 shows an example of the correlation between the Chaotic and Pleasant axis in the Soundscape Appraisal diagram of Fig. 3.

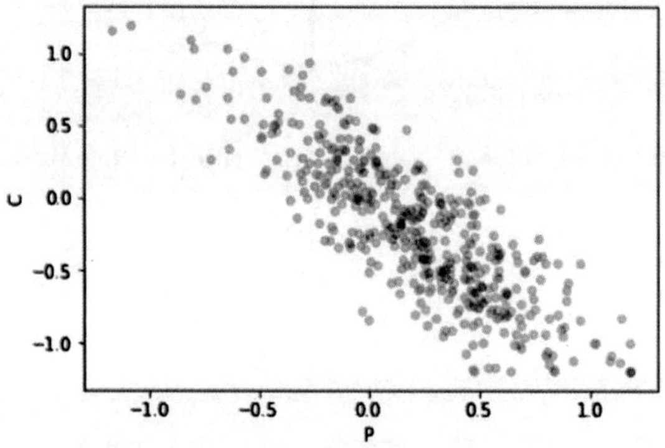

Fig. 4. Observed correlation and a negative slope between the Chaotic (C) and the Pleasant (P) soundscape rating. Lower chaos distinctly means more pleasant.

The analysis produced some interesting and relevant findings, for example.

- The A-weighted sound pressure level could explain the appraisal only very partially. The correlation between the passenger's judgement of the acoustic comfort and the sound pressure level at his seat was poor.
- The emotional indicators, viz. Pleasantness, eventfulness, chaotic vs. Calm (i.e. the horizontal, vertical and top left to downright axes), had a much better correlation to the appraisal than the sound pressure level, whereas boring vs. Lively (the top right to down left axe) did not show such good correlation.

- The qualification "calm" had a high response in general. Passengers judged the interior soundscape as being calm in all train types.

In a further analysis of the data, a distinction was made between sound emanating from the train and sound produced by the other passengers. Some clear findings that easily translate into recommendations for designers are:

- Complex sound is rated negatively whereas simple and easy to recognize sound is comforting and reliable.
- Particularly the alarming sounds (e.g. the door closing warning signal) stand out negatively. Warning sounds are perceived as alarming (as it is the intention) but do not contribute to pleasantness. The recommendation here would be to select sounds that are recognized easily but are experienced as pleasant at the same time.
- Conversation between passengers may become louder when more people are speaking at the same time (the so-called Lombard effect [15]). The strength of this effects is depending on the reverberation radius. There should not be too much reverberation in the train.
- It turns out to be useful to apply the Hi-Fi versus Low-Fi concept [16]. In a Hi-Fi soundscape, the ambient or background sound level is low enough so that distinct sound events can be heard clearly, whereas in a Low-Fi soundscape these sounds are lost. A hypothesis confirmed by the data analysis is that it appears to reassure people to perceive some sound produced by the train; it helps the brain to understand where the person is. The train sound should not affect the foreground to background ratio too much: the soundscape should be Hi-Fi.
- Passengers appreciate a calm atmosphere. In a boring atmosphere it is virtually impossible to conclude about one's "safety". Therefore, some liveliness is appreciated as well.

5 Recommendations

5.1 Recommendations for Further Work

The results of this first pilot allow to formulate some recommendations for the set-up of this type of passenger survey:

- During the survey, a certain time interval should be reserved to observe and perceive the sound in the train before one starts to fill out the questionnaire;
- It would be advantageous if one could measure the "mood" the particular passenger is in, as a consequence of the section of trip he or she has made already, as it is expected that the mood strongly influences the personal perception of the quality of sound.

With these and other recommendations, a second pilot could have a significantly improved set-up and therefore more reliable results.

5.2 Recommendations for Rolling Stock Design

Finally, a first - and very preliminary - list of recommendations for rolling stock design could be derived from this pilot with its limited scope, viz.

- In the design phase, speech and train noise should be distinguished and should be considered both, so special attention has to be paid to good acoustics (Hi-Fi soundscapes), but also to the interaction of train and passenger sounds.
- It should be prevented that people talk louder because they want to be louder than the other people (Lombard effect). This can be achieved particularly by reducing the reverberation radius, i.e. by adding sound absorption in the main frequency range of human speech (500–2000 Hz). Another approach is to shape the interior of the train such that it reminds people of a quiet surrounding, e.g. a library or a forest.
- Preferably some noise produced by the train should be audible as it is a comforting sort of background sound. But it should not be louder than foreground noise. So, a dB(A) based target, for train noise in particular, might be useful after all.
- Once on the train, alarming sounds shall be avoided. This equally applies to alarming sounds like curve squeal, but not to the impact noise in rail joints.

6 Conclusion

Measurement of the soundscape characteristics in terms of physical parameters combined with asking the emotional experience of that same soundscape appears to be a very promising way of discovering the most appropriate soundscape for passengers in a running train. In addition, the transfer functions between emotional and physical parameters could be derived from the data collected in the same experiment. These would be essential to decide on design requirements for new rolling stock. In this way, the soundscape approach presented in this paper is a promising method to predict and assess passenger comfort in a standardised way. The pilot carried out in the Destinate project produced some first interesting findings. These need to be confirmed by further surveys, preferably with a slightly improved method. Knowing the characteristics of all sources of airborne and structure borne noise combined with assessment of the emotional experience appears very promising in engineering the right soundscape for passengers in a running train.

Looking the other way around and starting from an "ideal" soundscape, in due course an "ideal" set of physical parameters could be defined, together with the values these parameters should assume in order to achieve a higher passenger satisfaction, an essential condition for the desired Shift to Rail.

References

1. www.destinate-project.tu-berlin.de
2. shift2rail.org
3. Delepaut, G., Dubois, D., Mzali, M., Guerrand, S.: Dénominations et représentations sémantiques du trajet en train, published in "L'acte de nommer, Une dynamique entre langue et discours," Presses Sorbonne, Sciences du language, Paris (2007)

4. van Hagen, M.: Waiting experience at train stations, thesis of the University of Twente, 1 April 2011. ISBN:98-90-5972-506-5
5. UIC 660, Measures to ensure the technical compatibility of high-speed trains (2002). ISBN:2-7461-0215-3
6. Schafer, R.M.: Tuning of the World, p. 43. Destiny Books (1993)
7. ISO 12913-1: Acoustics -- Soundscape -- Part 1: Definition and conceptual framework (2014)
8. Aumonda, P., Cana, A., De Coensel, B., Botteldooren, D., Ribeiroc, C., Lavandier, C.: Modeling soundscape pleasantness using perceptual assessments and acoustic measurements along paths in urban context, Acta Acustica (2017)
9. Andringa, T.C., Lanser, J.J.: How pleasant sounds promote and annoying sounds impede health: a cognitive approach. Int. J. Environ. Res. Publ. Health **10**(4), 1439–1461 (2013)
10. Apter, M.: Reversal theory: a new approach to motivation, emotion and personality, Anuario de psicología (1989)
11. Dubois, F., Meunier, S., Rabau, G., Poisson, F., Guyader, G.: Détection de sources émergentes au sein d'un bruit large bande: mécanismes perceptifs et applications, 10ème Congrès Français d'Acoustique, Lyon, France, April 2010
12. van den Bosch, K., Andringa, T.: The effect of sound sources on soundscape appraisal. In: 11th International Congress on Noise as a Public Health Problem (ICBEN) 2014, Nara, JAPAN (2014)
13. van den Bosch, K.A.M., Welch, D., Andringa, T.C.: The evolution of soundscape appraisal through enactive cognition. Front. Psychol. (2018)
14. Botteldooren, D., Andringa, T.C., Aspuru, I., Brown, A., Dubois, D., Guastavino, C., et al.: From sonic environment to soundscape. In: Ten Questions on the Soundscapes of the Built Environment, pp. 17–41. CRC Press (2015)
15. Lane, H., Tranel, B.: The Lombard sign and the role of hearing in speech. J. Speech Hear Res. **14**(4), 677–709 (1971)
16. Schafer, M.: Audio Culture: Readings in Modern Music. Continuum International Publishing Group, pp. 29–38 (2004)

Industrial Engineering Framework for Railway Interior Noise Predictions

Torsten Kohrs[1(✉)], Karl-Richard Kirchner[1], Haike Brick[1], Dennis Fast[1], and Ainara Guiral[2]

[1] Bombardier Transportation, Acoustic and Vibration, Am Rathenaupark, Hennigsdorf, Germany
torsten.kohrs@rail.bombardier.com
[2] CAF R&D, Noise and Vibration, J.M. Iturrioz 26, 20200 Beasain, Spain

Abstract. Within the European rail initiative Shift2Rail, funded by the European Union, the FINE1 project aims to improve state-of-the-art noise modelling and control for railway systems. To meet the increasing demands for acoustic comfort inside the rail vehicle while fulfilling more stringent targets, a framework for interior noise predictions is required which goes beyond the need for accuracy, the mere acceptance of calculation results or the intuitive noise control by design. In collaboration with other European railway vehicle manufacturers, a generic framework for interior noise predictions in an industrial context is elaborated to find a common basis for the definition and validation of the inherent acoustic processes. Air-borne, as well as structure-borne topics, are investigated including the relevant source and transmission characterizations, the necessary quantification and noise control activities. This paper presents an overview of the generic framework and includes selected examples for the airborne and structure-borne noise paths. These topics include the pressure field around a carbody of high speed and regional rail vehicle, the transmission loss of sub-assemblies, the interior sound distribution along an aisle and the structure-borne sound transmission and radiation of a roof-mounted HVAC system.

Keywords: Interior noise predictions · Railway vehicle · Measurements

1 Introduction

To meet the increasing demands for acoustic comfort inside the rail vehicle while fulfilling more stringent targets (e.g. mass reduction), a framework is required that allows systematic and comprehensive improvement of interior noise predictions.

In collaboration with other European railway vehicle manufacturers, a generic framework for interior noise predictions in an industrial context is elaborated to find a common basis for the definition and validation of the inherent acoustic processes. The generic framework (Fig. 1) is based on the generic acoustic source-path-receiver model and consists of two branches representing the airborne noise (ABN, direct and transmitted) and structure-borne noise (SBN) contributions to the interior noise. The SBN path follows

G. Degrande et al. (Eds.): *Noise and Vibration Mitigation for Rail Transportation Systems*, NNFM 150, pp. 158–166, 2021.
https://doi.org/10.1007/978-3-030-70289-2_15

the process sketched in [1]. For a review of interior noise prediction methods used in railway application see [2]. This paper presents an overview of the generic framework and includes selected examples for the airborne and structure-borne noise paths. Since the quantification of ABN sources by sound power level (SWL) and directivity is commonly used in the railway industry, framework module A is not treated here explicitly.

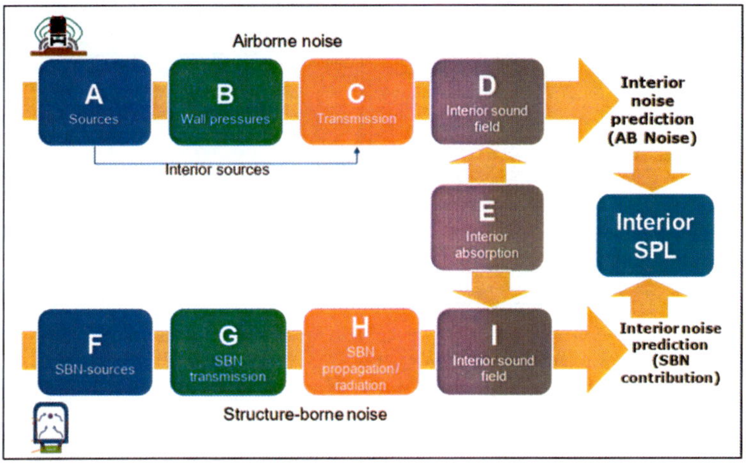

Fig. 1. Generic engineering framework for interior noise predictions

2 Framework Module: Wall Pressures (B)

The investigation of the exterior sound field around the carbody structure of a rail vehicle is selected as an approach for the wall pressure determination (Fig. 1, B). The relevant underframe and roof source positions are investigated. The BEM and the FEM simulations of a full-scale high-speed and regional railway vehicles are performed and validated in [3], using the measurements on the complete railway vehicles.

The sound pressure field around the carbody (Fig. 1, B) is considered essential information for the sound transmission calculation to the interior (Fig. 1, C) and tailored means of noise control. This approach is also validated by full-scale measurements (Fig. 3, left). The sound pressure level (SPL) distribution is computed with a sufficiently detailed model and evaluated as a specific transfer function (TF):

$$TF = SPL_{Spatial\ Distribution} - SWL_{Sound\ Source} \tag{1}$$

Figure 2 displays selected field points at different surfaces, for which frequency-dependent values of TF are demonstrated. If the measured or calculated SWL of the sources of interest are inserted into these TFs, the resulting frequency-dependent spatial SPL distribution can be quantified and visualized immediately and, even more important, can be implemented in overall prediction tools. In addition, simulation and measurement results of the pressure field distribution along selected paths enables the application of further simplifications, modelling and parametrization techniques.

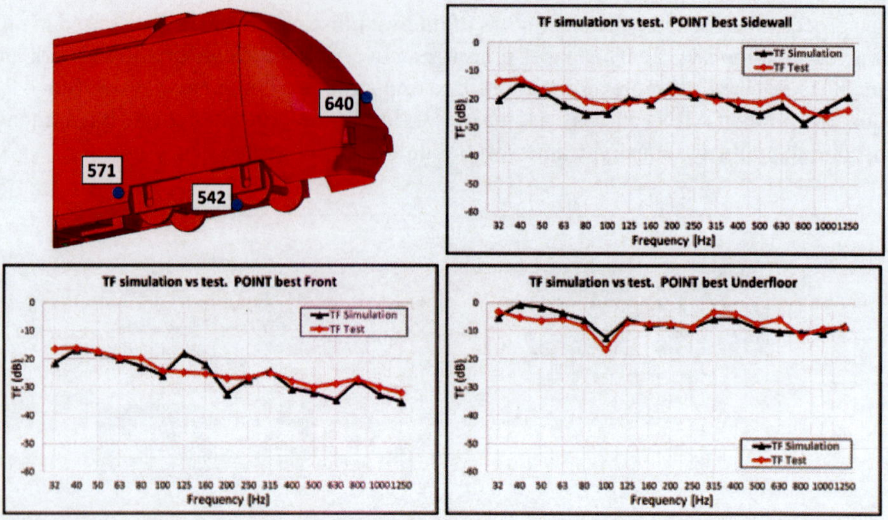

Fig. 2. FEM model of a regional railway vehicle with the receiver positions, measured and calculated TF at the sidewall, front and underfloor

In Fig. 3 (right), the measured (using microphone array) and the computed sound pressure field around the carbody of the regional vehicle at the field points (Fig. 3, left) at 630 Hz for different ballast absorptions are plotted. The measuring points corresponds to the sidewall, underfloor or front, therefore only the curve behavior of the same measuring points should be compared. The curve characteristics reveal the importance of accurate characterisation of the ballast absorption properties.

The absorption properties of the ballast can be measured and provided either as a frequency dependent absorption coefficient or as a specific flow resistivity [4, 5]. The variation of the acoustic performance of track ballast seems enormous in terms of quantified values or frequency dependence. The prediction of the pressure field around

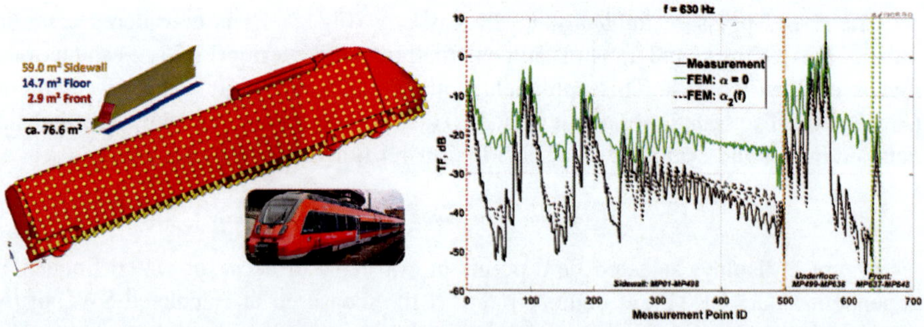

Fig. 3. Left – field points for the sound pressure distribution around carbody. Right – comparison of measured and calculated transfer functions (649 positions) for one-third octave at 630 Hz, where α is the ballast absorption coefficient

the carbody for selected underfloor sources could be done either with the fully reflective ground or with a defined ballast absorption.

3 Framework Module: Transmission (C)

Airborne transfer path from the exterior pressure field around the carbody (Fig. 1, B) to the vehicle interior is defined by the sound transmission loss (TL) of the vehicle subassemblies (Fig. 1, C) in SEA models. TL of subassemblies is usually measured either in a laboratory or directly when mounted on the train (in-situ). However, results differ depending on the test method applied. In the transmission module, the methodology for the use of effective TL is evaluated and improved where necessary.

The sound transmission loss of the subassemblies are measured in laboratory conditions and in-situ tests (Fig. 4, top). Tunnel and free-field interior noise calculations are carried out (Fig. 4, bottom). The ongoing work evaluates the differences in TL results depending on the method applied and evaluates the impact of the measurement uncertainty in the interior noise prediction.

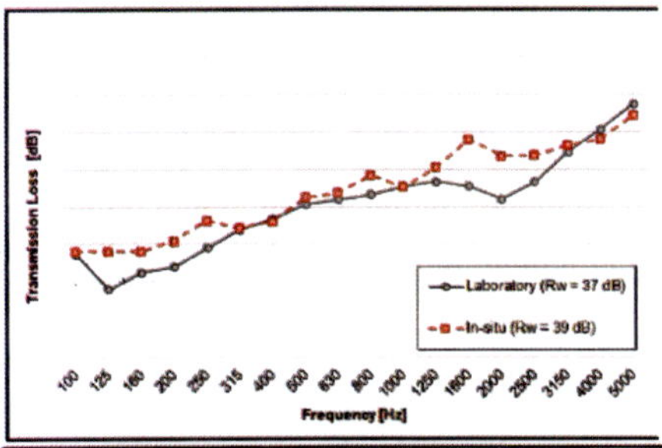

Subassembly	$TL_{in\text{-}situ} - TL_{lab}$
Floor	1 dB
Sidewall	1 dB
Window	2 dB
Door	3 dB
Roof	3 dB
Bellow	-5 dB

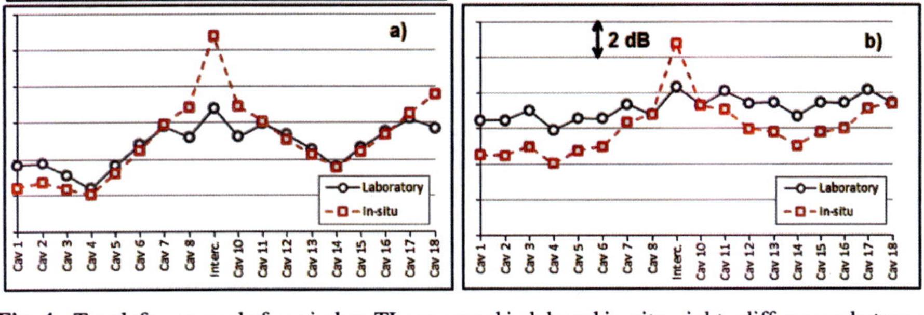

Fig. 4. Top: left – example for window TL measured in lab and in-situ, right - differences between TL results. Bottom - interior noise at 80 km/h running a) in free-field and b) in tunnel

Significant differences are found, especially for the tunnel predictions where all subassemblies are of importance for the sound transmission. Highest deviations occur in the gangway area as expected from the difference in measured TL.

These results raise several technical points of interest: The changing exterior sound field (or operation condition) changes the transmission relevance of selected subassemblies. Both TL measurements have their advantages and disadvantages regarding the sound field, flanking transmission, mounting conditions and dimensions. Interior noise measurement could support the discussion but are not available for this example.

4 Framework Module: Interior Sound Field (D) and Interior Absorption (E)

This section presents few results related to the framework modules 'Interior absorption' and 'Interior sound field'. Existing methodologies for calculation of SPL distribution in train interior were assessed and refined. The interior SPL distribution is modelled with two different methods: Ray tracing (Odeon) and SEA (BRAINS [6]). Sound absorption, the spatial decay of the SPL along the train interior, as well as drop in SPL due to area constrictions, i.e. a gangway opening, are calculated. Validation is carried out with a loudspeaker excitation (pink noise). The agreement between measurements and simulation calculations is good, except in the low-frequency bands (125–250 Hz), due to a low modal density and modal behavior of the sound field. However, at higher frequencies the prediction accuracy is excellent. The global total discrepancy between simulations and calculations is about 1 dB. In Fig. 5 an Odeon model is shown (left), as well as a BRAINS model (right) consisting of coupled sub-systems (cavities). The lower plots show Odeon (left) and BRAINS (right) predicted (o) in comparison with measured (x) A-weighted overall SPL along the train interior as the sound passes a gangway with an opening of 0.65 m (cross-section: h = 2.2 m, w = 2.6 m). The main advantage of the method is that this setup can be easily modified using results of characterization (Fig. 1, A) of a real sound source.

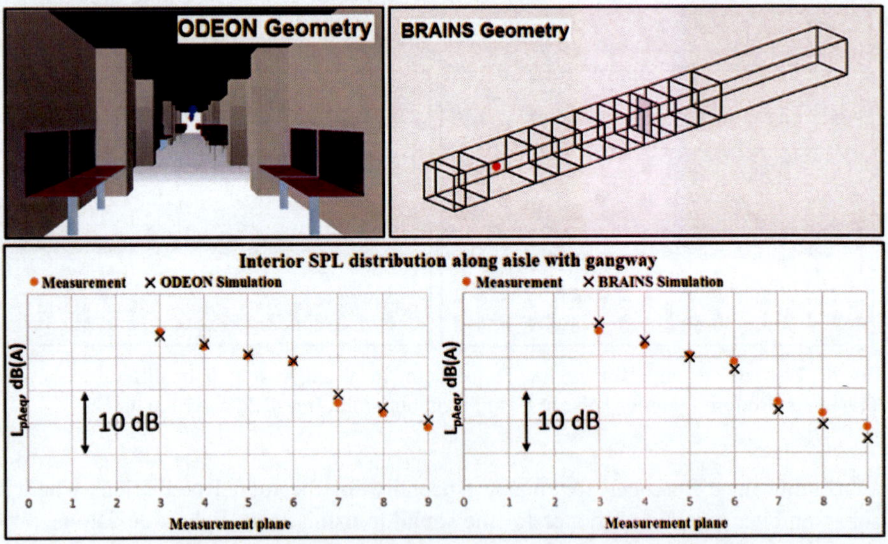

Fig. 5. Interior SPL distribution along aisle with gangway

5 Framework Module: SBN Sources (F)

To demonstrate the second major path of the framework, the SBN-contribution, a roof-mounted HVAC system is selected. In order to predict the SBN contribution to the overall interior SPL, the SB characteristics of main sources have to be quantified. The blocked force (F_{BL}) is considered the most relevant measurable quantity for a comparison of different SBN sources. According to [7], the F_{BL} vector can be obtained from measurements of coupled mobility matrix $Y_{C,ic}$ of the source-receiver system when the source switched off and rigidly installed on a test rig or vehicle and from measurements of the velocity operating response v_i at the coupling points (CPs).

To ensure the quality of measured blocked forces and cross-check the obtained results the on-board validation (OBV) is used according to [8]. The OBV approach requires that the transfer mobility vector $Y_{C,vc}$ between the CPs of source/receiver (c, Acc01-Acc06) and the validation point (v, AccVPZ) has been determined (Fig. 6). This can be practically done during the measurement of the coupled mobility matrix $Y_{C,ic}$. Using this data, the velocity at the validation point can be calculated according to:

$$\{F_{BL}\} = \left[Y_{C,ic}\right]^{-1}\{v_i\} \tag{2}$$

$$\left\{v_{v,calc}\right\} = \left[Y_{C,vc}\right]\{F_{BL}\} \tag{3}$$

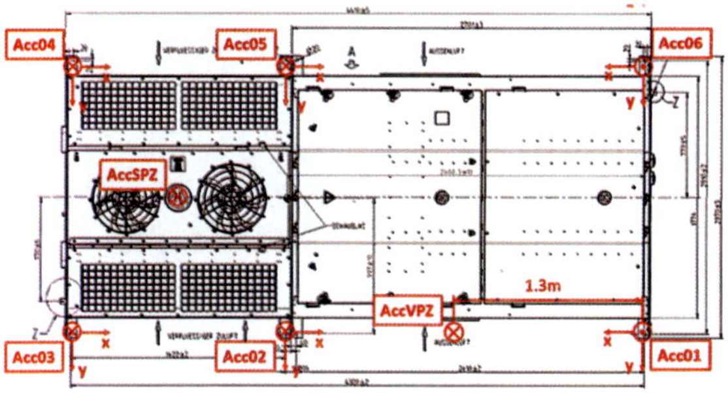

Fig. 6. Drawing of roof mounted HVAC unit incl. positions of 6 triaxial accelerometers (Acc01 – Acc06) and two single axis accelerometers (AccSPZ and AccVPZ)

Figure 7 (left) presents a good agreement of laboratory and vehicle measurement results in all translational directions when using only the point mobilities. Thus, test rig measurement can be used to get F_{BL} during the vehicle design phase. As shown in Fig. 7, (right), the complex operating response leads to greater prediction accuracy.

6 Framework Modules: Transmission (G), Propagation (H) and Interior Sound Field (I)

The SB source framework module (Fig. 1, F) provides F_{BL} of the HVAC system on the test rig as well as from the vehicle measurements. A vehicle measurement includes the

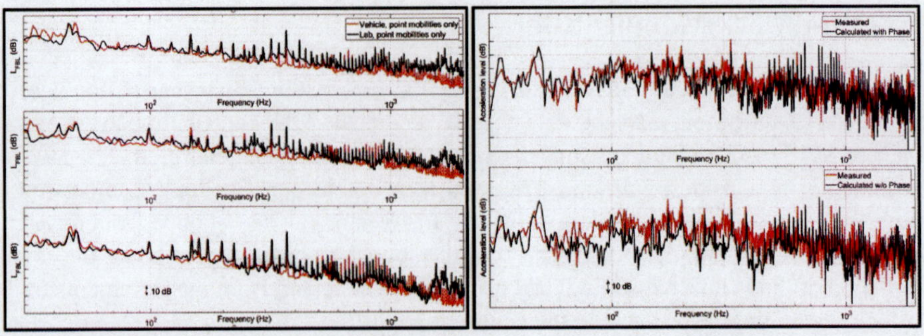

Fig. 7. Left - F_{BL} lab and in-situ measurement, top (x), middle (y) and bottom (z) at Acc02. Right - measured and predicted acceleration level (OBV) in z-direction with and w/o phase

vibroacoustic TF of the considered CPs of HVAC inside the railway vehicle (18 DoFs). The SB propagation and radiation were investigated with a method adopted from the automotive industry, coupling structural vibrations of the Body in White (BiW, Fig. 8) to the interior air volume, which is an acoustic cavity between exterior structure (Fig. 8, left) and interior structure (Fig. 8, right). BiW refers to the joined carbody (e.g. welding, gluing, riveting) before final painting and attachment of sub-assemblies, trim and drive components.

Fig. 8. Left – exterior FEM mesh of BiW with roof mounted HVAC system (magenta), right – interior FEM mesh of passenger compartment from the entrance area along the vehicle

The vibroacoustic response was computed for selected field point positions of the compartment and was compared to the measurement results. In total, 52 frequencies were computed. Figure 9 shows the comparison of the SBN spectra using the direct-coupled vibro-acoustic frequency response (BiW method) with the SBN spectra using the measured coupled vibro-acoustic transfer functions scaled with the blocked forces. In the frequency range where highest excitation from the HVAC vibration sources is expected (e.g. compressor at about 50 Hz), both methods lead to similar results.

This FEM-based simulation and validation scheme provided first good results with the combined amplitude/phase excitation and transforms a structure-borne characterisation based on blocked force into a quantitative estimate of the finally targeted sound pressure spectra inside the vehicle.

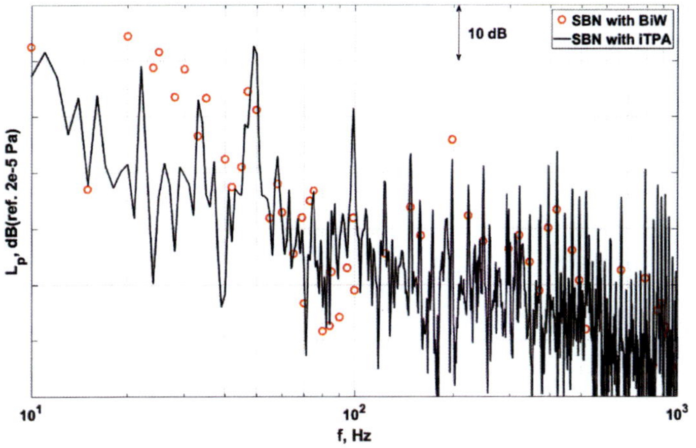

Fig. 9. Comparison of SBN calculation results using BiW method and iTPA method

In contrast to the experimental techniques, this procedure investigates the transmission of SBN exclusively and directly and allows an estimation of the resulting radiated airborne sound. Therefore, quantified SBN characterisation and specification can be validated and adapted. The effect of changing the carbody structure or the source interface points can easily be investigated by simulations and can lead to improved passenger comfort and less costs and weight for secondary passive noise control measures to mitigate SBN-transmission and radiation. The effects of the carbody interior (e.g. insulation and paneling) is the scope of further investigations of the applied BiW method.

7 Summary

The complete process defined in calculation framework is manageable but weaknesses and open points identified in this paper need to be improved in future refinements of interior noise predictions for railway vehicles.

Acknowledgement. The authors express their acknowledgement to the Shift2Rail Joint Undertaking for financing the projects FINE1 (GA 730818) related to the work presented here. This contribution reflects the views of the authors.

References

1. Petersson, B.A.T., Gibbs, B.M.: Towards a structure-borne sound source characterization. Appl. Acoust. **61**, 325–343 (2000)
2. Shift2Rail – FINE1 – Deliverable 7.1: Review of state of the art for industrial interior noise predictions, Grant Agreement Number: 730818, July 2017
3. Kohrs, T., Kirchner, K.R., Fast, D., Vallespín, A., Sapena, J., Guiral, A., Martner, O.: Sound propagation and distribution around typical train carbody structures, EURONOISE (2018)

4. Broadbent, R.A., Thompson, D.J., Jones, C.J.C.: The acoustic properties of railway ballast, EURONOISE (2009)
5. Betgen, B., Orrenius, U., Brunstrom, S., Vincent, N.: D 3.7 Model for geometrical integration transfer functions to be integrated in the global simulation model, Run2Rail (2014)
6. Frid, A., Orrenius, U., Kohrs, T., Leth, S.: BRAINS the concepts behind a quick and efficient tool for prediction of exterior and interior railway vehicle noise. In: Acoustics (2012)
7. Fehse, K.-R., Kohrs, T.: Der Einsatz von (in-)direkten Messverfahren und Berechnungen zur Charakterisierung und Quantifizierung von Köperschallquellen im Design-Prozeß von Schienenfahrzeugen, DAGA (2005)
8. Alber, T., Sturm, M., Moorhouse, A.: Independent characterization of structure-borne sound sources using the in-situ blocked force method, INTERNOISE (2016)

Structure-Borne Noise Contributions on Interior Noise in Terms of Car Design and Vehicle Type

Alex Sievi[1(✉)], Haike Brick[2(✉)], and Philipp Rüst[3(✉)]

[1] Müller-BBM, 82152 Planegg, Germany
Alex.Sievi@mbbm.com
[2] Bombardier Transportation, 16761 Hennigsdorf, Germany
Haike.Brick@rail.bombardier.com
[3] Bombardier Transportation, 8050 Zurich, Switzerland
Philipp.Ruest@rail.bombardier.com

Abstract. The aim of the investigation at hand was to elaborate a systemized view on the importance of structure-borne bogie noise on the interior noise by a comparative analysis of six different trains of different vehicle classes, construction methods and bogie designs. In doing so, it is intended to deepen the understanding of noise transmission into the passenger compartments of railway vehicles. For that purpose, the data sets of the different vehicles were analyzed similarly with the OTPA technique. This comparative analysis at typical train speeds clearly shows that there are agreements within the vehicle class and differences between the vehicle classes. The higher the maximum vehicle speed, the higher is the importance of structure-borne noise. This is caused firstly by the different bogie designs and second by the fact that the airborne transmission loss is improved for trains with a higher maximum speed, but the structure-borne transmission loss not to the same extent.

Keywords: Structure-borne noise · Bogie · Interior noise · Light rail vehicle · Electrical multiple unit · High speed train · OTPA

1 Introduction

The noise of the trailer and motor bogies determines the interior noise levels in passenger vehicles in the entire range of vehicle speed except at standstill. Beside the noise of auxiliary equipment and aero-dynamic noise, they are the most important sound sources concerning interior noise. During the design phase, limiting its impact on the interior noise is the main task of the acoustic engineering team. Therefore, a lot of effort is spent on designing quiet bogies and drives and on designing cars with a high sound transmission loss in order to meet the customers' requirements on the interior sound pressure.

The resulting interior sound pressure is always a combination of airborne noise contributions resulting from outside sound pressure excitations and the structure-borne noise contributions, which are generated by the forces acting on and in the bogie equipment

G. Degrande et al. (Eds.): *Noise and Vibration Mitigation for Rail Transportation Systems*, NNFM 150, pp. 167–175, 2021.
https://doi.org/10.1007/978-3-030-70289-2_16

and which are transmitted via the mechanical structure of the vehicle into the interior of the car. Individual investigations on different trains show that structure-borne noise normally dominates the interior noise in the lower frequency range [2]. The frequency limits depend on the investigated bogie and train type.

The work presented here gives a comparative view on the importance of the impact of structure-borne noise on the interior noise using the measurement data of six different vehicle types and using the same analysis method, the operational transfer path analysis.

2 Vehicle Classes from an Acoustic Point of View

From a constructional point of view, railway vehicles for passenger transport differ very widely. This is true not only for the car body, but also for the nature and type of interior fittings and the construction of drives and bogies, respectively.

With regard to the construction of the carbody a basic distinction is made between differential and integral construction. Differential construction consists of a framework which is filled with metal sheets. It is also referred to as steel-frame construction. The integral design is based on the use of rigid aluminium extrusion with hollow chambers that are as long as a carriage and welded together longitudinally to form a vehicle tube. From an acoustic point of view the carbody design is decisive both for the airborne sound insulation of the carriage and for its sensitivity to structure-borne sound excitation. Due to the clear homogeneity of the integral design, the attenuation properties with regard to structure-borne sound are very poor in longitudinal direction. Furthermore, the high bending stiffness of the extrusions results in even worse airborne sound insulation properties compared to components with the same mass, but designed with the differential construction method [3]. In addition, with the differential construction method usually a higher gap is achievable between the shell and the floor slab or interior fittings, respectively, leading to better airborne sound insulation. Looking only at the above-mentioned qualitative evaluations, without really having compared two equivalent vehicles with different car body structures, it is clear that the integral construction method is more crucial than the differential construction method, considering both airborne sound and structure-borne sound.

Due to the lower acoustic requirements and the lower speeds relevant in tram construction, it is quite common to have a single-shell vehicle floor only, while in the main-line sector and in high-speed trains double-shell floor superstructures are used without exception.

From an acoustic point of view bogies of tram vehicles and underground trains differ significantly from those of main-line vehicles or high-speed vehicles. The bogies for LRV (Light Rail Vehicles) generally have fewer dampers or coupling points between the car body and the bogie. Because of the lower maximum damping forces, LRV bogies can also be equipped with softer rubber mounts for connecting the dampers. The usually softer axle guidance also helps. On the other hand, the lighter and structurally less stiff bogie frames offer less impedance.

The aim of the investigation at hand is to present and systemize the impacts of different types of construction on the structure-borne noise contributed by the bogie on the basis of available measurement data for the different vehicle classes. In doing so, it is intended to deepen the understanding of noise transmission into the passenger compartments of railway vehicles. Data sets of the following vehicles were used for this purpose (Table 1):

Table 1. Vehicles under investigation, DCM: differential construction method, ICM: integral construction method, MB: motor bogie, TB: trailer bogie, MFR: manufacturer.

No	Vehicle type	v_{max}	Car body type	Bogie type	Floor construction	MFR
1	LRV/Tram	80 kph	DCM	MB	Single wall	1
2	LRV/Metro	100 kph	DCM	MB	Double wall	4
3	EMU/Inter-city	200 kph	DCM	TB	Double wall	2
4	EMU/Regional	160 kph	IDM	MB	Double wall	1
5	HST/Inter-city	300 kph	IDM	TB	Double wall	3
6	HST/Inter-city	250 kph	DCM	TB	Double wall	1

3 Analysis Approach

3.1 Operational Transfer Path Analysis

For the performed investigation the method of operational transfer path analysis (OTPA) was used in order to separate the sound contributions of airborne (abn) and structure-borne noise (sbn) to the sound in the passenger areas. Based on the operational data (time row data) of acceleration (a_i) and sound pressure (p_j) near the sound source or on certain points along the sound transfer paths (references) and the sound pressure inside (p) the passenger compartment (response), the OTPA technique is capable to determine transmissibilities $H_{p/aj}$ and $H_{p/pi}$, which are a kind of transfer function of interior sound pressure p related to the references a_j and p_i. The contribution of each reference to the response is determined by filtering the reference signals according to the corresponding transmissibility (Fig. 1).

By adding up the contributions of the acceleration reference signals and the sound pressure reference signals, the overall contributions of structure-borne noise and airborne noise can be determined. A network for this transfer path synthesis is displayed in Fig. 2. This analysis was performed with the measurement and analysis software package PAK®.

Two relations were determined for all data sets. The ratio p_{snb}/p_{bogie} was calculated to identify the proportion of structure-borne noise from the bogie to the overall contribution of the bogie. Since the interior noise may also be determined by other sources, the structure-borne noise contributions were additionally related to the actual measured interior noise: $p_{sbn}/p_{measured}$. With the second ratio it is possible to evaluate the significance of structure-borne noise in relation to the overall noise. The ratios were

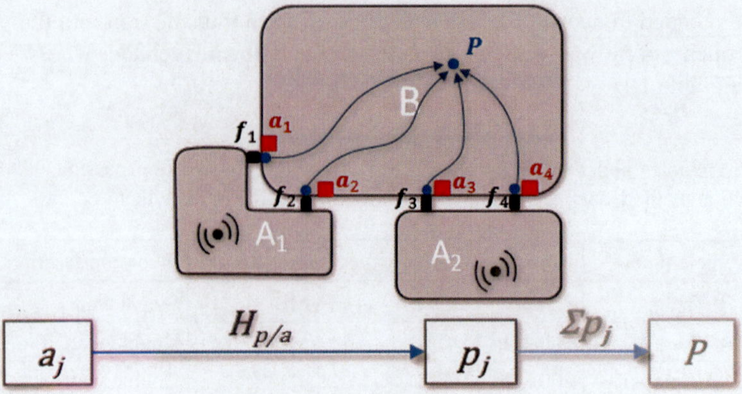

Fig. 1. Operational transfer path analysis [4].

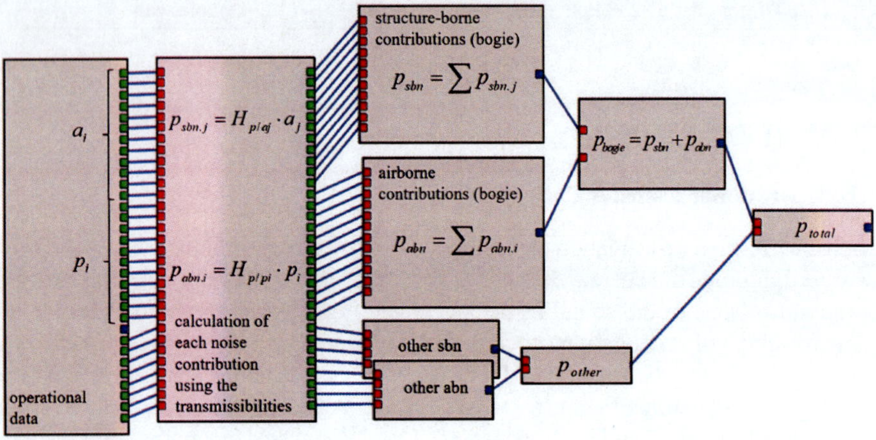

Fig. 2. Transfer path network for calculating the contributions of sbn and abn.

converted into dB-values with $20 \cdot \log_{10}\left(p_{sbn}/p_{bogie}\right)$ and $20 \cdot \log_{10}(p_{sbn}/p_{measured})$ and displayed as spectrograms over driving speed. Each ratio was formed by averaging five measuring steps. By selecting the scaling between 0 dB and -12 dB the relevance of the structure-borne sound proportions can be derived approximately as follows, depending on the colour (Table 2):

Table 2. Colour-coding, approximate dB-value range and meaning, sbn: structure-borne noise.

Colour	dB-values	Significance
Magenta	0 dB	sbn completely determines the internal sound level
Red-orange	-3 dB < X < 0 dB	sbn determines the level
Yellow	-3 dB	Half of the interior level is due to sbn
Green	-3 dB < X < -7.5 dB	sbn increases the level by 0.9 dB to 3 dB
Turquoise blue	-7.5 dB < X < -9 dB	sbn increases the level by 0.6 dB to 0.9 dB

4 Results and Discussion

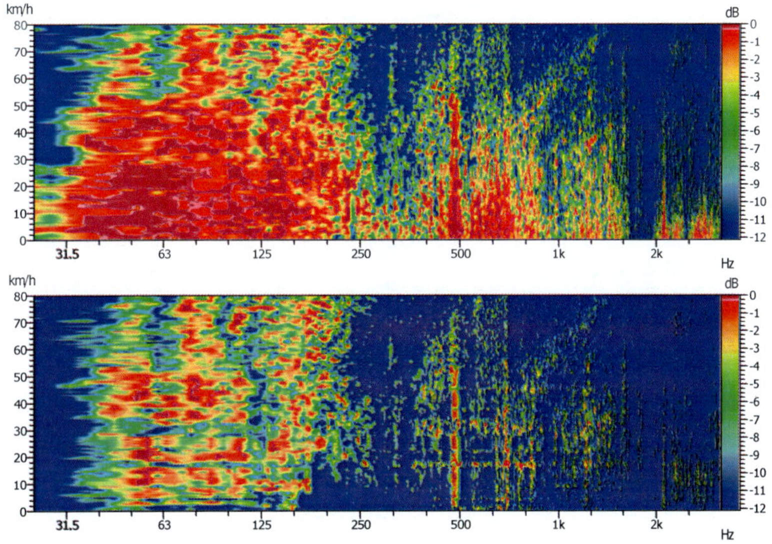

Fig. 3. Vehicle No. 1: LRV, top: $20 \cdot \log_{10}\left(p_{sbn}/p_{bogie}\right)$, bottom: $20 \cdot \log_{10}\left(p_{sbn}/p_{measured}\right)$.

The colored spectrograms of the other vehicles are displayed in [1] which also provides a detailed analysis for each train investigated. For presenting a more general and comprehensive analysis, the ratios were displayed comparatively and only for typical speeds for 1/12 octave bands.

4.1 Comparison of the Vehicle Classes at Certain Speeds

At 80 km/h for both investigated LRVs (see Fig. 3 and Fig. 5) structure-borne sound below 250 Hz is of great importance. This applies both to the total contribution of the bogies and to the total noise (measured in real terms) in the passenger areas. On the other hand, for both EMUs the structure-borne noise is relevant up to 2.5 kHz in terms

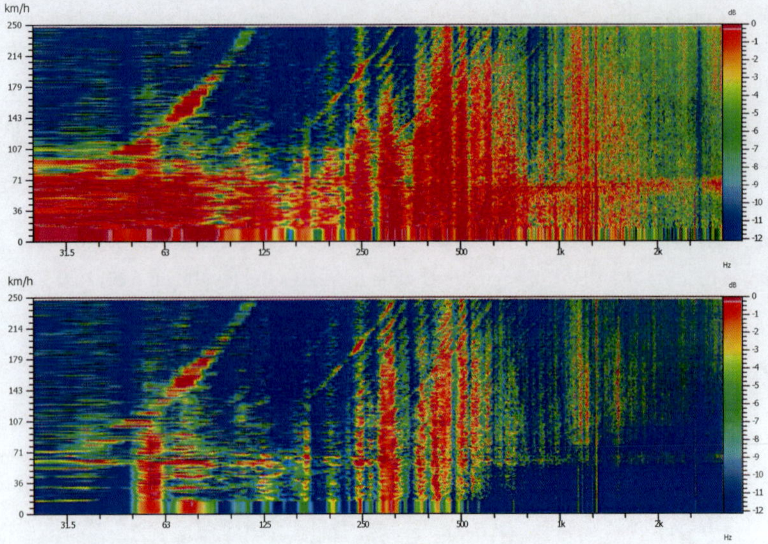

Fig. 4. Vehicle No. 6: HST, top: $20 \cdot \log_{10}\left(p_{sbn}/p_{bogie}\right)$, bottom: $20 \cdot \log_{10}\left(p_{sbn}/p_{measured}\right)$.

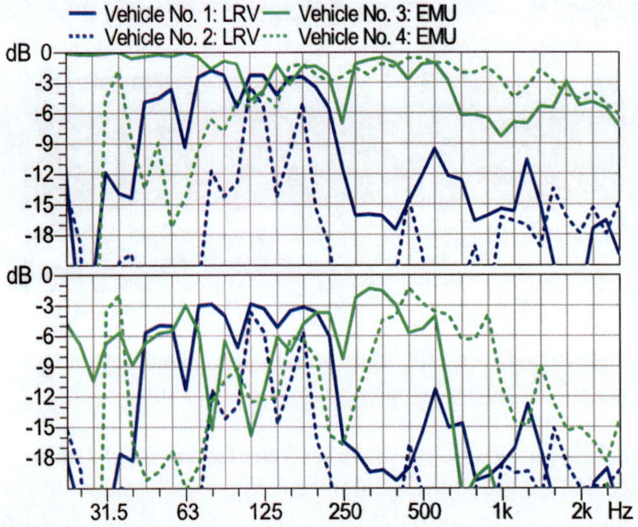

Fig. 5. LRVs (blue curves) and EMUs (green curves) for 80 kph, above: $20 \cdot \log_{10}\left(p_{sbn}/p_{bogie}\right)$, below: $20 \cdot \log_{10}\left(p_{sbn}/p_{measured}\right)$.

of total bogie noise and approximately up to 600 Hz and 800 Hz in terms of total noise at 80 km/h.

At 160 kph (see Fig. 4 and Fig. 6) the structure-borne noise of the bogie is relevant for the total bogie noise up to 600 Hz to 1000 Hz, depending on the type of the investigated

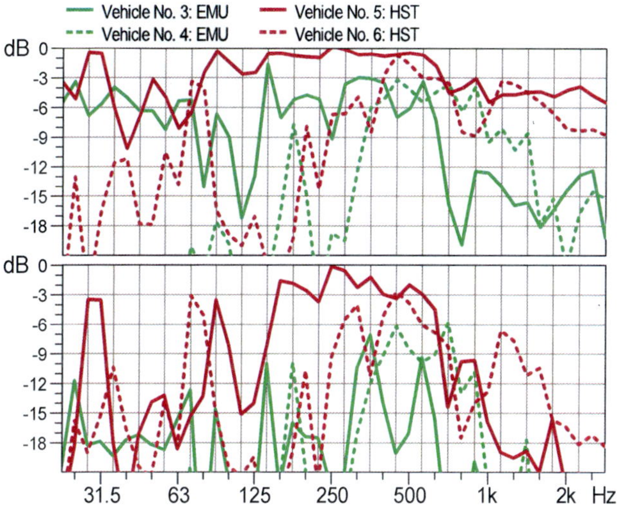

Fig. 6. EMUs (green curves) and HSTs (magenta curves) for 160 kph, above: $20 \cdot \log_{10}\left(p_{sbn}/p_{bogie}\right)$, below: $20 \cdot \log_{10}(p_{sbn}/p_{measured})$.

EMU. In terms of total noise, the structure-borne noise is only of marginal importance in the frequency range of 250 Hz to 1000 Hz for the EMU. In case of the HSTs, the structure-borne noise is much more important. At 160 kph the structure-borne noise is of high importance for the total bogie noise in the whole frequency range. For the total noise in the vehicle, the structure-borne noise of the bogie is relevant up to the frequency of 600 Hz. It is also interesting that the structure-borne noise at the sleeper pass-by frequency between 63 Hz and 125 Hz is pronounced for both HSTs.

4.2 Comparison of the Vehicle Classes Based on the Transmissibilities

Structure-borne proportions within the individual vehicle classes match better than those of different classes, although the car body construction methods differ as well as the floor structures in some cases (LRV). The type of car body construction method (differential or integral construction method) thus seems to be of secondary importance for the proportion of structure-borne noise in the total contribution of the bogie. The train type itself implies a certain concept for the transmission loss, because with increasing train speed the sound power levels of the exterior sound sources like aerodynamic and rolling noise sources increase and the vehicle's floor and sidewalls need a higher transmission loss for a sufficient isolation of the interior. That this is the case can roughly be seen in Fig. 7, which displays the average airborne transmissibilities $H_{p/pj}$ of the investigated trains for the floor.

The construction of the bogie also seems to have a greater impact on the structure-borne noise contribution. The SBN proportion increases over the categories LRV, Electrical Multiple Unit and High-Speed Trains. The higher structure-borne noise proportions are certainly caused by a larger number of coupling points between bogie and car body and stiffer rubber connections as well as increasing damping forces.

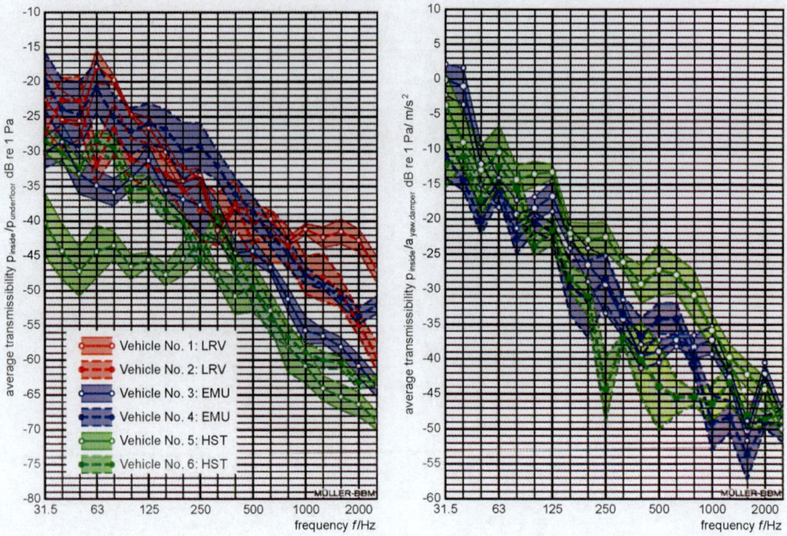

Fig. 7. Average airborne $H_{p/pi}$ (left) and structure-borne $H_{p/aj}$ (right) transmissibilities.

The higher contribution of structure-borne sound to the overall noise in high-speed trains is also caused by a high shielding of the exterior airborne sound by carbody and interior design. This hypothesis is assisted by the displayed average structure-borne transmissibilities of the jaw damper of the EMUs and HSTs in Fig. 7, where no systematic trend can be seen for the vehicle classes.

5 Summary

The aim of the investigation at hand was to present and systemize the impacts of different types of construction on the structure-borne noise contributed by the bogie on the basis of available measurement data for the different vehicle classes. In doing so, it is intended to deepen the understanding of noise transmission into the passenger compartments of railway vehicles. For this purpose, the OTPA method was applied.

With the aid of two transfer functions, the proportion of structure-borne bogie noise to the total noise of the bogie and the proportion of structure-borne bogie noise in the actually measured total noise in the vehicle interior were evaluated.

This comparative analysis at typical train speeds clearly shows that there are clear agreements within the vehicle class and clear differences between the vehicle classes. The higher the maximum vehicle speed, the higher is the importance of structure-borne noise. This is caused firstly by the different bogie designs and second by the fact that the airborne transmission loss is improved for trains with a higher maximum speed, but the structure-borne transmission loss not to the same extent.

References

1. Sievi, A.: Bedeutung des Körperschalls im Schienenfahrzeugbau, Bahnakustik Workshop – Infrastruktur, Fahrzeuge, Betrieb, Planegg (2018)
2. Sievi, A., Tao, G., Li, J.: The application of OTPA as a multi-dimensional tool for gaining a deep understanding of the sound transmission of railway-bound vehicles, Internoise Hongkong (2017)
3. Kohrs, T., Peterson, B.: Transmission loss of extruded aluminium panels with orthotropic cores. In: Proceedings of Forum Acusticum, 3rd European Congress on Acoustics, Sevilla (2002)
4. Lohrmann, M.: Einführung in die TPA-Methoden, TPA-Workshop am 22./23.3., Müller-BBM, Planegg (2017)

Shift2Rail Research Project DESTINATE Interior Railway Noise Prediction Based on OTPA

Nathan Isert[1](✉) and Otto Martner[2]

[1] Müller-BBM Rail Technologies GmbH, Robert-Koch Str. 11, 82152 Planegg, Germany
Nathan.Isert@mbbm.com
[2] Müller-BBM GmbH, Robert-Koch Str. 11, 82152 Planegg, Germany

Abstract. Noise abatement generally comes at a cost while offering the benefit of an acoustically preferable solution. In the concept or (re-)design phase for a train, both costs and benefits must be accessible to take informed decisions. Various prediction methods are currently used in the rail industry, that all come with their own advantages and limitations. The Shift2Rail research project DESTINATE aims to support this by developing and advancing acoustic prediction methods. Within this project, an OTPA model was developed to predict interior noise in a light rail vehicle. Various options to utilize the OTPA model in predicting design changes were explored, the most promising one being the integration of FIR filters in the sound synthesis. OTPA results produce rankings for the main noise contributors as well as audible simulations of interior sound signals.

Keywords: Interior noise prediction · Transfer path analysis · Railway noise

1 Introduction

With the acceptance that noise has become a major issue at many places in our modern society, raises the question on which steps we take to mitigate or prevent noise and who holds responsible. The operator of a passenger train could be held responsible for the acoustic comfort of his passengers, only that he himself has limited options to implement noise abatement measures. The interior noise in a traveling train is impacted by various factors, such as the vehicles acoustic isolation from exterior noise sources, the track condition but also the passenger's activities. Acoustic comfort may additionally relate to more than just actual noise levels.

A noise abatement process requires a holistic evaluation of the situation, considering all target groups, determining the main noise contributors and assessing cost and benefits for the available mitigation options, both with the immediate effect as well as their long-term impact in mind. The Shift2Rail research and innovation project decision supporting tools for implementation of cost-efficient railway noise abatement measures (DESTINATE) aims at contributing to this by developing and advancing acoustic prediction methods and facilitating informed decisions on railway noise mitigation, options and acoustic comfort. Deliverables will be available through [7].

G. Degrande et al. (Eds.): *Noise and Vibration Mitigation for Rail Transportation Systems*, NNFM 150, pp. 176–185, 2021.
https://doi.org/10.1007/978-3-030-70289-2_17

The presented paper focuses on interior noise prediction methods to support identifying dominant noise contributors and for making predictions. Within DESTINATE an operational transfer path analysis (OTPA) is assessed as a method to build a model upon for interior noise prediction. To develop the model, measurement tests were performed on a light rail tram vehicle.

2 Noise Prediction Methods

How sound generated at a rail-bound source finds its way to a passenger seated inside the train is linked to the transmission paths the sound takes travelling in the air, through walls and in the train body. The situation is depicted on the left hand side in Fig. 1, where two independent sources contribute to the interior sound field. Noise prediction often utilizes one of the following methods:

- Finite element method (FEM)/boundary element method (BEM) divides a large structure into small elements, assigns physical properties, formulates restraints and solves equations of motion for each such element to observe displacements in the global structure. This method is particularly efficient at lower frequencies, as higher frequencies require smaller elements and hence more computational effort [2].
- Statistical energy analysis (SEA) looks at the average transfer of energy from one part of the system into another. It divides a large structure into subsystems (cavities) which can all contain vibrational modes (energy). Coupling (transfer) functions allow for energy transfer from one subsystem to another. Feeding energy into one subsystem leads to a relaxation of the entire system to a minimum energy state. SEA models are quick to build but the results are averages only with high uncertainties at lower frequencies [3].
- Ray tracing models compose of 3-dimensional structures in a space with sources at distinct locations emitting sound. Rays sent out from the sources in random directions are reflected off boundaries and absorbed in the transmitting media or at the boundary interfaces. The sound pressure at any given point is determined from all passing rays [5].
- Transfer path analysis (TPA)/operational transfer path analysis (OTPA) require an actual test object. No 3-dimensional model is used; instead, the exchange of vibrational energy is determined from transfer functions. The transfer characteristics follow from excitation at all source positions and recording the response at receiver position. TPA typically uses impact excitation, while OTPA deducts transfer characteristics from operational excitations. Classical TPA requires the dismounting of sources to determine the transfer functions and an independent characterization of the sources airborne and structure-bore emissions. In an OTPA model, acceleration and sound pressure at the source-structure interface represent the sources structure-borne and airborne excitations. All source representations are measured in parallel and transfer characteristics are obtained from decomposing the signals for their individual contributions [1, 6].
- Additionally analytical solutions can be implemented or multibody oscillator models applied (i.e. TWINS/RIM) [8, 9].

The center part of Fig. 1 [3] depicts frequency limitations and complexity of some of these the models. Hybrid models combine several methods to overcome limitations [10]. The right hand side of Fig. 1 shows the driver's cab in a FE-SEA model [2] to cover a larger frequency range. Coupling functions required in SEA can be obtained from TPA measurements. A prediction model for rail noise that utilizes most of the described methods is BRAINS (Bombardier RAIlway Noise Software) [4].

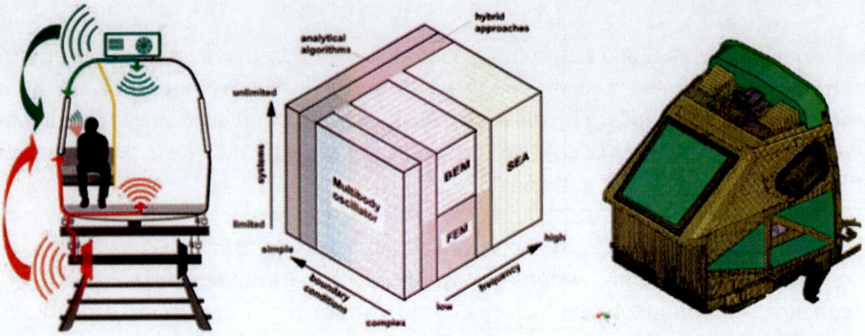

Fig. 1. Left: Schematic drawing of two independent sound sources (HVAC and rolling wheel) contributing to the interior noise inside a train. Center: Frequency limitations and complexity of commonly used methods for interior noise predictions taken from [3]. Right: Hybrid structure of a driver's cab in a FE-SEA model [2]

3 Operational Transfer Path Analysis (OTPA)

The OTPA approach considers paths instead of a geometrical model. It does so by assuming a fixed (linear) relation between a source and a receiver (i.e. passenger in Fig. 1 on the left hand side). Sources are represented by the responses at the source-structure interface; structure-borne noise and airborne noise are treated separately. **Fehler! Verweisquelle konnte nicht gefunden werden.** Schematically captures by displaying microphone (yellow circles) and accelerometer (red/green squares) placements in transmission paths on one of the test vehicles. Mathematically the source-receiver relation is [1].

$$\mathbf{V}_R(f) = \mathbf{T}_{RI}(f) \cdot \mathbf{V}_I(f) \tag{1}$$

Where $\mathbf{V}_I$ stands for the input signal, $\mathbf{V}_R$ for the signal at the receiver position and matrix $\mathbf{T}_{RI}$ for the transmissibility function connecting these two. The equation holds true for input and receiver signals in the frequency domain. In the time-domain, the simple pointwise multiplication turns into a convolution.

Equation (**Fehler! Verweisquelle konnte nicht gefunden werden.**) suggest that in order to predict the interior noise signal $\mathbf{V}_R$ at a receiver position we need input signals $\mathbf{V}_I$ from all contributing sources (at all interfaces) and the transfer matrix (transmissibility functions) containing the information of how vibrational energy induced at the interfaces is transmitted to the receiver position. We assume the transfer characteristics are fixed, such that determining them once for a given structure allows predicting the signal at the receiver position from any set of input signals. Input signals change dependent on operational conditions (i.e. speed, HVAC operation), boundary conditions (i.e. rail roughness) or the surroundings (i.e. tunnel); likewise the predicted signal at the receiver condition changes (Fig. 2).

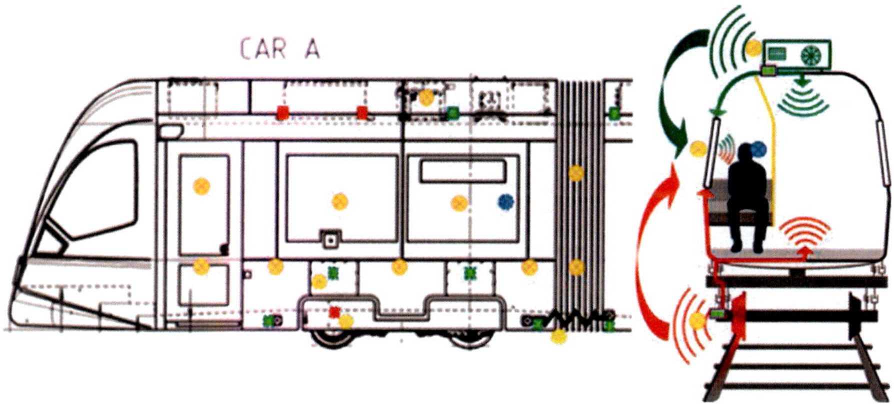

Fig. 2. Schematic drawing of the test vehicle (TramLink Prototype by Stadler Valencia). Microphones and acceleration sensors placed at source-structure interfaces in structure-born and airborne noise transmission paths.

The fundamental task of an OTPA is determining the transfer matrix (transmissibility functions). This is done by inversion of $\mathbf{V}_I$ in Equation **(Fehler! Verweisquelle konnte nicht gefunden werden.)**. However, since the receiver signal $\mathbf{V}_R$ has a relation with all sources that must hold true at various excitations, the variables in Equation **(Fehler! Verweisquelle konnte nicht gefunden werden.)** have matrix form. Inversion of $\mathbf{V}_I$ requires a mathematical procedure referred to as single value decomposition (SVD). This procedure identifies the strongest vibration mode in the receiver signal, assigns it to a source/path, and then proceeds finding the second strongest from the residue and so forth. The process of assigning modes is called principal component analysis (PCA).

PCA is useful for the later analysis of measurement results, as it allows separating the receiver signal into contributions from all sources and thus makes a ranking of the dominant sources and most contributing paths possible. It also provides an evaluation of airborne and structure-borne noise contributions. Finally, the procedure cancels crosstalk contributions from one source at the interface representation position of another. It is vital to have representations for all relevant sources.

For processing time-domain input (source) signals into time-domain output (receiver) signals, corresponding impulse response functions (IRFs) are generated for all transmissibility functions by fitting them with FIR filters and applied to the input signals by convolution [1, 11]. This is generally referred to as Transfer Path Synthesis (TPS) [1] or time-domain Independent Component Analysis (time-domain ICA) [11]. Alternatively, the measured time data of all input and output channels is transformed and divided into frequency blocks first and then the resulting responses of the output channels that fit the

OTPA model may be computed by multiplication of the input channel frequency blocks and OTPA Transfer function matrix [12].

4 Predicting Design Changes Within an OTPA Model

Often new trains are derivatives of prior series, with redesigned or optimized parts and features. The necessity for a redesign can stem i.e. from customer requirements, the need for modernization, improved operations or from cost saving. A redesign can take various forms, ranging from cosmetic changes, to the exchange of auxiliaries, and major redesigning of parts of the structure. Knowing the acoustic effect of a redesign is important even if the target has nothing to do with noise abatement. Interior noise prediction models help to estimate the acoustic effect of design changes. In conjunction with cost estimates, this supports informed decision processes.

Which noise prediction method is best used depends on a number of factors, such as resources like budget, time or the availability of test measurements. OTPA models require actual measurement recordings on a comparable structure. OTPA mainly qualifies if the redesign process leaves large parts of the train body structure unchanged. An OTPA model is relatively fast to build and further offers the benefit of producing audible outputs. Auralizations help evaluating noise perception. We should note that using any noise prediction method without proficient knowledge of the effect an exchange or redesign has locally raises the uncertainty of prediction results. FINE1, the complementary research project to DESTINATE in Shift2Rail, attempts to standardize characterization procedures for railway sources and subassemblies.

To build the OTPA model, a train was equipped with accelerometers and microphones for taking measurements under various operational conditions. For the measurement recordings, the crosstalk cancelling and the subsequent sound synthesis, we used the software PAK. MATLAB was used to alter time signals prior to the OTPA synthesis. Individual contributions from input signals belonging to interfaces of the same source or path were aggregated.

A number of options exist that allow simulating design changes in an OTPA based model. At first, the effect a redesign or change in excitation has on the source-structure interfaces and on the associated transfer path has to be evaluated. Estimates can be found from measurement on a mockup, from simulations, calculation or from experience. For each source, a source-receiver relationship as in equation (**Fehler! Verweisquelle konnte nicht gefunden werden.**) can be formulated and then modified in response to a design change. Mathematically, this equals a pre- or post-multiplication of equation (**Fehler! Verweisquelle konnte nicht gefunden werden.**) with appropriate matrices containing the scaling values for altering the source-receiver relations.

$$L \cdot \mathbf{V}_R \cdot R = L \cdot \mathbf{T}_{RI} \cdot \mathbf{V}_I \cdot R \tag{2}$$

Within the TPS, modifications were introduced from altering signals and/or transfer functions individually. Modifications were generally designed in the frequency-domain and subsequently transformed within the software PAK into corresponding impulse responses to be convoluted with the time-domain signals. Mainly the following modification approaches were explored [7]:

- Weighting factors: all instances of a signal are multiplied with a scalar factor. By doing so one can easily assess what the result would be like if individual input signals were enhanced or toned down. This approach alters only the amplitude of the signals.
- FIR filter method: frequency response functions containing the frequency information, i.e. of adding a damping for a certain frequency band, are applied to input signals. These are transformed into corresponding impulse response filters (IRFs) in the synthesis of time signals. This was the method applied to incorporate most of the effects of design changes within this research project. In example, the exchange from a single layer bellow to double layer system was derived from insertion loss measurement in acceptance tests for which the difference was formulated into a FIR filter function. A corresponding impulse response would then alter the airborne sound contributions from the bellow measurement positions at the receiver point and hence its time signal.
- Response modification analysis allows modifying the receiver signal as desired and therefrom calculates a set of input signals satisfying the source-receiver relation. This is useful to identify the main contributors to the altered signal and approximate targets for noise abatement.

5 Measurements and Results

Two independent measurement campaigns collected input data to build the OTPA model and to validate it. The first campaign used the TramLink prototype manufactured by Stadler Valencia to raise all necessary input data for building an OTPA model. The second measurement campaign took a derivative vehicle, the Gmunden TramLink, to test and validate the applicability of the OTPA model for predicting interior noise in a redesigned environment. The Gmunden TramLink had undergone some significant design changes, such that it was difficult to account for changes specifically. Only the effects of structural stiffening at joints and the transition from a single layer bellow to a double layer were addressed, as those were considered to have the largest impact on interior noise going by the results of the initial OTPA.

OTPA based interior noise prediction was shown to work well on the vehicles it was established for. A separation and ranking of the dominant noise sources and paths showed the structure-borne vibrations from engine and rolling contact excitations to be determining most of the interior noise levels on both vehicles. For confidentiality reasons, all results publically available were offset by an arbitrary dB value.

Structure-borne rolling noise was dominant in the interior noise on the prototype. Adopting the model to account for design changes therefore focused mainly on the structure and the lower tram sections. No information was attainable for the exchanged HVAC unit. The information available but for schematic drawings was that many structural joints on the Gmunden TramLink had been stiffened in comparison to the Prototype TramLink, the bellow was changed from single to double layer and absorption material was removed from the underfloor. For the presented results in Fig. 3 the following modelling assumptions were made:

- Stiffening benefits the structure-borne noise damping at joints and the added structural damping have a larger impact on higher frequencies. An approximating FIR filter was

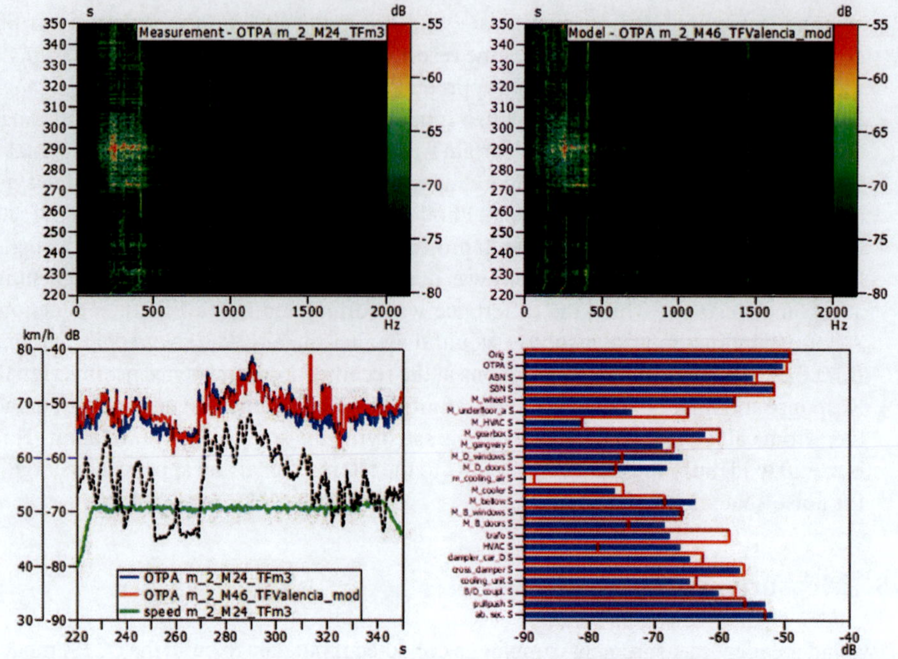

Fig. 3. Frequency-time plots (upper graphs), sound level (bottom left) and source contributions (bottom right) at an interior receiver position for an OTPA predictions using transfer characteristics from the measurement vehicle (upper left and blue) and for an OTPA model of a different vehicle where design changes are modelled from FIR filters approximating the impact of the change (upper right and red). The sound levels of both predictions match well. The frequency-time plots differ slightly and some of the contribution calculations do not match well but the dominant contributors yield similar impacts. The lower left curve also displays a roughness indication curve (dotted black) as extracted from the indirect roughness measurements (see Fig. 4).

designed for the increased dynamic stiffness from a logarithmically growing loss term and a damping of about 10 dB at 1 kHz.

- From acceptance tests for bellows the insertion losses were used to calculate a FIR filter for the exchange of single layer for a double-layer bellow.
- As the absorption characteristics of the material removed in the underfloor was unknown, its was assumed to approximately cancel out to the fact that the reference measurement was run on a ballasted track with higher absorptions than the modelling measurement on embedded tracks.

Corresponding FIR filters were designed and incorporated into the synthesis which then would transform them into IRFs and apply them to the time-domain input signals. The already complex validation procedure was further complicated by the differences in boundary conditions (i.e. embedded track with grooved rails in urban surrounding versus ballasted track, in a rural area with a partially unwelded rail) which further increases the uncertainties in the prediction results.

Overall, the interior sound prediction, using the extended OTPA model with FIR filter representations for design changes, matches the expected sound levels sufficiently well. This is showcased from the noise level plots on the bottom left of **Fehler! Verweisquelle konnte nicht gefunden werden**. The source and path ranking suggests that the most important design changes were captured in our approach (bottom right in **Fehler! Verweisquelle konnte nicht gefunden werden.**). The audible interior sound signals (visualized through frequency-time plots in the upper part of **Fehler! Verweisquelle konnte nicht gefunden werden.**), display overall similarities but also some distinct differences (mainly modes at certain frequencies). They were obtained through applying the TPS approach outlined in Sect. 3 and 4. Their differences already suggest that the listening experience of the two is different, which could easily be verified playing it to some test persons. No full listening experiments were conducted, as the sound signals were too distinct and due to the auralization from the modified setup sounding more artificial.

The frequency-time plots (APS) in Fig. 3 both are obtained from TPS models in conjunction with input signals recorded on the Gmunden TramLink. The one referred to as Measurement–OTPA corresponds to the blue noise levels and was obtained using transfer functions from the Gmunden TramLink. The other referred to as Model–OTPA corresponds to the red noise levels and results from using the transfer functions from the TramLink prototype (Valencia) OTPA in conjunction with FIR filters to account for the differences in the structure and operation of the two TramLink vehicles. To further improve the prediction and reduce the uncertainty, more detailed information would have to be available for the changes, i.e. from separate tests or simulations.

High interior contribution from rolling noise can be the cause of either a bad track or a low transmission loss. To separate and assess the impact of infrastructure on the OTPA result, additional information is required. Therein the rail surface condition has often the largest impact. In **Fehler! Verweisquelle konnte nicht gefunden werden.** The high interior noise at around 290 s into the measurement are the result of a severely damaged rail running surface. This is displayed as roughness indication (dotted black curve) in the bottom left with an arbitrary scaling such that it may be plotted alongside the sound levels and tram speed. Clearly the sound levels are strongly correlated to rail roughness. Rail roughness was assessed from applying a track monitoring. The results of the track monitoring take the form displayed in Fig. 4 with green indicating a low rail roughness whereas red stands for a high rail roughness or damaged rail surface.

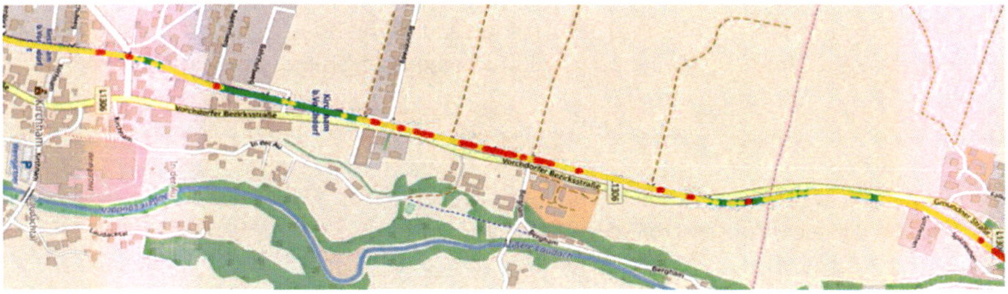

Fig. 4. Rail roughness information gained from applying the concept of track monitoring. Green indicates a low rail roughness whereas red stands for a severely damaged track.

6 Conclusion

OTPA is a useful tool for interior noise predictions. An OTPA model can be extended to account for design changes within some limits. In general, OTPA allows for fast identification and ranking of the main contributors to the interior sound. Frequency and time selective filtering can be applied. Estimated design changes are easily incorporated using weighting factors or FIR filter functions, whereas a response modification analysis offers support on selecting the most promising contributors for noise abatement. Results are audible and thus usable to assess perception in addition to mere sound levels. Finding appropriate filter functions to describe design changes can be challenging. This limits the approach to situations where the system can be assumed to be linear in its responses and the FIR filter fitting of the TPS is appropriate. For fast noise level predictions and an approximate impact analysis it may be helpful none the less.

Acknowledgements. The DESTINATE project has received funding from the Shift2Rail Joint Undertaking under the European Union's Horizon 2020 research and innovation program under grant agreement No 730829.

References

1. Van der Seijs, M., de Klerk, D., Rixen, D.: General framework for transfer path analysis: history, theory and classification of techniques. MSSP, Issue **68–69**, 217–244 (2016)
2. Sapena, J., Tabbal, A., Jové, J., Guerville, F.: Interior noise prediction in high-speed rolling stock driver's cab: focus on structure-borne paths (mechanical and aero sources). In: Maeda, T., et al. (eds.) Noise and Vibration Mitigation for Rail Transportation Systems. Notes on Numerical Fluid Mechanics and Multidisciplinary Design, vol. 118, pp. 445–452. Springer, Tokyo (2012)
3. Fischer, M.: Statistical energy analysis. In: Seminar Notes from Joint Advanced Student School (JASS), St. Petersburg (2006)
4. Frid, A., Orrenius, U., Kohrs, T., Leth, S., BRAINS - the concepts behind a quick and efficient tool for prediction of exterior and interior railway vehicle sound. In: Proceedings of the Acoustics, Nantes, France (2012)
5. Gabet, P., Drobecq, V., Noé, N., Jean, P.: Prediction of acoustic comfort of a trainset using ray-tracing. In: Proceedings of SIA (2009)
6. Sievi, A., Martner, O., Lutzenberger, S.: Noise reduction of trains using the operational transfer path analysis. In: Maeda, T., et al. (eds.) Noise and Vibration Mitigation for Rail Transportation Systems. Notes on Numerical Fluid Mechanics and Multidisciplinary Design, vol. 118, pp. 453–461. Springer, Tokyo (2012)
7. DESTINATE. https://projects.shift2rail.org/s2r_ipcc_n.aspx?p=DESTINATE, Accessed 11 June 2019
8. Thompson, D.: Railway Sound and Vibration - Mechanisms, Modelling and Means of Control. Elsevier Science, Amsterdam (2009)
9. Diehl, R., Hoelzl, G.: Prediction of wheel/rail sound and vibration - validation of RIM. EuroSound98, Munich (1998)
10. Sapena, J., Blanchet, D.: Interior Sound Structure Borne Path Prediction in a High Speed Train using "FE/SEA" Hybrid Modelling Methodologies. England, NOVEM, Oxford (2009)

11. Junji, Y., Hikaru, I.: Method for the separation of vehicle interior noise contributions using only response signals. J. Syst. Des. Dyn. **7**(4), 416–427 (2013). https://doi.org/10.1299/jsdd.7.416
12. de Klerk, D., Ossipov, A.: Operational transfer path analysis: theory, guidelines and tire noise application. MSSP Issue **24**, 1950–1962 (2010)

Virtual Test Method of Structure-Borne Sound for a Metro Bogie

Gang Xie[1](✉), Martin Rissmann[2], Pascal Bouvet[2], Xiaowan Liu[3], David J. Thompson[3], Luis Baeza[3], Juan Moreno[4], Julián Martín Jarillo[4], Francisco D. Denia[5], Juan Giner-Navarro[5], and Ines Lopez Arteaga[6]

[1] CDH-AG, Competence Center Vibration and Acoustics, 70565 Stuttgart, Germany
gang.xie@cdh-ag.com
[2] Vibratec, Chemin du Petit Bois BP 36, 69131 Ecully Cedex, France
[3] Institute of Sound and Vibration Research, University of Southampton, Southampton SO17 1BJ, United Kingdom
[4] Servicio de Ingeniería de Material Móvil, Área de Ingeniería, Metro de Madrid, Madrid, Spain
[5] Centro de Investigación en Ingeniería Mecánica (CIIM), Universitat Politècnica de València, València, Spain
[6] Department of Aeronautical and Vehicle Engineering, KTH Royal Institute of Technology, 100 44 Stockholm, Sweden

Abstract. This paper presents a virtual test method for structure-borne noise generated from railway running gear. This method combines a number of existing tools to form a system approach. The wheelset and bogie frame are modelled using FEM software Nastran to include details of their construction. The primary springs are simplified to standard CBUSH elements in Nastran with point and transfer stiffness modelled by frequency-dependent complex stiffness, which are tuned against measurements. The wheel-rail contact forces due to roughness excitation are obtained by the wheel-rail interaction tool TWINS. The vibration of the full running gear is simulated in Nastran by applying the wheel-rail contact forces. The forces transmitted to the vehicle body through traction bars and dampers are calculated for predicting structure-borne noise.

Keywords: Bogie · Structure-borne sound

1 Introduction

The concern of environment, climate change and increased living standards have driven road, railway and air industries to provide innovative technologies to develop better and more sustainable transport systems. The need to design quieter and more comfortable trains has never been higher than in the present time.

The most important source of noise from the rail transport system is rolling noise caused by wheel and rail vibrations induced at the wheel/rail contact [1]. This is particularly the case for low-speed metros and light rail systems. The mechanism of rolling

G. Degrande et al. (Eds.): *Noise and Vibration Mitigation for Rail Transportation Systems*, NNFM 150, pp. 186–193, 2021.
https://doi.org/10.1007/978-3-030-70289-2_18

noise has been well understood through many years of research and a validated modelling tool, TWINS was also developed by Thompson et al. [2] to predict wheel/rail contact forces and radiated noise of wheel and rail. Rolling noise is transmitted to the interior of vehicles through not only airborne paths, but also structure-borne paths, which include the whole running gear system. Airborne transmission tends to be dominant at higher frequencies while structure-borne transmission at lower frequencies. The difference between the mechanisms of airborne and structure-borne transmission also leads to different modelling approaches [1]. This paper focuses on the structure-borne noise. To control the structure-borne noise and vibration inside the vehicle, the whole running gear system including the bogie, suspensions and wheelsets must also be considered. However, a whole-system approach intended for use in design has not been reported before.

In the automotive industry, the full vehicle structure is digitized and its operational conditions are simulated by computers to predict interior acoustic responses and to approve vehicle designs. The modelling approach is mainly based on approaches such as the finite element method (FEM). As many similarities for the noise and vibration problems are shared between railway and automobile industries, this type of virtual test process used in the automobile industry can also be applied to rail vehicles for structure-borne noise. The key to build a good virtual test process is to have a validated engineering method and efficient computational techniques. This paper presents a comprehensive virtual test method for the structure-borne noise inside railway vehicles.

2 Methodology

The structure-borne path starts at the wheel/rail contact and propagates through the suspensions and bogie frame into the car body. The proposed method combines FEM for the running gear and the established TWINS method for the wheel/track interaction, which together lead to a whole system approach. A trailer bogie of a metro train is used as an example for the proposed method. Structural components of the running gear, including the bogie frame, traction bars, dampers, wheelsets and axle boxes are modelled using FEM. Special attention must be given to modelling suspension elements as they often show frequency-dependent characteristics. The rubber bushings of the traction bar and dampers are modelled with complex stiffnesses, which have been obtained from laboratory measurements. The primary springs have been modelled as a series of lumped masses and springs. The stiffness of the springs also takes account of the internal mass effect of the rubber elements and has been tuned with the measured dynamic stiffness. The wheel/rail contact forces can be calculated using TWINS with measured wheel and rail roughness. For the full bogie system, each wheel is excited by the wheel/rail forces and frequency response analyses are then carried out in FE software. Dynamic forces transmitted to the train body through the connection points obtained from frequency response analyses can then be used for transfer path analysis (TPA). The structure-borne noise can then be predicted by using these forces together with noise transfer functions from forces at these positions to the interior sound pressure. These could be obtained either through simulations or measurements. For frequencies up to 1 kHz, the number of acoustic modes is very high due to the size of the interior cavity of the railway vehicle. It is therefore very rare that the full vehicle body is modelled using FEM. In the present project measured transfer functions have been used. The model has been validated against field-test data obtained from an experimental campaign.

3 Structure-Borne Noise Transmission Model

3.1 FE Model

In the present study, a trailer bogie of a Metro de Madrid vehicle is modelled using the FE package Nastran. The full FE model includes the bogie frame, a front wheelset, a rear wheelset, axle boxes, primary suspension springs, lateral dampers and traction bars, as shown in Fig. 1. The wheelsets, bogie frame and axle boxes are modelled using solid elements. The structural part of dampers and traction bars could also be modelled using solid elements or appropriate one-dimensional beam elements; the latter approach is used here. They are connected by bushing elements to the bogie. The primary suspension spring is modelled using spring elements with frequency-dependent complex stiffness without FE representation of its structure. The whole FE model consists of over 1.36 million nodes.

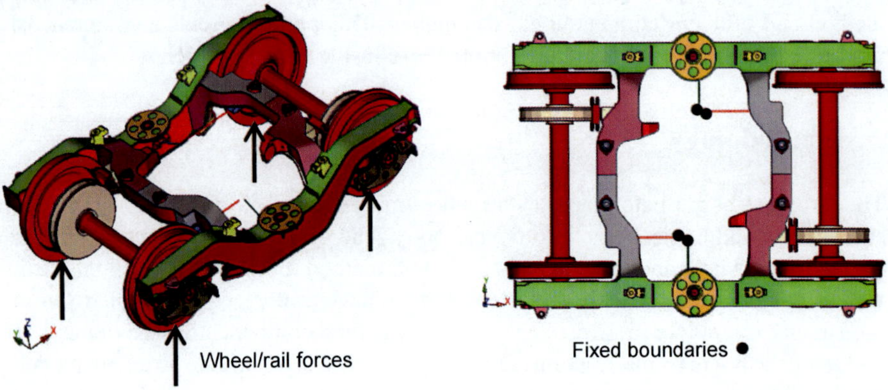

Fig. 1. The FE model for the bogie with dampers and traction bars modelled by 1D bar elements.

3.2 Frequency-Dependent Stiffness

The primary suspension spring is built with four layers of metal and rubber, as shown in Fig. 2. The dynamic model is composed of a series of masses and springs [3]. The mass-spring model was extended to include wave motion of rubber elements. As result, each spring has frequency-dependent point and transfer stiffness and this model gave the best correlation with dynamic tests carried out in laboratory. To simplify the global FE model, the series of masses and springs was reduced to a single frequency-dependent spring for each direction, connected between wheelset and bogie frame. The stiffness matrix is expressed as

$$K = \begin{bmatrix} k_{\mathrm{po\_b}} & -k_{\mathrm{tr}} \\ -k_{\mathrm{tr}} & k_{\mathrm{po\_w}} \end{bmatrix} \tag{1}$$

Note that the point stiffness $k_{\mathrm{po\_b}}$ on the bogie side is different from the wheelset side $k_{\mathrm{po\_w}}$ due to the asymmetric build of primary spring whereas the transfer stiffnesses k_{tr} are equal by reciprocity. To represent this stiffness matrix in Nastran, CBUSH element was used [4]. CBUSH element is a general spring and damper element with the capability of including frequency dependent stiffness and damping. A conventional CBUSH element, connecting bogie and wheelset, and two grounded springs on both bogie and wheelset sides have to be combined. These three springs together give the stiffness matrix as

$$K = \begin{bmatrix} k_{\mathrm{tr}} & -k_{\mathrm{tr}} \\ -k_{\mathrm{tr}} & k_{\mathrm{tr}} \end{bmatrix} + \begin{bmatrix} k_{\mathrm{po\_b}} - k_{\mathrm{tr}} & 0 \\ 0 & 0 \end{bmatrix} + \begin{bmatrix} 0 & 0 \\ 0 & k_{\mathrm{po\_w}} - k_{\mathrm{tr}} \end{bmatrix} \tag{2}$$

where the first part is a normal CBUSH and other two are grounded springs.

For the traction bar and lateral damper, the measured dynamic stiffnesses of the bushing are applied to local coordinates which are defined to be the same as that used in the measurements.

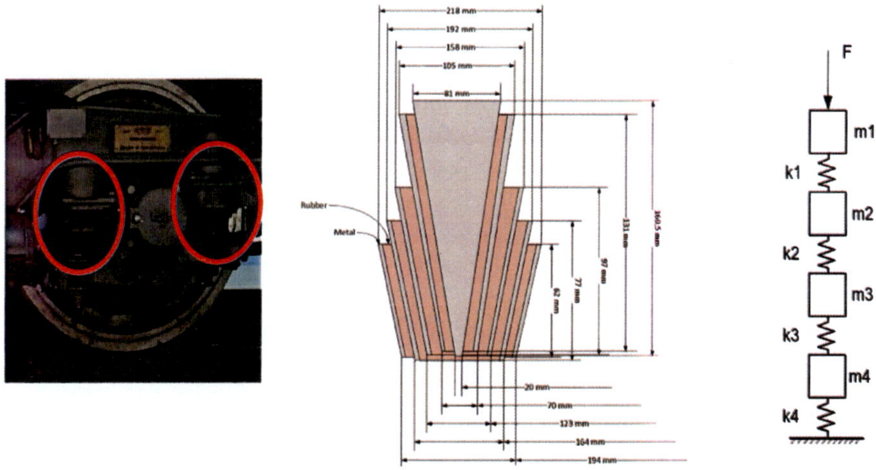

Fig. 2. Left, primary spring. Middle, cross section layout of rubber and metal. Right, the model.

The frequency response calculation in FEM is usually carried out using a modal summation approach to reduce the computational cost. The modal equation is given by

$$\left(-\omega^2[m] + i\omega[b] + [k]\right)\{q(\omega)\} = f(\omega) \tag{3}$$

where $[m]$, $[b]$ and $[k]$ are modal mass, damping and stiffness matrices of the FE model and $q(\omega)$ contains the modal generalised coordinates. This approach can also be used for a system which contains frequency-dependent stiffness and damping. In this case, the mode shape obtained from the constant stiffness is used and a modal correction to the modal damping and stiffness matrices is carried out to account for the frequency-dependent part of the stiffness. Equation (3) becomes

$$\left(-\omega^2[m] + i\omega[b + \Delta b(\omega)] + [k + \Delta k(\omega)]\right)\{q(\omega)\} = f(\omega) \tag{4}$$

where $\Delta k(\omega)$ and $\Delta b(\omega)$ are modal correction terms which are due to $\Delta K(\omega)$ and $\Delta B(\omega)$, the changes of stiffness and damping in physical coordinates, given by

$$\Delta b = [\phi]^T[\Delta B(\omega)][\phi] \tag{5}$$

$$\Delta k = [\phi]^T[\Delta K(\omega)][\phi] \tag{6}$$

where $[\phi]$ is the matrix of mode shapes for the system with constant stiffnesses.

3.3 Wheel/Rail Contact Force

The contact force $\vec{F}_{\text{contact}}$ resulting from the wheel-rail interaction is calculated using a TWINS-like approach [2] as follows. For this, the roughness excitation $\vec{R}$, and the receptance matrix $[A]$ of the wheel, rail and contact at the contact point are required. The contact force $\vec{F}_{\text{contact}}$ is then calculated according to

$$\vec{F}_{\text{contact}} = \begin{pmatrix} F_z \\ F_y \end{pmatrix} = ([A]_w + [A]_r + [A]_c)^{-1}\vec{R} \tag{7}$$

The roughness excitation $\vec{R}$ is actually an imposed displacement in the vertical direction, such that $\vec{R} = (R_z, 0)$. Furthermore, the input roughness is filtered by the contact patch. This filtering effect depends on speed and wheel load. A Remington contact filter will be used here [1].

The receptance matrices $[A]$ of the wheel, rail and contact are calculated by different means [3]. In the present paper, $[A]_w$ is calculated with the FE method; $[A]_c$ and $[A]_r$ are calculated with TWINS.

4 Results

4.1 Contact Forces

Figure 3 shows the lateral and vertical contact forces calculated from filtered roughness excitation. The wheel load is 6 tons and the vehicle speed is 60 km/h. The track dynamics was modelled with TWINS. Wheels 3 and 4 are the rear left and rear right wheels as shown in Fig. 1. Due to asymmetry of the wheelset, the left and right wheels of a wheelset have different contact forces.

4.2 Forces Acting on Vehicle Body

Since the vehicle body was not considered, the so called blocked-force method was used. Therefore the connecting points on the bolster, to which the traction bars and lateral dampers are attached, are fixed in the Nastran calculations. The forces acting on these four nodes are forces transmitted to the vehicle. For each wheel, vertical F_z and lateral F_y contact forces are correlated (longitudinal contact forces are negligible). However, contact forces on different wheels can be regarded as uncorrelated. Therefore, the forces acting on the vehicle body can be obtained by superposition of the force calculated for

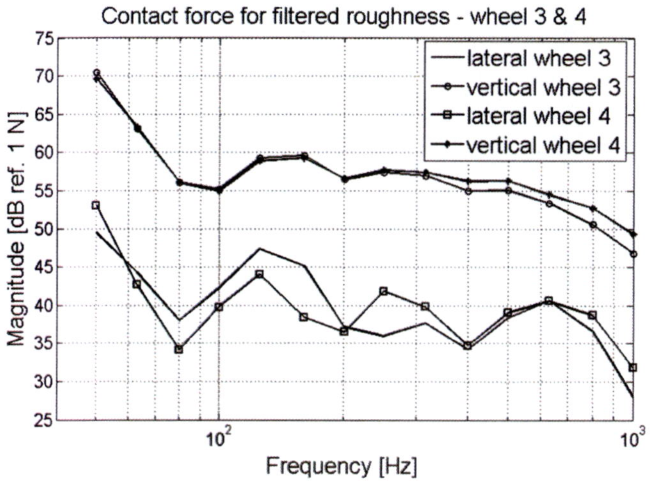

Fig. 3. Wheel/rail contact forces.

each of the four wheels individually. Such calculation can be readily carried out with the random analysis capability available in Nastran.

Figure 4 (top figure) shows the resultant forces acting on the vehicle body by the traction bars and lateral dampers. The lateral dampers contribute more to the resultant force than the traction bars, in particular between 200 and 500 Hz.

4.3 Interior Noise

The interior noise due to a single wheel is calculated by multiplying the transmitted forces obtained from above with transfer functions.

$$p_i = \sum\nolimits_{j,k} H_{kj} F_{kj} \tag{11}$$

where i is 1 to 4 wheels, k is 1 to 4 corresponding to four connection positions to the vehicle body and j is 1 to 3 DoFs. The total structure-borne noise is finally given by

$$p = \sqrt{\sum\nolimits_i |p_i|^2} \tag{12}$$

The transfer functions obtained through measurements and the predicted structure-borne noise are shown in Fig. 4 (middle and bottom figures). The main transfer paths can be identified as the longitudinal direction of the traction bar and the lateral direction of the lateral damper.

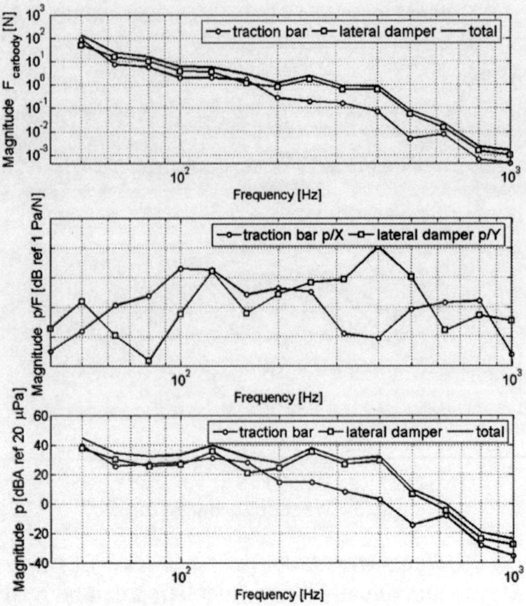

Fig. 4. Transmitted forces, measured noise transfer functions and predicted structure-borne noise from traction bars and lateral dampers

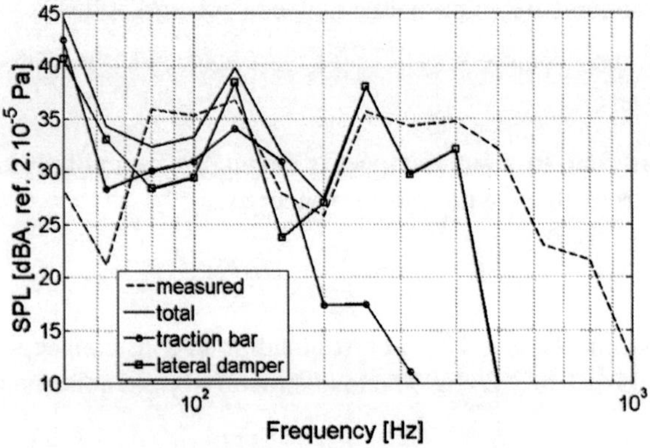

Fig. 5. Comparison of structure-borne noise levels inside the vehicle for a microphone above a trailer bogie.

Figure 5 compares the total structure-borne noise predicted for one microphone position inside the vehicle above a trailer bogie with the result from the experimental TPA as well as predicted contributions of the traction bar and the lateral damper. A very satisfactory agreement in the total structure-borne noise is observed between 80 and 400 Hz, which corresponds to the frequency range for which structure-borne noise was found to be significant during tests.

At low frequencies (below 50 Hz), discrepancies could be related to the measured rail roughness or track impedance. Above 400 Hz, the computations under-estimate the structure-borne noise contribution, which is likely to be due to inconsistent stiffness for the rubber bushings at the extremities of the lateral dampers and traction bars.

5 Conclusions

A virtual test method for the structure-borne noise of railway running gear has been presented in this paper. The bogie, wheelsets and suspensions have been modelled using FEM including dynamic properties measured in the laboratory. The excitation force on the wheel-rail contact has been calculated using TWINS. The blocked forces method was used to obtain the transmitted forces to the vehicle body. A case study implementing this method has been carried out on a metro vehicle using Nastran. The structure-borne noise inside the vehicle has been predicted using measured noise transfer functions. The developed model for the metro bogie requires validations against measurements, which will lead to further refinement and update to the model. Having validated the model, the developed virtual test model can be used to evaluate the impact of design changes at early stage, such as new material, component or noise and vibration countermeasure. Industry can potentially significantly shorten the design cycle and reduce costs of hardware tests.

Acknowledgement. The work presented in this paper has received funding from the Shift2Rail Joint Undertaking under the European Union's Horizon 2020 research and innovation programme under grant agreement no. 777564. The contents of this publication only reflect the authors' view and the Joint Undertaking is not responsible for any use that may be made of the information contained in the paper.

Provision of vehicle data from CAF Spain is gratefully acknowledged.

References

1. Thompson, D.: Railway Noise and Vibration: Mechanism, Modelling and Means, 1st edn. Elsevier, Oxford (2009)
2. Thompson, D., Hemsworth, B., Vincent, N.: Experimental validation of the TWINS prediction program for rolling noise, part 1: description of the model and method. J. Sound Vib. **193**, 123–135 (1996)
3. Report: Deliverable 4.2 – Complete virtual test method for structure-borne and airborne noise transmission, EU Shift2Rail/H2020 project RUN2rail (2019)
4. MSC NASTRAN Quick reference guide (2018)

Policy, Regulation and Perception

Railway Noise Mitigation Framework in Europe: Combining Policies with the Concerns of the Railways

Jakob Oertli(✉)

Swiss Federal Railways, Infrastructure, Hilfikerstrasse 3, 3000 Bern 65, Switzerland
jakob.oertli@sbb.ch

Abstract. Several European countries are implementing considerable railway noise mitigation programs, mostly consisting of noise barriers, retrofitting the freight fleet with composite brake blocks, and installing sound-proof windows. Nonetheless, both European and many national policies aim to further reduce railway noise. In addition, the WHO Noise Guidelines call for very low noise levels. Such additional noise mitigation is both costly and problematic for railways. Noise mitigation efforts must not undermine railway competitiveness, especially when considering that rail freight must cope with fierce intermodal competition and profit margins are low. Railways are a sustainable means of transport and must be promoted, if climate goals are to be achieved. It is therefore important that railway noise research resources are allocated in such a way as to take railway concerns into account. This can, for example, be achieved by focusing research efforts on the whole system including asset management and vibrations rather than only on noise. Also, it is useful to consider existing components of the track or the rolling stock, rather than developing additional ones. For trackside measures rare and ever shorter construction intervals, safety issues, track inspection requirements as well as life cycle costs must be considered additionally. Furthermore, annoyance studies must be critically appraised, and an evaluation is necessary to determine, if resources spent on noise mitigation are the most cost-effective method to achieve desired health goals.

Keywords: Railway noise policies · Railway concerns · Railway noise research

1 Introduction

Research on railway noise is both costly and complicated. Therefore, it is important that research efforts are aligned with the regulatory and policy framework of the European Union as well as with individual countries, take the guidelines of the World Health Organization into account and that they are consistent with railway concerns such as infrastructure and rolling stock life cycle costs. Recent years have shown significant developments on all levels. These are summarized, and conclusions are drawn for railway noise research.

G. Degrande et al. (Eds.): *Noise and Vibration Mitigation for Rail Transportation Systems*, NNFM 150, pp. 197–205, 2021.
https://doi.org/10.1007/978-3-030-70289-2_19

2 European Strategies and Policies

The EU considers railways a sustainable means of transport which should be supported [1]. Therefore, the EU's philosophy is to reduce noise without harming the competitiveness of the railways. The specific noise control strategy was formulated in a Commission Staff Working Document (SWD (2015) 300) in 2015, with a policy mix including the gradual applicability of noise limits set by the EU Technical Specification for Interoperability (TSI) to all freight wagons, the harmonization of noise-charging principles, a recommendation on financial support to help the sector make the fleet more silent, and the development of noise-related standards for railway infrastructure. The specific recent actions of the European Commission are described below:

Revision of the TSI Noise: The main effort to support the retrofitting of freight wagons with composite brake blocks is the revision of the TSI Noise by introducing so called Quieter Routes (QR) on which only silent freight wagons with composite brake blocks are allowed. On January 31, 2019 the TSI Noise was voted positively in the Rail Interoperability and Safety Committee (RISC). Starting December 8, 2024 noisy wagons with cast-iron brake blocks will be banned from Quieter Routes. These are defined as part of the railway network with a minimum length of 20 km, on which the annual averaged daily operated number of freight trains during night-time is higher than 12 trains. Specific cases or exemptions are possible. Examples include several countries such as Finland, track gauges that are different from the main rail network or specific situations such as the Channel Tunnel. Currently unresolved are reported problems with the braking performance of composite brake blocks during Nordic winter conditions. This issue is addressed with a backstop clause in the TSI Noise: If ongoing tests prove safety issues, the TSI Noise could be further amended to grant exemptions for a limited number of wagons in Europe. During 2019 the European Commission defined a procedure to address the issue and will issue a report in the second half of 2020 regarding operations with wagons with composite brake blocks in Nordic winter conditions.

Noise Differentiated Track Access Charges (NDTAC). The Directive on a single European railway area (Directive 2012/34/EU) enabled the Commission to develop rules for charging for railway noise. On this basis the Commission adopted the Implementing Regulation (EU) 2015/429 which provides the legal framework for NDTAC schemes within the European Union. A condition of NDTAC is that there is no influence on the competitiveness of the railways. To date, Germany, the Netherlands, Austria and the Czech Republic have implemented NDTAC (in addition to Switzerland, who is not a member of the EU and has adopted a different system). The European Commission is currently carrying out an evaluation of the implementation of the NDTAC schemes. The Commission is studying the costs (administrative and revenue related) of running NDTAC schemes to understand efficiency of the regulation. Results of the review are expected by mid-2020.

Connecting Europe Facility (CEF) [2]: The CEF provides funding of 20% of the costs for retrofitting and is formalized in Regulation (EU) No 1316/2013. A total of three calls were undertaken in the years 2014, 2016 and 2019. In total some 200′000 wagons will be retrofitted with these funds.

Environmental Noise Directive (END): The END 2002/49/EC4 obliges national authorities to draw up strategic noise maps and action plans for major railways and large agglomerations, allowing the most problematic "hot spots" to be identified and targeted. It is planned to add an additional annex providing a method to calculate the health effects of noise based on the WHO Guidelines (see below).

Shift2Rail: Shift2Rail is a European Public Private Partnership (PPP) for railways focused on research and innovation as well as market-driven solutions by accelerating the integration of new and advanced technologies into innovative rail product solutions. The issue of noise is addressed in several Innovation Programs (IPs) that require horizontal coordination, such as energy and noise management, safety, standardization, overall traffic management, maintenance and virtual certification, as well as long-term societal effects and human capital management projects [3]. The two most important noise projects are FINE 1 (Future Improvement for Energy and Noise) and DESTINATE (Decision supporting tools for implementation of cost-efficient railway noise abatement measures). The FINE 1 project aims to reduce operational costs of railways by a reduction of energy use and noise related to rail traffic [4]. DESTINATE aims to develop tools and methodologies for railway noise simulation and cost-benefit analysis of mitigation actions of interior and exterior noise [5].

3 The WHO Guidelines

In a report published in October 2018 [6] the WHO recognizes that noise is an important health issue. The organization therefore developed guidelines to provide recommendations on protecting human health from noise. For this purpose, so called Systematic Review Teams undertook an assessment of the overall quality of the evidence. Quality criteria included study limitations, inconsistency of results, indirectness of evidence, imprecision of effect estimates and publication bias. Based on these criteria the quality of the evidence was rated either high, moderate, low or very low. If the quality was high, further research is very unlikely to change the certainty of the effect estimate. With moderate quality, further research is likely to have an important impact on the certainty of the effect estimate and may change the estimate. With low quality, further research is very likely to have an important impact on the certainty of the effect estimate and is likely to change the estimate. Finally, if the evidence is of very low quality, any effect estimate is uncertain. The recommendations could be either strong or conditional based on the quality of evidence, the balance of benefits and harms, values and preferences and resource use.

For railways no studies were available for some high priority health outcomes such as incidence of ischemic heart disease (IHD), permanent hearing impairment, reading skills and oral comprehension in children. The quality of evidence for the incidence of hypertension was low, and the quality of the evidence for the prevalence of highly annoyed persons and sleep disturbance was moderate. Based on this evidence the WHO made a strong recommendation for noise levels below 54 dB L_{den}. For night exposure it was recommended that the levels are below 44 dB L_{night}. Furthermore, the WHO recommends that policy makers implement measures to reduce exposure from railways.

The recommendations do not include particular measures however, only stating that rail grinding and behavioral interventions such as communicating with the community may be beneficial. The recommendations were "strong" even though the evidence quality was non-existent, low (incidence of hypertension) to moderate (prevalence of highly annoyed population, sleep disturbance). There is no mention of the required costs or of the practicability of measures to meet the recommended levels in the guidelines or whether financial resources allocated to noise control are the most cost-effective method to improve the health of the general population. Furthermore, there is no mention of the effects of façade insulation when determining the recommended levels.

The dose-response relationships for the railway and road sector where found to be similar. This is supported by the SiRENE [7] study, where annoyance of noise was similar for rail and road traffic. This is bound to call railway noise bonuses into question in several European countries resulting in a de facto lowering of legislated limit values.

4 National Strategies and Policies

Several countries have adopted policies that influence the railway noise landscape [8]. Often the policies and strategies vary greatly throughout Europe depending on the local requirements. The variety of policies is shown in the following case examples:

Germany [9]: Germany has a comprehensive strategy to mitigate railway noise. Among other measures, legislation was enacted to ban noisy freight wagons on the whole network starting in December 2020. This national legal act is not in line with the revised TSI Noise, which bans noisy wagons only on Quieter Routes and in 2024 instead of 2020. Currently it is unclear how this inconsistency will be resolved. Also, in the past years noise limits have been reduced several times including abolishing a railway bonus which was initially introduced because noise from railways was considered less annoying than that of roads. Germany provides considerable financial means to mitigate noise along railway lines, mostly with noise barriers and sound proof windows. In addition, retrofitting the national freight rolling stock is also partially subsidized. Finally, Germany has a system of NDTACs.

Switzerland [10]: Switzerland has introduced a ban of noisy freight vehicles starting January 1, 2020. The country financed noise barriers, sound proof windows and the retrofitting of the entire national freight fleet. In a second phase, limit values on rail roughness have become effective starting 2020. A revised guideline calls for simultaneous noise control if noise increases by 1 dB, which can easily happen if the components are changed during track renewal. This in turn allows lineside inhabitants to register objections to simple track renewal projects thus reducing critical planning reliability.

Austria [11]: A considerable amount of noise barriers has been constructed and more than half of freight vehicles are silent. Austria has a system of NDTACs.

The Netherlands [12]: The Netherlands have defined permissible noise emission levels for each railway line. If a line no longer complies with these levels, the responsible authority (i.e. the infrastructure manager) must take measures to ensure that it does, for example by installing noise barriers or rail dampers. Also, the Netherlands have implemented a system of NDTACs to incentivize the retrofitting of freight rolling stock with composite brake blocks.

Italy [13]: In Italy an extensive program of noise barrier construction along railway lines is in progress, focusing on the reduction of existing noise levels. Also, in December 2018 the European Commission has approved under EU State Aid Rules Italy's € 20 million retrofitting aid scheme.

These countries are examples with considerable noise mitigation efforts. On the other hand, such policies are largely absent in countries such as Bulgaria or Romania where, given the economic situation in general and of rail freight in particular, railway noise does not seem to be such a large issue.

5 Railway Concerns

While the railways recognize the need to consider noise as an issue, they have several concerns that must be taken into account:

Competitiveness: The railway sector is in competition with other transport modes. Therefore, profit margins of rail freight are very low. While the sector promotes retrofitting of cast iron brake blocks of existing freight wagons with composite brake blocks, the construction of noise barriers as well as further measures on the infrastructure, these should not reduce the competitiveness of the railways [14]. To remain competitive, European policy makers must consider the costs of noise mitigation measures, the influence of noise mitigation measures on the traffic capacity of the network and the impact on intermodal competition. The costs stem not only from the investment and maintenance costs of the measure itself, but also from effects on other parts of the system such as the increased necessity for wheel reprofiling after retrofitting freight wagons or increased costs for maintenance and track diagnostics with the use of rail dampers. Keeping the railways competitive also has a greater societal benefit with regards to measures to counteract climate change. This is an important argument as it justifies the spending of public funds to wagon keepers to cover one-off costs and – at least for a transitional period – higher operational costs. The one-off costs are currently financed in part by the European Commission through the CEF (see above) and by national subsidies (see above), however there are currently no initiatives to cover the higher operational costs.

Trade-Offs: Aims for asset management, noise and vibrations can counteract one another. For example, soft rail pads are considered beneficial for asset management but lead to more noise in comparison to stiff rail pads [15]. In addition, there are indications that soft under sleeper pads are beneficial for vibrations but lead to more noise [15]. It must be noted that the frequencies relevant for asset management and vibrations are usually lower (below 400 Hz) than for noise (usually between 125 and 4000 Hz). Some railways have recognized the need for a whole system optimization such as SBB Infrastructure with their Go-Leise project [15].

Maintain Traffic Capacity of Network: In addition, noise policy should not reduce the traffic capacity of the network, for example by imposing operational restrictions.

Reduce Risk of Disruption of Service: As networks are being used more and more intensely, all additional elements that could lead to a disruption in service are problematic. In other words, every additional component that is added to the system can become a risk. It is therefore important that noise mitigation is undertaken with components of the system as such or are positioned outside of a critical area near the track (such as is usually the case with noise barriers or sound proof windows). Furthermore, available construction intervals are becoming shorter and rarer. Noise mitigation measures must therefore be installable in a minimum of time.

Compatibility with Other Elements of Infrastructure: Often when planning mitigation measures only acoustic criteria are considered. Because of the proximity to other elements of the track, possible interactions must be taken into account. Examples of such a framework are sufficient maneuvering room for maintenance crews or minimum distances to electrical installations.

Availability of Construction Crews and Safety Personnel: Currently in many European railways considerable construction work is in progress. Therefore, construction resources and work gang protection measures are often limiting factors. Noise mitigation measures and research efforts must take this into account and should therefore be planned in such a way so that they can be installed with a minimum of effort.

Access to Railway Lines is an Issue: Noise mitigation measures in built over areas often result in noise barriers. This makes access to the railway track difficult; not only for track workers but also for rescue organizations.

Damage and Vandalism Control: Many noise mitigation measures are prone to damage and vandalism, e.g. through graffiti on noise barriers, vehicles colliding with noise barriers and the like. Research efforts must focus on measures that take the potential for damage and vandalism into account.

Safety of Products: All new products used in railways must adhere to safety standards. It is best, if these are considered early in the product development process.

Product Requirements: Often noise control measures used in the track or on the railways must adhere to specific requirements. For example, they should exceed a minimum life span, be resistant to the effects of moisture, ozone, ultra-violet light, hydrocarbons and other railway related chemicals and be shock and abrasion resistant.

Minimal Maintenance and Replacement Costs: Many of the above concerns are also valid when the mitigation measure must be maintained or replaced. These costs must be included when choosing between different noise mitigation options. Often it is best to work with high quality products with long lifespans.

Although in many cases these concerns are met, this often occurs at a rather late stage in the process. Much can be gained, if they are considered as early as possible.

6 Conclusions

Both on the EU as well as on national levels railways are being promoted as an environmentally sound means of transportation. Therefore, if the competitiveness of the railways is to be maintained, noise mitigation must cost as little as possible and should not reduce traffic capacity. Nonetheless, it is expected that railway noise must be further reduced significantly to attain the guidelines currently proposed by the WHO as well as to adhere to European and national noise policies. The railway noise research challenge is to reduce noise without significantly increasing costs of railway operations.

It is therefore important to consider noise as well as the costs for railways simultaneously. The ideal approach is to gain a better understanding of track dynamics as well as train track interactions. These insights would then be useful both for noise and asset management, ideally reducing costs and noise at the same time and thus optimizing the whole system rather than considering noise and asset management separately. In short, railway noise research should not only include noise, but must be enlarged to include the whole system.

In this regard, the most promising research should focus on improving existing components or procedures (improved rail pads or optimized rail grinding) rather than adding components or procedures to the system (i.e. rail dampers, additional rail grinding). Additionally, research results should always be evaluated both in terms of noise and the effects on asset management as well as in terms of traffic capacity. This analysis should also consider the specific situation of each railway and country and therefore may lead to different conclusions in each case.

In sum, railway research efforts should include the following considerations:

Basic Philosophy: Develop noise control mechanisms without reducing the competitiveness of the railways.

Undertake a Whole System Optimization: The aims of noise, vibrations and maintenance are not necessarily the same. It is insufficient if only one of these is optimized. Cost-benefit calculations must therefore be undertaken to include all parts of the system.

Work with Existing Components of the Track: The track system is already complicated as is. Every additional component increases the complexity due to interactions between the components. It is therefore best to improve existing components rather than to add new ones. For example, it is better to improve the rail pad, the sleeper or the fastening system, than to add rail dampers or low height noise barriers close to the track.

Take the Whole Railway Framework into Account: The railway system is complex, and any new products must fit into the given framework. Operational, construction and safety considerations should be considered.

Understand the System: The above can best be achieved if the dynamics of the track, the rolling stock and the interaction are understood as well as possible.

Critical Appraisal of Promises by Manufactures of New Measures: Due to the noise reduction pressure, government ministries and railway companies are open to all sorts of innovative measure that might reduce noise. Also, lineside inhabitants often consider innovative projects to be more effective than traditional ones such as noise barriers. These leads manufacturers to develop noise mitigation measures with questionable effectiveness and which may hinder railway operations or maintenance. Furthermore, manufacturers have been known to take legal action if the railways are too critical of the effectiveness of their products. Therefore, all measures should be tested thoroughly before they are implemented.

Critical Appraisal of the Science Behind Annoyance Studies: The WHO Guidelines mention nonexistent or low to moderate quality of the studies as a basis to their suggestions. Further research on annoyance and a critical appraisal of the current studies are needed before costly decisions are made. All suggestions should be quantified in terms of the costs to achieve them.

Noise Mitigation Costs to Achieve Health Effects Should be Put into a Larger Context: Research is also needed to determine if the money spent on railway noise control is used in the most efficient area to achieve desired effects. The question is: Are the positive health outcomes best achieved it the resources are spent in railway noise control or could a larger effect be achieved if they are invested in other areas?

Railway noise research should focus on measures that do not reduce the competitiveness of the railways. For this purpose, the above concerns and suggestions should be taken into account. The importance of the concerns depends on the economic situation of every particular railway either as a private company striving for profit or as a state dependent party, where a deficit may be less critical.

References

1. European Commission: Roadmap to a Single European Transport Area – Towards a competitive and resource efficient transport system (2011)
2. CEF Transport Calls. https://ec.europa.eu/inea/en/news-events/newsroom/69-project-proposals-submitted-under-2019-cef-transport-calls. Accessed 13 May 2019
3. Shift2Rail. https://shift2rail.org. Accessed 13 May 2019
4. Burroughs, D.: The quiet revolution in noise abatement. Int. Railway J. **58**, 44–47 (2018)
5. DESTINATE. https://www.destinate-project.tu-berlin.de. Accessed 13 May 2019
6. World Health Organisation, WHO: Environmental Noise Guidelines for the European Region (2018)
7. Röösli, M., Wunderli, J.M., Brink, M., Cajochen, C., Probst-Hensch, N.: Die SiRENE Studie. In: Swiss Medical Forum, vol. 19, pp. 77–82 (2019)
8. International Union of Railways, UIC: State of the Art Report on Railway Noise (2016)
9. Deutsche Bahn Noise. https://www1.deutschebahn.com/laerm. Accessed 13 May 2019
10. Swiss Federal Office of Traffic. https://www.bav.admin.ch/bav/de/home/verkehrstraeger/eisenbahn/ausbauprogramme_bahninfrastruktur/laermsanierung.html. Accessed 13 May 2019
11. Bundesministerium für Nachhaltigkeit und Tourismus. https://www.laerminfo.at. Accessed 13 May 2019
12. Dutch Government. https://www.government.nl/topics/environment/noise-nuisance/noise-pollution-from-railways. Accessed 13 May 2019

13. Italian Railways. https://www.fsitaliane.it/content/fsitaliane/it/sostenibilita/tutelare-l-ambiente/emissioni-e-rifiuti/emissioni-acustiche.html. Accessed 13 May 2019
14. Community of European Railways and Infrastructure Companies, CER: Rail Freight Noise Strategy (2016)
15. Oertli, J., Hafner, M.: Ein neuer Ansatz zur Gesamtoptimierung der Fahrbahn. Der Eisenbahningenieur **68**(6), 32–37 (2017). ISSN 0013-2810

Short and Very Short Term Indicators to Characterize Train Pass-By Noise

Franck Poisson[1](✉) and Anne Guerrero[2]

[1] SNCF Mobilités Agence d'Essai Ferroviaire, 21, Avenue du Président Salvador Allende, 94407 Vitry sur Seine Cedex, France
franck.poisson@sncf.fr
[2] SNCF Réseau Direction du Développement durable, Campus Réseau, 15/17 rue Jean-Philippe Rameau, 93418 La Plaine Saint Denis, France
anne.guerrero@sncf.fr

Abstract. The environmental noise of the railway system can be quantified with a large number of different indicators. The main characteristic of these indicators is the integration time which varies from 125 ms to several hours. A measurement campaign of a large number of trains has been carried out and the most usual long term, short term and very short term indicators have been calculated ($L_{eq,Tp}$, $L_{Aeq,Tp}$, SEL, TEL, L_{max}, L_{Amax}, and L_{PAFmax}). A statistical analysis has been performed to evaluate the statistical characteristics of each indicator for each train type. Boxplot representations show that the range is higher for very short term indicators than for the others. Some representative traffic have been characterized at different locations along the French railway network. Noise indicators have been simulated by combination of the noise measurements of the previous measurement campaign. Then, another statistical analysis has been performed to characterize the noise indicators for each traffic type. For suburban and regional traffic, the correlation between all the indicators is high. This result is not valid for freight traffic. As a conclusion, the use of short and very short terms indicators is discussed.

Keywords: Environmental noise · Long term indicators · Short term indicators

1 Introduction

The environmental noise of the railway system is a key issue for its integration in the environment and its acceptance by residents. A combination of national [1] and European [2] laws limits the annoyance caused by the railway noise. Two noise limits are designed to control the noise produced by each train (train certification), and to control the noise of the train traffic at a given location within a certain time window.

When the traffic increases or a new infrastructure is built, environmental noise is evaluated and if noise limits are exceeded then, mitigation measures are installed.

Noise indicators are designed to be representative of the resident annoyance. If the noise indicators are not well suited for this purpose, then the mitigation measures can

G. Degrande et al. (Eds.): *Noise and Vibration Mitigation for Rail Transportation Systems*, NNFM 150, pp. 206–213, 2021.
https://doi.org/10.1007/978-3-030-70289-2_20

be inefficient to reduce the annoyance. Most of the noise indicators are based on the A-weighted sound pressure level integrated over a given period. According to this period, the indicator is considered as:

- A long term indicator: The sound pressure level is integrated over several hours, for example from 6 am to 22 pm, one day or one year
- A short term indicator: The sound pressure level is integrated during the duration of the train pass-by.
- A very short term indicator: The sound pressure level is integrated over 125 ms, 1 s...

In Europe, long term and short term indicators are used in the national and European laws. The noise emitted by each train is limited by the Technical Specifications for Operability based on the short term indicator $L_{Aeq,Tp}$. The traffic noise is limited at the receiver by long term indicators L_{DEN}, $L_{6h\text{-}22h}$, $L_{22h\text{-}6h}$.

In this context, some residents claim that the long term indicators are not sufficient to be representative of the noise annoyance. They want to add a new indicator to quantify the noise at the receiver based on short or very short term indicators.

Then, SNCF-Réseau, the infrastructure manager of the French railway network, has carried out a statistical analysis of short and very short term noise indicators. A dedicated measurement campaign and a statistical analysis have been performed. Main results are presented in the following paragraphs. The relevance of short and very short term indicators to represent the annoyance of residents is not treated in this study.

2 Methodology

2.1 Measurement Campaign

A dedicated measurement campaign has been carried out to characterize the pass-by noise of a large number of commercial trains.

High Speed Trains (HST) have been characterized on the Mediterranée high speed line (HSL) during 4 days. The high speed track is made of "UIC60" rail, "Fastclip Pandrol" fastening system and bi-bloc concrete sleepers. The commercial train speed was 300 km/h. During 3 days, freight trains, regional Diesel multiple units (DMU), regional electrical multiple units (EMU), intercity trains and HST have been measured on a classical line (CL) equipped with "UIC60" rail, "Nabla" fastening system, 9 mm thick rail pad and bi-bloc concrete sleepers. The commercial speed was around 160 km/h. During 3 days, suburban trains (EMU) have been measured along "RER-D" line made of "U33" rail, 4, 5 mm thick rail pad, "C165" fastening system and bi-bloc concrete sleepers.

The pass-by noise has been measured with a sound level meter at 150 m from the track to be representative of a house located close to a railway line in urban area. Simultaneously, the noise has been measured with another a sound level meter at 25 m as a reference. Time signals have been also recorded at the same locations with a microphone and a data acquisition system to perform post-treatment if needed. Optical train detectors have been used to detect the train pass-by, the train position and the train speed.

The following short and very short term indicators have been calculated:

- The equivalent sound pressure levels integrated over the duration of the pass-by: $L_{eq,Tp}$ and $L_{Aeq,Tp}$,
- The Sound Exposure Level: SEL,
- The Transient Equivalent Level: TEL,
- The maximum value of $L_{pAeq125ms}$ during the pass-by: L_{Amax},
- The maximum value of $L_{peq125ms}$ during the pass-by: L_{max},
- The maximum value of $L_{pAf(t)}$ during the pass-by: L_{PAFmax}.

For each location, the number of valid measurement is presented in Table 1.

Table 1. Number of valid measurements at each location.

Location	Train type	Number of pass-by
High speed line	HST	52
Classical line	Intercity train	29
	EMU	9
	DMU	23
	Freight	43
	HST	8
Classical line in urban area (RER-D)	EMU	18
	Freight	1
	Total	**183**

The traffic has been characterized at 22 locations representative of the French railway network for day, evening and night periods. The 183 sound samples have been used to reconstruct realistic measurements at the 22 locations. Sound samples have been randomly selected in the database using the bootstrap method.

Each of the 22 locations has a dominant traffic as it is presented in Table 2.

Table 2. Number of locations for each dominant traffic.

Dominant traffic	Sub urban trains	Regional trains and inter city trains	Freight trains	HST
Number of locations	4	6	9	3

2.2 Measurement Analysis

In the first part of the study, statistical analysis is performed for each train type: intercity, freight, suburban EMU, regional EMU, regional DMU, HST on HSL, HST on CL.

The main objective is to identify the statistical characteristics of short and very short term indicators for each train type to have a better understanding of the statistical characteristics of the measurement of the traffic.

The second part of the study is dedicated to the statistical analysis of mixed traffic at 22 locations of the French railway network. For each traffic, a particular train type is dominant (high speed trains, regional trains, suburban trains and freight trains) but some other train types are also included.

3 Results

3.1 Statistical Analysis for Each Train Type

Train Speed. For each train type, the variation of the train speed is characterized as it has a large influence on the railway noise. Box plot of the train speed is presented in Fig. 1.

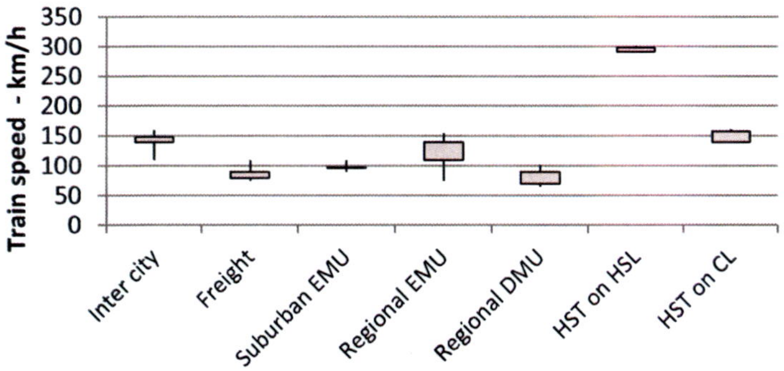

Fig. 1. Boxplots of the train speed during the pass-by noise measurement for each train family.

The range of the train speeds is limited for most of the given train types. This result is representative of commercial trains in operation. The higher range concerns the regional EMU because the measurement point was located in a very dense area of the railway network.

Short-Term Indicators. Boxplots are calculated for each short term and very short term indicator. Results for $L_{Aeq,Tp}$ are presented Fig. 2.

The range of the $L_{Aeq,Tp}$ between the first and the third quartile is between 1 dB for suburban EMU and 7 dB for regional EMU. The speed variations of regional EMU explain this result. The higher difference between the maximum and the minimum value is 13 dB for HST on HSL.

Boxplots for TEL are presented in Fig. 3.

The range of the TEL between the first and the third quartile is between 2 dB for Intercity trains and HST on CL and 8 dB for regional EMU. The speed variations of regional EMU explain this result. The higher difference between the maximum and the

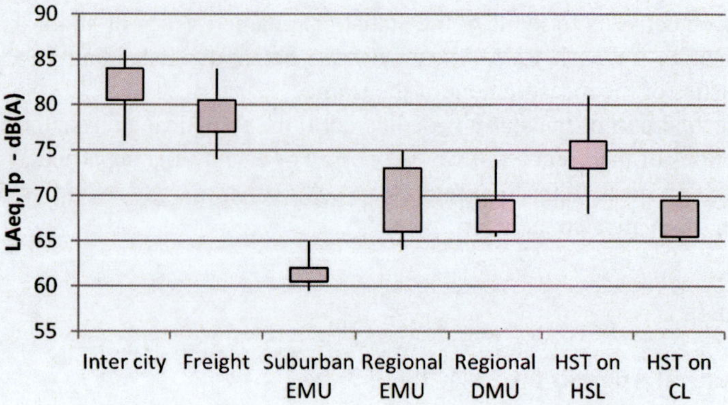

Fig. 2. Boxplots of the $L_{Aeq,Tp}$ according to the train type.

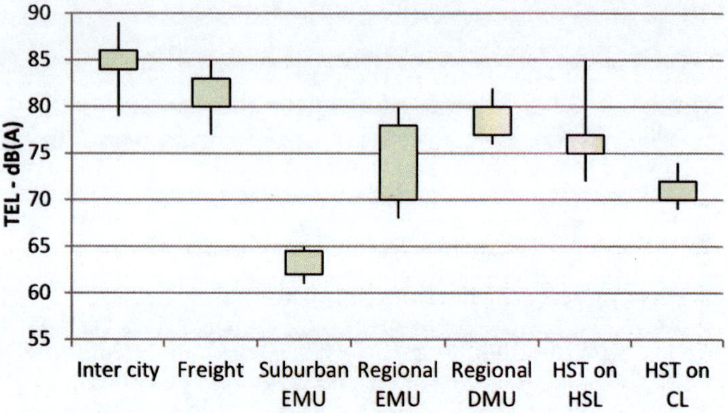

Fig. 3. Boxplots of the TEL according to the train type.

minimum value is 13 dB for HST on HSL. Results are similar for $L_{Aeq,Tp}$, TEL and SEL (not presented in this article). The same order of magnitude of ranges is obtained for the $L_{eq,Tp}$ but some high atypical points appear for suburban EMU and regional EMU. A frequency analysis of the EMU and HST sound samples shows a very high energy in the low frequency bands. For HST, aerodynamic noise variations [3] in the low frequency band are not enough to impact TEL. The conditions of propagation, and especially the wind, are probably responsible of these differences but the comparison between the spectra of measurement at 25 m and 150 m (strong effect of the wind) is not strong enough to make it clear. Further investigations are needed.

Very Short-Term Indicators. Boxplots of the very short term indicator $L_{PA,Fmax}$ are presented Fig. 4 for the different train types.

The range of the $L_{PA,Fmax}$ between the first and the third quartile is between 2 dB for suburban EMU and 7 dB for regional EMU. This range is between 0.5 and 2 dB higher than for short term indicators. The higher difference between the maximum and the minimum value can be also very high and reach 19 dB for HST on HSL.

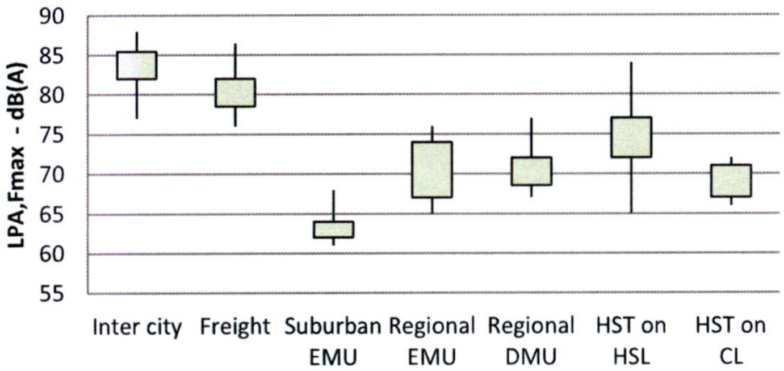

Fig. 4. Boxplots of the $L_{PA,Fmax}$ according to the train type.

The same order of magnitude is obtained for L_{Amax} (see Fig. 5). The higher difference between the maximum and the minimum value is also very high and reach 18 dB for HST on HSL.

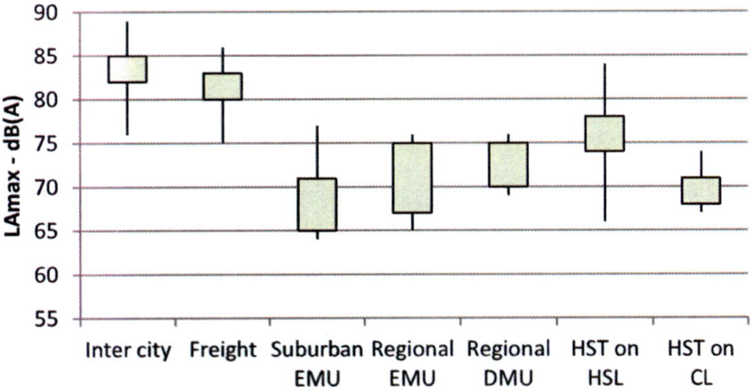

Fig. 5. Boxplots of the $L_{A,max}$ according to the train type.

3.2 Statistical Analysis for Traffic

As mentioned in Sect. 2.2, sound samples of the measurement campaign are combined according to representative traffic at 22 locations. Then, long, short and very short term indicators are calculated.

Correlation Between Indicators. The correlation between indicators is calculated for the different types of traffic.

For suburban and regional traffic, correlations between most of the indicators are high (0.6<–<0.8 – in italic in Table 3) or very high (>0.8 – in bold in Table 3). Results are presented in Table 3 for suburban traffic.

Freight traffic is the only one for which the correlation between indicators is low (see Table 4).

Table 3. Correlation between indicators for suburban traffic.

	L_{den}	$L_{eqA,6h\text{-}22h}$	$L_{eqA,22h\text{-}6h}$	Max ($L_{Aeq,Tp}$)	Max SEL	Max (L_{Amax})
L_{den}	**1**					
$L_{eqA,6h\text{-}22h}$	**0.99**	**1**				
$L_{eqA,22h\text{-}6h}$	**0.88**	**0.92**	**1**			
Max ($L_{Aeq,Tp}$)	**0.91**	**0.92**	*0.78*	**1**		
Max SEL	*0.68*	*0.68*	0.54	**0.82**	**1**	
Max (L_{Amax})	**0.90**	**0.91**	*0.77*	**0.99**	**0.84**	**1**

Table 4. Correlation between indicators for freight traffic.

	L_{den}	$L_{eqA,6h\text{-}22h}$	$L_{eqA,22h\text{-}6h}$	Max ($L_{Aeq,Tp}$)	Max SEL	Max (L_{Amax})
L_{den}	**1**					
$L_{eqA,6h\text{-}22h}$	**0.98**	**1**				
$L_{eqA,22h\text{-}6h}$	*0.60*	*0.68*	**1**			
Max ($L_{Aeq,Tp}$)	0.30	0.25	0.10	**1**		
Max SEL	0.19	0.25	0.36	0.11	**1**	
Max (L_{Amax})	0.30	0.25	0.09	**0.97**	0.02	**1**

Correlations between L_{den} and $L_{eqA,6h\text{-}22h}$ or $L_{eqA,22h\text{-}6h}$ can be explained by the quantity of trains running during each period “day”, “evening”, “night”, “6 h–22 h” or “22 h–6 h”. In Table 4, freight traffic is concentrated during the day “6 h–22 h” with less trains during the night “22 h–6 h”.

Nevertheless, the low correlations between indicators for freight traffic are not explained.

Results for HST traffic are not presented as there is no variation of Max SEL (variance = 0) and then the correlation cannot be calculated.

4 Conclusion

Long term indicators are used to limit the noise for the residents of railway infrastructures. Residents propose to add short or very short term indicator to improve the representativeness of indicators in terms of short term annoyance. This study presents a global view of ranges of short term indicators $L_{eq,Tp}$, $L_{Aeq,Tp}$, SEL, TEL, and very short term indicators L_{max}, L_{Amax}, L_{PAFmax}. The ranges of the very short term indicators between the first and the third quartile are higher than for short term indicators for all the train types. The differences between the maximum and the minimum value are also higher for very short term indicators. This difference reaches 19 dB. Therefore, it seems to be difficult to manage a limit value for very short term indicators considering this result.

From an acoustical point of view, an indicator measured in the surroundings of the track and integrated during the pass-by period like $L_{Aeq,Tp}$ is not appropriate. The acoustical energy of the approach and the escape periods is not taken into account. SEL and TEL are more representative of the train pass-by.

The correlation between all the indicators is high or very high for most of the train traffic. Nevertheless, for freight traffic, long term indicators and short/very short term indicators are not correlated. This result is not explained and must be investigated.

From a statistical and an acoustical point of view, it appears to be more relevant to use a short term indicator with a period of integration based on the complete train pass-by duration like the TEL or the SEL, than a very short term one.

After this statistical analysis, the relationship between short and very short term indicators on one hand, and short term annoyance in the other hand must be investigated by annoyance specialists. A survey must be carried out with residents living along the railway infrastructure in various locations (city, sub urban area, campaign) to quantify annoyance. High (acoustic hot spots) and low exposed locations must be selected. Locations including mitigation measures like noise barrier, façade insulation, rail dampers can also be added. In the same time, the long term, short term and very short term indicators must be recorded. The objective will be to identify the indicator the most correlated with annoyance. This step requires dedicated skills.

References

1. Loi bruit n°1444/92 (1992)
2. European Noise Directive 2002/49/CE (2002)
3. Poisson, F.: Railway noise generated by high-speed trains. In: Nielsen, J., et al. (eds.) Noise and Vibration Mitigation for Rail Transportation Systems. Notes on Numerical Fluid Mechanics and Multidisciplinary Design, vol. 126. Springer, Heidelberg (2015)

Cost Effectiveness of Noise and Vibration Mitigation Measures Using Life-Cycle Cost Analysis

Xiaohan Phrain Gu[1,2(✉)], Ziyan Ma[1,2], Anbin Wang[1,2], Longhua Ju[1], Xiaodong Chai[1], and Xiaogang Gao[1]

[1] School of Urban Railway Transportation, Shanghai University of Engineering Science, 333 Longteng Road, Shanghai 201621, People's Republic of China
evxxhpgu@sues.edu.cn
[2] Anhui Railway Technology Co. Ltd., 11 Chaoyang Road, Chahe EDZ, Lai'an 239236, Anhui, People's Republic of China

Abstract. This paper provides insights, through a case study, into the financial implications of using different noise and vibration mitigation measures in urban railway applications.

Noise complaints had been received at the S metro line 3 of S city in China. A comprehensive noise and vibration assessment was therefore conducted. Excessive levels of in-car noise and vibration, as well as rail corrugation were identified at 15 out of 44 track sections of the S metro line 3.

Three options of noise and vibration mitigation measures were proposed, with both technical feasibility and practicality properly examined. Long-term financial implications due to the selection of mitigation measures were evaluated using Life-cycle cost analysis. Cost profile and cost intensive year(s) throughout the life-cycle are valuable information, enabling a straight-forward and meaningful cost comparison between different mitigation options. The outputs of life-cycle cost analysis are of interests to asset owners and decision makers.

Furthermore, life-cycle cost analysis is useful at the design stage for new railway lines, as it facilitates early anticipation of the investment plan and overall expenditure.

Keywords: Noise and Vibration (N&V) mitigation measures · Financial implications · Life-Cycle Cost Analysis (LCCA)

1 Introduction

Railway transportation is undoubtedly the most sustainable mode of transportation for densely-populated urban environment. Comparing to other types of transportation, railway has the advantages of larger capacity, requiring less space, lower energy consumption and lower carbon-footprint.

The central government of China has granted substantial investment into urban railway development in the past decade or so. Up to the end of 2018, urban railway systems

G. Degrande et al. (Eds.): *Noise and Vibration Mitigation for Rail Transportation Systems*, NNFM 150, pp. 214–221, 2021.
https://doi.org/10.1007/978-3-030-70289-2_21

have been developed and are in operation in 35 cities in China, with a total network length of 5761 km.

Notwithstanding above strengths of urban rail systems, one of the vulnerabilities of current urban rail systems is Noise and Vibration (N&V) impact to passengers as well as to other parties along the railway line, whether underground, on the road surface or elevated above ground. Noise and vibration issues have become one of the focal points to railway operators in cities such as Shanghai, Guangzhou, Nanjing etc. where urban rail systems are considered relatively mature.

Research into noise and vibration control in China are primarily in areas of mitigation techniques, advanced noise and vibration products, rail corrugation and wheel polygonalisation etc. On the other hand, financial implications associated with different remedial schemes are yet to be examined in a systematic manner, considering that long-term operational costs are of a completely different order of magnitude from construction costs. Study on the cost-effectiveness of N&V mitigations could potentially lead to significant reduction of operational cost in the long-run.

The following sections demonstrate a systematic study on the use of different N&V mitigation measures and products, and corresponding life-cycle costs.

2 Noise and Vibration Assessment of the S Metro Line 3

The S Metro Line 3 (SML3) of S city in China was officially opened in December 2014. The underground line is approximately 23 km in total, with 21 stations.

Due to noise complaints, the operator of the SML3 implements routine rail grinding quarterly to give a reasonably even rail surface to lessen the effects of wheel-rail interaction, thus lower noise and vibration levels. However, excessive rail grinding inevitably causes additional wear of rails. According to statistics from urban railway lines in China, rail grinding quarterly to once every 4 months results in a significant reduction of rail life. Such intensive grinding consequently demands rerailing every 5 years; both rail grinding and rerailing are costly.

The team at Shanghai University of Engineering (SUES) and Anhui Railway Technology Co. Ltd. (ART) was commissioned by the operator of the SML3 for a N&V assessment and recommendation of mitigation measures, to provide a technically viable and more cost-effective solution to the N&V and rail wear problem.

The One-stop Noise and Vibration Assessment Method (ONVAM) of SUES and ART (per Fig. 1) was employed. In-car N&V levels were monitored and analysed to identify N&V-prone track sections. Rail profiles and corrugation characteristics [1], including rail vibration decay rates were measured, followed by real-time vibration monitoring for dynamic characteristics of vibration response as trains passing relevant track sections. Data analyses were performed, based on which N&V mitigation recommendation were proposed, reviewed and implemented. After the implementation of mitigation measures on a trial section, N&V levels and rail corrugation characteristics were monitored for at least 12 months; these indicators were then compared against their counterparts before the mitigation measures in place, to quantify the effectiveness of recommended mitigation measures.

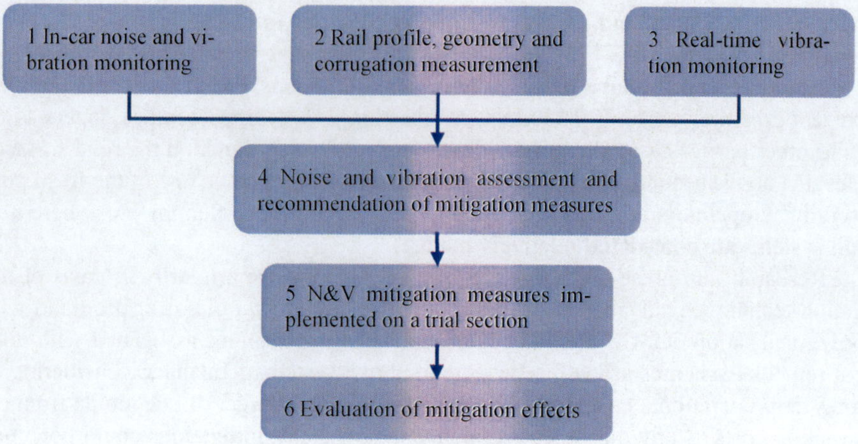

Fig. 1. The One-stop N&V Assessment Method (ONVAM)

In-car noise monitoring was performed according to Chinese National Standard GB 14892-2006 [2]. Rail corrugation and acoustic roughness were measured using Bi-CAT [3], and decay rate was measured according to BS EN 15461:2008+A1:2010 [4].

Measured in-car noise levels are summarised in Fig. 2. In-car vibration response was also measured at each track section. Vibration levels are in agreement with in-car noise levels. Vibration levels are not displayed herein due to the limitation of space.

N&V-prone sections were defined as in-car noise level exceeding 83 dB(A), namely the GB 14892-2006 noise limit line in Fig. 2. N&V-prone sections match well with track sections receiving most numbers of complaint records.

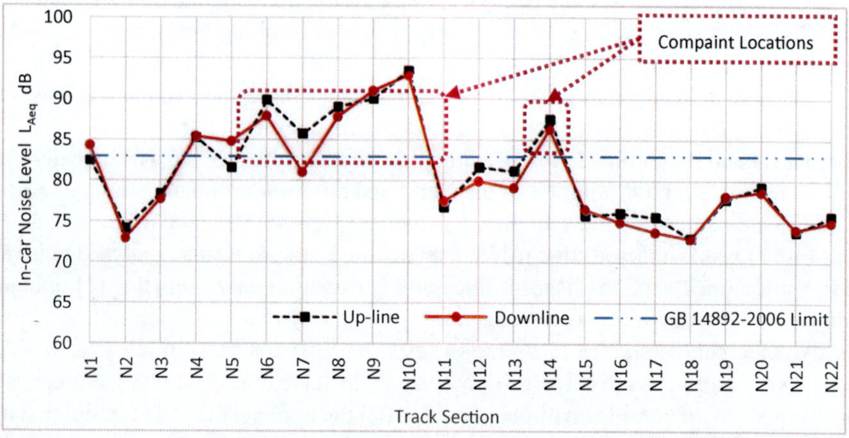

Fig. 2. Measurements of in-car noise level

At each N&V-prone track section, further investigations including rail profile; corrugation characteristics in terms of roughness and decay rate; as well as dynamic response as trains running through accelerometer locations, were all undertaken. A presentation of the frequency contents of rail roughness of a representative N&V-prone section is shown in Fig. 3. For all N&V-prone sections, 1/3 octave band central frequencies corresponding to peak roughness values are found to be at 315 Hz and 500 Hz, below the main noise frequency range of 1000 Hz for SML3.

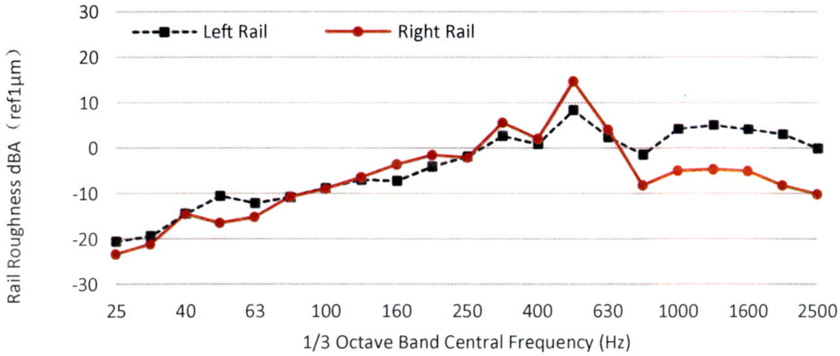

Fig. 3. Frequency contents of rail roughness

According to above noise, vibration and corrugation assessment, two technically feasible mitigation measures were proposed for N&V-prone track sections. Together with current routine rail grinding, treatment options with relevant information and assumptions are described below and summarised in Table 1. All three treatment options provide a satisfactory level of noise and vibration reduction, as shown in Fig. 4, which is a comparison of vibration level measured at the tunnel wall of trial sections after a month of implementation of treatment options.

With treatment option A, there is effectively no N&V mitigation measures, resulting in high maintenance cost, i.e. grinding cost, each year. As stated before, excessive rail grinding demands much more frequent rerailing. Material disposal cost is also a significant cost, financially and environmentally[1].

On the other hand, with option B, corrugation growth rate can be slowed down significantly using Tuned Rail Damper (TRD) [5], which has a life-span of at least 10 years. While Type I fastening, with a 5-years life-span, provides a satisfactory level of vibration insulation. With this mitigation option, the rail grinding interval can be increased to at least every 2 years; the rerailing interval is hence increased to every 15 years, which leads to a reduction of material disposal cost, thus less environmental and ecological impact.

Mitigation option C uses a versatile fastening system, offering the merits of both TRD and Type I fastening, but with much less expensive unit cost comparing to that of option B, the requirement of rail grinding and rerailing are the same as that of option B.

For mitigation option B and C, cost profile of maintenance costs related to TRD and fastening systems only are shown in Table 2. Maintenance costs for TRD and fastenings

[1] For various reasons, material disposal cost is not included herein. Inclusion of such cost would increase the cost difference between treatment option A and the other two.

Table 1. Current treatment vs. proposed N&V mitigation measures

Option ID	Treatment option	Implementation interval	Rate (RMB[a]/km) [b]	NB
A	Routing rail grinding only	Quarterly	250,000[c]	Rerailing every 5 yrs
B	Type I fastening System	Every 5 yrs	1,152,000	Grinding every 2 yrs; Rerailing every 15 yrs
	Tuned rail damper (TRD)	Every 10 yrs	3,118,050	
C	Type II fastening System	Every 10 yrs	2,160,000	Grinding every 2 yrs; Rerailing every 15 yrs

NB: [a]RMB is Chinese currency.
[b]Rerailing rate is RMB 1,680,000/km.
[c]Equivalent to RMB 1 m per year grinding cost.

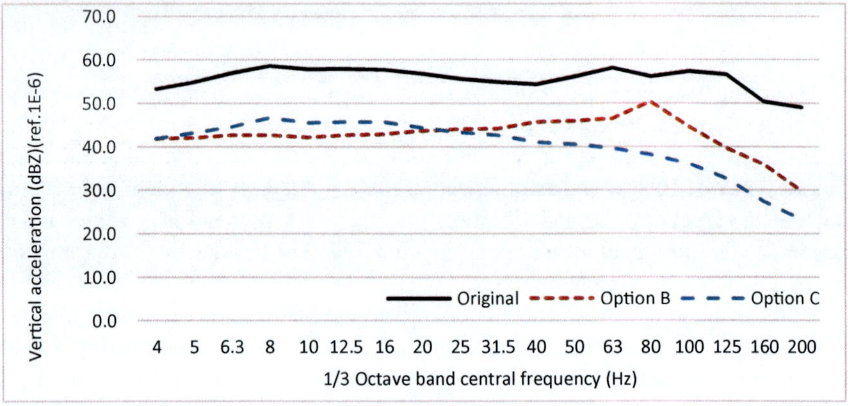

Fig. 4. Comparison of vibration levels at tunnel wall

typically include limited amount of material and labour costs only, and thus is independent of mileage. Total maintenance cost for option B and C, therefore, consists of mileage dependent biennial grinding cost, and mileage independent TRD- and fastenings-related maintenance costs.

3 Life-Cycle Cost Analysis

Life-Cycle Cost Analysis (LCCA)[2] is a tool to determine the most cost-effective option among different competing alternatives to, in this case, provide noise and vibration

[2] Life-cycle cost analysis in this particular study does NOT include environmental costs.

Table 2. Mileage independent annual maintenance cost profiles

Year	Fastening- and TRD-related maintenance cost (RMB)	
	Option B: type I fastening + TRD	Option C: type II fastening
0	0	0
1	180,000	180,000
2	180,000	180,000
3	180,000	180,000
4	180,000	240,000
5	240,000	240,000
6	240,000	240,000
7	360,000	270,000
8	390,000	270,000
9	420,000	300,000

mitigation measures, when each is equally appropriate to be implemented on technical grounds.

For each treatment option, intervention costs and associated maintenance costs are allocated to each year within the life-cycle, i.e. 60 years in this study. Discounted Cash Flow (DCF) analysis, based on concepts of time value of money, is used to give the present value of future costs (Eq. 1 below).

$$DCF = \frac{CF_1}{(1+r)^1} + \frac{CF_2}{(1+r)^2} + \frac{CF_3}{(1+r)^3} + \ldots + \frac{CF_n}{(1+r)^n} \quad (1)$$

Where CF_n is Cash Flow in each year; and r is the discount rate.

In this study, the discount rate is assumed to be 5%, based on which long-term cumulative costs, comprising intervention costs and maintenance costs were all computed and illustrated from Fig. 4, 5 and Fig. 6.

Compared to Option B and C, cumulative total cost covering 60 years life-cycle using option A is the most expensive means to achieve the required noise and vibration levels. Option B, using the combination of Type I Fastening and TRD, cost less in total from the 8th year onwards, compared to Option A. Option C is by far the most cost-effective means. Option B although is the most capital-intensive option compared to the other two, but is still overall a less expensive option comparison to Option A, due to the extensively higher maintenance cost of the latter.

Cost profiles below also allow a clear visualisation of cost intensive years, e.g. year 10 and year 30 of option B (Fig. 7).

A direct comparison of the total discounted cost of each treatment option is summarised in Table 3 below. In the long-run, Option C allows a massive saving of 42% compared to the current quarterly grinding scheme.

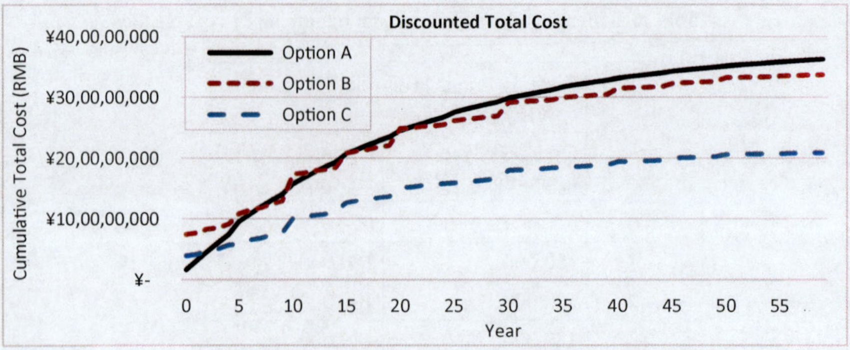

Fig. 5. Comparison of discounted total cost

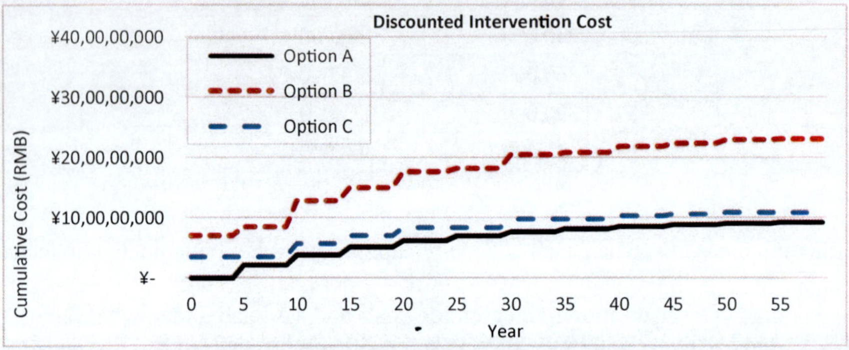

Fig. 6. Comparison of discounted intervention cost

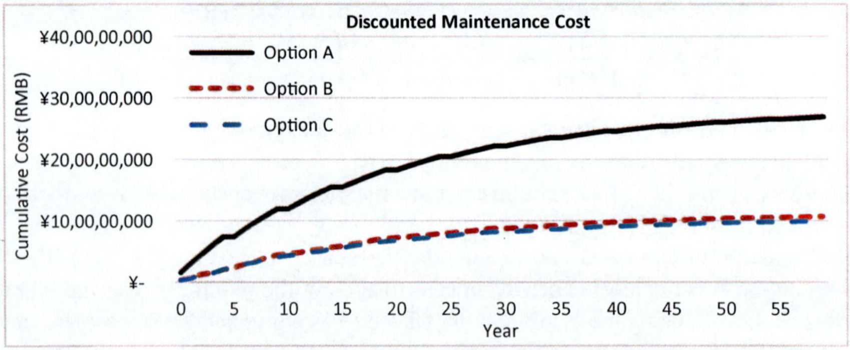

Fig. 7. Comparison of discounted maintenance cost

Table 3. Comparison of discounted total cost

	Option A	Option B	Option C
Savings compared to treatment option A	¥ 362,798,700	¥ 336,908,838	¥ 209,677,841
	NA	7%	42%

4 Conclusion

This case study applies the One-stop Noise and Vibration Assessment Method (ONVAM) of Shanghai University of Engineering Science (SUES) and Anhui Railway Technology Co. Ltd. (ART), to examine existing noise and vibration issues and to provide mitigation options to achieve required noise and vibration levels. Life-Cycle Cost Analysis was subsequently carried out for operators at S Metro Line 3, allowing a straight-forward comparison of applying different noise and vibration mitigation schemes and associated long-term financial implications.

Overall, the study shed insights for urban railway operators and relevant stakeholders to perceive the overall cost benefit by reducing long-term maintenance costs. Capital-intensive mitigation measures may not always be the seemingly worst option, as can be demonstrated from this study. Analysis outputs enable operators and property owners to envisage investment profile and cost-intensive years, to facilitate decision making.

References

1. Grassie, S., Kalousek, J.: Rail corrugation: characteristics, causes and treatments. Proc. Inst. Mech. Eng. Part F J. Rail Rapid Transit **207**(1), 57–68 (1993)
2. Noise limit and measurement for train of urban rail transit. GB 14892-2006, Standard fixed Institute of the Ministry of Construction PR China (2006)
3. Rail Measurement. https://www.railway-technology.com. Accessed 13 Nov 2018
4. Railway applications — Noise emission — Characterization of the dynamic properties of track selections for pass by noise measurements. BS EN 15461:2008+A1:2010, BSI (2008)
5. Wang, A., Wang, Z., Zhao Z., Zhang, Y., Duan, Y.: Effects of track stiffness and tuned rail damper on rail roughness growth and rail vibration levels on metro system. In: Nielsen, J.C.O., et al. (eds.) Noise and Vibration Mitigation for Rail Transportation Systems. Notes on Numerical Fluid Mechanics and Multidisciplinary Design, vol. 126, pp. 667–674. Springer, Heidelberg (2015)

Predictions, Measurements, Monitoring and Modelling

Application to Real Cases of a Methodology to Evaluate the Uncertainty of Train Exterior Noise Predictions

Eduardo Latorre Iglesias[1(✉)], A. L. Gomes Neves[3], A. Bistagnino[2], and Joan Sapena[2]

[1] Universidad Politecnica de Madrid, ETSIST Campus Sur Universidad Politecnica de Madrid, Crtra de Valencia, km 7, 28031 Madrid, Spain
eduardo.latorre.iglesias@upm.es
[2] Alstom, 48 Rue Albert Dhalenne, 93400 Saint-Ouen, France
[3] Ecole Nationale Supérieure d'Ingénieurs du Mans (ENSIM), 1 Rue Aristote, 72000 Le Mans, France

Abstract. Noise simulations are performed by train manufacturers during the development of new rolling stock. They are used to evaluate compliance with the customer's expectations and requirements demanded by train operators based on existing regulations. The success of the decisions taken during the project lifetime will depend on the accuracy of these simulations. Hence a methodology to assess the simulation uncertainty will be desirable to provide a robust framework for future virtual certification of new trains. Reducing the simulation uncertainty allows reducing risks and defining a more precise design-to-cost strategy; in addition a good understanding of the dependency of the uncertainty of the final predicted results on the uncertainty of the different input parameters will allow defining where to put the efforts to improve the prediction accuracy.

The authors have recently published a methodology to assess the uncertainty of exterior noise predictions. This methodology is applied in the present work to assess the uncertainty of the pass-by noise of a suburban rolling stock calculated using Alstom's tool SITARE according to the test conditions given in the standard ISO3095:2013. A parametric analysis is carried out showing the influence of the uncertainty of the different model inputs (sound power level, directivity, installation effects and ground acoustic resistivity) on the uncertainty of the predicted pass-by noise. Based on this analysis suggestions are given on how to reduce the uncertainty in simulations of this kind.

Keywords: Uncertainty · Simulation · Railway noise

1 Introduction

In the EU-funded ACOUTRAIN project, guidance was provided for the estimation of railway noise calculation uncertainty [1]. In addition, some recommendations for assessing the uncertainty of an exterior noise virtual testing procedure are given in [2].

G. Degrande et al. (Eds.): *Noise and Vibration Mitigation for Rail Transportation Systems*, NNFM 150, pp. 225–233, 2021.
https://doi.org/10.1007/978-3-030-70289-2_22

However, it is difficult to find in the literature a specific proposal of a framework for assessing the uncertainty of railway exterior noise calculations. In this regard, Latorre Iglesias et al. [3] have recently proposed a methodology for assessing railway noise simulation uncertainty, establishing the theoretical basis of a framework that could be used for train virtual certification.

In the present work, this framework is applied to assess the uncertainty of pass-by exterior noise calculations using the Alstom's internal tool SITARE [4] for a typical suburban rolling stock at constant speed of 80 km/h (hereafter called reference case). The standard uncertainty -including the Probability Density Function (PDF)- is defined for each model input based on measurements and guidelines given in international standards and literature, or arbitrary values are chosen when not enough data are available. The uncertainty for the overall pass-by noise is obtained and a parametric study is carried out to evaluate the impact on the uncertainty of the final result of the accuracy of the different inputs to the model: Sound Power Level (PWL), directivity, installation effects and ground properties. Based on the results some guidelines are given to improve the accuracy of this kind of calculations.

Section 2 presents SITARE's calculation model; the description of the framework for uncertainty assessment and the chosen input uncertainties is given in Sect. 3. Section 4 shows the results obtained for the reference case, Sect. 5 includes a parametric study of the final uncertainty calculation and conclusions are given in Sect. 6.

2 Prediction Model

Alstom's tool SITARE is used for noise prediction of a train during pass-by, at standstill conditions and during acceleration. SITARE is based on analytical models to predict the propagation of the noise from source to receiver. The different noise sources are defined in terms of the input PWL, source directivity, source geometry and position in the train. The noise from each source to one or several receivers is calculated considering the effects of installing the source in the train (diffraction, screening…) and the influence of the topography and ground properties. Figure 1 shows a sketch of the modeling in SITARE and the different transmission paths from the noise sources to the receiver. An input PWL L_W is defined for each noise source. In the case of equipment this can be estimated from laboratory noise measurements following the relevant ISO standards, or assessed using TWINS [5] in the case of the rolling noise. A source model must be chosen to represent the real source in the prediction model. These models can be as simple as a point source (monopole, dipole, combination of both…) or more complex such as a box source with a specific PWL assigned to each box surface. The source model will determine the directivity factor of the noise source, Q. The geometry of the source and its position in the train are also defined. The sources can be placed on the roof, underframe or in the bogie cavity. The relative position between the source and the diffraction edge of the train (edge between the train roof and sidewall, top of a roof barrier or bottom of a bogie skirt…) influences the correction applied to the transmitted noise due to diffraction. This correction depends on the difference between direct and diffracted path δ. Finally, the ground properties (i.e. ground resistivity R_g) and topography are modelled. The analytical models included in SITARE define the functional relation between the sound pressure level (SPL) produced by one source j at one receiver i, L_{PiSj}, and the different input parameters as follows:

$$L_{p_i s_j} = f\left(L_{w_j}, Q_j, \delta_{i,j}, R_g\right) \quad (1)$$

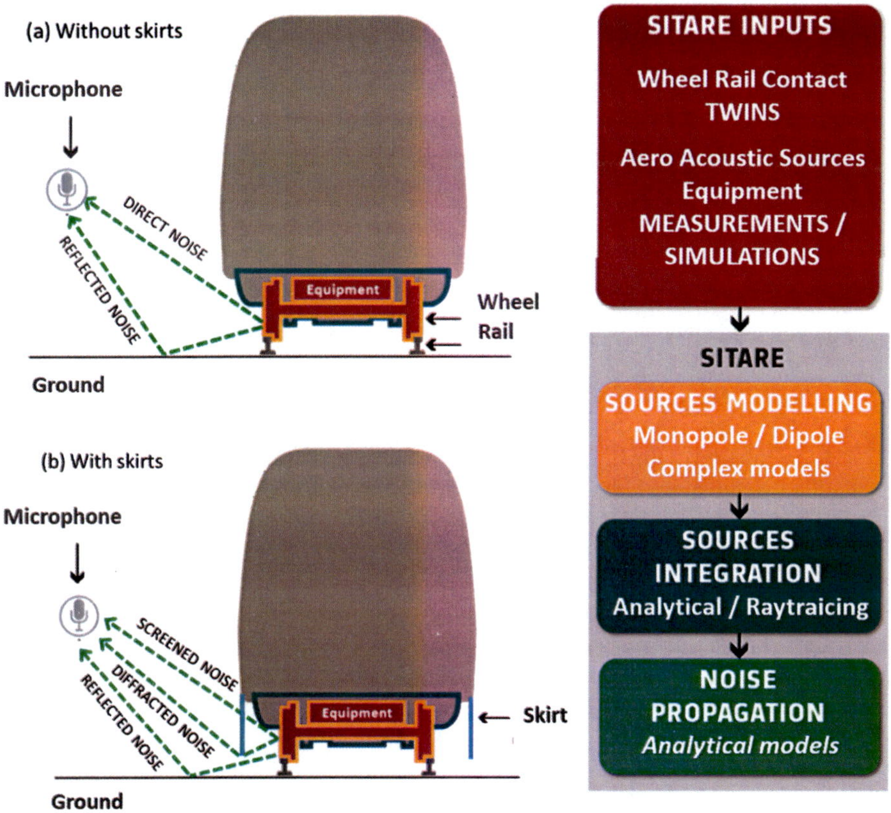

Fig. 1. Procedure applied for train exterior noise predictions using SITARE. An example is given of noise propagation from a source in the bogie cavity. (a) Without skirts. (b) With skirts.

3 Framework for Uncertainty Assessment

3.1 Methodology for Uncertainty Assessment

In the present study the uncertainty due to the chosen prediction model has been neglected. For simplicity, sources of different type are considered as statistically uncorrelated and sources of the same type are considered statistically correlated although in practice, these could be partially correlated due to installation and directivity effects in the noise propagation.

The methodology applied in the present work – described in detail in [3] – assumes that the uncertainty of the predicted parameter y, $u_c(y)$, can be assessed as the combined standard uncertainty of the input parameters x_k, $u(x_k)$. According with the guide to the expression of uncertainty in measurements (GUM) [6], if the input quantities are uncorrelated $u_c(y)$ is obtained by combining the standard uncertainties of the input estimates, $u(x_k)$ as shows in Eq. (2), while if the input quantities are correlated Eq. (3) shall be applied. The sensitivity coefficients $\partial f/\partial x_k$ define how the combine uncertainty $u_c(y)$ varies with the uncertainties of the different input parameters $u(x_k)$, where f is the functional relation given in Eq. (1).

$$u_c^2(y) = \sum\nolimits_{k=1}^{N} \left(\frac{\partial f}{\partial x_k}\right)^2 u^2(x_k) \tag{2}$$

$$u_c(y) = \sum\nolimits_{k=1}^{N} \left(\frac{\partial f}{\partial x_k}\right) u(x_k) \tag{3}$$

Three different steps can be identified in the proposed methodology:

1. For a given receiver i and source j the uncertainty of the predicted L_{PiSj}, $u(L_{PiSj})$, can be obtained using Eq. (2) and the input uncertainties defined in Table 1. The functional relation f is defined by the propagation and diffraction models used in SITARE.
2. The calculation of the $L_{pAeq,Tp}$, as defined in the ISO3095:2013, can be approximated as the averaged of all the samples recorded during the train pass-by using a sample interval $\Delta t = 125$ ms. For simplicity in the simulation, this is equivalent to consider the train at standstill and a number of microphones $N = L_{train}/(\Delta L)$ along the train length L_{train}, with a distance between the microphones $\Delta L = \Delta t \times V_{train}$, where V_{train} is the train speed. Therefore, the pass-by noise due to a single source j is given by the parameter $L_{pAeq,TP,j}$, defined as follows:

$$L_{pAeq,T_p,j} = 10log_{10}\left(1/N \sum\nolimits_{i=1}^{N} 10^{(Lpi,j/10)}\right) \tag{4}$$

The pass-by noise uncertainty for a single source j, $u(L_{pAeq,Tp,j})$, is given by Eq. (5), obtained by substituting in Eq. (3) the uncertainty of the predicted SPL at each point i due to source j, $u(L_{PiSj})$, which are assumed to be statistically correlated, and the sensitivity coefficients that account for the contribution of the noise at a given point i to the overall pass-by noise due to source j.

$$u\left(L_{pAeq,Tp,j}\right) = \sum\nolimits_{i=1}^{N} \left(10^{\frac{L_{PiSj}}{10}} / \sum\nolimits_{k=1}^{N} 10^{\frac{L_{PkSj}}{10}}\right) u\left(L_{PiSj}\right) \tag{5}$$

3. To obtain the pass-by noise uncertainty considering all the noise sources, $u(L_{pAeq,Tp})$, Eq. (2) is applied for noise sources of different types and Eq. (3) for those of the same type. The sensitivity coefficients are given by the contribution of each source j to the overall noise. Equation (6) gives an example of application for noise sources of different type, where S is the total number of sources.

$$u^2\left(L_{pAeq,Tp}\right) = \sum\nolimits_{j=1}^{S} \left(10^{\frac{L_{pAeq,Tp,j}}{10}} / \sum\nolimits_{k=1}^{S} 10^{\frac{L_{pAeq,Tp,k}}{10}}\right)^2 u^2\left(L_{pAeq,Tp,j}\right) \tag{6}$$

3.2 Definition of the Uncertainties

Table 1 shows the input uncertainties used for the reference calculation. The PWL of the equipment was measured according the standard ISO9614-2. In this case a standard uncertainty of 1.5 dB with a Gaussian distribution can be considered. The distribution of the rest of the input uncertainties is unknown. The estimate of the standard uncertainty is obtained by defining an upper and lower limit of the possible values of the input quantities x_k and considering a rectangular distribution. In the case of the rail and wheel PWL per unit roughness predicted using TWINS, the uncertainty was defined arbitrarily supposing accurate calculations for the wheel (i.e. correlated FEM models for the calculation of the modal basis) and for the rail (measured track decay rate, vibration response…). The standard uncertainty of the measured rail and wheel roughness was calculated using the input uncertainties suggested by the ISO standard prEN 15610:2017. It is recommended to carry out a dedicated uncertainty calculation when roughness measurements are performed. The uncertainty in the directivity factor Q (dependent on the precision of the source model) was defined arbitrarily considering a box source model with a given PWL per face measured as defined in the ISO9614-2. The uncertainty of the difference between direct and diffracted path δ is defined as function of the difference between the acoustic centre of the real source and that of the source model, defined in this case as 0.1 m. An uncertainty of 100 rayls/m for the ground flow resistivity was assumed.

Table 1. Definition of the input uncertainty used for the reference case prediction. W/R stands for the PWL normalized per unit roughness.

Parameter	Standard uncertainty, $u(x_k)$				PDF
	100–600 Hz	600–2k Hz	2k–5k Hz	Overall	
Equipment PWL, dB	1.5	1.5	1.5	*	Gaussian
Rail roughness, dB	3.6	3.6	3.6	*	Rectangular
Wheel roughness, dB	2.5	2.5	2.5	*	Rectangular
Rail W/R, dB re10^{-6}m	2.5	1.5	1.5	*	Rectangular
Wheel W/R, dB re10^{-6}m	1	1	1	*	Rectangular
Directivity factor Q, dB	1	1.5	2	*	Rectangular
Source geometry, m	*	*	*	0.1	Rectangular
Ground flow resistivity R_g, rayls/m	*	*	*	100	Rectangular

4 Results for the Reference Case

The rolling stock consist of 5 cars with length 18 m each and architecture M-M-T-M-M (80% of motorization), where M stands for motor and T for trailer car. This configuration gives a total of 16 motors and gear boxes for the whole train. The motor used is self-ventilated. Only motor, gear box and rolling noise were considered as they are normally the main contributors for pass-by noise at typical maximum metro speeds. The wheel used is monobloc with diameter 840 mm. The rail and wheel roughness and track decay rate (TDR) were measured during the test campaign. The rail pad stiffness was obtained by fitting the predicted TDR using TWINS to that measured, which might be an additional source of inaccuracy. Typical values of grassy ground are used for the ground flow resistivity. The calculation of the pass-by noise was made according to the guidelines given in the ISO3095:2013 for a microphone placed at 7.5 m from the centre of the track and a height of 1.2 m from the top of the rail.

A difference of 0.5 dB between the measured and predicted pass-by noise was obtained showing a very good correlation without applying any tuning of the inputs to the model. The uncertainty of the pass-by noise due to each noise source was calculated: 2.2 dB for the equipment, 3.9 dB for the wheel and 4.0 dB for the track, which gives an uncertainty of the overall pass-by noise of 1.9 dBA. A standard uncertainty of 1.7 dB was obtained during the pass-by measurements using the input uncertainties proposed in the Annex G of the standard ISO3095:2013 for an expanded uncertainty with a coverage factor $k = 2$, that is equal to two standard deviations.

5 Parametric Study

5.1 Uncertainty of the Rolling Noise PWL Due to Input Roughness

In TWINS calculations, the total roughness Rg_t is obtained as the sum in decibels of the rail Rg_r and wheel roughness Rg_w. Assuming that the uncertainty in the measurement of rail, $u(Rg_r)$, and wheel roughness, $u(Rg_w)$, are statistically uncorrelated the uncertainty of the total roughness, $u(Rg_t)$, can be obtained by applying Eq. (2) as follows:

$$u(Rg_t) = sqrt\left(\left(\frac{10^{Rg_w/10}}{10^{Rg_t/10}}\right)^2 u(Rg_w)^2 + \left(\frac{10^{Rg_r/10}}{10^{Rg_t/10}}\right)^2 u(Rg_r)^2\right) \quad (7)$$

In the reference case the measured rail roughness was significantly larger than the wheel roughness, so $Rg_r \approx Rg_t$. From Eq. (7) it can be inferred that $u(Rg_t) \approx u(Rg_r)$.

To assess the influence of the rail and wheel roughness on the final uncertainty, a measurement uncertainty of 1.5 dB was chosen assuming a high precision of the measured data. The results show an uncertainty of 2.4 dB of the pass-by noise calculation due to the wheels and 2.6 dB due to the track, which are significantly lower than those given in Sect. 4 for the reference case. The uncertainty of the overall pass-by noise is 1.6 dB, 0.3 dB lower than that of the reference case.

5.2 Uncertainty in the Source Directivity

Table 2 shows the variation of the uncertainty of the predicted pass-by noise for different input uncertainties $u(Q)$ of the directivity ratio Q. Different values of $u(Q)$ are proposed to consider different degrees of accuracy of the input Q. These values were chosen arbitrary due to the lack of measured data. The effect of the uncertainty in the input directivity ratio Q on the uncertainty of the result is moderate with an increase of 0.3 dB between the low and high input uncertainty $u(Q)$. The input uncertainties are assumed to be totally uncorrelated so the overall uncertainty can be obtained applying Eq. (2). As a result, the overall uncertainty is lower than the input uncertainties.

Table 2. Variation of the pass-by noise calculation uncertainty with $u(Q)$.

Accuracy	$u(Q)$, dB			$u(LpAeq,Tp)$, dB			
	100–600 Hz	600–2k Hz	2k–5k Hz	Equipment	Wheel	Track	Overall
Low	1.0	1.5	2.0	2.2	2.4	2.6	1.6
Middle	2.5	3.5	4.5	2.3	2.7	2.8	1.7
High	4.5	5.5	6.5	2.7	3.0	3.2	1.9

5.3 Uncertainty in the Source Integration

For simplicity, the assessment of the influence of the source model geometrical accuracy on the final uncertainty is carried only for equipment. Different lengths of the bogie skirts (this is shown in Fig. 1(b)) were used to assess the dependency on δ of the pass-by noise uncertainty. The uncertainty in the model geometry, $u(s_d)$, is defined as the difference between the position of the real source and the source model acoustic centres in the y-direction, $u(s_y)$, and in the z-direction, $u(s_z)$. It is difficult to define this parameter precisely so illustrative values are provided. This uncertainty is given by:

$$u^2(s_d) = \left(\frac{y}{\sqrt{y^2+z^2}}\right)^2 u(s_y)^2 + \left(\frac{z}{\sqrt{y^2+z^2}}\right)^2 u(s_Z)^2 \tag{8}$$

where y is the sideway coordinate of the source acoustic centre and z is its height. The noise reduction due to the screening effect and noise diffraction is calculated using Maekawa's model and it is a function of δ defined as:

$$\delta = r_{dif} - r_d \tag{9}$$

where $r_{\rm dif}$ is the length of the diffracted ray and $r_{\rm d}$ is the length of the direct ray. The uncertainty of the correction applied to account for the screening effect is proportional to δ and $u(s_{\rm d})$. Table 3 shows that the uncertainty of the predicted pass-by noise due to the equipment increases with the skirt height due to the increase of δ. Regarding the uncertainty in the source model geometry definition two different values are chosen: 0.1 m for detailed model (e.g. box model) and 0.3 if a simple source model -such as a point source- is chosen. Table 3 shows the resulting pass-by noise uncertainty using both values of the uncertainty in the model geometry. Although the results show a negligible impact on the overall pass-by noise uncertainty, this might be larger for a train at standstill where the point of maximum noise is evaluated. The results show that, for the same values of the input uncertainties, the overall uncertainty increases with δ. The value of δ depends on the source model geometry definition so it is important to be define it accurately to reduce the overall uncertainty.

Table 3. Variation of the pass-by noise calculation uncertainty with the skirt height and input uncertainty of the source model geometry.

		$u(L_{pAeq,Tp})$, dB							
		$u(s_{y,z}) = 0.1$ m				$u(s_{y,z}) = 0.3$ m			
Skirt height, m	δ, m	Equip.	Wheel	Track	Overall	Equip.	Wheel	Track	Overall
0.5	0.03	2.3	2.4	2.6	1.6	2.4	2.4	2.6	1.7
0.7	0.1	2.5	2.5	2.6	1.7	2.9	2.7	2.6	1.8
0.8	0.15	2.8	2.7	2.6	1.8	3.1	3.0	2.6	2.0

5.4 Uncertainty in the Ground Effect

Finally, the effect of the uncertainty in the ground flow resistivity, R_g, is studied. R_g might vary for porous grounds due to the user choice from 150 krayls/m for a grassy ground to 600 krayls/m for scrub. An uncertainty of 100 krayls/m was assumed in the reference case and lawn was considered as ground (typically with $R_g = 300$ krayls/m). The variation of the overall pass-by noise uncertainty in this case was 0.2 dB. This shows that, for soft grounds, there is no need to be precise in the ground resistivity definition. In the case of hard surfaces such as asphalt (typically with $R_g = 20000$ krayls/m) when R_g is unknown an uncertainty of 5000 krayls/m is considered. This gives a final uncertainty increase of 0.7 dB, which is not negligible.

6 Conclusions

A methodology for uncertainty assessment of train exterior noise calculations developed by the authors is applied in this work to assess the uncertainty of the exterior pass-by noise prediction for a typical suburban rolling stock at 80 km/h using Alstom's tool SITARE. Based on the methodology given in the GUM, the pass-by noise uncertainty is obtained for sources of different type: 2.2 dB for equipment, 3.9 dB for wheels and 4.0 dB for track. An uncertainty of 1.9 dB is obtained for the overall pass-by noise assuming that the previous sources are statistically uncorrelated. From the parametric study performed to understand the influence of the model input uncertainties in the final result the following conclusions are achieved:

- The uncertainty of the rolling noise depends largely on the uncertainty of the predominant roughness. It is recommended to perform accurate measurements or to use databases of tracks similar as the one previewed for the train certification tests.
- The uncertainty in the directivity of the chosen source model has not a significant impact on the pass-by noise uncertainty; however, it might not be the case for simulations of maximum noise at standstill.
- The accuracy on the source geometry definition becomes more important when the difference between diffracted and direct ray path increases. This is the case if skirts or barriers are used or if the microphone is placed close to the noise source.
- For grassy ground (low flow resistivity) the effect of the uncertainty on the definition of the ground resistivity on the final uncertainty is almost negligible. This is not the case for surfaces with large flow resistivity such as asphalt or concrete.

Future work is suggested in assessing the PDF of the final result using Monte Carlo simulations and obtaining larger samples of input data. The uncertainty due to the chosen prediction model could also be considered. In addition, the proposed methodology could be applied for a train at standstill and during acceleration and braking. Results obtained could be used as a starting point for virtual certification/acceptance of rolling stock by comparing noise targets with simulations plus its uncertainty.

References

1. Furio, N., Starnberg, M., Bongini, E., Thompson, D., Orrenius, U., Cuny, N.: ACOUTRAIN: virtual certification of acoustic performance for freight and passenger trains. Energy Environ. **1**, 491–500 (2016)
2. Bongini, E., Cordero, R., Bistagnino, A., Squicciarini, G.: Uncertainty and variability in acoustic virtual certification process of railway rolling stock. In: Proceedings of 21st ICSV, Beijing, China (2014)
3. Latorre Iglesias, E., Xia, J., Farroq, M.E., Bistagnino, A., Sapena, J.: Methodology to evaluate the uncertainty of train exterior noise prediction. Proc. Inst. Mech. E Part C J. Mech. Eng. Sci. (2019). https://doi.org/10.1177/0954406219853855
4. Gambard, N., Sapena, J., Planeau, V.: A methodology for exterior noise prediction of railways rolling stock. In: Proceeding of Euronoise, Edinburgh, United Kingdom (2009)
5. Track Wheel Interaction Noise Software (TWINS) User Manual (1996)
6. JCGM 100:2008: GUM 1995 with minor corrections. Evaluation of measurement data – Guide to the expression of uncertainty in measurement. International Organization for Standardization, Geneva, Switzerland (2008)

On Measurement of Railway Noise: Usability of Acoustic Camera

Günter Dinhobl[1](✉), Martin Petz[2], Sebastian Kümmritz[3], and Helmut Hutterer[4]

[1] ÖBB-Infrastruktur AG, Stab UE/Team Research and Development, 1020 Wien, Austria
guenter.dinhobl@oebb.at
[2] gfai tech GmbH, Branch Office Austria, 1220 Wien, Austria
[3] gfai tech GmbH, 12489 Berlin, Germany
[4] ÖBB-Produktion GmbH, 1100 Wien, Austria

Abstract. The paper gives an insight of the measurement campaign within the project 'LowNoiseTrain2', which took place in Austria in 2018. The acoustic measurement of this freight train was done with an acoustic camera and the results to identify hot-spots and best-practice design are promising: The results show the structure of railway noise emission and the location (height) of the noisiest areas for different wagon types depending on the frequency (range). For instance a promising design is to shield the wheels with the frame of the wagon, at least the upper wheel area; this results in a reduction of the noise of the wagon itself, but there are indications that more reflections occur from the surrounding area.

Keywords: Rolling noise · Acoustic camera · Track-vehicle interaction

1 Introduction

The European Commission identified the noise of railways as its environmental 'Achilles heel' [1]. Since then, theoretical and practical efforts aim to decrease railway noise. Theoretical, to understand the mechanism of generating railway noise [2], and practical, to get less noisy rail traffic [3]. Calculations and simulations predict e.g. acoustic modes of rail vehicle wheels or the spread of noise including a visualization on geographic maps. New materials and wagon layouts give the opportunity to reduce the pass-by sound level of new and existing rail vehicles. So far, the knowledge seems to be complete. However, in the future rail traffic noise will still be in focus: recommendations of WHO for decreasing values of noise indices might result in lower limit values for noise mapping. In addition, open questions between the theoretical approach and 'real' measurements, when rail-wheel-noise separation is measured, illustrate our incomplete understanding of railway noise quite well: "none of the methods were able to determine the wheel component in the current tests to sufficient accuracy" [4].

In 2018, ÖBB-Infrastruktur AG started a measurement campaign of systematic identification regarding noise and noise sources of freight wagons in Austria with a special test freight train (LNT2 – LowNoiseTrain2, see Fig. 1). Several test runs were done

G. Degrande et al. (Eds.): *Noise and Vibration Mitigation for Rail Transportation Systems*, NNFM 150, pp. 234–241, 2021.
https://doi.org/10.1007/978-3-030-70289-2_23

and the acoustic outcome was measured with different systems. Beside the conventional pass-by noise measurement, an acoustic camera with algorithms for the mapping of moving sources was used to visualize the sound sources. For the usage of an acoustic camera for a moving train, several essential aspects need to be considered: [5] gives hints on the dependence of the method and source scenario; [6] describes the importance to consider MSL (Maximum Side Lobe) in beamforming and [7] specifies the effects of Doppler shift on measurements of moving vehicles.

This paper presents the first results of the measurement campaign.

Fig. 1. Overall view of the test freight train 'LNT2' nearby the measurement site

2 Methodology

The test train consisted of 15 freight wagons of different types: flat wagons (Rns-z and Laaprs), container wagons (Sdggmrs), bulk freight wagons (Talns), tank wagons (Zans), high-sided open goods wagons (Eanos) and box wagons (Hbbills) - both loaded and unloaded. In addition, a wagon for electrical, magnetic and locomotive dynamics measurements was part of this test train. While most of the wagons were equipped with K-brake blocks, four wagons (Eanos and Hbbills) at one end of the train had conventional cast iron (CI) brake blocks. All wagons are part of the wagon pool of freight operators and were not specifically prepared. None of these wagons had wheels with significant out-of-roundness defects like wheel flat (Δr better than 0.1 mm), which was measured by an Argos® wayside-train-monitoring site.

The acoustic measurement campaign of the test train took place on a double track main line in the western part of Austria with less noise pollution and plane area alongside the tracks. Rail roughness and track decay rate were of good quality. The conventional pass-by noise measurement was at the position 'A' microphone position of EN3095 (7.5 / 1.2m) with a sound analyser Norsonic 'NOR-140'. The sound measurements with an acoustic camera of gfai tech GmbH were done at two distances from the track: at 18.6 m and at 14.6 m. A spiral array 'FlexStar' with 120 microphones enabled frequency dependent sound source localisation. The data recorder type 'mcdRec-721' collected all the data, which was analysed by the software NoiseImage 4.10.1. The test train passed the measurement site 24 times: on both tracks, in both directions and with different speeds (80 and 100 km/h). Sound from each pass-by run was recorded with a sampling frequency of 96 kHz. This result in raw data of 44 MB each second of a pass-by run, which is the basis of the post-processing analysis.

The analysis included the following steps: (1) the conventional measurement with a single microphone located at the EN3095-position 'A' gave a first and quick insight of repeatability and dependencies of speed, driving directions and chosen tracks; (2) the same was done with the data of the acoustic camera for each pass-by run. The result was the selection of a ‘typical’ test run for detailed analysis; (3) detailed analysis of (a) whole train, (b) individual wagons and (c) parts of wagons to identify general characteristics, frequency dependencies and level locations including split up of sources.

3 Results

The conventional pass-by noise measurement illustrates that the reproducibility is quite good (see Fig. 2). The graphs show that the running direction has very little effect. The biggest differences of the noise level are at the locomotive ‘Taurus’ (left) which might be caused by differences in traction power. The peak at the Zans wagon might be due to a selective out-of-roundness defect of one wheel with a size of $\Delta r = 0.1$ mm. However, the running line of the wheel on the rail of this single test run was just across the wheel defect and at the other test runs this acoustic peak was not measured.

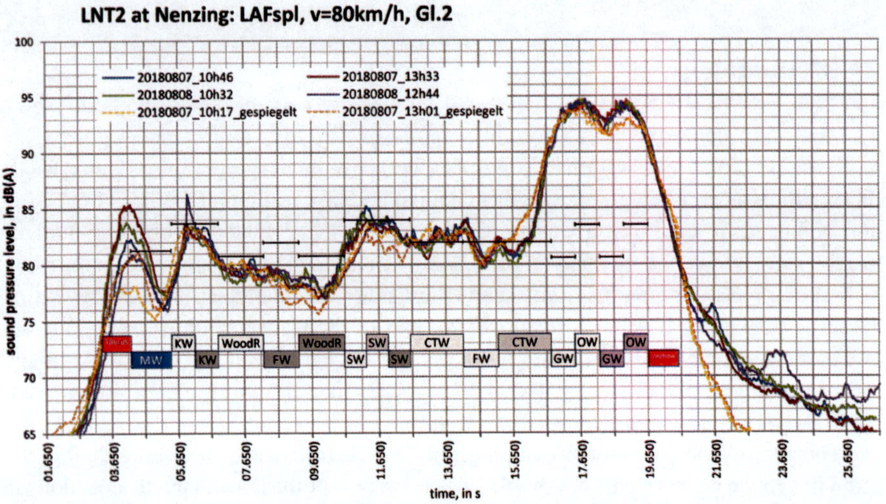

Fig. 2. Comparison of all LNT2 test runs with a speed of 80 km/h with conventional pass-by noise measurements at the microphone position ‘A’ of EN3095 (7.5/1.2 m). The lines with ‘gespiegelt’ show the test runs in the opposite direction. The short black lines represent the NOI TSI limit values of each wagon type. The CI-wagons are in the right area with around 93 dB(A).

All of the following analyses are based on the essential attribute of the acoustic camera, which uses beamforming as method to visualize sound sources. The assumption is that the emitted sound comes from (stationary) point sources. As this is not the case for the given set-up (moving: train, semi-linear: rail), the interpretation of the results becomes more complex. Due to this complexity, an automated analysis or interpretation for the measured data of the acoustic camera is not possible. All charts, images or videos have to be checked individually with a sensitivity of noise effects in combination with expertise in rail vehicle dynamics. While the data processing and analysing were done by a small team, the interpretation needed the experience of measurement experts like (railway) noise experts or technicians for freight wagons and running behaviour. The result indicates numerous noise issues which are presented in detail, e.g., general noise effects like directionality of pass-by noise of rail vehicles, individual effects of a certain wagon-type, of bogies, vehicle frames and/or wagon body depending on frequency range.

The acoustic camera generates conventional charts such as channel data graph and spectrograms (see Fig. 3). Nevertheless, of most interest are the videos with the passing train and the visualization of the sound pressure level, which can be switched between A-weighted, un-weighted, and individual frequency bands. In addition, during the analysis, a new tool enables the visualization of the complete train-pass-by in one image (see Fig. 4) [8]. This kind of visualization gives a quick overview of the sound pressure level of the complete passing train at a glance. In contrast, Fig. 5 shows the sound location of the complete train pass-by time period: it shows that the noisiest direction is not in the right angle to the track axis, but some degrees left and right of it. This is because of both the flexural wave of the line source 'rail' and -presumably- the appearance of geometric masking effects of the two wheels of one wheelset. In the case of the rail, this angle depends on the frequency – this rail sound emission was detected before the leading wheels passed the measurement point (unfortunately, there is no situation when wheel sound emissions occur without rail sound emissions).

The next stage of analysis was to generate an acoustic image of each wagon (see Fig. 6). Shown is a sequence of 120 ms in the middle of the wagon of a pass-by with a speed of 100 km/h – this implies that within this time frame the wagon moves forward for 3.3 m (that is why the image is representative and not accurate). The range of the noise source plots is 2.5 dB from the highest, noisiest direction of the chosen time frame. The extraordinary peak at 1300 Hz FFT-depiction and 1250 Hz-third-octave band is the result of a 0.1 mm wheel flat at one single wheel. It must be stated that at all frequencies the highest level is located either at the axle center height, the ballast embankment or the flat area in the foreground. We did not expect this result because there is a general expectation that an essential share of noise comes from the tank above the undercarriage.

By analysing the design of different wagons, 'good-practice' construction features are also identified: for example, the container wagon Sdggmrs has three bogies and the longitudinal frame covers two of them in the upper half of the wheel. This results in a lower noise peak of nearly 2.5 dB (see Fig. 7).

The same situation – a longitudinal frame, which hides large parts of the wheel – is given at the Laaprs wagon ('woodrailer'). The CLEAN-SC-tool (CLEAN Based on Spatial Source Coherence) enables to achieve a higher dynamic than conventional beamforming. With CLEAN-SC, a more detailed identification of the location of sources at a given frequency(-range) is possible. The example given shows that the source can also be an imaginary source (source on ballast), while the emitting, 'real' source is either covered by frame (wheel) or within the marking (rail). In addition, when looking for a precise location, it is essential to include the speed of the vehicle in the analysis (see Fig. 8).

During the runs of LNT2 conventional trains also passed by, and one of them was a block train. This train consists of 40 unloaded wagons of the same type (high-sided

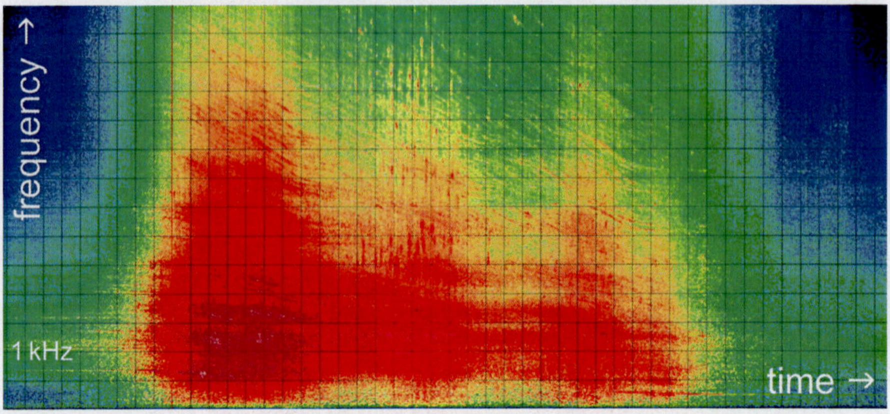

Fig. 3. Conventional spectrogram graph 0–7000 Hz of a train-pass-by acoustic measurement: channel data graph of LNT2 test run with a speed of 80 km/h. The noisy wagons with CI-brake blocks are on the left (pink color).

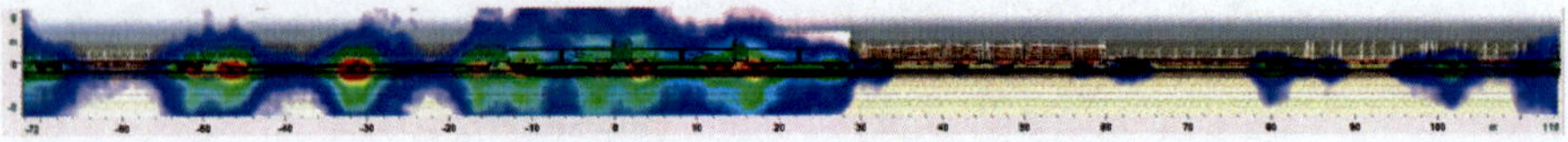

Fig. 4. Selection of the quieter wagons of LNT2 with K-brake blocks and the visualization of the A-weighted sound pressure level. While some wagons are very quiet, nearly no colored areas can be identified. The maximum difference between the highest and lowest level is 5 dB.

Fig. 5. Visualisation of the noise distribution of a pass-by for the complete test train LNT2 (pass-by from right to left): the localization of the A-weighted sound pressure level shows two eccentrically kidney-shaped peaks. The noisiest areas are pink coloured and the graphical spread up to the blue colored areas is 4 dB. The difference between the peaks and the middle area (right angle from track axis to acoustic camera) is about 2.6 dB (note that the image is a representative image of the middle of the train to visualize only the size and height of a single railway vehicle). The time frame of the image is 14.3 s.

open goods wagons 'Eaos'). All wagons were equipped with K-blocks, which enables a statistical analysis of a higher number of one wagon type. The wayside monitoring indicates that the 13th, 19th and 20th wagon had higher dynamic forces; the acoustic

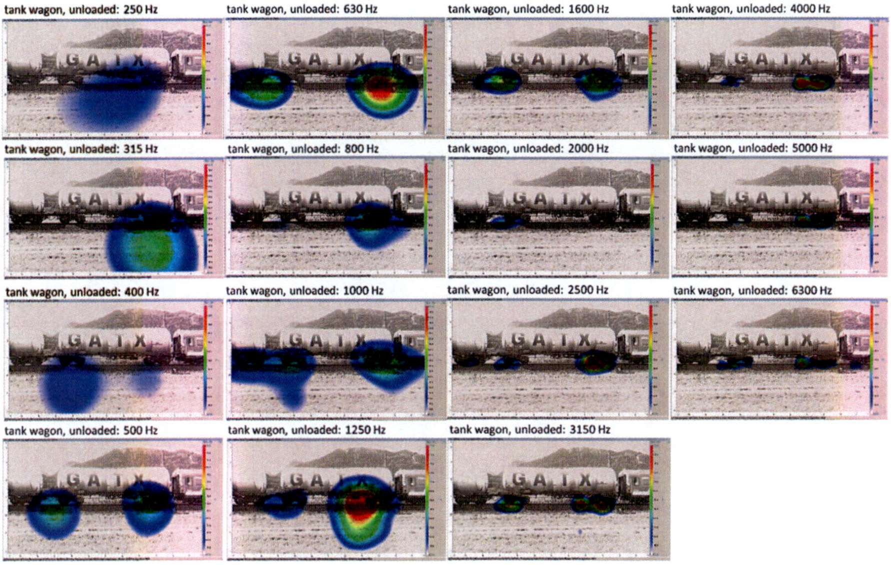

Fig. 6. Tank wagon and its location of noise sources in third-octave bands. The time frame of the images and the graph is 120 ms.

Fig. 7. Noise image of a container wagon (time frame 120 ms): the six-axle wagon (three bogies, two of them in the picture) shows the importance of construction details. While the bogie at the end of the wagon (left area) is partly covered by the body frame, the middle bogie (right area) is quite 'open' in terms of acoustics – and radiates around 2 dB more noise.

graph of the sound pressure level shows an increase in the first half of the train pass-by, probably because of an out-of-roundness effect like a wheel flat. That is why the sound analysis was limited to the second half of the train where no significant increase of the sound level occurs. The most important outcome of the analysis for our understanding of railway noise are:

Fig. 8. Illustration of the analysis-tool CLEAN-SC to identify and separate locations of single peaks of sources chosen from the FFT-chart at the wagon type Laaprs. The highest peaks of the frequencies 1148 to 1230Hz are at the ballast (see below the wheel, these might be reflections of the upper parts of the wheel which are behind the longitudinal frame of the wagon) and from the wheel tyre. Within the time frame of 40 ms the wagon moves around 1 m from right to left (diameter of wheel is about 1 m, image is from starting point of the movement within 40 ms).

- the center of the 315 Hz third-octave band is located at the height of the lower parts of the wagon body
- the center of the 500 Hz third-octave band is located at the height of the axle box
- the center of the 1250 Hz third-octave band is located at the height of the wheel center
- the center of the 2000 Hz third-octave band is located at the height of the lower part of the wheels.

In June 2019 another test train was put in operation, on the one hand flat wagons for container, equipped with disc brakes and K-brake blocks, and on the other hand, some wagon types of the LNT2-test train of 2018. The analysis of these data is on the way and sound reflections from the ballast (see [9]) and the grass will also be analysed.

Overall, the use of the acoustic camera shows that the measurement of several effects of good-practise-design of freight wagons is possible: a wagon frame, which covers at least the upper part of the wheel, result in a reduction of approx. 2 dB; a closed platform (container wagon platform is often open) will reduce the noise radiated from the wheels upwards. In addition, the measurement shows the frequency-dependent angle of noise radiation of the rail before the train passes by. Still an open point is the ballast and its share of radiation and reflection of noise, in particular when the frame covers the wheels; some small noise effects of the double suspension of freight wagons were also detected when the inner spring is unloaded.

4 Conclusion

The measurements of the LNT2-train show that the acoustic camera is useful to identify locations of loudest noise. Several wagon construction and design details are identified as acoustically effective, although it has to be considered that the beamforming method is based on singular point sources. In the case of railways, this is not always given: the rails are like a linear source while the wheels are something special like 'circular line sources' (this means that the circular sound source 'wheel' moves forward while running). Additionally, we expect in future to look more carefully to ground effects like reflections in certain (third-octave band) frequencies which are not typical for e.g. ballast which emits only in a certain (low) frequency range but is also a poor reflector. This indicates that the separation of track and vehicle, a separation of reflections and radiation from the ballast (and sleepers) is still an open point, which may be investigated in the next years.

References

1. European Commission, DG MOVE: Initiative and Roadmap 'Effective reduction of noise generated by rail freight wagons in the European Union'. Brussels, p.1 (2013)
2. Thompson, D.: Railway Noise and Vibration. Elsevier Science, Oxford (2008)
3. Practical effort: e.g. TSI NOI; END (environmental noise directive); legislation in Switzerland and Germany with a ban of Cast Iron Brake Blocks
4. Thompson, D., Wändell, J.: Roll2Rail Deliverable D7.4 - Assessment and validation of source separation methods, p. 105 (2017)
5. Sarradj, E., Herold G., Jekosch, S.: Array methods: Which one is the best? In: Proceedings of 7th Berlin beamforming Conference. Berlin (2018)
6. Ginn, B., Gomes, J., Hald, J.: Recent advances in Rail Vehicle Moving Source beamforming. In: Proceedings of Internoise 2013, paper No. 0443. Internoise, Innsbruck (2013).
7. Michel, U., Barsikow, B., Böhning, P., Hellmig, M.: Localisation of moving sound sources with phased microphone arrays. In: Internoise Proceedings 2004. Internoise, Prague (2004).
8. Meyer, A., Döbler, D.: Efficient methods for the analysis of moving sound sources with high channel microphone arrays. In: Proceedings of 5th Berlin beamforming Conference. Berlin (2014).
9. Sehu D., Wunderli, J.M., Heutschi, K.: SonRAIL-Projektdokumentation, p.42, Abb. 2.19. BAFU/EMPA, Bern (2010)

Acoustic Monitoring of Rail Faults in the German Railway Network

Astrid Pieringer[1,2(✉)], Matthias Stangl[1], Jörg Rothhämel[1], and Thorsten Tielkes[1]

[1] DB Systemtechnik GmbH, Department of Acoustics, Vibrations, Aerodynamics and HVAC (TT.TVE 34), Völckerstraße 5, Munich, Germany

[2] Chalmers University of Technology, Applied Acoustics, 41296 Gothenburg, Sweden

astrid.pieringer@chalmers.se

Abstract. The early detection of rail surface defects such as squats, poor welds, or wheel burns is important to prevent further rail deterioration. In this paper, a methodology for acoustic monitoring of squats in the German railway network is proposed based on the measurement of axle box acceleration (ABA) on the DB noise measurement car (SMW) and the previously developed numerical model WERAN for wheel/rail interaction. Specific characteristics of squats in the ABA signals are determined with the model and verified by pass-by measurements combined with direct geometry measurements of the squats. Based on these results, a logistic regression classifier is devised for the detection of squats in the measured ABA signals of the SMW. Trained with simulated and measured data, the classifier identifies all of the known severe squats and 87% of the known light squats in the measured test data.

Keywords: Acoustic monitoring · Squats · Axle box acceleration · Wheel/rail interaction · Time-domain modelling · Noise measurement car · Machine learning

1 Introduction

The early detection of rail surface defects such as squats, poor welds, or wheel burns is important to prevent further rail deterioration caused by large impact forces. Surface defects can be detected and measured optically by dedicated measurement trains, by ultrasound measurements or by visual inspection. These procedures give information about the geometry of the defect but cannot directly assess the severity in terms of the generated contact forces. Furthermore, early-stage defects are difficult to detect by these methods. An alternative to these direct measurements is the analysis of acoustic signals registered by monitoring systems on rolling stock. When the wheel rolls over a surface defect of the rail, the surface discontinuity leads to a relative motion of wheel and rail, causing vibrations and noise [1, 2]. The discrete nature of the event and the generated large amplitudes distinguish the phenomenon from rolling noise and provide opportunities for the detection of the surface defect in vibration or noise signals. Both

G. Degrande et al. (Eds.): *Noise and Vibration Mitigation for Rail Transportation Systems*, NNFM 150, pp. 242–250, 2021.
https://doi.org/10.1007/978-3-030-70289-2_24

axle box acceleration (ABA) and under-coach noise measurements can be used for this purpose. An advantage of such monitoring systems is that the measured quantity is directly related to the wheel/rail contact forces indicating the severity of the rail faults. A prerequisite for detection is, however, that the surface defect passes through the wheel/rail contact area. Defects located outside the running band cannot be captured.

Only a few publications deal with the detection of rail faults by acoustic monitoring on rolling stock. These include the detection of squats [3] and rail welds [4] in ABA signals and the detection of squats by under-coach noise measurements [5]. The most promising approaches apply detailed time-domain models of the rolling contact to determine the specific characteristics of the faults in the ABA signals and exploit the simulation results to devise a detection procedure. In the case of squats in the Dutch railway network, high hit rates could be achieved with this strategy even for light squats [3].

A similar methodology is proposed in this paper for the detection of rail faults in the German railway network with the main focus on squats. The methodology is based on ABA measurements on the noise measurement car (SMW) [6] of DB Systemtechnik presented in Sect. 2 and the numerical model WERAN for wheel/rail interaction and noise [7, 8] described in Sect. 3. After adaption to the SMW, WERAN is applied to calculate the ABA signals due to squats of different severity and for different track parameters to find characteristic signatures in the time or in the time-frequency domain allowing identification of the fault (Sect. 4). In addition, a measurement campaign on a particularly bad track with several squats has been carried out providing measured signatures of squats in combination with direct geometry measurements of the faults (Sect. 5). Based on the simulated and measured results, a logistic regression classifier is devised to detect squats in the measured ABA signals of the SMW (Sect. 6).

2 Noise Measurement Car “Schallmesswagen”

The noise measurement car SMW (Fig. 1a) is employed to guarantee low rail roughness and hence a reduction in rolling noise on track sections called “specially monitored track” [6]. For this purpose, the SMW measures its own rolling noise. When the rolling noise reaches the intervention limit, the rails are acoustically ground. The rolling noise is measured with a microphone mounted above a hole in a semi-anechoic chamber directly above the measurement bogie (Fig. 1b). The wheels of the measurement bogie are unbraked and specially maintained to ensure low wheel roughness and thereby eliminate the influence of the wheel roughness on the rolling noise. Recently, the wheelsets of the measurement bogie have been equipped with accelerometers on the axle boxes (Fig. 1c). Vertical ABA is measured on all four axle boxes (and lateral ABA on two of them) with a sampling rate of 20 kHz. As the SWM travels about 50,000 km a year on the German railway network, it provides a large database of recordings.

Fig. 1. Noise measurement car of DB Systemtechnik (a) equipped with a microphone above the measurement bogie (b) and accelerometers on the axle boxes of the measurement bogie (c).

3 The Model WERAN

WERAN (WhEel/RAil Noise) is a non-linear time-domain model for the calculation of wheel/rail interaction and noise [7, 8], which includes a detailed 3D submodel of the wheel/rail contact based on Kalker's variational method [9]. Nevertheless, it is computationally efficient and allows for large parameter studies. This is achieved by employing a Green's function approach. Vehicle and track are represented by Green's functions that are pre-calculated from the receptances at the contact point by inverse Fourier transform. Instead of performing a time integration, the wheel and rail displacements are obtained by convoluting the time series of contact forces with the Green's functions. The Green's function of the track is a moving Green's function that includes the motion of the contact point on the rail [7]. The ABA signal is determined in a post-processing step by convoluting the contact forces with the Green's function calculated from the transfer receptance between the contact point on the wheel and the axle box.

Due to the Green's function approach, WERAN is easily coupled to different vehicle and track models as long as they are linear and time-invariant. In this study, the track model is the finite element (FE) model accounting for discrete supports presented in [10]. The model comprises a 60E2 rail represented by Rayleigh–Timoshenko beam elements. The discrete supports consist of railpads and sleepers on ballast. The vehicle model includes a complete wheelset of the SMW comprising the axle, two wheels of type BA093 and two brake disks, see Fig. 2. The primary suspension is also included, while all vehicle components above the primary suspension are represented as a static preload. The axle boxes are not modelled. Instead, it is assumed that the acceleration at a point on the axle 8 cm from the axle end approximately corresponds to the axle box acceleration. The wheel radius is 469 mm corresponding to a reprofiled wheel (while a new wheel has a radius of 475 mm). The wheelset is modelled by 3165 axisymmetric finite elements using a commercial FE software. With this undamped FE model, the eigenfrequencies and eigenmodes have been calculated up to 10 kHz. Since a measurement of the transfer function from contact to axle box was not available, the modal damping has been determined from the measured radial and axial wheel receptances at the contact [11]. Modes that could not be identified in the measurements (involving in particular modes with axle bending which have higher damping) were assigned the approximate damping values proposed by Thompson [1]. The measured receptances were also used to update the eigenfrequencies while the eigenmodes were not changed. From this modal basis, the Green's functions of the wheelset were calculated.

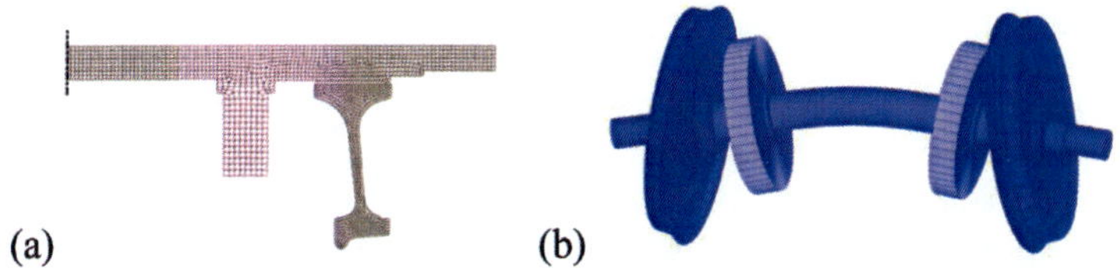

Fig. 2. Flexible wheelset model of the SMW: half of the 2D mesh of axisymmetric elements (a) and 3D visualization of the eigenmode at 244 Hz (b).

WERAN can consider both vertical and lateral wheel/rail interaction. In this study, only vertical wheel/rail interaction is included since this is considered sufficient for the excitation by rail surface defects.

4 Simulated Signatures of Squats

To determine the specific characteristics of squats in the ABA signals, a parameter study has been carried out with WERAN. The rail model has been adapted to the average measured track receptance from left and right rail at the shunting yard Munich North, see Sect. 5. This led to a rail pad stiffness $k_p = 400$ MN/m, rail pad damping 60 kNs/m, ballast stiffness $k_b = 200$ MN/m and ballast damping $c_b = 100$ kNs/m. The sleeper distance was 0.6 m. The center of the squats was placed at a quarter of the sleeper bay at the top of the rail. In all simulations, the train speed was 80 km/h.

4.1 Squat Modelling

The squats were modelled according to the following formula presented in [12]:

$$z_s(x, y) = -\frac{h}{2}\left(1 + \cos\left(\pi\sqrt{\left(\frac{x}{a}\right)^2 + \left(\frac{y}{b}\right)^2}\right)\right), x, y \in \left\{x, y : \sqrt{\left(\frac{x}{a}\right)^2 + \left(\frac{y}{b}\right)^2} \leq 1\right\} \tag{1}$$

where h is the squat depth, and a and b are the half-length and half-width, respectively. An example of a 3D squat shape is given in Fig. 3, along with different longitudinal profiles for the same squat depth.

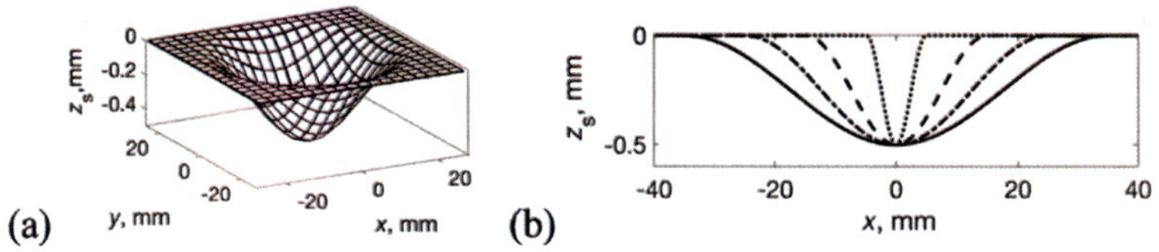

Fig. 3. Examples of squats according to Eq. (1): 3D shape for $h = 0.5$ mm and $a = b = 25$ mm (*a*) and longitudinal profile at y = 0 for $h = 0.5$ mm and $a = 5$, 15, 25 or 35 mm (*b*).

4.2 Influence of the Squat Dimensions

ABA signals for rolling over circular squats (i.e. $a = b$) with depths ranging from 0.1–1.4 mm and diameters ranging from 10 – 70 mm have been calculated. Except for the shallowest squats, the squat diameter has a larger influence on the maximum ABA and the shape of the ABA time signal than the squat depth. The influence of the squat diameter for a squat of depth 0.5 mm on the ABA time series is presented in Fig. 4a, where the center of the squat is located at position 0 m. When the wheel enters the squat zone, unloading occurs in the contact force, leading to a downward acceleration. The subsequent impact entails a strong positive peak in the acceleration. The larger the squat diameter, the larger the peak magnitude and the width of the peak. The corresponding power spectra in Fig. 5a reveal that the dominating frequencies in the signals are in the main independent of the squat diameter. Leaving aside the smallest squat whose length corresponds to the length of the contact area, the ABA levels are also similar at high frequencies above 700 Hz. Below larger differences occur.

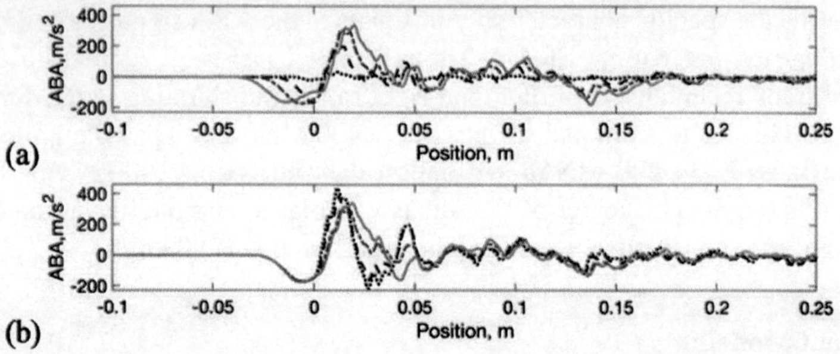

Fig. 4. Time series of ABA for rolling over a squat of depth 0.5 mm for different squat diameters: ..... 10 mm, --- 30 mm, -.-. 50 mm, — 70 mm (a); and for different rail pad stiffnesses: ..... 55 MN/m, --- 135 MN/m, -.-. 400 MN/m, — 800 MN/m, with squat diameter 50 mm (b).

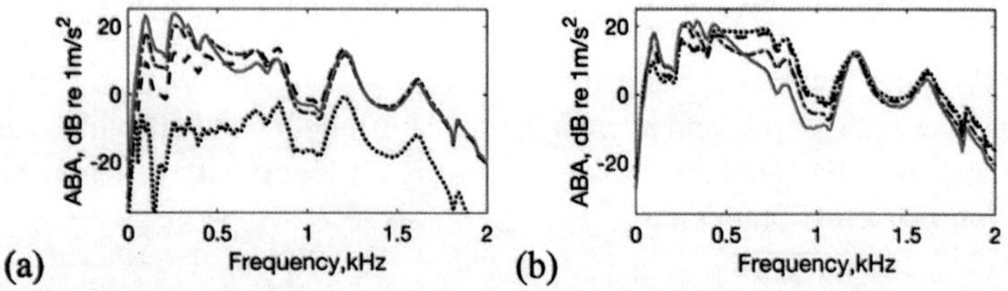

Fig. 5. Power spectrum of ABA for rolling over a squat of depth 0.5 mm for different squat diameters: ..... 10 mm, --- 30 mm, -.-. 50 mm, — 70 mm (a); and for different rail pad stiffnesses: ..... 55 MN/m, --- 135 MN/m, -.-. 400 MN/m, — 800 MN/m, with squat diameter 50 mm (b).

4.3 Influence of the Railpad Stiffness

The influence of the rail pad stiffness ranging from very soft pads (55 MN/m) to very stiff pads (800 MN/m) is shown in Figs. 4b and 5b for a squat of depth 0.5 mm and diameter 50 mm. The softer the pads, the higher and narrower the first peak in the ABA time series. In the spectrum of the contact force (not shown here), a change of the rail pad stiffness leads to a shift in frequency of some of the peaks. Due to the resonant behavior of the wheelset, these frequency shifts are not prominent in the spectrum of the ABA signal. The main response is still at the same wheel modes. The level of the ABA spectrum is influenced by the rail pad stiffness mainly below 1 kHz.

5 Measured Acoustic Signatures of Squats

To provide measured signatures of squats and data for validation of the model WERAN, a measurement campaign was carried out at the 5 km-long northern passing track at the shunting yard Munich-North in March 2019. The detailed geometry of three severe and four light squats was measured with the rail roughness measurement device RM1200E providing 1.2 m-long roughness lines with a longitudinal resolution of 0.5 mm. Up to 11 parallel roughness lines with a distance of 3 mm have been measured for each squat. A severe squat (A) and a light squat (B) that were located on the right rail are presented in Fig. 6, together with longitudinal roughness lines through the squat centers.

Fig. 6. Photograph of severe squat A (a) and light squat B (b) at shunting yard Munich North. Corresponding squat profiles at midline (c): — squat A, -.-. Squat B.

In addition, the vertical track receptance was measured on the right and left rail at a location 40 m from squat A and 1.3 km from squat B. It differed considerably between the two sides of the rail (not shown here) and is expected to vary along the track. Eight days later, pass-by measurements were carried out with the SMW. For four pass-bys in the main travelling direction and four in the opposite direction at 80 km/h, ABA signals were recorded on the SMW. At the time of the pass-by measurements, the rail at squat A had been equipped with an emergency fishplate connector for safety reasons.

As seen in Fig. 7, the measured ABA signals show good repeatability between pass-bys. Based on the measured squat geometries of squat A and B, WERAN has been used to predict the ABA signal. Here, the rail model was adapted to the track receptance measurement at the right rail. The comparison of simulated and measured signals is presented in Fig. 7 in terms of the ABA time signal and in Fig. 8 in terms of the spectrogram. The model predicts well the magnitude and width of the first ABA peak, but the simulated response is too highly damped. The measured and simulated spectrograms show similar patterns for the positions from −0.6 m to +0.6 m around the squat where roughness/geometry was measured. Some of the peaks appear, however, shifted in frequency, e.g. the measured peak at 1.4 kHz appears at 1.2 kHz in the simulated response. These differences are mainly attributed to the uncertainty about eigenfrequencies and damping

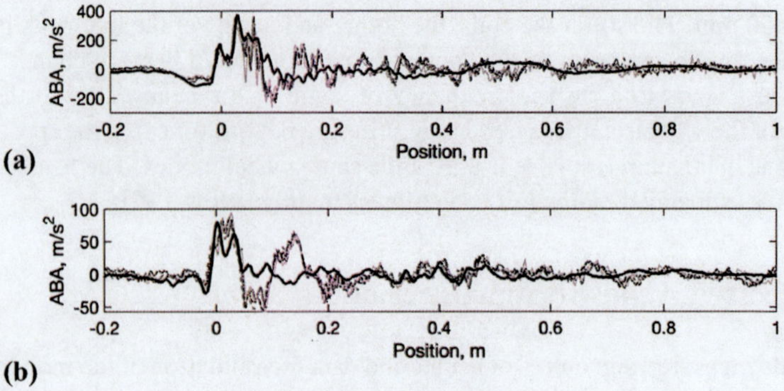

Fig. 7. Time series of ABA for rolling over squat A (*a*) and B (*b*): four measured pass-bys in the main travelling direction at 80 km/h (*grey, different line styles*) and simulation (*black line*).

of the wheelset modes involving bending of the axle, which could not be measured, see Sect. 3. Additionally, the rail receptance at the squat may have differed from the one at the measurement position, and the emergency fishplate connector at squat A has not been modelled.

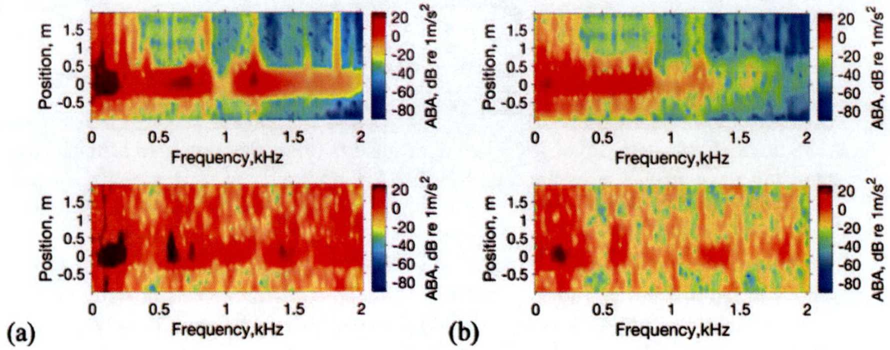

Fig. 8. Spectrograms of ABA for rolling over squat A (*a*) and B (*b*): simulation (*upper row*) and one selected measured pass-by in the main travelling direction (*lower row*) at 80 km/h.

6 Detection Algorithm

As seen in Sects. 4 and 5, the impact at the squat leads to a temporary increase in the ABA signal over a wide frequency range. For the same wheelset, the main excited frequencies are relatively independent of the squat dimensions and the rail support properties. Against this background, a pattern recognition algorithm based on machine learning seems a promising strategy to detect squats in the ABA signals of the SMW. A logistic regression classifier [13] predicting 'squat' or 'no squat' from the spectrogram of the ABA signal containing 93 x 29 acceleration values in dB (spatial resolution 1 mm, frequency resolution 22 Hz, overlap 95%) has been trained using in total 18,427 spectrograms of simulated and measured data. Applied on the measured ABA signals from the shunting yard Munich-Nord, the algorithm classifies spectrograms from all three severe squats correctly in all eight pass-bys. This corresponds to totally 28 squat passages since the signal from one axle box was corrupted and squat passages were discarded when the train velocity deviated more than 5% from the targeted 80 km/h. Concerning the four light squats, the spectrograms are classified correctly in 87% of the totally 31 squat passages. For the test of each squat, all data from this specific squat were removed from the training data. The algorithm also detects deteriorated insulated rail joints as 'squats'. Presumably, discrimination between squats and insulated rail joints would be possible when more data of insulated rail joints are available. Besides insulated rail joints, only a few (possibly) false alarms occurred. On two 250 m test sections of the measured signals, four 'squats' were detected at positions where no squats had been found in the visual inspection. Practically, these can all be discarded due to low peak accelerations below 40 m/s^2.

7 Conclusions

In this paper, the acoustic signatures of squats in the ABA signals of the DB noise measurement car SMW have been determined from measurements and simulations with the model WERAN. Good agreement has been found between both. Based on these results, a logistic regression classifier to detect squats in the ABA signals of the SMW has been proposed. It achieves high hit rates for both severe and light squats. More data must, however, be collected to discriminate squats from deteriorated insulated rail joints.

Acknowledgements. This work has been carried out in the project AMONTRACK that has received funding from the European Union's Horizon 2020 research and innovation programme under the Marie Skłodowska-Curie grant agreement No. 744605.

References

1. Thompson, D.: Railway Noise and Vibration. Elsevier, Oxford, UK (2009)
2. Nielsen, J.C.O., Pieringer, A., Thompson, D.J., Torstensson, P.T.: Wheel–rail impact loads, noise and vibration: a review of excitation mechanisms, prediction methods and mitigation measures. In: Proceedings of IWRN13, Ghent, Belgium (2019)
3. Molodova, M., Li, Z., Núñez, A., Dollevoet, R.: Monitoring the railway infrastructure: Detection of surface defects using wavelets. In: Proceedings of the 16th International IEEE Annual Conference on Intelligent Transportation Systems, The Hague, The Netherlands (2013)
4. An, B., Wang, P., Xu, J., Chen, R., Cui, D.: Observation and simulation of axle box acceleration in the presence of rail weld in high-speed monitoring. Appl. Sci. **7**(12), 1259 (2017)
5. Schwanen, W., Kuijpers, A.H.W.M.: Early rail defect detection using sound measurements. In: Pombo, J. (ed.) Proceedings of the Third International Conference on Railway Technology: Research, Development and Maintenance, Civil-Comp Press, Stirlingshire, UK (2016)
6. Asmussen, B., Onnich, H., Strube, R., Greven, L.M., Schröder, S., Jäger, K., Degen, K.G: Status and perspectives of the "specially monitored track". J. Sound Vib. **293**, 1070–1077 (2006).

7. Pieringer, A., Kropp, W., Nielsen, J.C.O.: The influence of contact modelling on simulated wheel/rail interaction due to wheel flats. Wear **314**(1–2), 273–281 (2014)
8. Pieringer, A.: A numerical investigation of curve squeal in the case of constant wheel/rail friction. J. Sound Vib. **333**(18), 4295–4313 (2014)
9. Kalker, J.J.: Three-Dimensional Elastic Bodies in Rolling Contact. Kluwer Academic Publishers, Dordrecht, Boston, London (1990)
10. Nielsen, J.C.O., Igeland, A.: Vertical dynamic interaction between train and track – influence of wheel and track imperfections. J. Sound Vib. **187**(5), 825–839 (1995)
11. Thompson, D., et al.: Assessment of measurement-based methods for separating wheel and track contributions to railway rolling noise. Appl. Acoust. **140**, 48–62 (2018)
12. Andersson, R., Torstensson, P.T., Kabo, E., Larsson, F.: The influence of rail surface irregularities on contact forces and local stresses. Vehicle Syst. Dyn. **53**(1), 68–87 (2015)
13. Ng, A.: Machine Learning. Online course at Coursera Platform, Stanford University. https://www.coursera.org/learn/machine-learning. Accessed 20 May 2019

Nonlinear Model of an Embedded Rail System for the Simulation of Train-Track Dynamic Interaction and the Analysis of Vibration Transmission

Andrea Collina, Roberto Corradi, Egidio Di Gialleonardo, and Qianqian Li(✉)

Politecnico di Milano, 20156 Milan, MI, Italy
qianqian.li@polimi.it

Abstract. Nonlinearities in the track system are important and sometimes necessary to be taken into consideration in the study of the train-track interaction and its relevant issues, especially when resilient materials are employed. In the present work, static and dynamic characterisations of an embedded rail system (ERS) were performed through laboratory tests. The dynamic behaviour of the ERS varies considerably with the test conditions. The collected data are then used to develop a nonlinear rheological model of unit length of the ERS. The identified model is in good agreement with the experimental data and reproduces effectively the dynamic behaviour of the ERS. The identified nonlinear model is then integrated with a 2D track model for the simulation of train-track dynamic interaction. Simulations of rail deflection time history corresponding to a train passage are performed employing both linear and nonlinear track models. The results demonstrate that the nonlinear model leads to a significantly different track response with respect to the one obtained with the linear track model and thus suggest that the nonlinear track behaviour needs to be carefully accounted for in train-track dynamic interaction simulation and that the nonlinear rheological model can be an efficient tool to achieve this objective.

Keywords: Track nonlinearities · Nonlinear track rheological model · Train-track dynamic interaction

1 Introduction

Theoretical modelling and numerical simulation are important tools for studying train-track dynamic interaction and its relevant issues. These tools usually include a track model, a vehicle model and a contact force model, the level of detail in each submodel being dependent on the phenomenon under study and on the frequency range of interest. Most of the train-track dynamic interaction models proposed in the literature (e.g. [1–3]) rely on linear track models. However, when dealing with railway noise and vibration issues, the nonlinearities in the track are sometimes necessary to be considered for properly assessing the system response. This is typically the case when resilient materials

G. Degrande et al. (Eds.): *Noise and Vibration Mitigation for Rail Transportation Systems*, NNFM 150, pp. 251–258, 2021.
https://doi.org/10.1007/978-3-030-70289-2_25

are employed to mitigate vibration transmission from the track to the supporting structure. In fact, resilient elements such as rail pads, sleeper pads or the continuous elastic support in embedded rail systems, show significant variations of their dynamic stiffness with load amplitude, frequency and preload [4–6]. When a train passage takes place, the generic track section is simultaneously subjected to the low frequency deformation associated with the moving axle loads and to the higher frequency vibrations due to the track/wheel geometrical irregularity. An accurate prediction of the vibrations transmitted from the track to the supporting structure (e.g. in case of bridge runnability or ground-borne vibration analysis) requires both phenomena to be accounted for at the same time.

This paper proposes a methodology to include track nonlinearities in the simulation of train-track interaction, by making reference to the test case of an embedded rail system (ERS) typically adopted on bridges, slab tracks or for installation in tunnels. The dynamic laboratory tests carried out on an ERS sample are first illustrated. The collected experimental data were then used to identify a nonlinear rheological model of unit length which reproduces the dynamic behaviour of the tested ERS sample. Finally, the identified model was integrated with a 2D track model reproducing the system's dynamics in the vertical plane for the simulation of train-track dynamic interaction.

2 Experimental Tests

The ERS experimental characterisation was performed in the labs of Politecnico di Milano, on a full-scale, 750 mm long sample. The experimental test setup is illustrated in Fig. 1.

The bottom plate of the test sample is fixed and the embedded rail is subjected to a uniform vertical displacement imposed by a hydraulic actuator which can be controlled either in displacement or in force. The rail displacement was measured by 4 laser sensors positioned at the corners of the test sample while the actuator force was measured by a load cell mounted between the actuator and a fixed rigid supporting structure. The quasi-static tests were carried out in accordance with the requirements of the reference technical standards [7]. The dynamic tests, where the actuator imposed a sinusoidal displacement, were performed based on different combinations of frequency (1, 3, 5, 10, 20 Hz) and preload (6, 18, 37, 55 kN where 55 kN corresponds to the maximum static deflection induced by a single bogie of a SCHIMMNS freight wagon) with a 5 kN amplitude.

The static stiffness of unit length of the tested ERS is characterised as 60 MN m^{-2} while the results of the dynamic tests, in terms of equivalent stiffness and damping, are reported in Fig. 2.

The dynamic behaviour of the tested ERS sample, especially in terms of the equivalent stiffness, varies substantially with the test condition, both in preload and frequency. With the preload fixed, the equivalent stiffness increases with frequency while there is no asymptotic trend with the increase of preload.

Fig. 1. Test bench for full-scale experiments on a 750 mm long ERS sample.

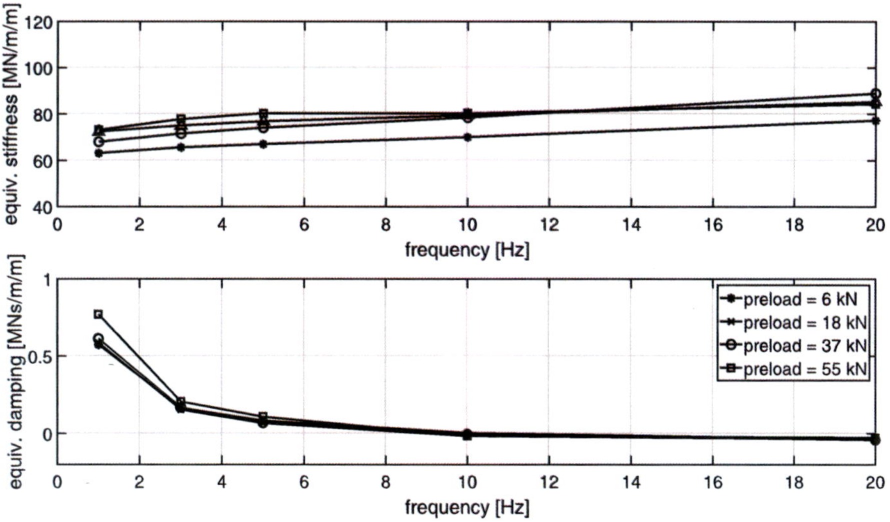

Fig. 2. Test results: equivalent stiffness and damping of the 750 mm ERS sample as a function of frequency.

3 Numerical Models

3.1 ERS Nonlinear Rheological Model

The nonlinear rheological model of unit length of the tested ERS is a Zener model with 3 parameters: a spring (k_1) in parallel with a combination of a spring (k_2) and a damper (r_2) in series [8]. All the three parameters are a function of preload (P) and are defined as:

$$\begin{aligned} k_1 &= k_{1,0}(1 + P/P_0)^x \\ k_2 &= k_{2,0}(1 + P/P_0)^x \\ r_2 &= r_{2,0}(1 + P/P_0)x \end{aligned} \tag{1}$$

where $k_{1,0}$, $k_{2,0}$, $r_{2,0}$, are constant reference values of k_1,k_2 and r_2; P is the preload applied on the ERS; P_0 is a constant reference preload; x is a non-dimensional constant number.

$k_{1,0}$, $k_{2,0}$, $r_{2,0}$ and x are determined through a minimisation procedure of the difference between the modelled and experimental equivalent stiffness and damping whose comparisons are reported in Fig. 3. The reference preload is defined as 64 kN and the identified values of the model parameters are the following: $k_{1,0}$ = 62 MN/m, $k_{2,0}$ = 9.6 MN/m, $r_{2,0}$ = 0.7 MN/m/s, x = 0.21.

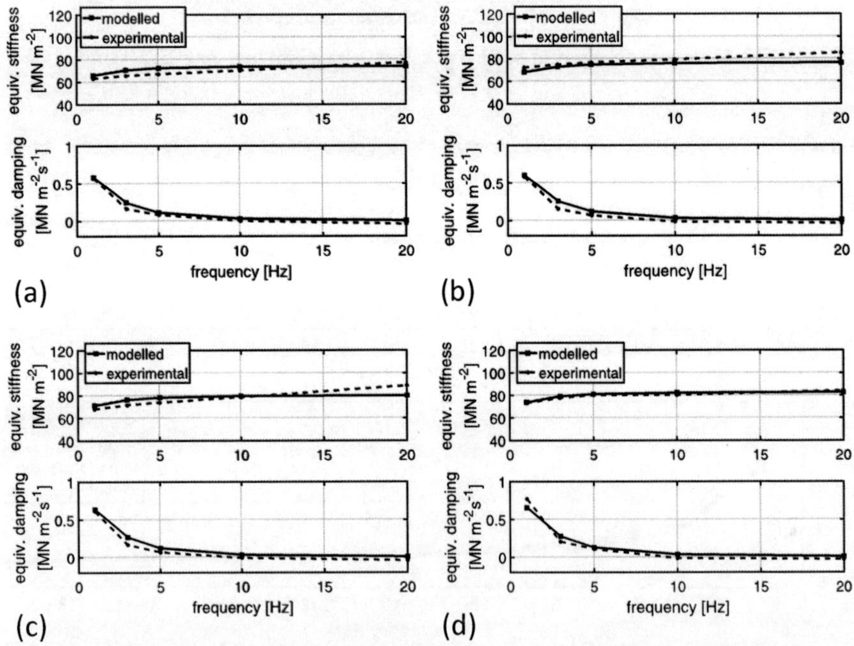

Fig. 3. Modelled and experimental equivalent stiffness and damping of the 750 mm ERS sample: (a) preload = 6 kN; (b) preload = 18 kN; (c) preload = 37 kN; (d) preload = 55 kN.

It can be observed from Fig. 3 that the modelled equivalent stiffness and damping are quite similar to the experimental ones, in terms of both the absolute values and the variation trend as a function of preload and frequency.

In order to verify whether the model can reproduce the dynamic behaviour of the tested ERS, a time domain simulation is performed and is compared to the corresponding experimental result. During the simulation, the lower point of the nonlinear Zener model is set fixed (corresponding to the fixed bottom plate of the ERS sample in laboratory test); the input of the simulation is a displacement time history at the upper end of the nonlinear Zener model (corresponding to the displacement of the embedded rail of the ERS sample imposed by the actuator in laboratory test) while the output is the force time history (corresponding to the force measurement in laboratory test). The input displacement time history is an approximated rail deflection time history corresponding to a passage of a train. This time history is obtained by converting the static deformation distribution along the track given by a single beam Winkler model, on which wheelset loads are modelled as a sequence of discrete loads, from space domain to time domain through the wheelset advancement velocity. The values of the wheelset loads (55 kN on each rail), wheelbase and bogie base refer to the passenger coach of train ETR 500 and the velocity is defined as 200 km/h.

The output force time histories obtained from the laboratory test and the numerical simulation are compared in Fig. 4. The visualization of the time history is limited to a time interval corresponding to the passage of a single bogie because it is verified that the portion of the time history related to the passage of a single bogie is independent from the other one.

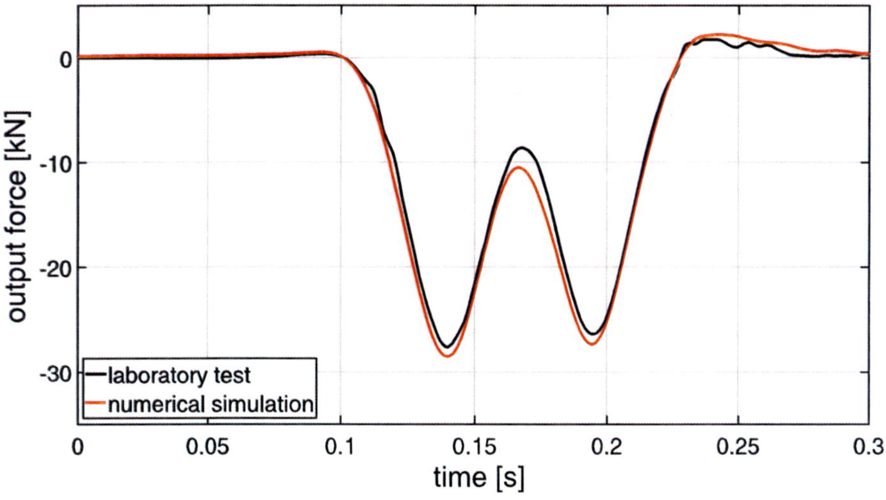

Fig. 4. Comparison of the output force time history obtained from the 750 mm ERS section subjected to an input displacement simulating train passage (ETR 500 train at 200 km/h): laboratory test and numerical simulation.

It can be observed that the output force simulated by the nonlinear Zener model is quite similar to the experimental one. Furthermore, the laboratory test result shows that the second peak of the force time history, corresponding to the passage of the posterior wheelset, has a lower absolute value than the first one and this phenomenon is reproduced by the numerical simulation.

3.2 Nonlinear Track Model and Train-Track Interaction Simulation

The nonlinear track model is obtained by substituting the linear foundation in a 2D track model with the identified nonlinear Zener model of the ERS. The original 2D track model (Fig. 5) consists in a finite element model of a Euler-Bernoulli beam of 30 m (representing the rail) on a one-layer foundation, which is formed by a continuous linear Kelvin-Voigt model. The train is modelled as a sequence of moving loads (Q, representing the wheelset loads) which is the input of the train-track interaction simulation. The properties of the rail refer to the rail type UIC 60; the values of the wheelset loads (Q), wheelbase (p) and bogie base (d) refer to the passenger coach of train ETR 500. The simulation input associated with the wheel/rail irregularity is not accounted for and will be considered in an updated version of the train-track interaction model.

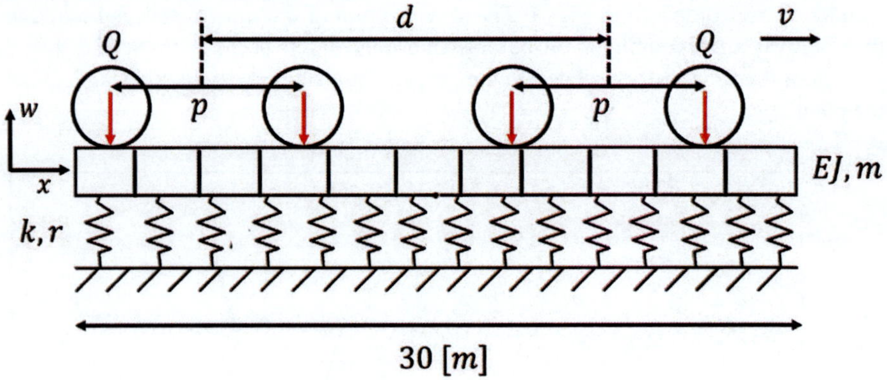

Fig. 5. 2D track model and moving loads input.

The simulation of a train passage is performed both with the linear and nonlinear track model. The linear Kelvin-Voigt model of the reference ERS is employed in the linear track model and the values of stiffness of unit length (80 MN m^{-2}) and damping of unit length (0.13 MN $m^{-2}s^{-1}$) are identified in the same way as the ones of the nonlinear Zener model. The simulated rail deflection time histories of a generic track section, corresponding to the passage of a train with 3 coaches, are reported in Fig. 6.

It can be observed from Fig. 6 that the two track models lead to several considerable differences. Firstly, the nonlinear track model results in different rail deflection values at the time instants corresponding to the passage of the two adjacent wheelsets while the linear track model does not. Secondly, the two models lead to a 0.05 mm (about 10%) difference of the maximum rail deflection. Thirdly, the two simulated time histories are obviously different in the time intervals soon after the passages of bogies.

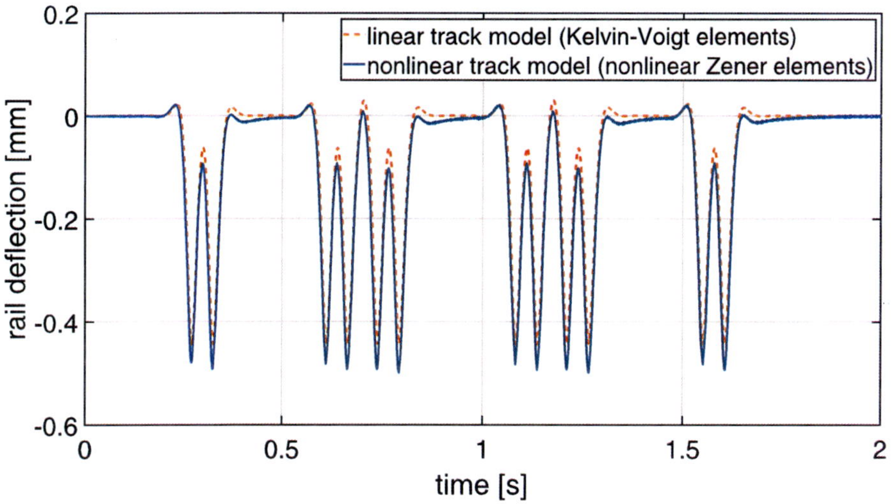

Fig. 6. Time history of the rail deflection corresponding to a train passage (ETR 500 at 200 km/h): simulation with linear and nonlinear track model.

4 Conclusion

In this paper, an ERS sample is tested with variable frequencies and preloads and shows considerable changes in its dynamic behaviour as a function of both test conditions. The identified nonlinear rheological model is in good agreement with the experimental data and reproduces effectively the dynamic behaviour of the ERS sample. When including, through the identified model, track nonlinearities in the simulation of train-track dynamic interaction, the system response shows considerable differences from the one obtained with a linear track model, in terms of rail deflection time history of a generic track section.

The results obtained suggest that the nonlinear track behaviour needs to be carefully accounted for, when accurate prediction of the track vibration mitigation performance is required, such as in the study of bridge runnability or ground-borne vibration in underground or at grade railway lines. To this end, rheological models of track resilient elements can be developed from lab tests (considering different combinations of the test parameters, so as to cover typical operating conditions) and then included in train-track dynamic interaction models.

References

1. Knothe, K.L., Grassie, S.L.: Modelling of railway track and vehicle/track interaction at high frequencies. Veh. Syst. Dyn. **22**, 209–262 (1993). https://doi.org/10.1080/00423119308969027
2. Sheng, X., Jones, C.J.C., Thompson, D.J.: A theoretical model for ground vibration from trains generated by vertical track irregularities. J. Sound Vib. **272**(3–5), 937–965 (2004). https://doi.org/10.1016/S0022-460X(03)00782-X

3. Lombaert, G., Degrande, G., Kogut, J., François, S.: The experimental validation of a numerical model for the prediction of railway induced vibrations. J. Sound Vib. **297**(3–5), 512–535 (2006). https://doi.org/10.1016/j.jsv.2006.03.048
4. Sjöberg, M., Kari, L.: Non-linear behavior of a rubber isolator system using fractional derivatives. Veh. Syst. Dyn. **37**(3), 217–236 (2002). https://doi.org/10.1076/vesd.37.3.217.3532
5. Thompson, D.J., Verheij, J.W.: The dynamic behaviour of rail fasteners at high frequencies. Appl. Acoust. **52**, 1–7 (1997). https://doi.org/10.1016/S0003-682X(97)00016-9
6. Bruni, S., Collina, A.: Modelling the viscoelastic behaviour of elastomeric components: an application to the simulation of train-track interaction. Vehicle Syst. Dyn. **34**, 283–301 (2000). https://doi.org/10.1076/vesd.34.4.283.2061
7. BS EN 13146–9:2009+A1:2011, Railway applications – Track – Test methods for fastening systems. Part 9: Determination of stiffness
8. De Man, A.P.: Dynatrack: a survey of dynamic railway track properties and their quality (2002)

Wheel-Rail Contact Analysis with Emphasis on Wear (Measurements/Simulation)

Kris Decroos[1], Jonathan Ceulemans[2], Bert Stallaert[1](✉), and Tom Vanhonacker[1]

[1] D2S International, Jules Vandenbemptlaan 71, 3001 Heverlee, Belgium
{kris.decroos,Bert.stallaert}@d2sint.com

[2] TEC Charleroi, Place des Tramways 9/1, 6000 Charleroi, Belgium
jonathan.ceulemans@tec-wl.be

Abstract. The contact between vehicle wheel and rail is a very complex phenomenon that can be considered as an imperfect link because of the high concentrated stresses that occur, the relative sliding of the wheelset with respect to the rail and the presence of dust, rain, sand or even ballast stones. Simulation of this dynamic changing contact and the consequences for wheel and rail wear of the tram on Pont de Sambre in Charleroi is the subject of this case study.

The software VI-Rail allows predicting wheel and rail wear with a very high precision during different operational conditions: new versus worn wheels and rails, influence of speeds, different curve radii and friction coefficients, etc.

First a measurement campaign was done to determine the rail profile, track gauge, speed of the trams and profiles of the wheels after different mileages.

The second phase of the study concerns the multibody dynamics analysis to determine by simulation wear profiles of wheel and rail in different conditions. Both the complete tram and the local track were precisely modelled.

Several scenarios were considered in which various parameters were modified (e.g. rail profile, track gauge, inclination of the rails, vehicle speed, etc.). The analysis results showed the location, shape and size of the damage on the rail and on the wheel, which allowed determining the cause of the measured wear and proposing an appropriate solution.

Keywords: Wheel-rail contact · Wear · Simulation

1 Introduction

In certain segments of the Belgian light rail network at Charleroi, significant rail wear was observed. The problem arises especially in curves with vertically installed grooved rails (no inclination). Pronounced wear on the wheels of the vehicles was also observed. The purpose of this study was to improve the vehicle track interaction to increase wheel and rail life and to be able to maintain the operational speed in tight curves, without increasing the rail and wheel wear.

G. Degrande et al. (Eds.): *Noise and Vibration Mitigation for Rail Transportation Systems*, NNFM 150, pp. 259–266, 2021.
https://doi.org/10.1007/978-3-030-70289-2_26

2 Methodology

The first phase of this project consisted of a measurement campaign to determine the rail profile, track gauge and speed of the trams. Also, the wheel profiles of the vehicle in new, average usage and completely worn state were measured.

The second phase of the study concerns the multi body dynamics analysis in order to determine wear profiles of wheel and rail in various conditions. VI-Rail [1] is used to perform these simulations. The software has a dedicated wear analysis module based on Archard's abrasive wear theory, that allows predicting wheel and rail wear with a very high precision during different operational conditions: new versus worn wheels and rails, different speeds, different curve radii and different friction coefficients.

In order to apply this theory, both the complete tram vehicle and the local track were precisely modeled, based on in situ measurements of wheel and rail profile and track gauge.

A sensitivity study was performed to assess the influence of different parameters in order to identify the cause of the high wear rate. Besides the rail profile (flat head EB50T, grooved 60Ri2 and Ri54G2) and the inclination of the rail, the influence of the vehicle speed was also investigated.

2.1 Measurement Campaign

Track. The first phase of this project consisted of a measurement campaign to determine the rail profile, track gauge and speed of the trams. This was done in three different locations. Here, the study will be focused on one location: the bridge 'Pont de Sambre' (see Fig. 1). The rails used in this segment of the track are grooved rails type Ri54G2 and the curve radius of the track is ± 40 m. Using a Miniprof, measuring device, profiles were obtained on the inner and outer tracks as shown in Fig. 2.

Fig. 1. Tracks and tram on Pont de Sambre in Charleroi

A clear worn section can be observed on both tracks: the high rail profile (high rail – red arrow) and a small worn impact can also be seen on the low rail profile (low rail – blue arrow). An S-shaped bead (false flange) is formed.

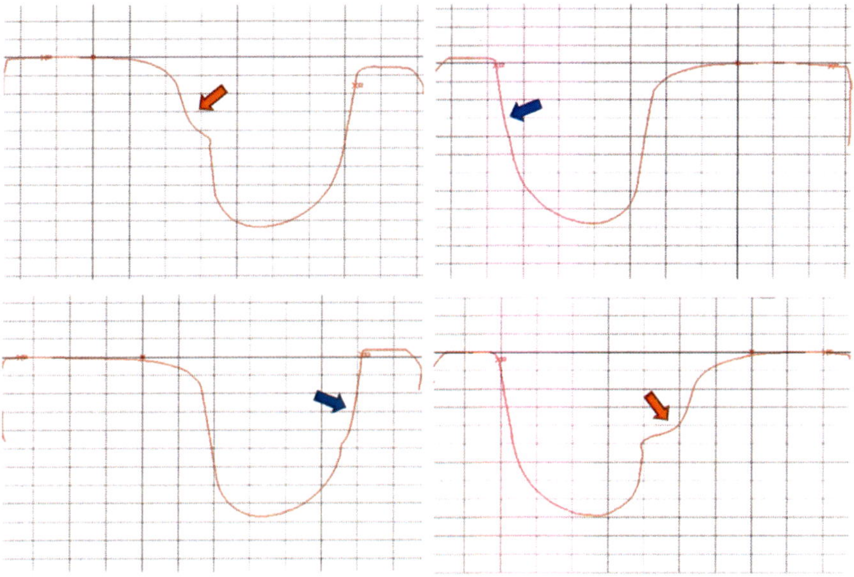

Fig. 2. Measured rail profiles on the inner (top) and outer track (bottom)

Wheels. Figure 3 shows the measured wheel profiles in new, average usage and completely worn state.

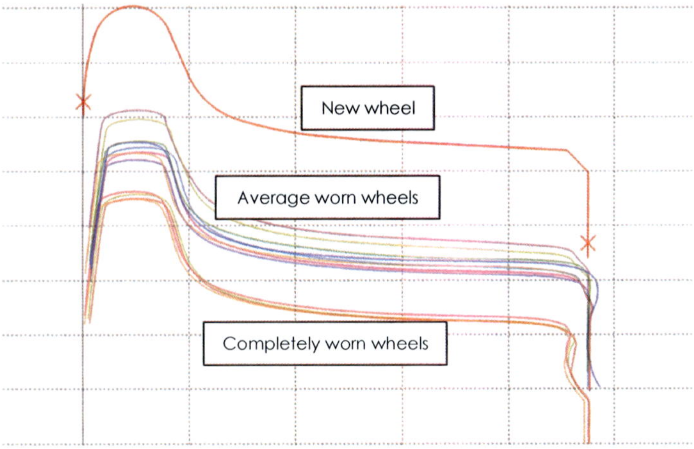

Fig. 3. Comparison of the different measured wheels

2.2 Preparation for Numerical Simulations

Purpose. As described before heavy wear occurs on wheels and rails in certain sections of the light rail network in Charleroi. With a measuring campaign, profiles have been analyzed. With a multibody dynamics analysis, the cause of this extreme wear behavior has to be found and an appropriate solution needs to be provided.

Background. For this wear analysis, a dedicated toolkit inside VI-Rail is used. This wear toolkit is based on abrasive wear theory according to Archard [2]. In this theory, the wear volume is proportional to frictional work. During the simulation, frictional power in wheel/rail contact is computed together with the changing of wheel and rail profiles due to the wear.

The term abrasive wear can be defined as the damage to sliding surfaces due to the action of solid particles present in the rubbing zone. Accordingly, abrasive wear includes wear produced by particles removed during the rubbing process.

The mechanical action of the abrasive particles on the wearing material strongly depends on:

- surface shape;
- strength of attachment;
- relation between the mechanical properties of the abrasive particles and the wearing surfaces;
- loads.

On this principle, the wear mechanism itself can change from elastic, pre-deformational to the most serious type of wear, that is micro-cutting. The basic assumption, which is used mostly for wheel/rail profile wear is that the abrasive wear is taken to be proportional to the frictional work (Archard, 1957). Frictional work is directly proportional to worn away mass.

Archard's wear equation:

$$Mw = k.P \tag{1}$$

with

M_w: wear mass [g]
k: wear coefficient [g/Nm]
P: frictional work [Nm]
N: normal force [N]
d: sliding distance [m]

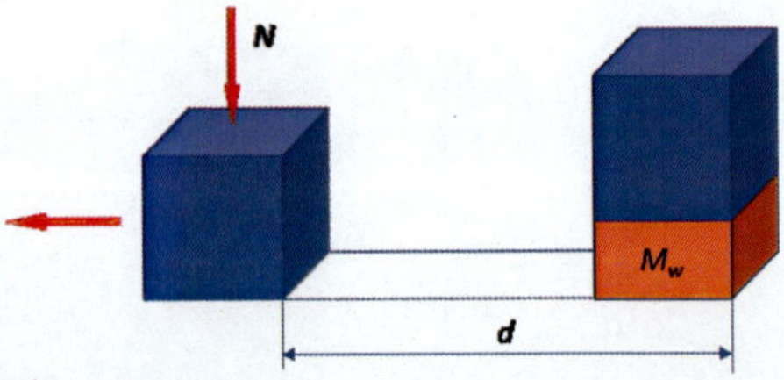

Wear coefficient is determined by experiment:

k_{mild} ~= 4.5 10-7 [g/Nm]
k_{heavy}~= 1.25 10-6 [g/Nm]

Transition between mild and heavy wear depends on friction power in contact patch. The wear coefficients are used for global and local description:

$$P_k = s_x . T_x + s_y . T_y \quad (2)$$

with
P_k: wear number [N]
s: translational creepage
T: translational creep forces

in the defined contact point or in the middle of meshes of the contact patch.

Transition between mild and heavy wear is computed by.

$$P_d = (P_k . v_0) / A \quad (3)$$

with
P_d: frictional power density [W/m^2]
v_0: mean velocity on surface of wheel and rail [m/s]
A: contact or mesh area [m^2]

Frictional work of a particle traveling in a segment of the profile is given by:

$$P = P_k \cdot L \quad (4)$$

with
L: loaded path length where the particle is traveling [m]

The depth of wear (δ) can be computed out of the wear volume divided by density of the material and contact patch or stripe area:

$$\delta = (k . P) / (A . \rho) \quad (5)$$

with
d: wear depth [m]
k: local wear coefficient [g/Nm]
r: density of the material [g/m^3]

Simulation wear depth is calculated for each integration step of the analysis.

Implementation. The use of the wear module in VI-Rail allows predicting the wear profiles of wheels and rail. The user has the capability to make successive runs in which the vehicle runs on the same or different tracks, using a certain velocity profile or running at different speeds, etc. In this way, wear can be accumulated to represent as much as possible the life of wheels and rails.

The wear module must combine the short term dynamics of the vehicle with the long-term dynamics responsible for the wear of wheel and rail profile.

2.3 Simulations

Modelling of the Vehicle. D2S International uses VI-Rail to analyze the interaction between the vehicle (tram) and the bridge (infrastructure). Figure 4 shows the vehicle model.

The tram is composed of 2 wagons and 3 bogies: 2 motor bogies and 1 trailer Jacob's bogie in the middle of the tram. All aspects were taken into account in order to approach as much as possible the real behavior: weight, inertia moments, locations of centers of gravity, suspension systems, axle boxes, wheelsets, etc.

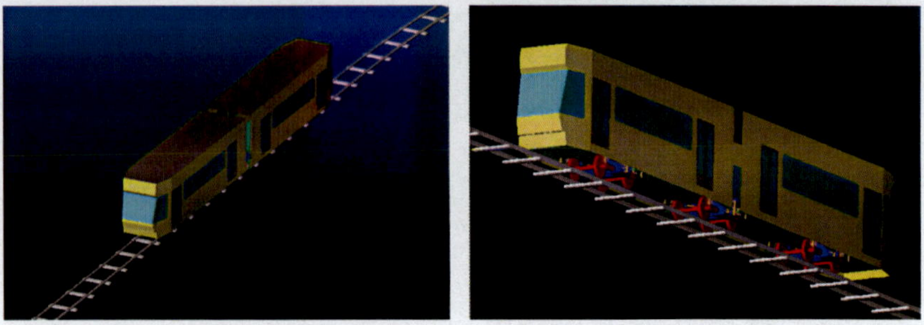

Fig. 4. Generated vehicle model (tram series 6100)

Modelling of the Track (Fig. 5). Each trajectory consists of straight sections and curved segments defined by the central curve of the track, the variation in track height, the type of rail, the inclination of the rails, the gauge of the track, etc. Different versions of the track have been created in order to compare running and wear behavior of the tram on those tracks. Three types of rails have been compared: flat head EB50T (as reference), grooved rails 60Ri2 and Ri54G2. Also, the inclination (0 or 1:20) and gauge widening (−2, −3, 0, + 5 mm) were modified parameters.

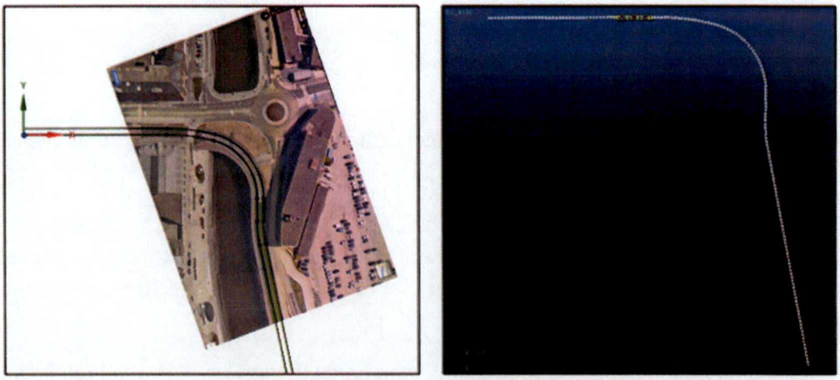

Fig. 5. Modelled track on the bridge

Simulations. Several situations were compared and analyzed, but only some results on the tightest curve on the bridge (Pont de Sambre) with a radius of ~40 m are presented. Figure 6 compares the results of the Ri54G2 rail type (with different inclinations) against the reference situation with an EB50T rail type inclined by 1:20.

A rail wear condition was calculated (right image) that corresponds very well with the measurements (left image). It is also noted that the guard rail is slightly affected - as measured. The effect of the inclination of the Ri54G2 rail is also noticeable. The inclined rails show a more even wear over the tread.

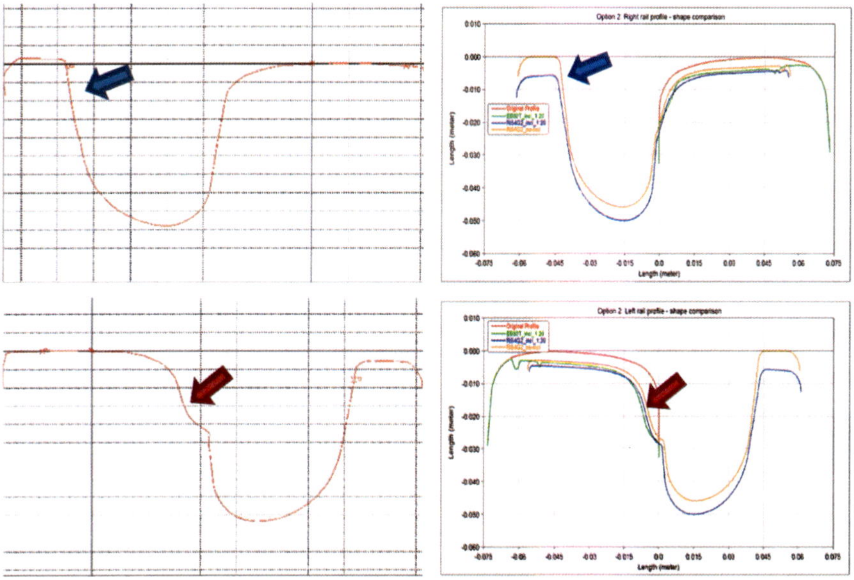

Fig. 6. Comparison left and right rail: Ri54G2 vs EB50T

Figure 7 presents the worn wheel profiles (left and right) for the leading axle. In reality, wear on those wheels will be averaged by the fact that the vehicle takes left and right turns. We also see that the wheel on the left rolls on the bead (false flange) formed on the left rail (red arrow).

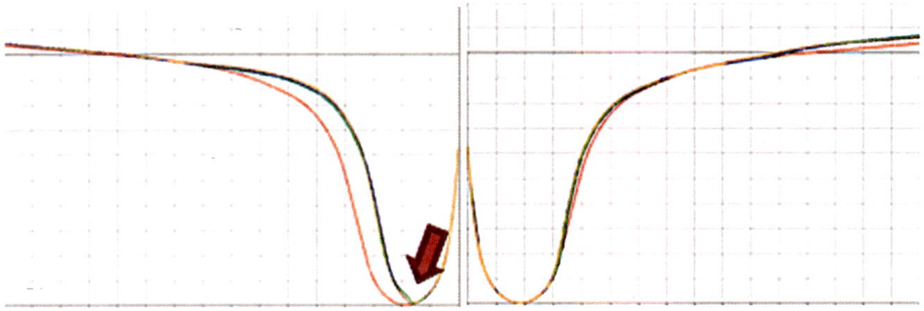

Fig. 7. Worn wheel shapes for the leading axle

3 Conclusions

In some segments of the Charleroi light rail network, high rail wear was observed. The problem is particularly acute in curves where vertical grooved rails were used. Also, pronounced wear on the wheels was observed.

In a first phase of the study, a measurement campaign was carried out at different sites, with different radii of the curve, to determine the wear state of the rail profiles. The profiles of the wheel of the vehicle in new condition, in average working condition and completely worn out were also measured.

In a second phase, a series of simulations was performed using the multibody dynamics software VI-Rail. In this software, the vehicle and track are accurately modelled.

For this study, several scenarios were considered where different parameters were modified (e.g. type of rail profile, rail spacing and inclination, influence of vehicle speed, influence of curve radius, etc.). This allowed showing the location, shape and size of the damage on the rail and wheel.

It was found that a 60Ri2 rail performs better in all the options investigated, both for wear on the two rails (external and internal to the curve) and for wheel wear. This rail contains a narrower groove which prevents greater lateral movement - resulting in very pronounced wear on the outer rail of the track. It was also found that it would be beneficial to tilt the rail (at 1:40 or 1:20). Since the EB50T rails are inclined 1:20 in the rest of the network, the same inclination of a 60Ri2 rail would be justified.

TEC Charleroi has chosen to use a 55R1 rail which has the same groove as the 60Ri2 (36 mm opening) but has a lower height. This was necessary because of space limitations on the bridge.

Combined with other measures, a significant improvement (more than a factor 2) was observed in the rail lifetime. The wheel lifetime was improved as well.

This study demonstrates the potential of the use of a multibody simulation tool, taking into account both vehicle and track parameters, to accurately predict wheel and rail wear. This is verified using experimental measurements. As such, it serves as a tool to optimize track and vehicle parameters.

References

1. VI-RAIL. www.vi-grade.com/en/products/vi-rail/
2. Archard, J.F.: Contact and rubbing of flat surfaces. J. Appl. Phys. **24**(8), 981–988 (1953)

Evaluation of BBI Performance Indicator in a Full-Scale Test Bench

Hamid Masoumi[1(✉)], Bram Veelhaver[1], Manthos Papadopoulos[1], Fülöp Augusztinovicz[2], and Patrick Carels[1]

[1] CDM Group, 3090 Overijse, Belgium
hamid.masoumi@cdm.be
[2] Budapest University of Technology and Economics Budapest, Budapest, Hungary
https://www.cdm.eu

Abstract. Prediction of the real performance of Building Base Isolation (BBI) systems to control ground borne vibration has been the subject of several research programs in recent years. Comparing the vibration level in the superstructure in an isolated setup with that obtained in a non-isolated setup, the performance of the BBI system can be directly calculated. However, in practice, a non-isolated building does not exist.

Recently, in the frame of the EUROSTAR BIOVIB project (Building Insulation against Outdoor Vibrations), based on Power Flow Insertion Gain (PFIG), a simplified performance indicator has been introduced. The new indicator is defined in terms of the measured transmissibility in the isolated building and the mobilities of the superstructure and substructure .

To validate the proposed performance indicator, a full-scale test bench has been developed. The test bench has three main parts: (1) a substructure (foundation), (2) an isolation bearing and (3) a superstructure. The superstructure consists of a concrete column and four steel blocks being hung from the column by four springs representing the effect of floor resonance frequencies. The substructure (foundation) is a thick steel plate supported by four thin pads.

The performance of a selected BBI system has been determined by the proposed indicator and then compared with that measured by the direct insertion loss in the test bench. The proposed indicator shows a reasonable agreement with the direct insertion gain up 100 Hz in the frequency range of ground borne vibration due to railway induced vibrations.

Keywords: Building Base Isolation (BBI) · Isolation performance · Ground borne vibration

1 Introduction

The buildings, near railway or subway networks, can be subjected to ground-borne vibrations resulting in an unacceptable level of noise (structural borne-noise) and vibration for the people inside the building. Despite different techniques to mitigate the vibrations at the source (the railway track) or along the

G. Degrande et al. (Eds.): *Noise and Vibration Mitigation for Rail Transportation Systems*, NNFM 150, pp. 267–274, 2021.
https://doi.org/10.1007/978-3-030-70289-2_27

propagation path (e.g. using isolating trenches), the building as the receiver can be decoupled from the surrounding ground using Building Base Isolation (BBI) techniques.

The building can be decoupled from the surrounding ground at the foundation level or at the columns or walls in a upper level using springs or elastomeric pads. Since the first known application in 1964, where the Royal Conservatory in Antwerp was decoupled from its surroundings by resilient bearings, the technology has evolved dramatically [1].

2 Performance of the Building Base Isolation System

Although the isolation efficiency of an individual isolation bearing can be obtained by simplifying the problem to a dynamic system with a single degree of freedom (SDOF), the dynamic behavior of a building with its flexible structural members is more complex, in practice. In addition, a simple SDOF model can not account for the multi-directional vibrations due to the rail traffic.

In fact, the assessment of real insertion gain requires a comparison between the transmitted vibration through the building with and without the isolation system. However, in practice, when a building is isolated, a non-isolated building does not exist, and the real insertion gain is not measurable. Determination of the real performance of BBI systems has been the subject of several research projects in recent years [2,3]. It has been shown that the real insertion gain is influenced not only by the dynamic characteristics of the isolation bearings but also by mobilities of the soil-foundation system and the superstructure including the flexibility of superstructure members like floors [2,4].

In the frame of the BIOVIB project, "Building Insulation against Outdoor Vibrations", to better express the performance of an isolated building, the performance indicator proposed by Villot et al. [4] has been has been validated. Based on the Power Flow Insertion Gain (PFIG) concept proposed by Talbot and Hunt [2], this BBI indicator has been defined in terms of the transmission loss "TL" (which can be measured in an isolated building) and the mobilities of the substructure (soil-foundation) and the superstructure (the structure above the vibration isolation cut) systems [4,5].

The isolation efficiency was determined by two different indicators: (1) Building Insertion Gain (BIG) that refers to the difference between the vibration level of the superstructure in the test setup with and without the isolation system, calculated as follows:

$$\mathrm{BIG}\,[\mathrm{dB}] = 10\log_{10}\left(\frac{v^2_{b,rms,iso}}{v^2_{b,rms,non-iso}}\right) \tag{1}$$

where $v_{b,rms,iso}$ and $v_{b,rms,non-iso}$ represent the building (superstructure) vibration levels in an isolated and non-isolated building, respectively.

(2) BBI indicator based on Power flow insertion gain (PFIG) that refers to a correction on the transmission loss and can be measured in an isolated building as follows:

$$\text{BBI indicator [dB]} = \text{TL [dB]} + 10\log_{10}\left(\frac{|Y_b|^2 + |Y_s|^2}{|Y_b|^2}\right) \quad (2)$$

where Y_s and Y_b are the mobilities of the substructure and the superstructure, respectively. The transmission loss (TL) is defined as the difference between the vibration level measured below and above the isolation bearing as follows:

$$\text{TL [dB]} = 10\log_{10}\left(\frac{\mathrm{v}^2_{\mathrm{b,rms,iso}}}{\mathrm{v}^2_{\mathrm{s,rms,iso}}}\right) \quad (3)$$

In Eq. (2), the first term (TL) depends on the resonance frequency of the system and is mostly influenced by the dynamic stiffness of the isolation bearing and the effective dynamic mass of the system. The second term acts as a modification factor and shows the effect of dynamic soil-structure interaction as well as the dynamic behavior of superstructure members. Note that, in practice, the mobility of the superstructure can only be calculated numerically using FEM.

To experimentally validate the proposed BBI indicator, a set of in-situ measurements has been performed [6,7]. In addition, a test bench has been developed in the CDM laboratory and the performance of full-scale BBI systems have determined by both the direct Building Insertion Gain (BIG) and the proposed BBI indicator.

3 Building Base Isolation Test Bench

The test bench has three main parts: (1) a superstructure, (2) isolation bearings and (3) a substructure (foundation), Fig. 1a. The superstructure consists of a concrete column and four steel blocks hung from the column by four springs to represent the effect of floor resonance frequencies. The substructure (foundation) is a thick steel plate supported by four thin pads adjusted to dynamically behave similar to a real soil-foundation system.

3.1 Measurement Configuration

The measurement configuration of the full-scale test setup for the evaluation of the isolation performance is given in Fig. 1b in which an isolation bearing is placed between the foundation plate and the concrete column. For the non-isolated setup, the isolation bearing was replaced by a rigid steel block. An isolation system type CDM-SEB (a solution 8 Hz) with four elastomeric pads 50 × 50 × 50 mm has been tested. The steel blocks on top of the column are supposed to show a resonance frequency at 20 Hz.

The foundation and the superstructure responses were measured at different points due to a hammer impact applied to the foundation: two on the foundation (CH2 and CH3), two on the end-plate bottom of the column (CH4 and CH5), and one on top of the column (CH6). In this work, the isolation efficiency of the isolation system consisting of four elastomeric discrete pads was determined as the difference between the vibration level at the super-structure in a non-isolated

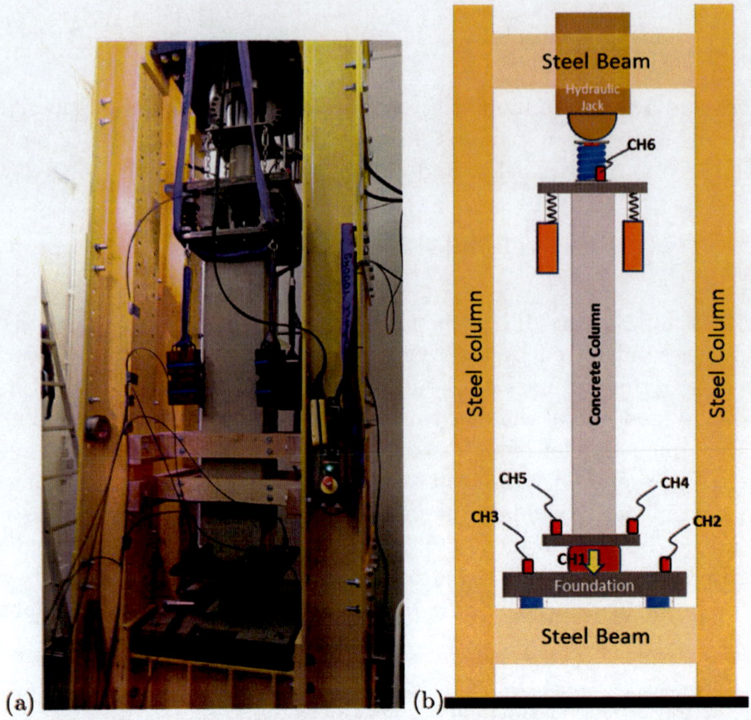

Fig. 1. Full-scale test setup for measuring the building isolation performance.

building and that obtained in an isolated building. The non-isolated building has been built by replacing the isolation bearing with a rigid steel block. The foundation is supported by four thin resilient pad with very high damping to represent the real soil-foundation dynamic behavior.

First, the mobilities of the foundation (substructure) and the superstructure were determined. The mobility of the superstructure was obtained by a hammer impact test. The impact was applied on the column foot when the superstructure was hanging over a soft spring. The column response was measured by means of two accelerometers mounted on the column end-plates. The mobility of the foundation was measured by the hammer impact applied vertically at the center of the steel plate and measuring the foundation response in the same direction. Then, to determine the performance of the isolation system, the full setup was excited by means of a hammer impact on the foundation.

3.2 Measurement Results

Figure 2 displays the mobilities of the substructure and the superstructure. The mobility of the foundation shows a highly damped resonance frequency 110 Hz. The column is relatively rigid and the anti-resonance at 22 Hz is related to the mass-spring system on top of the column which represents the effect of the resonance frequency of floors in a building.

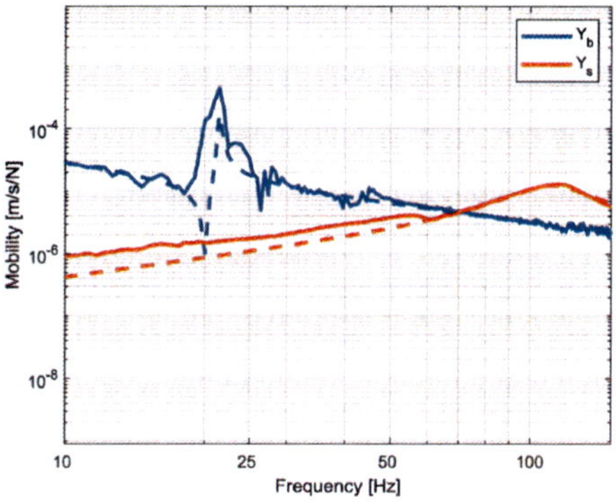

Fig. 2. Foundation and superstructure mobilities: the experimental result (solid line) compared with the numerical one (dashed line).

Figure 3 displays the responses of the foundation and the column (superstructure) in the full-scale test bench. Results are presented in terms of frequency response function (FRF's) at different measurement points. As expected, the column responses at the top and the bottom positions were quasi-identical. Furthermore, a peak at 8 Hz refers to the resonance frequency of the isolation system with four CDM-HR pads and two anti-resonance modes at 22 Hz 29 Hz reflect the effect of the steel blocks hanging on the column top plate. Note that because of weak coherency at low frequecies, the foundation response in the isolated setup has been modified based on that obtained in a tuned numerical model of 3DOFs.

The performance of the isolated system was calculated first by the BIG indicator as the difference between the vibration level of the superstructure in the isolated built-up and that in the non-isolated one, and then by the proposed BBI indicator in the isolated setup.

Figure 4 displays the performance of the isolation bearings obtained using BIG indicator. Results show a reasonable agreement between the numerical 3-DOFs model and those obtained directly from the test bench measurement.

It should be noted that the BIG indicator will be representative of the building performance, refering to the vibration reduction inside the rooms, and might vary from one place to another inside the building, since the floor vibration modes might be different in different rooms and locations. In the other hand, the BBI indicator would be more consistent and would depend on primarily on the performance of the isolators and would affect less by the floor response inside of the building. Note that the effect of the flloor resonance frequencies in the BBI indicator would be applied by the mobility correction factor (Eq. 2).

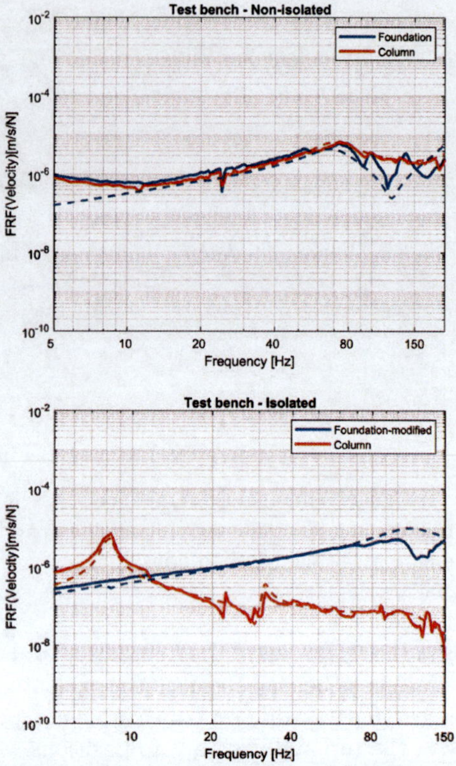

(a)

(b)

Fig. 3. The foundation and column responses (FRF's) measured in the test bench with (a) non-isolated setup and (b) isolated one with four CDM-HR elastomeric pads.

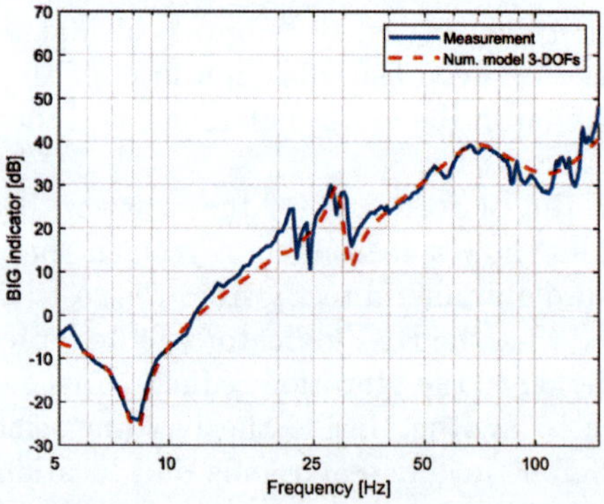

Fig. 4. The performance of full-scale system obtained by the BIG indicator: the measurement (solid line) compared with the numerical model (dashed line).

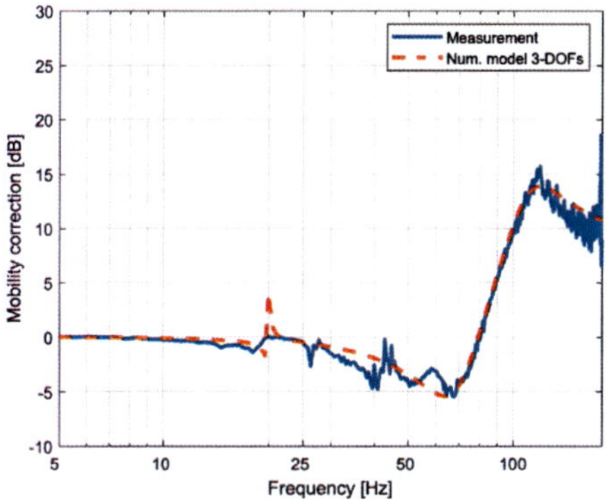

Fig. 5. Mobility correction factor: the measurement (solid line) compared with the numerical model (dashed line).

Figure 5 shows the mobility correction factor based on the measured mobilities of the substructure and the superstructure. It can be seen that the effect of the mobility correction factor becomes more pronounced at frequencies higher 30 Hz.

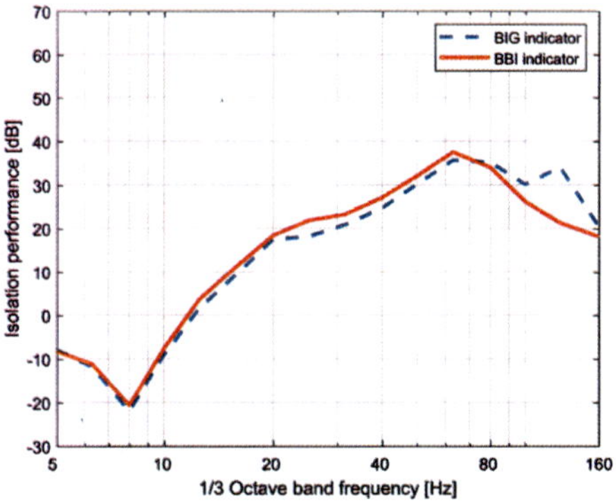

Fig. 6. Performance of the isolation system determined experimentally in terms of the BIG indicator (dashed line) and the BIOVIB BBI indicator (solid line).

Figure 6 displays the comparison between the isolation performance of the full-scale system obtained by the BIG indicator and that of the BBI indicator in one-third octave bands frequency. Although a good agreement was obtained between the two indicators up 100 Hz, at higher frequencies, the BBI indicator shows lower performance. Note that the same discrepancy at higher frequencies has been seen in the results of the numerical model.

4 Conclusions

A test bench with a full-scale BBI system has been built to evaluate the performance of isolation systems. The isolation performance was directly determined as the difference between the vibration level of the super-structure in a non-isolated setup and that obtained in an isolated one (BIG indicator). In this work, the efficiency of the CDM-SEB system has been examined with both the BIG indicator and BBI indicator implemented in BIOVIB. The resonable agreement between the two indicators confirms the validation of the BIOVIB indicator up 100 Hz. Both methods very well confirm the efficiency of the system with performance greater than 20 dB at frequencies 40 Hz.

Acknowledgment. The results presented in this paper have been obtained within the frame of the BIOVIB project funded by VLAIO and EUROSTARS. Their financial support is gratefully acknowledged.

References

1. Carels, P.: Building base isolation recent progress, feedback and practical hints. In: Proceeding of Acoustics Conference, Liverpool, UK, vol. 22, pp. 247–254 (2000)
2. Talbot, J.P., Hunt, H.E.H.: On the performance of base-isolated buildings. Build. Acoust. **7**(3), 163–178 (2000)
3. Wasmine, W., Bailhache, S., Villot, M., Philippe, J., Torbay, J., Bozzetto, D.: Projet DIAVIBAT : Prévision et contrôle de la performance sur site des dispositifs d'atténuation vibratoire dans les bâtiments. Acoustique et Techniques : trimestriel d'information des professionnels de l'acoustique", 79 / 80, August (2015)
4. Villot, M., Philippe, J.: Railway vibration: predicting the field performances of mitigation measures in buildings. In: Proceedings of the 9th International Conference on Structural Dynamics, EURODYN 2014, Porto, Portugal, (2014)
5. Trévisan, B., Grau, L., Bozzetto, D., Villot, M., Jean, P.: In-situ performance prediction of base-isolated buildings. In: Proceeding of EURONOISE conference, Crete, Greece (2018)
6. Bozzetto, D., Trévisan, B.: Validation of new methods by in situ experimental measurements. BIOVIB Progress report, November (2018)
7. Masoumi, H., Pinto, P., Carels, P.: In-situ investigation of building base isolation performance: construction details and bridging elements. In: Proceeding of INTER-NOISE conference, Madrid, Spain (2019)

Rail Roughness, Corrugation and Grinding

Tonal Noises and High-Frequency Oscillations of Rails Caused by Grinding Procedures

Jörg Rothhämel[1(✉)], F. Kendl[1], B. Lütke[2], S. Lange[3], T. Kempinger[4], and B. Asmussen[4]

[1] DB Systemtechnik GmbH, Völckerstraße 5, 80939 Munich, Germany
joerg.rothhaemel@deutschebahn.com
[2] Former DB Systemtechnik GmbH, Völckerstraße 5, 80939 Munich, Germany
[3] DB Netz AG, Pfarrer-Perrabo-Platz 4, 60326 Frankfurt am Main, Germany
[4] DB Netz AG, Theodor-Heuss-Allee 5-7, 60486 Frankfurt, Germany

Abstract. After certain rail grinding processes, the subsequent train traffic seems to cause high tonal noises. Since residents' complaints increased, the DB Netz AG set up the project noise-optimized rail grinding called "ZIP Lärmschleifen". Acoustic rail roughness criteria directly after grinding were developed and defined, based on annoyance and tonality. The optimization of grinding procedures has been evaluated. Narrow band spectra of airborne noise and rail roughness were evaluated and merged into one diagram based on the speed of the train. The development of the rail roughness over 2 Mega Gross Tons leveling the rail has been evaluated. An acoustic quality inspection of the rail grinding using the DB Systemtechnik noise measurement train is possible. The acoustic grinding process optimization has been successful. The acoustic grinding process optimization has been successful. The annoyance of residents due to tonal noise after grinding can be reduced significantly.

Keywords: Acoustic grinding · Tonal noise · Rail roughness criteria

1 Introduction

After certain rail grinding processes, the subsequent train traffic seems to cause high tonal noises. Since resident's complaints increased, the DB Netz AG set up the project noise-optimized rail grinding. In the framework of this project, six differently ground test sections were acoustically investigated. Public rail tracks have been used for the pass-by noise measurements of the regular track. In addition to microphone measurements, surface roughness and track decay rate measurements were carried out at all test sections over a period of four weeks. The aim of the project is the acoustical optimization of the grinding processes.

Within the project "ZIP Lärmschleifen" acoustic critical values for the rail roughness directly after grinding were developed and defined. The critical values are based on annoyance and tonality. This paper summarizes first results of the investigations recently performed at Deutsche Bahn.

G. Degrande et al. (Eds.): *Noise and Vibration Mitigation for Rail Transportation Systems*, NNFM 150, pp. 277–285, 2021.
https://doi.org/10.1007/978-3-030-70289-2_28

2 Procedure and Results

In sect. 2.1 the test track is defined and sect. 2.2 describes the test set-up. Section 2.3 gives an introduction into the special diagram design. Limiting curves and tonality criteria are defined in sect. 2.4. Section 2.5 presents the test results. Finally sect. 2.6 gives an overview about the results with the noise monitoring train.

2.1 Test Track

The test track was chosen based on the following criteria: straight lined, low ambient noise, mixed rail traffic (intercity transport, commuter trains and freight trains), same railway parameters over all test sections. The line 5300 north of Augsburg, Germany, was chosen, see Fig. 1.

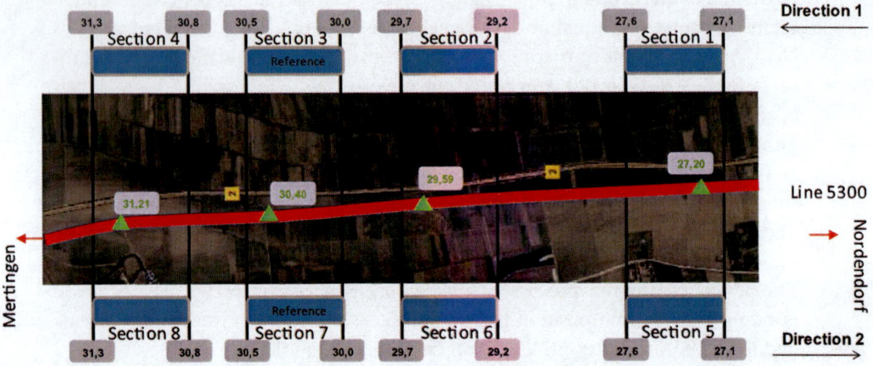

Fig. 1. Schematic overview of the test section, Line number 5300 (Photo: Google earth)

2.2 Test Design

The project started with a survey of the acoustic performance after grinding of 6 test sections and two reference sections, see Fig. 1, in 2017. In two further tests in 2018 the optimization of the grinding processes has been evaluated.

The acoustic rail roughness was measured with m|rail trolley (Müller-BBM) in accordance with EN 15610 [1] over a length of 60 m. The air borne noise was measured following ISO 3095 [2]. Three microphones were positioned in a distance of 15 m between each microphone (Fig. 2).

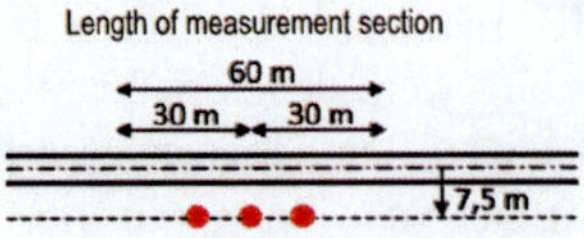

Fig. 2. Schematic overview of one test track section, legend: ● microphone position

The evolution of the acoustic roughness over gross tons and time was monitored directly after grinding, after 0.7 and after approx. 2 Mega Gross Tons (MGT). The reference sections (with a typical low roughness without grinding effects and same track design) are used to check validity of the measurements.

2.3 Diagram Design

To identify the high tonal noises that partly sound like a merger of a whistle and a circular saw the narrow band spectra of airborne noise and rail roughness were evaluated. Based on the speed of the train the two diagrams (Fig. 3) were merged into one diagram (Fig. 4). By merging the annoying frequency a clear correlation between rail roughness and airborne sound can be identified. The reduction of the frequency peaks after leveling by train with 2 Mega Gross Tons is shown in Fig. 4.

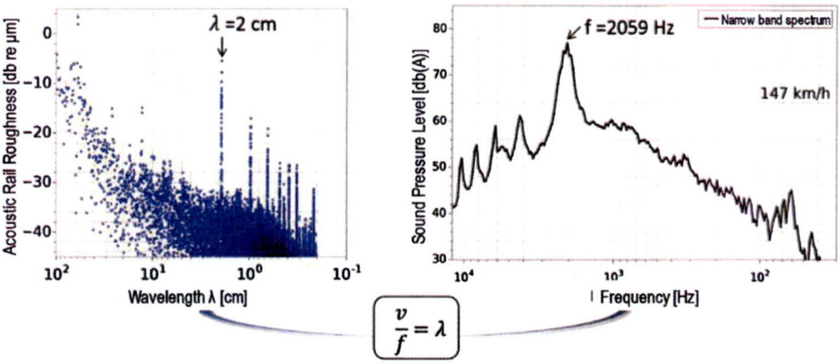

Fig. 3. Correlation between narrow band roughness spectrum (left), narrow band airborne sound spectrum (right) and speed of the train

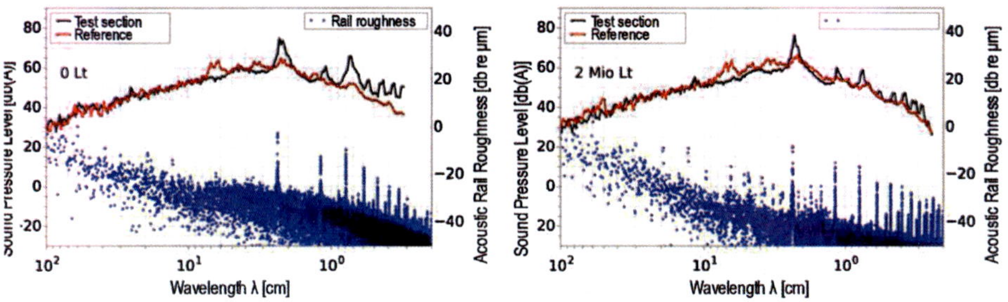

Fig. 4. Exemplary roughness spectrum and airborne spectrum after grinding (left) and after 2 MGT leveling (right)

2.4 Limiting Curves and Tonality Criteria

The limiting curve based on EN 3095 [4] is not designed for ambient noise, it is designed for train homologation. So, based on psychoacoustic listening tests, analysis of acoustic rail roughness on the different test sections and TWINS modeling new limiting curve for less acoustical annoyance of residents has been developed, see Fig. 5.

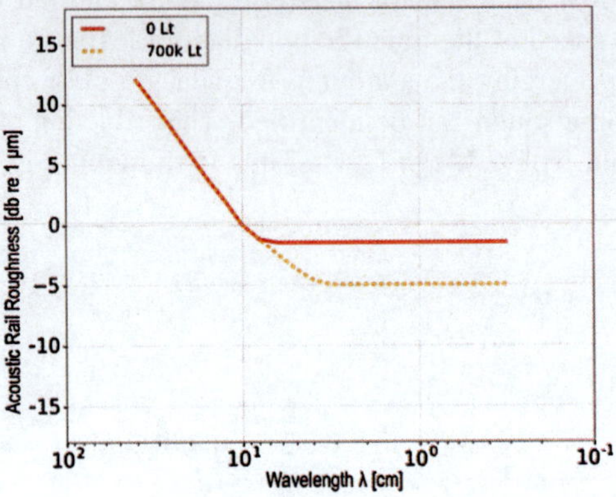

Fig. 5. Acoustic rail roughness limiting curves directly after grinding (solid line) and after 0.7 MGT (dashed line)

This limiting curve is defined for two different measurement times: Directly after grinding (red) and after 0.7 MGT leveling the rails. Only the value in one third-octave band is allowed to be over the limiting curve directly after grinding.

Based on the experiences of the listening tests it became clear that nevertheless tonal annoyance can happen. So a second criterion had to be defined:

Rule 1
The level in a third-octave band must not exceed the arithmetic average of the neighboring bands by more than 5 dB for wavelengths $\lambda \leq 50$ mm and more than 8 dB for wavelengths $\lambda > 50$ mm.

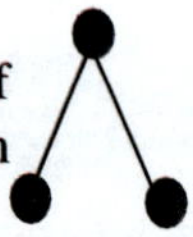

Rule 2
The arithmetic average of two neighboring third-octave bands must not exceed the arithmetic average of the neighboring bands by more than 5 dB for wavelengths $\lambda \leq 50$ mm and more than 7 dB for wavelengths $\lambda > 50$ mm.

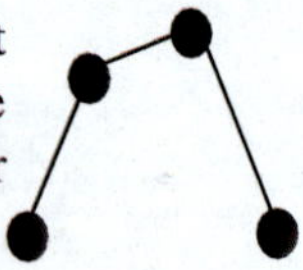

Both criteria have to be fulfilled after grinding.

2.5 Test Results

Six grinding companies participated in this project. But only three finalized the complete program for different reasons. So here only the results are shown of the three companies that finished the program. This does not say anything about the grinding quality of the other three companies.

2.5.1 Test Results After the Second Optimization Phase

The following figures show the results of three grinding sections as the final result of the acoustic optimization. First the roughness and airborne spectrum are shown directly after grinding. Second on the left hand side the acoustic roughness directly after grinding is shown and on the right hand side the development of the roughness after 0.7 and after 2 MGT.

Grinding Section A

In the short wavelength range ($\lambda < 2$ cm) a significant reduction of the roughness peaks is shown after 2 MGT, see Fig. 6. In comparison to the reference section a significant difference in airborne noise of SPL $\Delta L_{pA,max} = 8$ dB is found at a wavelength of $\lambda = 0.5$ cm. However, the noise does not show tonality.

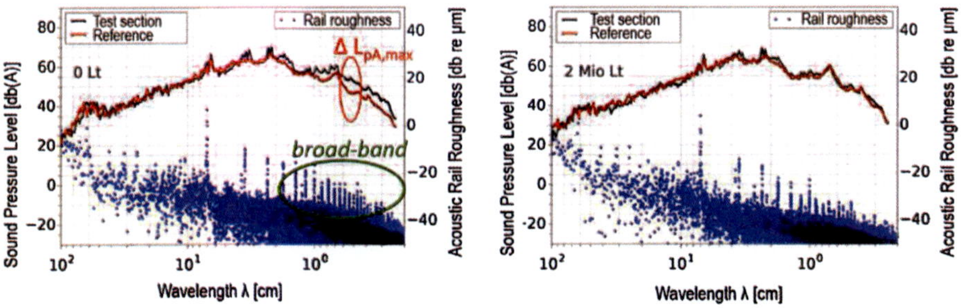

Fig. 6. Roughness spectrum and airborne spectrum after grinding (left) and after 2 MGT leveling

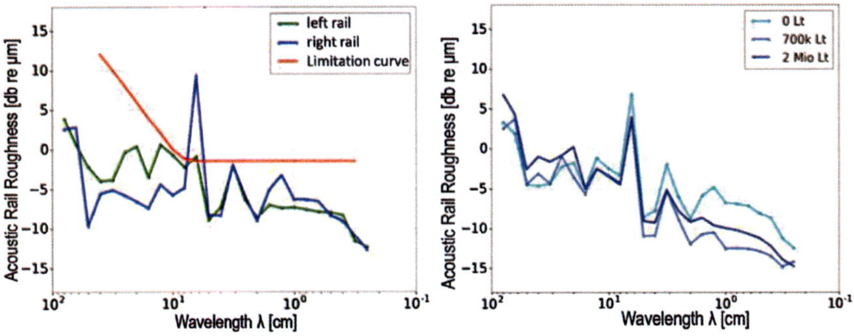

Fig. 7. Roughness (third-octave band) after grinding (left) and development of the rail roughness (right) after 0.7 and 2 MGT leveling

The limiting curve criterion is fulfilled for both rails. The tonality criterion is fulfilled on the left rail and three times not fulfilled on the right rail (Fig. 7).

Grinding Section B

In the short wavelength ($\lambda < 2$ cm) a significant reduction of the roughness peaks is shown after 2 MGT, see Fig. 8. The airborne tonality matches to the acoustic roughness spectrum very well. In comparison to the reference section a significant difference in airborne noise of SPL $\Delta L_{pA,max} = 11$ dB is found at a wavelength of $\lambda = 0.8$ cm directly after grinding. The tonality is given for wavelength $\lambda \leq 0{,}5$ cm.

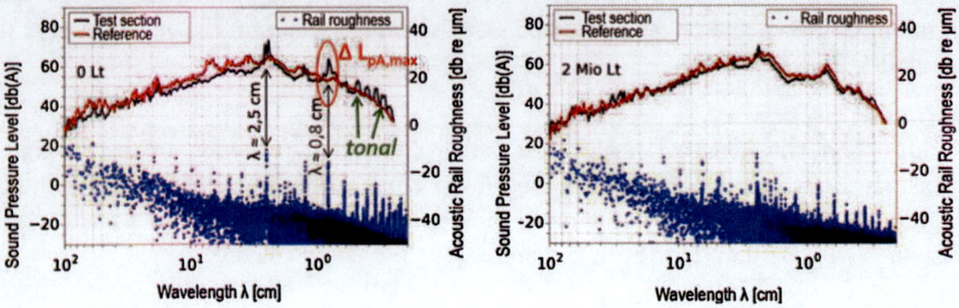

Fig. 8. Roughness spectrum and airborne spectrum after grinding (left) and after 2 MGT leveling

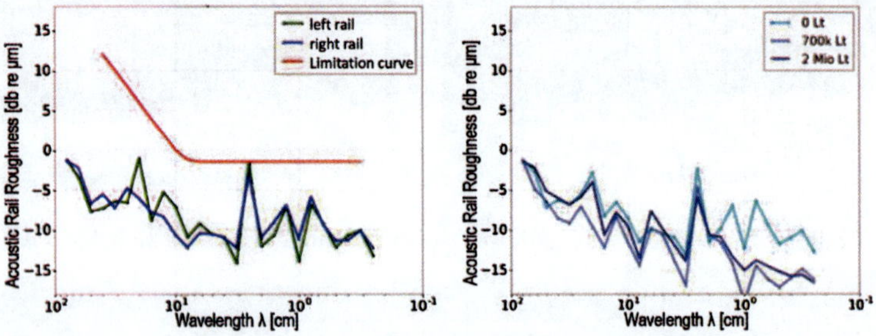

Fig. 9. Roughness (third-octave band) after grinding (left) and development of the rail roughness (average over both rails, right) after 0.7 and 2 MGT leveling

The limiting curve criterion is fulfilled. The tonality criterion is two times not fulfilled on the left rail and one time not fulfilled on the right rail (Fig. 9).

Grinding Section C

In the short wavelength ($\lambda < 2$ cm) a significant reduction of the roughness peaks is shown after 2 MGT, see Fig. 10. The airborne tonality matches the acoustic roughness spectrum very well. In comparison to the reference section a significant difference in airborne noise of SPL $\Delta L_{pA,max} = 8$ dB is found at a wavelength of $\lambda = 0.3$ cm. However, the noise does not show tonality.

The limiting curve criterion is fulfilled.

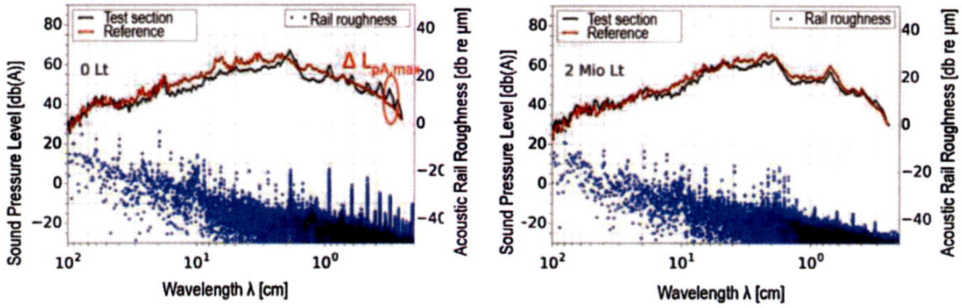

Fig. 10. Roughness spectrum and airborne spectrum after grinding (left) and after 2 MGT leveling (right)

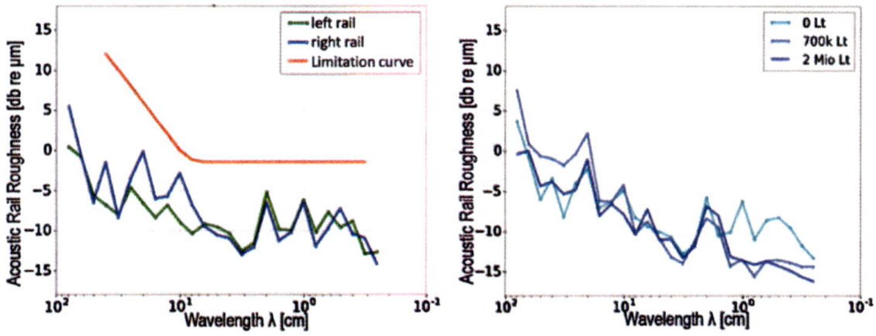

Fig. 11. Roughness (third-octave band) after grinding (left) and development of the rail roughness (average over both rails, right) after 0.7 and 2 MGT leveling

Eye-catching is the difference of the acoustic roughness of the left and the right trail (Fig. 11). The reason for the difference was not investigated in this program, but it should not be forgotten.

A strong improvement of the acoustic grinding quality was shown by three grinding companies. The following section shows the development results.

2.5.2 Development of the Acoustic Rail Roughness

At the beginning of the program each grinding company showed the grinding results as they thought it would give the best acoustic results. Based on the results of the measurements they optimized their grinding process. Figure 12 shows the significant results of the optimization.

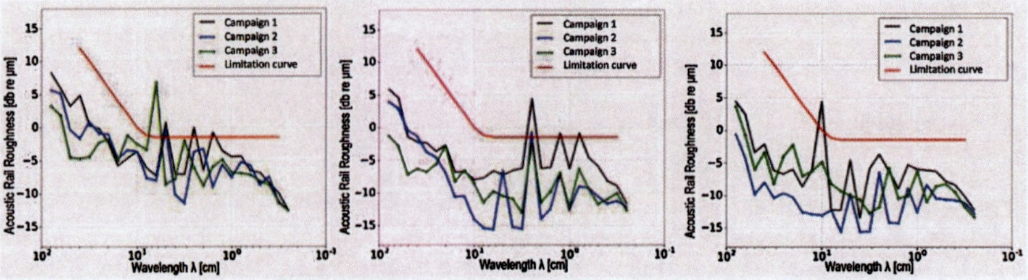

Fig. 12. Results of the grinding process development over two optimization phases (campaign 1 – 3), grinding sections A – C. The sections are the same that were ground in each campaign

2.6 Inspection with the DB Noise Measurement Train (Schallmesswagen)

Parallel to the acoustic rail roughness and air borne way side airborne noise measurements measurement the noise measurement train of DB Systemtechnik, see also [3] and [4], for periodic track continuous noise measurement has been used to identify if the grinding quality can also be measured with this train. We identified that the acoustic grinding quality over the complete grinding section was not always homogeneous, as shown in Fig. 13 in the left spectrogram. The acoustic roughness of the track can be identified in the spectral acoustic signal of the Schallmesswagen. So a track-continuous acoustic quality control of grinding would be possible.

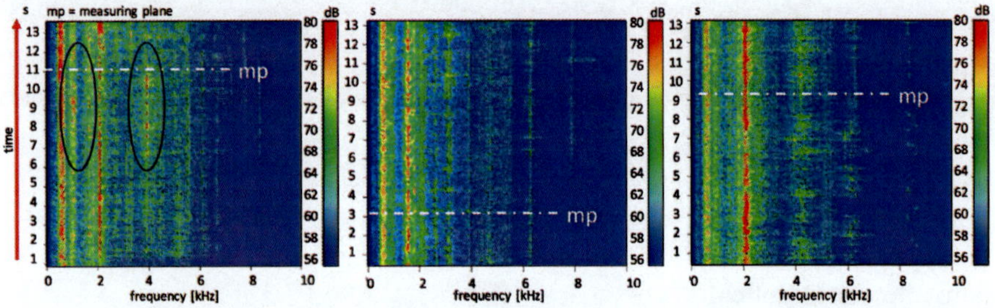

Fig. 13. Spectrogram of the noise measurement train signal over three grinding sections

3 Conclusion

Within the project, measures to reduce the acoustic annoyance of residents close to railways after rail grinding were compiled and evaluated. The results can be used to define acoustic grinding standards (Acoustic rail roughness limiting curves, see Fig. 5, and roughness tonality criteria (rule 1 &2)) in noise sensitive areas and to evaluate the acoustic quality of a local used grinding process. If the standard will be not fulfilled the grinding process can be optimized. This is independent of the grinding procedures. Furthermore, criteria of rail roughness are given to identify in the future less annoying rail grinding processes without airborne acoustic measurements. A strong evaluated

TWINS model is available at DB Systemtechnik considering the measured rail of 8 sections and train parameters. At the end of the project DB Netz AG is able to set up a guideline, which grinding procedures should be used in noise sensitive environments.

Acknowledgements. The authors would like to thank the German Federal Ministry of Transport and Digital Infrastructure (BMVI) for funding the project.

References

1. DIN EN 15610, Bahnanwendungen - Geräuschemission - Messung der Schienenrauheit im Hinblick auf die Entstehung von Rollgeräusch (2009)
2. DIN EN ISO 3095: Messung der Geräuschemission von spurgebundenen Fahrzeugen (2013)
3. Pieringer, A., et al.: Acoustic monitoring of rail faults in the German railway network. In: Noise and Vibration Mitigation for Rail Transportation Systems - Proceedings of the 13th International Workshop on Railway Noise, Ghent, Belgium, 16–20 September 2019 (2019)
4. Rothhämel, J., Schröder, S., Koch, B.: Akustischer Fahrflächenzustand im Netz der DB Netz AG, ZEVrail 139 (2015)

Extracting Information from Axle-Box Acceleration: On the Derivation of Rail Roughness Spectra in the Presence of Wheel Roughness

Tobias D. Carrigan and James P. Talbot(✉)

Department of Engineering, University of Cambridge, Cambridge CB2 1PZ, UK
{tdc41,jpt1000}@cam.ac.uk

Abstract. Railhead roughness increases over time, leading to increased environmental noise and vibration. The use of axle-box acceleration (ABA) measurements on in-service railway vehicles to monitor rail roughness is potentially more cost-effective than other techniques. The measured acceleration requires signal processing to derive suitable metrics of railhead condition. A transfer function may be calibrated with direct roughness and ABA measurements made on a reference track, which may then be used to derive roughness spectra from subsequent ABA measurements. However, this approach is affected by variations in track dynamic behaviour, as well as variations in wheel roughness, which is inherently combined with rail roughness in the ABA measurement. This paper proposes an improved approach that (i) extracts the track's dynamic stiffness parameters from the ABA measurements, enabling the derivation of the roughness-ABA transfer function for each section of track, and (ii) separates the wheel and rail roughness by synchronous averaging over several wheel revolutions. By accounting for variations in track properties and removing the influence of wheel roughness, initial modelling indicates that reliable measurements of rail roughness spectra can be obtained in practice.

Keywords: Axle-box acceleration · Rail roughness · Corrugation · Wheel roughness · Track stiffness

1 Introduction

The potential of axle-box accelerometers, mounted on in-service railway vehicles, has long been recognised as a cost-effective means of continuously monitoring the condition of railway track, without the need for dedicated measurement trains [1, 2]. An accelerometer measures the vertical motion of the axle-box as excited by irregularities on the running surfaces of the wheel and track. Axle-box acceleration (ABA) measurements have previously been analyzed through a variety of techniques, ranging from simple root-mean-square analyses [3] to time-frequency analyses [4], in order to detect

G. Degrande et al. (Eds.): *Noise and Vibration Mitigation for Rail Transportation Systems*, NNFM 150, pp. 286–294, 2021.
https://doi.org/10.1007/978-3-030-70289-2_29

corrugation and short track defects [5], for example. However, ABA alone cannot quantify rail roughness in absolute terms as it depends on vehicle speed and the dynamic properties of the vehicle and track, as well as wheel roughness.

This paper reports on progress towards deriving rail roughness spectra from ABA that are independent of vehicle speed, track dynamics and wheel roughness. The derivation requires knowledge of the transfer function from the roughness excitation to the axle-box response. This transfer function can be determined by comparing direct roughness and ABA measurements made on a reference track, as done for the High-Speed Rail Corrugation Analyzer (HSRCA) [6], Banverket's ABA system [7] and by Németh and Schleinzer [8], but the result is accurate only on track that is dynamically similar to the reference track at the time of calibration. In practice, the track's stiffness can vary significantly with location and time due to variations in track construction, soil properties and environmental conditions: trial HSRCA measurements [6] were affected by varying track properties at different locations, and ABA measurements on the UK West Midlands Metro have been found to vary seasonally with environmental conditions [9]. There is a clear need for improved methods.

2 Deriving Roughness Spectra from Axle-Box Acceleration

By considering force equilibrium and compatibility of displacements at the wheel-rail interface, and assuming stationary random signals of rail roughness and ABA, the roughness power spectral density (PSD) $S_{\delta\delta}(\gamma)$ in the wavenumber (γ) domain may be derived from the frequency (ω)-domain PSD of ABA $S_{aa}(\omega)$ using random process theory [9, 10]:

$$S_{\delta\delta}\left(\gamma = \frac{\omega}{v}\right) = vS_{\delta\delta}(\omega) = \frac{v}{|H(\omega)|^2} S_{aa}(\omega) \tag{1}$$

where v is the vehicle speed and $H(\omega)$ is the transfer function from roughness excitation to ABA, which depends on the point receptances of the wheel $H_w(\omega)$ and rail $H_r(\omega)$, and the contact spring stiffness k_H:

$$H(\omega) = -\omega^2 \left(\frac{H_w(\omega)}{H_w(\omega) + H_r(\omega) + k_H^{-1}} \right) \tag{2}$$

An estimate of the rail roughness PSD may therefore be derived directly from the ABA PSD, given the vehicle speed and estimates of the wheel and rail receptances.

2.1 PSD Calculation and Roughness Derivation Procedure

Here, PSDs are calculated using Welch's method of averaging periodograms (square-magnitude discrete Fourier transforms) of overlapping segments of the signal. The length and number of segments are selected based on recommendations in EN 15610 [11] for plotting rail roughness spectra at wavelengths up to 1 m: 30 segments of length 4 m are used to attain the required accuracy [9]. The time-histories acquired over the respective segments are Hann-windowed and overlapped with each other by 50%, so the spectra are taken over 62 m of track.

The basic procedure for deriving the rail roughness PSD from an ABA signal recorded on a section of track is as follows. The track is split into 30 overlapping segments and the time-domain ABA signal is aligned with the segment boundaries according to vehicle position and speed. Hann-windowed periodograms $S_{aa}(\omega)$ of the 30 ABA signal portions between segment boundaries are calculated and then converted into wavelength-domain roughness periodograms $S_{\delta\delta}(\lambda)$ using Eq. (1), substituting $\lambda = 2\pi/\gamma$. These are then averaged together to give the final roughness PSD. Applying Eq. (1) to the individual periodograms, rather than their average, reduces the effect of any variation in vehicle speed in the frequency-to-wavelength conversion [9]. This procedure can then be repeated on consecutive sections of track.

3 Simulator

A vehicle-track simulator has been developed to test the proposed algorithm. The simulator models an idealized vehicle-track system to calculate the ABA signal that would be measured as a wheel rolls along a given rail roughness profile. The ABA signal is fed into the roughness derivation algorithm under test, and the derived roughness spectrum is then compared to the actual roughness spectrum.

The simulator is based on the vehicle-track model illustrated in Fig. 1(a), which represents the track as an Euler beam on a continuous viscoelastic foundation that is discretized into finite elements [9] using the 'moving element' method developed by Koh *et al.* [12].

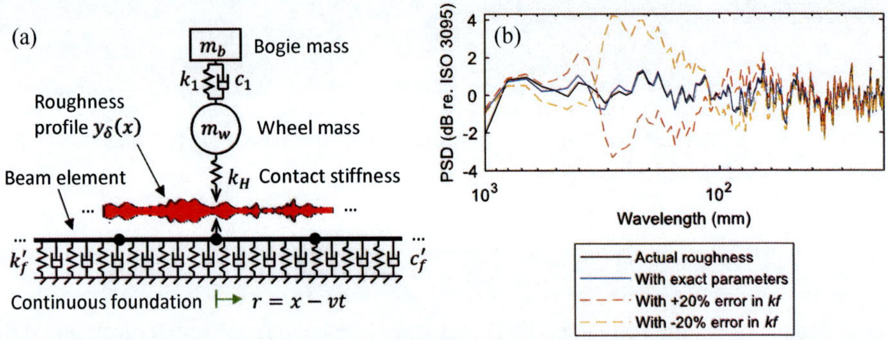

Fig. 1. (a) 2-DoF vehicle model on a moving-element beam on viscoelastic foundation with moving roughness profile. (b) PSDs of actual and derived roughness, relative to the ISO 3095:2013 limit spectrum, with exact model parameters and with ±20% error in track foundation stiffness k_f'.

3.1 Results for Known Vehicle-Track Parameters

The simulator is run along a random rail roughness profile with the vehicle and track parameters given in Table 1, which represent directly-fastened underground track [13]. The roughness profile is generated according to the roughness limit spectrum in ISO

3095:2013 [14]. Roughness PSDs are taken over 30 segments of length 4 m, overlapped by 50%, starting 5 m from the start of the simulation to avoid the initial transient in the model response. The roughness-to-ABA transfer function (Eq. (2)) is calculated using analytical point-receptance functions of the wheel and the rail, using the same vehicle and track parameters as the simulator (Table 1). The wheel receptance $H_w(\omega)$ corresponds to the mass-spring vehicle model of Fig. 1(a), and the rail receptance $H_r(\omega)$ is the frequency-domain solution for an Euler beam on viscoelastic foundation [9]. The actual roughness spectrum and three derived spectra (from the procedure outlined in Sect. 2.1) are plotted relative to the ISO 3095:2013 limit spectrum in Fig. 1(b); one of the derived spectra is obtained with the exact track parameters and two are obtained with ±20% error introduced into the track foundation stiffness k_f' used in the transfer function.

Table 1. Vehicle and track parameters used in the simulations [13].

Track parameter	Value	Vehicle parameter	Value
Wheel-rail contact stiffness k_H	1.35 GN m^{-1}	Vehicle speed V	20.0 m s^{-1}
Rail mass per length m_r'	56.0 kg m^{-1}	Wheel mass m_w	600 kg
Rail bending stiffness EI	4.86 MN m^2	Suspension stiffness k_1	1.40 MN m^{-1}
Foundation stiffness per length k_f'	77.0 MN m^{-2}	Suspension damping c_1	30.0 kN s m^{-1}
Foundation damping per length c_f'	32.8 kN s m^{-2}	Bogie mass m_b	1000 kg

If the exact parameters are known, the derived spectrum is very close to the actual spectrum. With the 20% deviation in the track stiffness, errors of up to 4 dB can be seen in Fig. 1(b) at wavelengths from 70 to 400 mm. This error is amplified because the P2 resonance frequency (near 62 Hz; wavelength of 0.32 m at 20 m/s) is shifted by the parameter deviation; this amplification increases with lower damping. This result indicates the importance of using an in-situ measurement of track stiffness where possible: errors can be reduced by ensuring that the natural frequencies of the transfer function match those of the actual vehicle-track dynamics.

4 Extracting Foundation Stiffness and Damping

To improve the accuracy of the roughness derivation, it is clearly desirable to compensate for variations in track stiffness. At least three techniques are reported in the literature to extract track dynamic stiffness from vehicle acceleration measurements with the purpose of monitoring track for stiffness-related defects. Cano *et al.* [15] measured the frequency of the P2 resonance peak in the ABA spectrum and used a single mass-spring model to derive the associated track stiffness, knowing the combined mass of the unsprung vehicle components, rail and sleeper. Cano *et al.* report results within 16% of direct track stiffness measurements. Quirke *et al.* [16] and Zhu *et al.* [17] devised model-based optimization methods to identify track stiffness variations from bogie acceleration and ABA respectively. These methods fit the time-domain vehicle acceleration calculated by a vehicle-track model to measurements by setting the model's foundation stiffness parameters. The accuracy of the results presented by Zhu *et al.* [17] is initially near-perfect but is affected by measurement noise and the presence of a wheel flat. The stiffness identified by Quirke *et al.* [16] is within 6% of the actual stiffness in a simulated environment with 3% measurement noise but without rail roughness.

The method proposed here uses the present vehicle-track model (Fig. 1(a)) to identify the track's foundation stiffness and damping parameters from the P2 resonance, knowing all other model parameters. Here, a least-squares optimization is used to curve-fit the model's transfer function $H(\omega)$ (Eq. (2)) to the square-root of the ABA PSD in the vicinity of the P2 resonance by optimizing the function's foundation stiffness and damping parameters, as well as its scaling. This method (along with the method of Cano *et al.* [15]) does not require the roughness to be known but assumes that it does not have features that alter the shape of the resonance peak in the ABA spectrum.

4.1 Results

The ABA spectrum resulting from the simulation in Sect. 3.1 is shown in Fig. 2(a). Here, spectra are calculated with an increased segment length of 8 m (covering 122 m of track) to improve the resolution of the resonance peak. The transfer function is assigned with the vehicle and rail beam parameters in Table 1 and is then curve-fitted to the ABA spectrum to identify the foundation stiffness k_f' and damping c_f'. The resulting values are given in Table 2, along with deviations from the actual simulated values. To examine the effect of errors in the wheel mass parameter m_w, the curve fitting is repeated with the wheel mass increased by 20%, and the resulting foundation parameters are given in Table 2 alongside the values with the actual wheel mass.

The result of deriving the roughness spectrum with the identified foundation parameters is plotted in Fig. 2(b), and is within 1 dB of the actual spectrum. The spectrum derived with the 20%-increased wheel mass is also plotted in Fig. 2(b), for both the actual and identified foundation parameters. The use of the identified, rather than actual, foundation parameters reduces the impact of the 20% error in wheel mass on the spectrum at longer wavelengths because the former corrects the transfer function's resonance frequency to match that of the vehicle-track system.

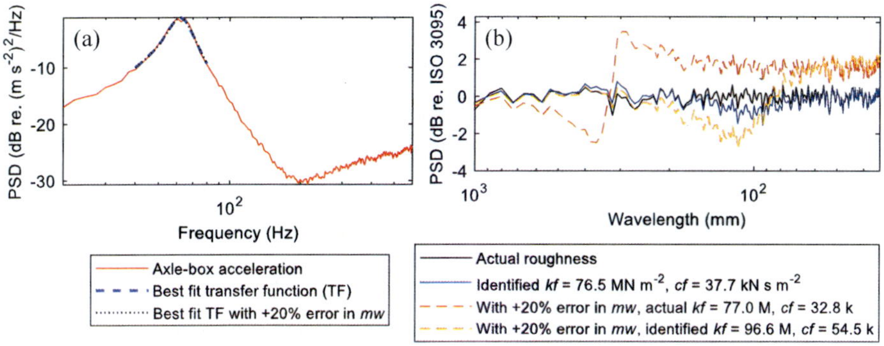

Fig. 2. PSDs of (a) axle-box acceleration and (b) actual and derived roughness relative to the ISO 3095:2013 limit spectrum, with identified track stiffness k_f and damping c_f parameters. Note that the three curves in plot (a) effectively overlap.

Table 2. Identified track foundation stiffness and damping compared to actual values.

Parameter	Actual value	Identified value for actual m_w	Identified value for 20%-increased m_w
Foundation stiffness per length $k_f^{\prime}$ (MN m^{-2})	77.0	76.5	96.6
Deviation in identified foundation stiffness		– 0.7%	+25%
Foundation damping per length $c_f^{\prime}$ (kN s m^{-2})	32.8	37.7	54.5
Deviation in identified foundation damping		+15%	+66%

5 Separating Wheel and Rail Roughness

The excitations associated with wheel and rail roughness are inherently combined in the response of the vehicle-track system. Unlike rail roughness, excitation by the wheel roughness is periodic with the rotation of the wheel. This makes it possible to separate the wheel- and rail-roughness components of an ABA signal by averaging together portions of the signal associated with each of a number of wheel revolutions, a process known as synchronous averaging [18].

The synchronous averaging technique was evaluated by Németh and Schleinzer [8] to extract the wheel component of ABA measurements before applying a transfer function to derive the wheel roughness spectrum, which agreed to within 1–5 dB of direct measurements depending on wavelength. Here, we evaluate this technique in a simulated environment with the aim of extracting rail roughness.

5.1 Method and Results

To test the synchronous averaging technique, a wheel roughness profile is generated according to roughness measurements presented by Johansson [19] for powered wheelsets of the X2 locomotive. A rail roughness profile is generated according to the limit spectrum in ISO 3095:2013 [14] plus two sinusoidal corrugation components at wavelengths of 40 mm and 200 mm. The (actual) PSDs of wheel, rail and combined roughness are shown in Fig. 3. The simulator is run with the combined roughness profile at a vehicle speed of 20 m/s and with the model parameters in Table 1. The wheel component of the resulting ABA signal is extracted by averaging over 18 wheel revolutions (62.3 m of track as the wheel diameter is 1.1 m). The rail component is then derived by subtracting the wheel component from the combined ABA signal. The combined signal and both separated components are processed by Eq. (1) and the procedure in Sect. 2.1 to derive the wheel, rail and combined roughness PSDs. All PSDs are computed using Welch's method over 62 m of track, split into 30 segments of length 4 m overlapped by 50%. This includes the wheel PSDs, which are facilitated by repeat-extending the wheel profile/component to cover 62 m of track. The derived wheel and rail roughness PSDs are also plotted in Fig. 3.

The derived rail roughness PSD (labelled "Derived rail (time domain sub.)") is close to the actual rail PSD at all wavelengths, including around the corrugation peaks and where wheel roughness exceeds the rail roughness by up to 20 dB. The derived wheel PSD tends to be less accurate where it is less than the rail PSD, with errors of up to 5 dB.

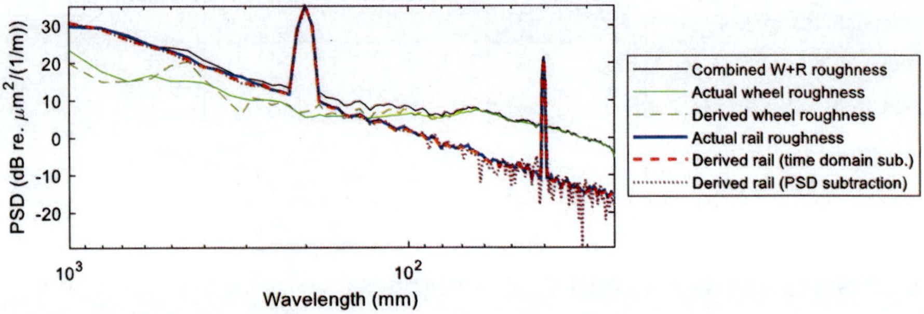

Fig. 3. Welch PSDs of derived wheel and rail roughness profiles compared to actual profiles.

An alternative rail roughness PSD is calculated by energy-subtracting the wheel roughness PSD from the combined PSD (labelled "Derived rail (PSD subtraction)" in Fig. 3). In contrast to subtracting in the time domain, the process of subtracting PSDs is affected by their statistical accuracy. The variance of the derived rail PSD increases as the wheel roughness PSD increases above the rail PSD. However, presenting the spectra in third-octave bands (not shown) averages out the variance, especially at shorter wavelengths where more PSD data points fall within a third-octave band, resulting in derived spectra that are within 1.5 dB of the actual ones. In most practical cases, wheel roughness does not exceed rail roughness at wavelengths longer than 300 mm, so energy subtraction may be sufficiently accurate for deriving the third-octave spectrum of rail roughness. This means that the wheel spectrum can be derived and updated on sections of track with conditions preferable for synchronous averaging (e.g. straight, invariant track) and then energy-subtracted from subsequent ABA-derived roughness spectra to obtain the rail roughness spectra.

6 Conclusions

By accounting for variations in track properties and removing the influence of wheel roughness, idealized vehicle-track modelling has indicated that reliable measurements of rail roughness spectra can be obtained from axle-box acceleration (ABA). Track dynamic stiffness parameters are extracted by curve-fitting a transfer function to the P2 resonance in the ABA spectrum. This method relies on sufficient spectral accuracy and resolution, and assumes that the roughness spectrum does not contain features near the P2 resonance that distort the ABA spectrum. Highly-damped resonances may also reduce accuracy. The synchronous averaging method of separating wheel and rail roughness assumes that the wheel excitation is perfectly periodic, which may not be true if the variation in lateral contact position on the wheel tread is considered.

Further work will develop the vehicle-track model and techniques, as necessary, to match the dynamics of real railway track. This will include considering discretely-supported track and distributed mass and stiffness within the track slab/ballast, as well as variations in the lateral contact position. The techniques will be tested using real measurements of rail roughness and ABA, with the overall aim of developing an autonomous ABA-based measurement system that continuously 'maps' railhead roughness, enabling more efficient and proactive scheduling of rail maintenance.

References

1. Grassie, S.L.: Measurement of railhead longitudinal profiles. Wear **191**, 245–251 (1996)
2. Talbot, J.P., Hunt, H.E.M., Hussein, M.F.M.: A prediction tool for the optimisation of maintenance activity to reduce disturbance due to ground-borne vibration from underground railways. In: Proceedings of the 8th International Workshop on Railway Noise, Buxton, UK (2004)
3. Bocciolone, M., et al.: A measurement system for quick rail inspection and effective track maintenance strategy. Mech. Sys. Sig. Process. **21**(3), 1242–1254 (2007)
4. Salvador, P., et al.: Axlebox accelerations: their acquisition and time-frequency characterisation for railway track monitoring purposes. Measurement **82**(C), 301–312 (2016)
5. Li, Z., Molodova, M., Núñez, A., Dollevoet, R.: Improvements in axle box acceleration measurements for the detection of light squats in railway infrastructure. IEEE Trans. Ind. Elec. **62**(7), 4385–4397 (2015)
6. Bongini, E., Grassie, S.L., Saxon, M.J.: 'Noise mapping' of a railway network: validation and use of a system based on measurement of axlebox vibration. In: Maeda, T. et al. (eds.) Noise and Vibration Mitigation, NNFM, vol. 118, 505–513. Springer, Tokyo (2012)
7. Berggren, E.G., Li, M.X.D., Spännar, J.: A new approach to the analysis and presentation of vertical track geometry quality and rail roughness. Wear **265**(9), 1488–1496 (2008)
8. Németh, I., Schleinzer, G.: Investigation into the indirect determination of wheel-rail surface roughness. In: Proceedings of the 11th Mini Conference on Vehicle System Dynamics, Budapest, pp. 135–146 (2008)

9. Carrigan, T.D., Fidler, P.R.A., Talbot, J.P.: On the derivation of rail roughness spectra from axle-box vibration: development of a new technique. In: Proceedings of the International Conference on Smart Infrastructure and Construction, Cambridge, UK (2019)
10. Newland, D.E.: An Introduction to Random Vibrations, Spectral and Wavelet Analysis, 3rd edn. Longman Scientific & Technical, Essex (1993)
11. EN 15610:2009: Railway applications – Noise emission – Rail roughness measurement related to rolling noise generation. CEN, Brussels (2009)
12. Koh, C.G., Ong, J.S.Y., Chua, D.K.H., Feng, J.: Moving element method for train-track dynamics. Int. J. Numer. Meth. Eng. **56**(11), 1549–1567 (2003)
13. Grassie, S.L., Edwards, J.W.: Development of corrugation as a result of varying normal load. Wear **265**, 1150–1155 (2008)
14. ISO 3095:2013: Acoustics – Railway applications – Measurement of noise emitted by railbound vehicles. International Standards Organisation (2013)
15. Cano, M.J., Fernández, P.M., Franco, R.I.: Measuring track vertical stiffness through dynamic monitoring. Proc. Inst. Civil Eng. Transp. **169**(1), 3–11 (2015)
16. Quirke, P., Cantero, D., OBrien, E.J., Bowe, C.: Drive by detection of railway track stiffness variation using in-service vehicles. Proc. Inst. Mech. Eng. F **231**(4), 498–514 (2017)
17. Zhu, Q.X., Law, S.S., Huang, L.: Identification of railway ballasted track systems from dynamic responses of in-service trains. J. Aerosp. Eng. **31**(5), 1–7 (2018)
18. Antoni, J.: Cyclostationarity by examples. Mech. Sys. Sig. Proces. **23**(4), 987–1036 (2009)
19. Johansson, A.: Out-of-round railway wheels – assessment of wheel tread irregularities in train traffic. J. Sound Vib. **293**(3), 795–806 (2006)

Numerical and Experimental Analysis of Transfer Functions for On-Board Indirect Measurements of Rail Acoustic Roughness

Anna Rita Tufano[1,3(✉)], Olivier Chiello[1], Marie-Agnès Pallas[1], Baldrik Faure[2], Claire Chaufour[2], Emanuel Reynaud[3], and Nicolas Vincent[3]

[1] UMRAE, Univ Gustave Eiffel, IFSTTAR, CEREMA, Univ Lyon, 69675 Lyon, France
olivier.chiello@univ-eiffel.fr

[2] SNCF Innovation & Research, La Plaine, 93212 St Denis Cedex, France

[3] Vibratec, 28 Chemin du Petit Bois, 69131 Ecully, France
anna-rita.tufano@vibratec.fr

Abstract. The paper presents some results concerning the development of a multi-sensor hybrid approach for measuring rail roughness using acoustic and vibration signals measured on-board. The main objective of the system is to provide better input data for noise mapping and maintenance strategies. Several improvements to existing systems are proposed, based on the analysis of transfer functions. First numerical and experimental results are discussed.

Keywords: Rolling noise · Monitoring system · Rail roughness · Transfer functions · Wheel/rail contact · Vibro-acoustic model

1 Introduction

Environmental railway noise is a critical public health issue which is subject to national and international legislative frameworks. At the European level, the Environmental Noise Directive 2002/49/EC requires Member States to produce strategic noise mapping using the CNOSSOS-EU (Common NOise aSSessment methOdS) method from 2019 onwards. This methodology requires input data that are characteristic of each type of source, instead of equivalent sound power levels. Accordingly, railway rolling noise should be described through a set of track and train parameters. Rail roughness and track decay rate (TDR) are the most influential parameters concerning the track contribution to rolling noise.

Nowadays, direct systems represent the standard for rail roughness measurement. Roughness direct measurements consist in estimating the vertical irregularities of the rail surface through fixed devices or mobile trolleys, that use displacement or acceleration sensors. The methodology is governed by norm EN 15610 [1]. Direct measurements guarantee absolute roughness estimations in controlled

G. Degrande et al. (Eds.): *Noise and Vibration Mitigation for Rail Transportation Systems*, NNFM 150, pp. 295–302, 2021.
https://doi.org/10.1007/978-3-030-70289-2_30

conditions. Moreover, direct techniques are free from perturbations coming from the railway environment, and enable separation of rail and wheel contributions. On the other hand, in order to perform direct measurements, it is required to stop train traffic in the measurement zone. This, and the operational constraints due to direct measuring devices, limit the acquisitions in time and space.

Indirect measurements focus on quantities resulting from wheel-rail interaction, such as noise in the bogie area or wheelset vibrations, and from which the actual combined roughness is deduced. Indirect measurement systems help overcoming the limitations of direct systems in terms of measured track length, and potentially enable measurements on the entire national railway network. On-board indirect measurement systems not only help in complying better with actual noise requirements, but are also crucial in maintenance and grinding policies. Among the existing indirect measurement systems, two techniques are found: by axle-box accelerometers or by microphones. The axle-box accelerometer method was first proposed in [2] and is currently exploited in some widely used systems [3,4]. Early systems employing microphone signals were based on a comparison of global measured noise levels, and did not provide an estimation of absolute roughness spectra [5–7]. More recently, it has been proposed to use frequency-dependent transfer functions, in order to estimate roughness spectra from on-board acoustic measurements. Different measurement systems or methods are available, they differ in the estimation technique for transfer functions, and in the position and number of sensors [8–10]. Other approaches exist, which are based on optical techniques, or sensors directly fixed on wheels. However, in their current state of progress [11], they are not fitted to vehicles in operational conditions.

Existing indirect measurement systems suffer from some limitations. First, the frequency range in which they are accurate is limited: for instance, in the case of axle-box accelerometers, the signal could be masked by other vibratory sources [12], that make roughness estimations unreliable at frequencies higher than the axle-box decoupling frequency, which depends on bearing properties. The second limitation is linked to the precision of transfer functions measured during calibration tests. Calibration takes into account neither the variability of parameters influencing track dynamics, nor the variability of kinematic parameters, which massively affect transfer functions. Further constraints are related to acoustic or vibration perturbations, and coupling between the two rails.

The French project MEEQUAI aims at developing an indirect measurement system for roughness, which overcomes the previously mentioned limitations. This paper presents the adopted strategy for indirect roughness measurements. Section 2 explains the proposed approach, Sect. 3 illustrates numerical results concerning the transfer functions, Sect. 4 contains experimental results needed for the first validation and calibration of the method; in the end, some conclusive statements, as well as future perspectives are given.

2 Overview of the Multi-sensor Hybrid Strategy

The principle of rail roughness indirect measurement is illustrated in Fig. 1. Two important original contributions of the system under development lie in the estimation of the effective combined roughness from vibro-acoustic signals.

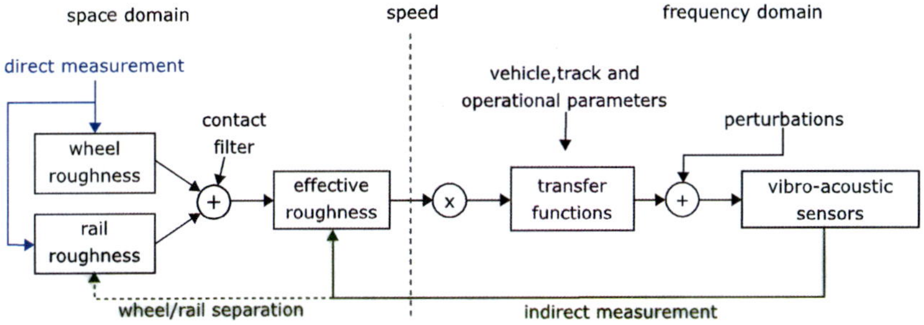

Fig. 1. Schematics of indirect measurement systems.

First, a *multi-sensor* approach is proposed for the measurement system. Indeed, the frequency domain limitation can be overcome by the redundant combination of different types of sensors, whose positions can be optimised in order to improve the robustness of the system. Moreover, sensor redundancy contributes to remove noise, in the context of inverse methods techniques. Table 1 summarizes the classes of sensors and their frequency range of applicability: the techniques and sensors found in the literature are used as a reference and enhanced.

Table 1. Summary of sensors used in the proposed approach.

Frequency range	Sensors
F100 Hz 1000 Hz	Axle-box accelerometers
From 500 / 1000 Hz 1500 Hz	Rail near-field microphones
F1500 Hz 5000 Hz	Wheel near-field microphones
From 500 / 1000 Hz 5000 Hz	Under coach microphones

Secondly, a *hybrid partial calibration* is proposed. Combining numerical and experimental methods for the estimation of transfer functions is a way to take into account the variation of track dynamic and kinematic parameters (speed, position of the contact point, wheel load).

3 Numerical Analysis

In order to develop and validate the approach, several numerical models have been set up, each one specific to a class of sensors listed in Table 1:

- Finite Element (F.E.) structural model of the track: solid elements for rail and sleepers, rail-pads and ballast modelled as discrete springs, anechoic terminations (no wave reflection at rail ends)

- F.E. structural model of the wheelset: solid elements, except for bearings, which are represented by a set of discrete springs
- Analytical model of wheel/rail interaction using point receptances (cf. Sect. 4.1 and [13])
- Exterior radiation F.E. models of wheel and track, including reflections in the bogie area: acoustic boundary condition enforced as a displacement field on the wheel and track surfaces, far-field conditions simulated through Infinite Elements, baffle plane with ballast impedance boundary conditions

Through these simulations it is possible to evaluate the influence of track parameters and optimize the position of sensors. Their advantage consists in using complete models, that include the behaviour of all components. Two examples of use of these models are presented below.

3.1 Track Influence on Roughness/Axle-box Transfer

F.E. models of several track architectures have been realized, in order to take into account the variability of track dynamics and kinematics. Thanks to a sensitivity analysis, we have determined the most influential parameters for the track dynamic behaviour and quantified the variation in track receptance for each parameter. Based on the actual track designs in the French network, a track database has been built in order to adjust the track dynamics in the wheel/rail interaction scheme. Figure 2 shows the axle-box displacement for unit roughness, calculated with different track receptance values. The comparison with an interaction scheme including a rigid wheelset is also included. Track dynamic properties affect the axle-box vibration at the first order. Therefore, a track correction is deemed necessary.

The same investigation on the influence of rail has been proposed in [4], but qualitative estimations of the track influence, based on analytical rail models, were shown. Here numerical full track models are used, and the influence of track parameters is quantified.

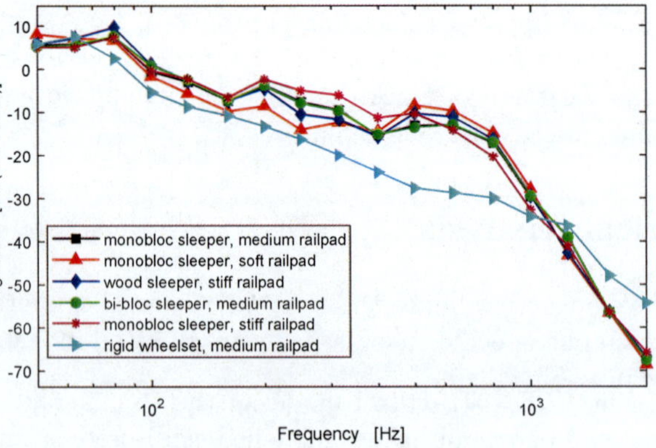

Fig. 2. Vertical axle-box displacement for unit roughness: influence of the track configuration.

3.2 Sensor Layout for Roughness/Wheel Near-Field Transfer

The optimal position and number of microphones have been selected after an analysis of the acoustic field around the wheel. The identification of an optimal position of a microphone is based on two requirements on the acoustic near-field impedance $Z' = p/\rho_0 c_0 V_n$: it should be stable along a line that runs out of the wheel, and should be as close as possible to unity, for the sake of robustness on acoustic transfer functions. Application of these criteria allows the definition of 2 optimal positions for wheel near-field microphones. Figure 3 shows the variation of the acoustic impedance along the directions indicated with red lines on the left. It should be noted that in [9] the condition $Z' = 1$ has been assumed as an hypothesis, based on observations on analytical models of wheel radiation, with the wheel simplified as a plane disc. Here, instead, the actual acoustic impedance is calculated with a model that includes all the geometric complexity of the wheel. More interestingly, a different acoustic model has been developed, that includes the effect of the surfaces surrounding the wheel (axle-box, brake discs, car-body). It is observed that the near-field is influenced by reflections on these surfaces (cf. the difference between red crosses and black circles in Fig. 3), and that the calculation of acoustic transfer functions should take them into account through calibration.

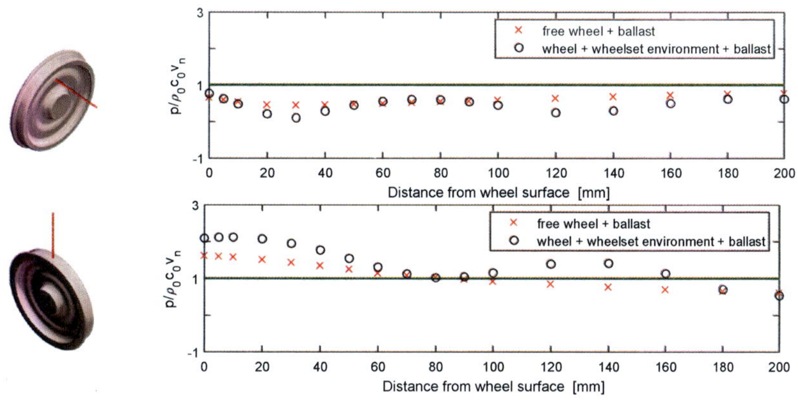

Fig. 3. Variation of the acoustic impedance in the wheel near-field.

4 Experimental Analysis and Partial Calibration

For the sake of illustration, we will focus here on the axle-box acceleration methodology; the other methods follow similarly.

4.1 Wheel/Rail Interaction and Force/Axle-box Transfer

In the wheel-rail interaction scheme that is considered here, contact receptance $\mathbf{A}_C$ is obtained analytically, based on the Hertz contact theory [13], while rail and wheel receptances at contact are calculated through F.E. models. Rail receptance $\mathbf{A}_R$ is estimated for different track architectures. Wheel receptance $\mathbf{A}_W$ is calculated by mode superposition, with mode shapes issued from numerical models, and other modal parameters obtained experimentally. Forces $\mathbf{F}$ at contact are obtained by imposing the continuity of displacements at the wheel-rail contact, and by considering imposed roughness $\mathbf{r}_{eff}$:

$$\mathbf{F} = \left(\mathbf{A}_R + \mathbf{A}_W + \mathbf{A}_C\right)^{-1} \mathbf{r}_{eff} \tag{1}$$

Axle-box accelerations are obtained by introducing the force/axle-box transfer function $\mathbf{B}_W$:

$$\mathbf{\Gamma}_{axle-box} = \mathbf{B}_W \mathbf{F} = \mathbf{B}_W \left(\mathbf{A}_R + \mathbf{A}_W + \mathbf{A}_C\right)^{-1} \mathbf{r}_{eff} \tag{2}$$

In these equations, vertical and lateral degrees of freedom are taken into account, as well as cross vertical-lateral terms. Coupling between both sides of the wheelset is also considered. Accordingly, forces and displacements are vectors of size 4×1 and transfer functions are matrices of size 4×4.

4.2 Measurement of Force/Axle-box Transfer on a Free Wheelset

The estimation of transfer function matrix $\mathbf{B}_W$ is a critical issue. At first, this function has been measured on a free wheelset, in the framework of an Experimental Modal Analysis aimed at validating the wheelset F.E. model in the frequency range [50 Hz : 2000 Hz]. The aim of the campaign was to state whether $\mathbf{B}_W$ should be calculated or measured. The validation of the axle-box components happened to be challenging, due to the complexity in characterizing bearings. Correlation between numerical and experimental transfer functions is shown in Fig. 4, where the first 2 curves represent the simulated and measured results for the transfer, respectively. Curves are shown in narrow- and 1/3 octave bands. Due to the significant discrepancies obtained, it was decided to estimate experimentally the force/axle-box transfer function, thus leading to a hybrid numerical-experimental methodology.

4.3 Measurement of Force/Axle-box Transfer on a Loaded Wheelset

Measurements on a free wheelset do not correspond to operational conditions, where the wheelset is installed on a vehicle. A second experimental campaign has been performed, on a wheelset standing still in real conditions. In this situation, a more realistic behaviour of axle-box bearings is expected, since they are thought to have a non-linear force-strain dependency. Some differences are seen between the transfer functions measured in the two conditions (the curve with triangles in

Fig. 4 refers to the loaded wheelset). These could be due to several factors: non-linearity of bearings and differences in wheelset architecture (some equipments are added when the wheelset is mounted on a vehicle), influence of rail stiffness on the wheelset dynamic behaviour, difference in excitation position (due to the presence of the wheel on rail, the excitation is applied at a point 180° away from the contact). Measurements proved that exciting the wheelset at a point 180° away from the contact does not significantly influence the transfer function, at least up 1200 Hz. Figure 4 shows that the influence of the rail on force/axle-box transfer is limited to the low frequency resonances. On the other side, non-linearities of the bearings and the dynamic behaviour of the attached components have an effect on the wheelset transfer at higher frequencies. These effects need to be taken into account in $\mathbf{B}_W$. Consequently, the recommended calibration measurement of $\mathbf{B}_W$ should be performed on a loaded wheelset. To improve the accuracy of the method a correction for the influence of rail can be applied (Eq. 3). The latter is obtained by analysing the impedance matching at contact.

$$\mathbf{B}_W \approx \mathbf{B}_{W_{on\ rail}} \left(\mathbf{A}_R + \mathbf{A}_C\right)^{-1} \left(\mathbf{A}_R + \mathbf{A}_W + \mathbf{A}_C\right) \qquad (3)$$

This approach will be tested in future rolling measurements.

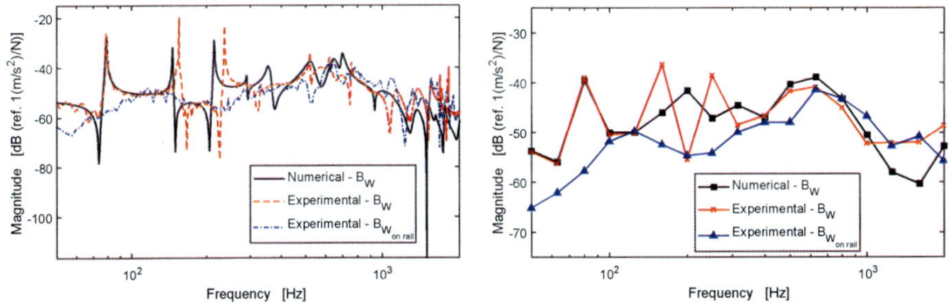

Fig. 4. Force/axle-box vertical transfer function. Numerical-experimental comparison.

5 Conclusions and Perspectives

The article presents a brand-new methodology for on-board indirect roughness measurements, which is based on a multi-sensor and hybrid calibration approach. Numerical analyses are used to quantify the dispersion due to track dynamic and kinematic parameters, and to optimize the number and position of sensors. The introduction of numerical approaches is a heavy innovation for indirect roughness measurement. Experimental transfer functions are employed for calibration purposes, in cases where numerical models prove to be not accurate enough.

A design for the measurement system has been set up, and several tests have been performed. A first campaign has been carried out in controlled conditions, with the vehicle standing still, and the rolling tests followed. Post-processing of the experimental tests is in progress.

Acknowledgements. The MEEQUAI project has been selected in the framework of the call for projects FUI-AAP23. It is funded by BPI France and by the AURA Region via FEDER. This project is co-financed by the European Union. This work was supported by the LABEX CeLyA (ANR-10-LABX-0060) of Université de Lyon, within the program "Investissements d'Avenir" (ANR-16-IDEX-0005) operated by the French National Research Agency (ANR). This support is greatly appreciated.

References

1. Technical Committee CEN TC 256 : Railway applications - Noise emission - Rail roughness measurement related to rolling noise generation EN 15610 :2009
2. Lewis, R.B., Richards, A.W.: A new method for the routine measurement of rail head corrugations. Rail Int. **17**, 37–41 (1986)
3. Spannar, J. : A new approach of assessing rail roughness. In: 4th IET International Conference on Railway Condition Monitoring, pp. 1–5, Derby (2008)
4. Bongini, E., Grassie, S.L., Saxon, M.J.: "Noise Mapping" of a railway network: validation and use of a system based on measurement of axlebox vibration. In: Noise and Vibration Mitigation for Rail Transportation Systems, pp. 505–513 (2012)
5. Bracciali, A., Ciuffi, L., Ciuffi, R., Rissone, P.: Continuous external train noise measurements through an on-board device. Proc. Inst. Mech. Eng. Part F: J. Rail Rapid Transit. **208**, 23–31 (1994)
6. Hardy, A.E.J., Jones, R.R.K., Turner, S.: The influence of real-world rail head roughness on railway noise prediction. J. Sound Vib. **293**, 965–974 (2006)
7. Asmussen, B.: Status and perspectives of the "Specially Monitored Track". J. Sound Vib. **293**, 1070–1077 (2006)
8. Kuijpers, A.H.W.M., Schwanen, W., Bongini, E.: In: Maeda, T., et al. (eds.) Noise and Vibration Mitigation for Rail Transportation Systems. Notes on Numerical Fluid Mechanics and Multidisciplinary Design. Indirect rail roughness measurement: the arrow system within the lecav project. Springer, Tokyo (2012). https://doi.org/10.1007/978-4-431-53927-8_67
9. Chartrain, P.-E. : Lecture acoustique de la voie ferrée. PhD Thesis (2013)
10. Höjer, M., Almgren, M. : Monitoring system for track roughness. In: Proceedings of EuroNoise (2015)
11. Fidecaro, F., Licitra, G., Bertolini, A., Maccioni, E., Paviotti, M.: Interferometric rail roughness measurement at train operational speed. J. Sound Vib. **293**, 856–864 (2006)
12. Phamová, L., Bauer, P., Malinský, J., Richter, M.: Indirect Method of Rail Roughness Measurement – VUKV Implementation and Initial Results. In: Nielsen, Jens C.O., et al. (eds.) Noise and Vibration Mitigation for Rail Transportation Systems. NNFMMD, vol. 126, pp. 189–196. Springer, Heidelberg (2015). https://doi.org/10.1007/978-3-662-44832-8_25
13. Thompson, D.J. : Railway Noise and Vibration: Mechanisms, Modelling and Means of Control. Elsevier (2009)

Analysis of the Effect of Running Speed and Bogie Attitude on Rail Corrugation Growth in Sharp Curves

Andrea Collina, Roberto Corradi, Egidio Di Gialleonardo(✉), and Qianqian Li

Politecnico di Milano, 20156 Milan, MI, Italy
egidio.digialleonardo@polimi.it

Abstract. In this paper an experimental and numerical analysis of the corrugation detected in a very sharp curve (radius equal to 110 m) and caused by a vehicle equipped with resilient wheels is shown. The results of the experimental tests carried out on the vehicle demonstrate that different values of the speed are able to trigger corrugation wavelengths in different ranges (30–40 mm and 150 mm). Additional tests are performed in order to make a modal analysis of the track and the wheelset with the aim of setting up a simple mathematical model used for the understanding of the phenomenon. The model for the interaction between the wheel and the rail in frequency domain is capable of predicting the frequency ranges where corrugation is most likely to occur.

Keywords: Corrugation · Rutting · Wheel-rail interaction

1 Introduction

Rail corrugation is a major issue for railway infrastructure owners and operators since it leads to additional cost due to the additional maintenance operations, a significant wear of the contacting surfaces, high vibration levels that can result in the degradation and also the damage of the components of vehicles and railway track. Additionally, the increased level of vibration is the source of significant rolling noise and ground-borne vibration. Many studies are present in literature addressing causes and mitigation actions that can be used to limit the consequences [1, 2]. Typical methods used to mitigate the effects of rail corrugation are the frequent rail grinding and the use of friction modifiers [3, 4], both having implications on maintenance cost. Resilient wheels are proven to be effective in reducing rolling noise and vibration transmitted to the track [5, 6] and, thus, may represent a mitigation measure. From the phenomenological point of view, a number of models to simulate the rail corrugation growth process have been developed going from simplified analytical models in frequency domain [7–9] to more complex numerical models in time domain [2]. However, there are still many open points, accounting for the fact that many factors can play a significant role in the corrugation growth process. The authors in [10] proved that this process is significantly affected by environmental

G. Degrande et al. (Eds.): *Noise and Vibration Mitigation for Rail Transportation Systems*, NNFM 150, pp. 303–311, 2021.
https://doi.org/10.1007/978-3-030-70289-2_31

factors and a non-uniform speed distribution in curves could diminish the growth rate of corrugation by 30% [11].

This paper presents a case study in which rail corrugation was observed on the low rail of a sharp curve (radius equal to 110 m) of an underground line. The causes of the corrugation may be found in the interaction between the track and the vehicle which is equipped with resilient wheels. A different operating speed was able to trigger different corrugation wavelengths.

The paper is organised as follows: in Sect. 2 the phenomenon is described, while in Sect. 3 the experimental campaign and its results are presented; Sect. 4 reports the findings obtained using models in the frequency domain and finally some concluding remarks are provided in Sect. 5.

2 Description of the Phenomenon

The phenomenon analysed in this paper is associated to the formation of rail corrugation in sharp curves. The phenomenon was observed since the opening of the underground line and got worse after the introduction of a speed limit.

The focus is on a track section characterized by the presence of a very sharp curve (radius of 110 m). The railway track in that section is made of a floating slab, elastically suspended with respect to the tunnel. Wooden sleepers are embedded into the slab and elastically supported on a continuous rubber layer. The vehicle is an automatic metro with four carbodies and five bogies (four motor bogies and one trailer bogie), each one is equipped with two wheelsets having resilient wheels.

Figure 1 synthetizes the corrugation wavelength (left) and peak-to-peak amplitude (right) found in the considered track section as a function of time.

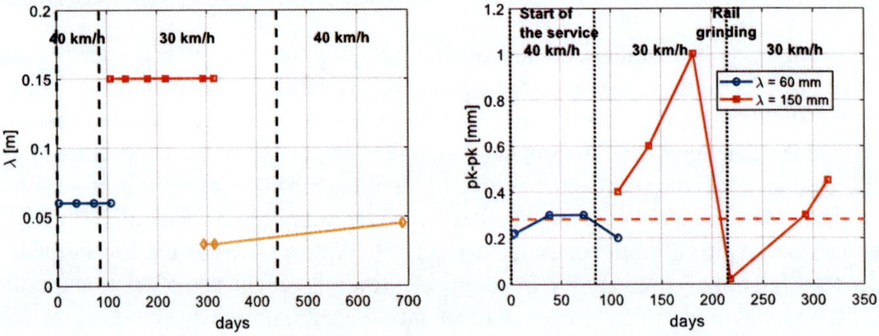

Fig. 1. Wavelength (left) and peak-to-peak amplitude (right) of rail corrugation as a function of time, since the opening of the line.

Since the start of the service, when the track section was negotiated at approximately 40 km/h, a short wavelength corrugation was found (around 60 mm). The growth rate was quite small, and the amplitude of the corrugation was around 0.3 mm. After less than 100 days after the opening of the line, the infrastructure owner introduced a 30 km/h limit for operative reasons. The speed limit triggered a corrugation with longer wavelength (around 150 mm) and with an extremely large growth rate. However, as shown in Fig. 1 left, when the speed limit was established, also a corrugation with shorter wavelength (around 30 mm) was found in the same curve. Finally, when the limit was cancelled, corrugation with longer wavelength was not found any longer while the shorter wavelength moved from 30 mm to approximately 45 mm.

3 Experimental Campaign

An extensive experimental campaign was planned and designed in order to better understand the corrugation process in the considered track section.

The investigation of the phenomenon was carried out by means of two series of tests:

- two tests (denominated as "test 1a" and "test 1b") repeated after 34 days when the track was operated at 30 km/h; "test 1a" was carried out some days after a rail grinding operation was performed on the considered track section;
- one test (denominated as "test 2") carried out when the track was operated at 40 km/h.

Among the different signals acquired during the experimental campaign, in this paper, the attention is focused only on the wheel acceleration measured on the inner wheel (considering the curve under analysis) of the trailing wheelset. This is due to the fact that this wheel is the one which negotiates the low rail of the sharp curves, where the corrugation was more evident.

Figure 2 shows the comparison between the tests 1a (in red dashed line) and 1b (in blue continuous line) that is the more recent one. On the top of the figure the acceleration is depicted as a function of the vehicle position along the track, while on the bottom, the moving RMS value over a 10 m window is reported. The yellow line on the top sub-figure represents a signal proportional to the track curvature (positive value means a left-handed curve).

The length of the curve under analysis is approximately 300 m. It is observed that when the vehicle enters the curve, especially the sharpest section, the acceleration level increases. Peak values of the acceleration are found at an almost constant distance between each other and correspond to the passage over rail joints. Additionally, looking at the moving RMS values and comparing tests 1a and 1b, it is clearly visible that the more recent test shows higher accelerations in the sharpest section with respect to the first test proving a degradation of the track.

In order to better clarify the effect of the corrugation on the acceleration of the wheel two zoomed views for the same wheel are presented in Fig. 3. It is observed that the corrugation evolved in time in two different sections along the curve (the distance between the sections in the figures is approximately 20 m) with two different wavelengths. The effect of a long wavelength corrugation is observed on the left figure, whereas a shorter one is presented in the right one. A reference dimension is highlighted in the diagrams so that it is possible to obtain the wavelength of the corresponding corrugation that is around 150 mm for the longer one and around 30 mm for the shorter one.

As far as the frequency content of the acceleration signal is concerned, the longer corrugation wavelength (around 150 mm) generates a vibration of the wheel at around

45 Hz, on the contrary, the shorter wavelength (around 30 mm) generates a vibration at higher frequencies around 250 Hz.

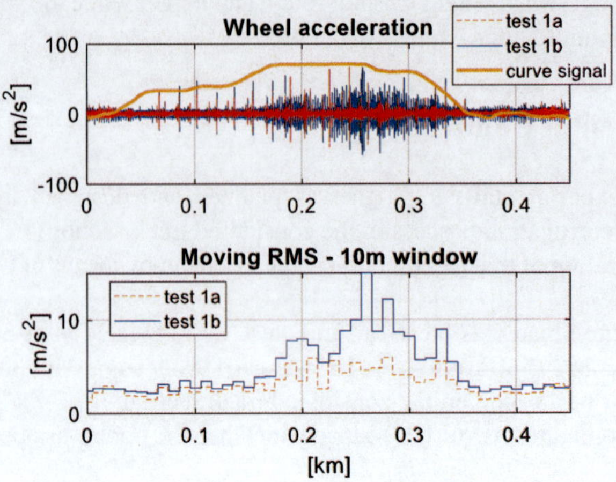

Fig. 2. Wheel acceleration measured in tests 1a and 1b as a function of the vehicle position along the track: time histories (top) and corresponding moving RMS computed over a 10 m window (bottom).

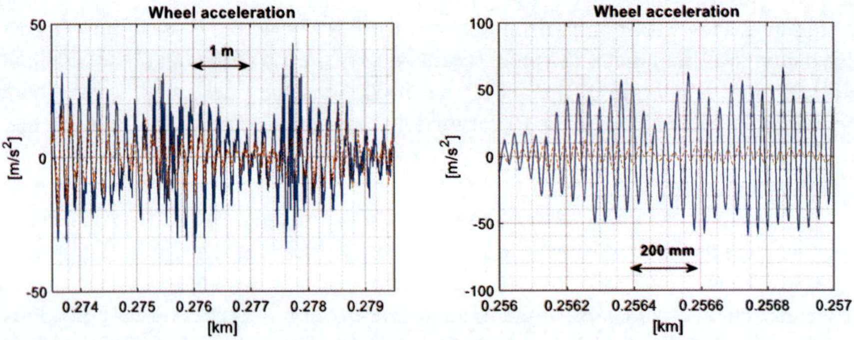

Fig. 3. Zoomed views of the wheel acceleration showing the effect of the evolution of a corrugation with long (left) and short (right) wavelength.

Some months later, the infrastructure owner decided to cancel the speed limit so to restore the original train speed along the track section, which was established back to approximately 40 km/h. The experimental tests were repeated also in this condition in order to investigate the effect of the speed change on the corrugation.

Figure 4 reports on the left the wheel acceleration measured on the inner wheel of the trailing wheelset as a function of the vehicle position along the track together with the corresponding moving RMS over a 10 m window. A zoomed view is shown on the right. As already observed for tests 1a and 1b, the acceleration levels increase when the vehicle negotiates the curve under analysis, especially in the sharpest section. Considering the identification of the wavelength of the corrugation a zoom is presented in Fig. 5 (right) and, in this case, it is possible to observe an almost mono-harmonic signal where 8 or 9 periods are found in approximately 400 mm, thus determining a wavelength of the corrugation of around 45 mm.

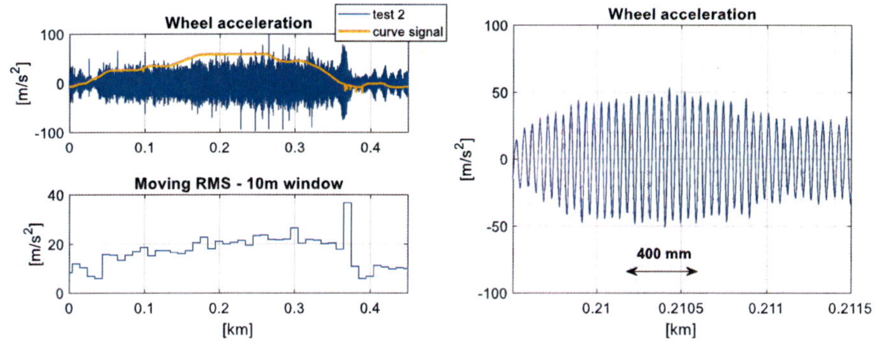

Fig. 4. Wheel acceleration as a function of the vehicle position along the track (top left) and in RMS of moving RMS computed over a 10 m window (bottom left) and a zoomed view (right).

Two important aspects must be highlighted: the first, the shorter wavelength moved from 30 mm up to around 45 mm and this is coherent with the increased speed, since the corrugation is a frequency driven mechanism; the second, and maybe a more important fact is that the longer wavelength was not present anymore on the low rail. A possible justification is associated to the different bogie attitude caused by the increased speed. In particular, the reduction of the cant excess allows the trailing wheelset to move into a more centred position avoiding flange contact, which seems to be crucial to avoid the formation of the long wavelength corrugation.

4 Mathematical Models in Frequency Domain

4.1 Track and Wheelset Dynamic Response

As far as the dynamic response of the track and the wheelset are concerned, hammer tests were carried out.

The structure of the track is of slab type: the rails are connected to the wooden sleepers using soft rubber pads and the sleepers are then suspended with respect to the slab by means of a sleeper pad. A group of three sleepers are connected to a single slab thus forming a block that is further suspended with respect to the substructure. Two different track sections were tested: one in straight track and the other one in the sharp curve previously shown (with radius equal to 110 m).

The measured track inertance[1] is reported in Fig. 5 (left), distinguishing between amplitude and phase. Three different zones can be identified in each curve: the first one

[1] The inertance is the frequency response function defined having as input the force applied to the system and as output the acceleration of one point [12].

in the range around 20 Hz, the second between 50–70 Hz, and then the band of high frequencies (larger than 200 Hz). In the first zone, the mode of vibration corresponds to the motion of the entire slab over the elastic supports that is found for all the track sections. In the second zone, where the mode corresponding to the in-phase motion of the rail and the sleeper over the slab is exhibited, important differences are found for the low rail in curve with respect to the other two. In fact, in this case, the frequency of the resonance is lower (52 Hz versus 70 Hz approximately). An important difference exists also in the third zone where a large dynamic amplification is found only in the low rail considering frequencies around 340 Hz. These differences are mainly due to the fact that the check rail, mounted on the track only in very sharp curves, is connected to the sleeper and not to the slab, thus increasing the mass resting on the sleeper and reducing the benefits (in terms of mass) introduced by the use of wood instead of concrete for the sleeper construction.

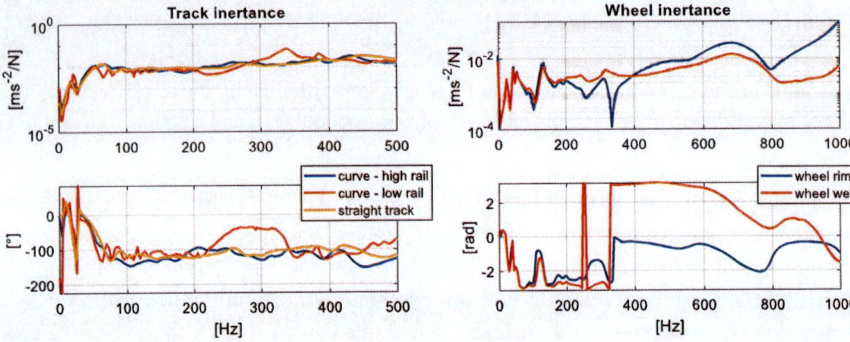

Fig. 5. Measured track inertance in different track sections (left) and wheel inertance (right).

An analogous procedure was performed with the aim of defining wheel inertance. The wheel and the axle were instrumented with a number of accelerometers placed in different positions on the wheel tread, on the wheel web and on the axle. A modal identification was performed permitting to identify the modes where the deformation of the axle gives the major contribution and modes where the relative motion between the wheel tread and the wheel web. Wheel inertance was measured when the hammer was positioned on the wheel rim. Figure 5 (right) depicts the inertance considering as output the acceleration measured on the wheel-rim (blue line) and on the wheel-web (red line): a number of resonances can be observed. These resonances are due to the rigid modes of the wheelset with respect to the bogie (around 20 Hz), the bending modes of the wheelset (with the first bending mode found at 70 Hz approximately) and the radial modes of the wheel tread with respect to the wheel web (with the first radial mode found at 350 Hz).

4.2 Dynamic Response of the Coupled Wheel-Rail System

A simple mathematical model in the frequency domain, developed by Grassie et al. [7] is used for the determination of the frequency ranges within which the probability for the corrugation to occur is higher. In particular, the contact force frequency response function of the coupled wheel-rail system is computed according to the following equations accounting for the contact stiffness:

$$x_W = A_{WV} W \tag{1}$$

$$x_R = -A_{RV} W \tag{2}$$

$$W = k_H (x_R - x_W) \tag{3}$$

$$\frac{W}{x_{irr}} = \frac{1}{\frac{1}{k_H} + A_{RV} + A_{WV}} = \frac{1}{T} \tag{4}$$

where A_{WV} and A_{RV} are respectively the wheel and rail receptances, obtained using the experimental results described in the previous subsections; T is the total receptance of the coupled system; k_H is the stiffness of the Hertzian spring; W is the vertical component of the contact force and x_{irr} is the vertical track irregularity. As far as the rail receptance is concerned, the analysis is performed considering the response measured on the low rail.

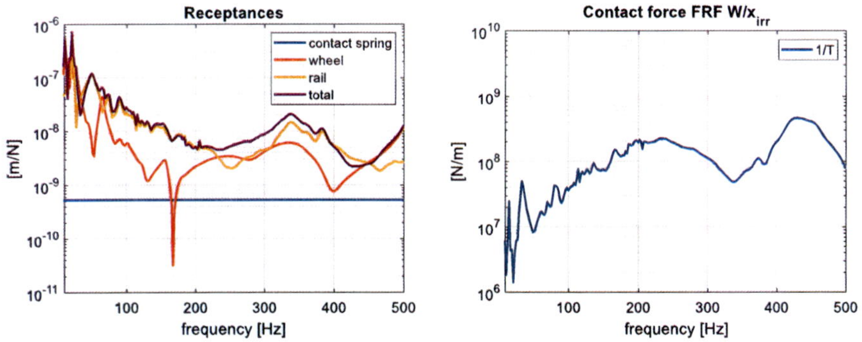

Fig. 6. Contributions of the single subsystems to the total receptance (left) and contact force FRF W/x_{irr} of the coupled wheel-rail system (right).

The combination of different frequency response functions defining T is shown in Fig. 6 left, while on the right the magnitude of the contact force frequency response function W/x_{irr} is depicted. A smooth shape of the function would lead to a broadband variation of the vertical force, while large peaks would suggest a dynamic amplification of the force corresponding to specific frequency ranges which may trigger a consequent corrugation. In this case, neglecting the contribution below 20 Hz, two major contributions may be detected, one around 35 Hz and a second one around 240 Hz. The variation of the vertical force will determine a consequent variation of the contact force in the tangential plane. Providing that the wear is the principle mechanism and low-damped modes of vibration exist in these two frequency ranges, then the corrugation may occur. This is because that the low-damped modes of vibration can determine large oscillations of the creepages, and thus the frictional power can be very high at the specific frequency. In this case, the longer corrugation (approximately 150 mm, see Fig. 1) wavelength should be associated to the low frequency contribution, whereas the shorter corrugation wavelength (30 mm when the speed was approximately 30 km/h and 45 mm when the speed was increased to 40 km/h, see Fig. 1) is associated to the high frequency one. Of course, this simple model is not able to justify why the change in speed is able to avoid the formation of the corrugation with long wavelength. More sophisticated models in the frequency domain like the ones by Frederick [8] and Tassilly [9] can be used, since they take into account the effect of the steady state values of creepages and creep forces in the curve.

5 Conclusions

In this paper an experimental and numerical analysis of the corrugation detected in a very sharp curve (radius equal to 110 m) and caused by a vehicle equipped with resilient wheels was shown. The results of the experimental tests carried out on the vehicle demonstrate that different values of the speed were able to trigger corrugation wavelengths in different ranges (30–40 mm and 150 mm). Additional tests were performed in order to make a modal analysis of the track and the wheelset with the aim of setting up a simple mathematical model used for the understanding of the phenomenon. The model helped to prove that the interaction between the resilient wheel and the low rail is causing the short wavelength corrugation. Another influencing fact is that the check rail was fixed to the sleeper instead of to the slab.

References

1. Grassie, S.L., Kalousek, J.: Rail corrugation: characteristics, causes and treatments. Proc. Inst. Mech. Eng. Part F: J. Rail Rapid Transit. **207**(1), 57–68 (1993)
2. Torstensson, P.T., Nielsen, J.C.O.: Simulation of dynamic vehicle-track interaction on small radius curves. Veh. Syst. Dyn. **49**(11), 1711–1732 (2011). https://doi.org/10.1080/00423114.2010.499468
3. Suda, Y., Hanawa, M., Okumuru, M., Iwasa, T.: Study on rail corrugation in sharp curves of commuter line. Wear **253**(1–2), 193–198 (2002)
4. Eadie, D.T., Kalousek, J., Chiddick, K.C.: The role of positive friction (HPF) modifier in the control of short pitch corrugations and related phenomenon. Wear **253**(1–2), 185–192 (2002)
5. Suarez, B., Chover, J.A., Rodriguez, P., Gonzalez, F.J.: Effectiveness of resilient wheels in reducing noise and vibrations. Proc. Inst. Mech. Eng. Part F: J. Rail Rapid Transit. **225**(6), 545–565 (2011). https://doi.org/10.1177/0954409711404104
6. Kouroussis, G., Verlinden, O., Conti, C.: Efficiency of resilient wheels on the alleviation of railway ground vibrations. Proc. Ins. Mech. Eng. Part F: J. Rail Rapid Transit. **226**(4), 381–396 (2012). https://doi.org/10.1177/0954409711429210
7. Grassie, S.L., Gregory, R.W., Harrison, D., Johnson, K.L.: The dynamic response of railway track to high frequency vertical excitation. J. Mech. Eng. Sci. **24**(2), 77–90 (1982)
8. Frederick, C.O.: A rail corrugation theory. In: Proceedings of the 2nd International Conference on Contact Mechanics of Rail-Wheel Systems, University of Rhode Island, pp. 181–211. University of Waterloo Press (1986)

9. Tassilly, E., Vincent, N.: A linear model for the corrugation of rails. J. Sound Vib. **150**(1), 25–45 (1991). https://doi.org/10.1016/0022-460X(91)90400-E
10. Meehan, P.A., Bellette, P.A., Horwood, R.J.: "Does god play dice with corrugations?": environmental effects on growth. Wear **314**(1–2), 254–260 (2014). https://doi.org/10.1016/j.wear.2013.11.027
11. Meehan, P.A., Batten, R.D., Bellette, P.A.: The effect of non-uniform train speed distribution on rail corrugation growth in curves/corners. Wear **366–367**, 27–37 (2016). https://doi.org/10.1016/j.wear.2016.05.009
12. Ewins D.J.: Modal testing: Theory, Practice and Application, 2nd edition, Research Studies Press Ltd, Boston (2009)

Rail Grinding and Track Vibration Resonances

Shankar Rajaram[1](✉), James Tuman Nelson[2], and Hugh Saurenman[3]

[1] Sound Transit, 401 S Jackson Street, Seattle, WA 94608, USA
Shankar.rajaram@soundtransit.org
[2] Wilson Ihrig and Associates, Emeryville, CA 94608, USA
jnelson@wiai.com
[3] ATS Consulting, Pasadena, CA 91101, USA
hsaurenman@atsconsulting.com

Abstract. Transit agencies in North America typically specify the ISO 3095:2005 rail roughness limits as an acceptance requirement in the grinding specifications. However, in practice, this requirement is rarely achieved, and the rail grinding industry typically proposes the EN 13231–3:2012 limits as alternate criteria for rail roughness. Rail grinding leaves grinding marks that show up as spectral roughness peaks at approximately 30 mm, 23 mm and 15 mm wavelengths. The roughness levels of these peaks are typically 5 to 15 decibels above the 2005 ISO curve. At 56 km/h train speeds, these peaks tend to increase the wayside noise by 1 or 2 decibels, or more where grinding marks are excessive. Measurements of track acceleration have shown that these roughness peaks generate vibration that coincides with the resilient fastener component resonance frequencies (800–900 Hz) at 89 km/h train speeds. These peaks are also close to the pin-pin vertical vibration mode of the rail. This paper focuses on an extreme case of this roughness peak exceeding the ISO curves and causing premature fatigue-induced failure of the rail clips installed on direct fixation fasteners.

Keywords: Rail · Vibration · Direct fixation · Transit · Fatigue

1 Introduction

The Sound Transit University Link extension in Seattle opened for revenue service in March 2016. The University Link comprises twin bore tunnels connecting Downtown Seattle to the University of Washington. Direct fixation track with continuous welded rail is used in the tunnels. To control groundborne noise emissions from the tunnels, the rails are supported on high compliance direct fixation (HCDF) fasteners with standard "e" shaped rail clips. These rail clips are designed to withstand high acceleration loads in demanding environments and are expected to perform under nominal light rail vehicle axle loads at speeds approaching 100 km/h in different grades and curves. The primary functions of these clips are to provide high clamping force for the rail, resist rail rollover, and provide controlled longitudinal restraint to the rail.

Approximately six months after the opening of revenue service in this extension, the rail clips started breaking at the rate of about a dozen clips a month. The majority

G. Degrande et al. (Eds.): *Noise and Vibration Mitigation for Rail Transportation Systems*, NNFM 150, pp. 312–319, 2021.
https://doi.org/10.1007/978-3-030-70289-2_32

of the failing clips showed signs of fatigue failure and cracks at the center leg of the "e" shaped clip where the tongue enters the top plate shoulder. The clip failures were concentrated in a few track sections. Further, in some of these sections a predominant number of failures were concentrated either on the left or right rail. Several fasteners showed repeat clip failures after broken clips were replaced with new clips. The clip failure rate did not rise to the level of being a safety concern. However, Sound Transit determined that it was necessary to understand the root cause and initiated a detailed investigation.

The investigation included the following:

- A review of track design elements
- Manufacturing quality of the clips
- Construction discrepancies
- Clip response under dynamic loading, vibration acceleration, and deflection of the tracks during train passage,
- Dynamic response of the clips to excitation in a lab setup,
- Onboard noise testing inside test trains, and
- Rail roughness measurements.

The focus of the current paper is on the noise, vibration, rail grinding, and rail roughness parts of the clip failure investigation [1, 2].

2 Background

The "e" shaped rail clip referred to as Clip-E in this paper is a lightweight clip with a bar diameter of 20 mm and a clamping deflection of 11 mm. A second type of clip referred to as Clip-P in this paper was installed in small test sections to evaluate its relative performance compared to Clip-E. Clip-P is a heavier and stronger clip with a bar diameter of 22 mm and a clamping deflection of 14 mm.

The commercial rail grinding services offered to transit properties in North America typically use 16 grit medium coarse grinding stones and 3600-rpm motors in grinding cars operated at 6.4 km/h for both profile grinding and polishing (finishing) of the rail surface. This method is attractive from a productivity point of view as it offers a reasonable means of grinding the rails within the limited non-revenue windows available for maintenance activities. However, this emphasis on productivity prevents meeting ISO 3095: 2005 rail roughness limits. The grinding service providers typically propose EN 13231–3:2012 as an alternate criterion for rail roughness that is 10 to 25 dB higher than the ISO 3095:2005 limits at critical wavelengths that influence wheel-rail noise and track vibration.

In 2017, Sound Transit developed rail grinding criteria that separated the grinding and finishing steps. The approach was similar to the strategy used in the Epping to Chatswood Rail in Sydney, Australia, and the Metro Network in Rotterdam. Sound Transit's finishing limit is 3 to 5 dB less stringent than the ISO 3095:2005 limits. Sound Transit performed a systemwide rail grinding program between November 2018 and April 2019 using the acceptance limits presented in Fig. 1. This paper includes rail roughness, clip acceleration and onboard noise data from 2017 and measurements repeated at the same location after the completion of the 2018–2019 rail grinding.

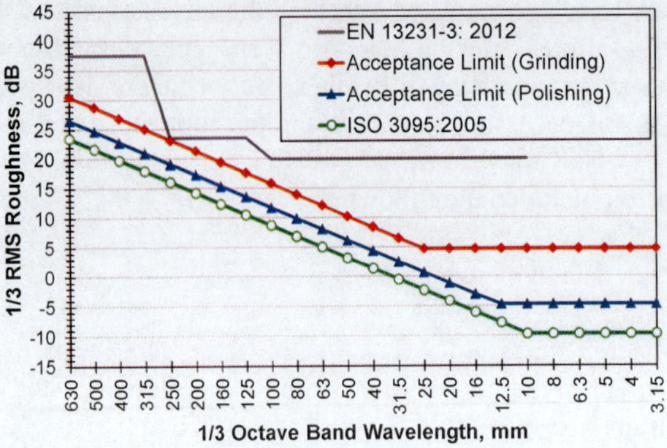

Fig. 1. Rail roughness acceptance limits for grinding and polishing.

Sound Transit has 62 KinkiSharyo LRVs that are a 70% low floor design. Each train in revenue service typically operates with three-LRV consists. Each LRV has three trucks: an idling center truck with independently rotating wheels and two powered trucks at each end with solid axle wheelsets. The wheelsets are fitted with resilient wheels.

3 Test Methodology

This section discusses the testing methodology used for the test results presented in this paper. The onboard noise measurements were performed with a custom tool developed for Sound Transit based on the Corrtracker® platform. The tool measures the noise levels inside the trailing cab area and identifies the location of the train with a 1.5 m accuracy. The rail roughness was measured using a Corrugation Analysis Trolley® (CAT).

Clip resonances were measured under laboratory conditions by clipping a short section of RE115 rail to a HCDF fastener with the clips under test. The clipped rail and fastener were placed on an approximately 500 kg solid steel inertia base with machined flat surface. Piezo-electric accelerometers were attached to the clip under test in lateral, longitudinal and vertical directions. The rail was excited by striking it with an instrumented hammer. The transfer function between clip acceleration and input force was measured in the laboratory with a fast-Fourier transform analyzer.

In situ acceleration measurements on the clips were performed at two locations on the rail alignment. Piezo-electric accelerometers were attached to the clips under test on the mainline rails in lateral, longitudinal and vertical directions. A triaxial accelerometer arrangement was also installed on the rail base (see Fig. 2). This measurement was performed on a southbound (SB) track fastener that had repeated clip failures. Trains on this section of track climb a nominal 4% grade at nominally 88 kph, so that tractive effort is high. The measurements at this location were performed in 2017 and repeated after rail grinding and finishing in 2019.

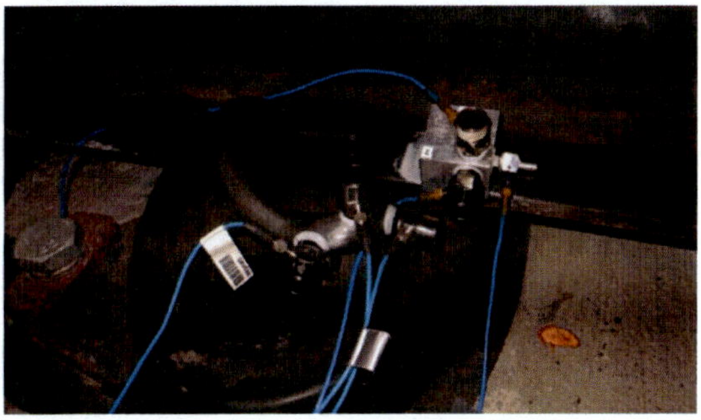

Fig. 2. Photograph of Accelerometers Installed in Three Axes on the Clip and on the Rail Base.

4 Results and Discussion

4.1 Onboard Noise Measurements

A review of the onboard noise data showed that the loudest section of the tunnels was between mileposts 5.1 and 5.2, which coincided with areas with concentrated clip failures. The noise data indicated a spectral peak at 870 Hz that reached a maximum level at this location (see Fig. 3). Discussions with the clip manufacturer indicated that Clip-E might have a natural resonance frequency around 850 Hz. Based on this information, an extreme resonance condition was shortlisted as a plausible cause for clip failure. Measurements in 2019 after rail polishing showed that the 870 Hz noise level peaks at this location were reduced by 8 to 10 dB.

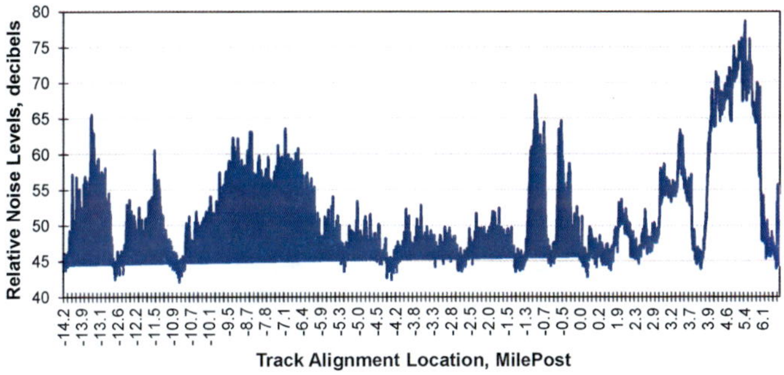

Fig. 3. Relative Onboard Noise Level at 870 Hz along the Train Alignment.

4.2 Rail Roughness

The roughness of both northbound (NB) and southbound (SB) track rails were measured at representative locations between milepost 5.1 and 5.2. Both tracks have comparable operating speeds and grades, though NB trains descend the grade under braking conditions, while SB trains climb the grade, presumably with similar tractive effort if constant speeds are maintained. The clip failures were concentrated on the right-hand rail of the

SB track in this section that was tested for roughness. The results from 2017 are shown in Fig. 4. The roughness spectra are 1/24 octave band data, while the criteria are for 1/3 octave band. The roughness data would be significantly higher if summed into 1/3 octave bands, though the discrete spectral peaks may dominate the 1/3 octave result in any case. The rail roughness spectra of all four rails exhibit a peak at 30 mm wavelength and harmonics thereof. The 2017 data also show that the SB right rail roughness was approximately 10 dB higher at 30 mm wavelength compared to the other three rails, consistent with the visual inspection. At train speeds of 88 ± 5 kph, the 30 mm peak would excite the track in the frequency range of 780 Hz to 870 Hz.

The rail roughness data measured after grinding and finishing in 2019 showed that the 30 mm peak has moved to 50 mm and its first harmonic at 25 mm was at least 15 dB less than the 30 mm peak in 2017.

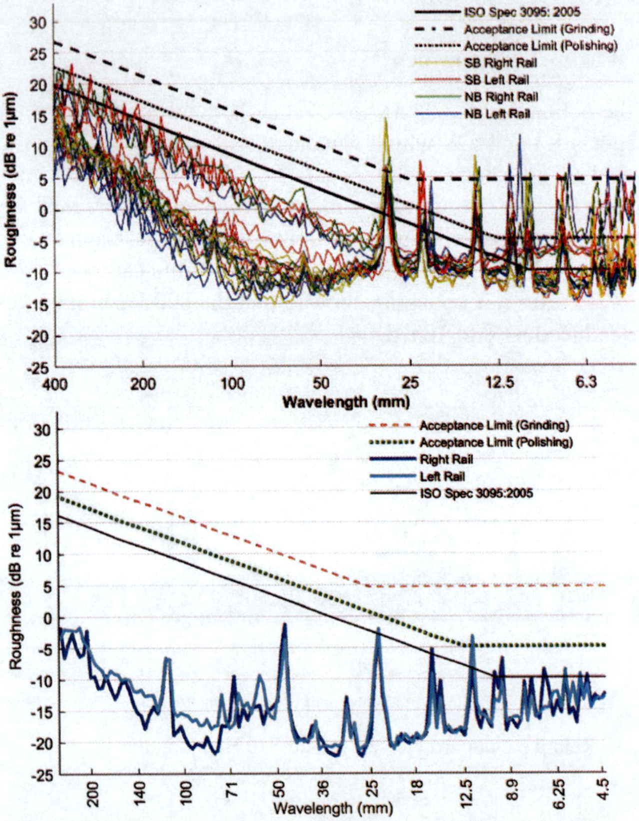

Fig. 4. Measured 1/24th Rail Roughness Levels in 2017 (Top) and in 2019 (Bottom).

4.3 Lab Testing of Clip Resonances

The resonance frequencies of the combined new Clip-E, high compliance fastener, and rail were measured in the laboratory. The results indicated sharp resonances at 500 Hz, 894 Hz, and 1400 Hz, with minor resonances at 1000 Hz. Damped resonances occur at 1810 Hz, 2000 Hz, and 2210 Hz. The peak at 894 Hz was pronounced and lightly damped. The resonance amplification of this peak was 40 to 60 dB. If grinding marks in the rail head are such that the rail head were sinusoidally excited at a discrete frequency exactly coincident with an installed clip resonance between 800 to 950 Hz of the type observed in the lab, the acceleration of the clip would be expected to be substantially amplified by the resonance.

4.4 Clip Acceleration

Figure 5 and Fig. 6 show results of measurements that were performed at a fastener with a history of clip failures. The figures show the acceleration levels in different 1/3 octave bands. The 2017 Clip-E data were collected before the 2019 grinding program. Then Clip-E was replaced with Clip-P in this section, including at this test fastener. The 2019 Clip-P data were collected after rail grinding. Finally, Clip-P was again replaced with new Clip-E (2019 Clip-E data) to understand the effects of the new rail grinding on Clip-E acceleration.

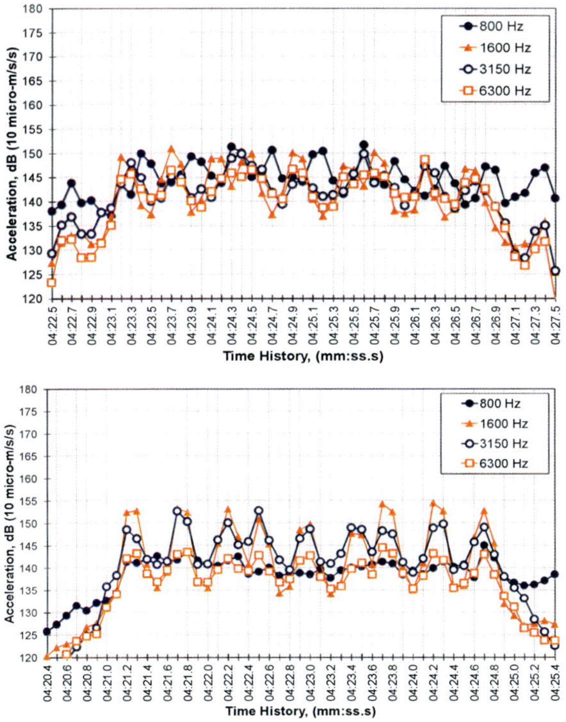

Fig. 5. Acceleration Time History (X-Axis) at a fastener in 2017 (Top) and 2019 (Bottom) from a 3-car LRV Consist Passing.

Figure 5 shows time history plots for transverse acceleration of Clip-E in 2017 and 2019. The time history plots represent a three-LRV consist revenue service train with a total of 9 bogies. The 1600 Hz time histories very clearly show vibration with each passing bogie with a slight increase in acceleration after grinding. The 800 Hz plot indicates a 10 dB reduction after grinding. An interesting feature of the 800 Hz acceleration in

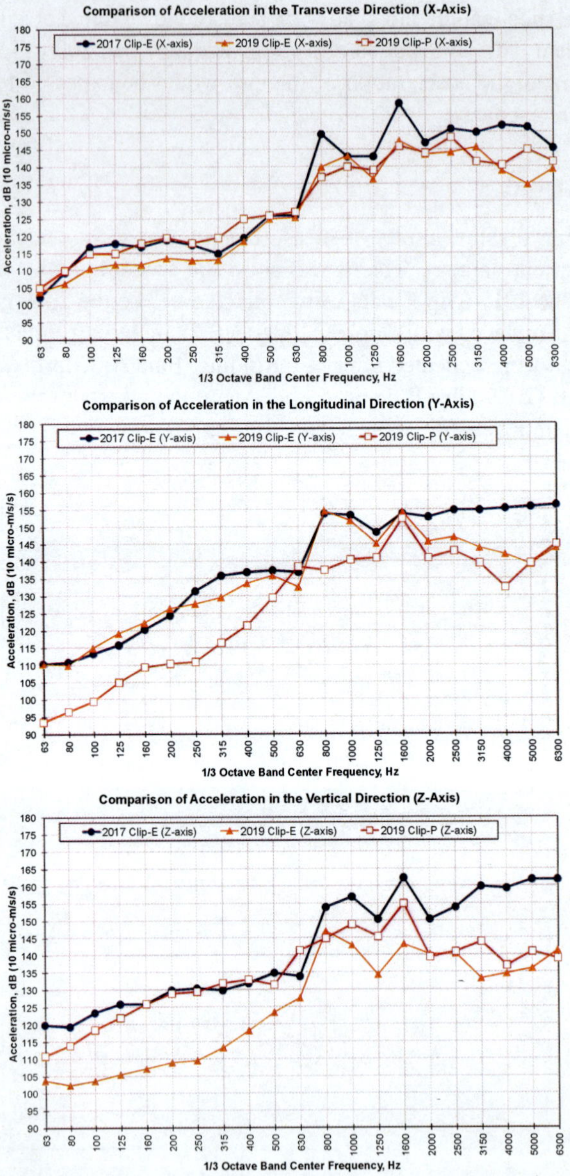

Fig. 6. Comparison of Clip Acceleration in the 800 Hz 1/3 Octave Band in 2017 and 2019 (Before and After Grinding).

2017 was the strong acceleration occuring on the fastener between the passing bogies. Although the mechanism was not clear, this may indicate a strong resonance condition near 800 Hz with deep grinding marks as the root cause.

Figure 6 shows a comparison of clip acceleration for Clip-E and Clip-P in three axes for passing revenue trains. The transverse X-axis data indicate that the acceleration of Clip-E was reduced by approximately 10 dB at 800 Hz and 1600 Hz after rail grinding. There was 5 to 20 dB reduction between 2000 Hz and 6300 Hz and a 2 to 5 dB reduction between 80 Hz and 250 Hz. The longitudinal Y-axis data shows that there was no noticeable reduction in Clip-E acceleration after rail grinding in most 1/3 octave bands. The vertical Z-axis data show a 10 dB reduction in Clip-E acceleration at 800 Hz and a 15 to 20 dB reduction at most other frequencies after grinding.

Rail grinding significantly reduced vertical acceleration in the 63 Hz to 500 Hz frequency bands which affect ground borne noise. In this immediate vicinity, ground-borne noise was audible at 800 Hz in a residential basement located roughly 60 m from the tunnel. The over-consolidated glacial tils, sands and gravels exhibited very low attenuation factor, so that ground borne vibration is propagated very efficiently.

The Clip-P data contained no peak at 800 Hz for all three axes. For similar excitation forces in the post-grinding scenario, only Clip-E has a noticeable peak at 800 Hz, which may be an indication that Clip-E combined with the top plate has a natural resonance that was excited by the track roughness. Most importantly, the rate of clip failures has dropped substantially in 2019, and field reports in the five months after grinding showed negligible clip breaks. This alone indicates the importance of rail grinding surface finish in controlling clip fatigue failure.

5 Conclusions

The failure of rail clips on direct fixation tracks on a light rail system was investigated. The vibration acceleration spectra of the clips and rails during passing trains contain a strong peak at 800 Hz. Although the pin-pin mode of rail vibration at about 800 Hz was expected, the clip breaks indicated something unusual. Investigation revealed that rail grinding marks excited rail vibration that was amplified by clip resonances, creating an unusual clip fatigue failure phenomenon. The grinding marks were removed by a grinding and polishing program, arresting the high rate of rail clip failures. The grinding marks apparently created a strong case of coincidence with system resonances that result in component fatigue failure.

Acknowledgements. The authors are grateful for many valuable contributions and support provided to this investigation by ST staff and management, Wilson Ihrig staff, ATS Consulting staff, HNTB, Stacey Witbeck Inc., Progress Rail, and Pandrol USA. The authors also acknowledge Steve Cox of Pandrol UK and Anthony Bohara of HDR for their insights.

References

1. Nelson, J.T.: U-Link E-Clip Resonance Tests. Prepared for Sound Transit, September 13 (2017)
2. Bergen, T.F., Hays, T.: Sound Transit U-Link Rail Clip Vibration Acceleration Measurement Results (Draft). Prepared for Sound Transit, May 24 (2019)

Measurement of Longitudinal Irregularities on Rails Using an Axlebox Accelerometer System

Stuart L. Grassie(✉)

RailMeasurement Ltd, 79 River Lane, Cambridge CB5 8HP, UK
stuart.grassie@railmeasurement.com

Abstract. Axlebox accelerometer (ABA) systems offer a means of measuring or monitoring the longitudinal profile of track at a relatively high speed over the wavelength range associated with corrugation: typically 10–1000 mm. There is little or no published data on the use of ABA systems for measuring longer wavelengths, in particular 1000–3000 mm. This paper provides data from one ABA system which was mounted in a road-rail vehicle that was used as a track recording car. Repeatability and accuracy are assessed in a similar and novel way that has not previously been presented in the literature. Both are shown to be quite acceptable throughout the wavelength range 10-3000 mm, with a correlation coefficient exceeding 0.9 for both repeatability and accuracy.

Keywords: Rail corrugation · Rail corrugation measurement · Acoustic roughness

1 Introduction

Measurement of railhead irregularities, in particular corrugation and acoustic roughness, is of particular interest to railway acousticians and track engineers. These irregularities can be measured using one of at least three different techniques: chord-based measuring systems, inertial systems and axlebox accelerometer systems. Some basic characteristics of these three techniques, including attractions and limitations, are examined in ref [1]. Chord-based systems suffer from having a very variable transfer function between input and output [1] and are discussed no further in this paper. These limitations, and a "decolouring" procedure to deal with the variable transfer function, are discussed in ref [2].

There are several references in which inertial systems have been used to measure corrugation and acoustic roughness e.g. [3–9]. References [5] and [7–9] consider measurement of relatively long wavelength irregularities using inertial equipment. References [8] and [9] present data from train-based systems showing measurements to wavelengths of 5 m and 10 m respectively.

G. Degrande et al. (Eds.): *Noise and Vibration Mitigation for Rail Transportation Systems*, NNFM 150, pp. 320–328, 2021.
https://doi.org/10.1007/978-3-030-70289-2_33

Axlebox acceleration (ABA) systems have existed since at least the early 1980s, when an analogue system was developed by British Rail Research and used thereafter to monitor corrugation on the UK network [10]. ABA systems have subsequently been used in Sweden [11], Australia [12] and elsewhere to monitor corrugation, primarily for purposes of rail maintenance and in particular reprofiling. ABA systems have also been used to detect and rank the severity of discrete, relatively short wavelength defects, including welds and insulated joints [12–14], while trials have been undertaken with an ABA system to assess its merit in producing a "noise map" of a railway system [15]. The paper by Berggren et al. [11] shows one-third octave spectra found using an ABA system to 1.28 m, but there is very little published work to show the performance of ABA systems for monitoring relatively long wavelength irregularities, which contribute to ground-borne noise and vibration (GBN&V).

The Crossrail system in the UK has adopted a specification to limit relatively long wavelength irregularities and thereby GBN&V. This IP D10 specification states that the combined wheel/rail roughness spectrum should not exceed a limit of 30dB re 1 micron centred on a wavelength of 2 m, decreasing by 15dB per tenfold reduction in wavelength [7]. Equipment to measure long wavelength irregularities is also of more general interest in framing an International Standard to deal with railhead irregularities pertinent to GBN&V. This work is being undertaken by WG8 of ISO/TC108/SC2.

Fig. 1. Axlebox accelerometer (ABA) system mounted on a road-rail track recording vehicle

Inertial and ABA systems have been developed by RailMeasurement Ltd (RML) to measure longitudinal irregularities on rails. The ABA systems developed by RML (known as the HSRCA or High Speed RCA) have been used primarily to identify the location and severity of corrugation on an entire railway network, and thereby help schedule rail grinding. One such ABA system was mounted on an unusual road-rail track recording vehicle (Fig. 1). One of the requirements of this system was to measure corrugation to about 1.5 m wavelength, which was known to be a problem on the network. Measurements from this HSRCA, have not previously been presented in the open literature.

Measurements which were obtained during validation testing of this equipment are presented here. A novel and concise method of analysis and display is proposed and examples are presented.

2 Axlebox Accelerometer System (HSRCA)

2.1 Principle of Operation and Analysis of Data

The ABA system was mounted as in Fig. 1. Data are produced in a format that is appropriate for surveying corrugation on the railway network.

Analysis is made in the frequency domain to find RMS values of railhead irregularities in different wavelength ranges. It can be deduced from refs [1] or [16, 17] that the transfer function between the railhead irregularity Y and axlebox acceleration $\ddot{Z}$ is

$$H(\omega) = \frac{Y}{\ddot{Z}} = -\frac{M_w}{k_H}(1 + k_H(\alpha_r + \alpha_w)) \tag{1}$$

where $\omega = \frac{2\pi v}{\lambda}$, v is the speed; λ is the wavelength of sinusoidal irregularity on the rail; k_H is the Hertzian contact stiffness; M_w is the wheel mass; α_r and α_w are respectively the rail and wheel receptances as seen from the point of contact. If the wheelset is modelled as a rigid body,

$$\alpha_w = {}^{-1}\big/_{M_w\omega^2} \tag{2}$$

In general, both α_r and α_w vary with frequency, so the transfer function varies with frequency. The spectral density S_y of railhead irregularity is related directly to the spectral density of axlebox acceleration by the transfer function:

$$S_y(\omega) = |H(\omega)|^2 S_z(\omega) \tag{3}$$

The mean square response is found from the spectral density using standard techniques e.g. ref [18]. Similar equations relating the roughness to the axlebox acceleration are found by Berggren et al. [11]. This technique is also used for the ABA systems described in refs [12] and [15]. In refs [11, 15] and for the ABA system described here, one-third octave spectra of the railhead roughness were calculated from the spectral density of railhead irregularity using the measured spectral density of axlebox acceleration and an assumed transfer function. For the current vehicle and equipment, it was assumed that the irregularities measured came entirely from the rail i.e. that the wheel irregularities were negligible. The instrumented axle was neither braked not driven: it was used for measurement alone. Although the wheels on this axle would therefore be relatively smooth, it would usually be advisable to correct for irregularities on the wheel, as in ref [15].

Axlebox accelerations cannot simply be integrated twice to produce output that is equivalent to irregularities on the rail, except for those conditions in which the wheel can be considered as isolated from the vehicle and also to be rolling up and down over irregularities that exist on a rigid rail that does not deflect under the measuring axle. For this reason, ABA systems cannot be used to provide data in the format required for EN13231–3:2012 [19], although they can be used to provide the RMS magnitudes of railhead irregularity that are permitted in the earlier version of this Standard [20].

2.2 Validation of the Equipment

Validation tests were undertaken at several sites to assess repeatability, effects of speed and direction of travel on reproducibility, and accuracy of the HSRCA equipment. Results are shown here from only one site of 400m length over which the test vehicle passed at speeds shown in Fig. 2. Runs 03, 05 and 07 were made in direction of decreasing km; runs 04 and 06 were made in the opposite direction.

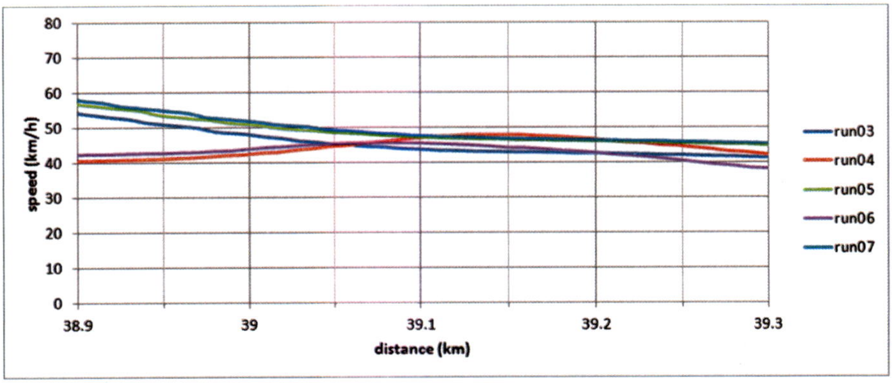

Fig. 2. Speeds for five runs with HSRCA over the test site

The validation was undertaken by comparing RMS amplitudes of railhead irregularity for left and right rails in the 10–30 mm, 30–100 mm, 100–300 mm, 300–1000 mm and 1000–3000 mm wavelength ranges. This calculation was made for a 100 m length of track every 10 m. These values for the 100–300 mm wavelength range are shown in Fig. 3 as functions of distance through the test site. Accuracy of the test equipment was assessed by comparing measurements of railhead irregularity made according to the method outlined in Sect. 2.1 with measurements made using the CAT instrument, for which measurements are also shown in Fig. 3. The CAT has been widely used to measure railhead irregularities e.g. refs [3–9, 11, 15, 20].

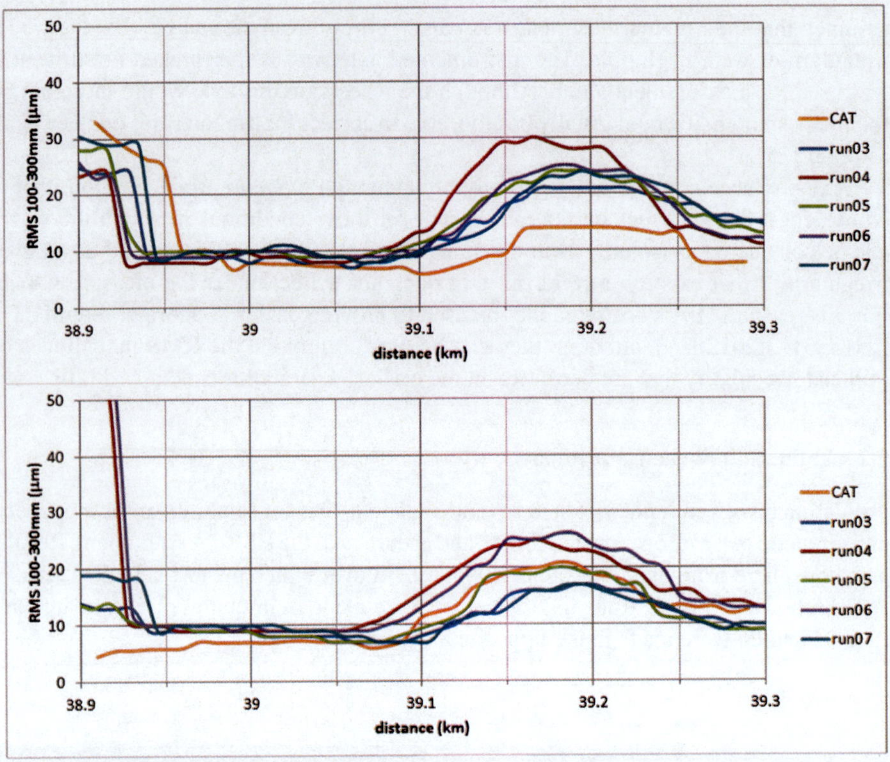

Fig. 3. RMS magnitude of corrugation in 100–300 mm wavelength range, calculated in 100 m window every 10 m, left rail (above), right rail (below)

The main method that is used here to assess repeatability and accuracy of the test equipment is directly to compare RMS values of irregularities in the five wavelength ranges mentioned above at increments of 10m throughout the 400 m test site. The results for repeatability, using data for runs 05 and 07, which were made in the same direction at roughly the same speed (Fig. 2), are shown in Fig. 4.

The accuracy of the test equipment is assessed similarly by comparing RMS values of irregularities calculated from the HSRCA in the five wavelength ranges mentioned above at increments of 10 m throughout the 400 m test site with corresponding values calculated from the CAT (Fig. 5). In this case the comparison is made using data for all five measuring runs shown in Fig. 2.

Summary statistical data for the two correlations are given in Table 1.

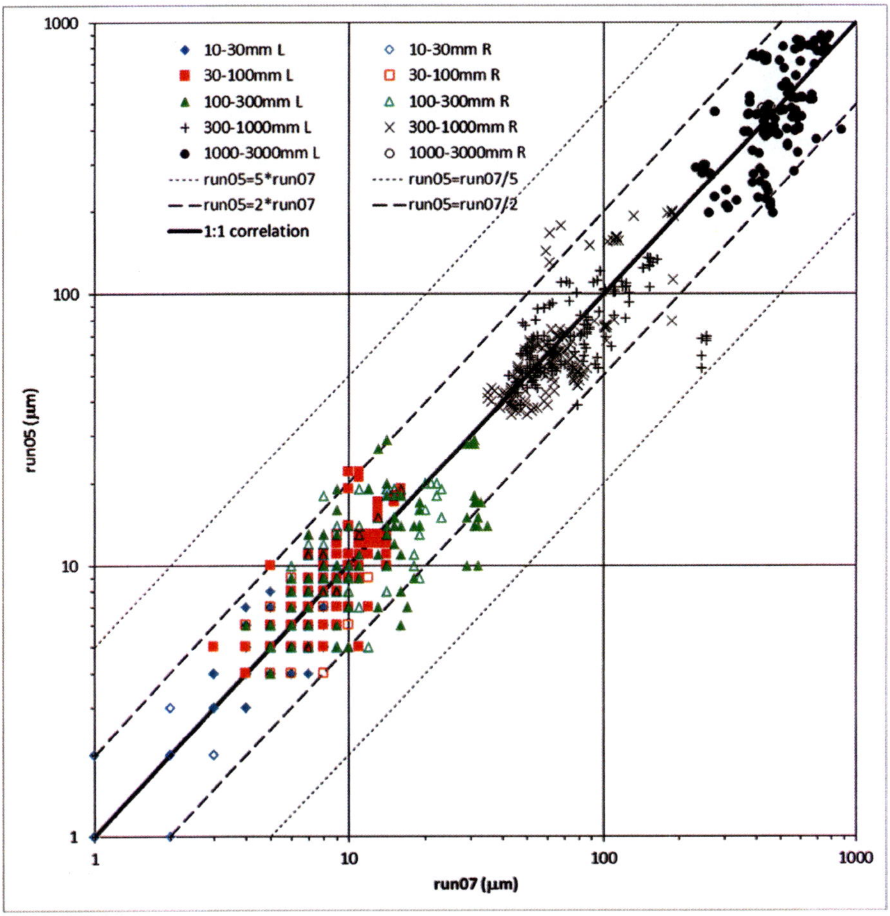

Fig. 4. Repeatability of HSRCA assessed from RMS magnitude of corrugation in five wavelength ranges over 10–3000 mm range at 10 m intervals through 400 m test site

Table 1. Statistical data for validation of HSRCA (repeatability and accuracy)

	No. of points	Slope	Intercept (dB)	Correlation coefficient, r^2	Standard error, slope
Repeatability	390	0.964	0.731 [1]	0.956	0.011
Accuracy	1950	0.867	6.659 [2]	0.920	0.006

[1] equivalent to intercept at 0dB (1 μm) of 1.088 μm.
[2] equivalent to intercept at 0dB (1 μm) of 2.153 μm.

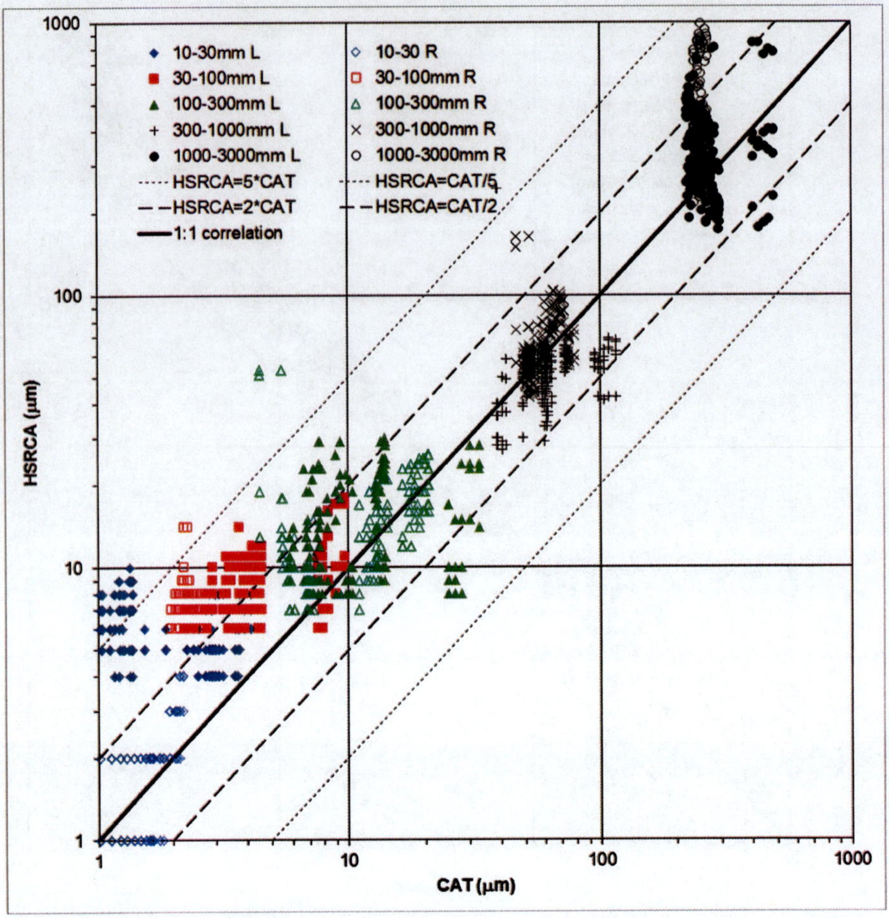

Fig. 5. Accuracy of HSRCA assessed by comparing RMS magnitudes of irregularity on different rails and in different wavelength ranges with corresponding magnitudes calculated from CAT

3 Discussion and Conclusions

Axlebox accelerometer (ABA) systems offer an attractive and versatile method of surveying corrugation and other longitudinal irregularities on a railway system at close to line speed. Measurements can in practice be produced only as RMS amplitudes of irregularity or the equivalent, such as one-third octave spectra.

This paper presents a novel method of demonstrating repeatability and accuracy of a system for measuring railhead irregularities. The method is based on correlation of RMS amplitudes of corrugation calculated over a test site. Measurements are taken from a particular ABA system developed by RailMeasurement Ltd, which had been developed for use at relatively low speeds: 40–60 km/h for the data presented here. For the longest wavelength range this corresponds to a frequency of less than 10 Hz.

Repeatability is assessed by comparing measurements for two separate measuring runs at the same conditions. There are 390 individual data points: 400 m at 10 m intervals, two rails and five wavelength ranges from 10–30mm to 1000–3000 mm. The correlation coefficient is 0.954 with a slope of the line of correlation of 0.964 c.f. an ideal of 1.0. Accuracy is assessed by comparing measurements from 5 measurements runs with the ABA system with corresponding data from the hand-held CAT instrument. For the 1950 data points, the correlation coefficient is 0.920 with a slope of the line of correlation of 0.867. Larger values of RMS irregularity are given by the HSRCA than the CAT at short wavelengths (10–30 mm and 30–100 mm), in part because the contribution of wheel roughness to the overall roughness is neglected. In previous work e.g. [15] the wheel roughness was measured and a correction made for this in calculating irregularities on the rail. No measurement of wheel irregularities was made on the measuring wheelset used here: indeed the tests reported here were undertaken before equipment to measure the circumferential profile of wheels was widely available.

This ABA system offers a reliable means of surveying corrugation on a railway system, provided that this can be characterized satisfactorily by RMS levels of irregularity or one-third octave spectra. This may not be a guide to the performance of ABA systems in general as this depends significantly on how reliably the transfer function between axlebox acceleration and railhead irregularity does indeed represent the system performance.

References

1. Grassie, S.L.: Measurement of railhead profiles: a comparison of different techniques. Wear **191**, 245–251 (1996)
2. Railway applications – track – track geometry quality – Part 1 Characterisation of track geometry quality, European Standard BS EN 13848–1:2019, March 2019, CEN, rue de la Science 23, B-1040 Brussels, Belgium
3. Grassie, S.L., Saxon, M.J., Smith, J.D.: Measurement of longitudinal rail irregularities and criteria for acceptable grinding. J. Sound Vib. **227**, 949–964 (1999)
4. Jiang, S., Meehan, P.A., Bellette, P.A., Thompson, D.J., Jones, C.J.C.: Validation of a prediction model for tangent rail roughness and noise growth. Wear **314**, 261–272 (2014)
5. Triepaischajonsak, N., Thompson, D.J., Jones, C.J.C., Ryue, J., Priest, J.A.: Ground vibration from trains: experimental parameters and validation of a numerical model. J. Rail Rapid Transit. Proc. Inst. Mech. Eng. **225F**, 140–153 (2011)
6. Grassie, S.L.: Rail irregularities, corrugation and acoustic roughness: characteristics, significance and effects of reprofiling on different types of railway system. J. Rail Rapid Transit. Proc. Inst. Mech. Eng. **226F**, 542–557 (2012)
7. Methold, R.H., Jones, C.J.C., Cobbing, C., Cronje, J.: The factors associated with the management of combined Rail/Wheel roughness to control groundborne noise and vibration from the UK's crossrail project. In: Anderson, D., Gautier, P.-E., Iida, M., Nelson, J.T., Thompson, D.J., Tielkes, T., Towers, D.A., de Vos, P., Nielsen, J.C.O. (eds.) Noise and Vibration Mitigation for Rail Transportation Systems. NNFMMD, vol. 139, pp. 671–682. Springer, Cham (2018)
8. Grassie, S.L.: Routine measurement of long wavelength irregularities from vehicle-based equipment. In: Anderson, D., Gautier, P.-E., Iida, M., Nelson, J.T., Thompson, D.J., Tielkes, T., Towers, D.A., de Vos, P., Nielsen, J.C.O. (eds.) Noise and Vibration Mitigation for Rail Transportation Systems, pp. 333–342. Springer, Cham (2018)

9. Grassie, S.L.: Measurement of long wave irregularities on rails. In: Nielsen, J.C.O., Anderson, D., Gautier, P.-E., Iida, M., Nelson, J.T., Thompson, D.J., Tielkes, T., Towers, D.A., de Vos, P. (eds.), Proceedings of the 11th International Workshop on Railway Noise, volume 126 of Notes on Numerical Fluid Mechanics and Multidisciplinary Design, pp. 643–649, Uddevalla, Sweden, 9–13 September 2013
10. Lewis, R.B., Richards, A.N.: A new method for the routine measurement of railhead corrugations. Rail International, pp. 37–41 February (1986)
11. Berggren, E.G., Li, M.X.D., Spännar, J.: A new approach to the analysis and presentation of vertical track geometry quality and rail roughness. Wear **265**, 1488–1496 (2008)
12. Grassie, S.L.: Corrugation on Australian national: cause, measurement and rectification. In: Proceeding of 4th International Heavy Haul Railways Conference, Brisbane, Institution of Engineers, Australia (1989)
13. Moldova, M., Li, Z., Dollevoet, R.: Axlebox acceleration: measurement and simulation for detection of short track defects. Wear **271**, 349–356 (2011)
14. Moldova, M., Oregui, M., Nūnez, A., Li, Z., Dollevoet, R.: Health condition monitoring of insulated joints based on axlebox acceleration: measurements. Eng. Struct. **123**, 225–235 (2016)
15. Grassie, S.L., Bongini, E., Saxon, M.: Noise mapping' of a railway network: validation and use of a system based on measurement of axlebox vibration. NNFM. In: Proceedings of IWRN10, (Nagahama, Japan, 18–22 October 2010), vol. 118, pp. 511–517 (2011)
16. Knothe, K., Grassie, S.L.: Modelling of railway track and of vehicle/track interaction at high frequencies. Veh. Syst. Dyn. **22**, 209–262 (1993)
17. Grassie, S.L., Gregory, R.W., Harrison, D., Johnson, K.L.: The dynamic response of railway track to high frequency vertical excitation. J. Mech. Eng. Sci. **24**, 77–90 (1982)
18. Newland, D.E.: An introduction to random vibration and spectral and wavelet analysis. 3rd edition, Chapter 7, Dover Publications Inc, New York (2012)
19. Railway applications – track – acceptance of works – Part 3: Acceptance of reprofiling rails in track, European Standard EN 1323, January 2012, CEN, rue de Stassart 36, B-1050 Brussels, Belgium, pp. 1–3 (2012)
20. Railway applications – track – acceptance of works – Part 3: Acceptance of rail grinding, milling and planing works in track, European Standard EN 1323, May 2006, CEN, rue de Stassart 36, B-1050 Brussels, Belgium, pp. 1–3 (2006)
21. Daniel, W.J.T., Horwood, R.J., Meehan, P.A., Wheatley, N.: Analysis of rail corrugation in cornering. Wear **265**, 1183–1192 (2008)

Squeal Noise

Investigation of Railway Curve Squeal Using Roller Rig and Running Tests

Takeshi Sueki(✉), Tsugutoshi Kawaguchi, Hiroyuki Kanemoto, Masahito Kuzuta, Tatsuya Inoue, and Toshiki Kitagawa

Railway Technical Research Institute, 2-8-38 Hikari-cho, Kokubunji-shi, Tokyo 185-8540, Japan
sueki.takeshi.99@rtri.or.jp

Abstract. The roller rig and the running tests were performed in this study to characterize curve squeal and its relation to wheel vibration. The results of and the differences between the tests are presented. The tests show consistent wheel vibration for rolling noise. The wheel vibration associated with curve squeal in the roller rig test was generated even at 5 km/h, with the greatest wheel vibration observed at 2 kHz. The wheel vibration in the running test was not generated clearly at 5 km/h, and the curve squeal at 2 kHz resembled that in the roller rig test above 15 km/h. Parameters related to the wheel and rail contact and the vibration characteristics of the wheel and the rail influence the curve squeal. The roller of the rig exhibited specific natural frequencies, and coincidence with several frequencies of the wheel response increased wheel vibration. Although the roller rig potentially simulates actual situations, the parameters and the vibration characteristics require appropriate tuning to correctly evaluate squeal.

Keywords: Railway curve squeal · Roller rig test · Running test · Wheel vibration

1 Introduction

Curve squeal is an important railway noise source that sometimes causes people to complain. To investigate the mechanism of and to counteract this squeal, running tests involving actual trains are imperative. Field investigations of wheel vibration characteristics and noise provide an avenue for understanding the mechanism of the noise.

Roller rig tests associated with curve squeal also exist (e.g. [1]). Since running tests with an actual train are difficult to manage, the roller rig test provides complementary information to better understand the characteristics of curve squeal and related vibration. The roller rig test is cheaper, quicker, and easier to replicate under uniform condition than a running test. However, the consistency of wheel vibration and noise from the roller rig test relative to the running test remains unclear from previous studies.

In this study, the roller rig and the running tests were conducted to understand the characteristics of curve squeal generation. Comparison of results clarified key factors for the relationship between the tests.

G. Degrande et al. (Eds.): *Noise and Vibration Mitigation for Rail Transportation Systems*, NNFM 150, pp. 331–338, 2021.
https://doi.org/10.1007/978-3-030-70289-2_34

2 Test Methods and Sensors

2.1 Running Test

Running tests were conducted at a test line of the Railway Technical Research Institute shown in Fig. 1. The test train consisted of one 20-m-long trailer vehicle weighing about 22,100 kg and one locomotive. The train's velocity ranged from about 5 to 33 km/h, during the tests. The study focused on the leading axle of the bogie, and the wheels termed "C-type wheel" in Japan had a straight web. Measurements involved the wheel vibration and the noise close to the wheel.

The track conditions varied along the test line, and the gauge was 1,067 mm. The data were obtained as the train traversed the curve characterized by curve squeal [2]. The curve involved a ballasted track with a curve radius of 160 m, a superelevation of 105 mm, a Japanese 50N-type rail, and wooden sleepers. The results when the train passed through a straight section are also discussed. The straight section contained a ballasted track with a Japanese 50N-type rail and concrete sleepers.

2.2 The Roller Rig Test

The roller rig shown in Fig. 2 measures the creep force of the railway vehicle, with a wheelset and rollers corresponding to the rail. The rig is capable of setting the angle between a wheel and rails known as the yaw angle. Setting the angle simulates the passing of trains through curves. The outer peripheral shape of the rollers is a railhead shape of a Japanese 60-type rail.

In this study, the roller rig was used for investigation of curve squeal. The wheels in the roller rig test were the same type used in the running test. Velocities of 5, 15, and 30 km/h and yaw angles of – 1, – 0.5, 0, 0.5, and 1° served as the test conditions. A load of 30 kN was applied to each axle box as a static wheel load, with the wheel for measurements on the left side (Fig. 2). Positive yaw angles indicated the wheel for measurements corresponded to the outer wheel in real situations and vice versa.

2.3 Sensors

Accelerometers were attached to the wheels as shown in Fig. 3. The signals of wheel vibrations were transmitted through the slip rings set at the axle boxes. The positions VS3/VS7 in the running test and VA2 in the roller rig test were slightly different. However, the wheel vibrations are expected to exhibit the same trend because almost all the responses were similar for both tests (see Fig. 5 in Sect. 3.1).

Microphones were installed as shown in Figs. 1, 2, and 3, with omnidirectional microphones in close to the wheels in the running tests, while those in the roller rig test were directional. The directional microphone reduces the impact of noise from the opposite wheel and the rollers. It consists of an omnidirectional microphone and a parabolic reflector. When the sound plane wave enters the parabolic reflector, the wave is concentrated at the focus of the reflector [3].

Fig. 1. Running test

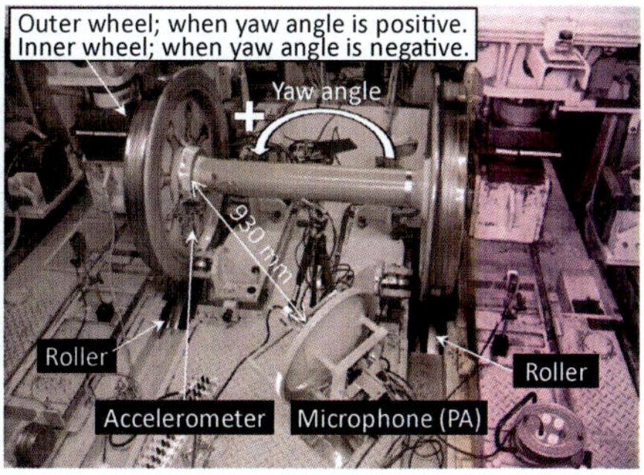

Fig. 2. Roller rig test

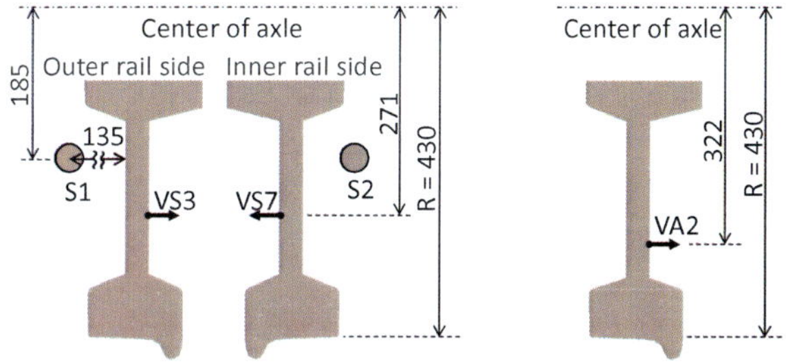

(a) Running test (b) Roller rig test

Fig. 3. Illustrations of the position of the sensors

3 Results

3.1 Frequency Analysis

Figure 4 displays frequency analysis data for the tests at a 15 km/h running and rolling velocity. The traces for the running test were obtained by averaging about 5 test train trips. The measurement time covered about 5 m length of train movement in the curve section, while the traces for the roller rig test were averaged over 30 s of measurements.

A direct comparison of the noise in the running test with that in the roller rig test has limitations because of differences in the distances from the wheel to the microphone for the two tests and the use of the directional microphone in the roller rig test. However, a comparison of trends from both tests are acceptable.

Rolling Noise. Figure 4(a) shows a comparison of results from a roller rig test at 0°s to those of a running test in the straight section. In this situation, the results correspond to rolling noise generation. For wheel vibration, the traces for the running tests and for the roller rig test were consistent. These tests displayed peaks at the same frequencies, suggesting that the roller rig test simulates running situations for rolling noise generation.

Curve Squeal Noise. Figure 4(b) shows a comparison of the results of the roller rig test at a positive angle with those of the running test in the curve section. The positive angle

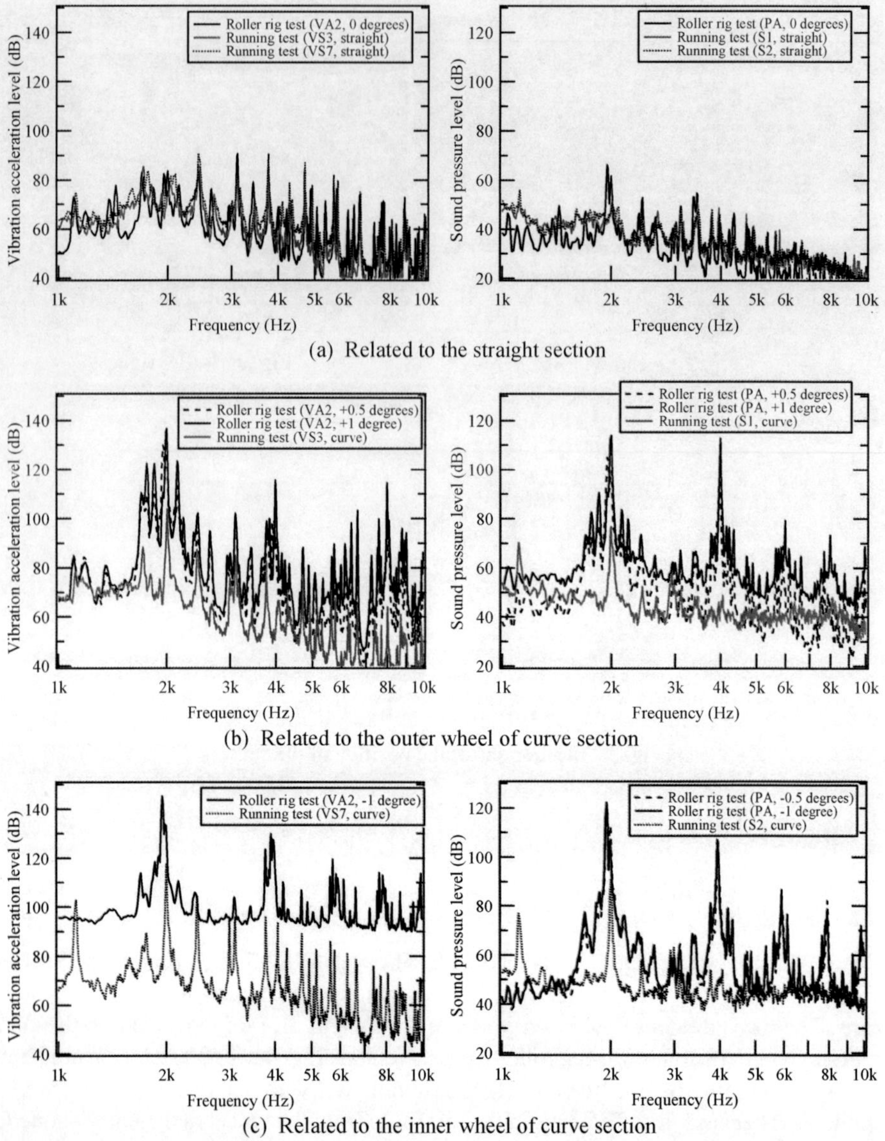

(a) Related to the straight section

(b) Related to the outer wheel of curve section

(c) Related to the inner wheel of curve section

Fig. 4. Frequency analysis at 15 km/h of train and rolling velocity. The left side represents wheel vibrations while the right side displays noise.

in the rig test corresponds to the outer wheel in the running test. Evidently, the wheel vibration and the noise in the roller rig test with angles were significantly above that without an angle, and these therefore correspond to curve squeal. The noise measured in the roller rig test displayed peaks mainly at 2, 4, 6, and 8 kHz. A comparison of the wheel vibration with noise in the roller rig test reveals that not all peaks in the wheel vibration were observed for the noise.

The frequency responses of the wheels during excitation of the wheel tread in the axial direction are shown in Fig. 5. The indexes (*m*, *n*) represent vibration modes; *n* is the number of nodal lines and *m* is the number of nodal circle ($m = R$ in radial mode). Figure 5 only partially displays the indexes of wheel vibrations. From the frequency responses, the peaks at 2, 4, 6, and 8 kHz correspond to the 4th, 6th, 8th, and 10th modes of the zero-nodal-circle. The vibrations, except for these modes of zero-nodal-circle, failed to generate greater noise. This response ressembles that of the inner wheel shown in Fig. 4(c).

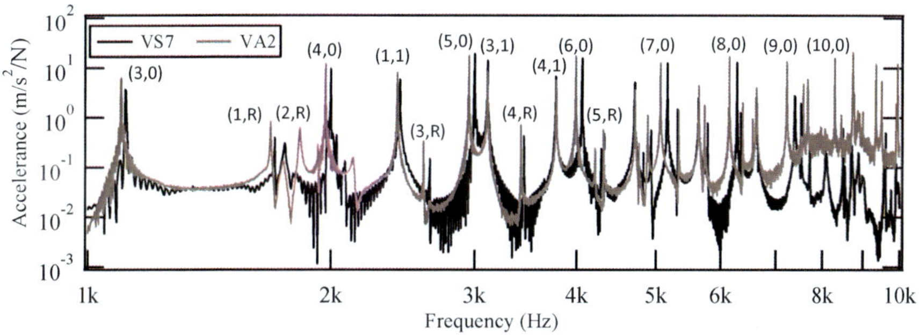

Fig. 5. Frequency responses of respective wheel (axial direction excitation).

Comparing the roller rig test and the running test in Figs. 4(b) and (c), the peak frequencies for wheel vibrations are similar, despite different vibration levels. Furthermore, the curve squeal from the inner wheel was above that from the outer wheel in the roller rig tests in Fig. 4(b) and (c).

3.2 Velocity Dependence of Peaks

This section focuses on two frequencies at which the wheel vibration and the noise were high and their relation to the velocity. Figure 6 shows plots of various peak levels at 2 and 4 kHz against the train and rolling velocity. The peak levels were obtained by summing the power within $\pm$ 50 Hz of each focused frequency. In Fig. 6, the respective measured results are displayed for the running test, whereas the average results of about 5 tests were shown in Fig. 4.

The wheel vibration at both frequencies yield similar levels for the running and the roller rig tests under conditions of the straight section in the running test and 0° in the roller rig test.

The trends for the running tests at the curve and the roller rig tests with yaw angles are different. The large wheel vibrations associated with curve squeal were generated even at 5 km/h in the roller rig tests but not clearly in the running tests. The velocity dependence of the wheel vibration for the two tests is obviously different, with the wheel vibrations at 2 and 4 kHz also drastically increasing above 15 km/h in the running tests.

The agreement of some high vibration levels in the running tests above 15 km/h with those of the roller rig tests indicates possibly equal contact forces between the wheels and the rails in the two tests. The noise trend was also like that of the wheel vibrations.

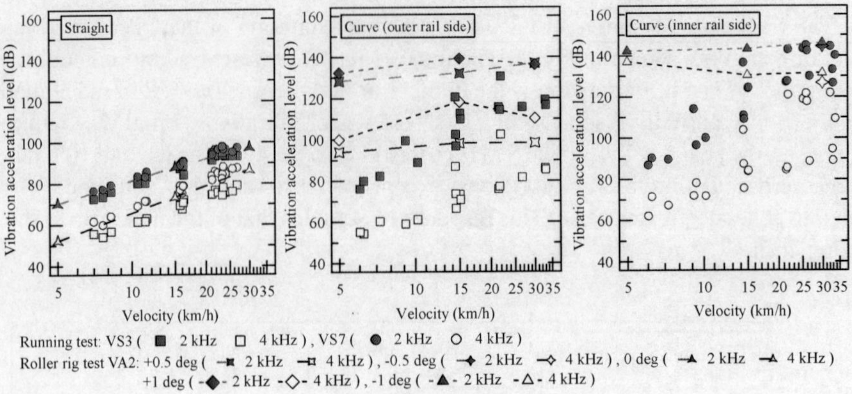

(a) Wheel vibration

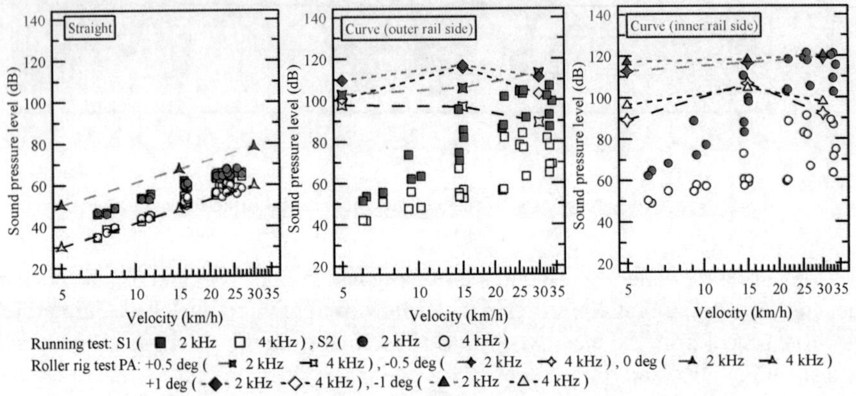

(b) Noise

Fig. 6. Velocity dependence of the wheel vibration and noise.

4 Discussion

The differences between the two tests are likely associated with the rail vibration behavior, the friction coefficient between the wheel and the rail, the contact position, and the size of the contact patch.

Analytical models for squeal have been proposed in previous studies. An example is a study that proposed excitation equations with wheel and rail dynamics described as follows [4]:

$$\dot{x}_2 = \left(Y_{22} - \frac{Y_{23}Y_{32}}{Y_{33}} \right) f_2 \tag{1}$$

$$f_2 = \frac{N}{V}\frac{d\mu_2}{d\gamma_2}\dot{x}_2 - \left(\mu_2 + \frac{N}{V}\frac{d\mu_2}{df_3} \right) \frac{Y_{32}}{Y_{33}} f_2 \tag{2}$$

$$Y_{ij} = Y_{ij}^{w} + Y_{ij}^{r} + Y_{ij}^{c} \tag{3}$$

where $\dot{x}_2$ is the wheel vibration velocity in an axial direction, f_2 and f_3 are the lateral and the vertical fluctuating forces at a contact point, N is the normal force, V is the rolling velocity, μ_2 is the friction coefficient in the lateral direction, γ_2 is the lateral creepage. Y_{ij}^{w}, Y_{ij}^{r}, and Y_{ij}^{c} are the responses (mobilities) in the i direction of a wheel, rail, and contact spring when excited in the j direction, respectively. Y_{ij} is the sum of mobilities for wheel, rail and contact spring. These equations constitute the feedback loop and the reasons advanced above may affect the stability of the loop. Differences in the friction coefficient and the vibration characteristics most likely explain the divergent squeal from the two tests. However, the train velocity also influences the stability of the loop.

Unlike the roller, the rail vibration displays no specific natural frequency above 1 kHz (Fig. 7). The peak at 2 kHz for the roller suggests the radial direction mode. This can cause a coupled motion of the wheel and roller and influence the stability as described in the Eqs. (1) and (2). Instability of the feedback loop due to usage of the roller easily generates squeal noise at a low rolling velocity in the roller rig test. We understand this partly accounts for differences between the two tests. A similar tendency is apparent at about 1.8 kHz as well as 2 kHz.

The friction coefficient also affects the stability of the loop. However, the friction coefficient showed an unclear influence in the two tests. Both tests were conducted in dry railhead conditions, although the rail and roller temperature and the air humidity were unknown; thus, further investigations are needed.

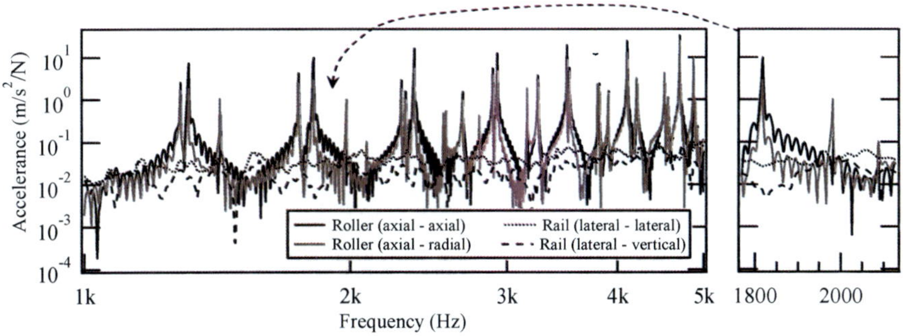

Fig. 7. Frequency responses of the rail and the roller. (The legend highlights the excitation and the response directions.)

The contact position and the size of the contact patch influence the creep force and the friction coefficient between the wheel and the rail. The attack angle of the leading wheel when the test vehicle passed through the curve of the test line was from 0.5 to 1°, independent of the train velocity at velocities < 30 km/h. The yaw angle in the roller rig test almost corresponds to the attack angle. The contact position and the size of the contact patch in the rig are reported to be like those of the running tests [5].

It is therefore presumed that combinations of wheel and rail responses, the friction coefficient, and the train velocity control the generation of curve squeal and account for differences between the two tests. The combinations for each test can be similarly stated for the train velocity at which similar curve squeal occurs. If the parameters related to stability are well-tuned for the two tests, the roller rig test simulates actual situations.

Although the roller rig test potentially simulates actual situations, the friction coefficient between the wheel and rail, the rolling velocity, and the vibration characteristics of the rail require consideration to correctly evaluate curve squeal.

5 Conclusions

The roller rig and the running tests were performed in this study to ascertain the characteristics of curve squeal and its relation to wheel vibration. The results were used to clarify differences between two tests. The wheel vibration associated with curve squeal in the roller rig test was generated even at 5 km/h, and the greatest wheel vibration occurred at 2 kHz. The vibration in the running test was not generated clearly at 5 km/h, and the curve squeal at 2 kHz resembled that in the roller rig test above 15 km/h. The parameters between the wheel and the rail and the vibration characteristics of the wheel and the rail influence curve squeal. Although the roller rig potentially simulates actual situations, the parameters related to stability and the vibration characteristics necessitate adequate tuning to correctly evaluate squeal. We will investigate the details of these parameters in future studies.

References

1. Thompson, D. J., Monk-Steel, A.D., Jones, C.J.C., Allen, P.D., Hsu, S.S., Iwnicki, S.D.: Railway Noise: Curve Squeal, Roughness Growth, Friction and Wear, Rail Research UK, Report: RRUK/A3/1 (2003)
2. Sueki, T., Kawaguchi, T., Kanemoto, H., Kuzuta, M., Inoue, T., Kitagawa, T.: Experimental Study on Curve Squeal Noise with the Actual Test Train, In: Proceedings of WCRR2019, No. OP-39–3, Tokyo (2019)
3. Kawaguchi, T., Sueki, T, Kitagawa, T., Nishimura, M., Abe, H.: Railway Noise Above 10 kHz Generated on a Curved Section of High-Speed Railway Line. In: Proceedings of inter-noise 2018, No. 1712, Chicago (2018)
4. Thompson, D. J.: Railway Noise and Vibration: Mechanisms, Modelling and Means of Control, Elsevier, Oxford (2009)
5. Satoh, E., Miyamoto, M.: Geometrical Contact between wheel and roller rig, Trans. Japan Soc. Mech. Eng. Ser. C, **59**(562), 1–4 (1993) (in Japanese)

A Full Finite Element Model for the Simulation of Friction-Induced Vibrations of Wheel/Rail Systems: Application to Curve Squeal Noise

Van Vuong Lai[1,2], Olivier Chiello[2(✉)], Jean-François Brunel[1], and Philippe Dufrénoy[1]

[1] Laboratoire de Mécanique Multiphysique Multiéchelle, UMR CNRS 9013, Centrale Lille, Univ. Lille, 59000 Lille, France
van-vuong.lai@polytech-lille.fr
[2] UMRAE, Univ Gustave Eiffel, IFSTTAR, CEREMA, Univ Lyon, 69675 Lyon, France
olivier.chiello@univ-eiffel.fr

Abstract. In this paper, a method is proposed for the modeling of dynamic wheel/rail frictional rolling contact. A full finite element formulation around the stationary position in an Eulerian reference frame is derived with a fine discretization of the contact surface combined with unilateral and Coulomb friction laws. Appropriate numerical techniques and reduction strategies are used in order to solve the non linear discrete equations in dynamic self-sustained conditions. In addition to the transient approach, a stability analysis allows the determination of unstable modes. This methodology is currently used in realistic wheel/rail contact in curves to simulate curve squeal.

Keywords: Curve squeal · Wheel/rail contact · Friction-induced vibrations · Finite element method · Model reduction · Non linear dynamics

1 Introduction

Squeal noise of rail-bound vehicles emitted in tight curves is characterized by high sound pressure levels at pure medium and high frequencies [1]. State-of-the-art abounds with models for curve squeal simulation [2]. They can be distinguished according to the mechanisms leading to instability but also wheel/rail contact models and solution types (time or frequency domain, linear or non-linear analysis). Although there are still many questions about the generation mechanisms [3–5], this paper focuses on the contact modelling issue.

There are mainly three types of model used for wheel/rail rolling contact. The point-contact models are based on analytical formulas, which are macroscopic laws based on simplified assumptions such as Hertz theory for the normal

G. Degrande et al. (Eds.): *Noise and Vibration Mitigation for Rail Transportation Systems*, NNFM 150, pp. 339–347, 2021.
https://doi.org/10.1007/978-3-030-70289-2_35

problem and Kalker's linear theory for the tangential problem [6]. They are better adapted to quasi-static cases. In slightly nonlinear dynamic cases (close to pure rolling), the combination of creep viscous damping and equivalent stiffness are used and give good results, especially for rolling noise modelling [7]. In other cases (large sliding with dynamic contributions), surface-contact models, where the contact zone is discretized, are more suitable. The variational theory of Kalker implemented in the CONTACT software is the most used model of this type [6]. However, a number of simplifications are generally performed in these models (elastic half-space assumption, contact and friction decoupling). The impact of these simplifications is unknown in the case of vibrations in curves where instability due to friction and excitation by surface irregularities may occur simultaneously. The third option is to use full finite element strategies in order to obtain reference solutions of the problem [8]. However, their application to 3D high frequency dynamics of rolling systems with friction is quite rare, essentially due to the computational challenge of combining rolling phenomena, nonlinear frictional contact and time integration schemes.

In [9], Lai *et al.* develop a full Finite Element (FE) computational method for the dynamics of frictional rolling contact systems. The validation of the whole approach is achieved in the case of two annular cylinders. In this paper, the method is applied to a realistic wheel/rail model in order to simulate curve squeal. In the first section, the main features of the numerical formulation are explained whereas the second section is dedicated to the results obtained for the wheel/rail model.

2 Numerical Method

2.1 Finite Element Formulation

Two bodies in rolling contact are considered. The conventions frequently used in rolling contact problems are adopted (see for instance [10]). In the absence of deformation and sliding, material particles of each surface move through the contact region in a direction parallel to $x-$axis with rolling speed V. In the case of sliding, relative tangential velocities $\boldsymbol{\Delta V} = \mathbf{V_1} - \mathbf{V_2}$ have to be considered in lateral and longitudinal directions. Spin can also occur but is not considered here for the sake of simplicity. In the Eulerian frame which moves with the point of contact, the relative sliding instantaneous velocities (or creep velocities) between the two deformable bodies at a fixed point of the potential contact interface S_c are given by:

$$\dot{\mathbf{s}}_\mathbf{t} = \mathbf{v_{t1}} - \mathbf{v_{t2}} = \Delta\mathbf{V} + V\left(\frac{\partial \mathbf{u_{t2}}}{\partial x} - \frac{\partial \mathbf{u_{t2}}}{\partial x}\right) + (\dot{\mathbf{u}}_\mathbf{t1} - \dot{\mathbf{u}}_\mathbf{t2}) \qquad (1)$$

where $\mathbf{u_t}$ and $\mathbf{v_t}$ denote respectively the displacement and velocity fields of the structure in the Eulerian frame.

To deal with frictional contact on the interface, unilateral (Signorini) and Coulomb laws with a constant friction coefficient law are chosen. Equivalent semi-regularized forms of these laws use the projection on the negative real set and the Coulomb cone (see for instance [11]):

$$\begin{aligned} r_n &= \mathrm{Proj}_{\mathbb{R}^-}\left(r_n - \rho_n \Delta u_n\right) & \forall \rho_n > 0 \\ \mathbf{r}_t &= \mathrm{Proj}_{C}\left(\mathbf{r}_t - \rho_t \dot{\mathbf{s}}_\mathbf{t}\right) & \forall \rho_t > 0 \end{aligned} \tag{2}$$

where $\Delta u_n = u_{n1} - u_{n2}$ is the normal relative displacement, r_n and $\mathbf{r}_t$ denote the normal and tangential contact stress whereas $\mathrm{Proj}_{\mathbb{R}^-}(x) = \min(x, 0)$ and $\mathrm{Proj}_C(\mathbf{x}) = \min(\frac{\mu|r_n|}{\|\mathbf{x}\|}, 1)\mathbf{x}$.

In order to introduce FE approximations, the principle of virtual power is used for the system dynamics with contact laws written in weak forms. Convective terms are neglected and finite element discretization directly gives [9]:

$$\begin{aligned} \mathbf{M}\ddot{\mathbf{U}} + \mathbf{C}\dot{\mathbf{U}} + \mathbf{K}\mathbf{U} &= \mathbf{F} + \mathbf{P_n}^\mathrm{T}\mathbf{R_n} + \mathbf{P_t}^\mathrm{T}\mathbf{R_t} \\ \mathbf{R_n} &= \int_{S_c} \mathbf{N_n}^\mathrm{T}\mathrm{Proj}_{\mathbb{R}^-}\left(\mathbf{N_n}\left(\mathbf{H_n^{-1}}\mathbf{R_n} - \rho_n(\mathbf{P_n}\mathbf{U} - \mathbf{G})\right)\right)\mathrm{d}s \\ \mathbf{R_t} &= \int_{S_c} \mathbf{N_t}^\mathrm{T}\mathrm{Proj}_C\left(\mathbf{N_t}\left(\mathbf{H_t^{-1}}\mathbf{R_t} - \rho_t\dot{\mathbf{S}}_\mathbf{t}(\mathbf{P_t}\mathbf{U}, \mathbf{P_t}\dot{\mathbf{U}})\right)\right)\mathrm{d}s \end{aligned} \tag{3}$$

where $\mathbf{U}$, $\mathbf{R_n}$ and $\mathbf{R_t}$ denote respectively the vectors of nodal displacements, equivalent normal reactions and equivalent friction forces, $\mathbf{M}, \mathbf{C}, \mathbf{K}$ are respectively the mass, damping and stiffness matrices of the structure without contact, $\mathbf{G}$ is the vector of nodal initial gaps and $\mathbf{N_n}, \mathbf{N_t}$ are the shape function vectors on the contact interface. In addition, $\mathbf{P_n}, \mathbf{P_t}$ are matrices allowing to pass the contact reactions from the local relative frame to the global frame whereas $\mathbf{H_n}$ and $\mathbf{H_t}$ are transformation matrices from nodal to equivalent forces. Finally, $\dot{\mathbf{S}}_\mathbf{t}(\mathbf{P_t}\mathbf{U}, \mathbf{P_t}\dot{\mathbf{U}})$ denotes the vector of nodal creep velocities which can be determined linearly using Eq. (1).

The computation of transient solutions from Eq. (3) is achieved using the modified θ-method. The detail of this integration scheme with an application to friction-induced vibrations can be found for instance in Loyer's work [12]. For the non linear resolutions for quasi-static and dynamic solutions at each time step, an iterative fixed point algorithm on equivalent contact reactions and friction forces is used with a stop criterion based on force convergence.

2.2 Stability Analysis

The aim of the stability analysis is to address the mechanism of self-excited vibration due to frictional contact through the determination of the evolution of small perturbations around the steady sliding equilibrium. In stable cases, the perturbations vanish and no vibration occurs. In unstable cases, some perturbations tend to diverge which can lead to self-sustained vibrations. Such an analysis is performed by a linearization of the non-linear equations around the equilibrium.

The quasi-static equilibrium is first obtained by neglecting the dynamic terms in Eq. (3) to provide the status of the nodes on the contact interface (no contact, sticking or sliding). Stability analysis is then carried out in the case of full steady sliding (no sticking region) and maintained contact configuration. Searching a discrete solution of the form $\mathbf{U^e} + \tilde{\mathbf{U}} \exp(\lambda t)$ where $\tilde{\mathbf{U}}$ stands for the complex displacement vector corresponding to small harmonic perturbations around the equilibrium, the linearized form of equation Eq. (3) leads to a constrained non symmetric eigenvalues problem:

$$\begin{aligned} (\lambda^2\mathbf{M} + \lambda(\mathbf{C} + \mathbf{C_b}) + \mathbf{K})\,\tilde{\mathbf{U}} &= (\tilde{\mathbf{P}}_\mathbf{n}^\mathrm{T} + \mu\tilde{\mathbf{P}}_\mathbf{t}^\mathrm{T})\tilde{\mathbf{R}}_\mathbf{n} \\ \tilde{\mathbf{P}}_\mathbf{n}\tilde{\mathbf{U}} &= 0 \end{aligned} \tag{4}$$

where $\mathbf{C_b}$ is the damping matrix provided by the linearisation of the sliding direction of friction force [12] and $\tilde{\mathbf{P}}_\mathbf{n}, \tilde{\mathbf{P}}_\mathbf{t}$ are new projection matrices corresponding to the effective contact region at equilibrium [9]. Complex modes and eigenvalues of the problem are then calculated. Modes corresponding to eigenvalues with positive real part are unstable. The divergence rate of a mode is notably defined as $D_v = \mathrm{Re}(\lambda)/\mathrm{Im}(\lambda)$ where $(\mathrm{Re}(\lambda), \mathrm{Im}(\lambda))$ are respectively the real and imaginary parts (pulsation) of the mode.

2.3 Reduction Strategy

As the size of the system is often large and the nonlinear solving process implies several resolutions of a linear system at each time step, reducing the size of the system is necessary to obtain reasonable computation times. The principle is to search an approximated solution $\mathbf{U} = \mathbf{B}\mathbf{q_r}$ of the problem spanned by a reduced basis $\mathbf{B}$, leading to a reduced dynamics equation. The approach proposed in [9] is here adopted with $\mathbf{B}$ composed of free-interface normal modes $\boldsymbol{\Phi}$ enriched by static residual attachment modes $\tilde{\boldsymbol{\Phi}}_s$. The residual modes are defined as the static displacement responses to unit contact reactions, after the elimination of the contribution of normal modes. They can be written:

$$\tilde{\boldsymbol{\Phi}}_s = \boldsymbol{\Phi}_s - \boldsymbol{\Phi}(\boldsymbol{\Phi}^\mathrm{T}\mathbf{K}\boldsymbol{\Phi})^{-1}\boldsymbol{\Phi}^\mathrm{T}\mathbf{P_c}^\mathrm{T} \tag{5}$$

where $\mathbf{P_c}^\mathrm{T} = [\mathbf{P_n}^\mathrm{T} \quad \mathbf{P_t}^\mathrm{T}]$ and $\boldsymbol{\Phi}_s$ are attachment modes verifying $\mathbf{K}\boldsymbol{\Phi}_s = \mathbf{P_c}^\mathrm{T}$. A stronger approximation proposed in [9] is also used which consists in neglecting the dynamic terms in the reduced equations corresponding to attachment modes. The use of basis $\mathbf{B} = [\boldsymbol{\Phi} \quad \tilde{\boldsymbol{\Phi}}_s]$ in Eq. (3) together with the elimination of the dynamic terms relating to the residual attachment modes gives [9]:

$$\begin{aligned} \boldsymbol{\Phi}^\mathrm{T}\mathbf{M}\boldsymbol{\Phi}\ddot{\mathbf{q}} + \boldsymbol{\Phi}^\mathrm{T}\mathbf{C}\boldsymbol{\Phi}\dot{\mathbf{q}} + \boldsymbol{\Phi}^\mathrm{T}\mathbf{K}\boldsymbol{\Phi}\mathbf{q} &= \boldsymbol{\Phi}^\mathrm{T}(\mathbf{F} + \mathbf{P_n}^\mathrm{T}\mathbf{R_n} + \mathbf{P_t}^\mathrm{T}\mathbf{R_t}) \\ \mathbf{P_c}\mathbf{U} &= \tilde{\boldsymbol{\Phi}}_s^\mathrm{T}(\mathbf{F} + \mathbf{P_n}^\mathrm{T}\mathbf{R_n} + \mathbf{P_t}^\mathrm{T}\mathbf{R_t}) + \mathbf{P_c}\boldsymbol{\Phi}\mathbf{q} \end{aligned} \tag{6}$$

This reduction strategy consists in solving the global dynamics using free-interface normal modes and adding a local static residual flexibility controlled by matrix $\tilde{\boldsymbol{\Phi}}_s^\mathrm{T}$ in the expression of the contact displacements.

The same reduction technique is used for the stability analysis but considering residual attachment modes $\tilde{\boldsymbol{\Phi}}_s$ defined on the effective contact region at the quasi-static equilibrium, leading to the following eigenvalue problem [9]:

$$\begin{gathered}\left(\lambda^2\boldsymbol{\Phi}^{\mathrm{T}}\mathbf{M}\boldsymbol{\Phi}+\lambda\boldsymbol{\Phi}^{\mathrm{T}}(\mathbf{C}+\mathbf{C_b})\boldsymbol{\Phi}+\boldsymbol{\Phi}^{\mathrm{T}}(\mathbf{K}+\mathbf{K_c})\boldsymbol{\Phi}\right)\tilde{\mathbf{q}}=\mathbf{0}\\ \mathbf{K_c}=(\tilde{\mathbf{P}}_{\mathbf{n}}^{\mathrm{T}}+\mu\tilde{\mathbf{P}}_{\mathbf{t}}^{\mathrm{T}})\left(\mathbf{I}_n\tilde{\boldsymbol{\Phi}}_s^{\mathrm{T}}(\tilde{\mathbf{P}}_{\mathbf{n}}^{\mathrm{T}}+\mu\tilde{\mathbf{P}}_{\mathbf{t}}^{\mathrm{T}})\right)^{-1}\tilde{\mathbf{P}}_{\mathbf{n}}\end{gathered}\tag{7}$$

This reduction strategy consists in solving the global dynamics using free-interface modes and adding a local static residual stiffness controlled by matrix $\mathbf{K_c}$.

3 Application to Wheel/Rail Dynamics in a Curve

In this section, an application of the numerical method presented in the previous sections is proposed for a wheel/rail model in rolling contact with imposed lateral creepages as in a curve squeal situation.

3.1 Description of the Model

The wheel FE model corresponds to a standard steel monobloc wheel (Fig. 1) with a nominal rolling diameter of 920 mm and a mass of 314 kg. From this model, 100 free-interface modes have been calculated up 8000 Hz considering clamped boundary conditions at the hub. 3 kinds of mode may be distinguished: the radial modes R_{mn}, the axial modes A_{mn} and the circumferential modes C_{mn} where n is the number of nodal diameters and m is the number of nodal circles [7]. Modal damping factors are chosen depending on the nodal diameters (1, 0.1 and 0.01 % for $n = 0$, $n = 1$ and $n \geq 2$ respectively [7]).

Fig. 1. Wheel model and contact zone

The track FE model consists of a periodically supported UIC60 rail on monobloc sleepers and ballast. The dynamics of the sleepers and ballast is neglected for this application. The rail has 48 m length but ends with 2 anechoic terminations of 6 m composed of 5 rail portions of length $L = (0.6; 0.6; 0.6; 1.2; 3.0)$ m with increasing damping, avoiding the return of waves. Pads that connect the rail and the sleepers are modelled by nodal springs of equivalent stiffness $k_x = k_y = 36$ MN/m and $k_z = 180$ MN/m. An equivalent viscous damping model is derived from structural damping factors of 0.02 % for the rail, 1 % for the pad and 0.1 to 1 % for the rail anechoic terminations.

Compatible meshes on wheel/rail interface are considered. The potential contact area is meshed with elements of length 1 mm as shown in Fig. 1. A vertical relative displacement is applied leading to a resultant contact force of 70 kN. A rolling velocity $V = 10$ m/s and a friction coefficient $\mu = 0.3$ are considered. Quasi-static solutions of Eq. (3) have been compared with those obtained with CONTACT software for several creepages $\Delta \mathbf{V}/V$ showing a excellent agreement for both partial and full sliding contact.

3.2 Stability Results

Stability analysis is carried out in a case of full sliding contact using a high creepage $V_y/V = 1\%$. In order to solve the large non symmetric eigenvalue problem resulting from the wheel/rail interaction, the reduction strategy presented in Sect. 2.2 (i.e. Eq. (7)) is adopted. 4 unstable modes are obtained (Fig. 2).

The first 3 unstable modes are very close to free-interface axial wheel modes without nodal circle A_{02}, A_{03} and A_{04} both in terms of mode shapes and frequencies. For these unstable modes, coupling between wheel modes does not seem to occur. The instability is mainly due to the rail effect as already pointed out in [4,5]. It is also important to note that these modes are often found experimentally to be the origin of curve squeal.

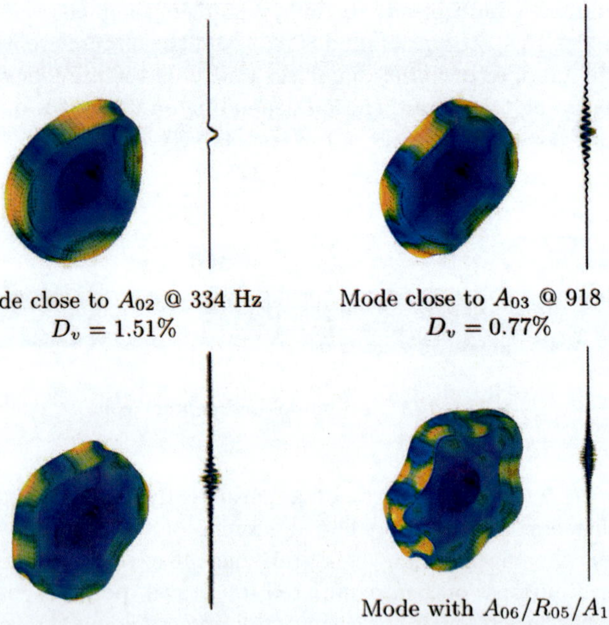

Mode close to A_{02} @ 334 Hz
$D_v = 1.51\%$

Mode close to A_{03} @ 918 Hz
$D_v = 0.77\%$

Mode close to A_{04} @ 1670 Hz
$D_v = 0.09\%$

Mode with $A_{06}/R_{05}/A_{11}$ contributions @ 3417 Hz
$D_v = 0.11\%$

Fig. 2. Unstable modes. Wheel and (vertical) rail deformation

The fourth unstable mode is quite different. Several wheel modes with close natural frequencies are involved in the corresponding instability and mode coupling can be clearly highlighted by looking at the bifurcations curves.

3.3 Transient Results

Transient non linear results corresponding to the previous stability analysis are determined using a numerical time integration from the quasi-static equilibrium, using the reduction strategy presented in Sect. 2.2 (i.e. Eq. (6)). The time step for the integration is $\Delta t = 1\ \mu s$. Figure 3 shows the evolution of the lateral contact resultant forces. It increases until a pronounced creep/slip oscillation builds up. When the tangential contact resultant force is smaller than the traction bound, a transient stick zone appears at the leading edge of the effective contact region.

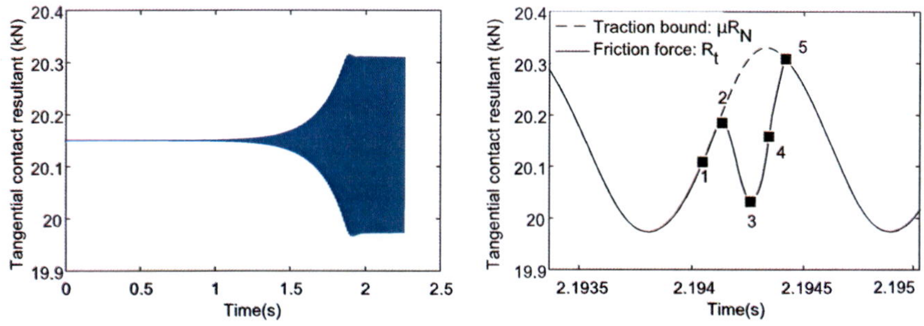

Fig. 3. Evolution of the tangential contact resultant force and zoom in the steady state

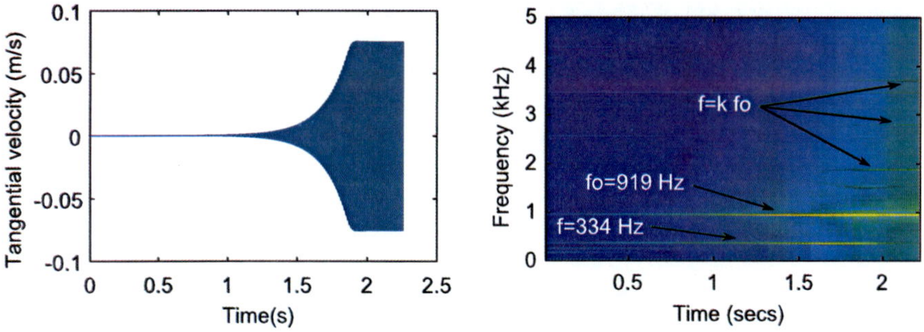

Fig. 4. Tangential velocity of a point outside the contact zone. Time history and spectrogram

The time series and the spectrogram of the wheel lateral velocity on a point outside of the contact zone are presented in Fig. 4. At the beginning of the simulation, there are two main peaks in the spectrum (334 919 Hz) which are very close to the natural frequencies of the two first unstable modes provided by the stability analysis. In the steaty state, where nonlinear events are present, the fundamental frequency $f_0 = 919$ Hz is clearly dominant and harmonic frequencies $f_k = kf_0$ appear. These results show that the transient calculations are consistent with the stability analysis and exhibit localized stick/slip oscillations even with a constant friction coefficient as already pointed out in [3].

4 Conclusion

In this paper, a numerical method for the analysis of friction-induced vibrations of rolling contact systems is applied to a wheel/rail dynamic model in order to compute reference solutions for curve squeal simulations. A reduction basis including free-interface modes and static residual flexibility is used allowing the reduction of the computation loads and times. Both stability and transient analysis are performed in order to predict unstable modes and determine the amplitude and full spectrum of the self-sustained vibrations.

References

1. Vincent, N., Koch, J.R., Chollet, H., Guerder, J.Y.: Curve squeal of urban rolling stock–Part 1: State of the art and field measurements. J. Sound Vib. **293**(3), 691–700 (2006). https://doi.org/10.1016/j.jsv.2005.12.008
2. Thompson, D.J., Squicciarini, G., Ding, B., Baeza, L.: A state-of-the-art review of curve squeal noise: phenomena, mechanisms, modelling and mitigation. In: Anderson, D., Gautier, P.-E., Iida, M., Nelson, J.T., Thompson, D.J., Tielkes, T., Towers, D.A., Vos, P., Nielsen, J.C.O. (eds.) Noise and Vibration Mitigation for Rail Transportation Systems. NNFMMD, vol. 139, pp. 3–41. Springer, Cham (2018). https://doi.org/10.1007/978-3-319-73411-8_1
3. Pieringer, A.: A numerical investigation of curve squeal in the case of constant wheel/rail friction. J. Sound Vib. **333**(18), 4295–4313 (2014). https://doi.org/10.1016/j.jsv.2014.04.024
4. Ding, B., Squicciarini, G., Thompson, D.J., Corradi, R.: An assessment of mode-coupling and falling-friction mechanisms in railway curve squeal through a simplified approach. J. Sound Vib. **423**, 126–140 (2018). https://doi.org/10.1016/j.jsv.2018.02.048
5. Lai, V.V., Chiello, O., Brunel, J.-F., Dufrénoy, P.: The critical effect of rail vertical phase response in railway curve squeal generation. Int. J. Mech. Sci. **167**, 105281 (2020). https://doi.org/10.1016/j.ijmecsci.2019.105281
6. Jacobson, B., Kalker, J.J. (eds.): Rolling Contact Phenomena. ICMS, vol. 411. Springer, Vienna (2000). https://doi.org/10.1007/978-3-7091-2782-7
7. Thompson, D.J.: Railway Noise and Vibration: Mechanisms. Modelling and Means of Control, Elsevier, Amsterdam (2009)
8. Wriggers, P.: Computational Contact Mechanics, 2nd edn. Springer-Verlag, Berlin Heidelberg (2006). https://www.springer.com/us/book/9783540326083
9. Lai, V.V., Chiello, O., Brunel, J.-F., Dufrénoy, P.: Full finite element models and reduction strategies for the simulation of friction-induced vibrations of rolling contact systems. J. Sound Vib. **444**, 197–215 (2019). https://doi.org/10.1016/j.jsv.2018.12.024

10. Johnson, K.L.: Contact Mechanics. Cambridge University Press, Cambridge (1985). https://doi.org/10.1017/CBO9781139171731
11. Jean, M.: The non-smooth contact dynamics method. Comput. Methods Appl. Mech. Eng. **177**(3), 235–257 (1999)
12. Loyer, A., Sinou, J.J., Chiello, O., Lorang, X.: Study of nonlinear behaviors and modal reductions for friction destabilized systems. application to an elastic layer. J. Sound Vib. **331**(5), 1011–1041 (2012)

Structure-Borne Noise and Ground-Borne Vibration

Importance of a Detailed Vibratory Characterization of a Railway Line for the Propagation of Vibrations in an Eco-Neighborhood

Guillaume Coquel[1(✉)], Catherine Guigou-Carter[2], and Philippe Jean[2]

[1] Régie Autonome des Transports Parisiens, 54, Quai de la Rapée, 75012 Paris, France
guillaume.coquel@ratp.fr

[2] Centre Scientifique et Technique du Bâtiment, Direction Santé et Confort, 24, Rue Joseph Fourier, 38400 Saint-Martin d'Hères, France

Abstract. The vibration field of a railway line in a 180 m linear cutting was measured and modeled to predict its impact on a future real estate program, as a work package of ECOCITE II, an urban innovation program. For decades, vibration levels due to rolling stock have been evaluated in the vicinity of tracks on the vertical z axis, at one or a few points, supposed to be representative of the entire vibration field around the tracks. When the size of the project is rather large, this assumption is incorrect, especially if the site is located near a train station, where stresses in the tracks can vary a lot depending on the direction of the train, i.e. depending on acceleration or deceleration of the train. This paper presents vibration measurements performed in different directions at six positions before the construction of any building (free field conditions) along the tracks. Steel spikes were used to mount vibration sensors and results suggest that resonnances can appear at some measurement points. In this paper, a procedure similar to the FTA method is used to further investigate this behavior. Furthermore, the attenuation of ground vibrations as a function of distance from the track can be used to evaluate the loss factor in the ground based on a comparison with a 2.5D BEM model of the ground including the different layers. The model is then used to identify equivalent forces in different directions (i.e. vertical, perpendicular and parallel to track directions) associated to train passage to lead to similar vibration transmission behavior in the ground.

Keywords: Vibration · Ground borne noise · Environmental · Mobility

1 Context, Site and Future Buildings

The vibration field of a railway line in a 180 m linear cutting was measured and modeled to predict its impact on a future real estate program, as a work package of ECOCITE II. This urban innovation program, from the conception to the execution, aims at improving environmental aspects (eco-neighborhood). Considering the short distance between the

G. Degrande et al. (Eds.): *Noise and Vibration Mitigation for Rail Transportation Systems*, NNFM 150, pp. 351–359, 2021.
https://doi.org/10.1007/978-3-030-70289-2_36

tracks and the future buildings and the size of the project, a detailed evaluation of vibration emission of the trains is performed. Part of the project is to evaluate which buildings need mitigation measures (on the tracks or in the buildings) and the associated required reduction. Vibrations levels were evaluated at six positions (three on each side) at the limit of the railway domain since previous studies [1] have shown that geological conditions can vary considerably in the same test site.

The site, which is 180 m long, is located on both sides of the RER A line 20 km east of Paris, close to a train station (140 m to the east). The tracks follow a very light curve (radius 1500 m). It is bordered by a parking (south) and a vegetated area (north) (Fig. 1). The north-bound trains coming from Paris (track 1) are accelerating. The south-bound trains (track 2) are slowly decelerating towards the train station. The train speed within the site varies from 60 and 70 km/h. The real estate project plans to erect housing and office buildings with two underground parking levels and up to 9 floors. These buildings will be located 12 m away from the existing tracks.

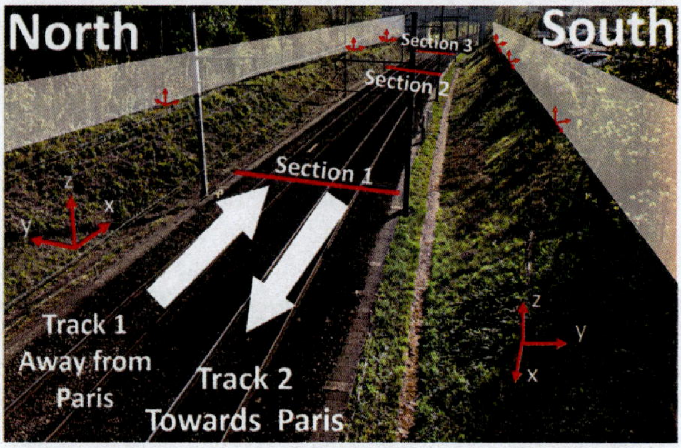

Fig. 1. Test site description. Three sections on both sides of the tracks. White areas: future building locations; position of measurement points and coordinate axes.

2 Rolling-Stock and Track

The track is used only by double decked, coupled trains (224 m long) similar one from the other in terms of vibration emission. The track is a ballasted track, with bi-block sleepers and a UIC-60 rail profile. The rail surface of both tracks was ground four months before measurements and no specific defects were seen on the rail surface. The purpose of the measurements was to evaluate an average behavior of the soil when the trains were passing-by so no detailed measurements of the track quality such as roughness or dynamic behavior at low frequencies were performed.

3 Measurements

3.1 Procedure and Measurement Systems

Different kinds of sensor coupling with the ground can be used and different authors reported difficulties to compare measurements for different sensor couplings [2–4]. As, for this actual site, trains are accelerating or decelerating, it was thought that measuring in the horizontal (x and y) directions might be important. Some authors have highlighted the difficulties to correctly measure vibrations in free field along the horizontal x and y axes [5]. For practical reasons (flexibility, good coupling along z axis, lack of structure on the test site), measurements were performed on top of steel spikes (50 cm long, X cross-section) previously tested [4]. Distances from the track to the measurement positions are of the same order for all sections (about 10 m). Vibration levels are averaged over at least 19 pass-by on each track. All sensors are 10 V/g, low noise PCB accelerometers (393B12 type). They are firmly screwed on top and on the sides of the spikes. The signals were analyzed using the $Lv_{eq,Tp}$ indicator in one-third octave bands. A 7 Hz high pass filter is applied directly to the signals during measurement and the sampling frequency is fixed at 8192 Hz. All pass-by measurements were performed over a two-day period; velocity levels on each section were evaluated during the same period of the day, one day apart for the north and for the south (Sect. 1: 10h30 to 12 h am; Sect. 2: 1 h/2h30 pm; Sect. 3: 3 h/4h30 pm). The load of the trains might therefore not be responsible for the measurement differences between the two sides presented below. In order to test the sensibility of the accelerometer mounting method a measurement similar to the FTA procedure was tested [6–8] (Fig. 2). The objective is to evaluate the response of the soil for a source of vibration applied at increasing distance from the measurement section. In this study, an instrumented impact hammer (B&K type 8210) was used (all other parameters are identical) due to its flexibility but another source of vibration could have been used [4]. Hits are applied (vertically) on the sleeper in order to obtain good reproducibility. The number of hits is proportional to the distance of the measurement section in order to improve the signal-to-noise ratio (Fig. 2). The complete setup (spikes included) is removed after the pass-by measurements and replaced for impact hammer tests. Furthermore, in order to evaluate the soil propagation attenuation between the tracks and future buildings, two extra accelerometers (y and z direction) also mounted on steel spikes were placed near (3 m away from the nearest rail) to the tracks when performing FTA-like measurements.

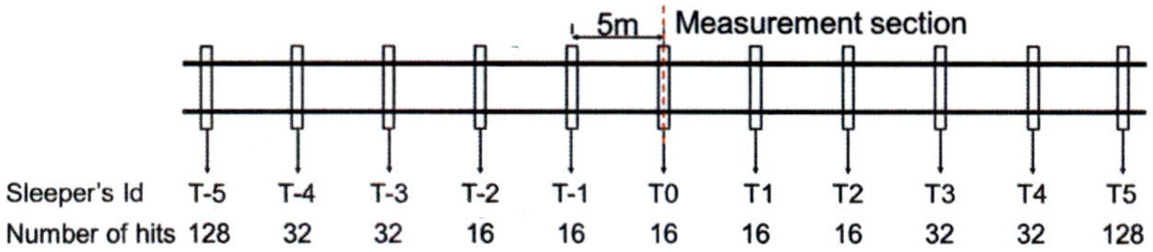

Fig. 2. Identification of source positions along the tracks (Sect. 2 – North side) and number of hits used to compute averaged responses.

3.2 Results

All measurement results were obviously calibrated, but full-scale values are not given in this paper for confidentiality reasons.

Pass-by

The first general remark is that for each measurement position, the velocity levels are very reproducible. For clarity, standard deviations cannot be displayed on the figures but their range in terms of global value from 6.3 Hz to 200 Hz varies between 0.2 and 1.8 dB on all measurements presented. All the results for the three sections and for all directions are given in Fig. 3. A general and logical trend is that the greater the distance from the measurement point (track 2 for north side and track 1 for south side) the smaller the vibration levels. On the north side, differences between trains passing on the two tracks tend to be lower than on the south side. This might be due to different soil propagation properties and terrain configurations on both sides. Figure 3 clearly shows that velocity levels are not constant along the test site and are related to train velocities; the train station is located 170 m away from Sect. 1. Train velocities are then quite low for this section. When trains are moving away from the station, their speed increases, and vibration velocity levels increase. An important reduction of vibration levels appears for all measurements for the 160 Hz frequency one-third octave band and at higher frequencies. The significant vibration levels observed in the very low frequency range (i.e. below 12 Hz), is not associated to background vibration levels (about 10–15 dB lower) and could be related to the moving load and parametric excitation associated to the train pass-by (even though measurement distance about 10 m from track), and/or the ground profile around the tracks (canyon type).

At Sect. 1, measurements show greater levels (around 5 dB for the averaged global level) for the horizontal directions. Compared to Sect. 1, levels measured at Sect. 2 tend to be higher. The North sensor position at Sect. 2 seems to be a special point of the site because velocity levels in horizontal directions are particularly high (10 dB) with a peak in the 80–100 Hz one-third octave bands. This distinctive behavior may be due to a bad coupling of the spike in the soil as no such resonance frequency appears at this level on the other side of the track. The measurements performed at Sect. 3 show no resonance frequency, but greater levels than at Sect. 1. This is probably due to the stabilized velocities of trains on both tracks.

Impact Hammer

Mobilities next to the future buildings are given for five source positions in Fig. 4 (narrow bands) and for all source positions in Fig. 5 (one-third octave bands). Both show greater values on the y axis (perpendicular to the track), a maximum of energy at 50 Hz (maximum of force injected in the track) band but no special amplification in the 80–100 Hz frequency bands. This is not due to a lack of energy transfer by the impact hammer to the sleeper since an analysis in one-third octave bands of the force spectrum (not presented here) shows differences less than 10 dB over the entire frequency range. Maximum levels on the y axis (perpendicular to the track) might be due to the difficulty to evaluate correctly the velocity response of the soil in the horizontal directions. Measurement results show that, for distances greater than 15 m, the coherence

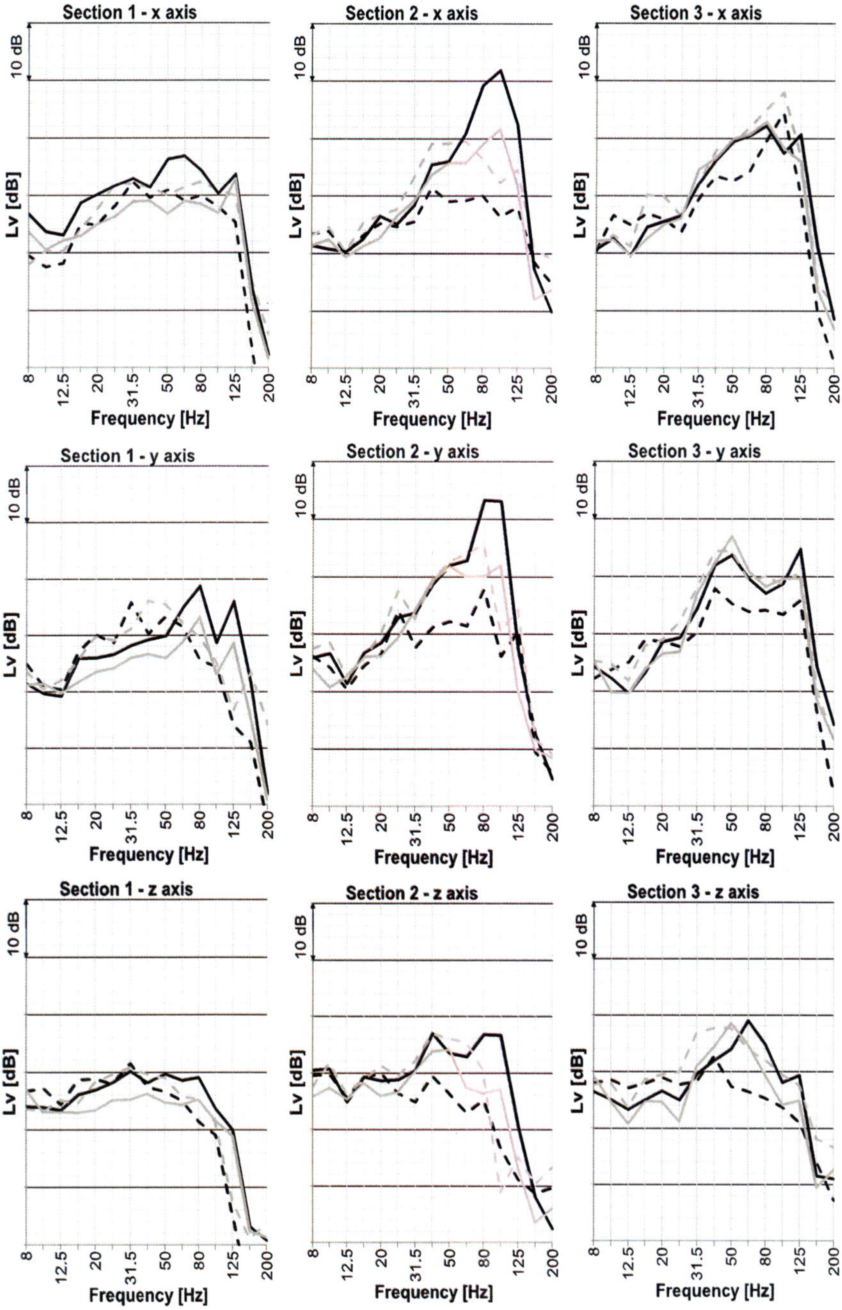

Fig. 3. Frequency content of mean pass-by in each section for x (parallel to the track), y (perpendicular to the track), and z (vertical) axis. All graphs are given with the same scale (black : trains running on track 1 – north side; grey: trains running on track 2 – south side). Plain lines: measurements performed on north side; dashed lines: measurements performed on south side.

decreases and the valid frequency range narrows. No specific behavior is visible in the 80–100 Hz frequency range of the narrow band analysis.

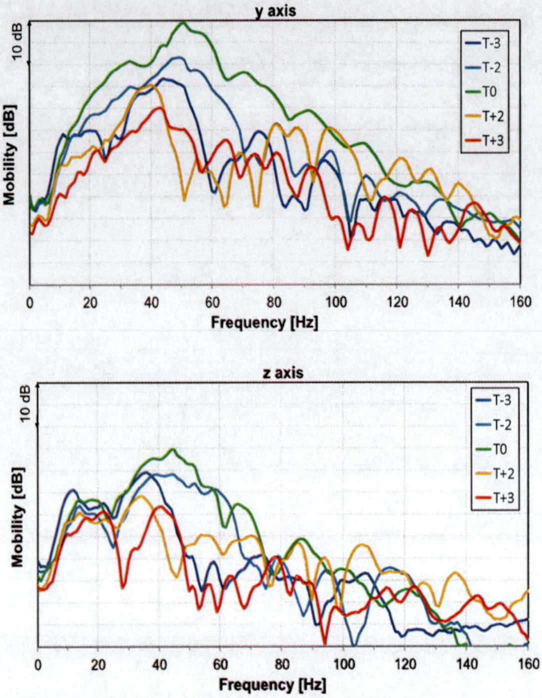

Fig. 4. Mobilities, on y (perpendicular to track) and z (vertical) axes for five source positions.

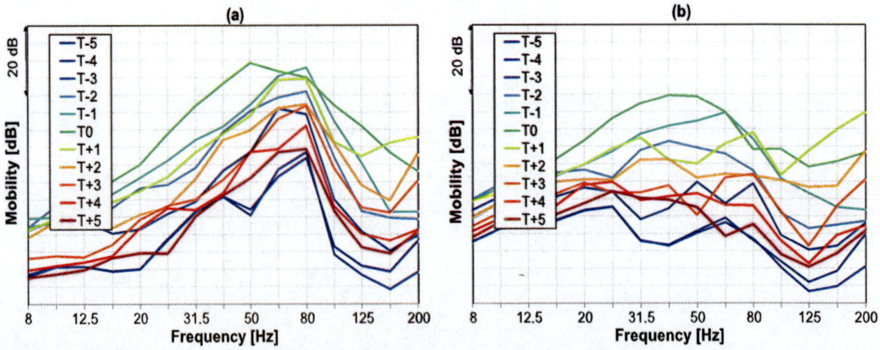

Fig. 5. Mobilities, on y and z axes (one-third octave band) for all source positions next to the track. (a) y axis (perpendicular to the track); (b) z axis (vertical). All graphs are given with the same scale.

In Fig. 5, near the track, a special behavior appears. At the 63–80 Hz one-third octave bands, there is an amplification of mobility. This behavior is not measured on the future building locations and appears only on the y axis. Therefore, it is believed that this amplification is due to a singular behavior of the spike anchored in the ground. As proposed by other authors [5], this kind of mounting setup can show specific frequency dependent behavior for horizontal sensors.

In the present case, it was impossible to perform measurements on the track as well as pass-by measurements during the same session, so accelerometer supporting spikes had to be removed and re-anchored, inducing potential differences in spike/ground coupling and behavior. A different method, maybe more optimized in terms of measurement equipment and time duration than the one used during this study could be implemented.

4 Comparison with a 2.5D Model

In order to evaluate vibration propagation in buildings some FEM/BEM 2.5D computations were performed with the MEFISSTO software [9]. From geotechnical characterizations at different positions on the site as well as bibliographical information, ground layer stratification and composition were proposed. Ground loss properties were evaluated according to other measurements described in [10].

Once the model was adjusted, computations in the wavenumber domain lead to results for a single force applied at the same source locations used in the FTA measurement procedure. In order to know if the results of the model are in good accordance to the measurements, a comparison of mobility difference between the point next to the tracks and the point next to the future building is proposed (Fig. 6). Differences appear between the computed and the measured results. This is probably due to the presence of the upper structure of the track (ballast, sleeper, rails) which was not modelled in the 2.5D model. Indeed, since the tracks were not properly characterized it was decided not to include them in the model, i.e. it did not seem appropriate to adjust the input data for the tracks (more specifically the ballast and ties) to match the measured behavior. Furthermore, this characterization of the tracks could also have helped to get a better insight of the ground characteristics under the tracks (maybe locally more compact).

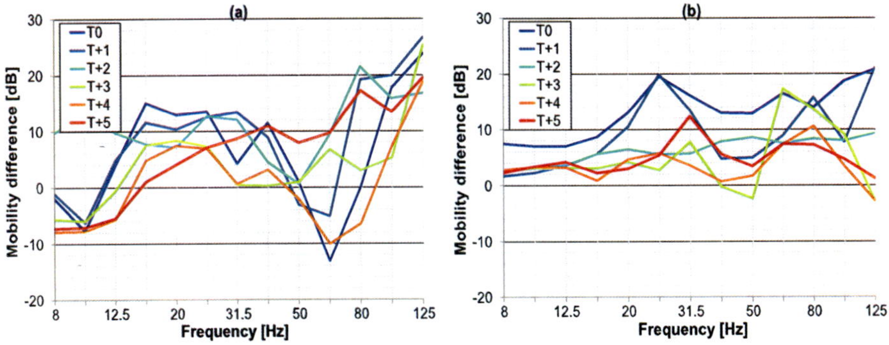

Fig. 6. Mobility differences (vertical axis only), sensors next to the tracks/future buildings for six source positions. (a) 2.5D computation; (b) measurements.

5 Conclusions

In order to assess existing environmental vibration field on a site dedicated to future buildings, a measurement campaign was undertaken. Measurements showed: i) very reproducible results for different pass-by; ii) an increase of vibration velocity levels for increasing train speed; iii) horizontal velocity levels greater than the vertical ones and iv) possible resonance for one measurement position in the 80–100 Hz one-third octave frequency bands probably due to a bad coupling between soil and spikes. A procedure similar to the FTA method showed no specific amplification at 80 or 100 Hz but higher mobilities for the horizontal axes than for the vertical axis for measurement positions near the future buildings. Accelerometers placed near the tracks showed resonant behavior in the 80 Hz one-third octave band, probably due to mounting conditions. Comparisons with a 2.5D model showed the complexity of making prognostics when the site is not fully characterized (ground, tracks, etc.).

Acknowledgements. The authors would like to thank LinkCity and Epamarne, as well as the Caisse des Dépôts et Consignations (Deposits and Consignments Fund) for supporting this work through the "Ecocité 2 – Ville de Demain" program. This French national program involves public and private actors in a territory to develop innovative projects to accelerate the energy and ecological transition of cities.

References

1. Müller, R., Brechbühl, Y.: Measurement report about a new under sleeper test track in a curve. Rivas project Deliverable 3.3; document code RIVAS_SBB_ WP3-3_D3_8_V07. (2013)
2. XP ISO/TS 14837-31, Mechanical vibration – ground-borne noise and vibration arising from rail systems – Part 31: Guideline on field measurement for the evaluation of human exposure in buildings. International Standard Organization (2018)
3. d'Avillez, J. et al.: The influence of vibration transducer mounting on the practical measurement of railway vibration. Internoise, Lisbon, Portugal, pp. 15–16 June (2010)
4. Coquel, G., Kengni Kengang, A.: Experimental comparison of maximum length sequence (MLS) and impact hammer methods to evaluate vibration transfer functions in soil. In: Nielsen, J. et al. (eds.) Noise and Vibration Mitigation for Rail Transportation Systems. Notes on Numerical Fluid Mechanics and Multidisciplinary Design, vol. 126, Springer, Berlin (2015)
5. Degrande, G., Van den Broeck, P., Clouteau, D.: A critical appraisal of in situ vibration measurements. European Association for Structural Dynamics European conference; 3rd, Structural dynamics; Florence; Italy in Structural Dynamics –Conference, vol. 2, pp. 1107–1116 (1996)
6. Hanson, C.E., Towers, D.A., Meister, L.D.: Transit noise and Vibration Impact Assessment, Federal Transit Administration, Office of planning and Environment, FTA-V1-90-1003-06, May (2006)
7. Kouroussis, G., Mouzakis, H.P., Vogiatzis, K.: Structural impact response for assessing railway vibration induced on buildings. Structural impact response for assessing railway vibration induced on buildings. Mech. Ind. **18**(8), 803 (2017)
8. Verbraken, H., Lombaert, G., Degrande, G.: Experimental and numerical determination of transfer functions along railway tracks. In: Proceedings of the 9th National Congress on Theoretical and Applied Mechanics, Brussels, 9-10-11 May (2012)

9. Jean, P., Guigou, C., Villot, M.: A 2D ½ BEM model for ground structure interaction. Building Acoustics **11**(3), 157–163 (2004)
10. Guigou-Carter, C., Coquel, G. Jean, P., Jolibois, A.: Reducing groundborne noise due to railways. Part 1 : assessing the problem. In: Proceedings of the 23rd International Congress on acoustics, Aachen. 9–13 September (2019)

An Efficient Three-Dimensional Track/Tunnel/Soil Interaction Analysis Method for Prediction of Vibration and Noise in a Building

Kazuhisa Abe[1(✉)], Koya Yamada[1], Sota Yamada[1], Masaru Furuta[2], Michiko Suehara[2], Sho Yoshitake[2], and Kazuhiro Koro[1]

[1] Niigata University, 8050, Igarashi 2-Nocho, Nishi-ku, Niigata 950-2181, Japan
abe@eng.niigata-u.ac.jp

[2] Nippon Koei Co., Ltd., 1-14-6, Kudan-kita, Chiyoda-ku, Tokyo 102-8539, Japan

Abstract. An efficient numerical method is developed for the evaluation of the influence of the track structure on vibration and noise in a neighboring building induced by a subway train. In order to save computational cost, the three-dimensional track/tunnel/soil interaction problem is formulated by virtue of the Floquet transform and then reduced to a quasi-two-dimensional problem. Furthermore, the vibration and noise levels in a neighboring building are predicted using the numerical result and empirical evaluation methods. As examples, two types of tracks with under-sleeper pad and with under-slab sheet are considered. Through numerical analyses, the influence of track structures on the vibration and noise in a building is examined.

Keywords: Train-induced vibration · Noise reduction · Floquet transform

1 Introduction

In the design of a subway track it is important to assess the train-induced ground-borne vibration and noise from the viewpoint of the residential environments. For this purpose, in the past few decades, different numerical methods have been developed for the interaction problems such as the tunnel-soil system subject to moving loads [1–3] and the dynamic reaction of a wheel-track-tunnel-soil coupling model [4]. Although a neighboring building can also be modeled, in the context of three-dimensional analysis, it leads to an increase of the computational cost and thus would be impractical. Consequently, in the development of a prediction method used for the track design, it is needed to balance accuracy and cost.

The authors have developed an efficient numerical method in the frequency domain [5]. In this method, to evaluate the influence of track structure on the

G. Degrande et al. (Eds.): *Noise and Vibration Mitigation for Rail Transportation Systems*, NNFM 150, pp. 360–367, 2021.
https://doi.org/10.1007/978-3-030-70289-2_37

dynamic behavior of tunnel, coupling problems consisting of track, tunnel and soil are considered. Due to the periodicity of the track characterized by the sleeper spacing, the Floquet transform [6,7] is applied to the interaction problem. Moreover, in order to reduce computational effort, the solution is represented by a Fourier series in the track direction. Since, in general, the speed of subway train, which is at most 30 m/s, is sufficiently slow compared to that of waves propagating through the tunnel and the surrounding soil, the effect of the moving load will be insignificant. Therefore, a stationary harmonic load acting on the rail is considered.

To predict the train-induced ground-borne vibration and noise in a nearby building, this paper presents an efficient evaluation method. The proposed method is composed of three procedures. The first one is the track/tunnel/soil interaction analysis using the numerical method developed in [5]. The second is the time-domain analysis of a wheel moving on a track supported by a rigid foundation. The tunnel vibration due to a running wheel is calculated approximately from both results. Once the tunnel reaction is obtained, in the final process the vibration and noise in a building are evaluated using empirical formulae [8,9] proposed for the vibration attenuation in the ground, and for wave transmission and noise radiation in a building. As examples, two types of tracks with under-sleeper pad and with under-slab sheet are considered. Based on numerical analyses, the influence of the track structure on the vibration and noise in a building is examined.

2 Outline of Numerical Method for Track/Tunnel/Soil Interaction Problem [5]

2.1 Track Modeling

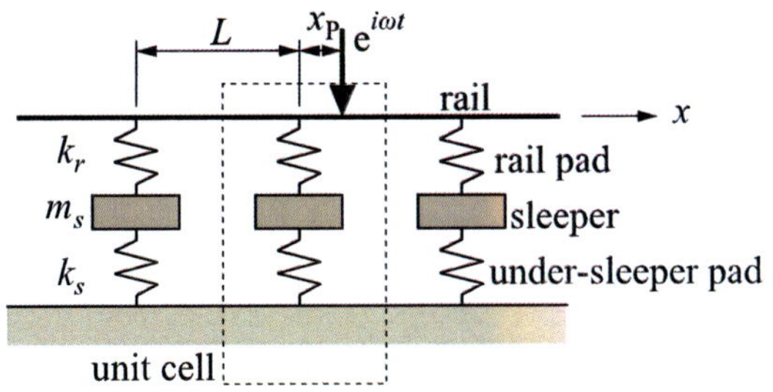

Fig. 1. Model of a track with under-sleeper pad.

In this analysis the interaction problem is decomposed into two sub-systems; the track and the tunnel-soil. The track consists of an infinite rail and discrete supports with a sleeper spacing L. As an example, the numerical model of the track with under-sleeper pad is illustrated in Fig. 1.

The Floquet transform and inverse transform [6,7] of the rail deflection w, associated with the sleeper spacing L, are defined as

$$\tilde{w}(\tilde{x},\kappa) = \sum_{n=-\infty}^{\infty} w(\tilde{x}+nL)e^{in\kappa L}, \quad w(\tilde{x}+nL) = \frac{L}{2\pi}\int_0^{2\pi/L} \tilde{w}(\tilde{x},\kappa)e^{-in\kappa L}\, d\kappa, \tag{1}$$

where $\tilde{w}$ is the Floquet transform of w, $\tilde{x}$ is the local coordinate defined in a unit cell $(-L/2, L/2)$ of the track and κ is the Floquet wavenumber.

The Floquet transform has the following periodicities:

$$\tilde{w}(\tilde{x}+L,\kappa) = e^{-i\kappa L}\tilde{w}(\tilde{x},\kappa), \qquad \tilde{w}(\tilde{x},\kappa+\frac{2\pi}{L}) = \tilde{w}(\tilde{x},\kappa). \tag{2}$$

With the aid of the Floquet transform, the dynamic problem of the infinite track can be reduced to that of a unit cell governed by the following equation of motion with periodic conditions (2),

$$EI\frac{\mathrm{d}^4\tilde{w}}{\mathrm{d}\tilde{x}^4} - \rho A\omega^2\tilde{w} = \delta(\tilde{x}-x_P) - k_e\tilde{w}\delta(\tilde{x}), \tag{3}$$

where the rail is modeled as a Euler–Bernoulli beam, EI is the bending stiffness, ρA is the mass per unit length, x_P is the position of the unit harmonic excitation and δ is the delta function, and ω is the circular frequency. The reaction force from the sleeper is given by $k_e\tilde{w}(0,\kappa)$ using the equivalent dynamic stiffness k_e which represents the sub-system consisting of the sleeper, tunnel and soil. The equivalent dynamic stiffness k_e of the track shown in Fig. 1 is expressed as

$$k_e = k_r\frac{k_s k_B - (k_s+k_B)m_s\omega^2}{k_s k_B + (k_s+k_B)(k_r - m_s\omega^2)}, \tag{4}$$

where k_B is the equivalent dynamic stiffness representing the tunnel-soil sub-system at the surface of the concrete slab. Its derivation will be described in Subsect. 2.3.

$\tilde{w}$ is expressed by a Fourier series, so that Eq. (2) can be satisfied, as

$$\tilde{w} = \sum_{n=-\infty}^{\infty} w_n(k_{xn})e^{-ik_{xn}\tilde{x}}, \quad k_{xn} = \frac{2\pi n}{L} + \kappa. \tag{5}$$

The n-th coefficient $w_n(z_n)$ which satisfies Eq. (3) is given by

$$w_n = \frac{1}{LR_n}\left(e^{-ik_{xn}x_P} - k_e Z\right), \quad Z = \sum_n w_n = \frac{1}{L + k_e\sum\frac{1}{R_n}}\sum_n\frac{e^{-ik_{xn}x_P}}{R_n}, \tag{6}$$

where $R_n = EIk_{xn}^4 - \rho A\omega^2$.

2.2 Tunnel-Soil Modeling

Floquet transform is applied to the tunnel-soil sub-system shown in Fig. 2(a). The transformed solution $\tilde{\mathbf{u}}$ in the unit cell is expanded in a Fourier series with

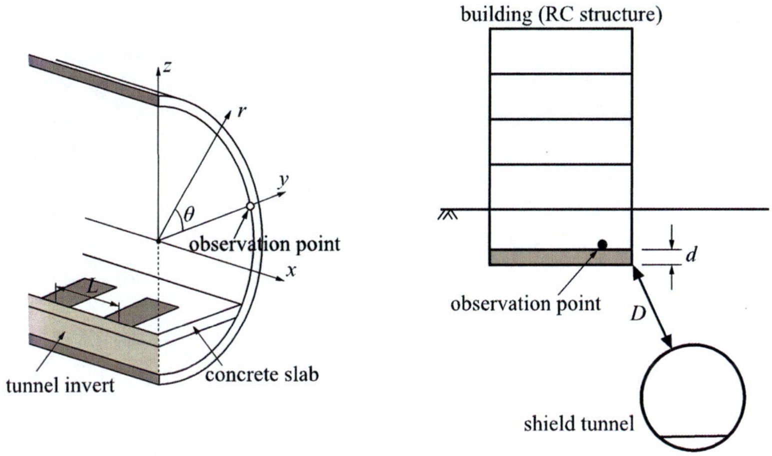

(a) Tunnel-soil sub-system (b) Path of wave transmission

Fig. 2. Tunnel-soil-building interaction problem.

respect to $\tilde{x}$, as the formulation in Subsect. 2. The present problem can then be reduced to a two-dimensional one. Since the tunnel section has a rather complicated shape, it is discretized by finite elements. In this case, the solution of tunnel sub-structure is expressed as

$$\tilde{\mathbf{u}} = [\mathbf{N}(y, z)] \sum_n \{\mathbf{U}_n\} e^{-ik_{xn}\tilde{x}}, \tag{7}$$

where $[\mathbf{N}]$ is a matrix composed of finite element shape functions, $\{\mathbf{U}_n\}$ is the nodal displacement vector of the n-th Fourier coefficient.

The finite element equation is given by

$$\begin{bmatrix} \mathbf{K}'_{naa} & \mathbf{K}'_{nab} \\ \mathbf{K}'_{nba} & \mathbf{K}'_{nbb} + \hat{\mathbf{K}}_n \end{bmatrix} \begin{Bmatrix} \mathbf{U}_{na} \\ \mathbf{U}_{nb} \end{Bmatrix} = \begin{Bmatrix} \mathbf{F}_{na} \\ \mathbf{0} \end{Bmatrix}, \quad [\mathbf{K}'_n] = [\mathbf{K}_n - \omega^2 \mathbf{M}], \tag{8}$$

where $(\ )_b$ stands for nodal components corresponding to the tunnel/soil interface, and $(\ )_a$ denotes other components. $[\mathbf{K}_n]$ and $[\mathbf{M}]$ are stiffness and mass matrices, respectively. $[\hat{\mathbf{K}}_n]$ is the equivalent stiffness of the soil on the tunnel/soil interface.

The soil region is modeled by a homogeneous infinite field. The Floquet transform of the soil displacement is expressed by [10]

$$\tilde{\mathbf{u}}_G = \nabla\phi + \nabla \times \{\psi \mathbf{e}_x + \ell \nabla \times (\chi \mathbf{e}_x)\}, \tag{9}$$

where $\mathbf{e}_x$ is the unit vector in the direction of x and ℓ is a parameter of any given value having the dimension of the length.

ϕ, ψ and χ in Eq. (9) are expressed by analytical forms as

$$g = \sum_{n,m} a_{nm}(\kappa) \mathrm{H}_m^{(2)}(k_n r) e^{im\theta} e^{-ik_{xn}\tilde{x}}, \qquad k_n^2 = \frac{\omega^2}{C^2} - k_{xn}^2, \tag{10}$$

where g is either ϕ, ψ and χ, a_{nm} is an expansion coefficient, and $\mathrm{H}_m^{(2)}$ is the m-th order Hankel function of the second kind. C is either the speed of the longitudinal wave C_L or the transversal wave C_T, and given by C_L for $g = \phi$ and C_T for $g = \psi$ or χ.

Imposing the compatibility and equilibrium conditions on the tunnel/soil interface S_R, the impedance matrix of the soil can be derived as

$$\hat{\mathbf{K}}_{n,jl} = -\frac{L}{2\pi R}[\mathbf{L}_j^T]\sum_m [\mathbf{Q}_{jm}^* \mathbf{S}_{nm} \mathbf{U}_{nm}^{-1} \mathbf{Q}_{lm}][\mathbf{L}_l], \quad [\mathbf{Q}_{jm}] = \int_{S_R} [\mathbf{N}_{sj}] e^{-im\theta}\, ds, \tag{11}$$

where $\hat{\mathbf{K}}_{n,jl}$ is a submatrix of $[\hat{\mathbf{K}}_n]$ associated with the j-th and l-th nodes, and $[\mathbf{L}_j]$ is the coordinate transformation matrix at the j-th node. $\mathbf{U}_{nm}$ and $\mathbf{S}_{nm}$ are matrices relevant to the displacement and traction, respectively.

2.3 Track/Tunnel/Soil Interaction Analysis

From Eq. (8), the solution is obtained for the unit harmonic excitation on the concrete slab at $\tilde{x} = 0$. k_B is then given by the vertical reaction $\tilde{u}_{B0}$ at the loading point as $1/\tilde{u}_{B0}$. Once k_B is obtained, k_e in Eq. (4) can be calculated. Finally, from Eqs. (1), (5) and (6), the present problem can be solved.

3 Tunnel Vibration Due to a Moving Wheel

The train-induced vibration is approximated by an interaction problem of a moving wheel and the track. To save computational cost, this interaction analysis is achieved separately from the aforementioned track/tunnel/soil interaction problem. The track is modeled by a finite rail with a random roughness generated by a power spectrum density and 80 sleepers resting on a rigid foundation. The wheel/track interaction is analyzed in time domain and the Fourier spectrum of the wheel/rail contact force is obtained. The tunnel vibration is evaluated by the multiplication of the contact force and the tunnel receptance.

Calculation of the vibration and noise in a building should be achieved for a series of axle loads. Although in this study the simplest model of a single moving wheel is used for the vibration evaluation, in general, the tunnel reaction induced by multiple axles can be approximated by superposing the solution of single wheel problem.

4 Prediction of Vibration and Noise in a Building

Decrease in the vibration acceleration level ΔL_{VA1} (dB re 1×10^{-5} m/s^2) due to wave propagation in the ground is evaluated by the following formula [8] which was constructed based on observation data obtained in the Tokyo area,

$$\begin{aligned} \Delta L_{VA1} &= -20q\log_{10} D - 8.68\alpha(f)D, \\ \alpha(f) &= 0.001f - 0.06, \end{aligned} \tag{12}$$

where D is the distance between the tunnel and the building, as depicted in Fig. 2(b). Since the length of a train is sufficiently long compared to D, in [8] the wave radiation from the tunnel was approximated by two-dimensional body waves, and the geometrical damping coefficient $q = 1/2$ was used. α is the material damping coefficient and f is a central frequency of a 1/3 octave band.

An RC structure near the subway tunnel is considered. The attenuation taking place between the foundation of the building and the observation point is predicted based on experimental data as [11]

$$\Delta L_{VA2} = -20\log_{10} d - \beta\sqrt{f}d, \tag{13}$$

where d is the distance between the base of the RC foundation and the observation point. β is a coefficient concerning the internal loss. Its value is set to 0.02, 0.03 and 0.04 in the frequency range of 10–63, 63–100 and larger 100 Hz, respectively.

The sound pressure level SPL (dB re 20μPa) is calculated by the following formula [9,12] for a diffuse sound field,

$$SPL = L_{VA0} + \Delta L_{VA1} + \Delta L_{VA2} - 20\log_{10} f + 10\log_{10}\frac{\kappa_0}{\bar{\alpha}} + 36 + 25\log_{10}\frac{V}{V_{60}}, \tag{14}$$

where L_{VA0} is the vibration acceleration level at the tunnel obtained by the interaction analysis. κ_0 is the radiation coefficient and $\bar{\alpha}$ is the sound absorption coefficient. The last term on the right-hand side is the velocity correction. V_{60} and V are the reference speed of 60 km/h and the present running speed, respectively. In the following analyses $\kappa_0 = 1$ and $\bar{\alpha} = 0.25$ are used.

5 Numerical Examples

5.1 Analysis Conditions

The track with under-sleeper pad is modeled as shown in Fig. 1, while the track with under-slab sheet and the direct fixation track have sleepers fixed on the slab. In case of the track with under-slab sheet, a mat made of microcellular polyurethane elastomer is placed under the concrete slab. The dynamic stiffness of the rail pad is $k_r = 83$ MN/m for the track with under-sleeper pad and 30 MN/m for other cases. The rail is modeled as UIC60. The semi-sleeper mass is $m_s = 100$ kg. The sleeper spacing is $L = 0.6$ m. Three types of under-sleeper pads of the dynamic stiffens $k_s = 7$, 10 and 17 MN/m are considered. The elastomer sheets have a dynamic stiffness of 7.5, 10, 12.5 and 15 MN/m^3. The complex stiffness of pad is expressed as $k(1 + ih)$ with $h = 0.14$.

The tunnel section shown in Fig. 2(a) is considered. Its inner radius and thickness are 3.25 m and 0.25 m, respectively. The concrete has a shear modulus $\mu_T = 13.5$ GPa, Poisson's ratio 0.2 and mass density 2400 kg/m^3. Its damping is represented by the complex shear stiffness as $\mu_T(1 + i\omega\eta)$, where $\eta = 1.27\times10^{-3}$ s/rad. The ground has a density 2200 kg/m^3, $C_L = 412$ m/s and $C_T = 220$ m/s.

The interaction analysis of a running wheel and track is performed for a wheel set with mass 500 kg and speed $V = 72$ km/h.

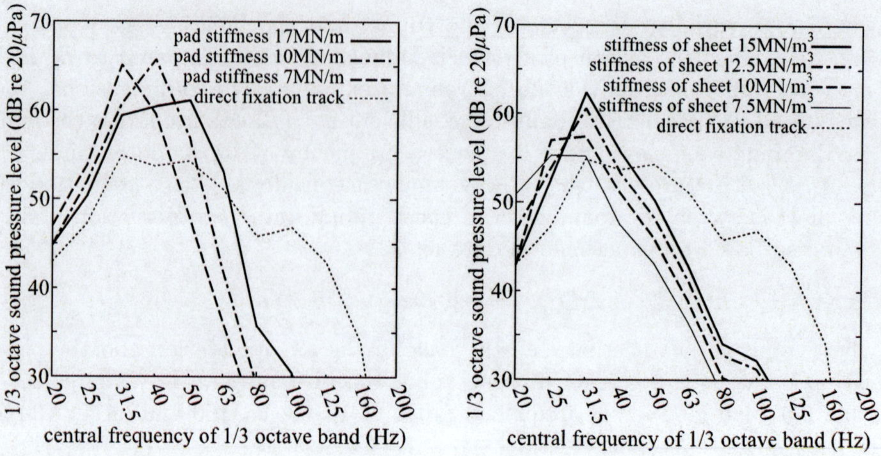

(a) Track with under-sleeper pad

(b) Track with under-slab sheet

Fig. 3. Sound pressure level in a nearby building.

5.2 Results

The sound pressure levels of both tracks are shown in Fig. 3 for the case of $D = 30\,\mathrm{m}$ and $d = 3\,\mathrm{m}$ (see Eq. (13)). The result for the direct fixation track is also drawn. The sound pressure levels of both tracks are lower than that of the direct fixation track at audible frequencies. The track with under-sleeper pad has the predominant frequency at 40 Hz, as shown in Fig. 3(a), corresponding to the resonance frequency of the wheel-track coupling system. Since this mode is characterized by the vertical vibration of the sleepers, its frequency is sensitive to the stiffness of the under-sleeper pad. The track with under-slab sheet has dominant frequency at around 31.5 Hz (Fig. 3(b)), determined by the mass-spring system consisting of the concrete slab and the elastomer sheet. Notice that the sleeper passing frequency $V/L = 33.3\,\mathrm{Hz}$ is approximately coincident with the resonance frequency of the wheel/track coupling system.

The peak at about 31.5 Hz observed for the track with under-slab sheet is lower than that of the track with under-sleeper pad. Because of this, the sound pressure level of the former is lower, in the range of 30–100 Hz. The noise reduction observed in the track with under-sleeper pad is sensitive to the pad stiffness compared to the case of elastomer sheet.

6 Conclusion

An efficient prediction method has been constructed for vibration and noise in a building nearby a subway tunnel. In order to save computational cost, a quasi-two-dimensional numerical method is employed for the three-dimensional track/tunnel/soil interaction problem. The damping of vibration transmitted through the soil and building and the sound radiation are predicted using empirical formulae. The developed method was applied to the track with under-sleeper pad and the track with under-slab sheet. The latter track is superior in reducing noise, whereas the noise originating from the track with under-sleeper pad is sensitive to the pad stiffness compared to the track with elastomer sheet.

References

1. Sheng, X., Jones, C.J.C., Thompson, D.J. : Ground vibration generated by a harmonic load moving in a circular tunnel in a layered ground. In: 10th International Meeting on Low Frequency Noise and Vibration and Its Control, pp. 161–176 (2002)
2. Abe, K., Kato, H., Furuta, M.: Tunnel and ground vibrations induced by moving train loads. Proc. of Railway Mech. JSCE. **8**, 47–52 (2004). (in Japanese)
3. Yang, Y.B., Hung, H.H.: Soil vibrations caused by underground moving trains. J. Geotech. Geoenv. Eng. ASCE. **134**(11), 1633–1644 (2008)
4. Abe, K., Satou, D., Suzuki, T., Furuta, M. : Three-dimensional analysis of subway track vibration due to running wheels. In: Chouw, N., Schmid, G. (eds.) Wave 2002, pp. 149–156. A.A. Balkema (2003)
5. Abe, K., Yamada, K., Furuta, M., Suehara, M., Koro, K. : A 3-D analysis method for dynamic soil/subway tunnel interaction problems. J. Appl. Mech. JSCE. **74**(2), 523–534 (2018). (in Japanese). https://doi.org/10.2208/jscejam.74.I_523
6. Clouteau, D., Elhabre, M.L., Aubry, D.: Periodic BEM and FEM-BEM coupling, application to seismic behaviour of very long structures. Comput. Mech. **25**, 567–577 (2000)
7. Clouteau, D., Arnst, M., Al-Hussaini, T.M., Degrande, G.: Freefield vibrations due to dynamic loading on a tunnel embedded in a stratified medium. J. Sound Vib. **283**, 173–199 (2005)
8. Tsuno, K., Furuta, M., Fujii, K. Nagashima, F., Kusakabe, O.: Wide range attenuation properties of ground vibration propagated from subway shield tunnel. J. JSCE. **792**(III-71), 185–197 (2005). (in Japanese)
9. Tsuno, K., Furuta, M., Fujii, K., Kusakabe, O.: Prediction of borne noise in buildings caused by subway-induced vibration. J-Rail **2005**, 501–504 (2005). (in Japanese)
10. Eringen, A.C. and Şuhubi, E. : Elastodynamics, vol. II. Academic Press, Inc., New York (1975)
11. Matsuda, Y., Tachibana, H., Ishii, S. : Propagation properties of ground-borne noise in a building structure. J. Acoust. Soc. Japan. **35**(11), 609–615 (1979). (in Japanese)
12. Ikeda, S. : Prediction and control of structure-borne sound in buildings adjacent to subway traffic. J. INCE. **21**(6), 397–400 (1997). (in Japanese)

Prediction of Structure-Borne Sound in Buildings Consisting of Various Building Elements, Generated by Underground Rail Traffic

Fülöp Augusztinovicz[1](✉), A. B. Nagy[2], P. Fiala[1], Hamid Masoumi[3], and Patrick Carels[3]

[1] Budapest University of Technology and Economics, Magyar tudósok krt. 2, Budapest 1117, Hungary
fulop@hit.bme.hu

[2] Budapest University of Technology and Economics, Stoczeku. 1, Budapest 1111, Hungary

[3] CDM NV, Reutenbeek 9-11, 3090 Overijse, Belgium

Abstract. The prediction of re-radiated structure-borne sound in large, complex buildings is a difficult task. Methods to do so exist from simple analytical formulae to full-size detailed numerical analysis of vibrations and re-radiated noise. Nevertheless, designers need an effective, practicable method to ensure that the interior noise of buildings planned above or in the vicinity of metro lines do meet the relevant limit values and to decide whether or not building base isolation is required. The paper summarizes a number of experiments performed in large portal frame buildings consisting of simultaneous vibration and noise measurements. It was established that the accuracy of the well-known Kurzweil formula [3] is highly limited: the difference between measured and predicted sound pressure values is frequency dependent with positive values for low and negative values for higher frequencies, and the standard deviation of the predictions is high. The experimental findings were pursued by numerical calculations for a very simple model, consisting of various typical wall and floor slab materials and dimensions. It was found that the positive difference for low frequencies is mainly caused by weak fluid-structure coupling. The large spread is related to the highly modal behavior of both the involved structures and the acoustic field, resulting in strongly frequency-dependent sound radiation from the surrounding surfaces. As a result, an improved prediction formula is proposed on the basis of the comparison of measured and simulated difference curves.

Keywords: Underground railway traffic · Structure-borne noise · Prediction

1 Introduction

Despite the fact that several EU projects, e.g. CONVURT [1] and RIVAS [2], have tackled the problem and further approaches have been published in the scientific literature (see

G. Degrande et al. (Eds.): *Noise and Vibration Mitigation for Rail Transportation Systems*, NNFM 150, pp. 368–376, 2021.
https://doi.org/10.1007/978-3-030-70289-2_38

e.g. [4] to [7]), the most frequently used method for prediction of re-radiated noise in the engineering practice is still based on the famous Kurzweil formula [3], combining the average floor vibration of a room and the resulting average sound pressure by

$$L_p = L_a - 20\,logf + 37 \tag{1}$$

where

$$L_a = 20\,log\frac{a}{10^{-6}[g]}\text{and}\,L_p = 20\,log\frac{p}{2\times 10^{-5}[Pa]} \tag{2}$$

It is easy to show that Eq. (1) is equivalent to the simple relation

$$L_p = L_v + 7,\,\text{where}\,L_v = 20\,log\frac{v}{5\times 10^{-8}[m/s]} \tag{3}$$

More recent publications have refined this relationship. Deliverable D1.6 of the RIVAS project [4] suggests that, while Eq. (3) is valid for concrete floors, a constant of –3 rather than + 7 dB should be used for lightweight wood floors. Annex A of ISO/TS 14837–31 [5] gives a formula which takes into account both the radiation efficiency σ of the radiating surfaces S and the acoustic behaviour of the room:

$$L_p = L_v + 10\,log\sigma + 10\,log\frac{4S}{A} \tag{4}$$

where A is total acoustic absorption. This equation is further simplified to equations similar to Eq. (3) again, but with different constants for various building structures.

Applying numerical techniques to take into account the typical radiation characteristics of the usual building elements and their interactions with the acoustic cavity, [6] and [7] present frequency dependent relationships instead of constants for some specific cases. Unfortunately, these detailed methods are not feasible for design purposes in the everyday engineering routine.

This paper, based on the results of the BIOVib EUROSTARS project [11], is devoted to tackle the problem from two aspects. Experimental results obtained from measurements on existing buildings are extended by a parametric study on a very simple numerical model, a rectangular room surrounded by various building structures and excited from below by point and distributed forces. It is shown that the prediction error lends itself to be explained by weak, highly modal and as such, strongly frequency dependent radiation efficiencies. Based on these findings, a frequency-dependent and still very simple prediction formula is proposed.

2 Experiments

2.1 Measuring Conditions

The correct evaluation of the accuracy of the investigated noise prediction method requires fulfilling quite a few conditions as much as possible. Just a few of them: simultaneous measurement of floor vibrations and the resulting re-radiated noise is necessary; the effect of any airborne noise should be excluded; the accelerometers should be attached to the walls and ceiling preferably in more than one point per building element; the measurements would last for hours, possibly in total silence, etc. All these are rarely tolerated by the owners or occupiers of the apartment or office room; hence it is rather difficult to perform useful and reliable measurements. Nonetheless, we have been able to perform appropriate measurements in a number of rooms of three portal frame buildings (i.e. buildings consisting of large rigid reinforced concrete frames) where ground-borne noise was an issue, and in rooms of two other, traditional brick buildings. In each case at least ten metro pass-bys were recorded and evaluated, because the technical conditions of the vehicles vary a lot in many cases. Floor vibrations and noise levels were recorded simultaneously and continuously. The recorded signals were processed subsequently, picking the peaks corresponding to the metro pass-bys and excluding background effects as much as possible.

2.2 Measured Results

Figure 1 shows the results of difference spectra between estimated and measured noise levels in the relevant frequency range of typical metro pass-bys. As can be seen, the prediction error is approx. + 10 dB or even more between 31.5 and 50 Hz, but sharply drops to below −10 dB up to 100 Hz. The standard deviation of the differences is 12–13 dB for portal frame buildings, and even higher if all measurements are considered. This frequency dependence poses a further problem, which is impossible to handle by using a constant such as in Eq. (3). Namely, if the dominant vibration components fall below 63 Hz, then the Kurzweil formula overestimates the resulting noise level, but the opposite can also happen if the dominant frequencies are high.

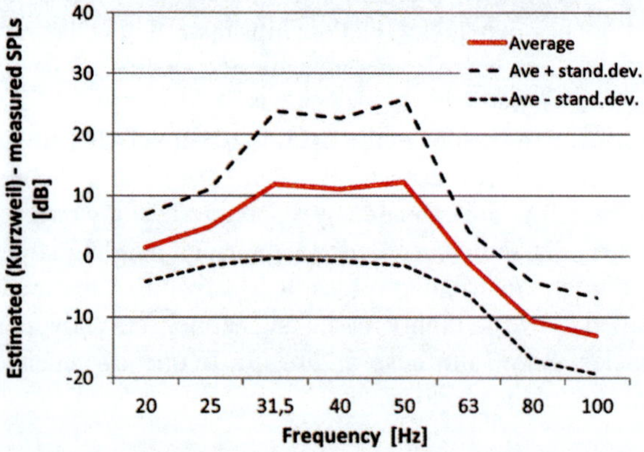

Fig. 1. Differences of the estimated and measured noise spectra, based on measurements performed on modern, portal-frame buildings.

3 Numerical Calculations

3.1 Aim and Approach of Calculations

The aim of the numerical calculations is to get a deeper insight into the vibro-acoustic phenomena of a closed acoustic space surrounded by walls and slabs of various thicknesses and materials by using numerical methods. Due to the extreme variety of the targeted building structures and possible modelling parameters one cannot expect good agreement between measured and calculated results. Nevertheless, the obtained results seem to explain the experimental findings and can serve as a basis to suggest an improved Kurzweil formula.

The calculations reported herein refer to a very simple model: a shoebox-shaped room in a hypothetical building. This model is used for a parametric study, varying the materials and thickness of the building structures, the boundary conditions (BCs) of the walls, and assuming various excitations (load cases).

The applied numerical approach is a weakly coupled structural FE – acoustic BE method, performed by using the numerical analysis software package SYSNOISE. Two sets of calculations were run. For the first, radiation efficiencies are calculated by assuming just one flexible wall/floor of the room excited by a point force. For the second, the vibration velocities of the floor and the sound pressure in the interior is calculated for various boundary conditions and load cases for the fully flexible model.

3.2 Model Parameters

The inner dimensions of the model were $5.1 \times 3.9 \times 2.7$ m. It was tested in the following boundary condition (BC) combinations:

1. All surrounding surfaces are assumed to be rigid, except for one long sidewall. This wall was assumed to be built of 14 cm concrete, 20 cm concrete, 12 cm and 17 cm brick and 5 cm gypsum, respectively.

2. All surrounding surfaces are assumed to be rigid, except for the floor slab of 20 cm concrete.

3. All surrounding surfaces are assumed to be flexible, made of 20 cm concrete, 12 cm brick and 5 cm gypsum, as descripted in Fig. 2 wherein the applied mesh of the mechanical structure (and at the same time the faces of the acoustical subsystem) is shown. The assumed material characteristics are collected in Table 1.

The assumed excitations were sinusoidal forces of swept frequency between 30 and 100 Hz, acting normally on one, randomly selected node of a wall or floor, and simultaneously on the nodes along the two longer edges of the floor slab (see Fig. 3). The amplitude and phase of these distributed forces were also varied:

1. 110 and 80 N vertical forces, identical phase (according to Fig. 3a)

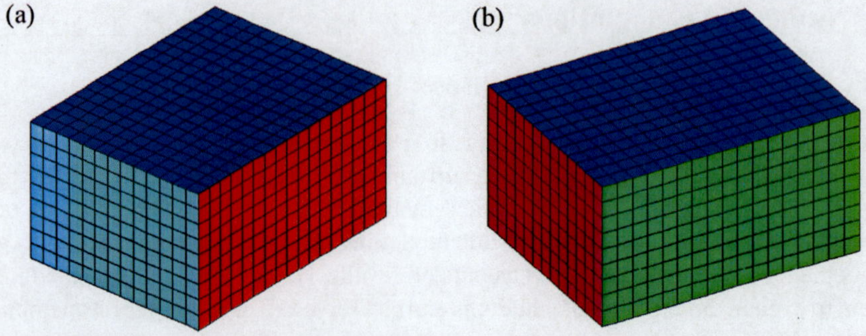

Fig. 2. Mesh and selected materials of the entirely flexible structural model, shown in both front and rear views. Floor and ceiling (blue): 20 cm concrete, sidewalls in front and on the right (cyan and red): 5 cm gypsum, left and back wall (green and magenta): 12 cm brick. (a): front side, (b): back side.

Table 1. Material parameters of the investigated model.

	Young-modulus [Pa]	Poisson coefficient [−]	Massdensity [kg/m^3]	Critical frequency of a hypothetical infinite plate [Hz]
Concrete, 14 cm	26 e9	0.2	2300	133
Concrete, 20 cm	26 e9	0.2	2300	93
Brick, 12 cm	9 e9	0.2	1700	226
Heavy brick, 17 cm	10 e9	0.2	2000	164
Gypsum, 5 cm	7e9	0.2	1200	517

2. 110 and 80 N vertical forces in varied phase (according to an assumed soil wave with propagation speed of 200 m/s; Fig. 3a)
3. both vertical and horizontal force components: $F_y = 55N$, $F_z = 110$ N along line 2, $F_y = 40N$, $F_z = 80$ N along line 1, equal phase (see Fig. 3b).

3.3 Results of Radiation Efficiency Calculations

The radiation efficiency is defined as the calculated sound power radiated from a wall of the model, referenced to the sound power radiated by a hypothetical rigid wall of equal size and RMS velocity. The results, demonstrated in Fig. 4, show that the radiation is rather weak, concordant with analytical and other numerical predictions [6, 7] and the theoretical critical frequencies of the assumed structural elements (see in Table 1). The values are far below unity, which are apparently caused by the excitations (point force) and by the fact that the critical frequency of the assumed materials refer to the radiation of infinite plates bearing ideal bending waves. One conclusion from the point

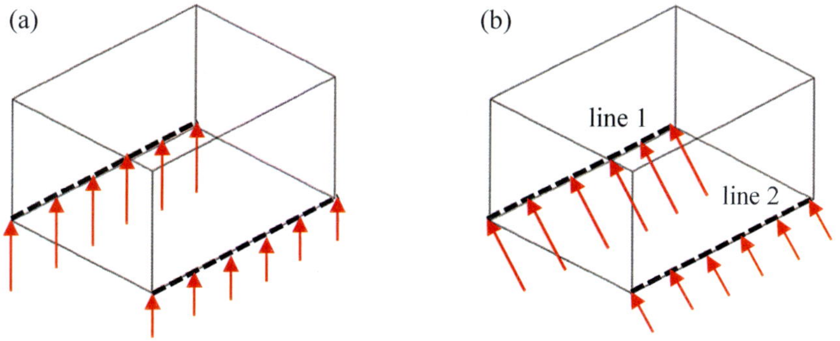

Fig. 3. Spatial distribution of exciting forces for the calculations

force excitation is that the strongest radiators are the thick, concrete walls and the floor, which seems to be supporting the Kurzweil approach.

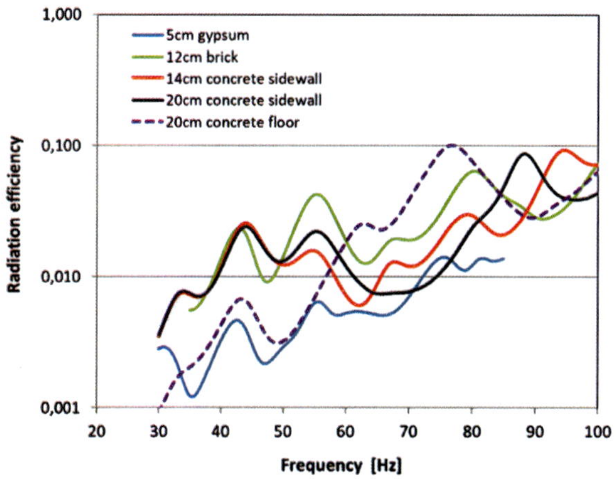

Fig. 4. Comparison of radiation efficiencies of single walls and floors for randomly selected point excitations.

3.4 Results of Radiated Noise Spectrum Calculations

The sound pressures, see Fig. 5, were calculated for six randomly selected points and energy averaged. They show similar trends as the radiation efficiency curves: increasing values vs. Frequency with many local maxima. If one compares the sound pressures numerically calculated and derived from floor vibrations in the same way as the experimental results above, Fig. 6 is obtained. The average values vs. Frequency curves show again that the traditional Kurzweil formula heavily overestimates the expected sound pressures below 50 Hz. Considering the tendencies, this finding is broadly in accordance with the experimental data (see Fig. 1): the average difference values are significantly higher but the trend is similar.

The sound radiation characteristics, and hence the final results of the noise prediction procedure, lend itself to be explained on the basis of the eigenfrequencies and mode shapes of the structural and acoustical modes and their coupling. The thick and rigid floor has only a few and rather simple normal modes in the relevant frequency range (mainly {1, 1}, {1, 2} and {2, 1} modes, where the numbers within brackets refer to the number of half wavelengths along the edges), while the walls show a number of rather complex mode shapes up to 100 Hz. The floor is set into motion with high amplitudes for a few frequencies, but due to the low radiation efficiencies only a small amount of vibration energy is converted into acoustical energy. On the contrary, the walls vibrate in more complex patterns and with varied amplitude. As a result, in the contribution of the surrounding surfaces the effect of the floor is relatively weaker than that of the walls, and the floor vibrations of high amplitude at lower frequencies are overestimated if the standard Kurzweil formula is applied.

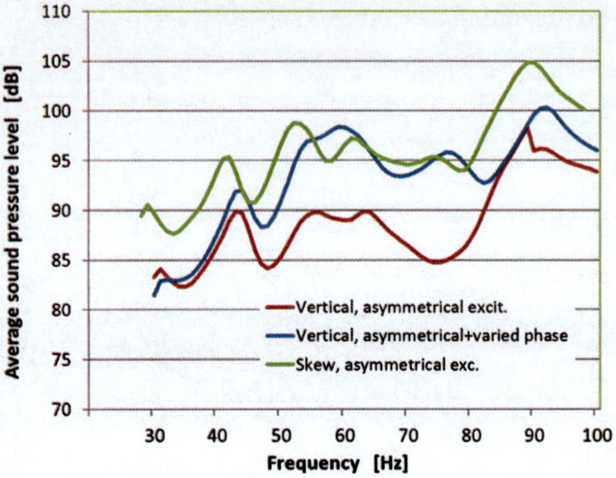

Fig. 5. Comparison of numerically calculated average sound pressure spectra for boundary conditions 3 and force configurations 1, 2 and 3 as detailed above.

4 Proposal for an Improved Estimation Formula

In order to introduce frequency dependence into the simple prediction formula that we aim at, we suggest to replace the constant of 7 dB in Eq. (3) by taking into account the overestimation between 25 and 63 Hz, and underestimation for 100 Hz. According to the statistical evaluation of the experimental data as shown in Fig. 1, we suggest to use the formula

$$L_p = L_v + C(f) \tag{5}$$

where $C(f)$ is given by the following table:

It is very important to note that this modified method can only be used for freely vibrating, rigid and heavy floor slabs. In case of floors embedded in soil or light-weight floor structures the formula is inappropriate and as such, cannot be used.

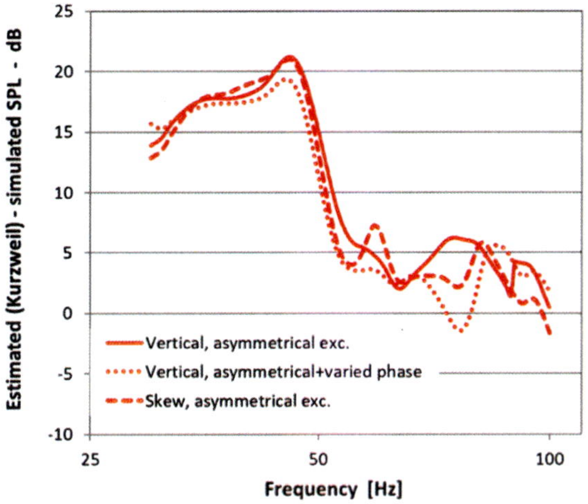

Fig. 6. Average differences of the estimated and numerically calculated noise spectra.

Table 2. Frequency dependent additive factor in Eq. (5)

f - Hz	20	25	31.5	40	50	63	80	100
$C(f)$–dB	7	2	−3	−3	−3	2	7	12

Acknowledgements. The results presented in this paper have been obtained within the frame of the BIOVIB project funded by VLAIO and EUROSTARS. Their financial support is gratefully acknowledged.

References

1. The Control of Vibration from Underground Rail Traffic (CONVURT). Growth 2000 EU RTD project.
2. Railway Induced Vibration Abatement Solutions (RIVAS). Collaborative EU project, SCP0-GA
3. Kurzweil, L.G.: Ground-borne noise and vibration from underground rail systems. J. Sound Vib. **66**(3), 363–370 (1979)
4. RIVAS project, WP 1 Deliverable D1.6: Definition of appropriate procedures to predict exposure in buildings and estimate annoyance. August 2012
5. ISO/TS 14837–31:2017 Mechanical vibration -- Ground-borne noise and vibration arising from rail systems -- *Part 31: Guideline on field measurements for the evaluation of human exposure in buildings*
6. Thornely-Taylor, R.: The relationship between floor vibration from an underground source and the airborne sound pressure level in the room. Int. J. Rail Transp. **4**(4), 247–255 (2016)
7. Villot, M., Jean, P., Grau, L., Bailhache, S.: Predicting railway-induced ground-borne noise from the vibration of radiating building elements using power-based building acoustics theory. Int. J Rail Transp. **6**(1), 38–54 (2018)

8. Dowell, E.H., Voss, H.M.: The effect of a cavity on panel vibration. AIAA J. **1**, 476–477 (1963)
9. Pan, J., Bies, D.A.: The effect of fluid-structural coupling on sound waves in an enclosure. Theor. Part. J. Acoust. Soc. Am. **87**(2), 691–707 (1990)
10. Pan, J., Bies, D.A.: The effect of fluid-structural coupling on sound waves in an enclosure. Exp. Part. J. Acoust. Soc. Am. **87**(2), 691–707 (1990)
11. Eureka Project BIOVib (Building Insulation against Outdoor Vibration)
12. Augusztinovicz, F., Fiala, P., Nagy, A.B.: Prediction of re-radiated noise from building structure elements. BIOVib Progress Reports TR 5.1 and 5.2, Budapest

Ground-Borne Noise and Vibration Propagation Measurements and Prediction Validations from an Australian Railway Tunnel Project

Peter Karantonis(✉), Conrad Weber, and Hayden Puckeridge

Renzo Tonin & Associates, PO Box 877, Strawberry Hills, NSW, Australia
peter.karantonis@renzotonin.com.au

Abstract. For a recent underground railway tunnel project in Australia, field tests were conducted to measure the vibration propagation from a hydraulic hammer operating inside partly constructed railway tunnels, to the outside and inside of buildings above the tunnels, with the aim of improving estimates of coupling loss and amplification factors and the conversion of vibration into noise inside buildings. Field measurements were used to reduce the uncertainty associated with this aspect of ground-borne noise and vibration predictions, and to ultimately inform the rail track design.

Keywords: Ground-borne noise · Ground-borne vibration · Building coupling loss · Building amplification · Railway tunnel · Propagation

1 Introduction

On projects where there is limited or only high-level information relating to how vibration propagates between an underground rail tunnel and receivers inside buildings, the uncertainty associated with ground-borne noise and vibration predictions can be large and therefore large prediction safety factors (engineering margins) are often used. During the detailed design stages of projects, such large safety factors can be costly in terms of the required mitigation measures.

On a recent underground railway tunnel project in Australia, a quantitative approach was applied with the aim of improving estimates of the combined coupling loss and amplification for typical building types and the conversion of vibration into noise inside buildings to better advise the design team of the level of design risk associated with predictions. Field measurements were used to reduce the uncertainty associated with this aspect of the ground-borne noise and vibration predictions and ultimately inform the rail track design.

2 Background

For underground railways, although vibration has the potential to be perceptible as tactile vibration, it is usually manifested as ground-borne noise (GBN), also referred to

G. Degrande et al. (Eds.): *Noise and Vibration Mitigation for Rail Transportation Systems*, NNFM 150, pp. 377–384, 2021.
https://doi.org/10.1007/978-3-030-70289-2_39

as regenerated, structure-borne or re-radiated noise. For most sensitive receiver types, assessment criteria relating to GBN are normally more stringent than the associated tactile vibration criteria. In rare cases, ground vibration (GBV) levels associated with train operations may impact the satisfactory operation of sensitive measurement equipment located within high technology facilities. At such locations, the associated GBV objectives may be more stringent than the associated GBN objectives. Typically, however mitigation measures for underground rail systems are controlled by GBN objectives. This study addresses both ground-borne noise and vibration (GBNV).

GBNV is influenced by several physical aspects relating to the train, ground conditions and receivers [1, 2]. The subject parameters investigated in this study are: (a) GBV propagation into receiver buildings (coupling loss and amplification), and (b) the conversion of floor GBV into GBN inside receiver buildings.

2.1 Coupling Loss and Amplification

Within a GBNV prediction model, each building can be assigned a frequency-dependent coupling loss based on a selected category [3]. Similarly, when modelling GBNV, allowances for amplifications (resonances) of building floors, walls and ceilings, must be made and there are a range of possible amplification values that can be applied. The FTA [4] recommends that an amplification value of 6 dB(A) should be utilized, whereas Nelson [3] notes that amplification is greatest in the 10 to 30 Hz frequency range corresponding with the natural frequency of floor structures, and amplification values may vary from 5 to 15 dB(A) over the 16 Hz to 80 Hz frequency range. Combining and applying such generic allowances for coupling losses and amplifications tends to provide large range estimates with large uncertainties. To reduce the range of estimates and degree of uncertainty, detailed assessment of foundations and building constructions is necessary. For large scale projects, it is not feasible to undertake detailed assessments of the foundation types and construction details of all buildings, so field vibration measurements were conducted at representative buildings nearest to the rail tunnel to reduce the uncertainty associated with these parameters.

2.2 Conversion of Floor Vibration to Audible Noise

Vibration of the main building elements (floor, walls and ceiling) may generate low-frequency GBN. The method that is most commonly applied to predicting GBN levels is based on the Kurzweil formula [3, 5, 6]. Utilizing this calculation method, the unweighted sound pressure level (dB re 20 x 10^{-6} Pa) is approximately equal to the rms vibration velocity level (dB re 10^{-9} m/s) of the floor, minus 27 dB for typical small rooms. In order to calculate the overall A-weighted noise levels, the 1/3 octave band noise levels are A-weighted and summed together over the 20 Hz to 250 Hz frequency range.

3 Methodology

GBNV measurements were undertaken at multiple sensitive receiver locations close to newly constructed underground tunnels for a recent railway tunnel project. The purpose of the measurements was to quantify the following parameters which present large uncertainties in relation to the GBNV predictions:

- Coupling loss and amplification (difference in ground borne vibration levels outside building and floor vibration levels inside building).
- Conversion of floor vibration levels to audible noise.

3.1 Vibration Source in Tunnel

The above parameters were investigated by generating vibration within the tunnel using an excavator with a hydraulic hammer attachment and measuring vibration levels in the adjacent tunnel, outside and inside several nearby buildings on the ground surface and noise levels inside buildings. A hydraulic hammer was utilized as it was not possible to use a train (or similar line source within the tunnels as the vibration source), because the invert and track form had not been constructed at that stage. The hydraulic hammer vibration source generated strong and steady vibration levels within the rail tunnels which were measurable inside nearby buildings on the surface whilst providing a strong signal-to-noise ratio that was greater than background noise and vibration levels in the important GBNV frequency range of 20 Hz to 250 Hz. For each measurement location, the hydraulic hammer was operated for a minimum of two 30-sec periods. The input force was not measured as this does not influence the coupling loss and amplification values; however the impact rate was at approx. 3–5 Hz.

A picture of the vibration source set up within the tunnel is shown in Fig. 1. To prevent damaging the tunnel, a concrete flood barrier was utilized between the hydraulic hammer tip and the tunnel rings which had the effect of distributing the vibration energy over a larger cross-sectional area than the tip alone. Conveyor belt material was placed between the flood barrier and the tunnel rings to assist in reducing the frequency content of the source vibration, consistent with the range applicable to GBNV predictions. A metal plate conveyor belt material was placed on top of the flood barrier to avoid damage over multiple tests. A portable crane was utilized to transport and position the flood barrier at the required measurement locations.

3.2 Measurements in Tunnels

Within the railway tunnels, vibration transducers were set up at two locations in the tunnel adjacent to where the hydraulic hammer was operating. The purpose of these measurements was to confirm that source vibration levels from the hydraulic hammer were consistent between test locations and measurement positions and make any necessary adjustments to the source levels.

3.3 Measurements on Surface

Above the tunnels, vibration monitoring was performed on the ground surface outside buildings and on the floor of habitable rooms inside buildings (near the centre of each room). Attended noise measurements were conducted concurrently with the vibration monitoring.

Fig. 1. Vibration source in one tunnel (left) and measurement transducers in adjacent tunnel (right)

Test locations were selected at eight representative buildings near the tunnel alignment where the tunnel was located 18 m to 42 m below the ground surface. Consideration was given to selecting a range of different building construction types in order to validate the coupling loss and amplification assumptions. Where possible, test locations were selected in areas with low background noise levels (away from major roads) to ensure that hydraulic hammer noise was audible within the buildings, providing a strong signal-to-noise ratio.

At each building location, source vibration levels were generated at five locations within the tunnel, identified as positions A to E, with each position offset by 22.5 m. Position C was always closest to each subject building.

Where the internal vibration measurements were undertaken on a solid floor (with tiles or floorboards), a metal plate was glued to the floor using epoxy glue (a thin layer of masking tape was used between the plate and floor surface for protection). Where the internal vibration measurements were undertaken on a carpeted floor, a carpet spike was used to provide a rigid connection with the underlying surface. For measurements on the ground surface outside the building, a 200 mm long metal spike was driven into the ground surface.

The noise and vibration measurement time histories were reviewed to assist in identifying periods during the 30 s hydraulic hammer events where noise/vibration levels were steady and not significantly influenced by extraneous events from road traffic or other sources. Average noise and vibration levels during typical 5 s periods were selected for detailed analysis. The vibration measurement results inside and outside each building during operation of the hydraulic hammer were typically 10 dB or more above the background levels within the 20 Hz to 250 Hz 1/3 octave frequency bands.

3.4 Calculations for Line Source Attenuation

Source vibration levels within the tunnel were based on hydraulic hammer vibration levels at five discrete locations within the tunnel. From a GBNV modelling perspective, the hydraulic hammers represent point source vibration levels, whereas the train is representative of a line source. The measurement results relating to the point source vibration levels have been converted into equivalent line source results using the methodology described in [3]. The 1/3 octave band point source vibration levels for each transducer location were energy summed following the trapezoidal rule for numerical integration to directly calculate the equivalent line-source vibration levels. This introduced some averaging into the results at each site and reduced the data spread. The multiple impact points within the tunnels excite the building structure over multiple paths and producing additional averaging. Conversion of discrete point sources to a line source was only relevant for the validation of ground vibration propagation models from the rail tunnel to receiver buildings, whereas this paper focuses on vibration at the receiver building only and how it transfers from outside to inside buildings and internal ground-borne noise.

4 Results

4.1 Coupling Loss/Amplification

Based on the measurement results, the combined coupling loss and amplification values were grouped into buildings with a concrete slab-on-ground and buildings with suspended timber floor constructions.

Figure 2 (a) shows a summary of the measured coupling loss and amplification values at each of five buildings and their average values, all being two-storey brick veneer buildings with concrete slabs on ground.

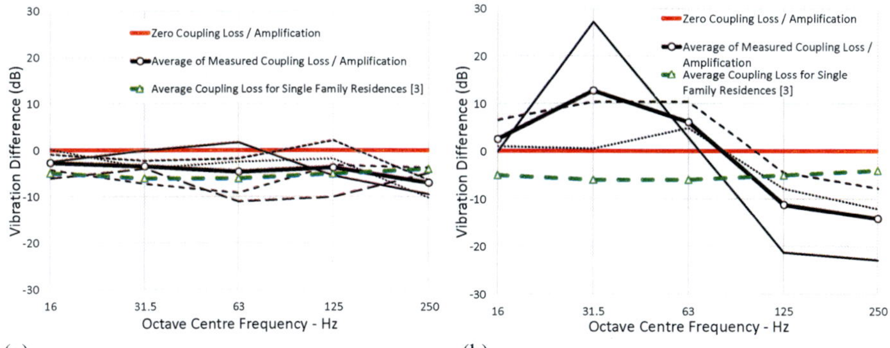

(a) (b)

Fig. 2. Summary of line source coupling loss/amplification values at test buildings (left: two-storey brick veneer buildings with concrete floor slabs on ground; right: single-storey brick veneer buildings with timber floors on piers). Vibration difference represents inside vibration levels minus outside vibration levels in dB.

Figure 2 (b) shows a summary of the measured coupling loss/amplification values at each of three buildings and their average values, all being single-storey brick veneer buildings with timber floors on piers. All building types are single family residences.

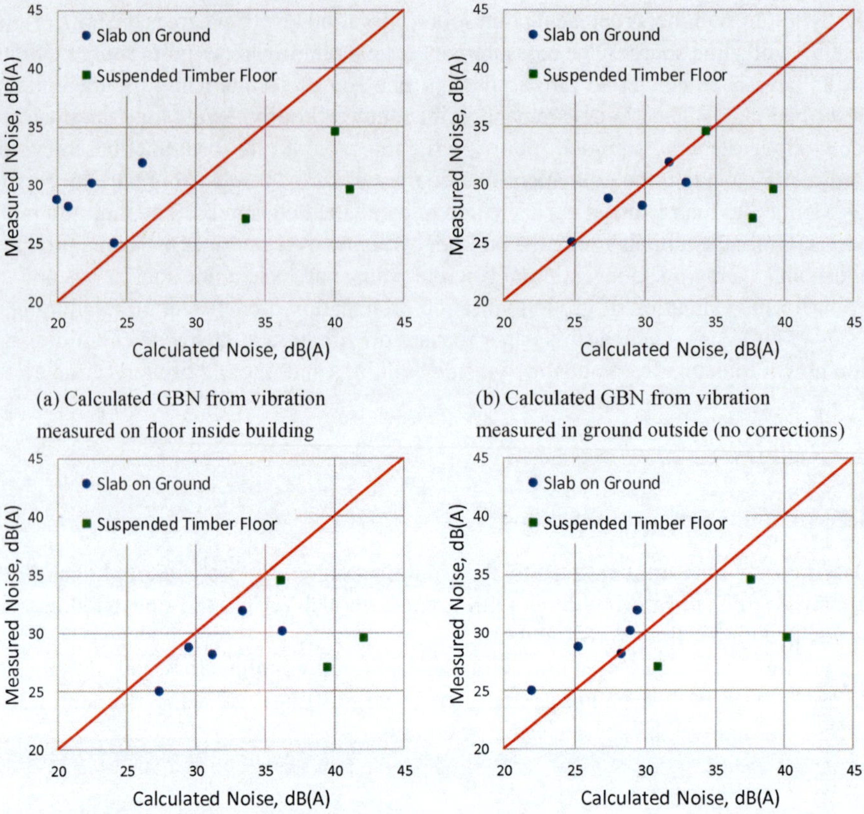

(a) Calculated GBN from vibration measured on floor inside building

(b) Calculated GBN from vibration measured in ground outside (no corrections)

(c) Calculated GBN from vibration measured in ground outside (with single family residence coupling loss and amplification as per [3])

(d) Calculated GBN from vibration measured in ground outside (with site measured combined coupling loss and amplification)

Fig. 3. Calculated A-weighted GBN levels using outside and inside vibration levels (z-axis) compared to measured indoor noise levels at test buildings (20 Hz to 250 Hz)

These coupling loss/amplification values were determined on a 1/3 octave frequency basis by subtracting the measured vibration levels outside the building from the measured vibration levels inside the building (inside minus outside). A positive value therefore represents higher vibration levels inside the building compared with vibration levels outside the building, implying an amplification, whilst a negative value implies a coupling loss. Also shown on the coupling loss/amplification plots are the default values for single family residences [3] and the zero-coupling loss/amplification line.

The measurement results show that the measured coupling loss/amplification values hover above and below zero. In Fig. 2 (a) the average coupling loss/amplification values are similar to the default assumptions for single family residences [3], but large variations occur for individual buildings. In Fig. 2 (b) the values are much higher than the default assumptions at low frequencies (63 Hz and below), implying the occurrence of amplification which most likely is caused by building and floor resonances.

4.2 Ground-Borne Noise

Measured vibration levels on the floor inside each building and the ground vibration outside each corresponding building are used to calculate internal A-weighted noise levels for each building using the Kurzweil formula. Figure 3 shows a summary of measured versus calculated noise levels, being the average of five measurements for each building. The 'blue' data points relate to the five buildings with concrete floor slabs and the 'green' data points relate to the three buildings with suspended timber floors. Figure 3 shows four different ways of calculating GBN from measured vibration levels. Where the points on the graphs are higher than the diagonal line, the measured noise levels are higher than the calculated noise levels, which indicates that the calculated noise levels underestimate the true GBN levels. Where the points on the graphs are lower than the diagonal line, the measured noise levels are lower than the calculated noise levels, which indicates that the calculated levels are conservatively high compared to true GBN levels.

Figure 3 (a) shows large scatter of data about the diagonal line, rendering the use of indoor floor vibration levels for calculating GBN as unreliable. One reason for this is that vibration levels vary significantly from one part of the floor to another and GBN is created from vibration of other surfaces as well as a room's floor [5].

Figure 3 (b) and Fig. 3 (c) show that using outdoor ground vibration levels in GBN calculations tends to overestimate GBN levels (i.e. data points tend to generally lie beneath the diagonal line).

Figure 3 (d) shows that using outdoor ground vibration levels and applying measured site-specific coupling loss/amplification corrections, provides a more even spread of data points about the diagonal line.

The data points relating to buildings with timber flooring tend to be below and furthest removed from the diagonal line, indicating calculations overestimate GBN levels. This ties in with the results in Fig. 2 (b) which whilst showing amplification at floor resonance frequencies, greater losses are shown for frequencies beyond 100 Hz that may be from more coupling losses and/or increased damping from timber floors.

Based on these results, the combined coupling loss and amplification values were grouped for buildings with suspended floor constructions and slab on ground constructions. A summary of this analysis was then determined for relevant building categories and predictions of internal noise levels were made for each building type. The predicted noise levels were then compared with the measured noise levels. Based on these results, a standard deviation was then determined which forms part of the design uncertainty analysis conducted for the rail tunnel project.

5 Conclusion

Field tests were conducted to measure the vibration propagation of a hydraulic hammer operating inside partly constructed railway tunnels, to the outside and inside of buildings above the tunnels. The measurements were conducted to quantify parameters which present large uncertainties in relation to ground-borne noise and vibration predictions for railway tunnel projects.

Vibration measurements were performed on the ground surface outside buildings and on the floor of habitable rooms inside buildings, whilst noise measurements were also conducted concurrently inside buildings.

The coupling loss/amplification values were measured at five two-storey brick veneer buildings with concrete slabs on ground, and at three single-storey brick veneer buildings with timber floors on piers.

Measured vibration levels on the floor inside each building and the ground vibration outside each corresponding building were used to calculate internal noise levels for each building. These noise levels were then compared to measured indoor noise levels and evaluated against known vibration-to-noise conversion factors (Kurzweil formula).

Based on these results, the combined coupling loss and amplification values were grouped for buildings with suspended floor constructions and slab on ground constructions. A summary of this analysis was then determined for relevant building categories and predictions of internal noise levels were made for each building type, starting with the predicted ground surface vibration levels. The large range of possible coupling loss and amplification categories that could be applied when calculating ground-borne noise and vibration is narrowed by measuring these values directly. The field measurements were therefore used to reduce the uncertainty associated with ground-borne noise and vibration predictions, with a focus on outside to inside building predictions, and to ultimately inform the rail track design for the rail tunnel project.

References

1. International Standards: ISO 14837-1:2005 - Mechanical vibration – Ground-borne noise and vibration arising from rail systems
2. Weber, C., Karantonis, P.: Rail ground-borne noise and vibration prediction uncertainties. In: Proceedings of the 12th International Workshop on Railway Noise, Australia, vol. 139 - Notes on Numerical Fluid Mechanics and Multidisciplinary Design, pp. 307–318 (2018)
3. Nelson, P.: Chapter 16 low frequency noise and vibration from trains (Remington, Kurzweil and Towers). Transp. Noise Ref. Book, Butterworths (1987)
4. U.S. Department of Transportation - Federal Transit Administration: Transit Noise and Vibration Impact Assessment, FTA Report no. 0123 (2018)
5. The Association of Noise Consultants: Measurement & Assessment of Groundborne Noise & Vibration (2001)
6. Thompson, D.: Chapter 13 Ground-borne Noise (by Chris Jones). In: Railway Noise and Vibration - Mechanisms, Modelling and Means of Control, Elsevier, Amsterdam (2009)

A Simplified Model for Calculating the Insertion Gain of Track Support Systems Using the Finite Difference Method

Rupert Thornely-Taylor[1](✉), Oliver Bewes[2], and Gennaro Sica[2]

[1] Rupert Taylor Ltd, Saxtead Hall, Saxtead, Woodbridge, Suffolk IP13 9QT, UK
rmtt@ruperttaylor.com

[2] High Speed Two Ltd, 1 Eversholt Street, Euston, London NW1 2DN, UK

Abstract. In the procurement of track support systems for railways planned in the vicinity of receptors sensitive to groundborne noise and vibration it can be necessary to specify parameters of components of the system in order to ensure that its in-service performance will deliver the required level of mitigation against the effects of groundborne noise and vibration. In previous comparable schemes in the UK it has been the practice to specify the performance of the track system in terms of Insertion Gain (IG) defined as the difference between the vibration spectrum at a reception point with the mitigating track in place, and with a hypothetically "very stiff" track support system. It is necessary to have a means of verifying that a proposed system is likely to satisfy the requirements and it is desirable to simplify the method of evaluation of predicted track performance based on the stated parameters. A simplified algorithm for rapidly calculating the insertion gain of a two-stage (or single-stage) resilient track support system has been developed using the finite-difference method. This provides a tool for evaluating the predicted performance of a variety of proposed track support systems in order to compare it against the required specification.

Keywords: Insertion gain · Finite difference · Vibration isolation · Rail · Track

1 Introduction

The High Speed 2 project in the UK will run in tunnels beneath residential receptors in a number of locations, in a variety of lithologies. Prediction of the likely level of groundborne noise and vibration has been made using an empirical model at the Environmental Assessment stage [1, 2], and a three-dimensional finite-difference-time-domain (FDTD) model in subsequent design stages [3, 4].

The FDTD model includes detailed representation of the vehicle, track support system, tunnel and surrounding lithology, and may also include buildings. Track support systems were considered for the purpose of determining the characteristics of the track system that would deliver the project's ground borne noise and vibration criteria. This paper focuses on how these characteristics are specified to a track system supplier.

G. Degrande et al. (Eds.): *Noise and Vibration Mitigation for Rail Transportation Systems*, NNFM 150, pp. 385–393, 2021.
https://doi.org/10.1007/978-3-030-70289-2_40

A rudimentary acoustic performance specification for track would be to place responsibility for achieving numerical commitments relating to groundborne noise and vibration on the track supplier. The prediction of ground-borne noise and vibration in buildings above the tunnels is however complex. Noise and vibration are affected by factors which are outside of the control of the track supplier, such as the characteristics of the rolling stock, tunnel, ground and receiving building.

An alternative is to separate the acoustic performance of the track from the source-path-receiver system. This was achieved on the High Speed 1 [5] and Crossrail [6] Projects in the UK by specifying the acoustic performance of the track as an 'insertion gain', the difference, in decibels, between the amplitude of vibration at a constant receptor when one track support system is replaced by another. The insertion gain enables the acoustic performance of the track to be characterized without the need to specify the performance of the individual components of the track which contribute to the overall performance. Specification as an insertion gain will lead a track system supplier to design and test the key parameters that determine the acoustic performance of the track and then calculate the insertion gain of the track system from these key parameters. Comparison with the insertion gain specification will demonstrate that the track will either achieve, worsen or improve on the ground borne noise performance that was modelled in the full detailed model of the source-path-receiver system.

While the insertion gain of a track support system can be calculated using the full FDTD model by comparing the output for the case in question with a hypothetical case involving a "very stiff" track, it is desirable to simplify the method to remove parameters which cannot be influenced by the track supplier (such as ground conditions or the tunnel design) and to create simplified code for rapid evaluation of predicted track performance.

Previous models have employed 2.5D or 3D numerical models [7] or have [8, 9] approximated the track with beam, spring and mass elements coupled to an elastic half-space, or [10] with the track system represented by its impedance per unit length of track. Software for the calculation of vibration from surface and underground railways has been developed [11] which is capable of computing vibration level differences for different track systems, assuming rail pads as continuous spring elements and sleepers as a continuous mass. The method described here models track support as discrete pads, baseplates, blocks or sleepers as appropriate using a simplified finite-difference algorithm and calculates the insertion gain from the difference in the dynamic forces below the track supports.

2 Definition of Insertion Gain

In the railway context, the insertion gain of a track support system is defined, for example in DIN 45673–4, as the difference, in decibels, between the transmitted vibration forces with the track components relative to those without the component. It may be expressed as a spectrum (e.g. in 1/3 octaves). While this may refer to an actual track replacement, it may also be used to express the predicted vibration that occurs with one track system in comparison with a hypothetical base case.

3 Principles for Calculating an Insertion Gain

3.1 General

A rail vehicle travelling on a track is a multi-degree-of-freedom dynamic system, the input signal to which occurs at the contact patch between the wheel and the rail. Looking upwards from the contact patch there is, in simple terms, a spring-mass-spring-mass-spring-mass system (if low frequency bending modes for the vehicle body are ignored). These springs and masses are: the Hertzian contact spring, the wheel and bogie unsprung mass, the primary suspension, the bogie sprung mass, the secondary suspension and the vehicle body. Looking downwards from the contact patch there is a complex mass-spring-mass-spring-mass/spring/resistance in which the components are the rail, the rail foot pad, the sleeper or block, the sleeper support (ballast or a resilient boot/pad) and the foundation in the form of tunnel invert/lining/soil/rock. In a mass-spring-mass-spring system, the result is two coupled natural frequencies, one of which is higher than the higher of the individual mass-spring frequencies and the other is lower than the lower of the individual mass-spring frequencies.

An important example of what happens when a sequence of masses and springs are coupled together is the behaviour of the Hertzian contact spring, which is further discussed in the next section.

3.2 Key Parameters

The key parameters that affect the insertion gain of a track system include the geometry of the track system (such as fastener spacing), the mass and bending stiffness of the rails, mass of the sleepers (if present) and the acoustic stiffness of the rail or sleeper supports.

Other key parameters will affect the insertion gain, such as the mass and suspension of the rolling stock, the Hertzian contact stiffness at the rail-wheel interface, the tunnel and the ground below the track. These are parameters that the track system supplier will have no control over. When comparing track systems these parameters should be kept constant or their effects removed from the insertion gain.

4 Methodology

The methodology started with a process of simplification of the full 3-dimensional model in order to observe the effect associated with a series of simplifications. These included (1) representing the track as a single rail, (2), replacing the line of axles with a single axle, (3) omitting the bogies and vehicle bodies, (4) removing the influence of the lithology (5) removal of the influence of the tunnel lining and (6) assuming frequency-independent damping. These simplifications are identified and illustrated in Fig. 1 which shows the insertion gain calculated using the full source-path-receiver model compared to the simplified models.

A difficulty encountered with the use of a fully detailed model of the system arose when modelling the “very stiff” track support system as a reference case. The effective rail mass, M_r, in a resiliently supported system is much less than the wheel mass, and the natural frequency, f_o, associated with the rail-contact-wheel system is close to the mass-spring natural frequency of the rail and the Hertzian spring, $f_o = 1/2\pi\sqrt{(k_h/M_r)}$ where k_h is the Hertzian spring stiffness, and is typically over 700 Hz. This is above the frequency range of interest. In the case of very stiff rail support the natural frequency becomes close to that of the wheel and the Hertzian spring, $f_o = 1/2\pi\sqrt{(k_h/M_w)}$ where M_w is the mass of the wheel, and is well within the frequency range of interest, typically about 180 Hz. Consequently the spectrum of vibration transmitted into the foundation by a train on a very stiff track will have a prominent peak within the frequency range of interest and reduced levels of vibration above that frequency. This peak gives a false indication of the performance of the track support, since it is not attributable to the behaviour of the track support system in a relevant way. The means of overcoming this problem, which occurs due to the choice of a reference system that is effectively rigid and does not represent any actual real-world track support system, was to raise the stiffness of the contact patch in the rigid system to a hypothetical value many times greater than its true value, so that the contact-wheel natural frequency is raised to a value above the frequency range of interest. (The stiffness has to remain finite in order to furnish the required input force through compression of the Hertzian spring).

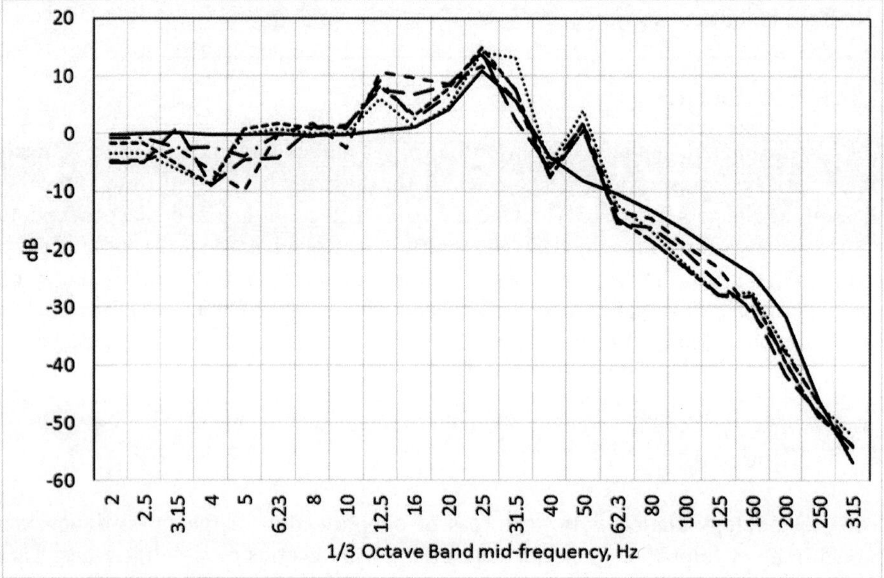

Fig. 1. The effect of simplification: — — Full model; – – one rail; — · — one rail, one axle; – – one rail, one axle, unsprung mass only; ···· one rail, one wheel, unsprung mass only, frequency-independent damping; —— one rail, one wheel, unsprung mass only, 1D model.

Having simplified the model in this way, it was also possibly to simplify the finite-difference code to fewer than 500 lines which could be ported to any convenient language. C++ was used in this case. The approach taken was to apply a relative displacement between the wheel and the rail and thus to compress the Hertzian contact spring at the wheel/rail contact patch, and to integrate the resulting force at the base of all the lowest resilient elements along the track, which was extended in length so as to be beyond the 10 dB down-points for the track decay rate. The relaxation damping algorithm used in the full 3D model was retained in order to be able to model frequency-dependent (or frequency-independent) damping.

4.1 The Simplified Finite-Difference Code

The finite difference method was employed in order to solve the bending wave equation for an Euler beam

$$EI\partial^4 u \Big/ \partial x^4 = -m'\partial^2 u \Big/ \partial t^2 \tag{1}$$

where E is the Young's modulus of the beam and I is the second moment of area of the section; m' is the mass per unit length; u is the vertical displacement; x is distance along the beam and t is time. This equation neglects rotary inertia and shear, which have an effect on the phase velocity of bending waves in a beam of less than 10% if the wavelength is greater than 6 h where h is the height of the beam.

The finite-difference approximation of Eq. (1) is obtained by dividing the rail into 2000 elements approximately 200 mm in length (the exact value depending on the fastener spacing to be studied) and for each element of the rail iterating the following two equations

```
ddx4=(((u[i+2]-u[i+1])-(u[i+1]-u[i]))-((u[i+1]-u[i])-(u[i]-u[i-1])))-
(((u[i+1]-u[i])-(u[i]-u[i-1]))-((u[i]-u[i-1])-(u[i-1]-u[i-2])))
```
(2)

```
v[i]=v[i]-ddx4*EIdx4
```
(3)

where `i` is the number of the rail element; `u[i]` is the vertical displacement of the rail and `v[i]` is the vertical velocity.

In the model described here, it is assumed that the rail is supported under every third element by a pad of stiffness `kpad`, below which is a block of mass `mblock`. Each block in turn is supported by a resilient boot of stiffness `kblock`.

Following the execution of lines (2) and (3) for each element `i`, the force exerted by the deflected rail pad at every third element is computed in terms of the incremental velocity as

```
dvr=-(u[i]-ub[i])*kpad/mrail/dx*dt
dvb=((u[i]-ub[i])*kpad-ub[i]*kblock)/mblock*dt
```
(4)

where `ub[i]` is the vertical displacement of the block, `dvr` is the incremental velocity of the rail and `dvb` is the incremental velocity of the block and `dt` is the finite difference time step.

Above the rail is the wheel, treated as a lumped mass on the Hertzian contact spring, with the lines

```
vw=vw-(uw-u[i])*kpatch/mwheel*dt
v[i]=v[i]+(uw-u[i])*kpatch/mrail/dx*dt
uw=uw+vw*dt
```
(5)

where `kpatch` is the stiffness of the Hertzian contact patch.

Damping is modelled using Boltzmann's relaxation model [12]

$$\sigma(t) = D_1\varepsilon(t) - \int_0^{\infty} \varepsilon(t - \Delta t)\varphi(\Delta t)d(\Delta t) \quad (6)$$

where $\sigma(t)$ is the stress at time t, $\varepsilon(t)$ is strain, $\varphi(\Delta t) = \frac{D_2}{\tau}e^{-\Delta t/\tau}$, D_1 and D_2 are constants and τ is a relaxation time.

Equation (6) may be replaced by a sum of three terms with different relaxation times, and by choice of values of D_1, D_2 and τ any desired frequency-dependence of the resulting damping effect can be achieved.

To implement Eq. (6) the lines in the code for the rail are

```
de=-(a[i]-ub[i])*kdamp/mrail/dx*dt
sp[i][0]=sp[i][0]*phi0+de*dt;
sp[i][1]=sp[i][1]*phi1+de*dt;                                    (7)
sp[i][2]=sp[i][2]*phi2+de*dt;
dv=dv-sp[i][0]*D2/tau0-sp[i][1]*D2/tau1-sp[i][2]*D2/tau2
```

```
v[i]=v[i]+dv                                                     (8)
```

with corresponding lines for the block.

The use of a length of rail of the order of 400 m means that the ends of the rail may be connected end-to-end to avoid reflections and maintain stability, and the length of the modelled rail is long enough for the bending waves to have decayed to insignificant amplitudes at the ends. The time step used was 1/131072 s, and the routine was run for 262144 iterations, i.e. 2 s. The run time is only a few seconds.

Excitation of the system is achieved by an initial condition at $t = 0$ consisting of a finite displacement of the Hertzian contact spring, i.e. a Heaviside step.

Output takes the form of the force under the boot below each block, as a time series. Each of these series is then subjected to FFT. As a result of the simplifications described above, the phase relationships between the forces below each block may differ from those in the full 3-dimensional case. To discard phase effects the squares of the forces at each block are summed to give a power-related total output as seen by the rigid support below the boots along the full length of the rail, expressed in decibels. This is subtracted from the corresponding curve for the rigid system to give the insertion gain (IG) in dB. The curve for the rigid system is found from the IG at frequencies well below the loaded track natural frequency, and has a straight line slope of-9 dB per octave.

5 Results

The results take the form of 1/3 octave band spectra of IG, which show the eigenmodes of a coupled two-stage system, taking account of the effect of damping, and the slopes to the insertion gain curves between and above the frequencies of the eigenmodes. For a single stage system, such as slab track with periodic rail supports, the code omits the lines for the blocks and boots. Examples are given in Fig. 2. In the results for two-stage systems, the characteristic loaded track resonance is clearly evident, and the coupled natural frequency of the blocks between the rail pads and the resilient boots also appears, reducing the insertion gain around 160 Hz–250 Hz in the examples shown in Fig. 2. This effect is absent in the case of the single stage slab track, but the enhanced attenuation that occurs above the coupled frequency in the two-stage systems is also absent.

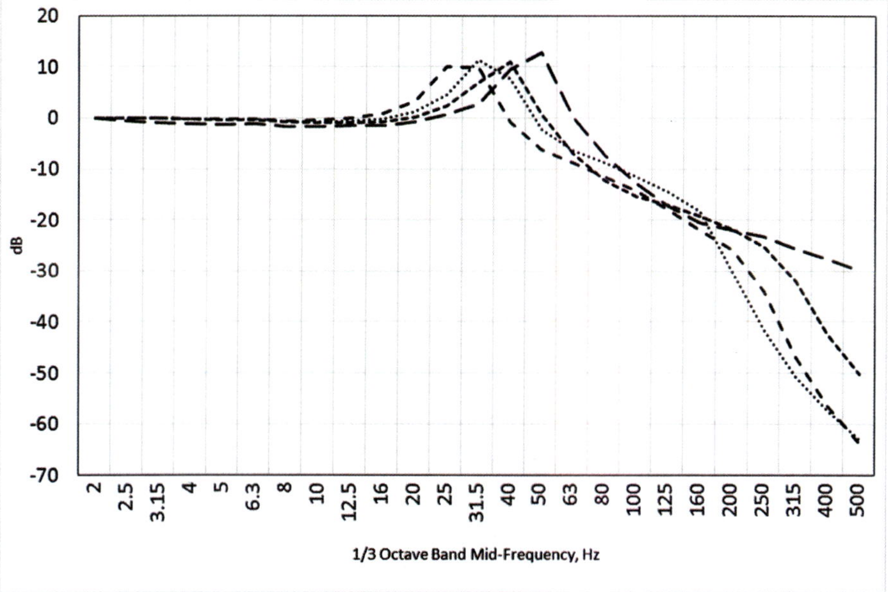

Fig. 2. Simplified IG of track support systems, dB re rigid track: – – slab track; - - booted block low mass; – – booted block low stiffness, high mass; ···· booted sleeper.

The properties assigned to each of the cases in Fig. 2 were as follows (Table 1).

Table 1. Input parameters for results in Fig. 2.

Track type	Rail pad stiffness MN/m	Rail pad damping (dimensionless)	Block mass kg	Boot stiffness MN/m	Boot damping (dimensionless)	Rail mass kg/m	Fastener spacing, m	Wheel mass kg
Slab track	40	0.15	–	–	–	60	0.65	950
Booted sleeper	150	0.2	210	31.5	0.2	60	0.75	950
Booted block low mass	100	0.2	45	28	0.2	60	0.6	950
Booted block high mass mass	200	0.2	120	14	0.2	60	0.6	950

6 Conclusion

The work has led to the development of an open-source computer code for rapidly calculating the insertion gain of a two-stage (or single-stage) resilient track support system. This provides a tool for evaluating the predicted performance of a proposed track support system in order to compare it against the required specification.

References

1. Bewes, O., et al.: Developing a prediction method for ground-borne sound and vibration from high speed trains operating at speeds in excess of 300 km/h. In: Proceedings of the 21st International Congress on Sound and Vibration, 13th–17th July 2014, Beijing, China
2. High Speed Two Ltd. Sound, noise and vibration: methodology, assumptions and assessment (route-wide), HS2 Phase One Environmental Statement, vol. 5 Appendix SV-001–000 (2013)
3. Thornely-Taylor, R.: The prediction of vibration, groundborne and structure-radiated noise from railways using finite difference methods–Part 1 –Theory. In: Proceedings of the Institute of Acoustics (2004)
4. Marshall, T., Sica, G, Fagan, N, Perez, D, Bewes, O, Thornely-Taylor, R: The predicted vibration and ground-borne noise performance of modern high speed railway tracks. In: proceedings, InterNoise 2015, San Francisco, California (2015)
5. Greer, R., et al.: Channel tunnel rail link, section 2. a review of innovative noise and vibration mitigation. In: Proceedings of the 8th International Workshop on Railway Noise, Buxton (2004)
6. Bewes, O., et al.: Track design to control railway induced groundborne noise and vibration from the UK's Crossrail project. In: Proceedings of the 23rd International Congress on Sound and Vibration, 10th – 14th July 2014, Athens, Greece
7. He, C., Zhou, S., Guo, P., Di, H., Zhang, X., Modelling of ground vibration from tunnels in a poro-elastic half-space using a 2.5D FE-BE formulation. Tunn. Undergr. Space Technol. **82**, 211–221 (2018)

8. Burgemeister, K., Greer, R.: Using insertion gains to evaluate railway vibration isolation systems. In: Proceedings, Acoustics, 3–5 November, Gold Coast, Australia (2004)
9. Jones, C.J.C.: Groundborne noise from new railway tunnels (Invited Paper). In: Proceedings of Internoise, Liverpool, UK, 30 July - 2 August 1996, Book 1, 421–426 (1996)
10. Guigou-Carter, C., Villot, M., Guillerme, B., Petit, C.: Analytical and experimental study of sleeper SAT S 312 in slab track Sateba System. J. Sound Vibr. **293**, 878–887 (2006)
11. MOTIV Homepage: https://motivproject.co.uk. Accessed 30 May 2019
12. Boltzmann, L.: Ann. Physik, Erg. Bd. 7, 624–654 (1876)

Evaluation of the Vibration Power Transmitted to Ground Due to Rolling Stock on Straight Tracks

Michel Villot[1(✉)], Catherine Guigou-Carter[2], Philippe Jean[2], and Roger Müller[3]

[1] MVexpert, 5bis Chemin Thiers, 38100 Grenoble, France
michel.jean.villot@gmail.com
[2] CSTB Health and Comfort Department, Acoustics, 24 Rue J. Fourier, 38400 St. Martin D'Hères, France
[3] Swiss Federal Railways, Infrastructure, Noise Abatement, Hilfikerstrasse 3, 3000 Bern 65, Switzerland

Abstract. In this paper, railway vibration emission is characterized from the vibration power transmitted to the ground. Train/track/ground systems can be considered as a source (train and track system) - receiver (ground) vibration system, where the power density transmitted is expressed from the force density applied to the ground and the ground mobility, both measurable on site. The power transmitted to the ground is first evaluated in a parametric study using a model coupling a numerical 2.5D ground model with a 2D model of the train/track system. The results obtained show the importance of ground properties on the force and power density transmitted. Finally, the approach is applied to an existing site where measurements have been carried out; however, the ground mobility as well as the combined rail-wheel unevenness, not measured, have been calculated in order to estimate force and power densities transmitted to the ground.

Keywords: Railway vibration sources · Ground-borne vibration · Vibration power

1 Introduction

An on-going project between CSTB and Swiss Federal Railways aims at understanding how and how much vibration power is transmitted to the ground due to train passby on straight ballasted track and identifying the most influential parameters (wheel/track unevenness, rolling stock, track system, ground type, …) through a parametric study.

The use of vibration power transmitted to the ground as the quantity characterizing railway vibration emission is something new; consequently, the first section of the paper defines the power transmitted from a vibration source to a receiver and applies this definition to railway lines. Now, understanding railway vibration emission, performing a parametric study or predicting a new situation from an existing one requires the use of a model which represents the source (train and track system here) and the receiver

G. Degrande et al. (Eds.): *Noise and Vibration Mitigation for Rail Transportation Systems*, NNFM 150, pp. 394–402, 2021.
https://doi.org/10.1007/978-3-030-70289-2_41

(ground under the track system). CSTB has developed such a model [1], consisting of a 2D source model (VibraFer software) coupled to a 2.5D ground model (MEFISSTO software); Sect. 3 of the paper describes these two models.

The power transmitted to the ground is then evaluated using the model in a parametric study (reduced here for the sake of brevity) where a common train/track system lies on typical homogeneous ground. Then the approach is applied to an existing site where measurements have been carried out. The parametric study and the application to a real site are presented in the last section (Sect. 4) of the paper and the results discussed.

2 Vibration Power

2.1 One DOF Source - Receiver System

In the case of a single DOF vibration source/receiver system as shown in Fig. 1, the power transmitted to the receiver is defined as:

$$W = \frac{1}{2}\,\mathrm{Re}\left(\tilde{F}_c \tilde{v}_c^*\right) \tag{1}$$

where $\tilde{F}_c$ and $\tilde{v}_c$ are respectively the frequency-dependent complex contact force and velocity (in narrow bands). A steady-state harmonic response is assumed in Eq. (1) and W represents the mean (Time-averaged) power.

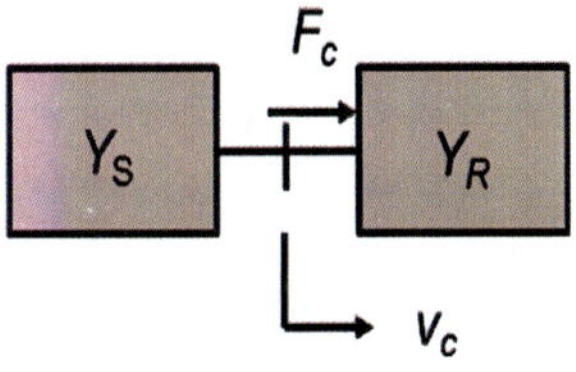

Fig. 1. Single DOF source/receiver system; contact force, contact velocity and mobility notation (YS and YR for the source and receiver mobilities respectively).

Introducing the receiver mobility $\tilde{Y}_R$ $\left(\text{ratio } \tilde{v}_c \middle/ \tilde{F}_c\right)$ and the RMS value of the contact force, Eq. (1) can be written as:

$$W = F_{c,rms}^2\,\mathrm{Re}\left(\tilde{Y}_R\right) \tag{2}$$

Both terms in Eq. (2) are real and can be expressed in 1/3 octave bands.

2.2 Application to Railway Lines

In the case of a railway line, the source can be represented as a line of uncorrelated excitation forces $\tilde{F}_{c,i}$, that are assumed to be equal in magnitude with random phase; the corresponding force density can be obtained on site, as suggested in [2], by measuring the ground velocity at a point away from the tracks during the train passby and the so-called ground line transfer mobility from the track to that point. Such source receiver configuration with N contacts and contact forces equal in magnitude but with random phase has already been theoretically studied [3]: the receiver line mobility at contact i is called effective mobility $\tilde{Y}_{R,i}^{\Sigma}$, includes the influence of the other contacts and is similar to the line ground transfer mobility used in [2] except that the ground response is measured on the excitation line as show in Fig. 2. $\tilde{Y}_{R,i}^{\Sigma}$ can also be calculated using a model (see below). The term effective mobility will be used throughout this paper.

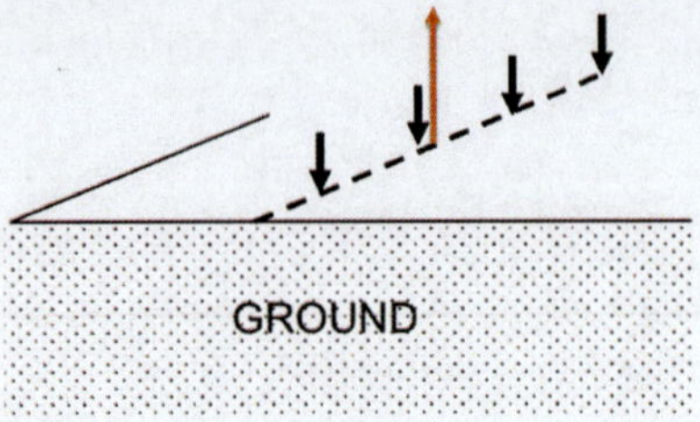

Fig. 2. Ground effective mobility obtained as velocity response at one contact (orange arrow) to a line of uncorrelated forces (black arrows); case of a half-space ground.

The transmitted power at each contact i can then be expressed as in Eq. (2), but using $\tilde{Y}_{R,i}^{\Sigma}$:

$$W_i = F_{c,rms}^2 \, Re\left(\tilde{Y}_{R,i}^{\Sigma}\right) \approx F_{c,rms}^2 \, Re\left(\tilde{Y}_{R,i}\right) \tag{3}$$

knowing from [3], that the real part of the effective mobility $\tilde{Y}_{R,i}^{\Sigma}$ can be approximated by the real part of the point mobility $\tilde{Y}_{R,i}$ at contact i. If the contact force is expressed as a squared force density per unit length (unit N^2/m) in Eq. (3), then the power obtained is also expressed as a power density (unit W/m).

3 Model Used

3.1 Ground Model

The ground model used, called MEFISSTO, is a 2.5D ground structure vibration interaction model [4] developed at CSTB, combining Boundary Elements (BE) for ground and Finite Elements (FE) for structures. The model is applied to different soil types homogeneous and infinite in the direction of the tracks. The model input parameters are the geometry of the configuration studied (layer interfaces and ground surface) and the material dynamic properties (Young's modulus, density, Poisson's ratio and loss factor); the excitation input is the force calculated by the source model (see section below). The model is used for calculating (i) the ground line mobility in the wave number domain (input data for the source model), (ii) the ground point mobility in the frequency domain (input data for calculating the power transmitted to ground), and (iii) the ground vibration field for comparison with measurements. However, the line of forces used in MEFISSTO does not correctly represent the real excitation, which is distributed over the width of the ballast. This width is taken into account by filtering the excitation force in the frequency/wavenumber domain as proposed by Auersch [5], the filter corresponding in the space domain to a convolution with a rectangular window of width equal to the width of the ballast. The power spectra therefore include this force filtering.

3.2 Railway Source Model

The railway source model, called VibraFer, is a 2D model based on the wave approach [6] also developed at CSTB. The rail is an infinite beam on a continuous multi-layered support representing rail pads, sleepers and ballast or other track components. The coupling to the ground is represented by the line mobility calculated for each frequency component in the wave number domain (ground surface velocity response to propagating stress waves of different wavelengths along a line) using MEFISSTO (see Fig. 3). The train is modelled as a set of springs and lumped masses representing the car body, bogie, wheel (unsprung mass) and connecting suspensions.

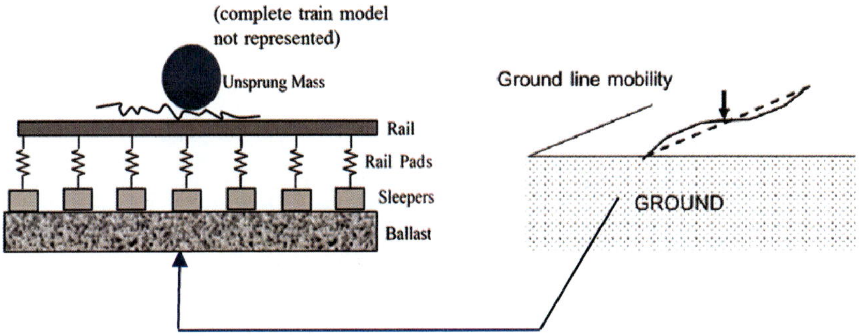

Fig. 3. VibraFer 2D model with ground line mobility calculated by MEFISSTO.

A known rail/wheel unevenness leads to the determination of the force applied to the rail, the contact force applied to ground, as well as the vibration velocity responses of the track components (rail, sleepers) and the ground surface. The moving load (quasi-static) excitation and the parametric excitation (discrete sleepers) are not modelled but can be taken into account through an equivalent combined rail-wheel unevenness back calculated using the model either from ground or axle box vibration measurements.

4 Applications

4.1 Parametric Study

Description of the Train/Ballasted Track/Ground Configuration: The source consists of Intercity (IC) and Regional (RZ) trains rolling on ballasted tracks as defined in the RIVAS project [7], deliverable D5.2 (properties not recalled here for the sake of brevity). Three speeds are considered (50, 100, 150 km/h). The ballast has the dynamic characteristics given in Table 1. Three half-space homogeneous grounds (stiff, medium and soft), with properties as indicated in Table 2, have been used.

Table 1. Track components characteristics (from RIVAS)

Component	Mass, kg/m	Dynamic stiffness, (MN/m)/m	Viscous damping, (MNs/m)/m
Ballast	520	1390	0.441

Table 2. Ground properties

	Young's modulus (MPa)	Loss factor (%)	Density (kg/m^3)	Poisson's ratio (-)	c_S (m/s)
Ground (soft)	50	10	1400	0.25	120
Ground (medium)	200	10	1600	0.25	224
Ground (stiff)	800	10	1800	0.25	422

The combined rail-wheel unevenness corresponds to an average curve suggested in RIVAS D5.2 in the case of relatively smooth wheel and rail (Fig. 4).

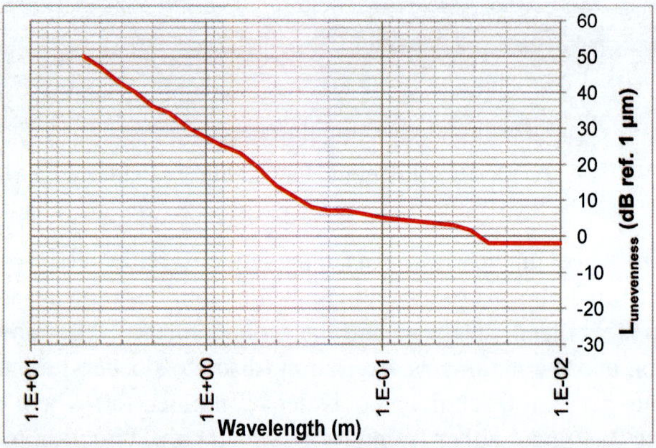

Fig. 4. Average combined rail-wheel unevenness (in 1/3 octave wavelength bands).

Contact Force and Power Densities Calculation: The ground mobility and contact force density have been calculated using MEFISSTO and VibraFer respectively and the power density has been obtained from Eq. (3). The power is given in Fig. 5, expressed in 1/3 octave band spectra.

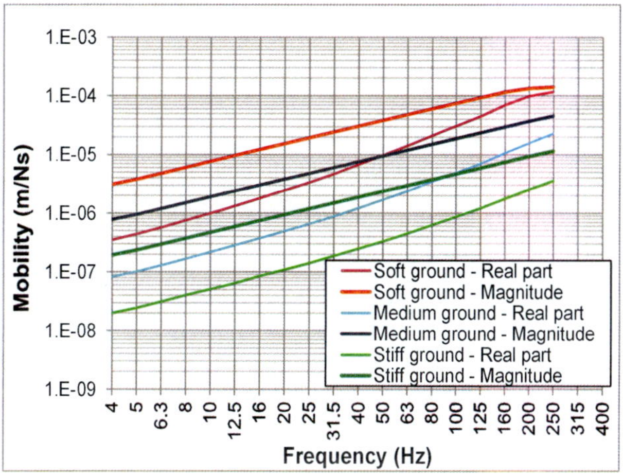

Fig. 5. Real part and magnitude of the ground input mobility (calculated).

For a given speed (100 km/h) the power density shows maxima at around 20, 40 and 80 Hz for the soft, medium and stiff grounds respectively, which corresponds to the track system - ground resonance. The train speed changes the power levels but does not affect their spectral shape. The force density (not represented for the sake of brevity) shows the same spectral shapes (Fig. 6).

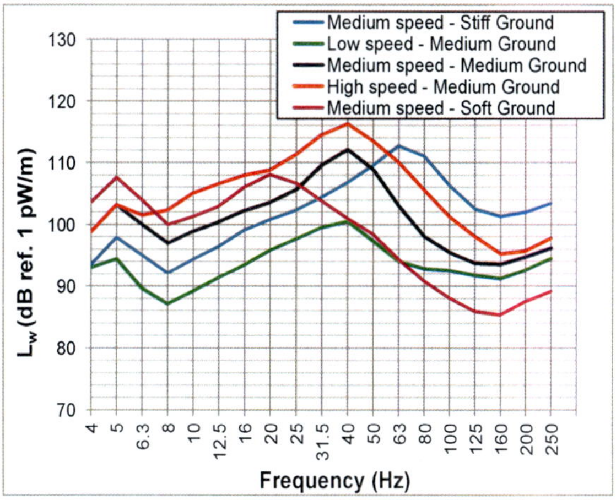

Fig. 6. Power density predicted using the model (ballasted track).

4.2 Application to an Existing Site (Pieterlen, Switzerland)

Description of The Train/Ballasted Track/Ground Configuration: Rolling stock and ballasted tracks were similar as in the parametric study. The ground consisted of a 3 m thick softer layer (Cs $\approx$ 150 m/s) on a stiffer ground (Cs $\approx$ 300 m/s). Ground vibration measurements at 5 m from the tracks and axle box measurements were available; the unknown combined rail-wheel unevenness has been back calculated using the source model from the measured axle box vibration levels. The ground mobility, not measured, has been calculated using MEFISSTO. Two situations without and with (soft) USP (dynamic stiffness around 0.15 N/mm^3 at 30 Hz) have been measured.

Ground Mobility: The calculated mobility (per 1/3 octave) is close to the one obtained for the medium ground used in the parametric study (see Fig. 5).

Combined Rail-wheel Unevenness: The unevenness has been back calculated from axle box measurements using the source model. The wavelength spectrum (per 1/3 octave) is given in Fig. 7 and compared to the average curve proposed in RIVAS.

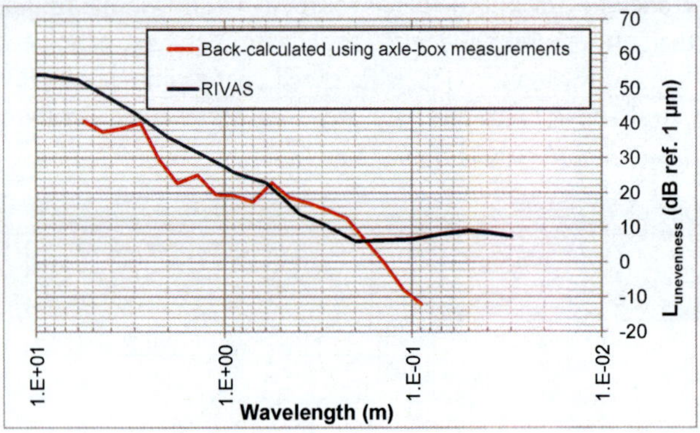

Fig. 7. Rail-wheel unevenness back calculated from axle box measurements.

Model Validation: The ground vibration levels at 5 m from the tracks and the USP insertion loss (in terms of ground velocity levels) have been calculated by the model. The results are good in terms of relative levels (insertion loss; see Fig. 8), but mediocre in terms of absolute vibration levels (correct in spectral shapes, but with discrepancies up to 10 dB; not shown for the sake of brevity), maybe due to the locally different properties of the soil directly under the tracks and the ballast.

Force and Power Density Levels: Contact force density and power density transmitted to the ground have then been calculated using the model. The results, given in Fig. 9, show that both the force spectrum and the power spectrum are monotonous.

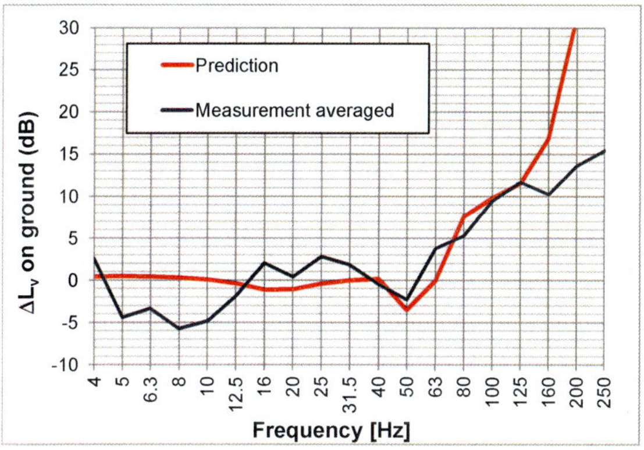

Fig. 8. Effect of under sleeper pads for the Pieterlen site; comparison between measured and calculated insertion losses (per 1/3 octave).

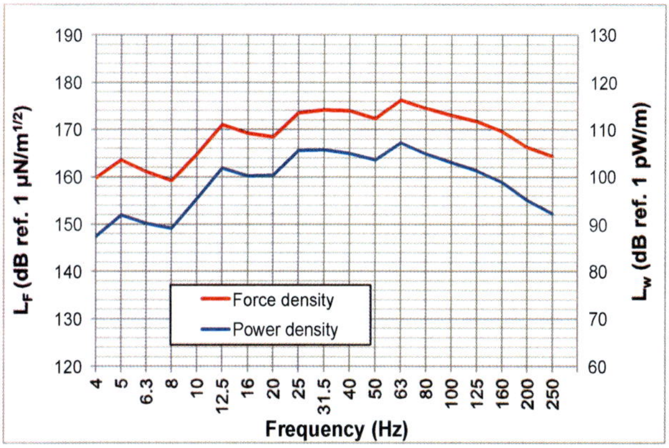

Fig. 9. Force and power densities (per 1/3 octave) predicted using the model and a wheel-rail unevenness deduced from axle box measurements (Pieterlen site).

5 Concluding Remarks

This paper shows the feasibility of estimating the vibration power density transmitted to the ground by a railway source from the contact force density and the ground input mobility, either measured on site or calculated using a train/track/ground model. The results show that:

- the power transmitted to the ground depends on the ground properties and is not, like for railway noise power, solely a characteristic of the railway source,
- combined rail-wheel unevenness can be back-calculated using the model from axle box (or ground) vibration measurements, thus indirectly taking into account moving load and parametric excitation,
- the model estimates well the effects of changes in the train/track/ground system considered, as shown in the model validation presented.

In the future, a better experimental validation of the approach could be undertaken through a dedicated field measurement campaign including ground vibration measurements close to the tracks. Further work on railway source characterization and transferability to other situations is under way, based on the same mobility approach.

References

1. Villot, M., et al.: Vibration emission of railway lines in tunnels – Characterization and prediction. Int. J. Railway Transp. **4**(4), 208–228 (2016)
2. US Department of Transportation, Federal Railroad Administration: High-speed ground transportation noise and vibration impact assessment (2012)
3. Petersson, B., Plunt, J.: On effective mobilities in the prediction of structure-borne sound transmission between a source and a receiving structure, Parts1 & 2, J. Sound Vib. **82** (4) 1982
4. Jean, P., Guigou-Carter, C., Villot, M.: A 2D ½ BEM model for ground structure interaction. Build. Acoust. **11**(3), 157–163 (2004)
5. Auersch, L.: The excitation of ground vibration by rail traffic: theory of vehicle-track-soil interaction and measurements on high-speed lines. J. Sound Vibr. **284**, 103–132 (2005)
6. Guigou-Carter, C., Villot, M., Guillerme, B., Petit, C.: Analytical and experimental study of sleeper SAT S 312 in slab track SATEBA system. J. Sound Vibr. **293**(3–5), 878–887 (2006)
7. RIVAS project. FP7 European project under grant agreement 265754 Nov 2013. www.rivas-project.eu

Vibration Excitation at Turnouts, Mechanism, Measurements and Mitigation Measures

Roger Müller[1(✉)], Yves Brechbühl[2], Stefan Lutzenberger[3], Samir Said[4], Lutz Auersch[4], Catherine Guigou-Carter[5], Michel Villot[5], and Roland Müller[6]

[1] Swiss Federal Railways, Infrastructure, Noise Abatement, Hilfikerstrasse 3, 3000 Bern 65, Switzerland
roger.ibmue.mueller@sbb.ch
[2] Swiss Federal Railways, Infrastructure, Project Engineering, Bahnhofstrasse 12, 4600 Olten, Switzerland
[3] Müller-BBM Rail Technologies GmbH, Robert-Koch-Str. 11, 82152 Planegg, Germany
[4] BAM, Federal Institute of Material Research and Testing, 12200 Berlin, Germany
[5] CSTB, 24 Rue Joseph Fourier, 38400 Saint-Martin-d'Hères, France
[6] Gleislauftechnik Müller, Riedlistrasse 33, 3123 Belp, Switzerland

Abstract. There is a strong need for cost-effective mitigation measures for turnouts. SBB has initiated a series of examinations using different methodologies to gain a deeper understanding of the excitation mechanisms at low frequencies, in addition to that obtained in the RIVAS project. To date it is not yet clear what constitutes a complete measurement data set that would enable understanding most of the vibration excitation mechanisms in turnouts. Increasing vibration at turnouts in comparison to normal track is observed for all measured frequencies. The different methodologies are presented in the paper. Under-sleeper pads (USP) are a cost-effective method to reduce vibration at frequencies above 63 Hz (1/3 octave), but there is probably no improvement for frequencies below 63 Hz. A first test of new frog geometry did not show relevant improvements in vibration emission in comparison to a reference frog geometry. Axle box acceleration measurements are an interesting method to identify defects in a turnout. A specialized measurement system of rail roughness could identify certain geometry problem areas for some frogs. Noise increases also are observed at turnouts for frequencies ranging between 80 to 1000 Hz. The use of railway source models to calculate contact forces for ballasted track and turnouts seems promising, in particular for understanding the influence of ground.

Keywords: Turnout · Switch · Vibration excitation · Vibration measurements · Noise measurements

1 Introduction and Methodology

A study of complaints at SBB showed that 32% of all complaints in terms of vibration are due to turnouts. Therefore, there is a strong need for cost-effective mitigation measures

G. Degrande et al. (Eds.): *Noise and Vibration Mitigation for Rail Transportation Systems*, NNFM 150, pp. 403–410, 2021.
https://doi.org/10.1007/978-3-030-70289-2_42

for turnouts in Switzerland. The mechanism for vibration excitation is not yet fully understood. Possible mitigation measures have been shown in a state-of-the art report in the RIVAS project [1]. A parametric study, also in RIVAS [2], about frog geometry improvements did not show relevant mitigation potential. A solution with stiff under-sleeper pads (USP) did not demonstrate relevant mitigation effects either [1, 3]. However, soft under-sleeper pads have shown promising improvements (Fig. 4) at high frequencies [4].

To date it is not yet clear which key parameters constitute a complete measurement data set in order to understand the predominant vibration excitation mechanisms at turnouts. This also means that it is difficult to measure the performance obtained [1].

Therefore, SBB initiated a series of examinations to gain a deeper understanding of the excitation mechanisms, especially for frequencies below 40 Hz. The strategy was first based on a vibration point of view by improving the measurement concept [5]. A validation measurement campaign [6] showed then that important effects in turnouts were still not understood sufficiently and that the most important parameters, such as the ground influence, were not controlled enough. However, axle box measurements could result in a better understanding of the dynamic effects in turnouts [7].

Next, the strategy was widened by applying a more holistic approach. This included learning from lifecycle-cost (LCC) improvements concerning dynamic defects in turnouts and deepening the understanding of turnout dynamics and measurements [8]. An important conclusion from the study was: There is a need for "cyclic interaction" of modelling, mitigation measures and measurements, see Fig. 1. With this procedure there is an increased chance to successfully develop mitigation measures. This idea is based on the standard EN50126-1 for innovation in Reliability, Availability, Maintainability and Safety (RAMS) for railways. To date a validated vibration model for turnout excitation is still lacking.

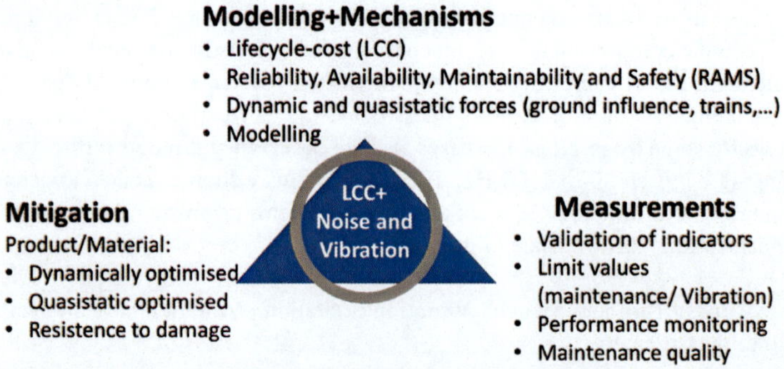

Fig. 1. Cyclic principle for developing innovations for Lifecycle cost (LCC), reliability, availability, maintainability and safety (RAMS) and vibration mitigation.

In this paper, the following methodologies are presented: state-of-the-art analysis of excitation mechanisms of turnouts, vibration measurements next to the track as well as in the track, noise measurements next to the track, frog geometry measurements, axle box acceleration measurements and modelling of vibration excitation.

2 Literature Study

The state-of-the-art was summarized in the RIVAS project [1]. The described studies contributed to a wider understanding of possible mechanisms. However, the wide range of experimentally determined insertion spectra from different turnouts could not be explained by means of statistical significance. It was then agreed that each turnout showed an individual behavior in terms of maintenance and ground conditions.

A further literature study demonstrated the importance of understanding dip mechanics (P1 and P2 forces [8]) to gain insight into the dynamics of the frog part of the turnout. The irregular dynamic forces resulting from the overrun of singular driving surfaces are determined not only by the driving speed but also to a large extent by the amplitude and length of the vertical wheel drop. To date it is not clear, if this concept is of use to define vibration excitation of the frog, how to measure P1 and P2 (onboard monitoring versus track measurements) or how to rate the results with respect to noise and vibration emissions. However, it is known from life-cycle-cost tests, that the forces in the frog are deduced by softer under-sleeper pads and by softer rail pads.

3 Measurement Results of Turnouts and Mitigation Measures

3.1 Turnout Measurement Techniques

A high-accurate innovative measurement of the vertical running geometry of turnouts was developed for SBB [5] and applied to a new or a worn wheel (Fig. 2) when passing over the turnout (especially the frog) [6]. This measurement procedure can contribute to detecting the maintenance status of the frog and defects (see e.g. in Fig. 2 at around 5.3 m a “defect” of 1 mm amplitude).

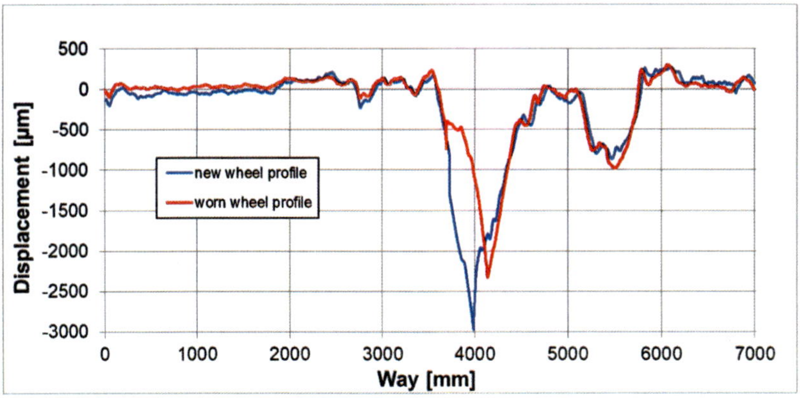

Fig. 2. Measurement of the trajectory of a new and hollow worn wheel at Hindelbank turnout 1 for the second new geometry of the frog [6].

Axle box acceleration (ABA) measurements have also been used to investigate the train-track interaction at turnouts and determine a relation between the impulse energy and defects within the turnout. The same information can be drawn from an analysis of vibration measurements next to the track [1]. ABA measurements as well as vibration measurements of the soil and the sleepers were performed for different velocities in Bülach [7]. ABA results are obviously correlated to defects within the turnout and therefore could contribute to the detection of defects in the turnout and in the frog.

The Federal Institute of Material Research and Testing (BAM) performed two campaigns in Switzerland: in 2011 at two sites with turnouts with and without stiff USPs (Le Landeron and Rubigen) and in 2013 at three sites (Selzach, Hindelbank, Berg) [9]. At some of these sites, ABA measurements were evaluated [10, 11]. The ABA measurements can generally be used to validate the prediction methods. Moreover, both ABA measurements and ground vibration measurements yield an increase of amplitudes with a factor of three for the turnouts in the range 15 to 80 Hz. There are indications (lower amplifications for softer soils) that the influence of soil and turnout must be considered together.

3.2 Vibration Measurements

For a total of 13 different turnout situations (10 turnouts, two of them with USPs) the vibration increase (frog versus normal track) was measured next to the track [6]. An increase of up to 10 dB or even more in the vicinity of turnouts can be seen over the whole frequency range (Fig. 3). It was concluded that the comparison of turnouts is still difficult and further measurement techniques must be developed [6].

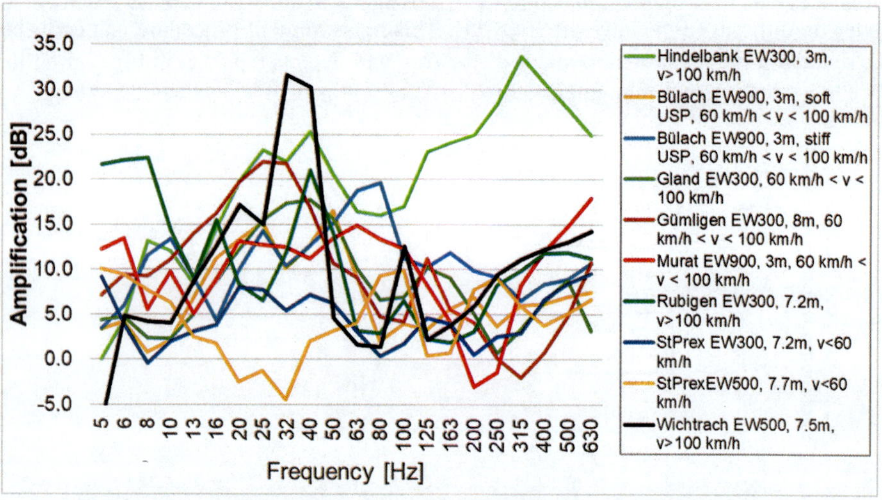

Fig. 3. Vibration measurements at turnouts and their amplification for different locations [6].

3.3 Vibration Mitigation Measures

SBB investigated a turnout with soft USPs showing promising results for high frequencies, but probably no improvement for frequencies below 63 Hz [4].

The vibration mitigation effect of soft USP in three turnouts was examined at two different sites in Switzerland (Bülach, Effretikon). The insertion loss for soft USPs (c_{stat} = 0.13 N/mm^3) in Effretikon [12] was examined by comparing vibration measurements before and after installing the soft USP, both in free field and in buildings. The results in the case of free field measurements are given in Table 1. These results suggest a substantial vibration reduction between 4 and 80 Hz. The ground influence was included by measuring the same turnout with and without mitigation measures.

Table 1 Vibration measurements of turnout T21 and T22 in Effretikon [12]: ratio (after substitution/before substitution) of the maximum rms vibration velocity (mean) for two free field points in direction orthogonal to the turnout's frogs.

Turnout	Distance from frog (m)	Ratio of max rms vibration velocity (mean)
T21	10.6	0.7
T22	6.7	0.6

The mitigation effect of soft USPs (c_{stat} = 0.11 N/mm^3) in Bülach was examined by comparison to a turnout equipped with semisoft USPs (c_{stat} = 0.21 N/mm^3). In Bülach soft USPs demonstrated a pronounced mitigation effect, as also demonstrated for normal track (RIVAS Deliverable D3.8), between 63 Hz and 250 Hz (Fig. 4). Considering the different maintenance status, as well as the fact that wooden sleepers were replaced with concrete sleepers, the mitigation effect at low frequencies (4 Hz–12.5 Hz) might be an artifact. The measurement campaign of Ziegler Consultant showed a different result, which is probably due to the measurement positions at a greater distance.

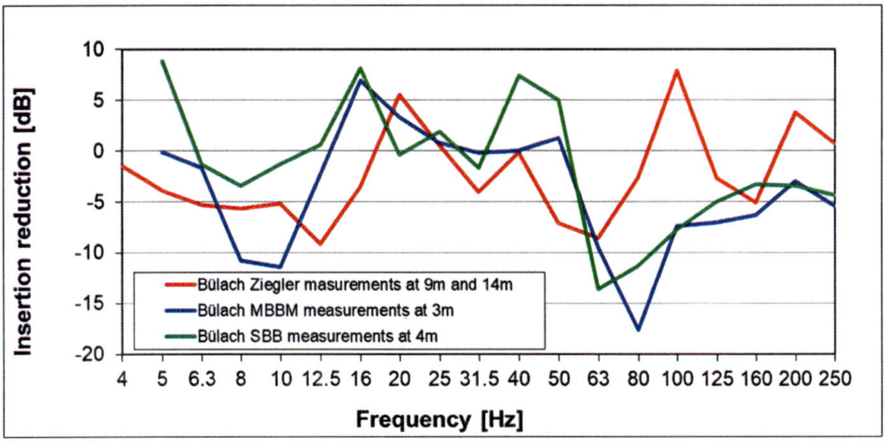

Fig. 4. Three measurement tests in Bülach for mitigation effect, reference turnout: stiff USP [4].

The vibration reduction of softer rail pads and two predefined frog geometries were studied in Hindelbank [6]. The new frog geometry 1 did not show relevant differences in vibration emissions to a reference frog geometry (Fig. 5). A mistake in the 2nd frog geometry happened because of the difficult manual grinding and is visible for 6–40 Hz. The soft rail pad resulted in a mitigation effect between 100–630 Hz.

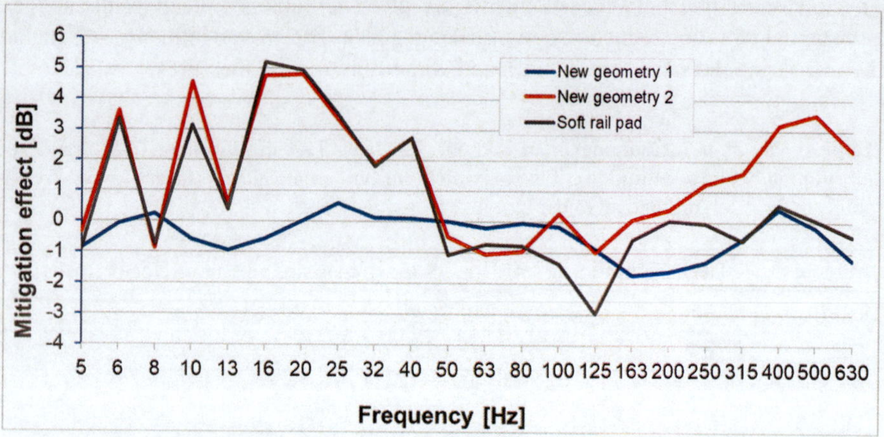

Fig. 5. Vibration reduction (new situations minus reference situation) in Hindelbank at frog in 3 m distance to middle of track [6].

3.4 Noise Measurements

Noise measurements showing additional noise emission for the turnout in comparison to the normal track for frequencies between 80 to 1000 Hz [6] are given in Fig. 6; the Murat and Wichtracht turnouts produced higher noise increases above 1000 Hz.

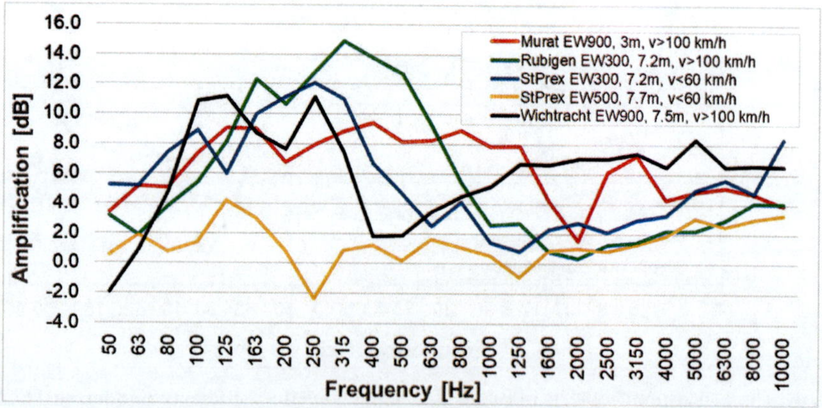

Fig. 6. Noise amplification of 5 turnouts in Switzerland [6] (measurement distances in legend).

4 Simulation of Vibration Emissions at Turnouts

The simulation consisted of evaluating a turnout singularity in terms wheel/rail unevenness and deducing the associated contact force using the same railway source models as the one presented in a companion paper [13]. Available measured data from commuter trains in a tunnel with slab track (STEDEF) passing over a turnout were used to calibrate the source model [14]. Then the corresponding unevennesses were back-calculated using the model from the measured tunnel invert vibration velocities without and with turnout. The unevennesses evaluated in the tunnel were then applied to a ballasted track placed on different ground types [13] and the contact forces calculated by the model. Figure 7 presents the calculated effects of three ground types on the contact force associated to standard ballast tracks and turnouts.

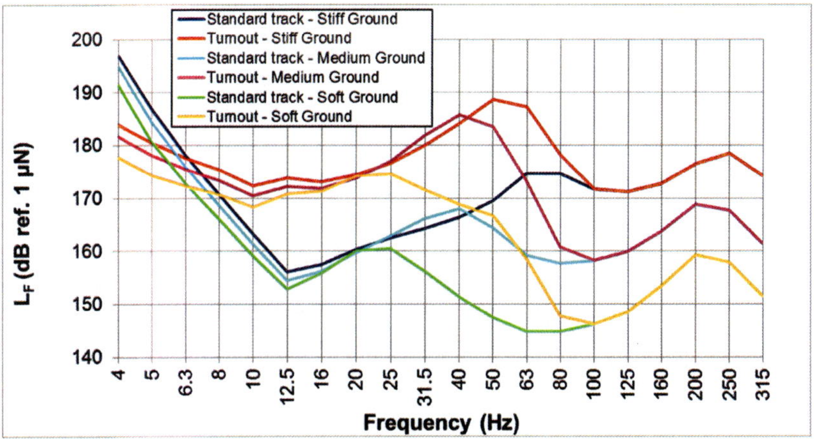

Fig. 7. Contact forces calculated for ballasted track and turnout for three ground types.

5 Conclusion and Outlook

It is not yet clear what constitutes a complete measurement data set that would enable understanding most of the vibration excitation mechanisms in turnouts. This furthermore means that it is difficult to measure and predict the performance of mitigation measures for turnouts. Vibration increase at turnouts in comparison to the normal track is observed for all frequencies measured. At frequencies above 63 Hz, USPs are a cost-effective method to reduce vibrations but there is probably no improvement for frequencies below 63 Hz. A first test of new frog geometries did not show relevant improvements in vibration emission in comparison to a reference frog geometry. Axle box acceleration measurements are an interesting method to identify defects in a turnout. A specialized measurement system of rail roughness could identify certain geometry problem areas for some frogs. Noise increases also are observed at turnouts for frequencies ranging between 80 to 1000 Hz. Simulations using a railway source model to express the effect of turnouts in terms of equivalent wheel-rail unevenness back-calculated from ground vibration measurements seem promising. Using the model allows then understanding for example the influence of the ground.

Outlook: the simulation approach [13] will be validated by SBB measured data (ground and axle box vibration) and extended with a wheel trajectory over the frog in the frame of an Austrian-Swiss project, which aims at developing cost-effective mitigation measures for turnouts for noise and low frequency vibration, based on measurements and modelling.

References

1. RIVAS Deliverable D3.6: Description of the vibration generation mechanism of turnouts and the development of cost-effective mitigation measures, 02.2013, Berne (Switzerland)
2. RIVAS Deliverable D3.12, Ground vibration from turnouts: numerical and experimental tests for identification of the main influencing sources/factors, 12.2013, Berne (Switzerland)
3. Müller, R., Nielsen, J.C.O., Nélain, B., Zemp, A.: Ground-borne vibration mitigation measures for turnouts: state-of-the-art and field tests, Notes on Numerical Fluid Mechanics and Multidisciplinary Design, vol. 126, pp. 547–554 (2015), Springer, Berlin Heidelberg (Germany)
4. Müller, R., Brechbühl, Y.: Eisenbahnerschütterungen - Neue Entwicklungen. Ziegler Symposium, Dübendorf (Switzerland), June 2016
5. Müller BBM: Erschütterungen bei Weichen, Mechanismen der Entstehung und Minderungsmaßnahmen, Report for SBB Nr. M122528/01, 31.3.2016, Munich (Germany)
6. Müller BBM Rail Technologies: Messtechnische Untersuchungen an Weichen SBB, Report for SBB Nr. M122528/03, 13.12. 2016, Munich (Germany)
7. Müller BBM: Achslagerbeschleunigungen, Untersuchungen und Auswertung der Messungen der SBB Mess- und Diagnosetechnik in Bülach, Munich (Germany) (2017)
8. Müller, R.: Übersicht zu Mechanismen, Diagnose, Abhilfemassnahmen gegen dynamische Schäden/Beanspruchung, Erschütterungen bei Weichenherzstück, Belp (Switzerland) (2018)
9. Auersch, L., Said, S.: Einfluss der Bodenparameter und von Weichen auf die Erschütterungsemission von Zügen. BAM report for SBB, Berlin (Germany) (2014)
10. Auersch, L.: Auswertung von Achslager-Beschleunigungs-Messungen zur Validierung der Erschütterungsprognose und zu den Auswirkungen von Weichen, Berlin (Germany) (2016)
11. Auersch, L., Said, S., Müller, R.: Measurements on the vehicle-track interaction and the excitation of railway-induced ground vibration. X International Conference on Structural Dynamics, EURODYN 2017, Procedia Eng. **199**, 2615–2620 (2017)
12. Basler & Hofmann: SBB Effretikon Nordkopf, Messungen zu Weichen mit Schwellenbesohlung, Report for SBB, 8.1.2018, Zürich (Switzerland)
13. Villot, M., Guigou-Carter, C., Jean, P., Müller, R.: Evaluation of the vibration power transmitted to ground due to rolling stock on straight tracks, IWRN13, Gent (Belgium) (2019)
14. Villot, M., et al.: Vibration emission of railway lines in tunnels – characterization and prediction. Int. J. Railway Transp. **4**(4), 208–228 (2016)

Predicted and Measured Amplitude-Speed Relations of Railway Ground Vibrations at Four German Sites with Different Test Trains

Lutz Auersch(✉)

Federal Institute of Material Research and Testing (BAM),
Unter den Eichen 87, 12200 Berlin, Germany
lutz.auersch-saworski@bam.de

Abstract. The present contribution evaluates four measuring series made by the Federal Institute of Material Research and Testing for the relations between train speed and ground vibration amplitudes. This experimental evaluation is supported by the simulation of the train passages at the different sites by using appropriate excitation mechanisms and forces as well as layered soil models which have been derived from impact measurements at each site.

Keywords: Train speed · Ground vibration · Excitation forces · Layered soils

1 Introduction

In many articles e.g. [1–5], an amplitude-speed relation for train-induced ground vibrations has been attempted. Generally, a single law is searched which can be used for many cases. This law is usually given as a power law $A \sim v_T^p$ between amplitude A (of particle velocity) and train speed v_T, and a power of $p \approx 1$ has often been stated.

Four measuring series have been performed by the Federal Institute of Material Research and Testing (BAM) which allow an evaluation for the amplitude-speed relation. These measurements are presented first before a theoretical analysis of the underlying mechanisms is undertaken.

2 Experimental Results

Site 1 has a homogeneous soil and two different trains have been measured. Sites 2, 3 and 4 have a layered soil. Sites 2 and 3 use the same train but have different soils whereas sites 3 and 4 have the same soil but different trains.

G. Degrande et al. (Eds.): *Noise and Vibration Mitigation for Rail Transportation Systems*, NNFM 150, pp. 411–419, 2021.
https://doi.org/10.1007/978-3-030-70289-2_43

2.1 ICE3 and Thalys Train on Homogeneous Soil (Site 1)

At site 1, an ICE3 train and a Thalys train have been measured with different speeds. The ICE3 run with 250 to 320 km/h while the Thalys run with 200 to 250 km/h. The corresponding ground vibrations are shown in Fig. 1 for the highest and lowest speed of each train. A certain mid-frequency range can be found in all spectra. For each increase of the train speed, this ground vibration component is shifted to the next one-third octave frequency, for the Thalys from 12–20 Hz to 16–25 Hz and for the ICE3 from 20–32 Hz to 25–40 Hz. At the same time the amplitudes of this ground vibration component increase, and the amplitude-speed relation can be evaluated for each of the two trains between $A \sim v_T^{1.5}$ and $A \sim v_T^3$. The same shift in frequency and increase of amplitude can be observed for the 8 Hz (10 Hz) peak of the ICE3 train. There are other ground vibration components at low and at high frequencies, but the mid-freqency component is dominant at the far field.

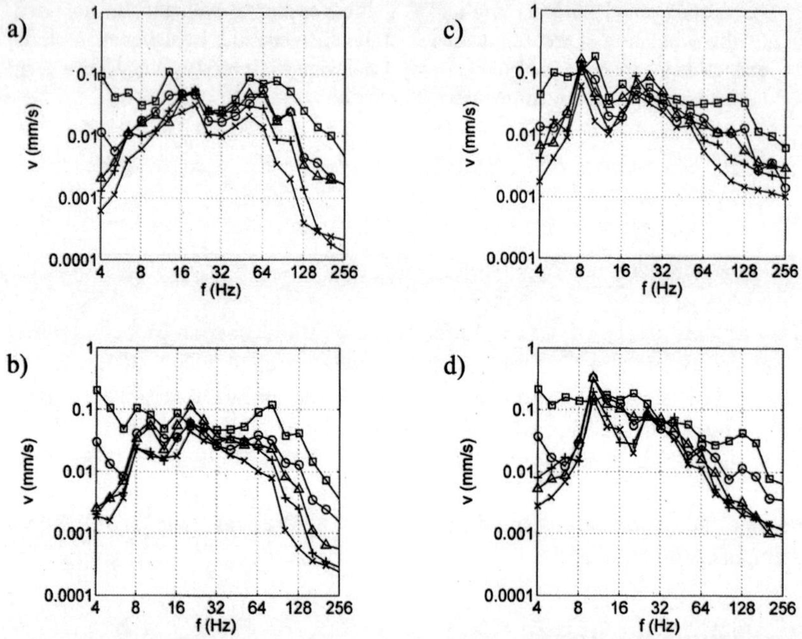

Fig. 1. Measured ground vibrations at site 1, one-third octave band spectra at distances $r = \square$ 4 m, ○ 8 m, △ 16 m, + 32 m, × 64 m, Thalys train with $v_T =$ a) 200, b) 250 km/h, ICE3 train with $v_T =$ c) 250, d) 320 km/h.

2.2 ICE Train on Two Layered Soils (Site 2 and 3)

A wider speed variation has been measured with the ICE (Intercity Experimental) train. Results are shown in Fig. 2 for site 2 and train speeds from 100 to 250 km/h and for site 3 and train speeds from 180 to 300 km/h.

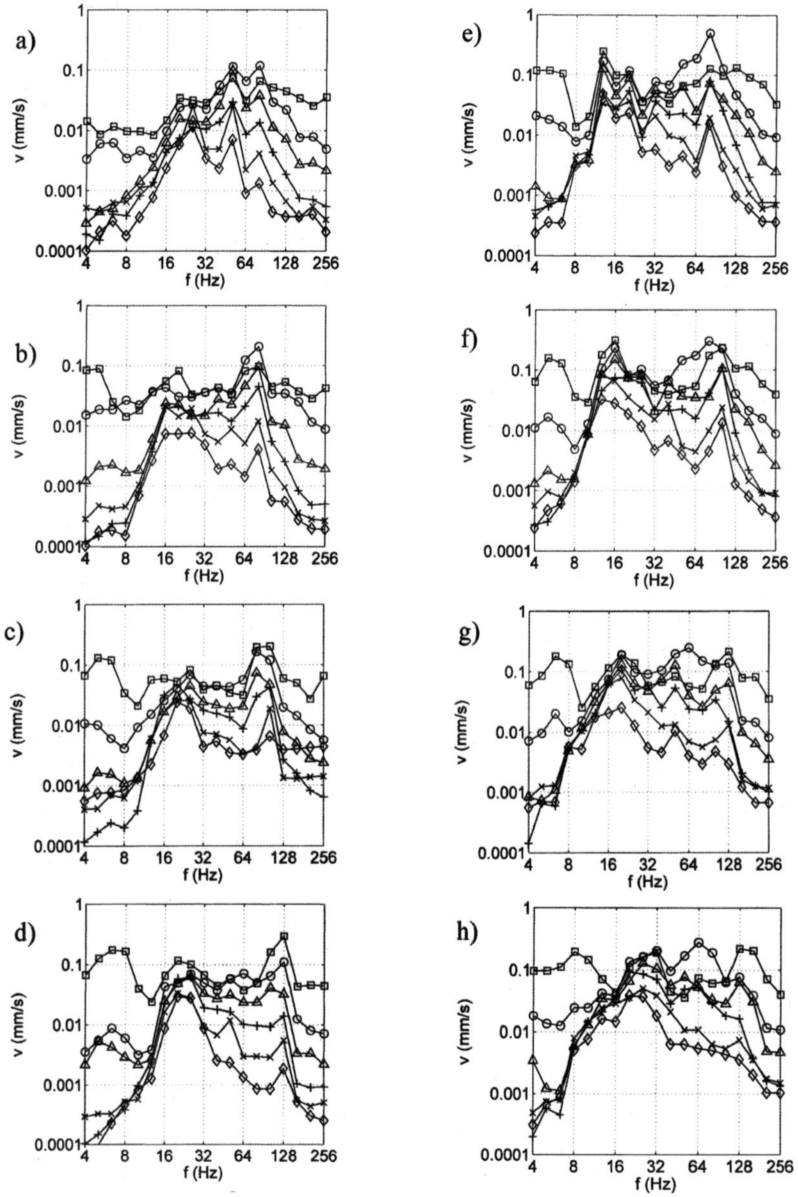

Fig. 2. Measured ground vibrations at site 2 and 3, one-third octave band spectra at $r = \square$ 2.5, $\bigcirc$ 5, $\triangle$ 10, + 20, $\times$ 30, and $\diamond$ 50 m, Intercity Experimental with v_T = a) 100, b) 160, c) 200, d) 250 km/h (site 2), and e) 180, f) 200, g) 250, h) 300 km/h (site 3).

There are a low-frequency component which is only observable in the near field, a high-frequency component which has a strong attenuation with distance, and a mid-frequency component which is dominant at the far field. The mid-frequency component shows different rules for the two measuring campaigns. For site 2 (Fig. 2 left), the mid-frequency amplitudes are increasing with train speed while the frequency is constant at 20–25 Hz. This is quite opposite for site 3 (Fig. 2 right), where the frequency is shifted from 12–20 Hz to 20–32 Hz and the amplitude is almost constant. So, the amplitude-speed relation is $A \sim v_T^0$ for site 3. The increase at site 2 can be measured in two ways. The amplitude at 20–25 Hz increases with $A \sim v_T^{1.0}$ whereas the specific component, which is shifted from 8–10 Hz to 20–25 Hz, increases faster with $A \sim v_T^{4.0}$.

The high-frequency component, which is due to the sleeper passage, is shifted from 50 to 128 Hz (160 Hz for 300 km/h) while the amplitudes increase for speeds from 100 to 160 km/h and only slightly at higher train speeds. The low-frequency component, which is the quasi-static response to the train passage observed in the near field, starts at 4–5 Hz for 160 km/h and widens to 4–8 Hz for 250 km/h and 4–10 Hz for 300 km/h. The maximum amplitude is almost constant, but the wider frequency band linearly increasing with train speed means also an increasing total quasi-static amplitude.

2.3 Passenger Train on Layered Soil (Site 4 ≈ Site 3)

Another measuring campaign has been performed at the same place (site 4 ≈ site 3) with a different train (a normal passenger train) and with lower train speeds from 63 to 160 km/h. Results have been evaluated for the locomotive and the passenger car in the mid of the train (Fig. 3).

At first, a similar mid-frequency component as for site 2 can be observed at a constant frequency of 12.5 Hz. The maximum amplitude can be found for 160 km/h and the increase/reduction follows the law $A \sim v_T^{2.0}$. The component can be followed with shifted frequency down to 5 Hz (4–6 Hz). These amplitudes increase more rapidly than $A \sim v_T^{4.0}$.

The sleeper passage component can also be observed at 32, 40 and 50 Hz for 63, 80 and 100 km/h (Figs. 3a, b, c, g, h). The quasi-static component appears at 4 Hz for 125 km/h (Figs. 3d, i) and at 5 Hz for 160 km/h (Figs. 3e, j). In general, the amplitudes of the locomotive are higher and more significant.

3 Interpretation and Analysis

3.1 Irregularities, Vehicle-Track Interaction and Excitation Forces

During the measurement campaign 4, wheelset accelerations have been measured in the test train. From these accelerations, amplitude-wavelength relations for the irregularities s have been found as $s \sim \lambda^{2.0}$ for the long wavelength alignment errors and $s \sim \lambda^{1.0}$ for the short wavelength roughness component [6]. The passage of the train over these irregularities generates forces that excite the ground vibration (Fig. 4a). In general, the irregularities $s(f = v_T/\lambda)$ must be multiplied by the vehicle-track transfer function $F/s(f)$ to get the excitation forces $F(f)$. At low frequencies (4–50 Hz), the transfer function is increasing with f^2 (proportional to the inertia of the wheelset). At higher frequencies (here at 90 Hz), there is a vehicle-track resonance where the transfer function has maximum amplitudes, and above, the amplitudes remain almost constant.

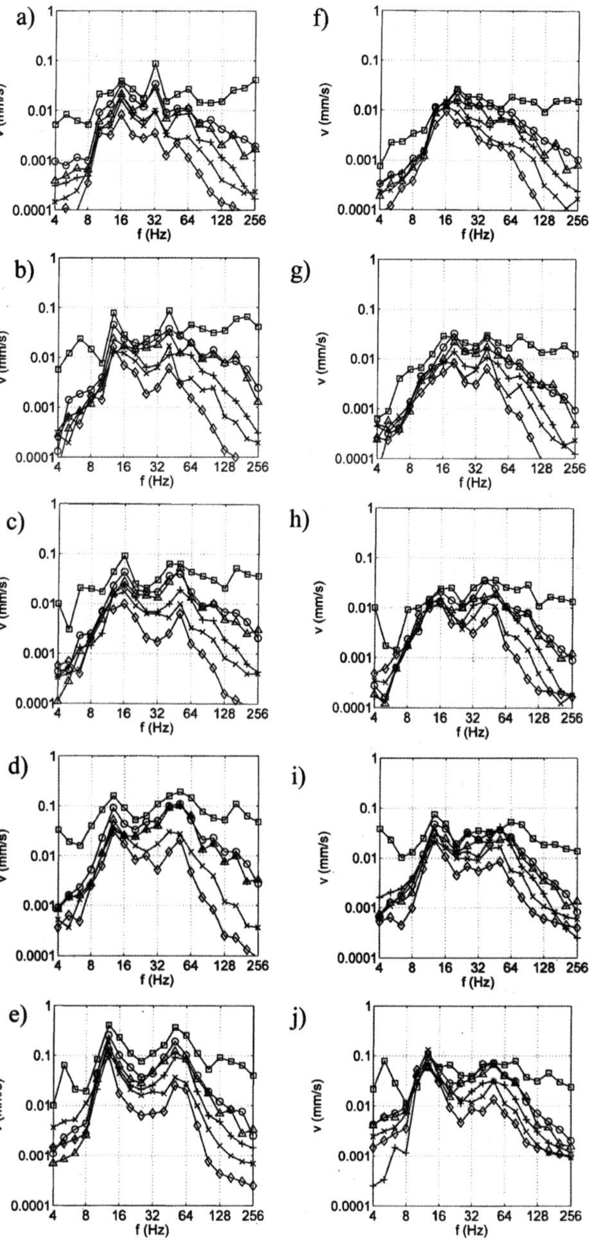

Fig. 3. Measured ground vibrations at site 4, one-third octave band spectra at $r = \square$ 2.5, $\bigcirc$ 7.5, $\triangle$ 12.5, $+$ 20, $\times$ 30, $\diamond$ 50 m, locomotive (left) and passenger car (right) with v_T = a,f) 63 km/h, b,g) 80 km/h, c,h) 100 km/h, d,i) 125 km/h, e,j) 160 km/h.

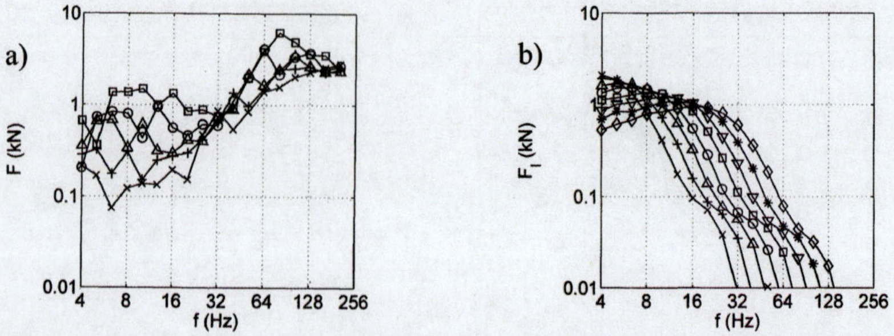

Fig. 4. Force spectra a) from irregularities, $v_T = \times$ 63, + 180, △ 100, ○ 125, □ 160 km/h, b) from axle impulses on the ground, $v_T =$ ◇ 100, × 125, + 160, △ 200, ○ 250, □ 320 km/h.

This can be observed for the sleeper-passage component s_S between 32 and 80 Hz where the corresponding forces $F_S(v_T) = F/s(f = v_T/d)s_S$ follow the transfer function [7].

The force spectra for each train speed (Fig. 4a) are rather constant because the increase of the transfer function and the decrease of the irregularities compensate. Nevertheless, the force amplitudes increase with train speed, either for a fixed frequency or for a fixed component, e.g. the sleeper passage between 32 and 80 Hz or the first wheel out-of-roundness between 8 and 16 Hz. The dependence on speed is quite different for the axle impulses which occur when a static axle load passes over the track (Fig. 4b) [8, 9]. With increasing train speed, the impulse spectrum, which has been calculated by the load distribution along the track [9], becomes wider, but the amplitude level is almost constant.

3.2 Transfer Functions of Homogeneous and Layered Soils

The transfer functions of the different sites have been measured by hammer impacts and are shown in Fig. 5. The homogeneous soil at site 1 (Fig. 5a) has a continuous increase of amplitudes $A \sim f^{1.0...1.5}$. At high frequency the transfer functions turn to constant and decreasing functions due to the material damping (and the track width).

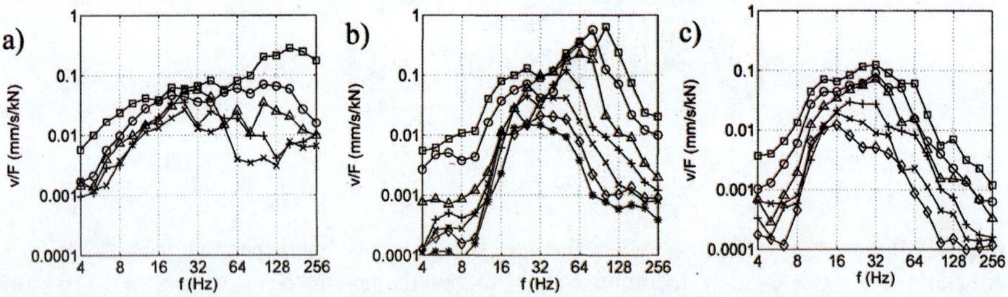

Fig. 5. Measured transfer functions at a) homogeneous site 1, $v_S = 170$ m/s, $r =$ □ 4 m, ○ 8 m, △ 16 m, + 32 m, × 64 m, b) at layered site 2 ($H = 5$ m) and c) site 3 ≈ site 4 ($H = 10$ m), $v_{S1} =$ 270 m/s, $v_{S2} = 1000$ m/s, $r =$ □ 2.5, ○ 5, △ 10, + 20, × 30, ◇ 40, ∗ 50 m.

In case of a layer over a stiffer half-space as at sites 2, 3, and 4, there is also a cut-off for the low frequencies. The cut-off or layer frequency is at $f_0 = v_{SI}/3H$ where v_{SI} is the shear wave velocity and H the thickness of the layer. Below the layer frequency, the amplitudes approach the lower amplitudes of the stiffer half-space. For frequencies above the layer frequency, the higher amplitudes of a softer half-space with v_{SI} are reached. On the other hand, a strong increase of amplitudes with frequency is found below the layer frequency. The layer frequency of site 2 is at 25 Hz (Fig. 5b) whereas it is at 12 Hz for site 3 and 4 (Fig. 5c).

3.3 Prediction of Train-Induced Ground Vibrations

A prediction of ground vibration and ground vibration components can now be done as the excitation forces and the transfer function of the soil are known [8–10]. Figure 6 shows examples for each site. The agreement of the prediction in Figs. 6a, b, c, d with the corresponding measurements in Figs. 2b, e, 3j, and 1c is quite good. The calculations have been done for different train speeds and different excitation components.

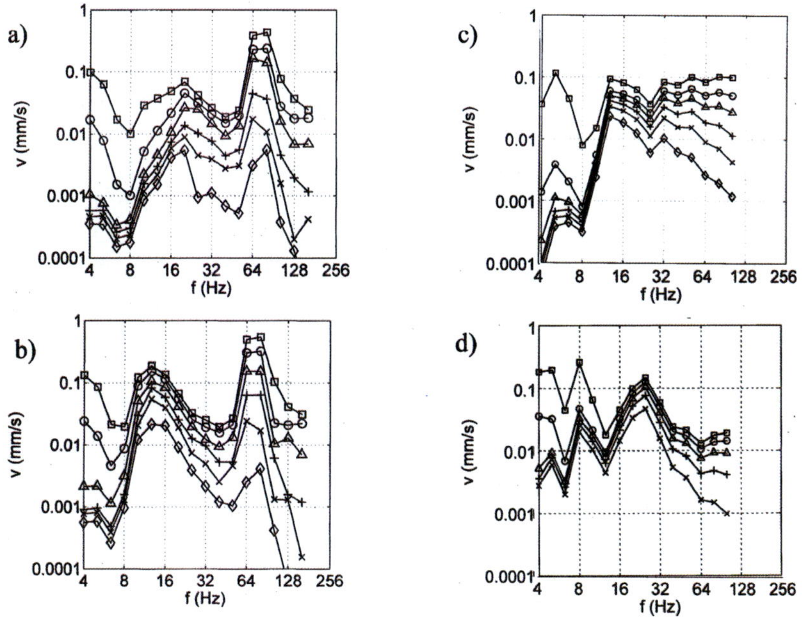

Fig. 6. Predicted ground vibrations for a) site 2, b) site 3, Intercity Experimental with v_T = 160 km/h c) site 4, passenger train with v_T = 160 km/h, r = □ 2.5, ○ 5, △ 10, + 20, × 30, and ◇ 50 m, d) site 1, ICE3 with v_T = 250 km/h, r = □ 4 m, ○ 8 m, △ 16 m, + 32 m, × 64 m.

If a low-frequency (long wavelength) irregularity is assumed and a fixed frequency is observed, it follows $s,F,A \sim v_T^{2.0}$ from $s \sim \lambda^{2.0}$ for all response quantities. This can be seen in Fig. 4a for the frequencies up to 20 Hz. For high frequencies with $s \sim \lambda^{1.0}$, the increase is also weaker as $s,F,A \sim v_T^{1.0}$. If a fixed component is considered, the whole transfer function from s to A must be regarded $A \sim A/F\ F/s\ s$ including the vehicle-track and the soil transfer function. At low frequencies usually both transfer functions are increasing so that the ground vibration amplitudes are strongly increasing with train speed.

The axle impulses yield the quasi-static component [8], but in a randomly inhomogeneous soil also a scattered part. As the force spectra are almost constant, the scattered response and its amplitude-speed relation is ruled directly by the transfer function of the soil. The strongest influence follows for site 2 where the axle-impulse component is below the layer frequency and therefore strongly increasing. A medium increase can be predicted for the homogeneous half-space at site 1 according to the medium and steady increase of the transfer function. Almost constant amplitudes follow for the increasing train speeds at site 3. The main part of the axle-impulse spectrum is above the layer frequency and follows the constant transfer function when the train speed is increased.

The passenger train at site 4 has a similar behaviour of the axle-impulse component, strongly increasing as at site 2, but it is at very low frequencies and not of importance. For low train speeds, other excitation components of shorter wavelengths are more important, for example the sleeper-passage component.

4 Conclusion

The different amplitude-speed relations of the different sites and soils clearly show that speed-dependency is not a single law. The amplitude-speed relation can be different for a certain frequency and for a certain component. It depends on the frequency (the distance), and most strongly on the soil at the measuring site. A homogeneous soil yields a different amplitude-speed law than a layered soil, and it is important if a vibration component is below or above the layer frequency. Scattered axle impulses yield power laws of $A \sim v_T^{0...2}$, whereas irregularities yield much stronger laws $A \sim v_T^{2...4}$ as the increasing transfer function of vehicle-track interaction are included.

References

1. Volberg, G.: Propagation of ground vibrations near railway lines. J. Sound Vib. **87**, 371–376 (1983)
2. Madshus, C., Bessason, B., Harvik, L.: Prediction model for low frequency vibration from high speed railways on soft ground. J. Sound Vib. **193**, 195–203 (1996)
3. Bahrekazemi, M.: Train-induced ground vibration and its prediction. PhD Thesis, KTH Stockholm (2004)
4. Lombaert, G., Degrande, G.: Ground-borne vibration due to static and dynamic axle loads of InterCity and high-speed trains. J. Sound Vib. **319**, 1036–1066 (2009)
5. Connolly, D., Alves Costa, P., Kouroussis, G., Galvin, P., Woodward, P., Laghrouche, O.: Large scale international testing of railway ground vibrations across Europe. Soil Dyn. Earthq. Eng. **71**, 1–12 (2015)
6. Auersch, L.: Theoretical and experimental excitation force spectra for railway induced ground vibration – vehicle-track soil interaction, irregularities and soil measurements. Veh. Syst. Dyn. **48**, 235–261 (2010)

7. Auersch, L.: The excitation of ground vibration by rail traffic: theory of vehicle-track-soil interaction and measurements on high-speed lines. J. Sound Vib. **284**(1–2), 103–132 (2005)
8. Auersch, L.: Ground vibration due to railway traffic – The calculation of the effects of moving static loads and their experimental verification. J. Sound Vib. **293**(3–5), 599–610 (2006)
9. Auersch, L.: Train induced ground vibrations: different amplitude-speed relations for two layered soils. J. Rail Rapid Transit **226**, 469–488 (2012)
10. Auersch, L.: Realistic axle-load spectra from ground vibrations measured near railway lines. Int. J. Rail Transp. **3**(4), 180–200 (2015)

Prediction of Ground-Borne Vibration Generated at Railway Crossings Using a Hybrid Model

Evangelos Ntotsios[1](✉), Dimitrios Kostovasilis[2], David J. Thompson[1], Giacomo Squicciarini[1], and Yann Bezin[2]

[1] Institute of Sound and Vibration Research, University of Southampton, Southampton SO17 1BJ, UK
e.ntotsios@gmail.com
[2] Institute of Railway Research, University of Huddersfield, Huddersfield HD1 3DH, UK

Abstract. Railway crossings are an important source of localized ground vibration. Models are required for identifying ways to tackle unacceptable levels of vibration from existing as well as future railway lines. Yet, the use of a prediction model that directly allows for the non-Hertzian wheel-rail contact dynamics and also includes three dimensional calculations of the ground response would be computationally expensive. In order to reduce the computational cost without affecting the accuracy of the predictions, a hybrid approach is proposed for the prediction of ground-borne vibration due to impacts at railway crossings. The approach combines the simulation of the vertical wheel-rail contact force in the time domain and the prediction of ground vibration levels in the far field using a linear wavenumber-frequency-domain approach. The proposed hybrid approach is used to investigate the influence of different vehicle speeds, crossing designs and wheel profiles on predicted ground vibration levels in the free field.

Keywords: Ground-borne vibration · Railway crossings · Hybrid model · Equivalent rail unevenness

1 Introduction

Turnouts (switches and crossings) are key elements of the railway system. A wheel passing over a railway crossing results in a high frequency impact load on the crossing nose (in the facing direction) or on the wing rail (in the trailing direction). The impact can cause wheel-rail contact fatigue and severe material degradation of the track components, but can also generate high levels of airborne noise and ground-borne vibration. This study addresses ground-borne vibration generated at a railway crossing that can cause annoyance (as either feelable vibration or re-radiated noise) or malfunctioning of sensitive equipment in nearby buildings.

In order to reduce the excessive levels of vibration that may be generated at railway crossings several empirical mitigation strategies have been proposed based on improved

G. Degrande et al. (Eds.): *Noise and Vibration Mitigation for Rail Transportation Systems*, NNFM 150, pp. 420–428, 2021.
https://doi.org/10.1007/978-3-030-70289-2_44

crossing designs [1], or installing vibration abatement measures on the track [2, 3]. Although the prediction of ground-borne vibration due to railway crossings is important for identifying ways to tackle unacceptable levels of vibration from existing as well as future railway lines, most studies that address wheel-rail interaction for complex rail profiles such as the crossing rail geometry are only focused on the prediction of wear and rolling-contact fatigue, rather than ground vibration. One reason is that the use of a prediction model that directly allows for the non-Hertzian wheel-rail contact and vehicle dynamics and also includes three dimensional calculations of the ground response would be highly computationally expensive [4]. In order to overcome the computational cost of such a complex model there have been a few studies in the literature that use a hybrid approach to predict ground-borne vibration due to impacts [5–7], but these focus mainly on singular defects such as joints and wheel flats.

A hybrid method has been proposed for the prediction of impact noise due to excitation by wheel flats [8], and this was recently extended for excitation due to wheel-rail contact at crossings [9]. Based on this, a similar approach is proposed here to address the ground-borne vibration from crossings. In the proposed approach the simulations are performed in two steps: (i) calculation of impact loads by a time-domain simulation of non-linear dynamic vehicle-track interaction, and (ii) prediction of ground-borne vibration in the far field using the MOTIV model [10] in the frequency domain based on excitation by an "equivalent roughness" spectrum. The proposed model is used to investigate the influence of different vehicle speeds, crossing designs and wheel profiles on predicted ground vibration levels in the free field.

2 Time-Domain Vehicle-Crossing Interaction Model

Simulation of the vibration due to the impacts requires accurate prediction of the magnitude and frequency content of the wheel-rail impact force. In the current work, dynamic vehicle-crossing interaction in the frequency interval up to 250 Hz is considered in the time domain using the VI-Rail software [11].

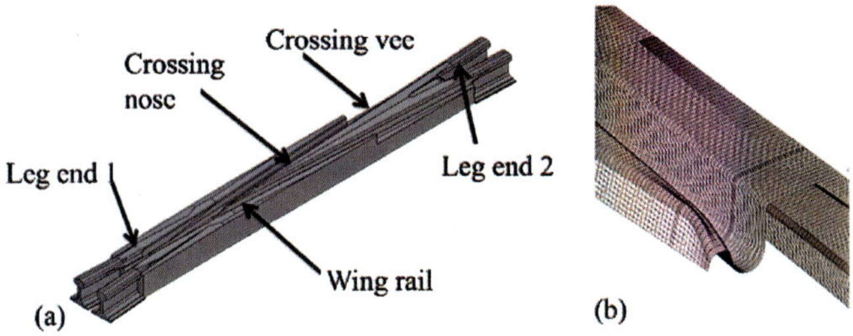

Fig. 1. The crossing geometry applied to the VI-Rail model: (a) overview; (b) crossing nose.

In VI-Rail the track is considered as multiple rigid masses supported by vertical, lateral and rotational springs. VI-Rail models the track as a series of 'flexible moving' subsystems, that represent the left/right rails and sleeper/ballast as sprung masses. Every axle of the vehicle is connected to such a subsystem, having homogeneous properties and being independent from one another. For the vehicle, the software can take into account a detailed model including flexible modes and non-linearities; however, for the purpose of this study the vehicle model is considered as linear. The model includes the detailed longitudinal and transverse rail profile at the crossing (Fig. 1) and calculates the wheel-rail contact forces along the rail due to the passage of the wheels.

3 Frequency-Wavenumber Domain Model for Ground Vibration

The MOTIV (Modelling Of Train Induced Vibration) model [10] is used for the calculations of the vibration levels of the ground. This is a semi-analytical linear model for calculating ground vibration from surface and underground railways and for assessing the performance of vibration countermeasures at the track and/or the train. For surface railways, that are studied here, the model uses the wavenumber-frequency domain formulation originally developed by Sheng et al. [12] and extended in [13] to take into account the traction variation across the track-ground interface and to allow the excitation and response of the two rails to be considered separately.

A longitudinally invariant ballasted track is coupled vertically to the surface of the ground. The track structure is modelled as multiple beams supported by vertical springs with consistent mass and the soil is modelled as a horizontally layered halfspace as can be seen in Fig. 2(a). Linear dynamic behaviour is assumed throughout. It is assumed that the wheels are always in contact with the rail and a linearized Hertzian contact spring is included between each wheel and the rail.

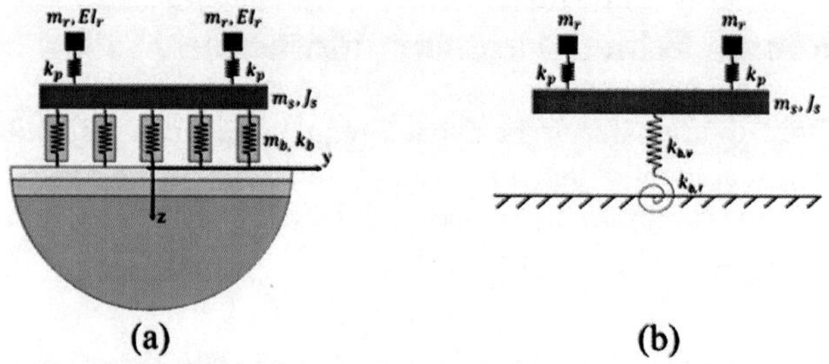

Fig. 2. Lateral view of the track multibody models for: (a) MOTIV; (b) equivalent VI-RAIL linear model.

For the ground vibration calculations only the vertical interaction forces are considered and only the vertical dynamics of the train are included. The vehicle is modelled as a linear 14-degree-of-freedom rigid-body system that includes the car, bogie and wheelset inertia and the primary and secondary suspension. The flexible modes of the car body are neglected. This is generally acceptable for ground vibration because the vehicle secondary suspension isolates the car body for frequencies above several times the suspension frequency (usually around 1–1.5 Hz). To enable analysis in the frequency domain each non-linear suspension is linearized, which is valid for small motion amplitudes. In general, the model assumes that the excitation is due to the passage of individual wheel loads along the track (quasi-static loading) and due to dynamic interaction forces caused by irregularities of the wheels and tracks (dynamic loading).

4 Hybrid Method for Predicting Ground Vibration at Crossings

Based on the equivalent roughness definition proposed in [8], a hybrid approach is developed to estimate the vibration levels of the ground from the contact forces calculated in the time domain. The idea is to find an equivalent roughness spectrum that, in a linear model, gives the same contact force spectrum as would be obtained with a non-linear model that includes the detailed rail profile and the discontinuity at the crossing. For this linear model, the receptances of the wheel and track must be the same as in the non-linear time-domain model but the non-linear properties of the contact are replaced by a linearized Hertzian contact spring. This roughness spectrum can then be used in a detailed linear vehicle-track-ground model, such as MOTIV [10], to predict the ground-borne vibration.

4.1 Equivalent Roughness

Within this framework, the first step is to calculate the transfer functions between roughness and contact forces given the vehicle, track and contact linear properties. By using a "moving roughness" approach, the contact force vector (at the two rails) $\mathbf{F}_0(\omega)$ for a unit roughness on one of the two rails $\mathbf{u}_0 = \left[\, 1\ 0 \,\right]^{\mathrm{T}}$ can be obtained as

$$\mathbf{F}_0(\omega) = [\mathbf{R}_{\mathrm{R}}(\omega) + \mathbf{R}_{\mathrm{W}}(\omega) + \mathbf{R}_{\mathrm{C}}(\omega)]^{-1}\mathbf{u}_0 \tag{1}$$

where $\mathbf{R}_{\mathrm{i}}(\omega)$ is a 2×2 matrix of receptances of the rail (i = R), wheel (i = W) and contact (i = C)

$$\mathbf{R}_{\mathrm{i}}(\omega) = \begin{bmatrix} R_{\mathrm{i},11}(\omega) & R_{\mathrm{i},12}(\omega) \\ R_{\mathrm{i},21}(\omega) & R_{\mathrm{i},22}(\omega) \end{bmatrix} \tag{2}$$

The rail and wheel receptances $R_{\mathrm{i},kk}$ are the point receptances and $R_{\mathrm{i},kl}$ with ($k = 1, 2$, $l = 1, 2$ and $k \neq l$) are transfer receptances from the contact point k on one rail to the contact point l on the other rail. These are calculated based on the properties of the time domain model presented in Sect. 2 by using a simplified linear track model to the VI-Rail track model that takes into account only the vertical dynamics of the track, as shown in Fig. 2(b), and vehicle. The contact receptance matrix $\mathbf{R}_{\mathrm{C}}(\omega)$ is a 2×2 diagonal matrix calculated from the axleload and the wheel radius that is constant for all frequencies ω. The inverse of the matrix in Eq. (1) can be considered as the dynamic stiffness matrix of the vehicle-track linear system at the contact points.

Next, the time histories of the vertical contact forces for the two wheel-rail contact points of a single wheelset are calculated using the non-linear VI-Rail model described in Sect. 2. These are transformed into the frequency domain and collected in the vector $\mathbf{F}_N(\omega)$ using the fast Fourier transform. The equivalent roughness component on the crossing rail $u_1(\omega)$ is obtained as

$$u_1(\omega) = \frac{F_{N,1}(\omega)}{F_{0,1}(\omega)} \tag{3}$$

where $F_{0,1}(\omega)$ is from Eq. (1). For the second rail, the rail profile is assumed smooth and the equivalent roughness component $u_2(\omega) = 0$.

4.2 Sensitivity of Equivalent Roughness Method to the Track Stiffness

The rail point receptance $R_{R,11}(\omega) = R_{R,22}(\omega)$ of the linear track model is shown in Fig. 3(a). This is calculated using the linear track model of Fig. 2(b) using the same properties as the VI-Rail model in Table 1. Three different cases of track stiffness are shown in Fig. 3(a); for the reference case the ballast stiffness has the value reported in Table 1, but for the other two cases it has half or double that value. The wheel point receptance $R_{W,11}(\omega) = R_{W,22}(\omega)$ is also shown. This is calculated using the linear model from [13] based on the vehicle properties reported in [14]. For validation purposes the rail receptance calculated with the MOTIV model (Fig. 2(a)) is also shown in Fig. 3(a). This uses the MOTIV track properties given in Table 1 that are given per unit length. It includes a ground with a 1 m deep soft soil surface layer with a shear wave (S-wave) speed of 190 m/s and dilatational wave (P-wave) speed 420 m/s and a stiffer half-space substratum with shear wave speed of 660 m/s and dilatational wave speed of 1350 m/s. Damping is included in both materials as a loss factor of 0.1. The receptance is similar to the one from VI-Rail up to 50 Hz.

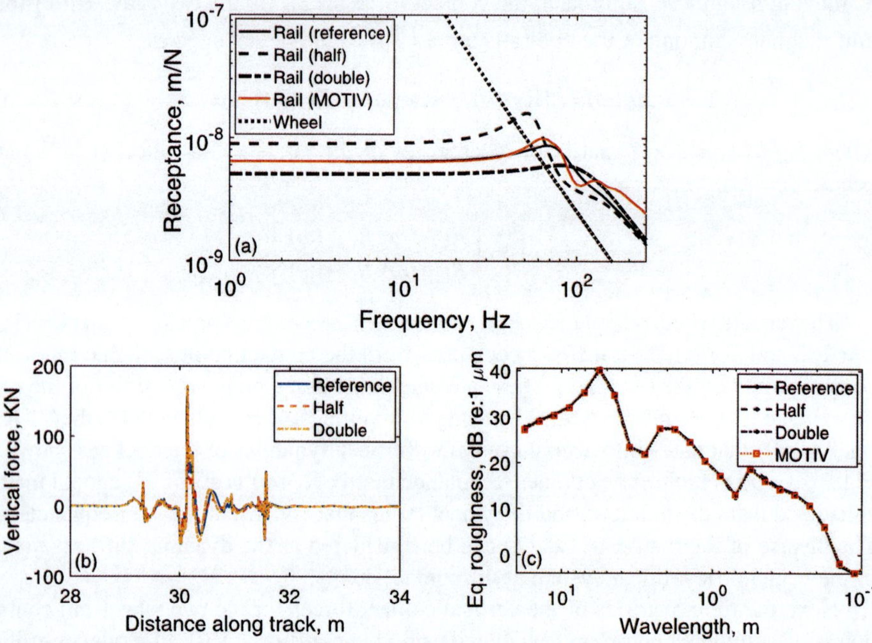

Fig. 3. (a) Rail and wheel receptance magnitudes; (b) contact forces; (c) equivalent roughness.

The contact forces at the crossing rail that are calculated with the VI-Rail model (using the properties reported in [14] for the vehicle and Table 1 for the track) and the equivalent roughness spectra in one-third octave band wavelengths calculated using Eq. (3) are shown in Fig. 3(b) and Fig. 3(c) respectively. The results are obtained for a train speed of 22.2 m/s (80 km/h) and for a contact spring stiffness of 1.11 GN/m.

Table 1. Track properties.

		VI-Rail	MOTIV
Rail	Bending stiffness per rail	–	1.5 MN/m^2
	Mass per unit length per rail	195 kg	195 kg/m
Rail pad	Vertical stiffness per rail	530 MN/m	530 MN/m^2
	Vertical damping coefficient	350 kN s/m	350 kN s/m^2
	Lateral stiffness per rail	430 MN/m	–
	Lateral damping coefficient	240 kN s/m	–
Sleeper	Mass	1200 kg	1200 kg/m
	Pitching moment of inertia	200 kg m^2	200 kg m
Ballast	Mass	0	0
	Vertical stiffness	390 MN/m	980 MN/m^2
	Vertical damping coefficient	480 kN s/m	480 kN s/m^2
	Lateral stiffness	37 MN/m	–
	Lateral damping coefficient	480 kN s/m	–
	Rotational stiffness	10 MN m/rad	–
	Rotational damping coefficient	10 kN m s/rad	–
	Ballast/ground interaction width	–	3.2 m

In Fig. 3(b) and Fig. 3(c) results are shown for the three different values of ballast stiffness; in Fig. 3(c) the case in which the MOTIV track model used for the calculation of the rail receptance is also shown. It can be seen that for the different cases of track stiffness, although the rail receptances in Fig. 3(a) are significantly different, the calculated contact forces are very similar and the estimated equivalent roughness spectra are almost identical. This suggests that the contact forces at the crossing contact points depend mostly on the stiff contact conditions and not on the less stiff track structure.

5 Results

The passage of the benchmark vehicle [14] is considered with speed 80 km/h in the facing move, from the switch panel towards the crossing panel of the through route. The track properties for the VI-Rail model and the MOTIV model are given in Table 1. For the ground model the properties that are used are the same as those reported in Sect. 4.2. Any further track irregularity outside the crossing region is not considered here even if it is allowed for by the simulation procedure.

Figure 4 compares the predicted ground vibration levels from the hybrid model for different speeds, worn wheel profiles and standardized crossing designs. The results are given in terms of the one-third octave spectrum of the predicted vertical velocity level of the ground surface, calculated at 16 m from the track due to the passage of the train over the crossing panel.

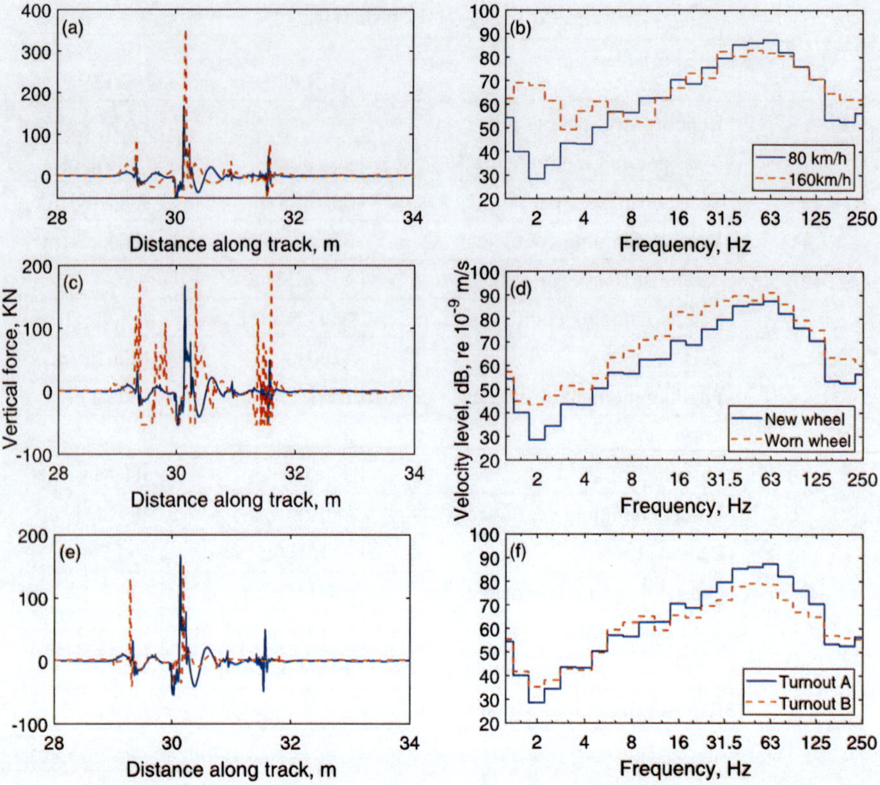

Fig. 4 Comparison of (a, c, e) contact forces and (b, d, f) ground vibration levels at 16 m from the track for different (a, b) train speeds; (c, d) wheel wear profiles; (e, f) crossing types.

The effect of the speed on the contact forces and the ground vibration levels shown in Figs. 4(a) and 4(b) is significant. An increase in vehicle speed from 80 to 160 km/h corresponds to an increase in the peak contact force. The vibration levels at low frequencies also increase due to the effect of the quasi-static response [13]. In addition, a decrease in the maximum vibration levels is observed at frequencies around the P2 resonance (at about 60 Hz). This is caused by the wheel having a smaller vertical drop at the crossing at the higher speed due to its inertia. Similar findings are reported in [6].

The influence of the worn wheel profile on the impact force magnitude and associated ground vibration level shown in Figs. 4(c) and 4(d) is also significant. For the selected worn wheel profile, it is shown that wheel wear can increase the contact forces and also the vibration levels in the free field for a wide range of frequencies.

The different crossing type considered in Figs. 4(e) and 4(f), also affects the contact forces and ground vibration level. However, although it seems to have increased the contact forces, the predicted vibration levels are lower between 16 and 125 Hz, where the dynamic contribution in the total response is significant.

6 Conclusions

A model for the prediction of ground-borne vibration at railway crossings has been developed. The hybrid approach based on the concept of an equivalent roughness spectrum is used in order to combine two already existing and validated prediction models: VI-Rail for numerical simulation of non-linear time-domain dynamic vehicle-track interaction, and the linear wavenumber-frequency domain model MOTIV for the prediction of ground vibration. The equivalent roughness concept can be properly applied to other wheel/rail interaction models (i.e. for rolling noise by using a flexible wheel model instead of a rigid one).

Based on the simulations it is concluded that the vibration levels are affected by train speed, wheel wear or crossing design in the higher frequencies of the spectrum, where the dynamic contribution in the total response is significant. At lower frequencies, the total vibration is influenced by the train speed due to its effect on the quasi-static component of the response. It is also affected by the selected worn wheel profile, which increases the low frequency components of the contact forces. These findings suggest that, when comparing predictions with measurements, uncertainties in wheel and rail profiles can influence the outcome considerably.

Acknowledgements. The financial support by the EPSRC under the grant EP/M025276/1 is gratefully acknowledged.

References

1. Kaewunruen, S.: Effectiveness of using elastomeric pads to mitigate impact vibration at an urban turnout crossing. In: Maeda, T. et al. (eds) Noise and Vibration Mitigation for Rail Transportation Systems. Notes on Numerical Fluid Mechanics and Multidisciplinary Design, vol. 118, pp. 357–365. Springer (2012)
2. Talbot, J.P.: Lift-over crossings as a solution to tram-generated ground-borne vibration and re-radiated noise. Proc. Inst. Mech. Eng. Part F J. Rail Rapid Transit **228**(8), 878–886 (2014)
3. Müller, R., Nielsen, J.C.O., Nélain, B., Zemp, A.: Ground-borne vibration mitigation measures for turnouts: state-of-the-art and field tests. In: Nielsen, J., et al. (eds.) Noise and Vibration Mitigation for Rail Transportation Systems. Notes on Numerical Fluid Mechanics and Multidisciplinary Design, vol. 126, pp. 547–554. Springer, Berlin, Heidelberg (2015)
4. Nielsen, J., Pieringer, A., Thompson, D., Torstensson, P.: Wheel-rail impact loads, noise and vibration: a review of excitation mechanisms, prediction methods and mitigation measures. In: Noise and Vibration Mitigation for Rail Transportation Systems - Proceedings of the 13th International Workshop on Railway Noise, pp. 16–20. Ghent, Belgium, September 2019
5. Nielsen, J.C.O., Lombaert, G., Francois, S.: A hybrid model for prediction of ground-borne vibration due to discrete wheel/rail irregularities. J. Sound Vib. **345**, 103–120 (2015)
6. Kouroussis, G., Connolly, D.P., Alexandrou, G., Vogiatzis, K.: Railway ground vibrations induced by wheel and rail singular defects. Veh. Sys. Dyn. **53**(10), 1500–1519 (2015)
7. Connolly, D.P., Galvín, P., Olivier, B., Romero, A., Kouroussis, G.: A 2.5D time-frequency domain model for railway induced soil-building vibration due to railway defects. Soil Dyn. Earthq. Eng. **120**, 332–344 (2019)
8. Wu, T.X., Thompson, D.J.: A hybrid model for the noise generation due to railway wheel flats. J. Sound Vib. **251**(1), 115–139 (2002)

9. Torstensson, P.T., Squicciarini, G., Krüger, M., Pålsson, B.A., Nielsen, J.C.O., Thompson, D.J.: Wheel–rail impact loads and noise generated at railway crossings - influence of vehicle speed and crossing dip angle. J. Sound Vib. **456**, 119–136 (2019)
10. MOTIV Software. https://motivproject.co.uk/motiv-software. Accessed 30 May 2019
11. VI-RAIL. https://www.vi-grade.com/en/products/VI-Rail. Accessed 30 May 2019
12. Sheng, X., Jones, C.J.C., Thompson, D.J.: A theoretical model for ground vibration from trains generated by vertical track irregularities. J. Sound Vib. **272**(3–5), 937–965 (2004)
13. Ntotsios, E., Thompson, D., Hussein, M.: The effect of track load correlation on ground-borne vibration from railways. J. Sound Vib. **402**, 142–163 (2017)
14. Iwnicki, S.: Manchester benchmarks for rail vehicle simulation. Veh. Syst. Dyn. **30**(3–4), 295–313 (1998)

FE Modelling as a Design Tool for Mitigation Measures for Railway Vibrations

Karin Norén-Cosgriff[✉], Birgitte M. Dahl, Joonsang Park, and Amir M. Kaynia

Norwegian Geotechnical Institute, NGI, P.O. Box 3930, Ullevål Stadion, 0806 Oslo, Norway
knc@ngi.no

Abstract. In Scandinavia, soft clay ground conditions often lead to complaints about annoying vibration in dwellings along railway lines. Since the effectiveness of vibration mitigation measures is highly dependent on the ground conditions, advanced methods such as FE-modelling usually need to be used when countermeasures are evaluated. However, for these methods to be suited for use as design tools, it is essential that the models are relatively easy to set up, have reasonably short computational times and do not require time-consuming post processing before the results can be obtained. It is therefore desirable to use 2D models to the greatest possible extent. In this paper we investigate the feasibility of using 2D models when calculating the effect of mitigation measures. We consider both low frequency vibrations from a railway line on soft ground with a screen as the mitigation measure, and more high frequency vibrations causing structure borne noise from a tunnel in clay with soft under-ballast mats as the mitigation measure. The results show that 2D models are suited as design tools as long as root mean square values of vibration velocity are considered, and the results are primarily used to compare different mitigation measures. Since geometrical attenuation is not correctly captured in 2D models, they should not be used to calculate absolute vibration values.

Keywords: Railway vibrations · Structure borne noise · Countermeasures · FE calculations

1 Introduction

In Scandinavia, soft clay ground conditions often lead to complaints about annoying vibration in dwellings along railway lines. Since the effectiveness of vibration mitigation measures is highly dependent on the ground conditions, more advanced methods usually need to be used when countermeasures are evaluated. However, for methods such as FE-modelling to be useful as design tools, it is essential that the models are relatively easy to set up, have reasonably short computational times and do not require time-consuming post processing before the results can be obtained. It is therefore desirable to use 2D FE-models to the greatest possible extent instead of e.g. full 3D or 2.5D.

In connection with planning for new infrastructure in Norway, $v_w = 0.3 - 0.6$ mm/s is usually suggested as limit values for vibration in new dwellings, where v_w is the

G. Degrande et al. (Eds.): *Noise and Vibration Mitigation for Rail Transportation Systems*, NNFM 150, pp. 429–436, 2021.
https://doi.org/10.1007/978-3-030-70289-2_45

frequency weighted RMS-1 s vibration velocity according to NS 8176 [1]. The lower value is used as a design goal, while the higher is the limit value. To meet this rather strict criterion, vibration mitigation measures with lime cement (LC) columns are often suggested in areas with soft ground conditions. Similarly, the Norwegian limit value for structure borne noise from traffic in tunnels, $L_{A,max,F} = 32$ dB, is strict compared to other countries [2]. Therefore, mitigation measures in the form of under-ballast mats are usually necessary in shallow tunnels in densely populated areas.

In this paper, we investigate the feasibility of using 2D models to calculate the effect of mitigation measures, both against low frequency vibration from a railway line on soft ground, and against high frequency vibration causing structure borne noise in dwellings above a railway tunnel in soft clay.

2 Methodology

The FE-analyses are performed in the frequency domain, using the software package COMSOL Multiphysics. The element types are quadratic Lagrange elements with the element size dependent on the frequency to maintain good resolution for higher frequencies. Table 1 shows the material properties of soil and track. The un-sprung mass of the wheel set is added as a weight on the rail, while the effect of the weight of the train is taken into account by increasing the elastic moduli of the ballast and the fill materials using the procedure described in [3].

Table 1. Material properties of soil and track used in the computations.

Material	Youngs module E (MPa)	Poisson's ratio ν	Density ρ (kg/m^3)	Loss factor η
Rail	205E3	0.28	49 (kg/m)	0.02
Sleeper	25E3	0.33	2300	0.02
Ballast	150–180	0.25	1900	0.3
Reinforcing layer	110–140	0.25	1900	0.3
Blasted rock	110–140	0.25	1900	0.3
Clay	55–330	0.49	1900	0.06
Lime cement columns	1670	0.33	2000	0.1
Soft under ballast mats (c_{63Hz} = 0.026 N/mm^3)	2.13	0.25	200	0.1

The models are excited by vertical dynamic unit forces at frequencies corresponding to the mid frequencies in the 1/3-octave frequency bands of interest. All models are equipped with specially-designed absorbing boundary domains, which prevent the waves from reflecting back from the outer boundaries of the numerical model [4].

Two different cases are studied. The first case is a railway line on soft ground with and without a 12 m deep lime cement (LC) screen between railway and neighbor dwellings. The results are compared with the results from a full 3D model describing a 50-meter long section of the track. In the 3D model, the forces are applied as point loads simultaneously at 8 contact surfaces corresponding to the positions of the wheels of a railway car with four axles on the left track. The root mean square value with 1 s integration time (RMS-1 s) is then estimated by moving the point loads with 1 m increments in the model, and calculating the average vertical vibration amplitudes from 30 consecutive results, corresponding to a train speed of 30 m/s. In the 2D models, which represents a cross-section of the 3D model, the forces are applied as line loads on both rails on the left track. The number of DOFs increases rapidly with the frequency, which limits the possibilities to perform calculations in full 3D for higher frequencies. Hence, the comparison has been limited to 8 Hz and 12.5 Hz in this study. Nevertheless, this can be defended by the fact that ground vibration from railway traffic on soft ground often displays a peak in the range 8 Hz–12.5 Hz. Further, the wavelength of the vibration wave in this frequency range is of the same order as the extent of mitigation measures, such as the depth of a LC-column screen.

The second case is a shallow concrete tunnel embedded in soft clay with and without ballast mats below the track. Vibrations from a tunnel subjected to dynamic loading (e.g. train-induced vibration) is a full 3D space problem. Nevertheless, in the higher frequency range which is of interest for structure borne noise, the use of 3D models is excluded because of computation time. In many practical cases, however, one can assume that the subsurface soil consists of homogeneous flat layers (isotropic or anisotropic). Furthermore, when some simplification is relevant e.g. ignoring the sleepers, and a tunnel being straight (linear), the mathematical description of the vibration phenomena can be simplified without any loss of generality, by using the so called 2.5D discrete formulation, e.g. [5]. Therefore, a 2.5D FE model is developed and implemented, the result of which is compared with the 2D results. In the 2.5D model the force is applied as two-point loads on the position of the two rails. The two models are shown in Fig. 1. The red lines illustrate the level where the results are compiled.

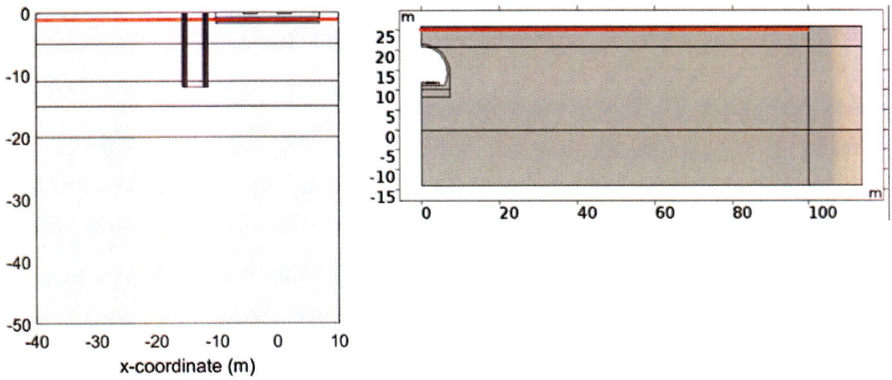

Fig. 1. Left: FE-model of railway line above ground with a 12 m deep LC-screen next to track. Right: FE-model of a shallow railway tunnel.

3 Comparison Between 2D and 3D Results

3.1 Effect of Poisson's Ratio

For soils with high Poisson's ratio, large oscillations in the vibration velocity amplitude may occur in 2D FE-models, which make it difficult to evaluate the effect of mitigation measures. This is especially true for low frequencies, for which the oscillations may extend into the area where the effect of mitigation measures is to be assessed. The oscillations are believed to be the effect of the vibration waves being confined to propagate only in two directions, causing near-field effects to persist longer than in reality. To verify this, FE-calculations are performed using a model with railway line on ground without any mitigation measures and the results are compared with calculations performed by use of an analytical model, where the ground is represented by the Green's functions of a layered half-space [6]. The analytical model describes the same soil profile as used in the FE-calculations but does not include the track bed and track. The comparison is performed both in 2D and 3D. Figure 2 shows the results from FE-analyses compared with results from the analytical model for different Poisson's ratios.

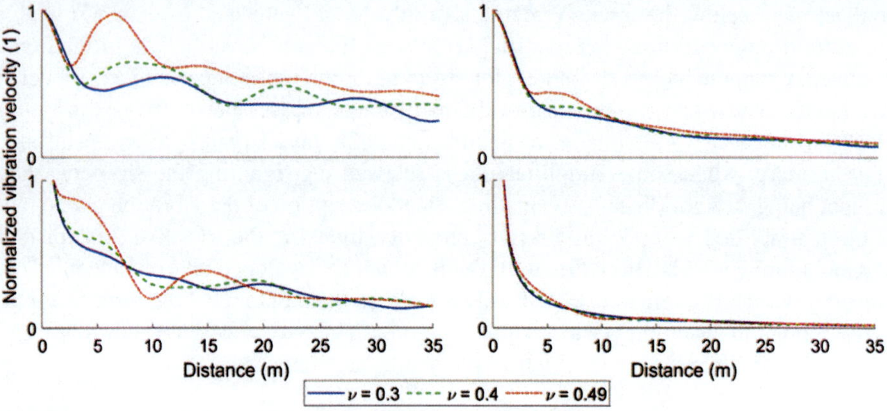

Fig. 2. Velocity at 12.5 Hz for different Poisson's ratio. Upper left: 2D FE-model, Upper right: 2.5D FE-model, Lower left 2D Analytical model, Lower right: 3D Analytical model.

The results are not identical since the FE-model and the analytical model are slightly different, but the resulting patterns are similar. It is clear from the figure that Poisson's ratio has a major impact on the oscillations, with a larger effect in the 2D model than in the 3D model. Further, Fig. 2 demonstrates that both for the FE-results and for the analytical solutions the oscillations persist for longer distances in the 2D models than in the 3D models. Therefore, the oscillations are due to differences in the extension of the near-field in the 2D and 3D models. This has also been reported in [7]. One practical solution to deal with the problem, which has been used in this study, is to average the calculation results by calculating the running average of the vibration velocity over the x-coordinate over a distance corresponding to the width of a typical building foundation, e.g. 8 m.

3.2 Effect of LC-Screen Against Low Frequency Vibrations

Comparisons are made between results from a 2D and a 3D model describing a situation with a 12 m deep LC-column screen at 7-meter distance from the track and a reference model without any mitigation measures. Figure 3 shows the ratio of the calculated vertical RMS-1 s vibration velocities for the models with a LC-screen to the reference model without mitigation measures as a function of the distance from the track. In Fig. 3 the vibration values from both the 2D and 3D analyses have been calculated as the running average over 8 m in the x-direction excluding the area in between the two LC-columns. Note that the 2D plots have no y-dimension.

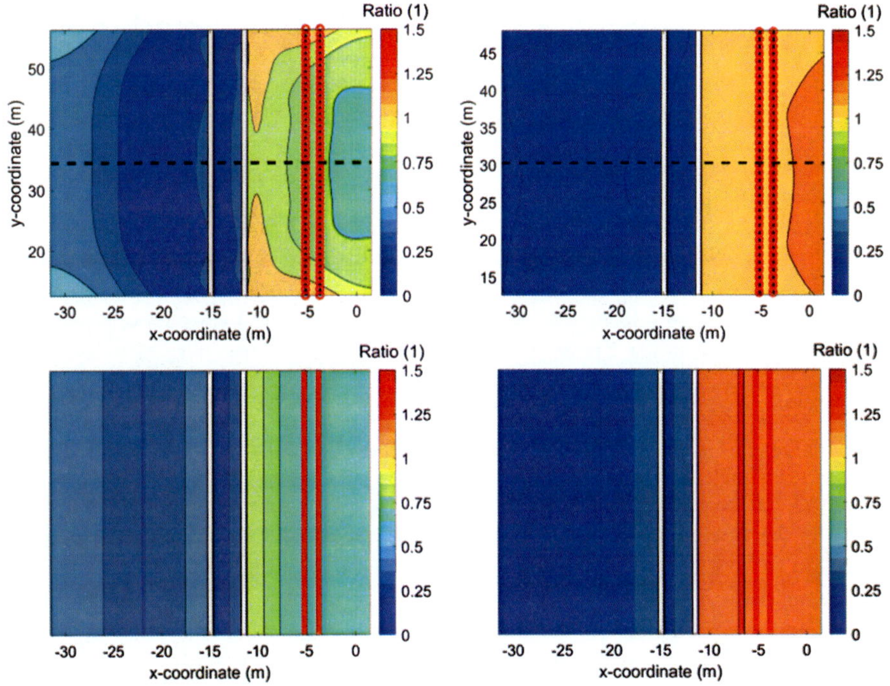

Fig. 3. Ratio of calculated vertical vibration velocities for models with LC-screens to the reference model without mitigation measures. Upper left: 3D model at 8 Hz, upper right: 3D model at 12.5 Hz, lower left: 2D model at 8 Hz. upper right: 2D model at 12.5 Hz. The position of the screen is shown with light grey lines and the two rails with red lines in 2D and red circles in 3D.

Results from the 2D and 3D models at the cross section marked with a black dashed line in Fig. 3 are compared in Fig. 4. The 3D results are shown as RMS-1 s values calculated as described in Sect. 2. To compare the results from the 2D and the 3D model, the results are normalized with the maximum vibration velocity for the reference models in 2D and 3D, respectively. The location of the LC column is marked with vertical grey lines. As expected, the difference in geometrical attenuation affects the results. However, when the aim of the study is to compare the effects of different mitigation measures, this difference is not of major concern as long as the models show similar effects of the evaluated mitigation measure. Figure 4 right shows the ratio of calculated vibration velocity for the models with LC-screens to the reference models without mitigation measures. The results in Fig. 3 and Fig. 4 show very good agreement between the 2D model and the RMS-1 s values from the 3D model. Note that the model describes a double track situation with traffic on the left track only (Fig. 1). Therefore, the results from the reference models are not perfectly symmetric along the x-axis.

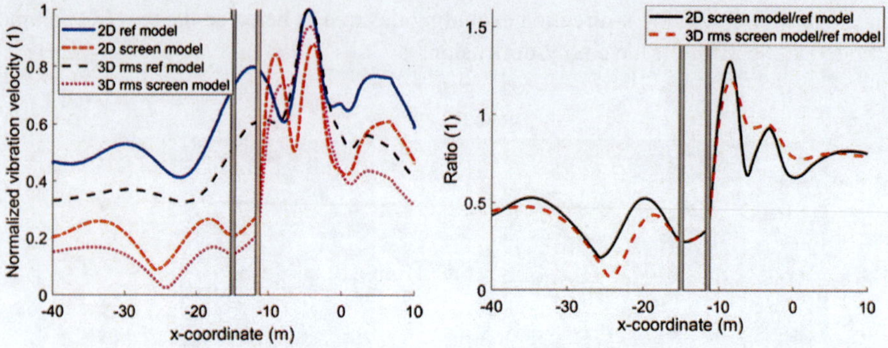

Fig. 4. Comparison between 2D and 3D results for the reference model and the model with a LC-column screen at 8 Hz. Left: Calculated vibration velocity normalized with maximum vibration velocity for the reference models. Right: Ratio of calculated vertical vibration velocities for the model with LC-screens to the reference model without mitigation measures.

3.3 Effect of Under-Ballast Mats Against Structure Borne Noise from Tunnel

Figure 5 left compares 2D and 2.5D models for a concrete tunnel embedded in clay (Fig. 1). Vertical vibration velocities are shown at 63 and 125 Hz. The results for a situation without tunnel and the source located on top of the soil is also shown at 125 Hz for comparison. The results are normalized with maximum vibration velocity for each model and averaged over the distance of a typical building foundation (8 m) to avoid local fluctuations. The results for the model with a concrete tunnel in clay show good similarity between 2D and 2.5D, and clearly better than the model without tunnel. This can be explained by the fact that the stiff concrete construction distributes the load over a larger area, hence acting more like a line source. Soft under-ballast mats are introduced under the ballast and Fig. 5 right compares the ratio of vibration velocity in a model with and without soft under ballast mats. The agreement is good between 2D and 2.5D.

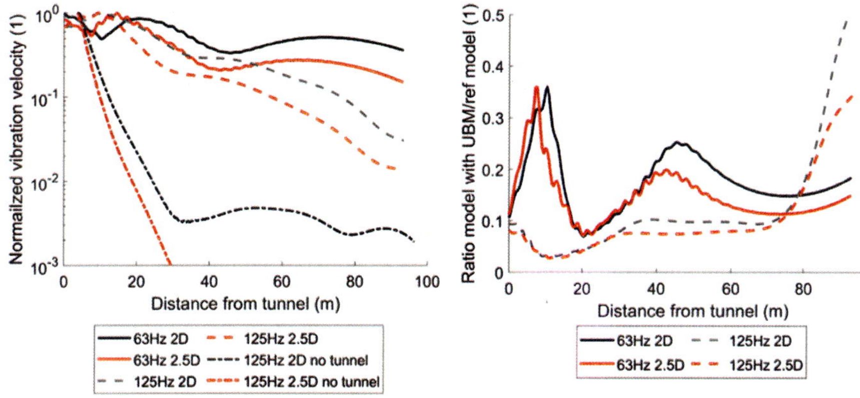

Fig. 5. Calculated vibration velocities 1 m below ground, red line in Fig. 1. Left: Normalized vibration velocity for model with tunnel and model with clay-only. Right: Ratio of vibration velocity in model with tunnel with and without under-ballast mats (UBM).

4 Calculations of Effect of Mitigation Measures in 2D

With the 2D model, calculations can easily be performed in the entire frequency range of interest. For low frequency vibration this is from about 4 Hz to 80 Hz and for structure borne noise from about 20 Hz to 250 Hz. The results of such calculations are presented in Fig. 6 as the ratio of the velocity in a model with mitigation measures to a model without mitigation measures. For low frequency vibration the applied mitigation measure is a LC-screen and for structure borne noise soft under ballast mats.

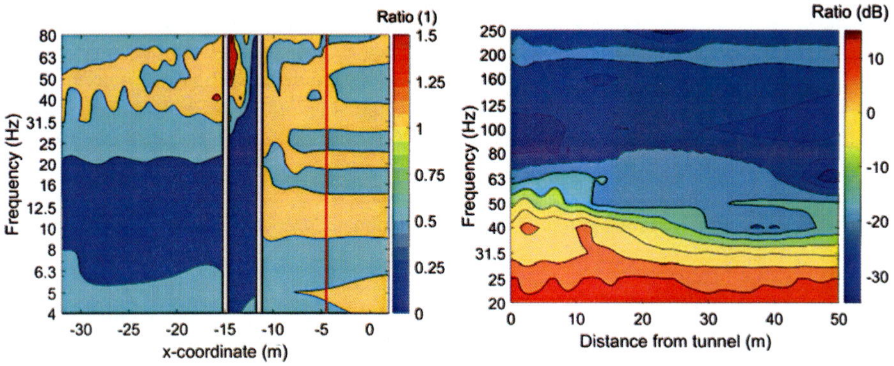

Fig. 6. Ratio of calculated vertical vibration velocities for models with and without mitigation measures. Left: LC-screen next to track, Right: Soft under-ballast mats in tunnel.

For low frequency vibrations, the calculations show a good effect of the LC-screen at this soft soil site, with at least 50% vibration reduction in the frequency range up to about 31.5 Hz. For structure borne noise the calculations show that use of soft under ballast mats increases the vibration values in the frequency range around 20 Hz. This is as expected since this is the frequency range where the combined system with under-ballast mat, ballast and the un-sprung mass of the wheel set has its resonance frequency. In the higher frequency range above 63 Hz, which is of most interest for structure borne noise, the under ballast mat gives about 20–25 dB reduction.

5 Conclusions

The study shows that 2D FE-models can be used as design tools for mitigation measures against vibration from railway lines. Both low frequency vibration from railway lines on soft clay and high frequency vibration from tunnels in clay causing structure borne noise can be handled. The premise is that the results are to be used primarily to compare different mitigation measures. Since geometrical attenuation is not correctly captured in 2D models, the models should not be used to calculate absolute vibration values. For soils with high Poisson's ratio, it may be necessary to average the results over for example the distance of a typical building foundation, to avoid unrealistic oscillations in the computed response.

Acknowledgements. This study was performed with support from the research project DESTination-Rail (Decision Support Tool for Rail Infrastructure Managers), funded by the European Commission, Grant Agreement 636285 (H2020-MG-2014-2015).

References

1. NS8176:2017: Vibration and shock - Measurement of vibration in buildings from land-based transport, vibration classification and guidance to evaluation of effects on human beings
2. Rivas SCP0-GA-2010-265754 deliverable D1.4: Review of existing standards, regulations and guidelines, as well as laboratory and field studies concerning human exposure to vibration
3. Cleave, R., Madshus, C., Grande, L., Brekke, A., Rothschild, K.: Mitigation of ground borne noise in rock railway tunnels-Part I: track design and simulation. In: Proceedings of the 7th International Conference on the Bearing Capacity of Roads, Railways and Airfields, BCRA05 (2005)
4. Park, J., Kaynia, A.M.: FE simulation of steady state wave motion in solids combined with a PML approach. In: Proceedings of the X International Conference on Structural Dynamics, EURODYN 2017, Procedia Engineering, vol. 199, pp. 1556–1561 (2017)
5. François, S., Galvin, P., Lombaert, G., Schevenels, M., Degrande, G.: A 2.5D coupled FE-BE methodology for the prediction of railway induced vibrations. In: Proceedings of the 8th International Conference on Structural Dynamics EURODYN 2011, pp. 799–803 (2001)
6. Kaynia, A.M., Madshus, C., Zackrisson, P.: Ground vibration from high-speed trains: prediction and countermeasure. J. Geot. Geoenv. Eng. **126**(6), 531–537 (2000)
7. Arcos, R., Romeu, J., Balastegui, A., Pàmies, T.: Determination of the near field distance for point and line sources acting on the surface of an homogeneous and viscoelastic half-space. Soil. Dyn. Earthq. Eng. **31**, 1072–1074 (2011)

Towards Hybrid Models for the Prediction of Railway Induced Vibration: Numerical Verification of Two Methodologies

Brice Nelain(✉), Nicolas Vincent, and Emanuel Reynaud

VIBRATEC, 28 Chemin du Petit Bois, 69130 Ecully, France
brice.nelain@vibratec.fr

Abstract. This paper investigates hybrid methodologies to predict railway ground-borne vibration as required at pre-project stages, for instance before a tramway network is constructed. In the hybrid approach, the train-track problem (source) is solved numerically and the soil transfer functions (propagation) are measured. The key point is the coupling between the two terms.

Two methodologies (force coupling and velocity coupling) are numerically tested and compared to a reference model. The obtained discrepancies are in the range of other existing methodologies for stiff and medium soils, but important errors are observed for softer soils. Corrections are tested numerically to explain the source of discrepancies and reduce them.

Keywords: Railway ground borne noise and vibration · Hybrid numerical and experimental models

1 Introduction

Current numerical models for ground vibration prediction assume that the soil can be represented by homogeneous layers [1]. However in urban areas, this assumption is no longer applicable, as the geometry is more complex. In this case, experimental transfer functions can be used to avoid more complex detailed 3D modelling.

The FRA guideline [2] proposes a full empirical way to assess train-induced ground vibration. The methodology relies on two measurements: force density L_F (source term) and line source transfer mobility TM_L (propagation term). The accuracy of this method was analysed using numerical models [3, 4], and the dependency upon the soil was highlighted, giving an uncertainty of ± 6 dB.

Kouroussis and Connolly [5, 6] modified this methodology to assess track singularities such as rail joints, using a numerical model and measured transfer functions. The excitation is seen as a point source with respect to wave propagation in the soil. Therefore, the force density of the FRA procedure was replaced by the wheel-rail contact force and the line source transfer mobility by a point source.

This article proposes a similar approach using a train-track interaction model to calculate the contact and transmitted forces, and measured transfer functions to cope

G. Degrande et al. (Eds.): *Noise and Vibration Mitigation for Rail Transportation Systems*, NNFM 150, pp. 437–444, 2021.
https://doi.org/10.1007/978-3-030-70289-2_46

with the complexity of urban areas. The discussion is focused on the physical quantities that are used to couple both terms. In this assessment study, the transfer functions are not measured, but simulated with layered soil models. All equations are written in the frequency domain. The proposed hybrid models are evaluated by comparison to a fully coupled train-track-soil numerical model. Two coupling approaches are assessed.

The first one (velocity coupling) combines a local train-track-soil model and soil transfer functions. The local model is used to compute the ground vibration velocity v_{ref} at a reference point close to the track. The transfer function $\left(v/v_{ref}\right)$ is then expressed as the ratio of the ground vibration velocity at the studied point divided by the vibration velocity at the reference point.

$$v = \left(\frac{v}{v_{ref}}\right) v_{ref} \tag{1}$$

The second one (force coupling) uses the force F_t transmitted to the ground below the track and the soil transfer function. The transmitted force is obtained with a track model. The transfer function (v/F_t) is the ratio of the ground vibration velocity at the studied point divided by the force applied to the soil.

$$v = \left(\frac{v}{F_t}\right) F_t \tag{2}$$

This work is performed using the TRAFFIC matlab toolbox, developed at KU Leuven [7, 8]. The article focuses on the vertical component, but the methodology can be extended to the other directions. All approaches assume wheel-rail roughness excitation in the frequency domain. Only at-grade traffic is considered.

2 Reference Model for Numerical Validation

The reference model is formulated in the wavenumber-frequency domain. In this approach, the track and the soil are represented in 2.5 D, assuming an invariant geometry in the track direction, which allows the use of a spatial Fourier transform.

The fully-coupled model computes the vibration in three steps. First, the wheel-rail contact force F_c is computed using the train-track interaction model (including the soil). Second, the transfer function is computed with the coupled track-soil model and used to compute the soil vibration level due to one wheel. Third, all the wheel contributions are energy-summed to obtain the train pass-by vibration level.

$$v = \sqrt{\sum_{i=1}^{N} v_i^2} \tag{3}$$

where N is the number of wheel-rail contacts in the train. The full mathematical development is presented in [9, 10].

A three-bogie tramway is represented by 6 independent 1200 kg unsprung mass axles. The track is a slab track with UIC54 rail and standard rail fixation (k = 150 kN/mm). The slab width is 2.5 m and its thickness is 40 cm.

Three grades of homogeneous soil are tested: shear wave velocity $C_s = 100$ m/s (soft soil), 300 m/s (medium soil) and 1000 m/s (stiff soil), compressive wave velocity $C_p = 2C_s$, mass density $\rho = 1945$ kg/m^3, shear and compressive damping $D_s = D_p = 0.025$.

3 Velocity Coupling Assessment

The hybrid model is made of a local numerical model coupled to measured soil transfer functions (simulated here), as illustrated in Fig. 1. The local model includes the track and its subgrade. A reference point (v_{ref}) is chosen for the coupling. This model should cover the area where the existing soil will be replaced by new track subgrades. The soil transfer function measurement will then cover the area where no change is expected.

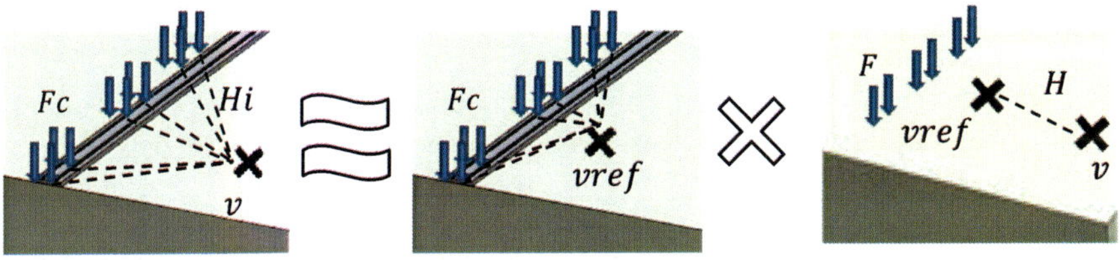

Fig. 1. Illustration of the procedure for velocity coupling

Figure 2 plots the ground velocity level at 16 m from the track center for the reference model and for the velocity coupling model, with a reference point at 4 m from the track.

The maximum difference is about 2 dB for stiff soil, 3 dB for medium soil, and 3.5 dB for soft soil. The discrepancy over the third octave spectrum is low, and is logically reduced when the reference and observation points are closer to each other.

Similar behavior has been observed on floating slab and ballasted tracks.

In practice, the soil will not be the same between the numerical model and the transfer function measurements, leading to probable higher errors than those computed here.

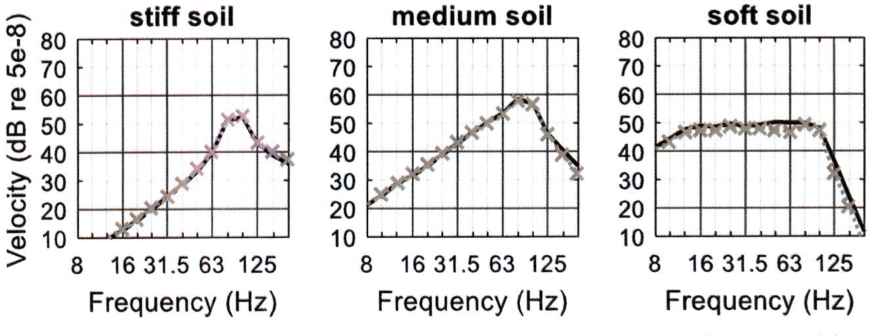

Fig. 2. Ground velocity (16 m from the track center) computed with the reference model (solid black line) and with velocity coupling (dotted line with crosses)

4 Force Coupling Assessment

4.1 Methodology and Main Approximations

As seen in Eq. (2), two physical quantities must be obtained: the force transmitted to the ground and the soil transfer function.

In the reference model, they are coupled in the wavenumber-frequency domain, while the hybrid coupling is performed in the spatial-frequency domain.

The proposed method is based on two steps. First, the contact force and the force transmitted to the ground are calculated using a train-track interaction model without soil. The absence of soil is one major approximation, but is convenient at the pre-project stage. Secondly, the transmitted force is multiplied by the measured soil transfer functions (simulated here) to obtain the vibration velocity level due to the train pass-by. In practice, and in this numerical study, the force is applied through a small interface (usually a metal disk), which constitutes an approximation with respect to the more complex real track-soil interface.

In the first step, the force transmitted to the ground is calculated as the integral of the force per unit length in the lowest resilient element of the track model.

$$F_t = \int_{-L_1}^{L_1} k\,u(x)dx \tag{4}$$

where k is the stiffness (complex) per unit length in the lowest resilient element of the track model, $u(x)$ is this element's relative displacement as a function of x (track direction) and L_1 is the integration length. The integration should go from minus to plus infinity, but as the energy is generally contained within ±20 m of the wheel position, this finite length L_1 is sufficient.

In this study of a standard slab track, the lowest resilient elements are the rail pads. For a floating slab track, the slab mat can be used, and for a ballast track, the ballast itself can be used (if modelled with spring elements).

In the second step, the force F_t computed for each axle is multiplied by the soil transfer function H_i to obtain the ground velocity v_i due to one axle.

$$v_i = F_t.H_i \tag{5}$$

The full train pass-by vibration velocity is obtained by the energy-sum of all axle contributions (Eq. (3)).

4.2 Results with Main Approximations

Figure 3 illustrates the procedure described in the previous section for a standard slab track. In the track model, the slab is modelled with a rigid cross-section and is clamped (which is equivalent to no slab). In the soil model, the impacts are applied with a rigid disk of 7 cm radius (usual dimension in practice) to obtain the transfer functions H_i.

Figure 4 plots the ground vibration levels at 16 m from the track center for the reference model and for the force coupling model. The best results are obtained with stiff soil (error of 3 dB at 100 Hz, 7 dB at 200 Hz), because the impedance difference between the track and the soil is the lowest, which minimizes the approximations. For medium soil, the error is 12 dB at 100 Hz, and 19 dB at 200 Hz. Soft soil presents a high error of more than 30 dB. The error is higher for observation points closer to the track.

4.3 Transmitted Force Corrected by the Slab Mass

One possibility to correct the approximation on the transmitted force is to compensate it with the slab mass. This requires that the soil be included in the track model.

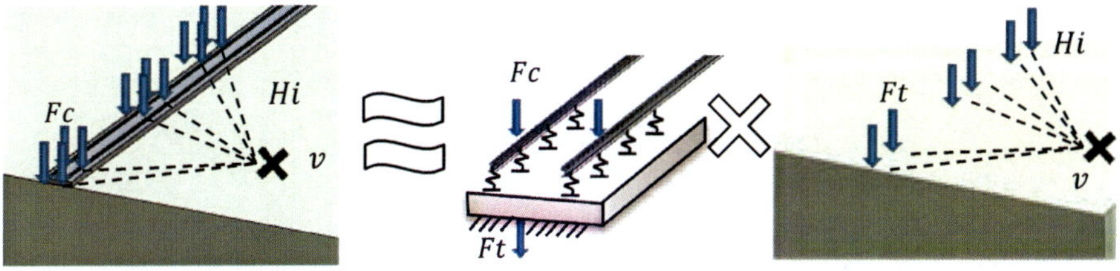

Fig. 3. Procedure for force coupling – track with rail, rail pad and slab (clamped), coupled with soil transfer functions

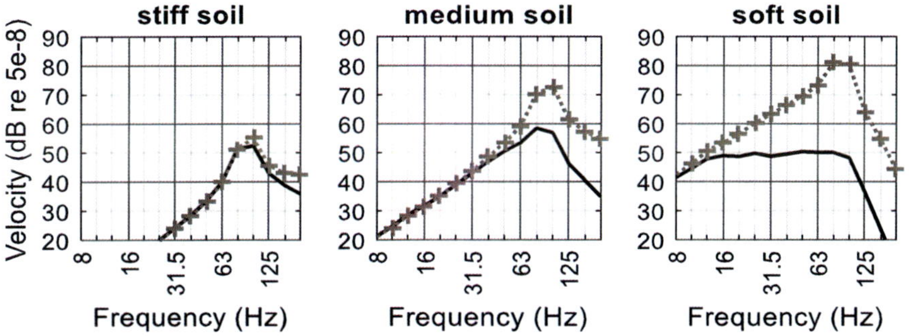

Fig. 4. Ground vibration levels (16 m from the track center) – track with rail, rail pad and slab (clamped), coupled with soil transfer functions (dotted line with crosses) – reference model (solid line)

The transmitted force is then obtained as:

$$F_t = F_{RP} - m\gamma \tag{6}$$

where F_{RP} is the rail pad force (F_t in Eq. (4)), and $m\gamma$ is the slab's change of momentum (mass multiplied by acceleration).

Figure 5 plots the ground vibration levels at 16 m from the track. The correction reduces the maximum error by up to 11 dB for soft soils, which is still not sufficient.

In the reference model, the force is applied to the ground through the width interface, while in the hybrid model, the force is applied through a small disk. For low frequencies, the characteristic dimension of the interface in both models is small compared to the soil wavelength, and hence they give the same results. When the soil wavelength is close to the slab width, the small disk interface approximation is no longer valid in the hybrid model, and discrepancies are observed. This divergence is observed at frequencies of ~160 Hz, ~50 Hz, and ~16 Hz for the soil for respectively 1000 m/s (stiff), 300 m/s (medium) and 100 m/s (soft), giving a wavelength of about 6 m for all three cases, which is about 2.4 times the slab width.

In practice, a correction to account for slab width can be applied to point source impacts, according to Auersch's investigations [11]. The ground vibration obtained with this method converges to the reference model (Fig. 6). The remaining discrepancy for soft soil is due to the difference between wavenumber coupling (reference model) and spatial coupling (force coupled model).

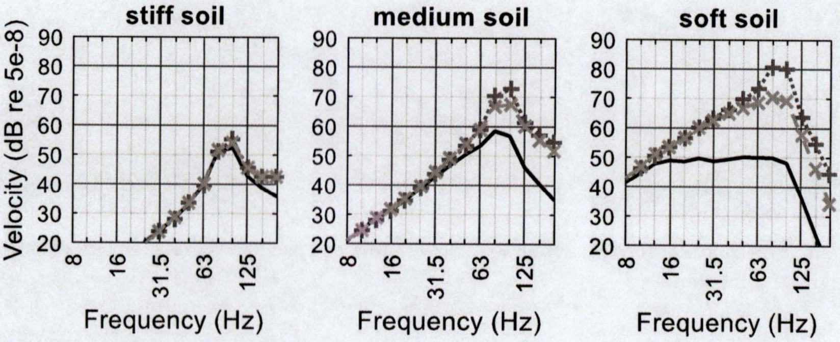

Fig. 5. Ground velocity (16 m from the track center): correction of the force coupling with slab mass (dashed line with exes) – force coupling (dotted line with crosses) – reference model (solid line)

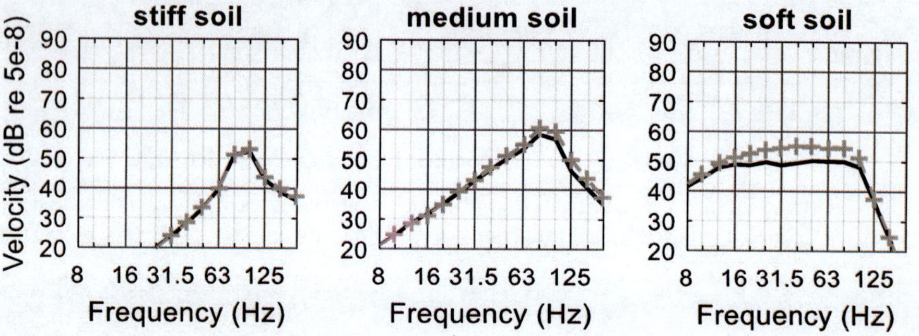

Fig. 6. Ground velocity (16 m from the track center): correction of the force coupling with slab mass and point force corrected by Auersch's formula (dotted line with crosses) – reference model (solid line)

4.4 Soil Transfer Function Including the Slab

Another way of correcting the assumption is to include the slab in the soil transfer function: the impacts are directly applied at the axle positions on the slab (see Fig. 7).

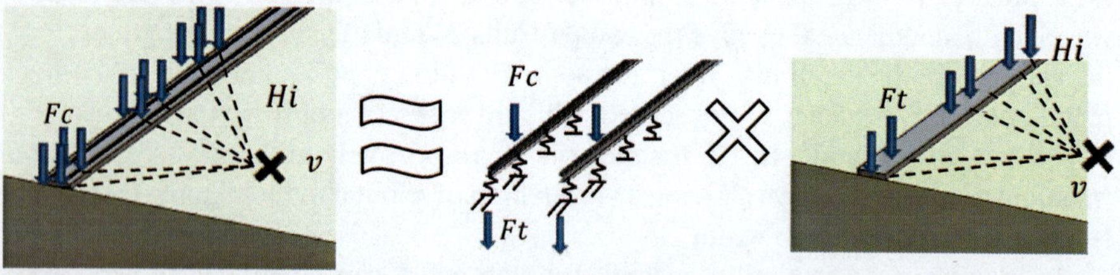

Fig. 7. Procedure for force coupling – track with rail and rail pads (clamped), coupled with slab on soil

Figure 8 plots the ground vibration levels with the correction, compared to the initial computation with main assumptions (Sect. 4.2) and to the reference model. The error is less than 6 dB for the soft soil, 4 dB for the medium soil, and 2 dB for the stiff soil.

This model has similar soil stiffness and wave propagation as the full track on soil model, explaining these lower discrepancies. However in practice, the slab is seldom built when the soil measurements are performed. The remaining error is due to point application of the axle forces on the slab, compared to longitudinally distributed forces under the rail pads in the reference model.

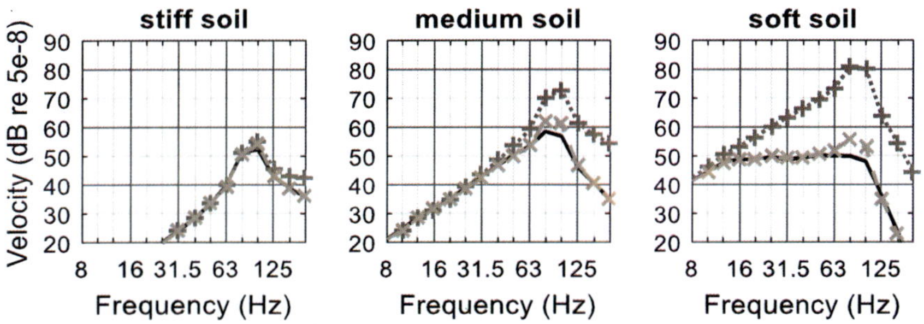

Fig. 8. Ground velocity (16 m from the track center) computed with the reference model (solid line), with force coupling (dotted line with crosses), and with force coupling with slab on soil (dashed line with exes)

5 Conclusion

Two hybrid models have been evaluated numerically: a velocity coupled model and a force coupled model. They have been compared to a reference model, in which the problem is solved in the frequency-wavenumber domain. The investigations showed the limits and advantages of each hybrid model.

The velocity coupled model is based on a local model which includes track and surrounding soil, coupled to soil transfer functions. The numerical study showed that the approach presents low error (<4 dB) compared to the reference model.

The force coupled model is valid for stiff soil (error ~5 dB, Cs > ~1000 m/s) and acceptable on medium soil (error ~12 dB, 100 m/s < Cs < ~300 m/s). However, this methodology is not recommended for soft soil (error > 15 dB, Cs < ~100 m/s).

Two approximations were highlighted. The main source of error is due to the application of point forces on the soil, instead of application through a slab width interface. A second source of error is the influence of the slab mass on the transmitted force. This hybrid approach is improved (error < 6 dB on soft soil) when the transfer function measurements are performed with the slab on soil, but this is rarely the case in practice.

The conclusions are drawn from investigations on homogeneous soils, but must be confirmed on heterogeneous soils. Investigations should also be carried out with measured soil transfer functions, including buildings.

The investigations focused on standard tramway slab tracks, and should be enlarged to other track types.

References

1. Lombaert, G., Degrande, G., François, S., Thompson, D.J.: Ground-borne vibration due to railway traffic. In: IWRN11, pp. 266–301 (2013)
2. Hanson, C.E., Towers, D.A., Meister, L.D.: Transit Noise and Vibration Impact Assessment -FTA Federal Transit Administration (2006)
3. Verbraken, H., Eysermans, H., Dechief, E., Franc, S.: Verification of an empirical prediction method for railway induced vibration
4. Verbraken, H., Lombaert, G., Degrande, G.: Verification of an empirical prediction method for railway induced vibrations by means of numerical simulations. J. Sound Vib. **330**(8), 1692–1703 (2011)
5. Connolly, D.P., Kouroussis, G.: A hybrid experimental-numerical approach to predict ground vibrations from localised railway defects. In: Railway Engineering, Edinburgh (2017)
6. Connolly, D.P., Kouroussis, G., Vogiatzis, K.: A hybrid numerical-experimental assessment of railway ground vibration in urban area. In: ICRT, Chengdu, no. 1987 (2017)
7. Lombaert, G., Degrande, G., Kogut, J., François, S.: The experimental validation of a numerical model for the prediction of railway induced vibrations. J. Sound Vib. **297**(3–5), 512–535 (2006)
8. Lombaert, G., François, S., Degrande, G.: TRAFFIC, matlab toolbox for traffic induced vibrations manual (2011)
9. Lombaert, G., Degrande, G., Bekaert, J.: The influence of the train speed on vibrations due to high speed trains, pp. 1–6 (2008)
10. Degrande, G., Schillemans, L.: Free field vibrations during the passage of a Thalys high-speed train at variable speed. J. Sound Vib. **247**(1), 131–144 (2001)
11. Auersch, L.: The excitation of ground vibration by rail traffic: theory of vehicle–track–soil interaction and measurements on high-speed lines. J. Sound Vib. **284**(1–2), 103–132 (2005)

A Methodology for the Assessment of Underground Railway-Induced Vibrations Based on Radiated Energy Flow Computed by Means of a 2.5D FEM-BEM Approach

Dhananjay Ghangale[1], Robert Arcos[1,2](✉), Arnau Clot[1,2], and Jordi Romeu[1]

[1] Acoustical and Mechanical Engineering Laboratory (LEAM), Universitat Politècnica de Catalunya (UPC), C/Colom, 11, 08222 Terrassa, Barcelona, Spain
{dhananjay.ghangale,robert.arcos}@upc.edu
[2] Universitat Politècnica de Catalunya (UPC), C/Colom, 11, 08222 Terrassa, Barcelona, Spain

Abstract. In this paper, a comprehensive numerical approach formulated in the two-and-a-half dimensional domain (2.5D) for modelling track/tunnel/soil systems in the context of ground-borne railway-induced vibration problems considering a full-space model of the soil is proposed. The approach consists of a coupled finite element-boundary element method (FEM-BEM) of the tunnel/soil system, a semi-analytical model of the track, a multibody model for the vehicle and a model for the vibration propagation in the soil based on semi-analytical solutions of a cylindrical cavity in a full-space. Since this methodology uses finite elements (FE) to model the tunnel structure, its modelling detail is higher than previously developed methodologies dedicated to computing the vibration energy flow radiated by underground railway infrastructures, as they are based on semi-analytical modelling of the tunnel structure. An application of the methodology for studying the efficiency of using one accelerometer for assessing vibration mitigation measures is presented.

Keywords: Railway-induced vibration · Ground-borne vibration · Vibration energy flow · 2.5D FEM-BEM

1 Introduction

Numerical methods have been used in the last few decades to study the ground-borne vibration induced by underground railway infrastructures. Typically in the modelling of these problems, the soil is modelled as a layered half-space, which is considered to be an acceptable approximation to the wave propagation characteristics between an underground tunnel and a nearby building. However, the consideration of a full-space model of the soil based on the locally surrounding

G. Degrande et al. (Eds.): *Noise and Vibration Mitigation for Rail Transportation Systems*, NNFM 150, pp. 445–453, 2021.
https://doi.org/10.1007/978-3-030-70289-2_47

soil condition is a computationally efficient approach especially for evaluating the efficiency of vibration mitigation measures applied on the tunnel or the track. In the context of this simplified modelling approach, vibration energy flow radiated upwards by the tunnel is proposed to be the indicator used for designing effective vibration mitigation measures.

Hussein et al. [1] proposed a power flow study based on the Pipe-in-Pipe (PiP) model [2,3]. The model was later improved by replacing the initial full-space model of the soil with a layered half-space model [4]. For the case of double-deck tunnels, Clot et al. [5,6] proposed a modified PiP model that considers an interior floor inside the tunnel to compute the induced vibration energy flow induced of such structures. A limitation of the above methodologies is that the geometrical configurations of the tunnel are limited due to the semi-analytical nature of the tunnel and track modelling considered. This paper presents a methodology to compute the vibration energy flow from underground railway tunnels that allows for a detailed modelling of tunnel structures. The developed methodology is then used for a preliminary study of the efficiency of using one accelerometer for assessing the effectiveness of vibration mitigation measures.

2 Numerical Methodology

The global computation scheme of the methodology used to obtain the vibration energy flow from underground railway tunnels consists of four different models: the track/tunnel/soil model, the train/track interaction model, the train pass-by response model and the vibration energy flow computation scheme. Details of each of these models can be found in [7]. Here, only the important steps of the methodology are described. Initially, the displacement Green's functions at the tunnel/soil interface (and other desired evaluation points along with the track/tunnel system) due to a set of vertical forces applied at the FE nodes in contact with the track model are computed using the 2.5D FEM-BEM approach presented in [7,8]. The track is then coupled to the tunnel/soil system and the response of the track/tunnel/soil model is obtained. The track is modelled by a semi-analytical model and consists of two rails coupled to the tunnel/soil model. In this model, the rails are assumed to be Euler-Bernoulli beams and the fasteners are modelled as longitudinally distributed linear viscous springs.

In the context of this coupled track/tunnel/soil model, the Green's functions of the vertical displacements of the rails $\tilde{\mathbf{H}}_r^r$ due vertical forces on them in the moving frame of reference are given by

$$\tilde{\mathbf{H}}_r^r = \left[\left[E_r I_r k_x^4 - \rho_r S_r \left(\tilde{\omega} + k_x v_t\right)^2\right]\mathbf{I} + \tilde{\mathbf{K}}_{ft}\right]^{-1}, \tag{1}$$

$$\bar{\mathbf{K}}_{ft} = \left(k_f + \mathrm{i}\omega c_f\right)\left[\mathbf{I} - \left(\frac{1}{k_f + \mathrm{i}\omega c_f}\mathbf{I} + \bar{\mathbf{H}}_{tr}^{tr}\right)^{-1}\bar{\mathbf{H}}_{tr}^{tr}\right], \tag{2}$$

where $\tilde{\mathbf{K}}_{ft} = \tilde{\mathbf{K}}_{ft}(k_x, \tilde{\omega}) = \bar{\mathbf{K}}_{ft}(k_x, \tilde{\omega} + k_x v_t)$, $\bar{\mathbf{H}}_{tr}^{tr}$ are the Green's functions that relate the equivalent vertical displacement of the tunnel below rails with equivalent vertical forces applied there, E_r is the Young's modulus, I_r is the second moment of area, ρ_r is the density, S_r cross-sectional area and k_f and c_f are the stiffness and viscous damping of the fasteners, respectively.

The Green's functions that relate the response of the tunnel/soil system coupled with rails at any arbitrary set of evaluation points due to the vertical forces on the rails in the moving frame of reference $\tilde{\mathbf{H}}_s^r$ are given by

$$\tilde{\mathbf{H}}_s^r = \tilde{\mathbf{H}}_s^{tr}\tilde{\mathbf{K}}_{ft}\tilde{\mathbf{H}}_r^r, \tag{3}$$

where $\tilde{\mathbf{H}}_s^{tr}$ are the Green's functions of the tunnel/soil system due to forces applied below the rails, obtained using the 2.5D FEM-BEM approach considering that $\tilde{\mathbf{H}}_s^{tr} = \tilde{\mathbf{H}}_s^{tr}(k_x, \tilde{\omega}) = \bar{\mathbf{H}}_s^{tr}(k_x, \tilde{\omega} + k_x v_t)$.

Equation (3) is used to obtain the 2.5D displacement Green's functions at the tunnel/soil interface of the track/tunnel/soil model, which can then be related to displacements and tractions at any arbitrary position in the soil by using the semi-analytical solutions of a cylindrical cavity in a full-space given in [2,3]. This results in the 2.5D displacement Green's functions that relate (in the moving frame of reference) the displacement and the traction response in the soil to the vertical forces applied on the rails.

The response of the coupled train/track/tunnel/soil model is then obtained in two steps. In the initial step, the track receptances in the moving frame of reference obtained from the 2.5D FEM-BEM model and the wheel receptances associated with the vehicle are used in the train/track interaction model to compute the wheel/rail interaction forces in the frequency domain. The train/track interaction model [7] mainly consists of a multibody vehicle model and a wheel/rail contact model based on the linearised Hertz contact theory. The train/track interaction model considers both the quasi-static and dynamic excitation mechanisms. In the next step, the wheel/rail interaction forces and the 2.5D Green's function of the system between the rails and the soil are used to obtain the vibration and/or traction response at selected evaluation points. Finally, the vibration energy flow in the soil can be obtained by using velocity and traction responses at a set of points located around the tunnel [9].

3 Calculation Example

A calculation example is presented in this section, where the previously described methodology is applied to study the correctness of using one accelerometer for assessing the efficiency of railway-induced vibration countermeasures applied on the track or the tunnel. The underground railway tunnel systems considered are a tunnel with a single track with direct fastening system (DFF) and a tunnel with a floating slab track (FST), consisting of a 3.2 m wide and 40 cm thick concrete slab supported over the invert by an elastomeric mat. The models for the DFF and FST systems are shown in Figs. 1 and 2, respectively. The mechanical properties of both systems are summarised in Table 1. Both tunnels have an inner radius of 3 m and a wall thickness of 0.25 m. Two soil cases are considered in this calculation example: stiff soil (Young's modulus of 480 MPa) and soft soil

(Young's modulus of 180 MPa), both of them have a density of 2100 kg/m^3, a Poisson's ratio of 0.3 and a damping ratio ($D_p = D_s$) of 0.25. The tunnels are excited by vertical 2.5D point loads at the points shown by blue star markers and the rails are coupled to the tunnel/soil system at these points. The response of the system during train pass-by is obtained on tunnel evaluation point A and on the evaluation points in the soil located above the tunnels at radial distance of 5 m from the outer boundary. A total of 37 evaluation points are placed in the soil covering a semi-circle of π rad. The vibration energy flow is obtained at these evaluation points by computing traction and velocity responses. A complete 2D vehicle model was used to obtain the train pass-by response. The model of the train and the train properties can be found in [7]. The train speed was assumed to be 25 m/s.

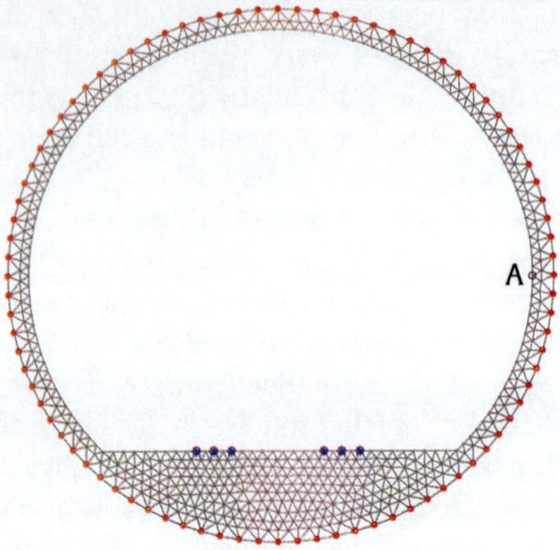

Fig. 1. DFF system modelled by 2.5D FEM-BEM. Red solid markers represent the BE nodes, blue star markers are the points where vertical rail/tunnel interaction forces are applied and pink circular markers denote the evaluation points, where the evaluation point in the tunnel wall is denoted by A.

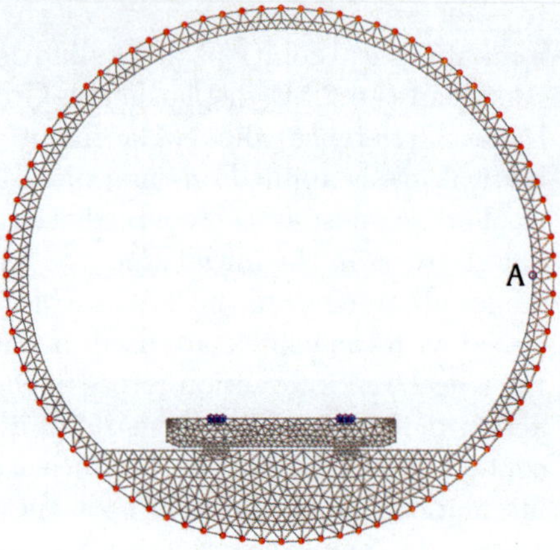

Fig. 2. FST system modelled by 2.5D FEM-BEM. Red solid markers represent the BE nodes, blue star markers are the points where vertical rail/tunnel interaction forces are applied and pink circular markers denote the evaluation points, where the evaluation point in the tunnel wall is denoted by A.

Figure 3 shows the frequency content of the vertical component of the vibration acceleration of rail and the tunnel evaluation point A for DFF and FST systems and for the two soil stiffness. From this figure, it can be seen that the FST is reducing the level of vibration at frequencies 30 Hz and amplifying the levels of vibration at low frequencies, for both soil stiffnesses considered. These results has been previously shown by other investigations [10] and experimental measurements [11].

Table 1. Properties of DFF and FST.

Subsystem	Parameters	Units	Value
Tunnel	Young's modulus	[GPa]	3.5
	Poisson's ratio	[–]	0.15
	Density	[kg/m^3]	2500
	Damping ratio	[–]	0.01
Floating slab	Young's modulus	[GPa]	3.5
	Poisson's ratio	[–]	0.15
	Density	[kg/m^3]	2500
	Damping ratio	[–]	0.01
Elastomeric mat	Young's modulus	[MPa]	2.73
	Poisson's ratio	[–]	0.35
	Density	[kg/m^3]	1328
	Damping ratio	[–]	0.05
Rails	Density	[kg/m^3]	7850
	Young's modulus	[GPa]	207
	Cross-sectional area	[m^2]	$23.5 \cdot 10^{-6}$
	Second moment of area	[m^4]	$6930 \cdot 10^{-6}$
Fasteners	Stiffness	[kN/mm]	35
	Viscous damping	[Ns/m]	$35{\cdot}10^3$

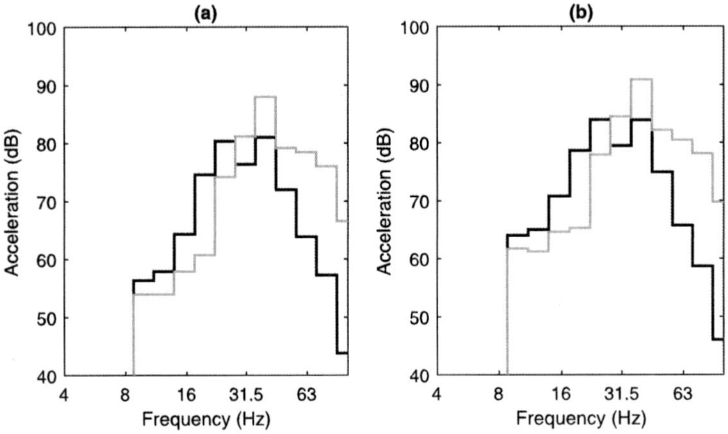

Fig. 3. One-third octave bands of the vertical acceleration spectrum at the evaluation point A for the stiff soil (a) and soft soil (b) cases. Solid black lines represent the response in the case of the FST system and solid grey lines represent the DFF case results. The reference for the dB is 10^{-6} m/s^2.

Table 2. Total energy flow from DFF and FST.

Track type	Soft soil	Stiff soil
DFF	0.2547 J/m	0.1743 J/m
FST	0.2227 J/m	0.1446 J/m

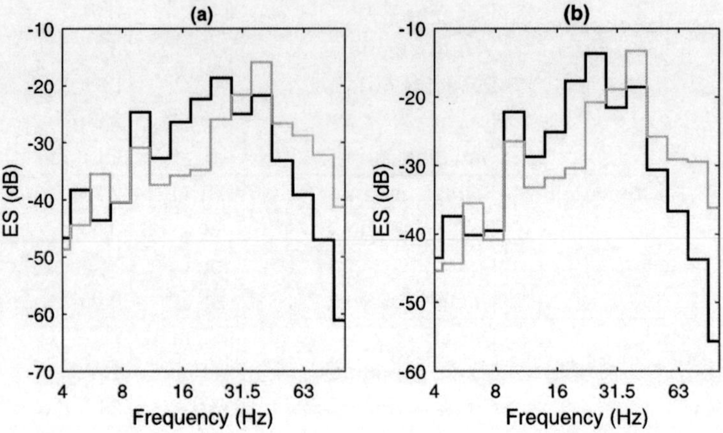

Fig. 4. Energy spectra (ES) in one-third octave bands of the vibration energy flow radiated upwards for stiff soil (a) and in soft soil (b) cases and for the cases of the DFF system (solid grey line) and FST system (solid black line). The reference for the dB is 1 J/m.

The vibration energy flow radiated upwards per meter by underground railway infrastructures can be obtained by using the expression

$$E = r_m \int_{\theta_1}^{\theta_2} \int_{-\infty}^{+\infty} \boldsymbol{v}(0, \theta, t) \cdot \boldsymbol{\tau}(0, \theta, t) \mathrm{d}t \mathrm{d}\theta. \tag{4}$$

Table 2 shows the total energy flow per meter radiated upwards by both tunnel systems and for the two soil cases. In terms of frequency content, energy spectra (ES) of the vibration energy flow radiated is obtained for the DFF and FST for the two cases of soil stiffnesses and is shown in Fig. 4. A similar behaviour to the one seen in the acceleration spectrum is observed for the ES, since the FST is shifting the energy content to lower frequencies.

In order to study the validity of using one accelerometer, the insertion loss between the DFF and FST for the tunnel evaluation point A is compared with the one that comes from the comparison between the ES associated with the vibration energy flow. The former is computed as

$$\mathrm{IL_a} = 20 \log_{10} \left(\frac{a_{\mathrm{FST}}}{a_{\mathrm{DFF}}} \right), \tag{5}$$

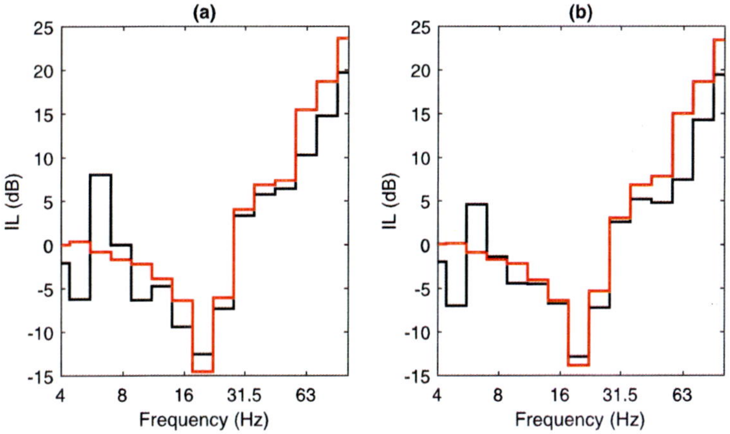

Fig. 5. Insertion loss obtained with ES (solid black line) and acceleration in vertical direction (soild red line) in stiff soil (a) and soft soil (b).

and the latter as

$$\mathrm{IL_{ES}} = 10\log_{10}\left(\frac{\mathrm{ES_{FST}}}{\mathrm{ES_{DFF}}}\right). \tag{6}$$

In Eqs. (5) and (6), a_{FST} represents the acceleration in vertical direction at the tunnel evaluation point A of the FST system, a_{DFF} represents the acceleration in vertical direction at the tunnel evaluation point A of the DFF system, $\mathrm{ES_{FST}}$ represents the ES associated to the FST system while $\mathrm{ES_{DFF}}$ represents the ES related to the DFF system.

In Fig. 5, the insertion loss $\mathrm{IL_a}$ for the tunnel evaluation point A is compared with the insertion loss $\mathrm{IL_{ES}}$ obtained by vibration energy flow, for the soft and stiff soil cases. It can be seen that IL_a approximately follows the $\mathrm{IL_{ES}}$ at frequencies between 10 and 50 Hz. For frequencies above 50 Hz, the IL_a associated to the tunnel evaluation point A is overestimated with respect to $\mathrm{IL_{ES}}$. Moreover, a large difference between IL_a and $\mathrm{IL_{ES}}$ appears at 6.3 Hz, especially for the case of stiff soil.

4 Conclusions

In this work, a methodology for computing the vibration energy flow from underground railway infrastructures is presented. The developed approach has been also used for a preliminary study on the validity of using one accelerometer for assessing the efficiency of vibration mitigation measures in the framework of railway tunnels. It is found that the differences between the insertion loss that comes from the vibration energy flow with respect to the insertion loss coming from one accelerometer in the tunnel wall are up to 8 dB at frequencies 63 Hz, 80 Hz and 100 Hz. Therefore, at these frequencies, the usage of only one accelerometer located at the tunnel for assessing the vibration induced by an underground railway infrastructure seems to be insufficient to catch the global vibration radiation behaviour of the tunnel. It is also found that the IL of a mitigation measure could be significantly dependent on the local subsoil surrounding the tunnel infrastructure. However, in order to achieve more rigorous conclusions about the validity of using one accelerometer in the tunnel for the assessment ground-borne railway-induced vibration in tunnels, a complete parametric study needs to be carried out accounting for a much more representative set of parameters of the system.

Acknowledgements. The present work is supported by the Ministerio de Economía y Competitividad under the research grant BES-2015-071453, individual research grant related to the project ISIBUR supported by the Ministerio de Ciencia e Innovación, Retos de Investigación 2014, with reference TRA2014-52718-R. The authors want to also acknowledge the financial support provided by the project VIBWAY supported by the Ministerio de Ciencia e Innovación, Retos de Investigación 2018, with reference RTI2018-096819-BI00. The second author also wants to acknowledge the funds provided by the NVTRail project with grant reference POCI-01-0145-FEDER-029577, funded by FEDER funds through COMPETE2020 and by national funds (PIDDAC) through FCT/MCTES.

References

1. Hussein, M.F.M., Hunt, H.E.M.: A power flow method for evaluating vibration from underground railways. J. Sound Vib. **293**(3), 667–679 (2006). Proceedings of the Eighth International Workshop on Railway Noise
2. Forrest, J.A., Hunt, H.E.M.: A three-dimensional tunnel model for calculation of train-induced ground vibration. J. Sound Vib. **294**(4), 678–705 (2006)
3. Hussein, M.F.M., Hunt, H.E.M.: A numerical model for calculating vibration from a railway tunnel embedded in a full-space. J. Sound Vib. **305**(3), 401–431 (2007)
4. Hussein, M.F.M., François, S., Schevenels, M., Hunt, H.E.M., Talbot, J.P., Degrande, G.: The fictitious force method for efficient calculation of vibration from a tunnel embedded in a multi-layered half-space. J. Sound Vib. **333**(25), 6996–7018 (2014)
5. Clot, A., Romeu, J., Arcos, R., Martín, S.R.: A power flow analysis of a double-deck circular tunnel embedded in a full-space. Soil Dyn. Earthq. Eng. **57**, 1–9 (2014)
6. Clot, A., Romeu, J., Arcos, R.: An energy flow study of a double-deck tunnel under quasi-static and harmonic excitations. Soil Dyn. Earthq. Eng. **89**, 1–4 (2016)
7. Ghangale, D., Arcos, R., Clot, A., Cayero, J., Romeu, J.: A methodology based on 2.5D FEM-BEM for the evaluation of the vibration energy flow radiated by underground railway infrastructures. Tunnelling and Underground Space Technology. Accepted for publication (2020)
8. Ghangale, D., Colaço, A., Alves Costa, P., Arcos, R.: A methodology based on structural finite element method-boundary element method and acoustic boundary element method models in 2.5D for the prediction of reradiated noise in railway-induced ground-borne vibration problems. J. Vib. Acoust. **141**, 031011 (2019)
9. Noori, B., Arcos, R., Clot, A., Romeu, J.: Control of ground-borne underground railway-induced vibration from double-deck tunnel infrastructures by means of dynamic vibration absorbers. J. Sound Vib. **461**, 114914 (2019)

10. Lombaert, G., Degrande, G., Vanhauwere, B., Vandeborght, B., François, S.: The control of ground-borne vibrations from railway traffic by means of continuous floating slabs. J. Sound Vib. **297**(3–5), 946–961 (2006)
11. Ding, D.Y., Liu, W.N., Li, K.F., Sun, X.J., Liu, W.F.: Low frequency vibration tests on a floating slab track in an underground laboratory. J. Zhejiang Univ. Sci. **12**(5), 345–359 (2011)

Assessment of Dynamic Vibration Absorbers Efficiency as a Countermeasure for Ground-Borne Vibrations Induced by Train Traffic in Double-Deck Tunnels Using an Energy Flow Criterion

Behshad Noori[1], Robert Arcos[1,2](✉), Arnau Clot[3], and Jordi Romeu[1]

[1] Acoustical and Mechanical Engineering Laboratory (LEAM), Universitat Politècnica de Catalunya (UPC), Colom 11, 08222 Terrassa, Barcelona, Spain
robert.arcos@upc.edu

[2] Serra Húnter Programme, Universitat Politècnica de Catalunya (UPC), Colom 11, 08222 Terrassa, Barcelona, Spain

[3] Department of Engineering, University of Cambridge, Trumpington Street, Cambridge CB2 1PZ, UK

Abstract. Double-deck tunnels are an innovative tunnel layout design useful to enhance the efficiency of underground railway transportation systems in which the tunnel is divided into two sections by an interior floor. However, vibration measurements and various recent investigations indicate that train traffic in the upper section of this type of tunnels could induce significantly larger levels of noise and vibration at the nearby residential buildings. This study aims to assess the efficiency of dynamic vibration absorbers (DVAs) as vibration mitigation measures for ground-borne vibrations induced by train traffic in the upper section of a double-deck tunnel. A previously developed semi-analytical model of a track-tunnel-ground system along with a two-dimensional multi-degree-of-freedom rigid body model of the vehicle is employed to compute the vibration energy flow radiated upwards due to a train pass-by. Considering the crucial role of DVA parameters in their efficiency, a global optimisation approach based on a genetic algorithm is used to obtain the optimum parameters of the set of DVAs. The performance of DVAs is assessed for two train speeds taking into account their efficiency in reducing the total vibration energy radiated from the tunnel. The results show more than 6 dB reduction in total radiated energy due to the use of the optimised DVAs.

Keywords: Traffic induced vibrations · Dynamic vibration absorbers · Ground-borne vibration · Double-deck tunnel dynamics

G. Degrande et al. (Eds.): *Noise and Vibration Mitigation for Rail Transportation Systems*, NNFM 150, pp. 454–461, 2021.
https://doi.org/10.1007/978-3-030-70289-2_48

1 Introduction

Double-deck circular tunnels are an innovative tunnel design in which the tunnel is divided into two sections by an interior floor. Tunnels of this type have been constructed in several cities for road and/or railway traffic. A design methodology of specific ground-borne vibration mitigation measures for this type of tunnels is necessary as they show a different vibration propagation pattern in comparison with simple circular tunnels for traffic in the upper section of the tunnel [1]. The modification of the rail pad stiffness and the implementation of an elastomeric mat between the interior floor and the tunnel structure are two mitigation measures studied by Clot et al. [2,3] using a two-and-a-half-dimensional (2.5D) semi-analytical model of the system [4].

Dynamic vibration absorbers (DVA) have been applied on several mechanical, civil and aerospace structures to attenuate their vibration either at a specific frequency or over a particular range of frequencies. Except for very simple cases, the design parameters of the set of DVAs needs to be optimised using an optimisation procedure. Among several optimisation approaches, the ones based on genetic algorithms (GA) have been extensively used to determine optimum parameters of DVAs [5,6].

Recently, DVAs have been used to address some problems in the field of railway-induced vibration. Zhu et al. [7] studied the potential of DVAs to deal with the low-frequency vibrations of discontinuous floating slab tracks. In another investigation, Wang et al. [8] showed that DVAs can be an effective countermeasure to control vertical vibration of car-bodies. Furthermore, DVAs have been found to be effective in reducing the radiated noise [9] and in decreasing rail corrugation growth [10].

In this paper, the methodology proposed in [11] is used to assess the efficiency of DVA as a countermeasure for railway-induced ground-borne vibration. Two energy flow criteria are considered in the assessment. An optimisation algorithm based on GA has been used to determine the optimum parameters of the DVAs set. The vibration energy flow radiated upwards before and after the application of the optimal set of DVAs has been used to evaluate the efficiency of this vibration abatement solution.

2 Modelling of the Train-Track-Tunnel-Soil System

The complete model of the train-track-tunnel-soil system used in the present work consists of a 2.5D model of track-tunnel-soil and a train-track interaction model. This model is detailed in [11] and it is briefly described in this section. The track-tunnel-soil model is used to compute the 2.5D Green's functions of the system which are required to couple the train and DVAs to the track and interior floor, respectively, and to compute the velocity and traction Green's functions in the soil. The train-track interaction model is used to obtain the contact forces caused by the wheel-rail interaction. These forces, along with the velocity and traction Green's functions, can be used to compute the velocity and traction vibration fields induced in the soil from which the energy flow radiated upwards due to a train pass-by can be computed.

In the 2.5D model of track-tunnel-soil, the track-tunnel system consists of two rails coupled to the interior floor of a double-deck tunnel via direct fixation fasteners. The rails, interior floor and the fasteners are modelled as infinite Euler-Bernoulli beams, as a homogeneous isotropic thin strip plate with a rectangular cross-section and as a continuous mass-less distribution of springs, respectively. Assuming that the tunnel and soil can be represented as an infinite thin cylindrical shell and as an infinite homogeneous elastic medium, respectively, the PiP model [12] is used to describe tunnel-soil subsystem. Figure 1 shows a schematic representation of the track-tunnel system embedded in a full-space as a model of the soil. In this work, the methodology for coupling a set of DVAs to a railway system presented in [11] is employed with the aim of coupling one longitudinal distribution of DVAs to the interior floor of a double-deck tunnel. This is also shown in Fig. 1, where k_d, c_d and m_d represent the stiffness, viscous damping and mass of each DVA, respectively. Since this method is based on substructuring techniques, it does not require a full system calculation each time that the response of the system in the presence of a new set of DVAs is computed, avoiding huge computational costs and allowing to apply optimisation procedures over the DVAs set parameters, as presented in the next section.

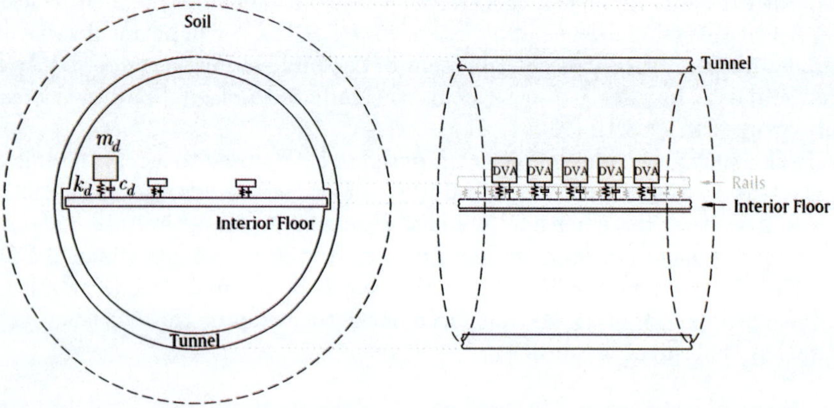

Fig. 1. A scheme of a double-deck tunnel with one longitudinal distribution of DVAs.

The wheel/rail interaction forces are computed with the train-track interaction model, which considers the rail unevenness as the main excitation source. The track receptances required for this calculation are computed in the absence of DVAs, since the effect of the DVAs on the resulting interaction forces has been found to be negligible. Once the contact forces are computed, the response at an arbitrary position of the railway infrastructure system due to the passage of the train can be found using the Green's functions of the system between the rails and the evaluation points.

3 Optimisation Approach

A global optimisation approach based on GA is used to obtain the optimum parameters of DVAs as countermeasures for underground railway-induced ground-borne vibration. In the optimisation process, the total vibration energy radiated upwards by the tunnel due to train pass-by is considered as the objective function to be minimised, leading to the optimum parameters of the DVAs set.

The vibration energy flow radiated upwards by the tunnel through a cylindrical strip with radius r_m at any arbitrary cross section x_e can be computed using the vibration velocity $\mathbf{v}$ and traction $\boldsymbol{\tau}$ fields along the surface as

$$E = r_m \, \Delta \mathrm{x} \int_0^{\pi} \int_{-\infty}^{+\infty} \mathbf{v}(x_e, \theta, t) \cdot \boldsymbol{\tau}(x_e, \theta, t) \mathrm{d}t \mathrm{d}\theta, \tag{1}$$

where θ is the angle that goes across the cylindrical strip, where the limits of integration for this angle are 0 and π in order to catch only the upwards vibration energy flow. The required velocity and traction fields induced by a train passage in the soil can be computed using the methodology presented previously.

To assess the efficiency of the DVAs as vibration abatement solutions, the total energy radiated upwards by the tunnel is computed in the absence (E) and in the presence (E') of the DVAs, and the insertion loss (IL) resulting from the expression $\mathrm{IL} = 10 \log_{10}(E/E')$ is used to quantify their efficiency. The design parameters of the DVA set are the position of the distribution y_d, the number of DVAs in a distribution N_d, the distance between two consecutive DVAs l_d, the natural frequency of the DVAs f_d in Hz, m_d and c_d. The parameters m_d and N_d are defined in the pre-design stage as 800 kg and 15 DVAs, respectively. The parameters l_d, f_d and c_d are assumed to be in the range between 1 m to 8 m, between 1 Hz and 80 Hz and between 5 kN s m^{-1} and 500 kN s m^{-1}, respectively. Possible values of y_d (the horizontal distance from the centre of the interior floor) range between 1.25 m and 4.5 m, at both sides of the interior floor.

4 Results and Discussion

In this section, optimised parameters of DVAs to minimise the energy flow radiated upward computed using the optimisation procedure explained previously are presented. Also, the efficiency of the application of the optimised DVAs on the interior floor of the double-deck tunnel is discussed. In the present study, the parameters used to model the soil are as follows: the density is 2000 kg m^{-3}, P- and S-wave velocities are 608.4 and 325.2 ms^{-1}, respectively, and damping ratios are 0.03. Tunnel and interior floor are assumed to be made of concrete with a Young's modulus of 50 GPa, a density of 3000 kg m^{-3} and a Poisson ratio of 0.175. Regarding the geometry of these systems, the thickness and the interior radius of the tunnel are 0.4 m and 5.56 m, respectively, and the thickness and the width of the interior floor are 0.7 m and 10.9 m, respectively. The two rails of the track are centred in the interior floor they are separated by 1.5 m. The rail mechanical parameters are: Young's modulus of 207 GPa, density of

7850 kg m^{-3}, cross-section area of 6.93·10^{-3} m^2 and the second moment of area of 23.5·10^{-6} m^4.

The considered train consists of two identical 2D models of the vehicle. The distance between the wheels of a bogie, bogies of the same car and bogies of two consecutive cars are 2.2 m, 15 m and 7 m, respectively. Regarding mechanical parameters of the vehicle, the mass of the combined wheel and half-axle system is 950 kg, the mass and inertia of mass of a half-bogie are 4700 kg and 1300 kg m^2, respectively; the stiffness and viscous damping of the primary suspension are $14 \cdot 10^5$ N m^{-1} and $9 \cdot 10^3$ N s m^{-1}, respectively; the mass and inertia of mass of the half-car body are 22500 kg and $55 \cdot 10^4$ kg m^2, respectively; and the stiffness and viscous damping of the secondary suspension are $6 \cdot 10^5$ N m^{-1} and $21 \cdot 10^3$ N s m^{-1}, respectively. The unevenness profiles of the rails are modelled using a stochastic random process characterised by its power spectral density (PSD) using class 3 track according to the Federal Railroad Administration (FRA) [13]. The profiles of the rails are assumed to be uncorrelated between themselves.

An optimisation process has been carried out for train speeds of $v_t = 20$ m s^{-1} (Case V20) and $v_t = 25$ m s^{-1} (Case V25), which are typical train speeds in urban railway lines. The resulting optimal values for the parameters of the DVA set and the associated IL regarding the total vibration energy radiated upwards are presented in Table 1. In this study, it is assumed that all DVAs in a distribution have the same properties.

Table 1. The optimum values of DVA parameters and the associated IL

Parameters	y_d (m)	l_d (m)	f_d (Hz)	c_d (kN s m^{-1})	IL (dB)
Case V20	3.55	7.5	31.52	14.09	6.2
Case V25	−2.45	4.5	28.83	27.07	6.6

Figure 2 shows energy spectrum in one-third octave bands radiated upwards through the cylindrical strip, before and after the application of optimised DVAs for both studied cases [11]. For both cases, most significant content of the vibration energy is concentrated in a frequency range between 25 and 35 Hz. It can be observed that the natural frequency of the optimal DVAs is also in the same frequency range. In other words, the natural frequencies of the DVAs have been set in this frequency range, which makes them effective in minimising the radiated energy.

Figure 3 shows the radiation pattern of the energy flow before the application of DVAs as a function of θ [11]. As expected, the total radiated energy is higher for Case V25 than for Case V20. Moreover, it can be seen that, in both cases, most of the energy is radiated in the directions of angles 0, $\pi/2$ rad and π rad. For both cases, the radiation pattern is significantly affected by the application of the DVAs. Considering the relation between the mode shapes of the interior floor and the radiation pattern of the energy, shown by Clot et al. [1], it can be inferred that the mode shapes of the interior floor are modified by the application of DVAs.

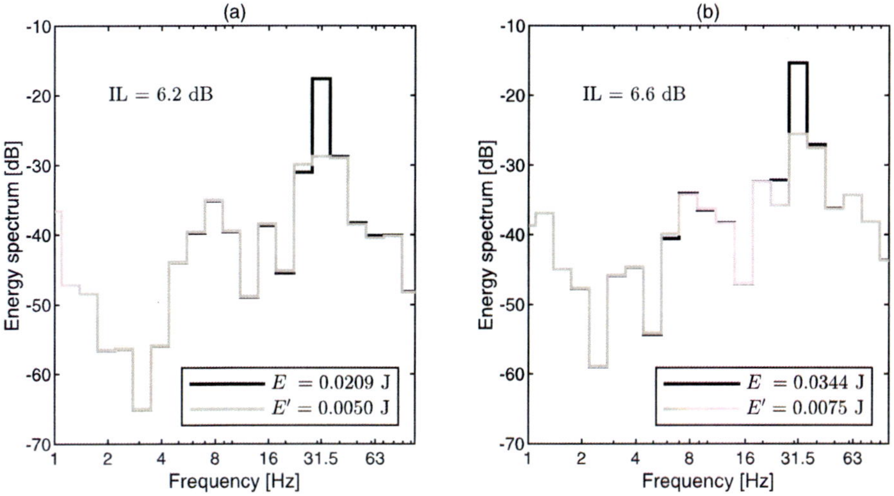

Fig. 2. One-third octave bands of the energy spectrum in dB (dB reference 1 J) radiated through the cylindrical strip for Case V20 (a) and Case V25 (b). The grey and black solid lines represent the results with and without the application of DVAs, respectively. The total radiated energy and associated IL are also presented for both cases.

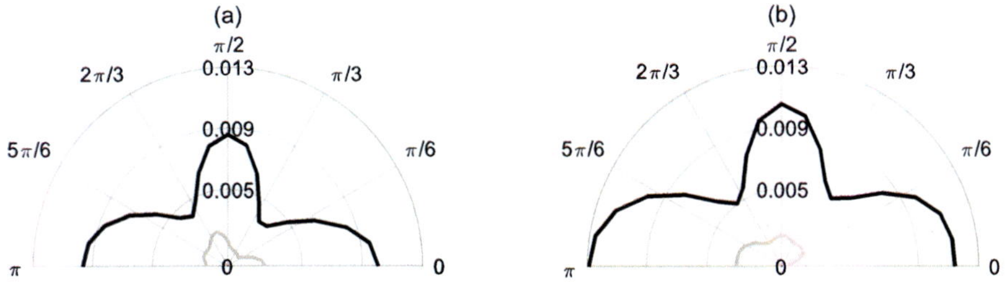

Fig. 3. Radiation pattern of the vibration energy flow in J/rad through the cylindrical strip before (black solid line) and after (grey solid lines) the application of DVAs for Case V20 (a) and Case V25 (b).

Figure 4 shows energy spectrum in one-third octave band radiated upwards through the cylindrical section between the θ angles of $\pi/3$ rad and $2\pi/3$ rad before and after the application of optimised DVAs. The total radiated energy with and without DVAs and the associated IL are also presented for each case. It is shown that the DVAs are slightly more effective in minimising the energy radiated through this specific section. In general, it is expected that by defining the objective function as the energy flow radiated upwards through a specific section rather than whole the complete strip, the optimisation process might reach larger reductions of the vibration energy radiated due to the optimal DVAs set application. This could be an interesting approach when the location of the target buildings with respect to the tunnel is known.

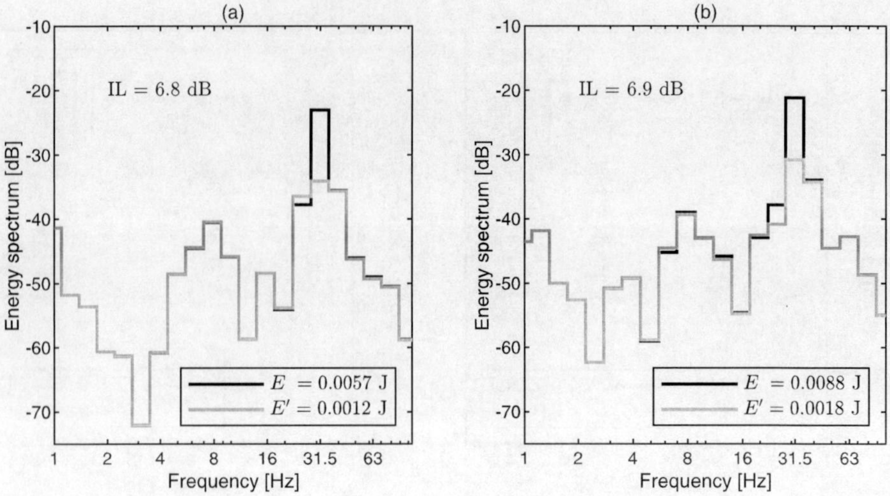

Fig. 4. One-third octave bands energy spectrum in dB (dB reference 1 J) radiated through the section $\theta = [\pi/3\ 2\pi/3]$ rad of the cylindrical strip for Case V20 (a) and Case V25 (b). The red and blue lines represent the results with and without the application of DVAs, respectively. The total radiated energy and associated IL are also presented.

5 Conclusions

The application of DVAs as an innovative countermeasure to mitigate the ground-borne vibration induced by train traffic over the interior floor of a double-deck tunnel is presented. The performance of one longitudinal distribution of DVAs in minimising the vibration energy flow radiated upwards by the tunnel has been evaluated for two different train speeds. The results show that by tuning the natural frequency of DVAs in the range of frequency at which most of the energy spectral content is concentrated, they can provide significant vibration mitigation benefits. Reductions of 6.2 dB and 6.6 dB in the total radiated energy has been achieved for train speeds of $v_t = 20$ and $v_t = 25\,\mathrm{m\ s^{-1}}$, respectively. It is expected that using more than one longitudinal distribution of DVAs or targeting a specific radiation path would result in larger reductions of the total vibration energy radiated. Noteworthy, DVAs would be a more cost-effective solution for double-deck tunnels in existing underground railway networks since their implementation would be cheaper than other vibration countermeasures, as for example vibration isolation screens, rail fastening system retrofitting or building base isolation.

Acknowledgements. This work has been carried out in the context of the Industrial Doctorates Plan, with the financial support of AV Ingenieros and AGAUR (Generalitat de Catalunya). The authors want to also acknowledge the financial support provided by the project VIBWAY: Fast computational tool for railway-induced vibrations and re-radiated noise assessment, supported by the Ministerio de Ciencia e Innovación,

Retos de Investigación 2018, with reference RTI2018-096819-B-I00. The second author also wants to acknowledge the funds provided by the NVTRail project, with grant reference POCI-01-0145-FEDER-029577, funded by FEDER funds through COMPETE2020 (Programa Operacional Competitividade e Internacionalização (POCI)) and by national funds (PIDDAC) through FCT/MCTES.

References

1. Clot, A., Romeu, J., Arcos, R., Martín, S.: A power flow analysis of a double-deck circular tunnel embedded in a full-space. Soil Dyn. Earthq. Eng. **57**, 1–9 (2014)
2. Clot, A., Arcos, R., Romeu, J., Noori, B.: Prediction of the isolation efficiency of vibration countermeasures for a double-deck tunnel. In: Proceedings of EuroRegio 2016 (2016)
3. Clot, A., Arcos, R., Noori, B., Romeu, J.: Isolation of vibrations induced by railway traffic in double-deck tunnels using elastomeric mats. In: 24th International Congress on Sound and Vibration, ICSV 2017 (2017)
4. Clot, A., Arcos, R., Romeu, J., Pàmies, T.: Dynamic response of a double-deck circular tunnel embedded in a full-space. Tunn. Undergr. Space Technol. **59**, 146–156 (2016)
5. Hadi, M., Arfiadi, Y.: Optimum design of absorber for MDOF structures. J. Struct. Eng. **124**(11), 1272–1280 (1998)
6. Arfiadi, Y., Hadi, M.: Optimum placement and properties of tuned mass dampers using hybrid genetic algorithms. Int. J. Optim. Civil Eng. **1**, 167–187 (2011)
7. Zhu, S., Yang, J., Yan, H., Zhang, L., Cai, C.: Low-frequency vibration control of floating slab tracks using dynamic vibration absorbers. Veh. Syst. Dyn. **53**(9), 1296–1314 (2015)
8. Wang, Q., Zeng, J., Wei, L., Zhou, C., Zhu, B.: Reduction of vertical abnormal vibration in carbodies of low-floor railway trains by using a dynamic vibration absorber. Proc. Inst. Mech. Eng. [F] J. Rail Rapid Transit **232**(5), 1437–1447 (2018)
9. Thompson, D., Jones, C., Waters, T., Farrington, D.: A tuned damping device for reducing noise from railway track. Appl. Acoust. **68**(1), 43–57 (2007)
10. Ho, W., Wong, B., Tsui, D., Kong, C.: Reducing rail corrugation growth by tuned mass damper. In: Proceedings of the 11th International Workshop of Railway Noise (2013)
11. Noori, B., Arcos, R., Clot, A., Romeu, J.: Control of ground-borne underground railway-induced vibration from double-deck tunnel infrastructures by means of dynamic vibration absorbers. J. Sound Vib. **461**, 114914 (2019)
12. Forrest, J., Hunt, H.: A three-dimensional tunnel model for calculation of train-induced ground vibration. J. Sound Vib. **294**, 678–705 (2006)
13. Gupta, S., Liu, W., Degrande, G., Lombaert, G., Liu, W.: Prediction of vibrations induced by underground railway traffic in Beijing. J. Sound Vib. **310**(3), 608–630 (2008)

Increase in Force Density Levels of Light Rail Vehicles

Shankar Rajaram[1](✉), James Tuman Nelson[2], and Hugh Saurenman[3]

[1] Sound Transit, 401 S Jackson Street, Seattle, WA 94608, USA
Shankar.rajaram@soundtransit.org
[2] Wilson Ihrig & Associates, Emeryville, CA 94608, USA
jnelson@wiai.com
[3] ATS Consulting, Pasadena, CA 91101, USA
hsaurenman@atsconsulting.com

Abstract. Ground vibration force generated by rail vehicles and track are usually characterized by one-third octave band force density levels (FDL) in North America. The FDLs depend on a wide variety of vehicle and track characteristics. FDL data measured over several years showed that the FDLs increased substantially between 40 and 100 Hz at two test sites at the Sound Transit system within a period of three years. FDLs at a third site did not change at those frequencies over a longer period. No noticeable differences in rail roughness were measured. However, differences in rail wear and lubrication existed between the three sites. The FDLs between 40 and 125 Hz were lower after grinding to correct the rail profile and remove grinding-induced corrugation (GIC) from earlier grinding. The track geometry of the test sites appeared to influence the speed dependencies of the FDLs.

Keywords: Force density level · Track · Rail grinding

1 Introduction

Force density levels (FDL) form the starting point for prediction of wayside vibration generated by rail transit vehicles running on railway tracks [1]. The FDL of light rail vehicles (LRVs) depend on a wide variety of factors including train speed, suspension stiffness, unsprung wheelset mass, wheel profile, condition of the wheel/rail interface, spacing of sleepers and rail fasteners, and rail support stiffness. Vibration studies in North America have shown that separating the factors that affect specific vibration peaks can be difficult.

Sound Transit LRVs are 70% low floor vehicles. Each LRV has three bogies: an idling center bogie with independently rotating wheels and two powered end bogies with solid axle wheelsets. The powered and unpowered bogies behave differently with different FDLs.

Sound Transit conducted detailed FDL measurement campaigns at different track types under different conditions over a decade. Test data in 2016 showed that Sound

G. Degrande et al. (Eds.): *Noise and Vibration Mitigation for Rail Transportation Systems*, NNFM 150, pp. 462–469, 2021.
https://doi.org/10.1007/978-3-030-70289-2_49

Transit LRVs have three distinct frequency regimes: low frequency, mid frequency, and high frequency regimes [2, 3]. The low frequency peaks between 8 Hz and 10 Hz are influenced predominantly by the primary suspension characteristics, driven by rail undulation and wheel run-out. The mid frequency FDL peaks, especially at 20 Hz and 25 Hz are influenced by the center truck characteristics, possibly related to lateral slip and axial wheel or suspension stiffness. The FDL peaks between 40 Hz and 100 Hz are influenced by rail roughness, the wheel-rail profile match, and the frictional forces at the wheel/rail interface [2].

In theory, FDLs between 40 Hz and 100 Hz may be affected by rail grinding and wheel truing, the rail fastener stiffness, vehicle primary suspension, wheelset characteristics, wheel stiffness, and damping of the bogie [3–6]. Factors such as roughness of the wheel and rail, unevenness of rail, and track type also influence FDLs [7]. FDLs at peak groundborne vibration and groundborne noise frequencies have shown 7 to 10 dB reduction at Sound Transit and Sydney Metro Line with rail profile grinding [2, 8 and 9].

The FDL measurements at Sound Transit indicate a substantial increase in FDLs at two sites between 2013 and 2016 [3, 10]. A review of rail surface conditions showed no noticeable difference between the 2013 and 2016 roughness data measured with a Corrugation Analysis Trolley (CAT). The condition of the vehicles and the wheel truing cycles were similar during this monitoring period.

Rail grinding was performed system wide at Sound Transit between 2018 and 2019. The rail grinding included profile correction of the rail to better match the wheels after extensive study of the rail and wheel interaction using vehicle dynamics models and field monitoring of rail wear. Studies are underway at Sound Transit to optimize the wheel profile for improved performance of both the powered and center car bogies. The rail grinding also included a polishing step to eliminate grinding-induced corrugation (GIC) by using softer grinding stones and higher grinder car speed compared to that used during the profile correction step.

FDLs were measured at the test sites after two to three weeks of revenue train service following grinding [11]. This paper presents FDL data from before and after grinding and discusses the results.

2 Background

2.1 Test Sites

The FDL measurements were performed at the following sites:

- **Site X (2013, 2016, and 2019):** This test site is located on 148th Street in Tukwila near Seattle. The rails are installed on natural rubber direct fixation fasteners with a nominal vertical static stiffness of 24 MN/m at 0.76 m pitch, giving a static rail support modulus of 33 MPa. The radius of curvature for this track section is 1524 m and the super-elevation is 32 mm. The allowed design speed for this track is 89 kph. The tests were performed at train speeds ranging from 40 kph to 89 kph.
- **Site Y (2013, 2016, and 2019):** This test site is a ballast-and-tie track section located south of the Rainier Beach Station on Martin Luther King Boulevard (MLK Site) in Seattle. The track section is approximately 300 m long and has two curves separated

by 60 m. The curve radii are 792 m and 1067 m, respectively. The design speed of this section is 64 kph and maximum revenue train speed is 56 kph. The tests in 2013 were performed over a range of 40 kph and 56 kph. The 2013 FDLs at higher speeds were extrapolated from Site X in 2013. The 2016 and 2019 FDL campaigns measured train vibration at speeds ranging from 40 kph to 89 kph.

- **Site Z (2012 and 2018):** This test was in the southbound Beacon Hill Tunnel with 4-car test trains and trued wheels. The rails are supported by resilient direct fixation fasteners with nominal stiffness 24 MN/m at 0.76 m pitch. Both measurements were performed before the rail grinding program. This section has a slight curve and is close to a wayside flange lubrication unit located at the southern end of the tunnel.
- **Site Z Station (2018 and 2019):** This test was performed at the Beacon Hill Station southbound platform with 4-car trains and trued wheels. The rails supported by resilient direct fixation fasteners of nominal stiffness 24 MN/m at 0.76 m pitch. Beyond the ends of the station platform are curves with 335 m radii. The allowed design speed for this track is 64 kph. The 2018 measurement was performed before grinding at the same time as Site Z. The 2019 measurement was performed after rail grinding.

The test campaign in 2013 was performed with a 3-car test train. All other tests were conducted with 4-car test trains. All test train wheels were visually checked for smoothness and none had any apparent wheel flats. The FDLs presented in this paper are based on test train vibration from northbound (NB) tracks.

2.2 Force Density Level

The FDLs were measured using the empirical method described in the FTA Guidance Manual [1]. The Line Source Transfer Mobilities (LSTM) were measured in 2013 and 2016 at two sites and the results are presented in Sect. 3. The differences in the two LSTM measurements were within typical uncertainties for one-third octave bands. For example, the uncertainties represented by the standard deviation for LSTMs between 31.5 Hz and 200 Hz were within 1 dB. However, at very low frequencies such as 5 Hz the uncertainties were closer to 2 dB. The train vibration levels (Lv) were measured at the same locations as the LSTMs at different train speeds. The FDLs were calculated with the following formula:

$$\mathrm{FDL} = \mathrm{Lv} - \mathrm{LSTM} \quad (\mathrm{dB}) \tag{1}$$

(All values are in decibels with consistent reference magnitudes and units. In the US, the force density level uses a force density reference magnitude of 1 lb/ft$^{1/2}$ and the vibration velocity reference magnitude is 1 micro-in/s).

2.3 Rail Wear

Rail wear data are collected systemwide with a track-geometry car with laser scanning technology on an annual basis and the data are stored in a database. The rail wear data presented in Table 1 are head loss and vertical rail wear. Both head loss and vertical rail wear represent the deviation of the respective metrics from the rail grinding template.

2.4 Rail Grinding

Sound Transit uses four rail profiles: two profiles for the tangent rail, and one profile each for the high rail and low rail on curves. The tangent rail profiles produce a contact band either closer to the gauge side or the field side. The two tangent rail profiles are used strategically to spread the wheel tread wear and mitigate wheel tread hollowing.

The rail grinding car uses a 3600-rpm vertical-axis grinding motor and grinding car speed of approximately 5.6 kph–6.4 kph (3.5 mph to 4 mph). To minimize GIC, a polishing step was implemented using softer stones with finer grit and higher grinding car translational speeds. The grinder speed for the polishing step was approximately 9.6 kph (6 mph).

3 FDL Results and Discussion

The focus of this section is to compare 1/3 octave band FDLs at ground-borne noise frequencies above 31.5 Hz and discuss their relationship to rail condition. The FDL differences below 31.5 Hz are not discussed here. Figure 1 shows that the third octave FDLs in 2017 increased at 63 Hz and 80 Hz. A 7-dB increase occurred at 63 Hz for 89 kph train speed at Site X. The increase at Site Y at 56 kph train speed was 3 dB. This vibration increase was remarkable and was the motivation for this study.

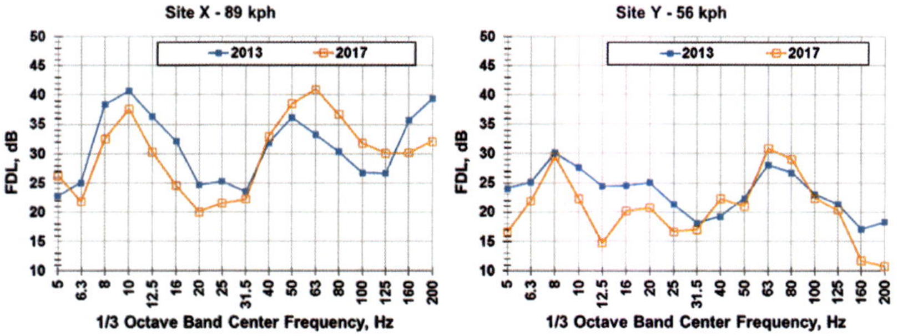

Fig. 1. FDL increase in 2017.

Figure 2 shows the FDL results at Site Z and Site Z Station at 56 kph. Site Z results for tangent track did not change between 2012 and 2018 above 31.5 Hz. This was unlike the vibration increase in 2017 at Sites X and Y. Site Z might have benefitted from the wayside lubrication system located about 600 m before this site and the tunnel preventing the lubrication coat on the rail from being washed away by the rain. If so, friction as well as roughness would appear to play a strong role in ground vibration generation.

Site Z Station data in Fig. 2 show the pre-grinding (2018) and post-grinding (2019) FDLs. The 2018 FDL data contain a sharper peak at 50 Hz and 63 Hz. This is believed to be the result of track geometry that has two 335 m-radius curves separated by a 122 m tangent section. The rail wear data from before grinding is shown in Table 1. These data show that the vertical wear of the high rail (East rail) at Site Z Station was the highest of the four test sites. The deceleration and acceleration of the trains in the station area and the track geometry would have contributed to this higher wear rate. The wear in the high rail of the curve south of the station could have resulted in increased lateral motion as the train entered the station during the tests. This could have contributed to the sharp peak at 50 and 63 Hz. The rail grinding in 2019 reduced the Site Z Station FDL by about 10 dB at 50 Hz indicating that the improved wheel rail profile match at the curve was favorable to vibration.

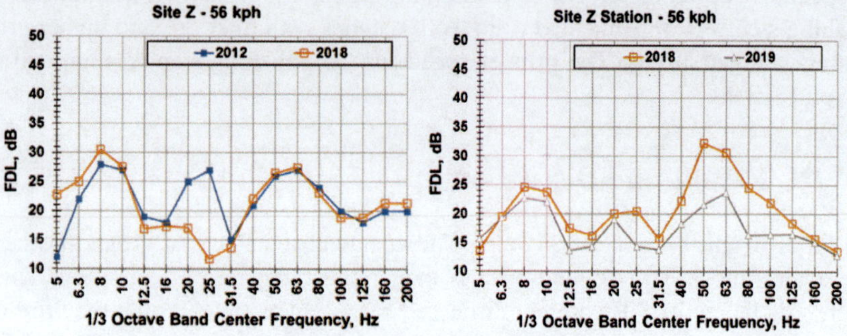

Fig. 2. Comparison of FDL at Site Z and Site Z Station

Table 1. Rail wear measured before rail grinding in 2018.

Site	Rail	Rail head loss	Vertical wear
Site X, NB track	West rail	3.3%	1.02 mm
	East rail	3.5%	1.27 mm
Site Y, NB track	West rail	2.6%	1.02 mm
	East rail	2.5%	1.02 mm
Site Z, NB track	West rail	0.9%	0.00 mm
	East rail	2.6%	1.02 mm
Site Z Station, NB track	West rail	1.6%	0.76 mm
	East rail	3.6%	1.78 mm

Figure 3 shows the FDL results at Site X from 2017 and 2019. FDL's between 31.5 Hz and 80 Hz were generally less in 2019 than in 2017 at all train speeds. Figure 4 shows Site Y to have more variation of vibration with speed between 31.5 Hz and 80 Hz. There is noticeable reduction in vibration at Site Y after grinding between 80 and 160 Hz.

Figure 5 shows the FDLs at Site Z Station. The pre-grinding data from 2018 show very limited speed dependency between 40 and 80 Hz. The post-grinding data show higher speed dependency in the same frequency range, which is difficult to explain. It is as if removing rail roughness and related vibration allows vibration from perhaps wheel roughness or other mechanisms to become more apparent, mechanisms that may be less predictable and even involve non-linear interactions of some sort.

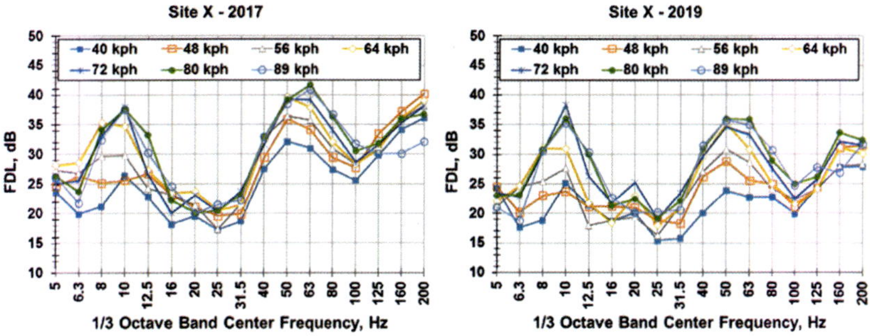

Fig. 3. FDL at Site X.

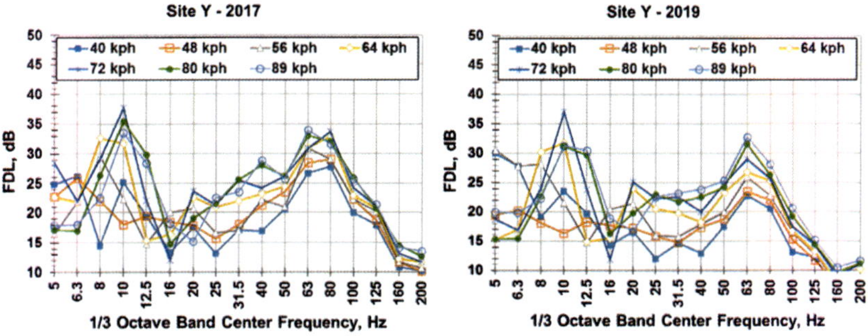

Fig. 4. FDL at Site Y.

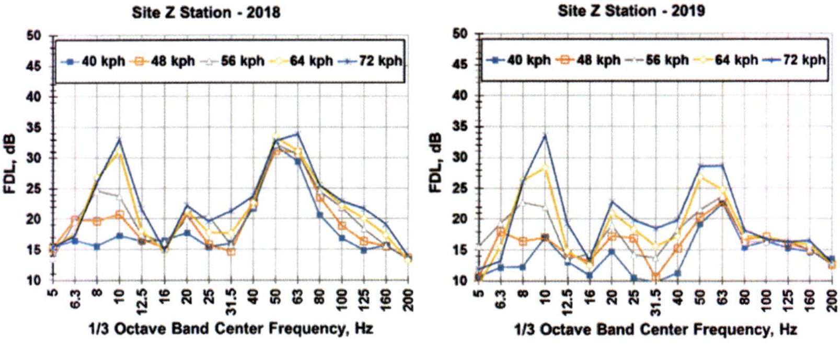

Fig. 5. FDL at Site Z Station.

Figure 6 shows the overall change in FDL at different speeds due to grinding at Sites X, Y and Z Station. In general, vibration above 31.5 Hz was reduced at all three sites. However, the level of reduction varied with speed and frequency. The vibration reduction at Site X showed the least speed dependency between 63 and 100 Hz, where the reduction ranged from 5 dB to 9 dB.

The vibration reduction at Site Y was at least 4 dB at all speeds at 40 Hz. At 50 Hz and 63 Hz the vibration reduction was greater at lower speeds. The least vibration reduction

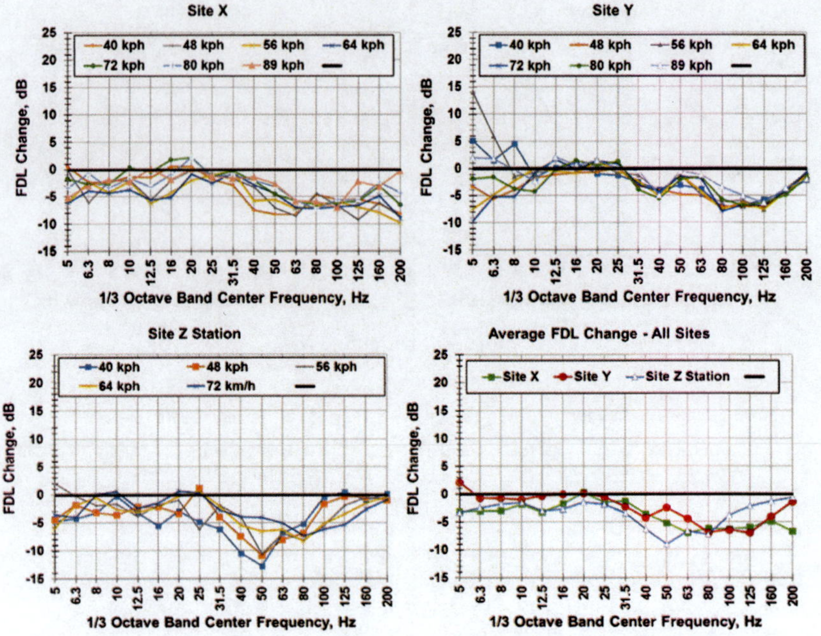

Fig. 6. FDL change at Sites X, Y and Z Station due to rail grinding.

above 31.5 Hz at Site Y was measured at 55 mph. The low frequency variation is likely due to differences of wheel rotation frequency rather than differences in roughness. As discussed earlier, this site has two curves that are within 60 m of each other. The ratio between the actual super-elevation (Ea) and the unbalance (Eu) at the curves is expected to move the center of gravity of the bogies more towards the high rail at speeds exceeding the 64 kph design speed. Also, the independently rotating wheels (IRW) of the center bogies at these speeds will flange through the first curve and may not recover in the short spiral before entering the second curve. Therefore, these results indicate that the vibration reduction from profile grinding was limited by the track geometry of this site at higher train speeds.

The vibration reduction at Site Z Station was most significant at 50 Hz for train speeds ranging from 40 kph to 56 kph, exceeding 10 dB. The vibration reduction from rail grinding was negligible above 80 Hz at lower train speeds.

The data from the three test sites show that rail grinding reduced vibration between 40 and 125 Hz by an average of 3 to 10 dB.

4 Conclusions

Light rail system FDLs were monitored at several test sites for about a decade. In 2017, there was a substantial increase in measured FDL at the track resonance frequency between 30 and 80 Hz. The only exception was at Site Z Station. The FDLs were measured at three sites after a systemwide rail grinding program designed to correct the rail profile and minimize GIC. The post-grinding data showed significant vibration reduction in the frequency ranges that are critical to groundborne noise and vibration. The FDL dependency on train speed varied with frequency and there was no linear correlation. A main conclusion is that ground vibration forces may vary considerably with train speed, rail profile and wear, corrugation, grinding procedures, bogie type, and other factors that may include non-linear lateral slip of IRWs.

References

1. Transit Noise and Vibration Impact Assessment Manual, report FTA Report No. 0123, Federal Transit Administration, Washington DC (2018)
2. Rajaram, S., Nelson, J.T., Saurenman, H.: A comprehensive review of force density levels from sound transit's light rail transit fleet. In: Anderson, D., Gautier, P.-E., Iida, M., Nelson, J.T., Thompson, D., Tielkes, T., Towers, D.A., de Vos, P., Nielsen, J.C.O. (eds.) Proceedings of 12th International Workshop on Railway Noise. Notes on Numerical Fluid Mechanics and Multidisciplinary Design, vol. 139, pp. 633–644. Springer, Heidelberg (2016)
3. Rajaram, S., Saurenman, H.J.: Evaluation of force density levels of light rail vehicles. Transp. Res. Rec. J. Transp. Res. Board **2302**, 21–29 (2015)
4. Nelson, J.T.: Tri-Met track vibration isolation tests. Technical report, Prepared for Tri-Met, 30 October 1998
5. Nelson, J.T., Watry, D.L.: Sound transit Kinkisharyo LRV force density level. Technical Memorandum, Prepared for Sound Transit, 12 December 2007
6. Rajaram, S., Saurenman, H.J.: Variation of force density levels of light rail vehicles. In: Burroughs, C.B., Mailing, G. (eds.) Proceedings of 24th National Conference on Noise Control Engineering (Baltimore), vol. 1 of 3, pp. 1554–1563. Noise Control Foundation, Ashland Ohio (2010)
7. Auersch, L.: Excitation of ground vibration due to the passage of trains over a track with trackbed irregularities and a varying support stiffness. Int. J. Veh. Mech. Mobility **53**, 1–29 (2015)
8. Nelson, J.T., Watry, D.L.: North link extension final design task 830.2. 90% ground borne vibration predictions. Technical Memorandum, Prepared for Sound Transit, 14 September 2012
9. Lawrence, B.: Effect of rail grinding on rail vibration and groundborne noise: results from controlled measurements. In: Proceedings of ACOUSTICS 2004, pp. 105–110 (2004)
10. Phillips, J., Nelson, J.T., Bergen, T.F.: Test report for LRV reference vibration measurements re: Lynnwood link extension final design. Prepared for Sound Transit, 28 July 2017
11. Hays, T., Bergen, T.F.: Sound transit LRV passby vibration measurements before and after rail grinding (draft). Prepared for Sound Transit, 30 May 2019

Predicting Vibration in University of Washington Research Facilities up to 500 m from a Light Rail Tunnel

Hugh Saurenman[1](✉), Roberto Della Neve Longo[1], and James Tuman Nelson[2]

[1] ATS Consulting, 130 West Bonita Avenue, Sierra Madre, CA 91024, USA
{hsaurenman,rdellaneve}@atsconsulting.com
[2] Wilson Ihrig, 6001 Shellmound Street, Suite 400, Emeryville, CA 94608, USA
jnelson@wiai.com

Abstract. The Northgate Link extension of the Sound Transit Link Light Rail system starts at the southern tip of the University of Washington (UW) Seattle campus and extends northward through the campus in twin-bore tunnels. The tunnels pass 24 UW research laboratories that are identified as being vibration sensitive. UW researchers are concerned that vibration and electro-magnetic interference generated by the light rail system would degrade the research environment and compromise the quality of UW research. A second concern is their ability to attract researchers to the campus if there is a perception that the transit system vibration degrades the research environment. The time frame of interest to the UW is the foreseeable future, perhaps a century or more. To address these concerns Sound Transit agreed to install a vibration monitoring system in the tunnels and use the measured train vibration to identify any trains that create vibration that exceeds the negotiated limits. This paper describes the equipment and procedures used to determine the relationships between the vibration in the tunnel and the vibration at the sensitive spaces.

Keywords: Groundborne vibration · Vibration sensitive laboratories · Transfer function

1 Introduction

The Northgate Link (N-Link) extension to the Sound Transit Light Rail Transit System starts at the southern tip of the Seattle campus of the University of Washington (UW) and extends approximately 2 km to the northern boundary of the campus. In response to the concerns of UW researchers about vibration, Sound Transit and UW negotiated a Master Implementation Agreement (MIA). Among other provisions, the MIA includes maximum allowable train vibration levels inside 24 buildings and requires Sound Transit to install a Vibration Monitoring System (VMS) that will continuously monitor train vibration at 28 locations between UW Station and the University District Station.

G. Degrande et al. (Eds.): *Noise and Vibration Mitigation for Rail Transportation Systems*, NNFM 150, pp. 470–477, 2021.
https://doi.org/10.1007/978-3-030-70289-2_50

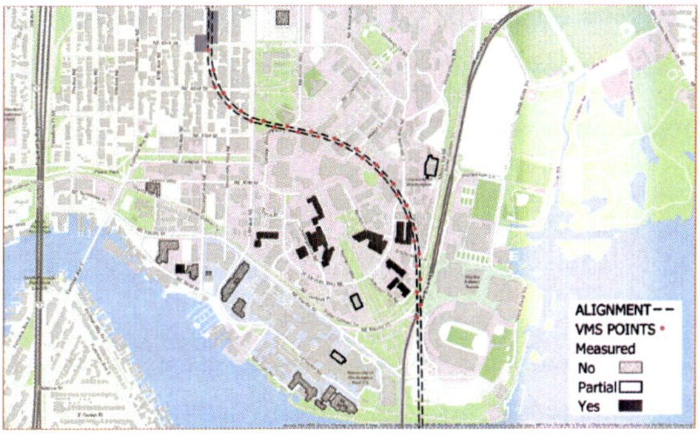

Fig. 1. Map of University of Washington Campus

Figure 1 shows the UW Campus with some of the buildings covered by the MIA. The straight line distance from where the N-Link route enters the southern tip of the UW campus to where it exits the northern boundary of the campus is 1.7 km.

To determine if and when the vibration level at one of the 28 VMS positions indicates that vibration at the research facility exceeds the MIA limit, relationships are needed between vibration at the VMS positions and the resulting vibration at each of the 24 research laboratories. This paper presents the procedures that were used to measure the Vibration Adjustment Estimates (VAE) from 18 of the VMS sites to 14 of the 24 buildings. This involved a total of 252 measurements that were performed to characterize 672 combinations of VMS positions and research facilities. The remaining 420 VAEs were estimated using the results from the 252 measurements.

A key point is that Sound Transit considered accurate VAEs to be critical to ensure that the UW will be able maintain its reputation for world class research facilities. Another factor is that the MIA includes provisions for substantial liquidated damages (financial penalties) if any combination of a VAE and the measured train vibration at a VMS site exceeds the MIA limit. To minimize the potential for exceedances of the MIA limits, a 5 Hz floating slab system was installed for the entire N-Link route from the UW Station at the southern tip of the campus to the western boundary of the campus.

2 Background

The Northgate extension to the Sound Transit Link light rail system is currently under construction with revenue service operations scheduled to start in 2021. The extension as it passes through the University of Washington Campus consists of dual subway tunnels. As the Northgate project went through the environmental assessment and the final design, considerable concern existed at the UW that vibration created by trains would adversely affect UW research facilities. To address these concerns, Sound Transit and the University developed a Master Implementation Agreement (MIA), dated 14 June 2007. The MIA lists the maximum allowable 1/3 octave band vibration velocity levels inside 24 facilities due to train operations. It also stipulates the potential for substantial financial penalties whenever train vibration exceeds the agreed upon limits. To check whether train vibration ever exceeds the MIA vibration limits, the MIA also stipulates that Sound Transit will install a Vibration Monitoring System (VMS) that will continuously measure the vibration at 28 locations in the tunnel between the UW Station and the University District Station.

The goal of this project was to perform measurements that allow accurately estimating the relationship between vibration measured at the 28 VMS positions and the resulting vibration at the 24 UW buildings listed in the MIA. These relationships are referred to as Vibration Adjustment Estimates (VAEs), which are defined as:

$$L_V(Build_i) = L_V(VMS_j) + VAE_{ij}$$

$L_V(VMS_j)$ is the measured vibration in the tunnel at VMS site j, VAE_{ij} is the VAE between VMS site j and Building(i) and $L_V(Build_i)$ is the predicted vibration level at Building(i). All these parameters are assumed to be in decibels with consistent reference values.

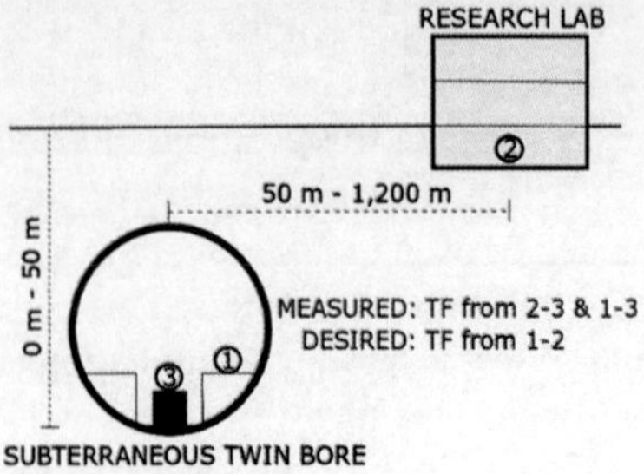

Fig. 2. Schematic of VAE measurement

Figure 2 is a schematic of the measurements used to measure the VAEs. The vibration source is at position 3 in the tunnel. The research laboratory is at position 2 in the building. The VMS site where train vibration will be monitored is position 1 on the safety walk. The desired measurement is 1 to 2 and the actual measurement is from 3 to 2. The 1 to 2 values were estimated by combining a measurement from 3 to 1 with the measurements from 3 to 2.

The VAE study for the Northgate Link is a follow-on to a study of the University Link (U-Link) extension (Ref. 1, 2), which opened in 2016. The experience developing VAE estimates for the U-Link extension provided important guidelines for developing VAE estimates for the N-Link project. Key guidelines are:

- A substantial vibration source is needed. The vibration source for the U-Link testing is an apparatus that dropped a 20 kg weight from a height of 1.2 m. The hammer is suitable for measuring at distances up to about 100 m. For this project measurements at distances up to 2,000 m were required.
- All of the data should be synchronized. In the U-Link analysis we found that obtaining transfer functions between any two channels was important for evaluating the measurement results.
- Testing should be performed during nighttime hours when ambient vibration is at a minimum. For the U-Link testing, vibration from traffic as much as 75 m from the tunnel created vibration in the tunnel at frequencies below 20 Hz that would sometimes exceed the vibration generated by trains that were less than 3 m from the measurement position. To minimize interference from other vibration sources, all N-Link testing was performed between 12 AM and 6 AM.

3 Instrumentation

The vibration sources investigated included various types of impact devices and shakers. Impact devices have the advantage of test efficiency because an impulse excites all frequencies at once. Impact devices mounted on pickup trucks would have provided sufficient force amplitudes, but it was not feasible to get a shaker truck into the tunnel. The best alternative was a portable shaker that could provide either swept-sine or sine-dwell testing. The equipment selected were two 181 kg (400 lbf) "Lightning" shakers from Seismic Mechatronics. Two shakers were used in synchrony to double the force signal. Figure 3 is a photograph of the shaker pair used for the testing.

Figure 4 shows the maximum force generated by a single shaker at low frequencies. The single shaker provided 181 kg (400 lb) force at frequencies greater than about 20 Hz. The maximum force decreased dramatically at lower frequencies. The decreased force at lower frequencies was a primary reason for deciding to use two synchronized shakers rather than a single shaker.

Fig. 3. Lightening Shakers used for vibration testing

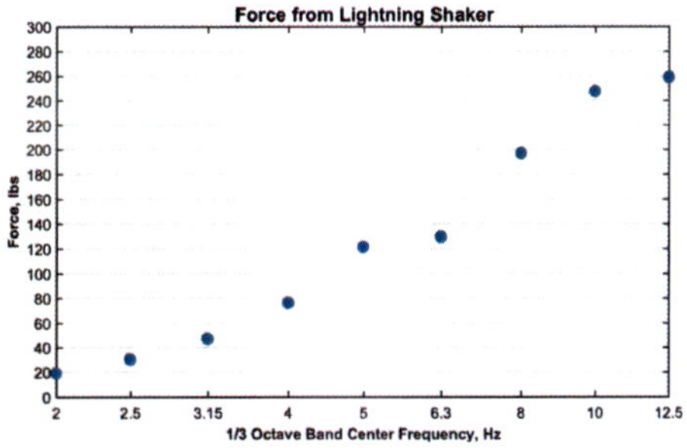

Fig. 4. Force generated by single Lightening Shaker

The vibration responses were measured with geophones mounted on small flat plates that were fixed to the test surface with beeswax. For outdoor measurements, the geophones were mounted on aluminum ground spikes. For the indoor measurements, the geophones were mounted on the bottom basement floor of the buildings, which was typically a slab-on-grade floor. The geophones have a corner frequency of 4.5 Hz that, with deconvolution, provided accurate results down to as low 1 Hz.

The signals from all the geophones were transmitted wirelessly to a central recording station using a Wireless Seismic RT2 system. Wireless Remote Units (WRU) communicated via radio telemetry using the 2.5 GHz band. Data from all WRUs were collected at a Line Interface Unit (LIU) and then transmitted to the Central Recording System (CRS), which was controlled by a laptop. Data from up to 20 WRUs were recorded by the CRS, both within the tunnel as well as buildings.

Data from all geophones and the force transducer were recorded into single Seg-Y files. The data collection system was set for a sampling frequency of 1000 Hz to avoid aliasing at frequencies up to 125 Hz.

4 Test Approach

Because of the concern of the UW administrators and researchers, performing measurements that would provide accurate estimates of all 672 VAEs was critical for the success of this project. The process to develop the successful test plan included:

1. Performing a preliminary test program to verify and refine the test procedures. Of particular concern was whether the selected vibration source would be of sufficient strength to obtain valid measurements at the more distant positions.
2. Selecting a suitable vibration source and test procedures.
3. Working closely with the UW staff to make sure that they understood the importance of the test program. The cooperation of the UW staff aided the on-time completion of this project.
4. Coordination with the ongoing construction. There were a number of construction activities occurring in parallel with our measurements. This process would have been much been more difficult if we had not had the cooperation of the contractors.

Following is an outline of the approach used to collect the vibration data:

1. Instrument 14 of the 24 buildings listed in the MIA with geophones and simultaneously record the vibration responses from all of the geophones. Using the Wireless Seismic RT2 system, there were no discernable instrumental phase drifting between the receiver channels and excitation forces.
2. Use the shaker vibration source to apply dynamic forces to the tunnel floor and simultaneously record the responses at the instrumented buildings, surface locations, and tunnel locations. Most of the tests were performed using a sine-dwell technique to overcome background vibration and achieve the best possible coherence between the excitation force and the resulting vibration velocity. The tests were performed at nominal 1/3 octave band center frequencies, with minor adjustments in some locations to avoid discrete spectral peaks in background vibration. This process was enhanced by the remarkably stable shaker vibration frequency over the duration of each test, which typically was 10 to 40 min.
3. The point source transfer mobility (PSTM) was calculated between the shaker force and the vibration velocity recorded at each geophone location at the UW buildings.

Because each measurement required testing at 10 frequencies and the test at each frequency would require 30 to 40 min, each night of testing was limited to testing at one VMS site. The testing was performed between December 10, 2018 and January 20, 2019, with a break in the testing between Christmas and New Year. The test schedule included a week of contingency days. Because the testing went smoothly, it was not necessary to use all of the contingency days.

In addition to the careful planning that went into developing the test plan, the cooperation of our UW contacts, the contacts from the three construction teams who were active during the measurements, and the Sound Transit Personnel responsible for this project, were important elements in the success of this test program.

5 Processing Raw Data

The initial step in the data processing is to obtain the transfer functions between the exciting force in the tunnel and the resulting vibration at each of the building measurement position. Using the terminology in the FTA Guidance manual [3], this transfer function is referred to as the Point Source Transfer Mobility (PSTM). Because the sine-dwell approach was used for most of the data collection, each measurement resulted in the transfer function and coherence at a single frequency. Figure 5 is an example of the transfer function between Wilcox Hall and the VMS position 7 at 16 Hz. The left plot is the PSTM in decibels and the right plot is the coherence. There is a distinct dip at 16 Hz in the PSTM and a corresponding peak in the coherence. The PSTM at 16 Hz is −48.54 dB and the coherence is 0.58. There also is a coherence peak at 60 Hz. This is an artifact of the 60-cycle power that was used for the measurement equipment. For this data point, the shaker was run for eight 30-s periods with a 15 s quiescent period between each 30 s period. These quiescent periods were used to avoid overheating the shakers.

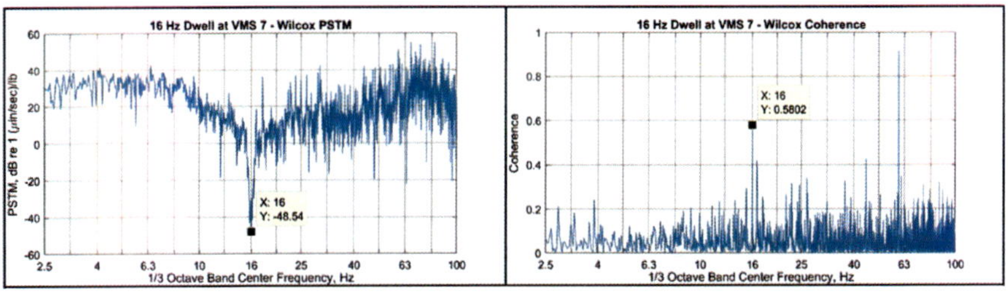

Fig. 5. Example PSTM Result from Sine-Dwell Test at 16 Hz

Figure 6 shows the PSTM magnitude and coherence for the frequencies between 4 and 100 Hz for the measurements at VMS sites 3 to 10 near the southern end of the N-Link tunnels. VMS 3, 4, and 6 are the closest to Wilcox Hall. The PSTM magnitude and the coherence are highest for these sites. The coherence data indicate very good data for these sites at frequencies of 6.3 Hz and greater. Coherence values range from 0 to 1. Coherence values close to 1 indicate strong confidence in the results and values less than 0.1 indicate the results may be strongly influenced by background vibration. Although the coherence values computed from the Phase 2 test data were often relatively low, they are almost always greater than 0.1, which shows that the results are valid for estimating VAEs.

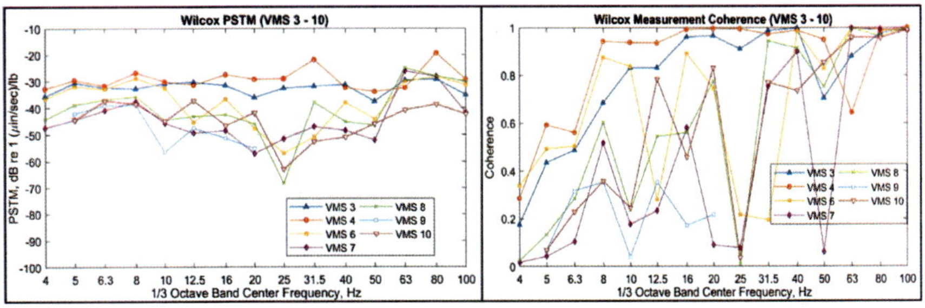

Fig. 6. Measured PSTM and Coherence at Wilcox

The PSTMs measured inside the UW buildings were used to develop level-versus-log10 (distance) regression curves for each 1/3 octave band from 4 Hz to 100 Hz. A variety of linear regression curves were investigated. The most appropriate regression curve was selected as a chi-square regression of the PSTM vs. Log distance. Both least-squares and chi-square minimizations were investigated. The three approaches to developing best fit curves were global using the data from all measurements, local regressions based on different grouping of building to see if any grouping reduced the spread, and building specific data using only the data applicable to each building. The conclusion of this effort was that the global regression was as good an approach as either of the alternatives.

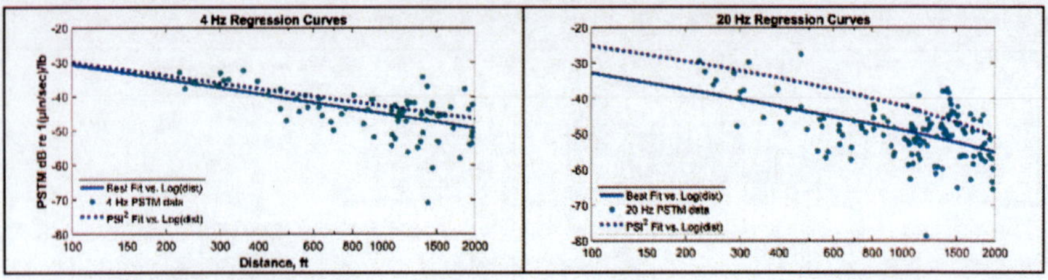

Fig. 7. Example Regression Curves for 4 Hz (left) and 20-Hz (right) Data.

Figure 7 shows examples of the regression lines for the 4 Hz and 20 Hz data that were used to derive the VAE values for all 672 building/VMS combinations. The data are the blue circles and two solid lines are the best fits for the data using the linear level vs. Log_{10}(dist). The broken lines are the linear level vs. Log_{10}(dist) derived using a chi-square fit. The advantage of the chi-square fit is that the variance derived from the signal coherence is used as a weighting that reduces the influence of values with low coherence. The least squares curves are lower than the chi-square curves, because the low coherence data points are below the curves and not above the curves.

At the buildings that are within 600 feet from the tunnel, the coherence at 10 Hz is generally greater than 0.2, which we considered satisfactory for including in the analysis. Also, the coherence tended to decrease at frequencies below 10 Hz because the input force of the shaker was limited at these frequencies. However, a coherence above 0.2 down to 4 Hz was achieved at the VMS positions closest to the buildings, which is significant considering the difficulty detecting the signal at low frequencies.

Chi-square regression curves similar to those shown in Fig. 7 were used to define PSTM vs. Distance curves for all frequencies. The steps from the PSTM for each VMS/building combination to the VAEs and then to the predicted vibration were:

1. Compare the PSTM from the PSTM vs. Distance curves to the measurement results for the specific buildings. Almost always the conclusion of this step was that the PSTM derived from the best fit curves were appropriate for the building. The chi-square fits were used in the analysis, although the chi-square and least squares fits usually were very similar.

2. Add an adjustment to approximate a measurement from the future VMS positions on the safety walk to the buildings. As shown in Fig. 2, the actual measurements were from the tunnel floor to the buildings.
3. Integrate the PSTMs over the train length to obtain the line source transfer mobility (LSTM), because trains are best modeled as line vibration sources. This adjustment was derived to be a 15 dB adjustment for the tunnel floor to each VMS position. These LSTMs are combined to obtain VAEs between VMS monitors and buildings.

6 Prognosis and Conclusion

The future train vibration levels at the 24 MIA buildings were estimated by adding the expected train force density level to the LSTMs between the tunnel floor and the buildings. The FDL curve used was derived from measurements at a section of the U-Link extension where a test version of the N-Link floating slab system is installed. The results are:

1. The estimated levels of train vibration at all 24 buildings listed in the MIA are well below the MIA limits.
2. The train vibration is predicted to be closest to the MIA limits at frequencies of 5 to 10 Hz, which range includes the floating slab and vehicle primary suspension resonance frequencies.
3. Comparing the predicted vibration levels from the closest VMS position to the MIA limit for each building, the predicted vibration levels are always well below the MIA limits. The smallest margin relative to the MIA limit is 9 dB at 5 Hz at the Mechanical Engineering Building Annex, which is directly above the subway and is the closest sensitive UW building to the Northgate Link tunnels.

References

1. ATS Consulting: Updated Vibration Attenuation Estimates for IEM Vibration Monitoring. Submitted to Sound Transit, 8 Jul 2016
2. Evans, A.L., Ono, C.G., Saurenman, H.J.: Estimating adjustment factors to predict vibration at research facilities based on measurements in a subway tunnel. Presented at IWRN12, Terrigal NSW, Australia, September 2016
3. Federal Transit Administration, Transit Noise and Vibration Impact Assessment Manual, FTA Report No. 0123, September 2018

Evaluation of the In-Situ Performance of Base Isolated Buildings

Raphaël Cettour-Janet[1](✉), Benjamin Trévisan[1], and Michel Villot[2]

[1] ACOUPHEN, 33 Route De Jonage, 69330 Pusignan, France
{raphael.cettour-janet,benjamin.trevisan}@acouphen.fr
[2] MV-Expert, 5b Chemin Thiers, 38100 Grenoble, France
michel.jean.villot@gmail.com

Abstract. Mitigation measures against railway vibration in buildings can include elastomeric mounts or springs inserted between building foundation and building upper structure. The method for predicting the performance is usually simplistic and based on the dynamic transmissibility of a single degree of freedom mass-spring oscillator on a rigid base. One method is to express the isolation performance as an insertion gain based on the building floor velocities. It is named the Building Insertion Gain Indicator (BIGI). Another method, called the Power Flow Insertion Gain (PFIG), has also been proposed in [1]. Then, the isolator performance is expressed as the difference of injected power to the building upper structure. In practice, both indicators need two identical buildings (with and without isolator) to be obtained automatically. A method to indirectly obtain the PFIG, using only an isolated building, has been developed and numerically validated for a 2D case in [2, 4]. It is based on the isolator transmissibility (ratio of velocities on both sides of the isolator) that is corrected with foundation and upper structure mobilities. In this paper, the theory is extended to a construction near railway tracks after being validated through comparisons between experiments and simulations. In this case, the isolator transmissibility and foundation mobility are measured. As the upper-structure mobility cannot be measured, it is obtained with a 2.5D simulation. Finally, it is compared to the other methods of evaluating isolator performance.

Keywords: PFIG · BIGI · Base isolation performance

1 Introduction

Usually, the mass-spring assumption is considered to estimate the performance of base isolated building. In practice at the end of the construction, the isolator transmissibility is measured to check its performance, without taking into account building behaviour. This article proposes to evaluate this performance using two other approaches: the BIGI and the PFIG, which take into account the structure behaviour.

G. Degrande et al. (Eds.): *Noise and Vibration Mitigation for Rail Transportation Systems*, NNFM 150, pp. 478–486, 2021.
https://doi.org/10.1007/978-3-030-70289-2_51

2 Indicators of Performance for Base Isolated Building

2.1 Theoretical Performances of Building and Isolator

The in-situ performance of base isolated buildings mainly depends on the ground and on the building's construction. Assuming the acoustic radiation is not impacted by the isolator insertion, the performance of such isolation can be given by the BIGI (Building Insertion Gain Indicator) using the ratio of floor velocities in the isolated and un-isolated cases [4]:

$$BIGI = 10log\left(\frac{v^2_{floor,rms,isol}}{v^2_{floor,rms,unisol}}\right) \tag{1}$$

It is related to inhabitant perception in a specific room. Its value will be different in each room. Consequently, it is a local indicator.

The velocity levels in the case of railway ground borne noise are compared on Fig. 1 [3]. Then, the isolator performance as a BIGI, represented by arrows (↓), leads to three main conclusions:

- The performance tends to an asymptotic value when the modal overlap is sufficient (above 80 Hz, in this example)
- The performance decreases at the floor eigen frequencies

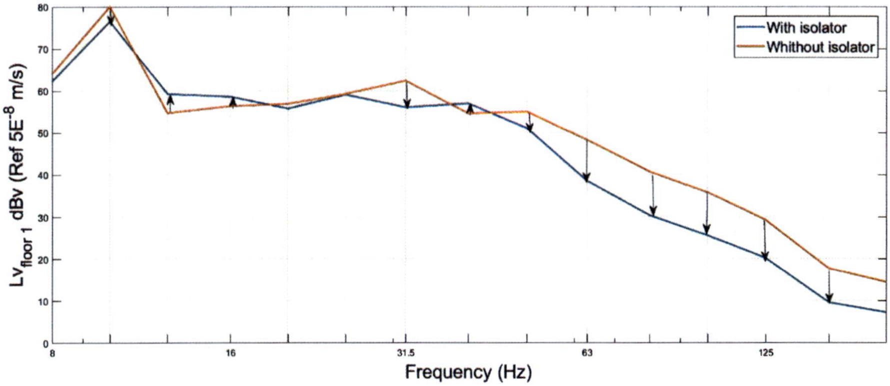

Fig. 1. Velocity levels of the first floor for isolated (blue) and un-isolated (red) cases. In situ performance (BIGI) represented as arrows (↓).

The performance can also be estimated with the PFIG (Power Flow Insertion Gain) allowing estimating the performance of the isolator. It is a global indicator. It is defined as follows [1, 2]:

$$PFIG = 10.log\left(\frac{\Pi_{isol}}{\Pi_{unisol}}\right) \tag{2}$$

Where Π_{isol} and Π_{unisol} are the power flow of the isolated and unisolated building.

2.2 Indirect Method for on-Site Performances

The indicators previously presented cannot be obtained by direct on –site measurements. However, considering that vertical waves are dominant, BIGI and PFIG [1] can be written differently leading to:

$$BIGI = \mathrm{TF}_{\mathrm{isolatedbuilding}} - \mathrm{TF}_{\mathrm{un-isolatedbuilding}} \tag{3}$$

$$PFIG \approx 10.log\left(\frac{v^2_{\substack{above\\ isolator}}}{v^2_{\substack{below\\ isolator}}}\right) + 20.log\left(\frac{|Yr| + |Ys|}{|Ys|}\right) \tag{4}$$

Where $\mathrm{TF}_{\mathrm{building}}$ refers to the transfer function from vibration measurement in free-field to the floor (can be defined as TF2+TF3 in ISO 14837-31:2007 [5, 7]) *Yr*; and *Ys* are respectively mobilities of the upper-structure and of the substructure. $\mathrm{TF}_{\mathrm{un-isolatedbuilding}}$ and *Yr* cannot be measured on site and have to be calculated.

3 Experiments and Comparison with 2.5D Simulation

Experiments have been done during the construction of an isolated building at different steps. These measurements will be compared to numerical models. The following measurements have been done:

- Mobility of the foundations *Ys* with impact excitation (see Fig. 2 at left)
- Isolator transmissibility with impact hammer (see Fig. 2 at the middle)
- Floor velocity with railway excitation (see Fig. 2 at right)
- $\mathrm{TF}_{\mathrm{isolatedbuilding}}$ with railway excitation according to ISO 14837–31:2007 [5] (see Fig. 3)

Fig. 2. Experiments on an isolated building at different steps. At left: foundations' mobility; at the middle: isolator transmissibility (end of construction with windows and doors); at right: floor velocity (end of construction with windows and doors).

In our case, the free field velocity could not be measured. The vibrations at a few meters from the building were measured and we assumed that they are the same whether the building is isolated or not. This assumption has been confirmed by the calculation.

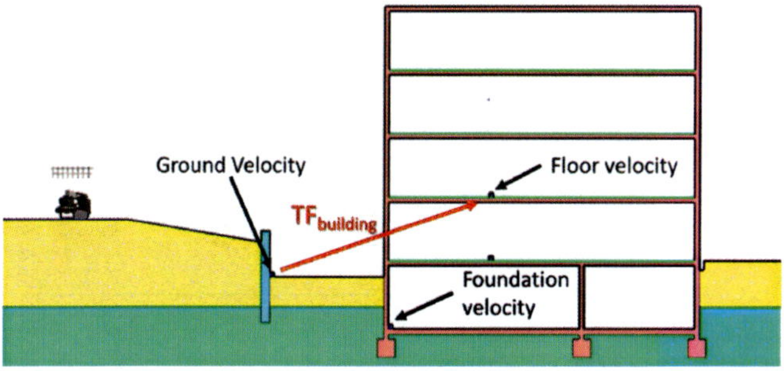

Fig. 3. Definition of $TF_{building}$ according to ISO 14837-31:2007 [5, 7]. Configuration of numerical model.

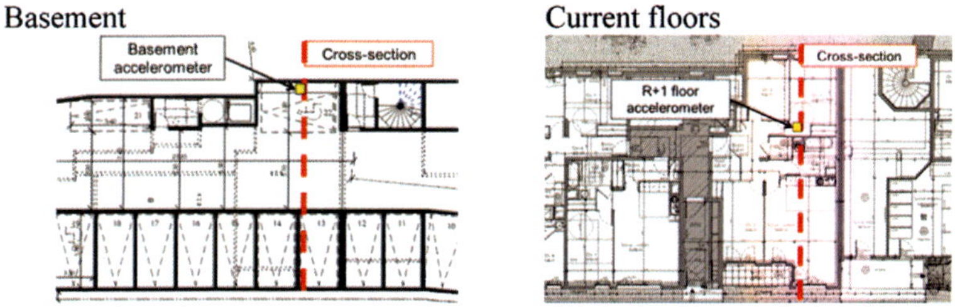

Fig. 4. Experimental setup labelled on building blueprints. Cross-section of numerical modelling represented in dashed red line.

3.1 Validation of Numerical Modelling

A FEM/BEM 2.5D model (MEFISSTO software) of the building is computed and compared to experiments. The cross section chosen for the simulation is shown on Fig. 3 and Fig. 4. Only the structural works of the section represented in red on Fig. 4 are modelled. Then, ground data are determined through MASW (Multichannel Analysis of Surface waves) and the building construction is based on blueprints (see Fig. 4). Moreover, the building is isolated with resilient elements for whose the resonant frequency is evaluated at 12 Hz using a mass-spring assumption.

Experiments are during train traffic and compared to model with a line source. Then the results are readjusted using the ground point velocity as the reference (see Fig. 3). Finally, experimental and model vibration levels are compared at the first floor and basement. Figure 5 presents the results for a specific train.

These results show a good agreement of the modelling with the experiments. On the first floor, a resonance peak in the simulation case does not appear in the experiment results. However, this frequency is out of the audible frequency range; consequently, this difference is not significant for ground borne noise (16–250 Hz according to ISO 14837-31:2007 [5]). In addition, at high frequency, the simulation underestimates the vibrational response. The 2.5D modelling that imposes simplifications in the third dimension can explain these gaps.

Beyond that, it is assumed that those variations are equivalent in the isolated and un-isolated case.

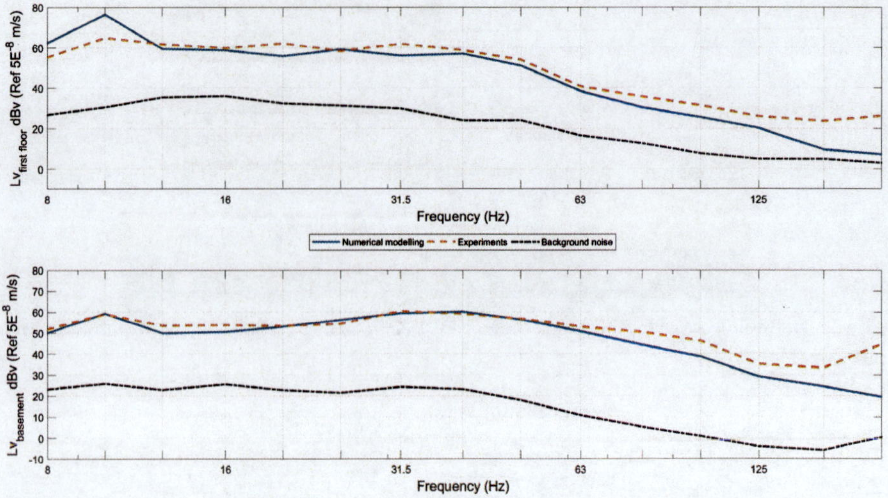

Fig. 5. Experiment vs modelling for a specific train: at first floor (top) and basement (bottom).

3.2 Building Insertion Gain Indicator

Figure 6 shows numerical and experimental $TF_{\text{isolated building}}$ and numercal $TF_{\text{unisolated building}}$, which are used to calculate the BIGI (Eq. 3).

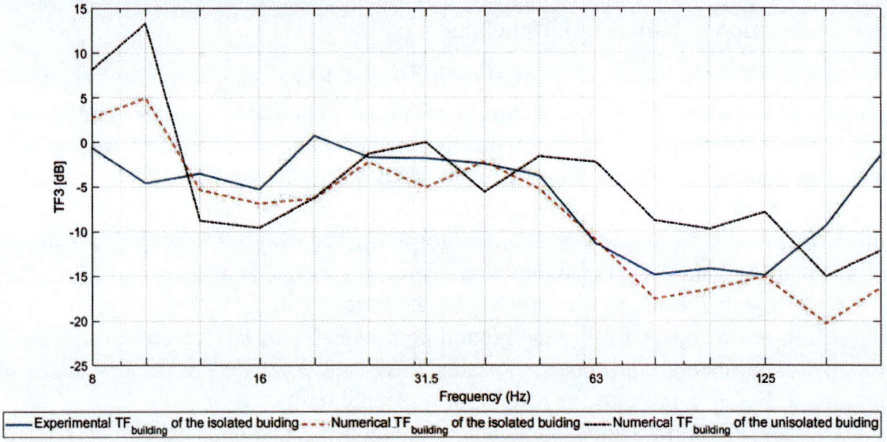

Fig. 6. Numerical and experimental $TF_{\text{isolated building}}$ and numerical $TF_{\text{unisolated building}}$

Numerical and experimental $TF_{\text{isolated building}}$ are similar above 12.5 Hz which is relevant in the case of ground borne noise [5]. The deviation at 200 Hz is due to the energy produced by the train becoming insufficient to provide sufficient excitation to the first floor. The resonance peak at 20 Hz is not calculated. This can be due to the 2.5D approximation.

BIGI (see Eq. (3)) is plotted (Fig. 7). Above 25 Hz, the most audible frequencies for ground borne noise [6], the curves are similar. Below this frequency, the greater difference is due to simplifications in the modelling (2.5D, infinity in the third direction) that not perfectly match with experimental as shown in Fig. 5.

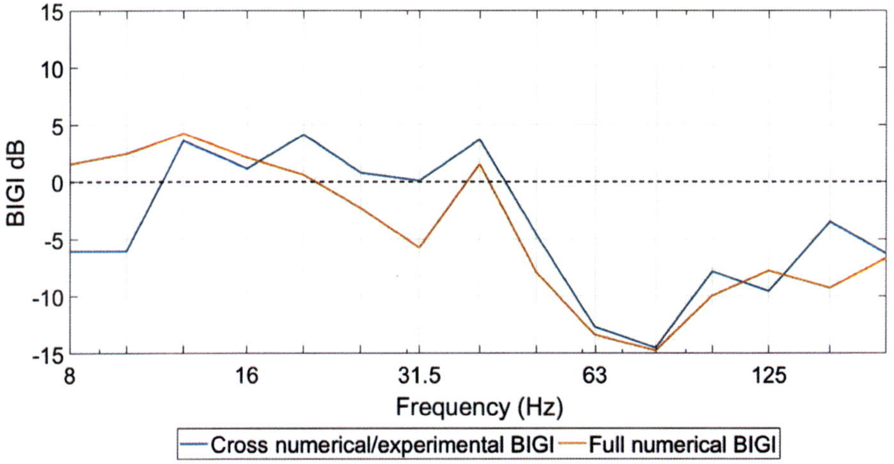

Fig. 7. BIGI evaluated through full numerical method and cross numeric/experimental method.

The isolator becomes efficient (i.e. the isolators allow to reduce the vibration level) at about twice the uncoupling frequency. Moreover, the isolation performance decreases at the main eigen-frequencies of the floor (40 Hz).

3.3 Transmissibility of the Isolator

The transmissibility is defined as follows:

$$TL = 10.log\left(\frac{v^2{}_{\substack{above\\isolator}}}{v^2{}_{\substack{below\\isolator}}}\right) \tag{5}$$

The transmissibility of the isolator has been measured using a hammer for a given isolator. (see Fig. 8). Using a 2.5D model, the transmissibility cannot be simulated. Consequently, in the PFIG calculation (Eq. (4)), the experimental result must be used.

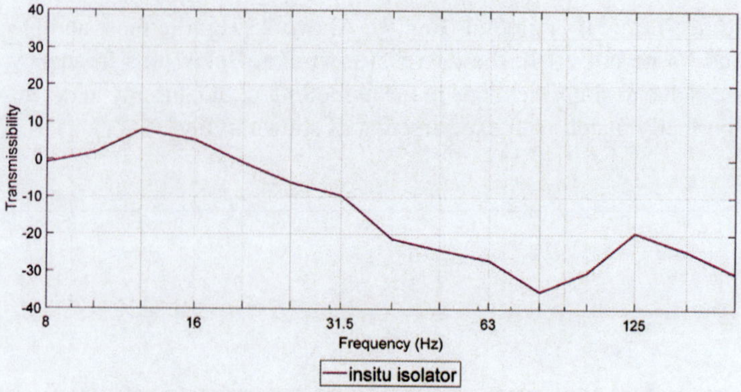

Fig. 8. Transmissibility of the isolator on a column.

3.4 Substructure and Superstructure Mobilities

Figure 9 compares the experimental mobilities of a column, a wall and a simulated wall.

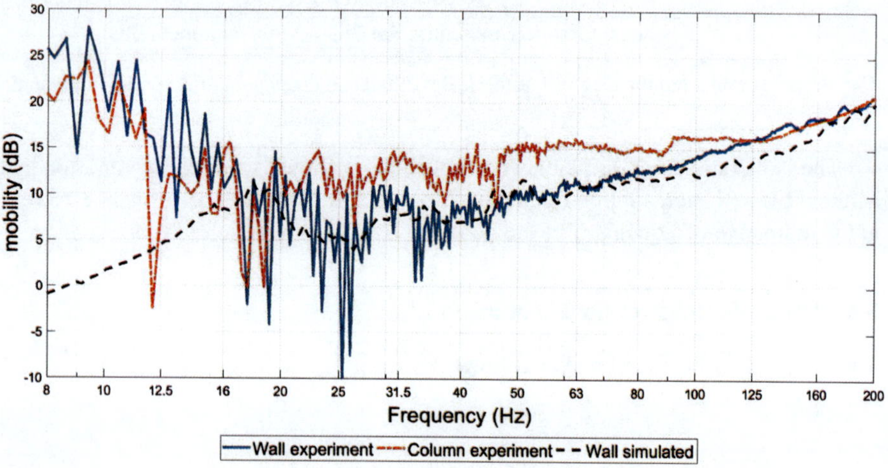

Fig. 9. Substructure mobilities.

These results show a good agreement between experimental modelling and simulation for walls. Moreover, the mobility of the column is more important than for the wall below 125 Hz, which confirms that the 2.5D modelling cannot simulate the column condition. In the following section, the experimental column mobility with a correction at low frequency is considered in the PFIG formula (Eq. 4).

Because the superstructure mobility cannot be measured, it is calculated with the 2.5D modelling and used in the PFIG formula (Eq. 4).

3.5 Comparison of Isolator Performance Indicators

In this section, comparison of the four isolator performance indicators is presented (see Fig. 10).

The PFIG is calculated with formula Eq. 4 using:

- The experimental transmissibility of the isolator
- The experimental substructure mobility
- The numerical superstructure mobility

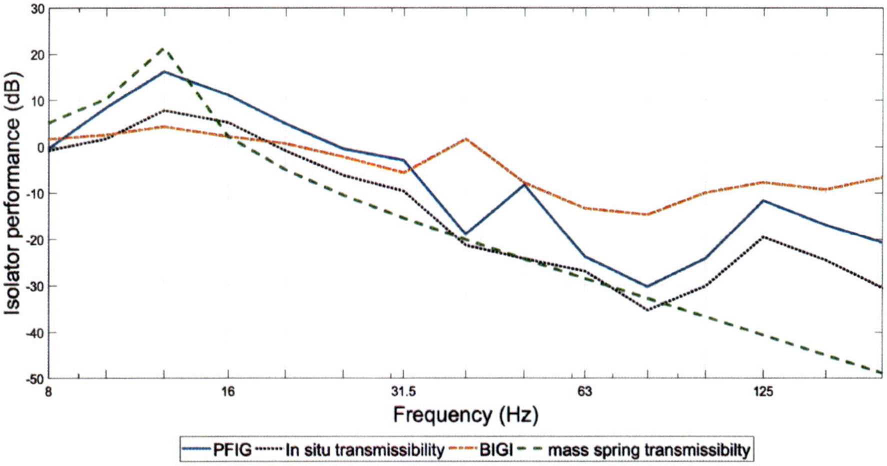

Fig. 10. Evaluation of isolator performance

The BIGI is numerically computed using the cross numeric/experimental method and the in-situ transmissibility is measured with a hammer.

These isolator performance indicators are compared with the performance given by usual mass-spring assumption which can be used to estimate the isolator performance.

4 Conclusion

In this article, the curve of a mass-spring system commonly used to estimate the performance of base isolated buildings has been compared to the isolator transmissibility, PFIG and BIGI (see Fig. 10). This comparison shows that the mass-spring assumption overestimates the performance of the isolator. The transmissibility allows full experimental evaluation but does not consider the building response.

On the contrary, the BIGI seems to be the most adapted indicator considering eigenfrequencies of the floor. However, it is a local indicator and it depends on the room measurement.

Finally, the PFIG used the transmissibility with a correction term considering mobilities in order to offer an intermediate approach. It is also possible to develop an adapted test bench to consider the superstructure's mobility through the addition of masses. This test bench has been built and tested in [8].

Acknowledgment. The authors thank BPI France and the Eurostar program for financially supporting the BIOVib project [3].

References

1. Talbot, J.P., Hunt, H.E.M.: On the performance of base-isolated building. Buil. Acoust. **7**(3), 163–178 (2000)
2. Villot, M., Trévisan, B., Grau, L., Jean, P.: Indirect method for evaluating the in-situ performance of building base isolation. Acta Acustica **105**, 630–637 (2019)
3. BIOVib project: Building Isolation against Outdoor Vibration. EUROSTARS call 17 number 11 420 - BIOVib agreed by the European Union. Partners: Belgian manufacturer CDM and ACOUPHEN (French consultancy company) (Sept. 2017 – March 2019).
4. Trévisan, B., Grau, L., Villot, M., Jean, P.: In-situ performance of base-isolated building. In: Euronoise 2018 Proceedings (May 2018)
5. ISO/TS 14837–31:2017: Mechanical vibration - Ground-borne noise and vibration arising from rail systems - Part 31: Guideline on field measurements for the evaluation of human exposure in buildings
6. ISO 226:2003 – Acoustique – Lignes isosoniques normales.
7. Villot, M., et al.: RIVAS – Procedures to predict exposure in buildings and estimate annoyance. CSTB. 2012. European report.
8. Masoumi, H., Veelhaver, B., Papadopoulos, M., Agusztinovicz, F., Carels, P.: Evaluation of BBI performance indicator in a full-scale test bench. In: Proceedings IWRN 13th, Ghent, Belgium (2019)

Ground and Building Vibration Estimation for Health Impact Research

Mikael Ögren[1(✉)], Alf Ekblad[2], Peter Johansson[2], Arnold Koopman[3], and Kerstin Persson Waye[1]

[1] Occupational and Envrionmental Medicine, Sahlgrenska Academy, University of Gothenburg, Box 414, 405 30 Gothenburg, Sweden
mikael.ogren@amm.gu.se
[2] Trafikverket - The Swedish Transport Administration, Gothenburg, Sweden
[3] Level Acoustics and Vibration, Eindhoven, The Netherlands
http://medicine.gu.se/english/phcm

Abstract. Building vibration from railway traffic is relatively common in Sweden. The vibration causes annoyance and interferes with sleep. Less is known about long term health effects of living in dwellings exposed to vibration, which is why a research project named EpiVib has been started where more than 6000 persons living close to railways have been recruited into a health cohort. Therefore we need to estimate the vibration exposure. However, the number of dwellings makes it difficult to use measurements for all, and limits how much data on geology can be collected. Instead, a semi-empirical model was developed based on 829 measurements in the area and geology data from official maps. The resulting method divides all soil types in the area into three classes and then estimates the vibration for different distances to the railway, and different main layer depths.

Keywords: Vibration from railways · Railway noise

1 Introduction

The health effects of noise exposure are relatively well researched; a recent report from WHO [1] has used the results of many studies when forming their guidelines. For vibration exposure there is not as much evidence. Therefore, a new study named EpiVib was started in 2016 with the aim of trying to link vibration exposure to health effects using similar methodology as previous research on noise exposure and health effects.

Three areas along railway lines where the ground is known to be sensitive to vibration propagation were selected in the south western part of Sweden. In these areas interviews were performed and a questionnaire was distributed; two papers have already been published, [2] analyzing the relation between annoyance and distance from the railway and [3] looking at interviews with a number of residents along the railway line.

G. Degrande et al. (Eds.): *Noise and Vibration Mitigation for Rail Transportation Systems*, NNFM 150, pp. 487–495, 2021.
https://doi.org/10.1007/978-3-030-70289-2_52

In a second step, those who answered the questionnaire and gave their consent will have registry based data on health and medication use linked to their response and the exposure to noise and vibration, and the relation between health outcomes and the exposure will be analyzed using epidemiological methods. Therefore, the vibration exposure of more than 6000 dwellings in the areas must be predicted.

In the vibration measurement database of Trafikverket (the Swedish traffic administration) there are 1447 measurements in the three areas. The measurements are performed either on the building foundation or indoors (typically at the center of the largest floor span), or both. A measurement series at a building often starts with a foundation measurement, and if the maximum level in the vertical direction measured at the foundation over 3–10 days is below 0.2 mm/s the results are considered to lie below the detection limit, and no indoor measurements is performed. If above 0.2 mm/s an indoor measurement is often, but not always, performed.

All measurement results and estimated vibration velocities in this paper are the maximum velocity using an exponential time weighting of 1 s, corresponding to the SLOW weighting used in acoustics, expressed in the unit mm/s. The measurements are band limited to the range 1–80 Hz according to the Swedish standard SS 4604861. Building foundation measurements are taken in the vertical direction only and are not frequency weighted. The indoor measurements are weighted according to the standard, and the direction which gives the highest maximum velocity is selected as the result.

The aim of this paper is to describe a method for predicting vibration exposures for the buildings in the study areas that have not been measured. The aim is not to develop a universally useful prediction scheme, but a method limited to the typical infrastructure, geology and buildings of the EpiVib study.

Prediction methods for noise from railways are typically official documents openly available, and the methods are also often implemented in free or commercially available software. There is even a common European method (CNOSSOS-EU). For vibration prediction the situation is different, but there is at least one openly available method in Denmark [4].

2 Method

2.1 Decay over Distance

Previously a simplified method of predicting vibration based on the Trafikverket database was developed [5]. The method used different distance decay functions for six classes of ground types under the receiving building to calculate a mean exposure at the corresponding distance. In order to estimate how many buildings were exposed to a certain level, the uncertainty of the estimation together with an assumed probability distribution function was used.

A similar approach was used in the Netherlands [6], where four different decay functions were used for distances between 25 and 200 m. However, the decay functions were exponential instead of power functions. Wave propagation in layered soils is a complex phenomenon involving several different wave types, including modal behavior [7]. For a very simplified case, a vertical point load generating Rayleigh waves in an infinite elastic half-space, the decay function is proportional to one over the square root of the distance if the material damping is insignificant. If the damping is high the decay function becomes exponential. The distance dependent part of the Greens function is approximately

$$v \propto \frac{\mathrm{e}^{-\alpha r}}{r^{1/2}}, \tag{1}$$

where v is the vertical surface velocity at a distance r from the point source, and α the material damping. It is difficult to draw more detailed conclusions from this simple case, but it illustrates that close to the source, when the damping is low, a power relation is relevant, and for longer ranges or when the deamping is high an exponential decay is more appropriate.

Determining a correct distance decay function for all soil classes in the database is difficult for several reasons. First and foremost there are limitations to what information is available. For example in the Trafikverket database only the soil type of the top layer close to the ground surface is known. Also, the soil type is not determined by measurements or inspection at each receiving building, but from a soil class map.

Fitting a statistical model of the vibration without a distance decay function, for example by introducing a number of free parameters that describes the vibration velocity by for example a spline function, is very challenging. Receiver points far from the railway line are more likely to have complexities such as varying soil class between the railway and receiving building, than receiver points close to the railway.

Also the database is not a complete survey of all buildings, but it is a compilation of measurements performed for various reasons; for example after receiving a complaint, when planning new buildings or when conducting surveys of areas with a known sensitive soil type.

During the work with the previous prediction method [5] this complexity was addressed by using inputation (addition of virtual measurements) of low values for long ranges in relation to building density, but still it was a challenge to get useful results. Here we instead start from a theoretical distance decay function including both a power function important for short ranges and a exponential decay relevant for long ranges.

2.2 Ground Layer Depth Dependence

This paper extended the method from [5] mainly by including one more important parameter, the depth of the main geological layer under the receiving building. For a typical high vibration situation in the areas of interest the ground is very soft; clay or silt, and the main frequency component of the ground and building vibration is in the 4–10 Hz region. Typically the soft layer is on top of a more rigid layer, and theoretically the depth or the softer layer should be important. For very shallow top layers, the rigid layer will determine the propagation, and for very deep layers it will be determined by the soft layer. At intermediate depths modal behaviour will start to appear, and certain wavelengths/frequencies will propagate better.

The main methodology used in this paper is a regression/optimization approach where different models of distance and depth dependance are compared to the measurements in the database. The goodness of fit, or ability to predict the mean vibration velocity, is used for the different mathematical models to get the final resulting prediction method.

Depth dependence is not important for soil types that do not transmit vibration very well, since the vibration velocity will be relatively low for all depths in that case, so we limited the depth analysis to those soil types that had significant measured vibration velocities. We also limited the analysis to a range span of 15 to 100 m. The lower limit is to remove cases where significant nearfield effects can occur. The upper limit is arbitrarily selected to reduce the number of measurements that are based on complaints only, and thus are skewed towards higher vibration values. Measurements below the detection threshold (0.2 mm/s for the foundation measurement) were also excluded.

3 Results

3.1 Distance Dependence

Using data from the Swedish geological survey (www.sgu.se/en), a total of 22 different geological classes for the main layer were present within the set of measurements within the three different research areas. In total there were 829 measurements on the building foundation, of which 643 were above the detection threshold.

As a first step a distance decay function was determined by looking at the decay functions estimated from measurements and calculations using numerical methods in [5,6]. The final result was obtained by trial and error and comparing the results visually, using a function of the form (1),

$$v = A\,\frac{\mathrm{e}^{-0.05r}}{r^{0.9}}, \tag{2}$$

where A is a constant dependent on the soil type and r the distance to the center line of the closest track.

3.2 Soil Classification

The sensitivity to vibration transmission of the different ground classes were determined using the following procedure. For each ground class only measurements within the distance range discussed above (15–100 m) and above the detection threshold (0.2 mm/s) were selected. Only classes with 10 or more measurements were analyzed, and in Table 1 the corresponding soil classes are presented together with the number of measurements for each class and the average vibration velocity (M1 in the table). Since different soils have measurements at different distances it is difficult to use this value to sort them in terms of vibration sensitivity, therefore two additional values were calculated where each measurement was normalized to a distance of 20 m using either the exponential decay (relevant for high damping) and power decay (relevant for low damping). The average of these values were then calculated and is presented in Table 1 in column M2 and M3.

Table 1. Mean vibration velocity of measurements for soil class. M1 is the mean over all measurements between 20 and 90 m, M2 is normalized to a distance of 20 m using a power decay function, M3 is normalized to 20 m using an exponential decay function.

Soil description	SGU code	Number of meas.	M1 mm/s	M2 mm/s	M3 mm/s
Postglacial clay	17	44	1.566	3.731	5.262
Fluvial sediment coarse silt - fine sand	8802	11	1.150	2.357	2.884
Postglacial fine sand	28	29	1.003	2.089	2.968
Postglacial sand	31	121	0.990	1.710	2.283
Postglacial silt	24	15	0.847	1.369	1.508
Clay-silt	86	17	0.765	1.001	1.189
Glaciofluvial sediment sand	55	28	0.744	1.704	2.516
Glacial clay	40	104	0.738	1.391	1.599
Glaciofluvial sediment	50	62	0.587	1.082	1.245
Sand till	95	27	0.531	0.888	1.171
Glacial silt	48	13	0.422	1.458	3.300

By investigating the results of the calculation, and then grouping soil types together that should have similar properties we classified all soil types present in our research areas in four classes, from no vibration transmission to very sensitive, according to Table 2. The classification is provided for all 31 soil types present in the EpiVib study, although measurements were only available for 22 of them.

3.3 Ground Layer Depth Dependence

The next step was to include the main layer depth in the vibration prediction scheme. The main layer depth parameter was available for each receiving point in the EpiVib study. When looking at the overall pattern in the measurements it was noted that the depth parameter was only important for vibration class 2 and 3. In order to maximize the data material for estimating the depth dependence,

Table 2. Soil classification from 0 (no vibration transmission) to 3 (very sensitive to vibration).

Soil type	SGU code	Vibration class
Rock	850,888,890	0
Till, peat	5,13,95,100,9147,9794	1
Glacial, glaciofluvial	31,33,39,40,43,48,50,55,57,86,200,9060	2
Postglacial, clay, fluvial	9,10,16,17,19,24,28,8802,8809	3

all measurements in vibration class 2 and 3 were grouped together and then normalized to a distance of 20 m using (2). Then a function was fitted using the least squares algorithm to the data, which gave a depth dependence

$$F(h) = 0.166h\,\mathrm{e}^{-0.0359\,h}, \tag{3}$$

where h is the main layer depth. The fitted function is plotted together with the normalized measurement points in Fig. 1.

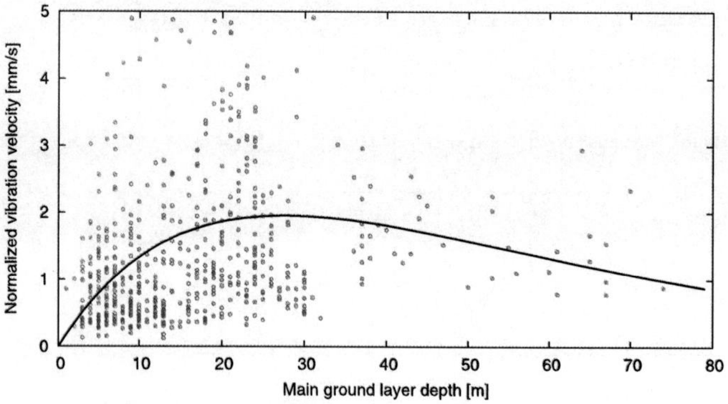

Fig. 1. Normalized maximum vibration velocity for measurements on sensitive ground versus the main layer depth. Solid line is the best fit for the predicted depth dependence.

3.4 Vibration Estimation

The final formula for predicting the maximum vibration velocity on the building foundation v in mm/s with time weighting SLOW becomes

$$v(r,c,h) = A(c)\frac{h\,\mathrm{e}^{-0.0359\,h}\,\mathrm{e}^{-0.05r}}{r^{0.9}}, \tag{4}$$

for the soil vibration class $c = 2$ and $c = 3$, where r is the distance in meters and h is the main layer depth in meters, and $A(c)$ is a factor given below.

For the class $c = 1$ the depth dependence is removed and the equation becomes

$$v(r, c) = A(c)\frac{e^{-0.05r}}{r^{0.9}}. \quad (5)$$

For the class $c = 0$ the estimated vibration is zero, or no vibration at all. The indoor weighted maximum vibration v_k can be estimated as $v_k = 0.83\,v$ [5].

The factor $A(c)$ for the different classes was determined using a global model fit using the least squares algorithm against all measurements in the range 15–100 m, and assuming that measurements below detection threshold is actually the detection threshold divided by $\sqrt{2}$. The resulting best fit factors for the three classes are $A(1) = 9.2477$, $A(2) = 3.4268$ and $A(3) = 6.2443$.

3.5 Estimation Error

The vibration velocity estimated using the procedure above gives large deviations compared to measured values. Figure 2 shows the difference (predicted-measured) as a function of distance for ground class 2 and 3.

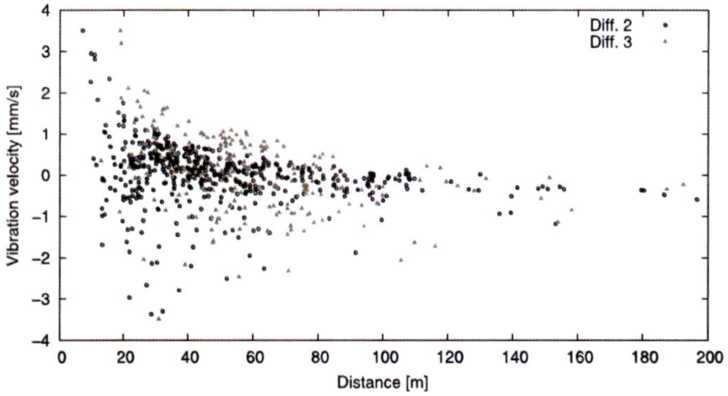

Fig. 2. Difference between measured and predicted maximum vibration velocity for soil vibration class 2 and 3 as a function of distance.

Deviations where the predicted vibration velocity is lower than the measured could be explained by a situation where the soil type dominating the propagation is more prone to vibration transmission than the soil type right below the house, and the other way around for higher predicted velocity. Also the source strength may be an explanation, for example if the rail is very smooth and well maintained.

4 Discussion

This paper introduces a method for estimating vibration from railways in a limited area in Sweden. The method extends previous simple methods [5] by including not only soil type but also the main layer depth. The method is needed since there are no official vibration prediction tools available, and the size of the area and number of locations (more than 6000) where predictions are needed prohibits a complete measurement campaign or using a numerical method that demands very detailed input.

There are many limitations to the method. The data included in the prediction is very limited, only the top layer of the ground is known, and only at the receiving position. The source strength of the railway is unknown, no details about the vehicles, the track or the ballast is included in the modeling of the vibrations. The method is more or less an empirical prediction based largely on the statistics of measurements in the area.

The strengths of the method are also related to the limitations; it can estimate vibration from data easily available from official sources such as Lantmäteriet, Trafikverket and SGU. It is also very fast in terms of calculation time and it is relatively easy to understand in principle.

The uncertainty of the predicted vibration velocity compared to measurements is very significant. In some ways it could be improved by using more available data, such as the geology information not just under the house, but for many points between the source and the receiver. In other cases it is very challenging to improve, such as trying to include very detailed information on ground properties that are not available except by making on-site geological measurements.

Compared to the previous method [5], the improved method shares the basic structure but introduces depth dependence and new distance decay functions. Unfortunately the improvements are not clearly visible in terms of better precision or less spread compared to measurements. As an example, for ground class 3 and distances between 15 and 100 m, the mean absolute difference between predicted and measured vibration velocities is 0.71 mm/s for the improved method and 0.73 mm/s for the previous method.

References

1. WHO: Environmental Noise Guidelines for the European Region. WHO (2018)
2. Maclachlan, L., Ögren, M., van Kempen, E., Hussain-Alkhateeb, L., Persson Waye, K.: Int. J. Environ. Res. Public Health **15**(9) 1887 (2018)
3. Maclachlan, L., Persson Waye, K., Pedersen, E.: Int. J. Environ. Res. Public Health **14**(11), 1303 (2017)
4. Banedanmark, New vibration model. Technical report A026780-006.1, COWI, Lyngby (2015)
5. Arnesson, M.: Analysis and estimation of residential vibration exposure from railway traffic in Sweden. Master's thesis, Department of Civil and Environmental Engineering, Chalmers (2016)

6. van Kamp, I., van Kempen, E., van Wijnen, H., Verheijen, E., Istamto, T., Breugelmans, O., Dirven, E., Koopman, A.: Wonen langs het spoor: Gezondheidseffecten van trillingen door treinen. RIVM raport 2014-0096, Bilthoven (2015)
7. Li, Y., Feng, S., Chen, H., Chen, Z., Zhang, D.: Random vibration of train-track-ground system with a poroelastic interlayer in the subsoil. Soil Dyn. Earthq. Eng. **120**, 1 (2019)

1 dB per Floor? How Does Noise and Vibration Propagate in High-Rise Buildings Near Railway Lines?

Dave Anderson(✉), Sav Shimada, and David Hanson

Acoustic Studio, Unit 27 43-53 Bridge Road, Stanmore, NSW 2048, Australia
Dave.anderson@acousticstudio.com.au

Abstract. This paper examines how vibration from rail operations propagates through high-rise buildings. Results are presented from four buildings of varying heights, ages and construction types. Each building sits either directly above or adjacent to a heavy rail corridor. The results suggest the traditional assumptions around vibration propagation through high-rise buildings may be non-conservative.

Keywords: Ground-borne noise · Vibration propagation · High-rise buildings

1 Introduction

New underground rail lines are often located in urban areas near to high-rise buildings. It is necessary to predict noise and vibration impacts in high-rise buildings to inform the design of track.

Often there is considerable focus placed on the calculation of the vibration level at the foundation of the building, but then less focus placed on how that vibration propagates through the building. A common rule of thumb is to assume 1 to 2 dB reduction in vibration for each floor of the building [1]. Work by Anderson [2] and Lurcock [3] raised doubts about the accuracy of this approach, indicating that amplification (rather than attenuation) can occur at some low frequencies, and that attenuation at higher frequencies may be less than indicated by the rule of thumb.

The previous work by Anderson and Lurcock covered buildings of up to 8 levels (plus a basement); this paper presents results from some taller buildings (up to 60 levels, plus basements) in Sydney.

2 Methodology

Tower A is a 60 floor commercial tower (plus 7 basement levels) constructed in the 1970s, comprised of a concrete core and eight concrete exterior columns that taper towards the top of the structure. The basement of the tower is within approximately 20 m of an

G. Degrande et al. (Eds.): *Noise and Vibration Mitigation for Rail Transportation Systems*, NNFM 150, pp. 496–503, 2021.
https://doi.org/10.1007/978-3-030-70289-2_53

underground rail tunnel carrying heavy-rail passenger services at 5-min intervals during peak times. Figure 1 shows the general arrangement of the site.

Vibration from train events was recorded inside the building, adjacent to the exterior structural column closest to the rail tunnel. Vibration was measured at the intersection of the exterior wall and floor, immediately adjacent to the structural column.

A vibration logger was deployed in the lowest basement level which recorded throughout the measurement exercise. Attended measurements were then undertaken on levels 11, 18, 40 and 62. The attended measurements also included vibration at mid-span and noise from train events.

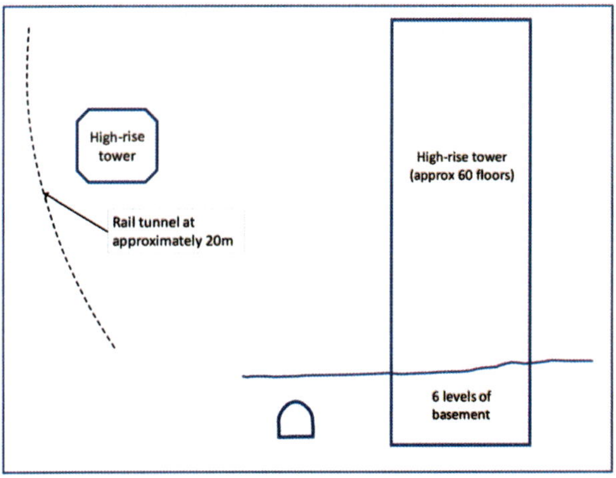

Fig. 1. Indicative site plan (left) and section (right) showing the location of the rail tunnel relative to the tower A

Supplementary measurements were made at three other concrete framed towers, referred to as towers B, C and D and described in Table 1. The methodology was similar to that for tower A except that it was not possible to access the occupied areas of upper floors, so measurements at upper levels were restricted to vibration in the lift lobby area.

Table 1. Details of Towers B, C and D.

Tower	Levels	Year constructed	Rail location
B	35	1988	Tunnel directly below
C	18	1996	Tunnel directly below
D	12	2001	Surface track at 20 m

3 Results

3.1 Tower A

Figure 2 shows the average one-third octave Lmax,F column vibration and internal noise spectra measured during train events at tower A. Average overall vibration ranged from 0.001–0.03 mm/s (rms) and the internal noise levels ranged from 40–49 dBA.

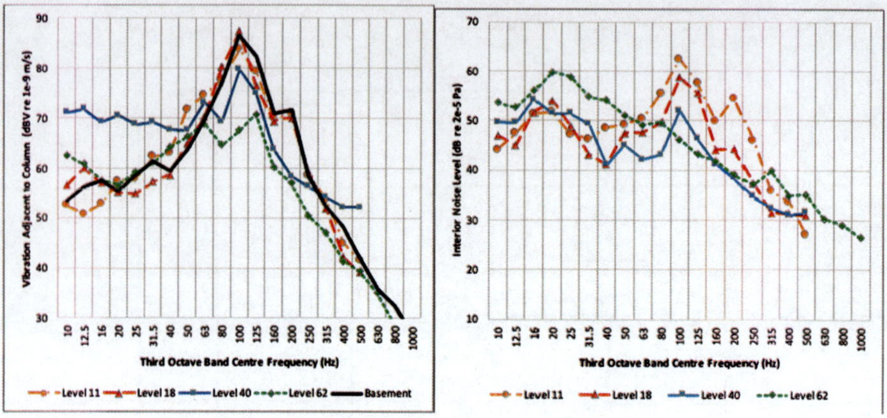

Fig. 2. Vibration levels (left) and noise levels (right) in tower A

As is evident in the noise spectra, the measured levels were influenced by ambient noise and vibration; the signal to noise ratio on level 62 was poor so the remainder of the analysis concentrates on measurements on levels 11, 18 and 40.

With the exception of level 62, all of the noise and vibration spectra show a strong peak in the 100 Hz third octave band. This is related to rail roughness in the tunnel due to severe corrugation. It is interesting to note that the 100 Hz vibration level is very similar at the basement, level 11 and level 18 (1 dB difference) and that there appears to be around 7 dB attenuation between basement and level 40.

The floor-to-floor vibration reduction during train passby was calculated with reference to vibration levels in the basement and is presented in Fig. 3. Very little reduction in vibration was recorded throughout the structure, with less than 0.25 dB per floor reduction occurring across the 80–400 Hz frequency range typically associated with ground-borne noise. Slight amplification was observed at some frequencies, such as 40–63 Hz.

The amplification of floor vibration between mid-span of the main structural members and the structural column is presented in Fig. 4. These results show amplification of up to 5 dB at frequencies below 100 Hz and attenuation of up to 5 dB for frequencies above 100 Hz. The large variance below 16 Hz suggests these frequencies may have been impacted by extraneous noise. Significant variations in amplification were noted between floors. This is likely to be due to differences in internal fit-out; levels 11 and 18 were occupied and fitted out with internal walls, partitions etc. whereas levels 40 and 62 were empty at the time of measurements and largely free of partitions or furniture.

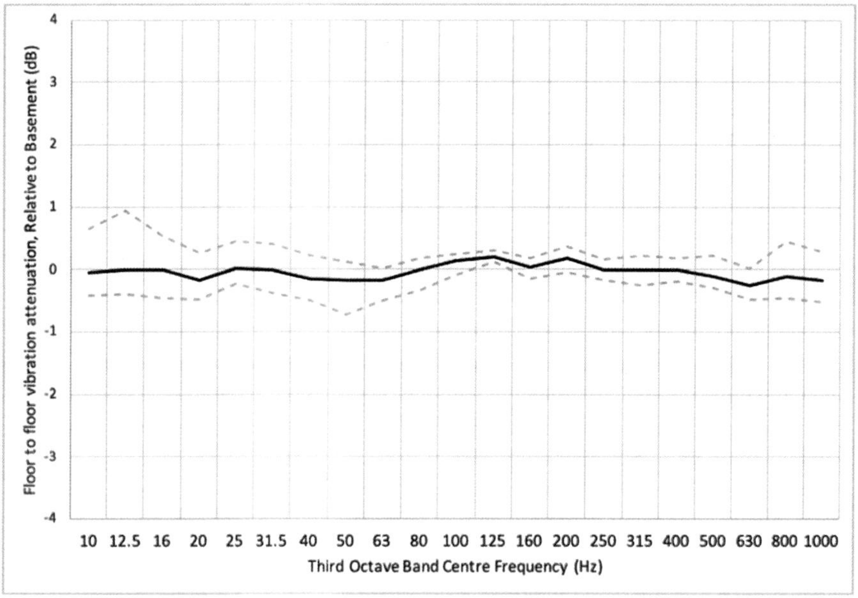

Fig. 3. Floor to floor vibration attenuation at external column of tower A– mean (solid), max and min (grey, dashed)

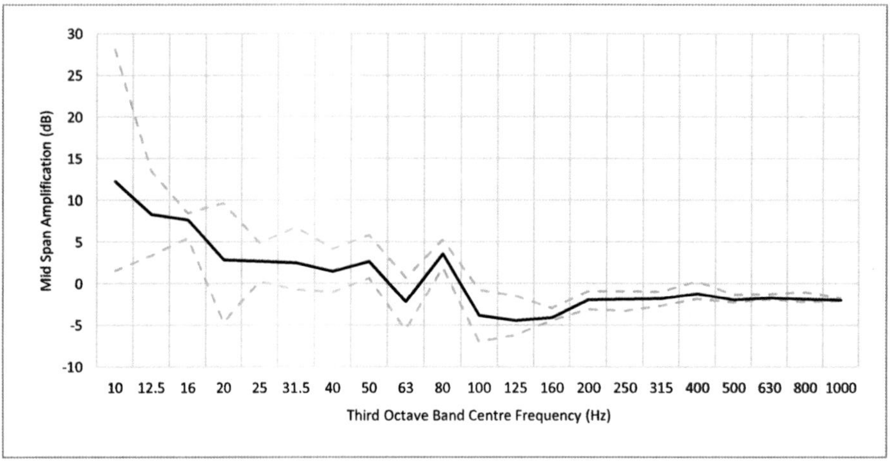

Fig. 4. Vibration amplification between mid-span and column, from level 40 at tower A

Average ground-borne noise levels were 49 dBA at level 11, 45 dBA at level 18 and 40 dBA at level 40. These levels suggest ground-borne noise attenuation ranging between 0.2 and 0.6 dB per floor. It is also notable that vibration levels at levels 11 and 18 were very similar to those in the basement, suggesting that ground-borne noise attenuation between basement and upper floors would be close to zero.

3.2 Towers B, C and D

Measurements in towers B, C and D were not as extensive as in tower A, but nevertheless provide insight into vibration transmission through the building from train vibration underneath or adjacent to the building foundations.

The vibration spectra measured within tower B, and the corresponding vibration reduction per floor, are presented in Fig. 5. Measurements were recorded for each train pass in the basement car park, and then in the lift shaft on various floors up the building. There was amplification between the basement and the level 4 car park location (directly above). Vibration at the level 4 lift shaft location was lower than in the car park; the vibration reduction results at upper floor lift lobbies are referenced to the vibration levels at the lift shaft location on Level 4. The spectra measured on each floor (left plot) show that vibration from train passes is prominent above the ambient vibration between around 20 Hz and 200 Hz. The vibration reduction per floor is therefore most valid within this frequency region.

The vibration spectra measured within tower C, and the corresponding vibration reduction per floor, are presented in Fig. 6. As for tower B, the vibration measured on the internal floors during train passes is similar to the ambient at low frequencies, and hence the reduction per floor (right plot) is most valid above 16 Hz.

The vibration spectra measured within tower D, and the corresponding vibration reduction per floor, are presented in Fig. 7. The ambient vibration on each floor was not recorded on the upper floors of tower D, but observations from the attended monitoring confirm the characteristic train vibration spectrum was clearly prominent between 16 Hz and 200 Hz, even on the highest floor recorded (Level 8, which is 13 levels above the basement reference measurement location).

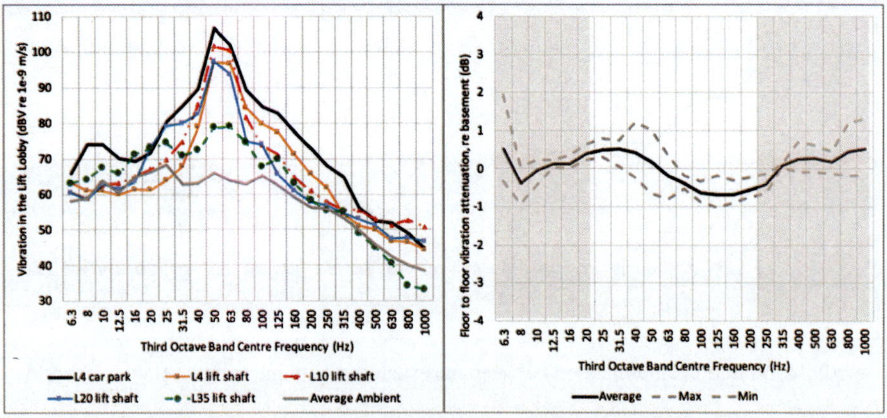

Fig. 5. Tower B: vibration spectra (left) and floor to floor vibration attenuation relative to basement (right). Shading indicates frequency regions with low signal-to-noise.

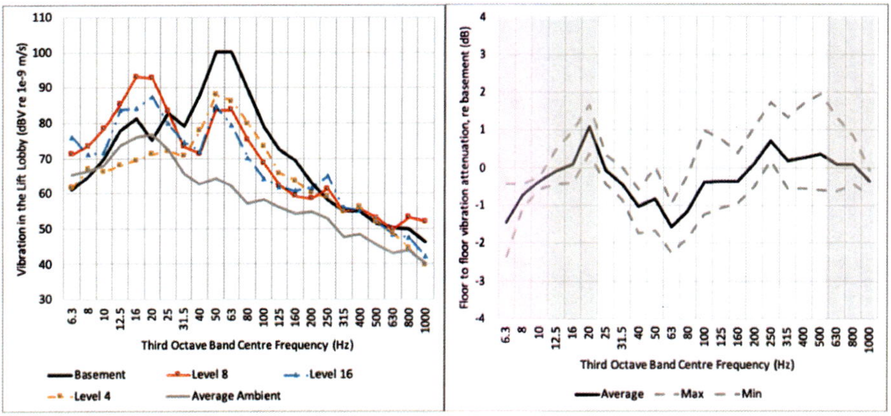

Fig. 6. Tower C: vibration spectra (left) and floor to floor vibration attenuation relative to basement (right). Shading indicates frequency regions with low signal-to-noise.

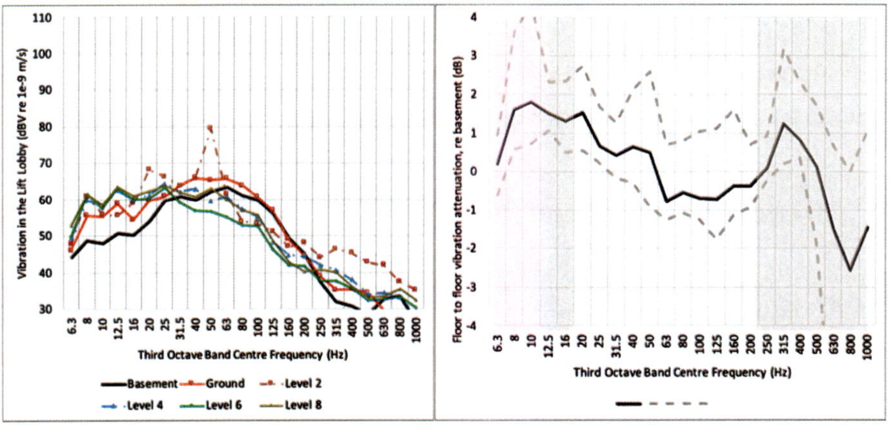

Fig. 7. Tower D: vibration spectra (left) where P5 is the lowest basement level, and floor to floor vibration attenuation relative to basement (right). Shading indicates frequency regions with low signal-to-noise.

4 Discussion

Figure 8 reproduces Table 6–13 from the FTA manual [1] showing the rule of thumb estimate of 1 to 2 dB reduction in vibration for each floor of the building.

The measurements at tower A showed that vibration levels in the frequency range of interest for ground-borne noise and vibration reduced by less than 0.25 dB per floor. This is significantly less than the rule of thumb estimate of 1 to 2 dB per floor. Wall to floor mid-span amplification was generally up to 5 dB at frequencies up to 100 Hz, which is within the 6 dB rule of thumb estimate.

Measurements in towers B and D show no attenuation per floor at frequencies below 63 Hz and less than 1 dB per floor at frequencies above 63 Hz. Tower C shows attenuation

Receiver Factor	Adjustment to Propagation Curve		Comment
Floor-to-floor attenuation	1 to 5 floors above grade	-2 dB/floor	This factor accounts for dispersion and attenuation of the vibration energy as it propagates through a building starting with the first suspended floor.*
	5 to 10 floors above grade	-1 dB/floor	
Amplification due to resonances of floors, walls, and ceilings	+6 dB		The actual amplification will vary greatly depending on the type of construction. The amplification is lower near the wall/floor and wall/ceiling intersections.

* Floor-to-floor attenuation adjustments for the first floor assume a basement.

Fig. 8. Table-6–13 from the FTA Manual

of approximately 1 dB per floor between 40 and 80 Hz (which is typically the most important frequency range for ground borne rail noisc), but less attenuation at other frequencies.

5 Conclusions

The low frequency amplification effect identified in [2, 3] was not evident in the high-rise towers assessed in this paper. However, the floor-to-floor attenuation of vibration was significantly lower than established rules of thumb, consistent with some of the results from [2, 3].

This has implications for accurate prediction and assessment of ground borne noise and vibration impact within high-rise buildings. It means that assessment of the impact at the lowest occupied floor of the building would not be representative in some cases, such as if ground borne noise-sensitive accommodation (such as residential use) or vibration-sensitive use (such as medical imaging or research laboratories) is at an upper floor of the building. This is often the case in practice; 3 of the 4 towers used in this study had the most sensitive accommodation at an upper floor, as follows:

- commercial office space and medical suites at level 7, in tower A,
- residential apartments at level 6, in tower B,
- court rooms at level 4, in tower C.

6 Future Work

There are two aspects that warrant further work following this study. The first is to consider more extensive measurements of noise and vibration throughout the buildings to improve signal to noise across the frequency range of interest; coverage of occupied areas (as well as lift lobbies); and insight into ambient vibration (from sources other than rail). The second is to compare measurement results with dynamic modelling using methods such as those in [4].

References

1. Federal Transit Administration. Transit Noise and Vibration Impact Assessment Manual. FTA Report No. 0123 (2018)
2. Anderson, D.C.: Engineering Prediction of Railway Vibration Transmission in Buildings. In: Proceedings of Euronoise 1992, London (1992)
3. Lurcock, D.E.J., Thompson, D.J., Bewes, O.G.: Attenuation of railway noise and vibration in two concrete frame multi-storey buildings. In: Nielsen, J.C.O., Anderson, D., Gautier, P.-E., Iida, M., Nelson, J.T., Thompson, D., Tielkes, T., Towers, D.A., de Vos, P. (eds.) Proceedings of the 11th International Workshop on Railway Noise. Notes on Numerical Fluid Mechanics and Multidisciplinary Design, Uddevalla, Sweden, September 2013, vol. 126, pp. 539–546. Springer, Heidelberg (2013)
4. Jin, Q., Thompson, D.J., Lurcock, D.E.J.: Prediction of building vibration induced by underground trains using a coupled waveguide finite/boundary element method. In: Anderson, D., Gautier, P.-E., Iida, M., Nelson, J.T., Thompson, D.J., Tielkes, T., Towers, D.A., de Vos, P., Nielsen, J.C.O. (eds.) Proceedings of the 11th International Workshop on Railway Noise. Notes on Numerical Fluid Mechanics and Multidisciplinary Design, vol. 139

Design of Railway-Induced Ground-Borne Vibration Abatement Solutions to be Applied in Railway Tunnels by Means of a Hybrid Modelling Approach

Robert Arcos[1,2](✉), Dhananjay Ghangale[1], Behshad Noori[1], Hassan Liravi[1], Arnau Clot[3], and Jordi Romeu[1]

[1] Acoustical and Mechanical Engineering Laboratory (LEAM), Universitat Politècnica de Catalunya (UPC), C/Colom, 11, 08222 Terrassa, Barcelona, Spain
robert.arcos@upc.edu
[2] Serra Húnter Programme, Universitat Politècnica de Catalunya (UPC), Colom 11, 08222 Terrassa, Barcelona, Spain
[3] Cambridge University, Trumpington Street, Cambridge CB2 1PZ, UK

Abstract. This paper presents a methodology for the design of vibration abatement solutions to be applied in railway tunnels in the context of problems where an already operative underground railway infrastructure is inducing excessive vibration levels to particular nearby buildings. In such cases, where some parts of the global system are already constructed, the inclusion of experimental data measured at the specific site to be studied increases significantly the accuracy and reduces the uncertainty of the predictive model. This paper proposes to use experimentally measured vibration transmissibility functions between the tunnel wall and the particular buildings to be studied as a model of the tunnel/soil/building system. This model, combined with a numerical model of the tunnel/track system, forms the global hybrid approach presented in this paper. Vibration abatement solutions can be included in the tunnel/track model in order to assess their efficiency. This methodology can be used for the study of any vibration abatement solution to be applied in the tunnel structure. In this paper, two types of these solutions are specifically studied: rail fastening system retrofitting in order to modify the rail fastener stiffness and the application of dynamic vibration absorbers.

Keywords: Railway-induced ground-borne vibration · Vibration mitigation measures · Experimental/numerical hybrid modelling · Optimisation

1 Introduction

Railway-induced ground-borne vibration is a major environmental concern in urban areas. When a new railway infrastructure becomes operational, it is common to have neighbourhood complaints about noise and vibration annoyance.

G. Degrande et al. (Eds.): *Noise and Vibration Mitigation for Rail Transportation Systems*, NNFM 150, pp. 504–511, 2021.
https://doi.org/10.1007/978-3-030-70289-2_54

Although a predictive study of railway-induced ground-borne noise and vibration levels of the new infrastructure was conducted and required vibration mitigation measures were designed and finally implemented, the noise and vibration levels could not accomplish the required law (or more restrictive vibration levels required) in some particular nearby buildings or facilities. These situations are normally due to particular conditions not accounted for in the general prediction study. Therefore, at least for particular sites where the legislation is not accomplished, vibration abatement measures should be applied in these cases.

In the present investigation, a methodology for the prediction of the efficiency of vibration mitigation measures applied in operational railway tunnels is presented. In such cases, vibration measurements on the rail, track, tunnel and surrounding buildings can be used to enhance the accuracy of the efficiency predictions for new vibration mitigation measures to be installed in the tunnel. This type of hybrid modelling has been found as an interesting option to enhance the accuracy of railway-induced vibration predictions in previous investigations. In [1], Verbraken et al. proposed to combine numerical and empirical predictions, aiming to construct a hybrid model that combines the advantages of both approaches: on one hand, numerical models allow for a great flexibility in dealing with different train/track models; on the other hand, empirical models based on experimental measurements of the particular site to be studied allow for an accurate assessment of the vibration propagation between the measuring points. They presented analytical expressions for the force density and the line source transfer mobility functions defined in the procedure stated by the Federal Railway Administration (FRA) of the U.S. Department of Transportation [2] for detailed vibration assessment [3]. These analytical expressions can be used instead of experimentally obtained ones in order to transform the empirical model into a hybrid approach. Degrande et al. [4] presented two hybrid models where the applicability of these analytical expressions is shown. Kuo et al. [5] presented a more detailed application example of these two hybrid models. More recently, Lopez-Mendoza et al. [6] have proposed a hybrid model to predict railway-induced vibrations in buildings in which they combine the building response, obtained from a FEM model of the isolated structure, and the ground surface response due to rail traffic, obtained from experimental measurements in the specific site. In [7], a hybrid model is stated in order to accurately deal with track singularities, such as switches and crossings, by combining experimental measurements of the vibration transfer functions between the rail and nearby structures with a theoretical model of the vehicle/track system that is able to account for track singularities. Recently, Mouzakis et al. [8] have proposed to use measured transmissibility functions between the vibration in nearby buildings and the vibration at the tunnel wall or invert induced by a muck track operator for the assessment of ground-borne noise and vibration of new underground railway lines.

The methodology presented in this paper combines transmissibility functions between the vibration of the floors of the studied building and the vibration in the tunnel wall, as a model of the tunnel/soil/building system with a numerical model for modelling the system consisting of the track, the tunnel and the locally surrounding soil. The term *transmissibility function* is used in this paper instead of *transfer function* in order to clearly specify that it is a transfer function between the vibration response at two points. Two applications of this methodology for assessing the efficiency of ground-borne vibration mitigation measures are presented in this paper: rail fastening system retrofitting (RFSR) and dynamic vibration absorbers (DVAs).

2 Methodology

2.1 Modelling Approach

Since the present model is created for optimization purposes, it has been developed to be highly computationally efficient in a particular task: obtaining the building responses associated to a large set of parameters of the vibration abatement solutions studied. The global hybrid modelling approach used to dynamically represent track/tunnel/ground/building systems consists of the following parts:

- Modelling of the tunnel and the locally surrounding ground: It is proposed to use a two-and-a-half-dimensional finite element method and boundary element method (2.5D FEM-BEM) coupled approach, where a FEM mesh is used to model the tunnel and a BEM mesh is used to account for the locally surrounding soil. This 2.5D FEM-BEM approach used in this investigation is outlined in [9].
- Railway track modelling: The track model consists of two Euler-Bernoulli beams, as a model of the rails, connected to the tunnel invert through continuously distributed springs, as a model of the rail fastening systems.
- Vibration abatement solutions: In order to design a modelling strategy able to efficiently deal with parametric studies of the vibration abatement solutions, a model of these solutions should be included in the track/tunnel system by dynamic substructuring techniques, always avoiding its inclusion as a part of the 2.5D FEM-BEM model. As already mentioned, two type of abatement solutions are studied in this paper. On one hand, for the RFSR, the semi-analytical model of the track allows for the calculation of the response of the system for various rail fastening system stiffnesses avoiding the recalculation of the 2.5D FEM-BEM model of the tunnel/soil system. On the other hand, for the application of DVAs, the methodology for coupling a distribution of DVAs over a model of a railway system presented in [10] is used. In the present study, only one longitudinal distribution of DVAs is considered. Moreover, all the DVAs of the distribution are assumed to be exactly equal to each other.
- Tunnel/soil/building model: This model consists on the transmissibility functions between the acceleration of vibration at the tunnel wall and at the specific buildings to be studied. These transmissibility functions should be obtained by simultaneous experimental measurements in the tunnel wall and the particular buildings due to normal operation train pass-bys. The influence of abatement solution on the measured transmissibility functions is discussed in Sects. 2.2 and 3.

- Train pass-by response: In this paper, the response due to the train passages is computed by means of the methodology presented in [11].

2.2 Tunnel/Soil/Building Model

To model the wave propagation from the tunnel to the building, the present approach takes advantage of the physical existence of the full transmission path between these two systems to use a fully experimental model. This model is based on the transmissibility functions that relates the acceleration of vibration in the tunnel wall of the tunnel with the acceleration of vibration at a particular point (or a set of points) in the studied building. To generate these transmissibility functions, the response of the tunnel wall and the building floor due to the train pass-by needs to be measured simultaneously using triaxial accelerometers. Then, a single input spectrum in the frequency domain $\ddot{U}_t(\omega)$ can be approximately obtained using triaxial information at the tunnel wall as

$$\ddot{U}_t = \sqrt{|\ddot{U}_{t_x}|^2 + |\ddot{U}_{t_y}|^2 + |\ddot{U}_{t_z}|^2}, \tag{1}$$

where $\ddot{U}_{t_x}$, $\ddot{U}_{t_y}$ and $\ddot{U}_{t_z}$ are the spectra of the vibration acceleration measured at the tunnel wall in x, y and z directions, respectively, in the frequency domain. A single output signal in the frequency domain $\ddot{U}_b(\omega)$ can also be approximately obtained using triaxial information measured at the building floor as

$$\ddot{U}_b = \sqrt{|\ddot{U}_{b_x}|^2 + |\ddot{U}_{b_y}|^2 + |\ddot{U}_{b_z}|^2}, \tag{2}$$

where $\ddot{U}_{b_x}$, $\ddot{U}_{b_y}$ and $\ddot{U}_{b_z}$ are the spectra of the vibration acceleration measured at the building floor in x, y and z directions, respectively, in the frequency domain. As both input and output signals are transient signals, the transfer function between the tunnel wall and the building can be obtained by

$$T_{tb}(\omega) = \frac{E_{tb}(\omega)}{E_{tt}(\omega)} \tag{3}$$

where $E_{tb}(\omega)$ is the cross-energy spectrum between $\ddot{U}_t$ and $\ddot{U}_b$, and E_{tt} is the energy spectrum of $\ddot{U}_t$. Thus, this T_{tb} can be used in the context of the global model as a zero-phase filter to be applied to the simulated time signal of the vibration acceleration at the tunnel wall to obtain the predicted response in the target building. Although this filter is simplified by avoiding the phase information of the exact transmissibility between the vibration at the tunnel wall and the target building floor, this assumption seems to be reasonable taking into account the stochastic nature of the rail unevenness profiles used as the input of the simulation process presented.

Noteworthy, it is assumed here that the application of DVAs would not affect the transfer function between tunnel wall and the building. So, the transfer function, obtained in the absence of DVAs, could be also used after the application of DVAs. This assumption is considered based on the work of Arcos et al. [12], in which a similar hybrid model approach was used to compute the optimal rail fastening system stiffness. The conclusions of that work arise that the transfer functions between tunnel wall and the building are not significantly affected by changing the stiffness of the rail fastening system.

3 Results

As already mentioned, the methodology proposed in this paper can be used for the optimization of vibration mitigation measures to be applied in railway tunnels. Application examples for RFSR and DVA mitigation measures are presented in this section.

3.1 Application for Rail Fastening System Retrofitting

For the example of application of this methodology for RFSR, a simple tunnel of 6 m of radius is considered. The evaluation point in the tunnel is placed at the tunnel wall, at 2 m from the tunnel invert. An evaluation point in the soil is also considered and it is placed at a distance of 10 m from the centre of the tunnel and at $\pi/4$ rad from the vertical. The vehicle model used in the simulation is the one presented in [13]. Experimental transmissibility functions between the tunnel wall and the targeted building floor are obtained from vibration measurements on the studied site. Left plot on Fig. 1 shows a prediction of the insertion loss (IL) associated to the maximum transient vibration value (MTVV) in the building floor as a function of the rail fastening system stiffness. Right plot on this figure shows the simulated transmissibility functions between the tunnel wall and the evaluation point in the soil for discrete values of the stiffness of the rail fastening system ranging between 5 kN/mm and 75 kN/mm. Maximum deviations with respect to the mean value of 2 dB are found, showing the validity of the hybrid methodology presented.

3.2 Application for Dynamic Vibration Absorbers

For the example of application of this methodology in the case of DVAs, a particular stretch of the double-deck tunnel of the Line 9 of Metro Barcelona and a nearby building with excessive vibration levels are considered in the study. Specific parameters of that case study are accounted for in the model. Experimental vibration measurements from hammer testing at the track and the tunnel and from train circulations are performed and used for tuning the vehicle/track/tunnel model and to obtain the transmissibility functions between the tunnel wall and the building floor. Figure 2 shows the response of the building floor with or without DVAs. The DVAs parameters used in this case are obtained applying a optimisation process based in genetic algorithm, applicable in this case due to the substructured nature of the present methodology.

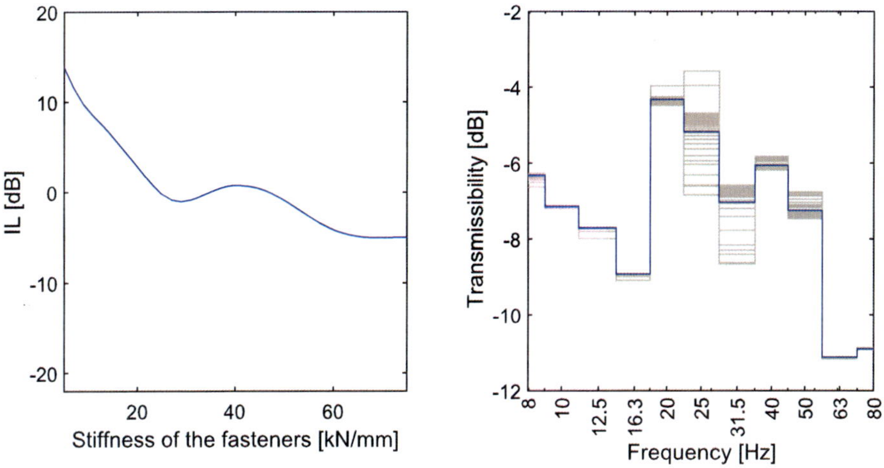

Fig. 1. Results of the application example for RFSR: Predicted IL in terms of MTVV of the building floor (left) and variability of the transmissibility functions in one-third octave bands between the tunnel wall and the evaluation point in the soil.

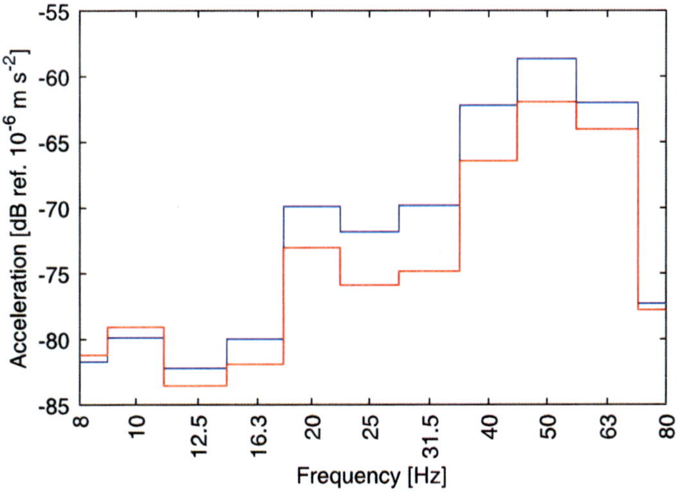

Fig. 2. Acceleration of building floor vibration in one-third octave bands with (red plot) and without DVAs (blue plot).

4 Conclusions

The methodology outlined in this paper allows for the designing of vibration abatement solutions to be applied in operational railway tunnels that are inducing excessive vibration levels to particular nearby buildings. In such cases, since most of the global system associated to the ground-borne vibration problem to be studied is already existing, the inclusion of experimental data measured at the specific site increases significantly the accuracy and reduces the uncertainty of the predictive model. Experimentally measured vibration transmissibility functions between the tunnel wall and the particular buildings and a numerical model of the vehicle/track/tunnel/system are the foundations of the present methodology. The most important benefits of the presented hybrid approach are the enhancement of the accuracy because of the inclusion of experimental data in the model and the suitability of the methodology for parametric studies of the vibration abatement solutions to be applied in the tunnel.

The applicability of the present methodology has been discussed for two kinds of vibration mitigation measures, showing its potential for the design of vibration countermeasures to be applied in the tunnel or the track. This methodology can be used for other kinds of vibration mitigation measures to be applied in the tunnel or the track as long as the inclusion of the vibration mitigation measure model on the global model is performed by using dynamic substructuring techniques.

Acknowledgements. The present work is funded by the project ISIBUR: Innovative Solutions for the Isolation of Buildings from Underground Railway-induced Vibrations supported by the Ministerio de Ciencia e Innovación, Retos de Investigación 2014, with reference TRA2014-52718-R. The authors want to also acknowledge the financial support provided by the project VIBWAY: Fast computational tool for railway-induced vibrations and re-radiated noise assessment, supported by the Ministerio de Ciencia e Innovación, Retos de Investigación 2018, with reference RTI2018-096819-BI00. The first author also wants to acknowledge the funds provided by the NVTRail project, Noise and Vibrations induced by railway traffic in tunnels: an integrated approach, with grant reference POCI-01-0145-FEDER-029577, funded by FEDER funds through COMPETE2020 (Programa Operacional Competitividade e Internacionalização (POCI)) and by national funds (PIDDAC) through FCT/MCTES.

References

1. Verbraken, H., Eysermans, H., Dechief, E., François, S., Lombaert, G., Degrande, G.: Verification of an empirical prediction method for railway induced vibration. Notes Numer. Fluid Mech. Multi. Des. **118**(8), 239–247 (2012)
2. Hanson, C.E., Towers, D.A., Meister, L.D.: Transit noise and vibration impact assessment (2006)
3. Nelson, J.T., Saurenman, H.J.: A prediction procedure for rail transportation groundborne noise and vibration. In Transportation Research Record 1143. In: Proceedings of A1F04 Committee Meetingon the Transportation Research Board, volume 1143 of Notes on numerical fluid mechanics and multidisciplinary design, pp. 26–35 (1987)
4. Degrande, G., Verbraken, H., Kuo, K., François, S., Dijckmans, A., Lombaert, G.: Numerical, experimental and hybrid predictions of ground vibration produced by high speed railway traffic. In: Proceedings of Inter-Noise 2015, the 2015 International Congress and Exposition on Noise Control Engineering, pp. 1–12, San Francisco, USA (2015)

5. Kuo, K.A., Verbraken, H., Degrande, G., Lombaert, G.: Hybrid predictions of railway induced ground vibration using a combination of experimental measurements and numerical modelling. J. Sound Vib. **373**, 263–284 (2016)
6. López-Mendoza, D., Romero, A., Connolly, D.P., Galvín, P.: Scoping assessment of building vibration induced by railway traffic. Soil Dyn. Earthq. Eng. **93**(June 2016), 147–161 (2017)
7. Kouroussis, G., Vogiatzis, K.E., Connolly, D.P.: A combined numerical/experimental prediction method for urban railway vibration. Soil Dyn. Earthq. Eng. **97**(March), 377–386 (2017)
8. Mouzakis, C., Vogiatzis, K., Zafiropoulou, V.: Assessing subway network ground borne noise and vibration using transfer function from tunnel wall to soil surface measured by muck train operation. Sci. Total Environ. **650**, 2888–2896 (2019)
9. Ghangale, D., Colaço, A., Alves Costa, P., Arcos, R.: A methodology based on structural finite element method-boundary element method and acoustic boundary element method models in 2.5D for the prediction of reradiated noise in railway-induced ground-borne vibration problems. J. Vib. Acoust. **141**(June), 031011 (2019)
10. Noori, B., Arcos, R., Clot, A., Romeu, J.: Control of ground-borne underground railway-induced vibration from double-deck tunnel infrastructures by means of dynamic vibration absorbers. J. Sound Vib. **461**, 114914 (2019)
11. Lombaert, G., Degrande, G.: Ground-borne vibration due to static and dynamic axle loads of InterCity and high-speed trains. J. Sound Vib. **319**(3–5), 1036–1066 (2009)
12. Arcos, R., Ghangale, D., Clot, A., Noori, B., Romeu, J.: Hybrid model for rail fasteners stiffness optimization in railway-induced ground-borne vibration problems. In: Proceedings of the 6th Conference on Noise and Vibration Emerging Methods (NOVEM), pp. 1–11 (2018)
13. Clot, A., Arcos, R., Noori, B., Romeu, J.: Isolation of vibrations induced by railway traffic in double-deck tunnels using elastomeric mats. In: 24th International Congress on Sound and Vibration, ICSV 2017 (2017)

Check for updates

Response of Periodic Railway Bridges Accounting for Dynamic Soil-Structure Interaction

Pieter Reumers[1(✉)], Kirsty Kuo[2], Geert Lombaert[1], and Geert Degrande[1]

[1] Department of Civil Engineering, Structural Mechanics Section, KU Leuven, Kasteelpark Arenberg 40 Box 2448, 3001 Leuven, Belgium
pieter.reumers@kuleuven.be
[2] Wood Technical Consulting Solutions, 240 St George's Terrace, Perth, Australia
https://bwk.kuleuven.be/bwm

Abstract. Predicting the dynamic response of long multi-span railway bridges with 3D element-based models is a computationally challenging task, especially if dynamic soil-structure interaction (SSI) is to be accounted for. In this paper, the wave finite element method (WFEM) is used to exploit periodicity. The methodology fully accounts for wave propagation in the superstructure as well as through-soil coupling between neighboring foundations. A simplified case study of an infinitely long viaduct founded on piled foundations is presented. The influence of the soil stiffness and span length on the response is investigated.

Keywords: Periodic railway bridges · Piled foundations · Dynamic soil-structure interaction

1 Introduction

Elevated railway bridges founded on piled foundations are frequently constructed to save space in densely populated areas, to span regions of changing topography, or to prevent track settlements in areas with soft soil. For cost-effectiveness, these bridges often consist of identical spans repeated over long distances.

Due to their length, the use of 3D element-based methods that account for dynamic soil-structure interaction (SSI) is challenging. In order to reduce the computational domain, periodic structure theory can be applied [1]. The Floquet transform is a useful tool to model infinitely long periodic structures represented by a single reference cell of length L. Clouteau et al. [2] presented a periodic finite element-boundary element (FE-BE) method based on Green-Floquet functions and applied it to e.g. an underground tunnel [3]. Alternatively, the wave finite element method (WFEM) can be used to efficiently compute the forced response of infinite or finite, periodic or quasi-periodic structures [4]. An FE model of the reference cell is used to compute free wave characteristics of positive- and

G. Degrande et al. (Eds.): *Noise and Vibration Mitigation for Rail Transportation Systems*, NNFM 150, pp. 512–520, 2021.
https://doi.org/10.1007/978-3-030-70289-2_55

negative-going waves. The forced response is obtained as a superposition of these free waves.

This paper is outlined as follows. Section 2 presents the WFEM-based methodology. In Sect. 3, a case study of a railway bridge on piled foundations is introduced. The receptance functions of a single piled foundation are computed and the influence of neighboring foundations on these functions is investigated. The frequency response functions of the railway bridge are computed and the influence of dynamic SSI is discussed. Section 4 concludes the paper.

2 Methodology

The railway bridge shown in Fig. 1 is assumed to be infinitely long and consisting of identical spans with length L. The dynamic stiffness matrix $\mathbf{D}$ of a reference cell k relates the forces $\mathbf{f}_k$ to the displacements $\mathbf{u}_k$:

$$\begin{Bmatrix} \mathbf{f}_{\mathrm{L}k} \\ \mathbf{f}_{\mathrm{I}k} \\ \mathbf{f}_{\mathrm{R}k} \end{Bmatrix} = \begin{bmatrix} \mathbf{D}_{\mathrm{LL}} & \mathbf{D}_{\mathrm{LI}} & \mathbf{D}_{\mathrm{LR}} \\ \mathbf{D}_{\mathrm{IL}} & \mathbf{D}_{\mathrm{II}} & \mathbf{D}_{\mathrm{IR}} \\ \mathbf{D}_{\mathrm{RL}} & \mathbf{D}_{\mathrm{RI}} & \mathbf{D}_{\mathrm{RR}} \end{bmatrix} \begin{Bmatrix} \mathbf{u}_{\mathrm{L}k} \\ \mathbf{u}_{\mathrm{I}k} \\ \mathbf{u}_{\mathrm{R}k} \end{Bmatrix}. \tag{1}$$

The subscripts L, I, and R indicate the degrees of freedom (DOFs) belonging to the left interface, the interior, and the right interface, respectively. If no external forces are applied to the interior DOFs (i.e. $\mathbf{f}_{\mathrm{I}k} = \mathbf{0}$), these DOFs can be eliminated resulting in the following condensed system of equations:

$$\begin{Bmatrix} \mathbf{f}_{\mathrm{L}k} \\ \mathbf{f}_{\mathrm{R}k} \end{Bmatrix} = \begin{bmatrix} \tilde{\mathbf{D}}_{\mathrm{LL}} & \tilde{\mathbf{D}}_{\mathrm{LR}} \\ \tilde{\mathbf{D}}_{\mathrm{RL}} & \tilde{\mathbf{D}}_{\mathrm{RR}} \end{bmatrix} \begin{Bmatrix} \mathbf{u}_{\mathrm{L}k} \\ \mathbf{u}_{\mathrm{R}k} \end{Bmatrix} \tag{2}$$

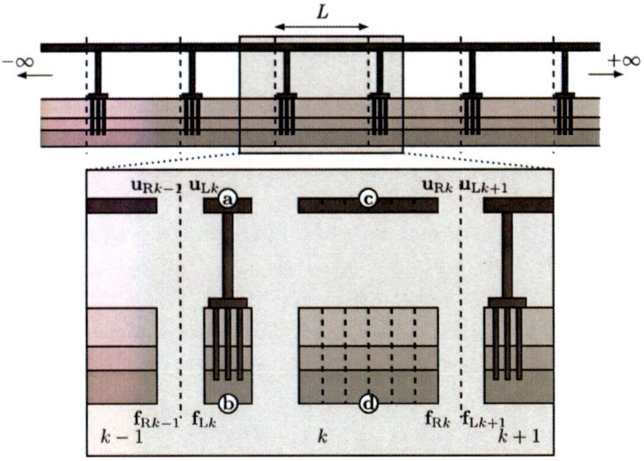

Fig. 1. Periodic elevated railway bridge consisting of identical cells of length L. The reference cell k consists of four parts: the pier (part a), the piled foundation (part b), the bridge deck between two piers (part c), and the soil between two foundations (part d).

where

$$\begin{bmatrix} \tilde{\mathbf{D}}_{\mathrm{LL}} & \tilde{\mathbf{D}}_{\mathrm{LR}} \\ \tilde{\mathbf{D}}_{\mathrm{RL}} & \tilde{\mathbf{D}}_{\mathrm{RR}} \end{bmatrix} = \begin{bmatrix} \mathbf{D}_{\mathrm{LL}} & \mathbf{D}_{\mathrm{LR}} \\ \mathbf{D}_{\mathrm{RL}} & \mathbf{D}_{\mathrm{RR}} \end{bmatrix} - \begin{bmatrix} \mathbf{D}_{\mathrm{LI}} \\ \mathbf{D}_{\mathrm{RI}} \end{bmatrix} \mathbf{D}_{\mathrm{II}}^{-1} \begin{bmatrix} \mathbf{D}_{\mathrm{IL}} & \mathbf{D}_{\mathrm{IR}} \end{bmatrix}. \tag{3}$$

A tilde is used to indicate variables after condensation of the interior DOFs. If the number of interior DOFs is large, the evaluation of Eq. (3) is computationally expensive. Hence, it is preferred to divide the reference cell into four parts (Fig. 1): the pier (part a), the piled foundation (part b), the bridge deck between two piers (part c), and the soil between two foundations (part d). It is assumed that parts c and d are periodic so that these parts are divided into N_{c} and N_{d} cells, respectively. This way, the FE modeling is limited to parts a and b, and the reference cells of parts c and d, drastically reducing the required computer memory. This approach is elaborated in the following paragraphs.

Assume first that the condensed dynamic stiffness matrix $\tilde{\mathbf{D}}$ of the reference cell k is known, and that no external forces are applied to the railway bridge. Free waves propagate through periodic structures with only a change in amplitude and phase, which is represented by the propagation constant μ of the free wave [1]:

$$\mathbf{u}_{\mathrm{L}k+1} = e^{-\mu}\mathbf{u}_{\mathrm{L}k}, \tag{4}$$

$$\mathbf{f}_{\mathrm{L}k+1} = e^{-\mu}\mathbf{f}_{\mathrm{L}k}. \tag{5}$$

Imposing continuity of displacements and equilibrium of forces between two neighboring cells k and $k+1$ results in:

$$\mathbf{u}_{\mathrm{R}k} = \mathbf{u}_{\mathrm{L}k+1} = e^{-\mu}\mathbf{u}_{\mathrm{L}k}, \tag{6}$$

$$\mathbf{f}_{\mathrm{R}k} = -\mathbf{f}_{\mathrm{L}k+1} = -e^{-\mu}\mathbf{f}_{\mathrm{L}k}. \tag{7}$$

The free wave propagation constants μ are obtained from the following eigenvalue problem, which originates from the combination of Eqs. (3), (6), (7):

$$\begin{bmatrix} \mathbf{0} & \mathbf{I} \\ -\tilde{\mathbf{D}}_{\mathrm{LR}}^{-1}\tilde{\mathbf{D}}_{\mathrm{RL}} & -\tilde{\mathbf{D}}_{\mathrm{LR}}^{-1}\left(\tilde{\mathbf{D}}_{\mathrm{LL}} + \tilde{\mathbf{D}}_{\mathrm{RR}}\right) \end{bmatrix} \begin{Bmatrix} \mathbf{u}_{\mathrm{L}k} \\ \mathbf{u}_{\mathrm{R}k} \end{Bmatrix} = e^{-\mu} \begin{Bmatrix} \mathbf{u}_{\mathrm{L}k} \\ \mathbf{u}_{\mathrm{R}k} \end{Bmatrix}. \tag{8}$$

The real part of μ determines the wave attenuation, while the imaginary part denotes its phase change over the cell length L. A free wave is called positive-going if its amplitude decreases towards the right ($\mathrm{Re}(\mu) > 0$) or, if power is flowing towards the right ($\mathrm{Im}(\mu) > 0$) when its amplitude is constant ($\mathrm{Re}(\mu) = 0$). All other waves are called negative-going. The propagation constants of positive- and negative-going free waves are indicated by μ_{p} and μ_{n}, respectively.

The eigenvectors in Eq. (8) contain the free wave modes $\boldsymbol{\psi}_{\mathrm{p}}$ and $\boldsymbol{\psi}_{\mathrm{n}}$ for the positive- and negative-going waves, respectively. Each mode $\boldsymbol{\psi}$ is associated to a force vector $\boldsymbol{\rho}$, which is directly obtained from Eq. (3). This yields:

$$\boldsymbol{\rho}_{\mathrm{p}} = \left(\tilde{\mathbf{D}}_{\mathrm{LL}} + e^{-\mu_{\mathrm{p}}}\tilde{\mathbf{D}}_{\mathrm{LR}}\right)\boldsymbol{\psi}_{\mathrm{p}} \quad \text{and} \quad \boldsymbol{\rho}_{\mathrm{n}} = \left(e^{+\mu_{\mathrm{n}}}\tilde{\mathbf{D}}_{\mathrm{RL}} + \tilde{\mathbf{D}}_{\mathrm{RR}}\right)\boldsymbol{\psi}_{\mathrm{n}}. \tag{9}$$

Now, consider a finite part consisting of N identical cells to which no external loads are applied. The displacements at both ends are written as a linear superposition of positive- and negative-going free waves:

$$\begin{Bmatrix} \mathbf{u}_{\mathrm{L}1} \\ \mathbf{u}_{\mathrm{R}N} \end{Bmatrix} = \begin{bmatrix} \boldsymbol{\Psi}_{\mathrm{p}} & \boldsymbol{\Psi}_{\mathrm{n}} e^{+N\mu_{\mathrm{n}}} \\ \boldsymbol{\Psi}_{\mathrm{p}} e^{-N\mu_{\mathrm{p}}} & \boldsymbol{\Psi}_{\mathrm{n}} \end{bmatrix} \begin{Bmatrix} \boldsymbol{\alpha}_{\mathrm{p}} \\ \boldsymbol{\alpha}_{\mathrm{n}} \end{Bmatrix} \tag{10}$$

where $\boldsymbol{\Psi}_{\mathrm{p}}$ and $\boldsymbol{\Psi}_{\mathrm{n}}$ collect the free wave modes $\boldsymbol{\psi}_{\mathrm{p}}$ and $\boldsymbol{\psi}_{\mathrm{n}}$, respectively. Similarly, the forces at both ends are written as a linear superposition of the force vectors:

$$\begin{Bmatrix} \mathbf{f}_{\mathrm{L}1} \\ \mathbf{f}_{\mathrm{R}N} \end{Bmatrix} = \begin{bmatrix} \mathbf{R}_{\mathrm{p}} & -\mathbf{R}_{\mathrm{n}} e^{+N\mu_{\mathrm{n}}} \\ -\mathbf{R}_{\mathrm{p}} e^{-N\mu_{\mathrm{p}}} & \mathbf{R}_{\mathrm{n}} \end{bmatrix} \begin{Bmatrix} \boldsymbol{\alpha}_{\mathrm{p}} \\ \boldsymbol{\alpha}_{\mathrm{n}} \end{Bmatrix} \tag{11}$$

where $\mathbf{R}_{\mathrm{p}}$ and $\mathbf{R}_{\mathrm{n}}$ collect the force vectors $\boldsymbol{\rho}_{\mathrm{p}}$ and $\boldsymbol{\rho}_{\mathrm{n}}$, respectively. Eliminating the modal coordinates $\boldsymbol{\alpha}_{\mathrm{p}}$ and $\boldsymbol{\alpha}_{\mathrm{n}}$ from Eqs. (10) and (11) results in the condensed dynamic stiffness matrix $\tilde{\mathbf{S}}$ of the finite part consisting of N cells:

$$\begin{bmatrix} \tilde{\mathbf{S}}_{\mathrm{LL}} & \tilde{\mathbf{S}}_{\mathrm{LR}} \\ \tilde{\mathbf{S}}_{\mathrm{RL}} & \tilde{\mathbf{S}}_{\mathrm{RR}} \end{bmatrix} = \begin{bmatrix} \mathbf{R}_{\mathrm{p}} & -\mathbf{R}_{\mathrm{n}} e^{+N\mu_{\mathrm{n}}} \\ -\mathbf{R}_{\mathrm{p}} e^{-N\mu_{\mathrm{p}}} & \mathbf{R}_{\mathrm{n}} \end{bmatrix} \begin{bmatrix} \boldsymbol{\Psi}_{\mathrm{p}} & \boldsymbol{\Psi}_{\mathrm{n}} e^{+N\mu_{\mathrm{n}}} \\ \boldsymbol{\Psi}_{\mathrm{p}} e^{-N\mu_{\mathrm{p}}} & \boldsymbol{\Psi}_{\mathrm{n}} \end{bmatrix}^{-1} . \tag{12}$$

For semi-infinite parts to which no external forces are applied, Eq. (12) simplifies to:

$$\tilde{\mathbf{S}}_{-\infty} = \mathbf{R}_{\mathrm{n}} \boldsymbol{\Psi}_{\mathrm{n}}^{-1} \quad \text{and} \quad \tilde{\mathbf{S}}_{+\infty} = \mathbf{R}_{\mathrm{p}} \boldsymbol{\Psi}_{\mathrm{p}}^{-1} \tag{13}$$

for a semi-infinite part towards the left and right, respectively.

Consider again the reference cell shown in Fig. 1. The condensed dynamic stiffness matrix of the union of parts a and b is indicated by $\tilde{\mathbf{D}}^{\mathrm{ab}}$, which contains coupling terms between parts a and b. Parts c and d are uncoupled, and their condensed dynamic stiffness matrices $\tilde{\mathbf{S}}^{\mathrm{c}}$ and $\tilde{\mathbf{S}}^{\mathrm{d}}$ are obtained from Eq. (12), after first computing the free wave characteristics of the reference cell in parts c and d. By imposing continuity of displacements and equilibrium of forces between parts a and c, and b and d, the following system of equations is obtained:

$$\begin{Bmatrix} \mathbf{f}^{\mathrm{a}}_{\mathrm{L}k} \\ \mathbf{f}^{\mathrm{b}}_{\mathrm{L}k} \\ \mathbf{f}^{\mathrm{c}}_{\mathrm{R}k} \\ \mathbf{f}^{\mathrm{d}}_{\mathrm{R}k} \\ \mathbf{0} \\ \mathbf{0} \end{Bmatrix} = \begin{bmatrix} \tilde{\mathbf{D}}^{\mathrm{aa}}_{\mathrm{LL}} & \tilde{\mathbf{D}}^{\mathrm{ab}}_{\mathrm{LL}} & \mathbf{0} & \mathbf{0} & \tilde{\mathbf{D}}^{\mathrm{aa}}_{\mathrm{LR}} & \tilde{\mathbf{D}}^{\mathrm{ab}}_{\mathrm{LR}} \\ \tilde{\mathbf{D}}^{\mathrm{ba}}_{\mathrm{LL}} & \tilde{\mathbf{D}}^{\mathrm{bb}}_{\mathrm{LL}} & \mathbf{0} & \mathbf{0} & \tilde{\mathbf{D}}^{\mathrm{ba}}_{\mathrm{LR}} & \tilde{\mathbf{D}}^{\mathrm{bb}}_{\mathrm{LR}} \\ \mathbf{0} & \mathbf{0} & \tilde{\mathbf{S}}^{\mathrm{cc}}_{\mathrm{RR}} & \mathbf{0} & \tilde{\mathbf{S}}^{\mathrm{cc}}_{\mathrm{RL}} & \mathbf{0} \\ \mathbf{0} & \mathbf{0} & \mathbf{0} & \tilde{\mathbf{S}}^{\mathrm{dd}}_{\mathrm{RR}} & \mathbf{0} & \tilde{\mathbf{S}}^{\mathrm{dd}}_{\mathrm{RL}} \\ \tilde{\mathbf{D}}^{\mathrm{aa}}_{\mathrm{RL}} & \tilde{\mathbf{D}}^{\mathrm{ab}}_{\mathrm{RL}} & \tilde{\mathbf{S}}^{\mathrm{cc}}_{\mathrm{LR}} & \mathbf{0} & \tilde{\mathbf{D}}^{\mathrm{aa}}_{\mathrm{RR}} + \tilde{\mathbf{S}}^{\mathrm{cc}}_{\mathrm{LL}} & \tilde{\mathbf{D}}^{\mathrm{ab}}_{\mathrm{RR}} \\ \tilde{\mathbf{D}}^{\mathrm{ba}}_{\mathrm{RL}} & \tilde{\mathbf{D}}^{\mathrm{bb}}_{\mathrm{RL}} & \mathbf{0} & \tilde{\mathbf{S}}^{\mathrm{dd}}_{\mathrm{LR}} & \tilde{\mathbf{D}}^{\mathrm{ba}}_{\mathrm{RR}} & \tilde{\mathbf{D}}^{\mathrm{bb}}_{\mathrm{RR}} + \tilde{\mathbf{S}}^{\mathrm{dd}}_{\mathrm{LL}} \end{bmatrix} \begin{Bmatrix} \mathbf{u}^{\mathrm{a}}_{\mathrm{L}k} \\ \mathbf{u}^{\mathrm{b}}_{\mathrm{L}k} \\ \mathbf{u}^{\mathrm{c}}_{\mathrm{R}k} \\ \mathbf{u}^{\mathrm{d}}_{\mathrm{R}k} \\ \mathbf{u}^{\mathrm{a}}_{\mathrm{R}k} \\ \mathbf{u}^{\mathrm{b}}_{\mathrm{R}k} \end{Bmatrix} . \tag{14}$$

The superscripts a, b, c, and d indicate the DOFs belonging to the interface of the respective parts. The displacements $\mathbf{u}^{\mathrm{a}}_{\mathrm{R}k}$ and $\mathbf{u}^{\mathrm{b}}_{\mathrm{R}k}$ are eliminated by taking the Schur complement of the coefficient matrix in Eq. (14). This finally results in the expression for the condensed dynamic stiffness matrix $\tilde{\mathbf{D}}$, which relates the forces to the displacements at both sides of the reference cell k.

From the condensed dynamic stiffness matrix $\tilde{\mathbf{D}}$, the free wave characteristics of the reference cell are computed. Using Eq. (13), the dynamic stiffness matrix of the semi-infinite parts on both ends of the reference cell are constructed. Finally, the forced vibration response is computed after assembling the dynamic stiffness matrices of the reference cell and the semi-infinite parts [5].

3 Case Study

This section presents a case study of a railway bridge founded on piles. The piled foundation consists of piles that are connected to a rigid pile cap. The piles are modeled with beam elements. The soil surrounding the piles is modeled with 20-node solid elements. The FE model contains cavities that correspond to the pile volumes (Fig. 2). The translational degrees of freedom of the pile elements are then coupled to the translational degrees of freedom on the soil-pile interface assuming perfect bonding conditions. The rotational degrees of freedom of the pile elements are left free. A perfectly matched layer (PML) surrounds the soil domain to prevent spurious wave reflections [6].

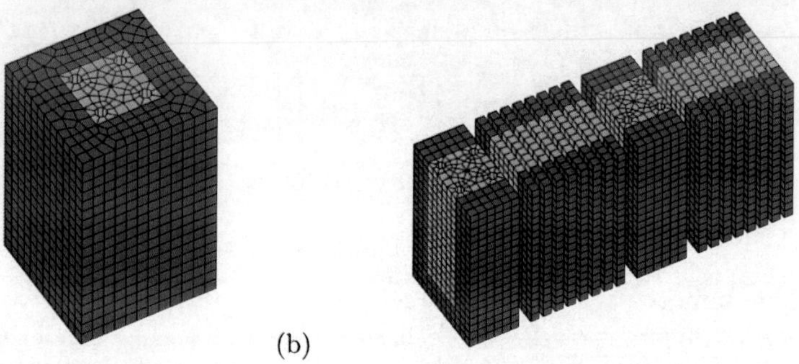

(a) (b)

Fig. 2. (a) 3D FE-PML model of a single piled foundation and (b) WFEM model of a row of piled foundations. It is composed of the physical soil domain (light grey) with cavities for the piles, and the perfectly matched layer (dark grey).

The piled foundation consists of four piles of length $L_\mathrm{p} = 7.5$ m, diameter $D_\mathrm{p} = 0.5$ m, and center-to-center distance $s = 2$ m. The size of the pile cap equals $3\,\mathrm{m} \times 3\,\mathrm{m} \times 1\,\mathrm{m}$. The piles and the pile cap are made of concrete with Young's modulus $E_\mathrm{p} = 30$ GPa, Poisson's ratio $\nu_\mathrm{p} = 0.25$, and density $\rho_\mathrm{p} = 2500$ kg/m^3. For the soil, two cases are considered: a soft soil with shear wave velocity $C_\mathrm{s} = 75$ m/s and a stiffer soil with $C_\mathrm{s} = 150$ m/s. The Poisson's ratio $\nu_\mathrm{s} = 1/3$, the density $\rho_\mathrm{s} = 1800$ kg/m^3, and the material damping ratio's $\beta_\mathrm{s} = \beta_\mathrm{p} = 0.02$ for shear and dilatational deformation are assumed constant for both soils.

The bridge consists of a deck that is periodically supported by cylindrical piers with a height of 5 m and a diameter of 1.2 m. The bridge deck and piers are modeled with beam elements. The pier end is rigidly connected to the pile cap. The bridge deck has a cross-sectional area of 4 m^2 and second moment of area of 1.5 m^4. The pier and deck are also made of concrete.

A floating slab track is installed on the bridge. The track is composed of two UIC60 rails and a concrete slab with cross-section $2.5\,\mathrm{m} \times 0.5\,\mathrm{m}$. Each rail is supported by rail pads with stiffness $k_\mathrm{rp} = 213.2 \times 10^6$ N/m and damping

$c_{rp} = 14.8 \times 10^3$ Ns/m. The slab is supported by a slab mat with stiffness $k_{sm} = 15 \times 10^6$ N/m^3 and damping $c_{sm} = 30 \times 10^3$ Ns/m^3. The slab resonance frequency 17 Hz.

3.1 Coupling Between Piled Foundations

Figure 3 shows the real and the imaginary part of the piled foundation receptance for the horizontal, vertical, and rocking motion of the pile cap. The 3D FE-PML model of a single foundation (Fig. 2a) is compared to the WFEM model of a row of foundations (Fig. 2b) with spacing $L = 12$ m and $L = 24$ m. It is observed that the stiffness of the piled foundation increases with increasing frequency. The piled foundation generally shows stiffer behavior when it is embedded in stiffer soils. The peaks and dips in the receptance curves are caused by pile-soil-pile interaction and depend on the pile spacing within a single foundation. For this particular case, the stiffness of the piled foundation is not affected by the presence of neighboring foundations.

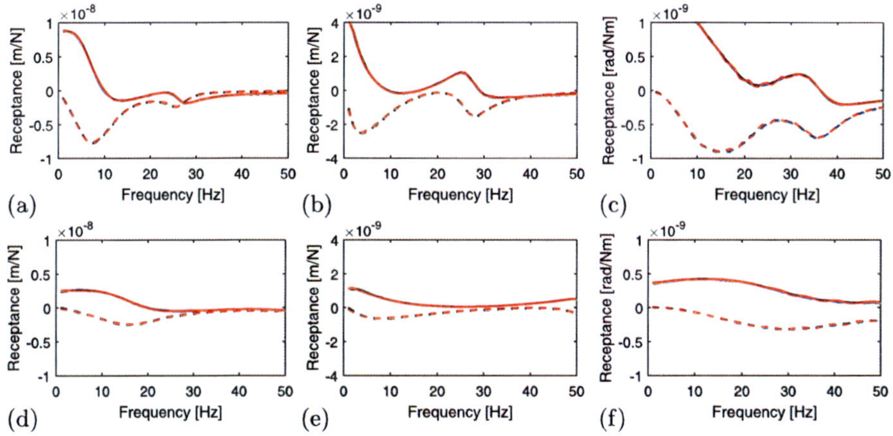

Fig. 3. Real (solid line) and imaginary (dashed line) part of the piled foundation receptance for (a-c) the soft soil, and (d-f) the stiffer soil. The receptance is shown for (a,d) the horizontal, (b,e) the vertical, and (c,f) the rocking motion of the pile cap. Three different cases are considered: a single foundation (black line), a foundation row with spacing $L = 12$ m (red line), and a foundation row with spacing $L = 24$ m (blue line).

3.2 Bridge Response

Figures 4 and 5 show the frequency response functions (FRFs) for a vertical unit harmonic load applied to the track above a pier and at midspan between two piers, for span lengths of $L = 12$ m and $L = 24$ m, respectively. The results are presented for the viaduct founded on soft soil and a stiffer soil. Three cases are considered: (1) dynamic SSI is disregarded (clamped piers), (2) dynamic

SSI is accounted for but through-soil coupling between foundations is neglected, and (3) dynamic SSI is accounted for including through-soil coupling between foundations. As already evidenced in Fig. 3, the through-soil coupling between the foundations is limited. Hence, cases (2) and (3) yield nearly identical results.

When the load is applied directly above the pier, the frequency response functions look similar for both soil conditions. Furthermore, they are only weakly influenced by the span length of the viaduct. It is observed that the response is generally more damped when dynamic SSI is accounted for, especially at the peak 17 Hz, which corresponds to the bending mode of the bridge and the slab track resonance.

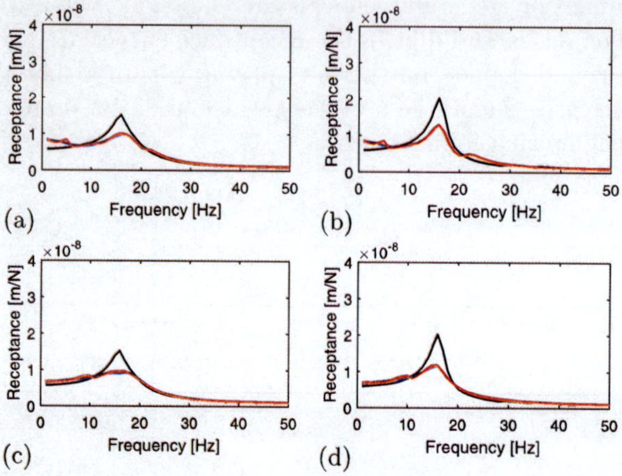

Fig. 4. Amplitude of the frequency response functions of the bridge with a span length $L = 12$ m that is founded on (a, b) soft soil, and (c, d) a stiffer soil. A vertical unit harmonic load is applied to the track, (a, c) above a pier and (b, d) at midspan between two piers. Three cases are considered: no dynamic SSI (black line), dynamic SSI neglecting through-soil coupling (blue line), and dynamic SSI accounting for through-soil coupling (red line).

When the load is applied at midspan, the influence of the span length becomes clear. For $L = 24$ m, a resonance peak is observed 7 Hz, which corresponds to a bending mode of the bridge deck. If the viaduct is founded on soft soil, this resonance peak is attenuated and shifted, while for the stiffer soil it is only slightly attenuated. Above 15 Hz, the influence of the soil conditions vanishes.

In the high frequency range, the influence of dynamic SSI can be neglected and the FRFs converge to a single curve. When studying e.g. moving load problems where high frequency range computations are required, this is beneficial because the computational domain can be reduced to only the superstructure, or even the track or rails.

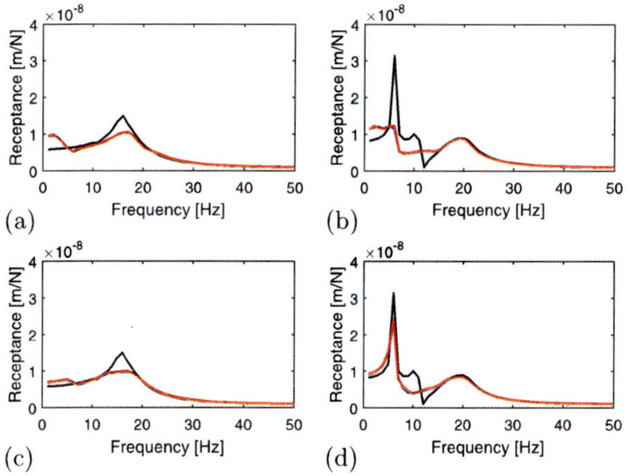

Fig. 5. Amplitude of the frequency response functions of the bridge with a span length $L = 24$ m that is founded on (a, b) soft soil, and (c, d) a stiffer soil; legend as for Fig. 4.

4 Conclusion

This paper presents a model to predict the response of periodic railway bridges accounting for dynamic SSI and through-soil coupling between piled foundations. The model exploits the repetitive nature of these bridges and is based on the WFEM. A simplified case study is presented which shows that through-soil coupling between foundations is limited. For the present case, dynamic SSI is important up 40 Hz, but can be neglected in the higher frequency range.

Acknowledgements. Results presented in this paper were obtained within the frame of the FWO project G.0903.17N "Dynamic response of repetitive elevated structures accounting for soil-structure interaction". Financial support by FWO Flanders is gratefully acknowledged.

References

1. Mead, D.: Wave propagation in continuous periodic structures: research contributions from Southampton. J. Sound Vib. **190**(3), 495–524 (1996)
2. Clouteau, D., Elhabre, M.L., Aubry, D.: Periodic BEM and FEM-BEM coupling: application to seismic behavior of very long structures. Comput. Mech. **25**, 567–577 (2000)
3. Degrande, G., Clouteau, D., Othman, R., Arnst, M., Chebli, H., Klein, R., Chatterjee, P., Janssens, B.: A numerical model for ground-borne vibrations from underground railway traffic based on a periodic finite element-boundary element formulation. J. Sound Vib. **293**(3–5), 645–666 (2006)
4. Mace, B., Duhamel, D., Brennan, M., Hinke, L.: Finite element prediction of wave motion in structural waveguides. J. Acoust. Soc. Am. **117**(5), 2835–2843 (2005)

5. Reumers, P., Kuo, K., Lombaert, G., Degrande, G.: Response of periodic elevated structures accounting for soil-structure interaction. In: Papadrakakis, M., Fragiadakis, M. (eds.) Proceedings of the 7th International Conference on Computational Methods in Structural Dynamics and Earthquake Engineering, COMPDYN 2019, Crete, Greece, June 2019
6. Basu, U., Chopra, A.K.: Perfectly matched layers for time-harmonic elastodynamics of unbounded domains: theory and finite-element implementation. Comput. Methods Appl. Mech. Eng. **192**(11–12), 1337–1375 (2003)

Identification of a Randomly-Fluctuating Continuous Model of the Ballasted Track Based on Measurements at the Pass-By of High-Speed Trains

Patryk Dec[1,2(✉)], Régis Cottereau[1], and Baldrik Faure[2]

[1] Aix-Marseille Univ., CNRS, Centrale Marseille, LMA, 4 impasse Nikola Tesla, 13013 Marseille, France
dec@lma.cnrs-mrs.fr

[2] SNCF, Innovation and Research Department, 1/3 Avenue François Mitterrand, 93212 La Plaine Saint Denis, France

Abstract. In this paper a randomly-fluctuating continuous model of ballasted railway tracks is studied. The ballast layer is considered as a linear, continuous, heterogeneous medium. Its properties are modeled as a sample of a random field, with a characteristic fluctuating length close to the average size of a ballast grain. The parameters of this model (mean wave velocities and coefficient of variation in the ballast layer) have to be identified from open-field measurements of the vertical acceleration in a ballasted railway track. The identification is performed for the mean model of the ballast and first steps in the identification process for the fluctuations are introduced. The geometry of the track (soil, ballast and sleepers) follows the real geometry where the experimental campaign was conducted.

Keywords: Railway track · Ballast · Wave propagation · Spectral element method · Inverse problem

1 Introduction

In the context of railway dynamics modeling, the consideration of the ballast layer is still a challenging issue. The following two approaches are mainly used for the ballast modeling. The first one, known as Discrete Element Method (DEM), considers every grain separately and takes into account the interactions between them (contact, friction, impact). However, the calculation time and resources needed are beyond the reach of current supercomputers for applications requiring large models such as wave propagation analyses. The second approach consists in modeling the ballast layer as a homogeneous medium and solving the problem with Finite (FEM) or Boundary (BEM) Element Methods. Despite the efficiency of these approaches in terms of computational costs, there seems to be a modeling ingredient missing because they are still limited in terms of

G. Degrande et al. (Eds.): *Noise and Vibration Mitigation for Rail Transportation Systems*, NNFM 150, pp. 521–528, 2021.
https://doi.org/10.1007/978-3-030-70289-2_56

prediction potential for track dynamics applications. Using a 3D-FEM model, Arlaud et al. [1] highlighted for instance this limitation on the track receptance (ratio of the resulting displacement of the rail to the exciting force), particularly in the 20–60 Hz frequency range.

An alternative to continuous homogeneous models has been recently introduced in [2]. The idea of this model consists in representing the ballast layer as a continuum still, but heterogeneous. Mechanical properties are obtained as samples of 3D random fields with specifically designed statistical properties. Approximation of the wave propagation problem is then performed with the Spectral Element Method (SEM), which is well-suited for large-scale problems, and widely used in seismic engineering for instance [3]. In the original paper, model parameters were identified on numerical DEM experiments in statics. This paper aims at presenting the identification of the statistical parameters based on dynamic ballasted railway track measurements.

First, the average properties of the ballast layer are identified. The model is therefore homogeneous. In a second part, fluctuations are identified, and the ballast is modeled as randomly-fluctuating. Ballast parameters are identified using dynamic acceleration measurements performed on a High-Speed Line (HSL) during train pass-by at commercial speed (300–320 km/h). The identification is performed for the average model of the ballast and first steps in the identification process for the fluctuations are introduced.

2 Acceleration Measurements on a HSL During the Passage of Trains

2.1 Description of the Measurement Campaign

The measurements have been performed in the context of a research project funded by SNCF Réseau and led by SNCF I&R. Its purpose is to characterize the mechanical behaviour of the railway track in the medium frequency range. The measurements were performed on the LGV-EE (East European High Speed Line) from Paris to Strasbourg. The track is used for commercial purposes, thus the ballast layer is well stabilized after a large number of pass-by, ensuring a high strength and stiffness [4].

Measurements were realized using triaxial accelerometers on 4 instrumented zones. Accelerometers were installed on each zone over two cross sections separated by 6 m. On each section, 4 accelerometers were placed as illustrated on Fig. 1: at the rail web, at the sleeper exterior part, in the upper layer of ballast at mid-distance between the two rail lines using piles driven halfway through the ballast and in the soil using piles driven at the foot of the ballast. Strain gages installed in Wheatstone bridge configuration at the rail web detected the bogie passages. More than 800 passages at commercial speed were recorded.

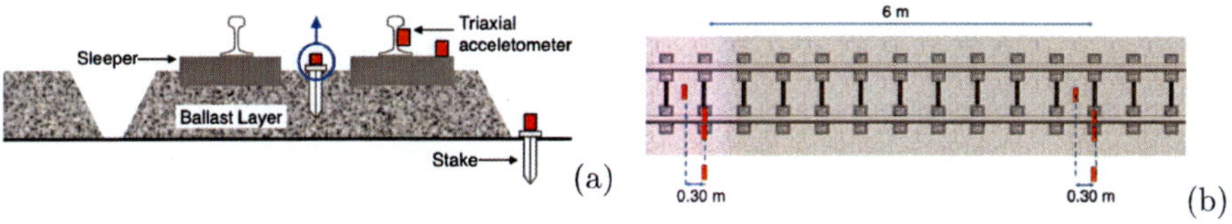

Fig. 1. Position of the sensors (in red) on the track during the measurement campaign. (a) cross-sectional view and (b) top view

2.2 Spectral Analysis of the Vertical Acceleration Inside the Ballast Layer

A Fourier transform of the measured vertical accelerations is realized to perform an analysis in the spectral domain. Amplitude of the frequency spectra for trains at various speed are presented on Fig. 2, as a function of characteristic length $\lambda = v/f$, where v is the velocity of the passing train and f is the frequency. Different peaks are noticeable and can be divided into two groups depending of their evolution in function of v.

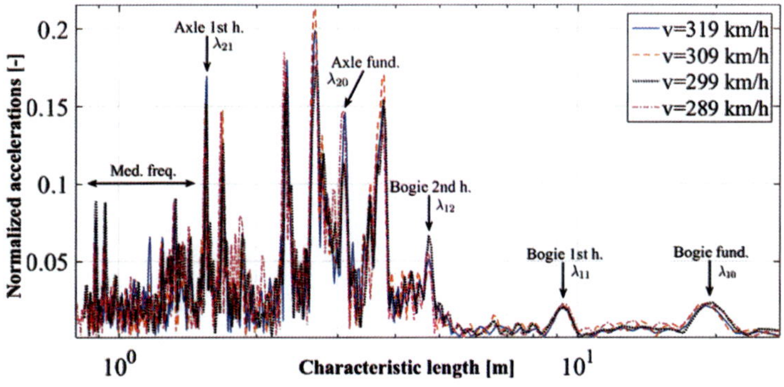

Fig. 2. Normalized frequency spectra for the vertical acceleration inside the ballast in function of the characteristic length $\lambda = v/f$, at the passage of TGVs from 289 to 319 km/h. Linearly-shifting peaks (denoted as *fund.*) and their harmonics (n-th *h.*) are designated, together with the medium frequency band (*med. freq.*).

For the first kind of peaks, their amplitude and/or frequency increases linearly with v. In the following they are denoted by λ_{ij}, with j being the j-th harmonic of the i-th peak. It appears clearly that the peaks of higher λ_{ij} match for identical trains, regardless of the speed. For example, λ_{1j} represents the typical length between two bogies (17.8 m) and λ_{2j} the space between two axles of a bogie (3 m). This shifting shows that those peaks are only due to the moving load, and not to the track components. Nevertheless there are some other peaks which do not follow this rule and depend non-linearly of the train speed v. Those frequency peaks are particularly numerous between $\lambda = 1$ m and 1.45 m, which correspond to frequencies between 50 and 80 Hz. This frequency range is likely to be above the frequency range of validity for many FE track models : for example in [1], the comparison of a numerical receptance test with experiments shows a significant error 50 Hz.

2.3 Identification of the Wave Velocity in the Soil

The wave velocity in the soil is obtained from measurements described in Sect. 2.1. A train pass-by induces vibrations that propagate in the soil as waves. The wave velocity in the soil is generally higher that the train speed. The wave that propagates in the soil is then detected before the wave corresponding to the train pass-by [5]. For both acquisitions in the soil a signal of duration 0.3 s corresponding to the propagation of the wave in the soil is selected. The cross-correlation is then calculated between these signals corresponding to two sensors in the soil with a spacing of 6 m, and the maximum yields an estimation of the Rayleigh wave velocity in the soil. Averaging these estimations over all recorded train passages, Rayleigh wave velocities are obtained and reported for each zone in Table 1. A similar operation on sensors positioned within the ballast layer yields an estimation of the Rayleigh wave velocity in the ballast, also reported in Table 1. The difference in wave velocities for different zones is due to experimental errors. Sensors in the soil and in the ballast are fixed on a metal pile which is planted in the soil and the ballast. The coupling between the sensor and the soil is questionable. Data from some sensors are very noisy and therefore they are difficult to exploit.

Table 1. Rayleigh wave velocities in the soil and the ballast

	Zone 1	Zone 2	Zone 3	Zone 4
C_R **Soil (m/s)**	248	192	270	250
C_R **Ballast(m/s)**	275	232	309	260

3 Construction of an Average Model of a Railway Track

In this section, an average model of a double track is constructed, based on the measurements discussed above, and a simulation of waves induced by train passages in this average model is presented.

3.1 Description of the Homogeneous Model

A simple model of the track is considered, where only the soil, the ballast layer and the sleepers are taken into account. The sleeper dimensions are

$2 \times 0.3 \times 0.4$ m^3, the sleeper spacing is $d = 0.6$ m, the distance between the tracks is 6 m, the upper ballast width is 8 m, the lower ballast width is 14 m and the soil is modeled in a box of dimensions $24.3 \times 18 \times 4$ m^3. All materials are assumed isotropic linear elastic. The sleepers are assumed homogeneous with pressure wave velocity $C_p = 3825$ m/s, shear wave velocity $C_s = 2236$ m/s and $\rho = 2400$ kg/m^3. The soil is assumed homogeneous with $C_p = 600$ m/s, $C_s = 350$ m/s and $\rho = 1800$ kg/m^3. The ballast is assumed homogeneous in this average model with $C_p = 500$ m/s, $C_s = 300$ m/s and $\rho = 1700$ kg/m^3.

The wave propagation is simulated using a spectral element [3] software co-developed by MSSMat Laboratory (CentraleSupélec, CNRS, Université Paris-Saclay), CEA (French Atomic Energy Commission) and IPGP (Paris Institute of Earth Physics). Perfectly Matched Layers are used to simulate wave propagation in an open space by absorbing waves exiting the computational domain. The hexahedral mesh was tailored so as to accurately transport frequencies up 100 Hz: the mean element size equals $h = 0.6$ m, with a minimal distance between nodes of 5 cm. There are around 24,000 hexahedral elements of 6th order in each direction, for a total of $1.1 \cdot 10^6$ nodes. The time step is about 10^{-6} s. To simulate 4 s of propagation, 36 days of CPU time on 144 processing units were used (6 h of real time).

3.2 Response to a Passing Train Obtained from the Response of a Single Bogie on a Single Sleeper

We show here that the acceleration induced by a passing train can be reconstructed from the acceleration of a single bogie on a single sleeper, under the simple hypotheses of periodicity (the period is the sleeper spacing d) and linearity of the model. We first discuss the transformation from a single passing bogie to a passing train, and then the transformation from a passing bogie to a bogie in a single sleeper.

By modeling the train as a set of $N + 1$ bogies with different weights, and under the assumptions above, the acceleration at point x due to a passing train is given by:

$$a_{\text{Train}}(x,t) = \sum_{k=0}^{N} a_{Bk}(x,t) = \sum_{k=0}^{N} \alpha_k a_B(x, t + \tau_k) \tag{1}$$

where a_{Bk} is the acceleration produced by the bogie B_k, τ_k the time delay of the k-th passing bogie pass-by with respect to the first one, α_k is the ratio of the weight of bogie k with respect to bogie 0, $\alpha_0 = 1$, and $a_B = a_{B0}$.

The acceleration at point x for a moving bogie can be reconstructed from the acceleration $a(x_\ell, t)$ measured in sensors distributed periodically along the track and excited by a non-moving loading applied on only one sleeper using formula

$$a_B(x,t) = \sum_{\ell=-\infty}^{+\infty} a_0 \left(x + \ell d, t + \ell \frac{d}{v} \right), \tag{2}$$

where $a_0(x,t)$ is the acceleration due to a non-moving bogie in x_0. The spectrum of a train pass-by is thus given by :

$$\hat{a}_{\text{Train}}(x,\omega) = \left(\sum_{\ell=-\infty}^{+\infty} \hat{a}_0(x+\ell d,\omega)\exp\left(i\omega\ell\frac{d}{v}\right)\right)\left(\sum_{k=0}^{N}\alpha_k \exp(i\omega\tau_k)\right). \quad (3)$$

The spectrum due to a full train can then be analytically calculated from the spectrum of a single passing bogie.

The input force representing the bogie loading is determined using the approach of Hoang et al. [6]. The rail is modeled as an Euler-Bernoulli beam, supported by a periodical elastic support modeling the rail pads on the sleepers. The total force applied on the beam can be expressed as a function of the moving load and reaction forces of sleepers. The Fourier transform of the reaction force of a support is obtained by performing a double Fourier transform in time and space of the Euler-Bernoulli beam equation that gives the vertical displacement u_z.

3.3 Results for a Passing Train

The displacement and acceleration fields induced by a train pass-by are reconstructed following the methodology described in Sect. 3.2. First, displacements in the soil are calculated for a limited size of the track with a bogie loading on only one sleeper. They are then extrapolated beyond the simulated physical domain (Fig. 3 (Left)). The reconstructed displacements for a moving bogie along the track are presented in the Fig. 3 (Right).

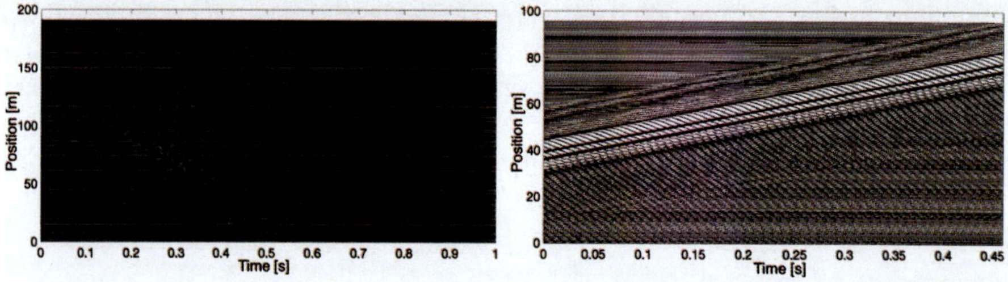

Fig. 3. (Left)Displacement ($\times 5 \cdot 10^4$) in the soil for a passing bogie on a single sleeper (Right) Reconstructed displacement ($\times 5 \cdot 10^5$) for a bogie pass-by through the simulated domain

4 Randomly-Fluctuating Heterogeneous Model of the Ballast

In this section, the randomly-fluctuating heterogeneous model of the ballast is presented. The future work of the identification process of the shear wave velocity, pressure wave velocity and coefficients of variation from experimental measurement is also introduced.

4.1 Description Randomly-Fluctuating Heterogeneous Model of the Ballast

The granular medium is represented by a randomly-fluctuating heterogeneous continuum model. A short description is given below whereas a detailed description can be found in de Abreu Corrêa et al. [7].

In order to build a heterogeneous layer for the granular medium, stochastic fields are generated. A log-normal distribution is assumed for the stochastic variables: the density $\rho(\mathbf{x})$, the pressure wave speed $c_p(\mathbf{x})$ and the shear wave speed $c_s(\mathbf{x})$. The mean values $\bar{c}_p$ and $\bar{c}_s$ are these determined from experimental data above. The random fields can be generated in space domain or in the spectral domain. In our case, random fields are generated using the spectral decomposition proposed by de Carvalho Paludo et al. [8]. The random field $g(\mathbf{x})$ in the spectral domain $\mathbf{k}$ discretized over a regular grid of size $\mathbf{N} = [N_x, N_y, N_z]$ and indexed by $\mathbf{n} = [n_x, n_y, n_z]$ can be expressed as:

$$g(\mathbf{x}) = \sum_{\mathbf{n} \leq \mathbf{N}} \sqrt{2\mathcal{S}(\mathbf{k_n})\bar{\Delta}\mathbf{k}} \cos(\mathbf{k_n} \times \mathbf{x} + \varPhi_\mathbf{n}) \tag{4}$$

where $\mathcal{S}(\mathbf{k_n})$ is the power spectral density of the random field, $\bar{\Delta}\mathbf{k}$ is the unit volume in the spectral domain and $\varPhi_\mathbf{n}$ are independent random variables of a N-dimensional variable, uniform over $[0, 2\pi]$. The first-order marginal density for each of these random fields is taken as gamma law. The standard deviations σ_{c_p} and σ_{c_s} are assumed to be equal and defined by the coefficient of variation : $CV = \sigma_{c_p}/\bar{c}_p = \sigma_{c_s}/\bar{c}_s$. The associated correlation lengths are set to $\lambda_{\text{corr}} = 7$ cm in the three directions of space, corresponding to the characteristic length of interaction between ballast grains. Because of the small correlation length, the mesh size inside the ballast has been reduced to $h = 0.07$ m (which corresponds to a minimal distance of 6 mm between two nodes) and the time step to 10^{-7} s. The computational cost rises up to 60 h of walltime (7 months of CPU time on 288 processing units).

4.2 Future Work: Identification Process of Heterogeneous Model of the Track

In order to identify ballast properties for the heterogeneous model, the normalized simulated spectrum of the vertical acceleration in the ballast $\hat{A}_z^{\text{sim}}$ will be compared to the normalized measured one $\hat{A}_z^{\text{data}}$. Spectrum are normalized with a quadratic norm. A normalization is necessary to compare the amplitude of frequency spectra because the amount of energy introduced in the model is different of that one given by the train. Spectra are compared in L^2 norm $\mathcal{N}(\hat{A}_z^{\text{sim}}) = \|\hat{A}_z^{\text{sim}} - \hat{A}_z^{\text{data}}\|_2$. The identification process consists in the minimization of the $\mathcal{N}$ function The density field is assumed to be known and homogeneous. The frequency range is limited to $R_f = [10, 40 \text{ Hz}]$ because of data quality issues. To ease the solution of this minimization problem, two successive steps are considered. First, the mean values $\bar{c}_p$ and $\bar{c}_s$ are fixed from experimental data. An approximate value CV^{pred} is determined. Then, the volume next to $(\bar{c}_p, \bar{c}_s, CV^{\text{pred}})$ is analyzed to search a more precise value of the optimum.

5 Conclusion

A linear randomly-fluctuating model of the ballast is constructed, along with its average. The pressure and shear wave velocities for the soil and the ballast are determined thanks to the cross correlation of vertical accelerations measured at two different position. In a future work, acceleration spectra will be compared to the measured one to identify the properties of the fluctuations of the ballast. Concerning numerical simulations, the use of a single bogie loading on only one sleeper allowed to reduce the costs. The response of the passing bogie and the spectrum of the train pass-by can then be reconstructed from the simulation results of only a standing bogie in one sleeper. Heterogeneity is expected to play a major role in the mechanical behavior of the railway track at medium frequencies which are not well represented by the average model.

References

1. Arlaud, E., Costa d'Aguiar, S., Balmes, E.: Receptance of railway tracks at low frequency: numerical and experimental approaches. Transp. Geotech. **9**, 1–16 (2016)
2. de Abreu Corrêa, L., Quezada, J.C., Cottereau, R., Costa d'Aguiar, S., Voivret, C.: Randomly-fluctuating heterogeneous continuum model of a ballasted railway track. Comput. Mech. **60**(5), 845–861 (2017)
3. Komatitsch, D., Tromp, J.: Introduction to the spectral element method for three-dimensional seismic wave propagation. Geophys. J. Int **139**(3), 806 (1999)
4. Suiker, A.S.J., Selig, E.T., Frenkel, R.: Static and cyclic triaxial testing of ballast and subballast. J. Geotech. Geoenviron. Eng. **131**(6), 771–782 (2005)
5. Ditzel, A., Herman, G.C., Drijkoningen, G.G.: Seismograms of moving trains: comparison of theory and measurements. J. Sound Vib. **248**(4), 635–652 (2001)
6. Hoang, T., Duhamel, D., Foret, G., Yin, H.P., Joyez, P., Caby, R.: Calculation of force distribution for a periodically supported beam subjected to moving loads. J. Sound Vib. **388**, 327–338 (2017)
7. de Abreu Corrêa, L., Quezada, J.C., Cottereau, R., d'Aguiar, S.C., Voivret, C.: Randomly-fluctuating heterogeneous continuum model of a ballasted railway track. Comput. Mech. **60**(5), 845–861 (2017)
8. de Carvalho Paludo, L., Bouvier, V., Cottereau, R.: Scalable parallel scheme for sampling of gaussian random fields over very large domains. Int. J. Numer. Methods Eng. **117**(8), 845–859 (2019)

Numerical Prediction and Experimental Validation of Railway Induced Vibration in a Multi-storey Office Building

Manthos Papadopoulos, Kirsty Kuo, Matthias Germonpré, Ramses Verachtert, Jie Zhang, Kristof Maes, Geert Lombaert, and Geert Degrande(✉)

Department of Civil Engineering, KU Leuven, 3001 Leuven, Belgium
geert.degrande@kuleuven.be
https://bwk.kuleuven.be/bwm

Abstract. This paper reports on an extensive measurement campaign in a three-storey office building close to a ballasted track on embankment. Dynamic soil characteristics are determined by means of in situ geophysical tests. A coupled Finite Element-Boundary Element (FE-BE) model of the reinforced concrete building, accounting for soil-structure interaction, is updated by means of modal characteristics that were identified using both ambient and forced excitation. The response of the track, the free field and the building is measured simultaneously during impact loading on the sleepers and the passage of freight and passenger trains. A 2.5D coupled FE-BE track model is calibrated based on the measured track receptance and transfer functions. The incident wave field due to impacts on the sleepers and train passages is very sensitive to uncertain dynamic soil properties. This uncertainty explains to a large degree the deviation between the predicted and measured response of the soil and the building.

Keywords: Railway induced vibration · Dynamic soil-structure interaction · System identification · Model calibration · Uncertainty

1 Introduction

Numerical prediction of railway induced vibration requires information on track, soil and building parameters which should be identified experimentally. Given the complexity and uncertainty of the problem, in combination with the wide frequency range of interest (1–80 Hz for vibration and 16–250 Hz for structure-borne noise), accurate prediction of railway induced vibration is challenging. This paper reports on an extensive measurement campaign and numerical investigation conducted at the Blok D building, a three-storey reinforced concrete building with below-ground basement located at 40 m from the railway line L1390 Leuven-Ottignies. The latter consists of two ballasted tracks on embankment and is operated by freight and passenger trains (Fig. 1).

G. Degrande et al. (Eds.): *Noise and Vibration Mitigation for Rail Transportation Systems*, NNFM 150, pp. 529–537, 2021.
https://doi.org/10.1007/978-3-030-70289-2_57

Vibration measurements were performed on the track (on the rail web, above 2 sleepers and at midspan between 2 sleepers; and at 10 consecutive sleeper positions), in the free field (9 locations on 3 parallel lines perpendicular to the track) and in the building (4 triaxial accelerometers on 4 floors), resulting in 84 measurement directions (Figs. 1b and 1c). Transfer functions were determined using impact hammer excitation at 17 sleeper positions equally spaced over a distance of 192 m [2]. The response of the track, free field and building was simultaneously measured during one week, for over 500 freight and passenger train passages.

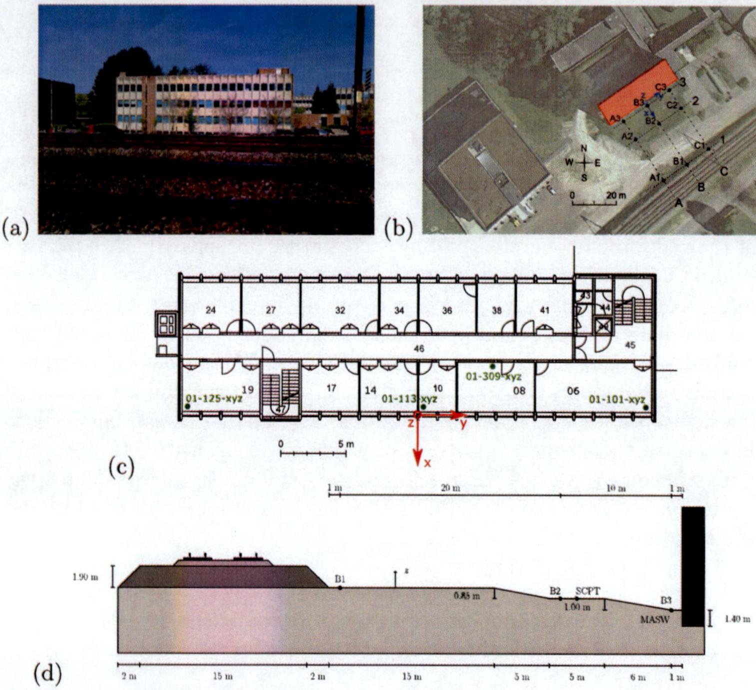

Fig. 1. (a) Blok D building, (b) measurement locations in the free field, (c) measurement locations on the first floor, and (d) cross section of the measurement site.

The measured transfer functions and response to passing trains are compared with numerical predictions and the influence of uncertain dynamic soil characteristics on prediction accuracy is discussed.

2 Dynamic Soil Characteristics

The Blok D building is located in the alluvial plain of the Dijle river consisting of an approximately 6 m thick quaterny layer of loose to dense sand (locally clayey) on top of a tertiary formation consisting of medium to dense sand with sand stone concretions. The ground water table is located at a depth of about 8 m. The dynamic soil characteristics were identified by means of Spectral Analysis of Surface Waves (SASW) tests and Seismic Cone Penetration Tests (SCPT). The data from these tests were combined in a probabilistic Bayesian inversion framework to identify a set of possible soil profiles [4]. Figure 2 shows the maximum a posteriori probability (MAP) soil profile along with 20 other possible realizations. The shear wave velocity of the shallow top layer and below a depth of 15 m, as well as the material damping ratio below 8 m, are highly uncertain.

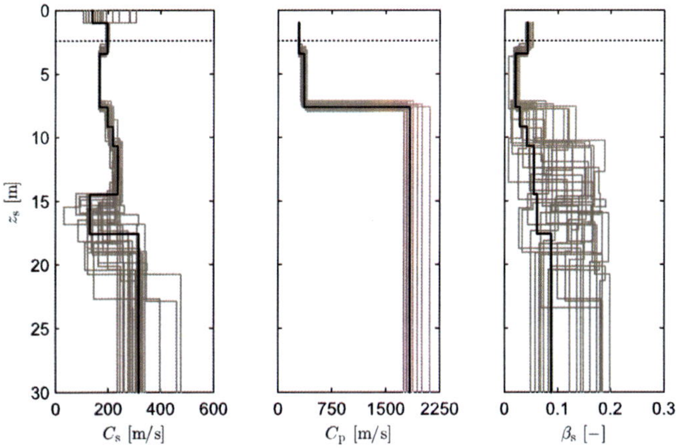

Fig. 2. Realizations of the shear wave velocity C_s, dilatational wave velocity C_p and shear material damping ratio β_s of the soil at the site of the Blok D building. The MAP soil profile is shown in black. The dotted line indicates the building foundation level.

3 Track and Embankment Characteristics

The ballasted tracks consist of UIC60 rails supported by resilient rubber rail pads and prestressed concrete monoblock sleepers. The rail pads have medium stiffness $k_{rp} = 150 \times 10^6$ N/m and damping $c_{rp} = 13.5 \times 10^3$ N/(m/s). The sleepers have a mass of 300 kg and spacing $L = 0.60$ m and are supported by a 0.40 m thick porphyry ballast layer, for which a density $\rho = 1800\,\text{kg/m}^3$, Poisson's ratio $\nu = 0.33$ and material damping ratio $\beta = 0.025$ are assumed. The track is located on top of a 1.90 m high embankment. The dynamic properties of the ballast layer, embankment and top soil layer are unknown.

The shear wave velocity C_s of the ballast and the width l_c of the sleeper-ballast contact area are tuned by comparing the measured sleeper mobility with the mobility computed with a periodic FE-BE model, incorporating the track, ballast, and a horizontally layered medium with the embankment and the soil

Table 1. Dynamic characteristics of the ballast, embankment and soil layer 1.

	h [m]	C_s [m/s]	C_p [m/s]	β_s [-]	β_p [-]	ρ [kg/m^3]
Ballast	0.40	120	240	0.025	0.025	1800
Embankment	1.90	130	260	0.044	0.044	1800
Soil layer 1	0.85	141	282	0.044	0.044	1800

layers [1]. A parametric study is performed and best agreement of the mobility is found for $C_s = 120\,\mathrm{m/s}$ and $l_c = 0.3\,\mathrm{m}$. The dynamic characteristics of the embankment and top soil layer are determined (Table 1) by comparing the measured transfer functions between the sleepers and the free field positions closest to the track (Fig. 1b) with the transfer functions computed with a 2.5D FE-BE model.

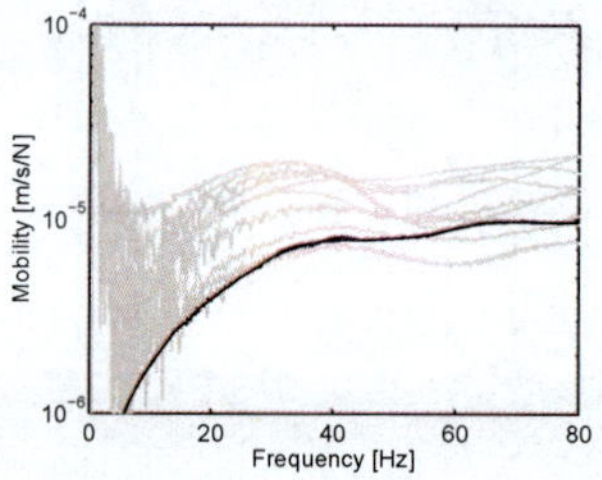

Fig. 3. Modulus of the average sleeper mobility (solid back line) and 90% confidence interval (grey shaded area) computed with the ensemble of 20 soil profiles. Comparison with the 10 measured sleeper mobilities (grey lines).

Figure 3 shows the vertical sleeper mobility measured on 10 consecutive sleepers, demonstrating large variability depending on contact conditions. The sleeper mobility is also computed with the identified characteristics of the ballast, embankment and top soil layer, and an ensemble of 20 soil profiles. A reasonably good correspondance with the experimental values is observed. The 90% confidence interval of the sleeper mobility is very small, revealing a negligibly small influence of the uncertain dynamic soil characteristics.

Figure 4 compares the measured averaged vertical transfer function between the sleeper and the free field at two distances from the track (Fig. 1d), with the transfer function computed with a 2.5D FE-BE model including a detailed topography of the embankment and the subsoil; computations are made for 20 identified soil profiles [1]. The measured transfer function lies almost entirely in the 90% confidence interval. The latter is wide, revealing a large influence of uncertain soil characteristics.

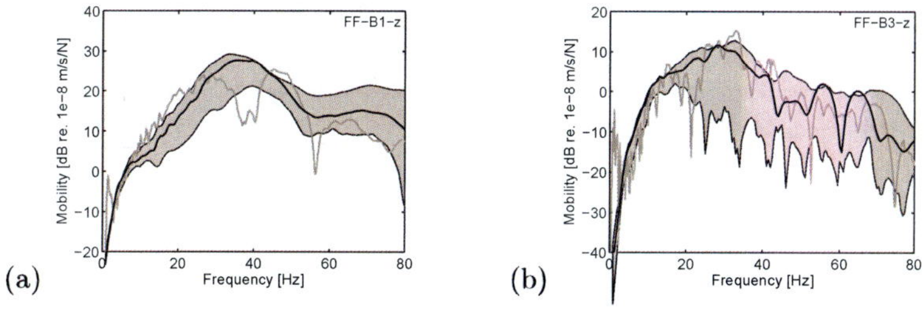

Fig. 4. Modulus of the predicted transfer function between a vertical force on the sleeper at $y = 0$ and the vertical velocity at (a) FF-B1-z and (b) FF-B3-z, using the MAP soil profile (solid black line). 90% confidence interval of the transfer functions based on 20 identified profiles (grey shaded area). Comparison with the measured (average) transfer function (grey line).

4 Dynamic System Identification of the Building

The primary structure of the building consists of concrete columns (every 1.5 m in the facades along the y-direction and every 3 m along the main corridor in the basement) and beams supporting the slabs (primary beams along the y-direction in the facades and main corridor; primary beams along the x-direction every 3 m; secondary beams along the x-direction every 1.5 m). This is complemented by concrete and masonry walls, elevator shafts and staircases. The basement and ground floor are connected to an adjacent building by means of walkways.

A FE model of the building is developed in SAP2000 using shell and beam elements [3]. Non-structural elements (floor coverings, masonry infill walls, plasterboard and plywood infill/partitioning walls) are incorporated as far as they contribute to the mass and/or stiffness. The dynamic stiffness of the footings and strip foundations, and the equivalent forces due to incident wave fields, are computed with the BE method. The soil properties under the building are modified to account for the increased stiffness under higher effective stress [3].

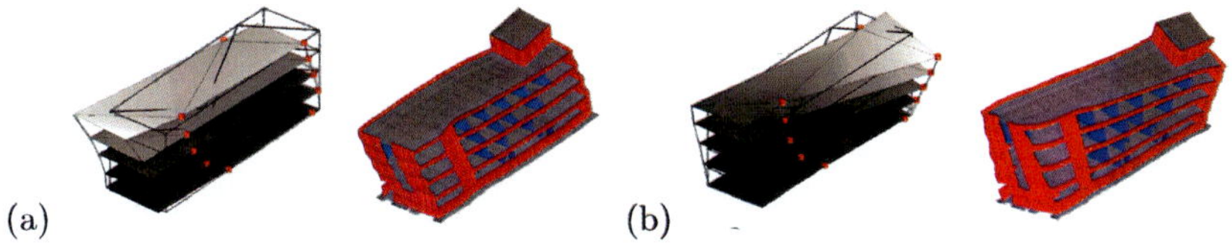

Fig. 5. Identified and computed global eigenmodes of the Blok D building: (a) first lateral mode with $f_1^{\text{exp}} = 3.37\,\text{Hz}$, $f_1^{\text{num}} = 3.38\,\text{Hz}$ and $\xi_1^{\text{exp}} = 0.032$, and (b) first torsional mode with $f_3^{\text{exp}} = 5.12\,\text{Hz}$, $f_3^{\text{num}} = 5.11\,\text{Hz}$ and $\xi_3^{\text{exp}} = 0.023$.

Ambient vibration measurements with 3 triaxial wireless accelerometers (Geo-SIG) on 5 floors (including basement and roof) were performed to identify the two lowest lateral modes and first torsional mode of the building [5].

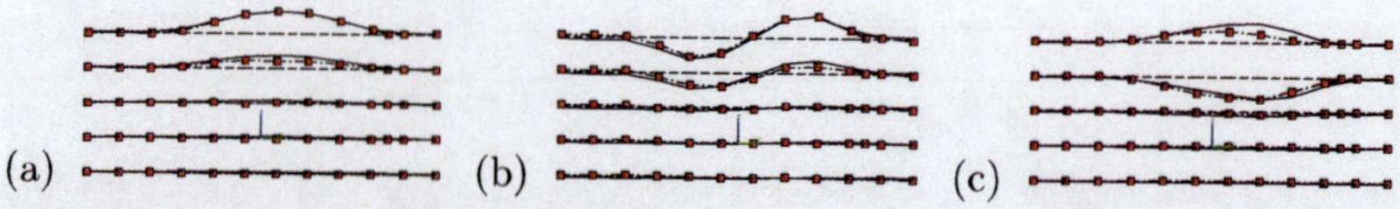

Fig. 6. Identified and computed vertical floor modes of the Blok D building: (a) first mode with $f_1^{\text{exp}} = 9.14\,\text{Hz}$, $f_1^{\text{num}} = 8.28\,\text{Hz}$ and $\xi_1^{\text{exp}} = 0.032$; (b) second mode with $f_2^{\text{exp}} = 10.70\,\text{Hz}$, $f_2^{\text{num}} = 10.22\,\text{Hz}$ and $\xi_2^{\text{exp}} = 0.013$; and (c) third mode with $f_3^{\text{exp}} = 11.18\,\text{Hz}$, $f_3^{\text{num}} = 11.02\,\text{Hz}$ and $\xi_3^{\text{exp}} = 0.027$.

Figure 5 shows the mode shape, eigenfrequency and modal damping ratio of the identified first lateral and first torsional mode, together with the mode shapes computed with the updated coupled FE-BE model. Additional ambient and forced vibration (using an impact hammer) tests with 15 GeoSIG units and 7 uniaxial accelerometers were performed to identify 8 vertical floor modes [5]. Figure 6 shows the mode shape (along the main corridor), eigenfrequency and modal damping ratio of the 3 lowest identified vertical floor modes, together with the mode shape computed with the updated FE-BE model. Vertical modes involve coupled motions of the floors. Model updating was applied to identify 15 uncertain building parameters (Young's moduli of structural materials; floor masses; springs representing connections with an adjacent building), in a multi-stage scheme, using vertical modes computed with a fixed base FE model as well as lateral and vertical modes computed with a coupled FE-BE model [3]. The modal characteristics of the building are significantly affected by dynamic soil-structure interaction.

Figure 7 shows the measured and predicted transfer functions between the force applied on a sleeper in front of the building and the displacements at points close to midspan of two floors in the building. The uncertainty on the measured transfer functions is relatively low. The predicted and measured FRFs have the same order of magnitude, but the accuracy of the predicted transfer functions decreases for increasing frequency. The realizations of the predicted transfer functions for different soil profiles suggest that a significant part of the

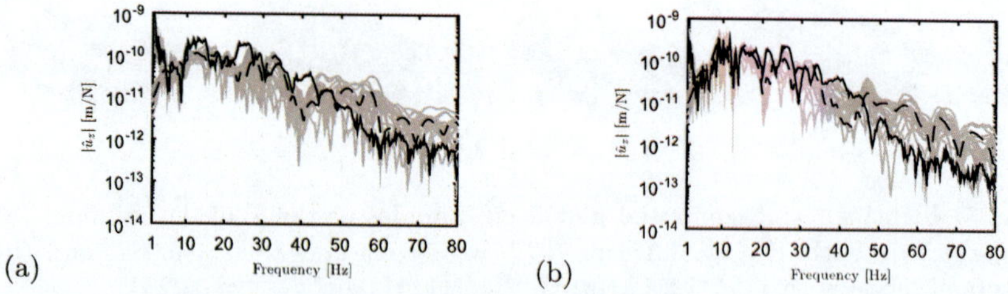

Fig. 7. Measured (solid line) and predicted (dashed line) vertical displacement at flexible points (a) V_1 (ground floor) and (b) V_3 (second floor) for vertical excitation on the sleeper. 90% confidence intervals of measured transfer functions and realizations of predictions for different soil profiles are shown in grey.

observed discrepancy can be explained by uncertain subsoil conditions; the latter affect both the dynamic stiffness of the coupled soil-structure system and the incident wave field that impinges on the building. Part of the uncertainty can also be attributed to modelling uncertainty.

5 Railway Induced Vibration

Most passenger trains are of the Desiro MS08 type and have 3 carriages. This train is modelled by means of 6 uncoupled bogie models, each consisting of 7 degrees of freedom: the vertical displacements of the car body, the bogie, two wheel sets and two wheel-rail contact points, and the pitch rotation of the bogie [1]. Rail unevenness was measured by Infrabel one week before the measurement campaign, revealing a track of moderate to poor quality [1]. Figure 8 shows the time history and frequency spectrum of the dynamic axle load at two bogies of a train running at 81 km/h, computed with a periodic track model coupled to the MAP soil profile [1]. The peak between 1 and 2 Hz corresponds to the resonance between the car body and the secondary suspensions, and the peak around 6 Hz due to the resonance of the bogie between the primary and secondary suspensions; the sleeper passing frequency is observed around 37.5 Hz.

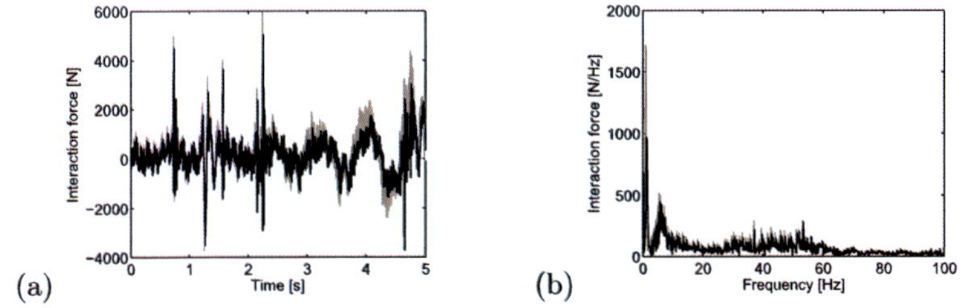

Fig. 8. (a) Time history and (b) frequency spectrum of the dynamic axle load of the first axle of the first (grey line) and third (black line) bogie of a Desiro MS08 train at 81 km/h.

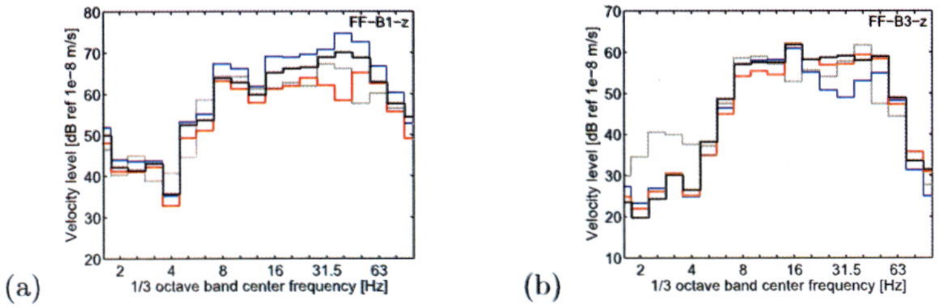

Fig. 9. One-third octave band spectrum of the vertical free field velocity at points (a) FF-B1-z and (b) FF-B3-z during the passage of a Desiro MS08 train at 81 km/h, predicted with the MAP soil profile (black), profile 1 (blue) and profile 6 (red), and compared with the measured response (grey).

Figure 9 compares the one-third octave band spectrum of the measured and predicted vertical velocity in the free field at two distances from the track during the passage of a Desiro MS08 train at 81 km/h; the measured vibration is very similar for different passages. Computations are made with a 2.5D coupled FE-BE model for the MAP soil profile, as well as for identified soil profiles 1 and 6, with a soft and stiff top layer, respectively [1]. The computed results differ by up to more than 10 dB for the different soil profiles, while the measured response lies between these results at most frequencies. Soil profile 1 generally results in higher free field vibration than soil profile 6. At low frequencies, the results computed with the three soil profiles underestimate the measured vibration. At the point B3, a larger discrepancy is found which is probably due to the presence of the building, which is not incorporated in the model. Figure 10 compares the one-third octave band spectra of the measured and predicted vertical velocity at two flexible locations in the building during the same passage. The uncertain soil properties result in a variation of the predicted response of up to 10 dB, particularly at higher frequencies.

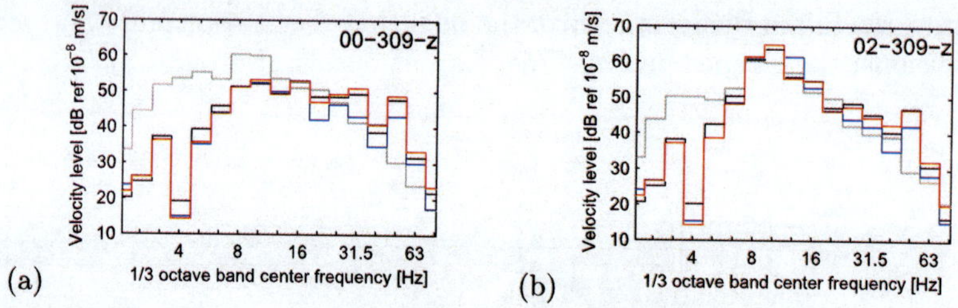

Fig. 10. One-third octave band spectrum of the vertical velocity at the receivers (a) V_1 (ground floor) and (b) V_3 (second floor) during the passage of a Desiro MS08 train at 81 km/h, predicted with the MAP soil profile (black), profile 1 (blue) and profile 6 (red), and compared with the measured response (grey).

6 Conclusion

Railway induced vibration was measured on the track, in the free field and in a building. These data were used to calibrate and validate numerical models for the railway induced vibration. The measured and predicted response show significant difference, which can partly be explained by uncertain soil properties.

Acknowledgements. This research was performed within the frame of the project OT/13/59 "Quantifying and reducing uncertainty in structural dynamics" funded by the Research Council of KU Leuven. KK was and KM is a postdoctoral researcher, while MG and RV were PhD students, all funded by FWO Flanders.

References

1. Germonpré, M.: The effect of parametric excitation on the prediction of railway induced vibration in the built environment. Ph.D thesis, Department of Civil Engineering, KU Leuven (2018)
2. Kuo, K., Germonpré, M., Maes, K., Degrande, G., Lombaert, G.: Processing of vibration measurements at the Blok D building of the administrative complex of KU Leuven. Report BWM-2017-20, Department of Civil Engineering, KU Leuven, Project OT/13/59 (2017)
3. Papadopoulos, M.: Influence of dynamic SSI on the building response to ground vibration. Ph.D thesis, Department of Civil Engineering, KU Leuven (2018)
4. Verachtert, R.: Deterministic and probabilistic determination of dynamic soil characteristcs. Ph.D thesis, Department of Civil Engineering, KU Leuven (2018)
5. Zhang, J.: Numerical analysis and experimental assessment of the dynamic characteristics of periodic structures. Ph.D thesis, Department of Civil Engineering, KU Leuven (2019)

Rail Roughness Evolution on a Curved Track and Its Impact on Induced Structure Borne Vibration

Vincent Jurdic[1(✉)], Rob Lever[2], Adrian Passmore[1], and Mark Scotter[3]

[1] Arup, 8 St. Thomas Street, Winchester SO23 9HE, UK
vincent.jurdic@arup.com

[2] Technical Services, Rail Vehicles Transport for London, Hearne House, 3 Museum Way, Acton W3 9BQ, UK

[3] Earls Court Partnership Limited, 15 Grosvenor Street, London W1K 4QZ, UK

Abstract. To assess the impact of rail grinding on structure borne vibration a study was undertaken on the track between the West Brompton (WB) and Earls Court (EC) stations. Before and after rail-grinding took place, rail roughness was measured at three different occasions and, in parallel to that, vibration due to train movements was measured at three locations inside the former Earls Court 1 exhibition centre (EC1) basement. This measurement process was undertaken on four separate occasions to assess the evolution of rail roughness/train vibration. Trains in the WB tunnel have generally been found to travel at lower speed than the track design speed, which suggests a higher wear rate of the outer rail (OR) when compared to the inside rail (IR). Contrary to the common approach of averaging the roughness of both rails, a better correlation between vibration levels and rail roughness for this curved track was found when considering the maximum roughness levels of both rails.

Keywords: Rail roughness · Structure borne vibration · Curved alignment

1 Introduction

The Earls Court redevelopment area is one of the highest profile mixed-use development sites in London. The site includes the area that was previously occupied by Earls Court 1 and Earls Court 2 Exhibition Centres (EC1 and EC2) and is traversed by London Underground Limited (LUL) tracks which are part of the District Line (DL) and Piccadilly Line systems and the Transport for London Overground Network between West Brompton and Olympia.

During the early design stages, preliminary structure borne noise predictions indicated a potential need for mitigation measures to control vibration from the underground train lines in order to achieve the requirements mandated by the local planning authority and the aspirations of the developer (*Earls Court Partnership Limited (ECPL)*). As part of the possible mitigation strategies, the potential benefit of controlling rail roughness

G. Degrande et al. (Eds.): *Noise and Vibration Mitigation for Rail Transportation Systems*, NNFM 150, pp. 538–545, 2021.
https://doi.org/10.1007/978-3-030-70289-2_58

on the DL tracks was considered. To evaluate this, rail grinding was carried-out on the DL track in the WB-EC tunnel. A series of rail roughness and vibration surveys were undertaken.

This paper presents the results of these surveys, and through the analysis considers whether the common approach of using the average roughness of both rails to predict structure borne vibration is appropriate where curved track is installed.

2 Surveys Description

2.1 Site Description

EC1 was built in the 1930s above and around a major rail network serving South West London, with the building foundation structure juxtaposed with two Piccadilly tunnels and four District line tunnels, see Fig. 1. During the time of this assessment, there have been three types of rolling stock operating on the DL: C stock, which operated until June 2014, D stock[1], and S7 stock which have operated since February 2014.

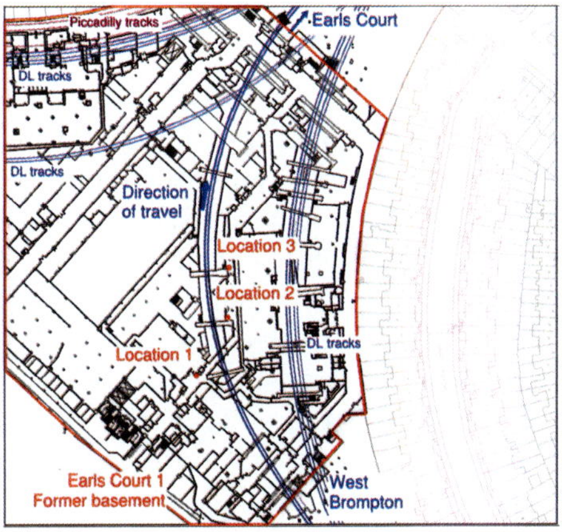

Fig. 1. Former EC1 basement and vibration measurement locations

This study was based on a circa 400 m long section of ballasted track, with a tight curve of 160 m radius on the WB-EC tunnel. When trains travel on such a curve they are subject to a centrifugal force which can induce an uneven distribution of the wheel loads on the rails. To counteract this effect, 'cant' (ie height difference between both rails) is commonly introduced in the rail track design, which also facilitates train steer and helps to reduce friction and wear. To achieve equal distribution of the wheel loads, the equilibrium cant value is defined for the radius of curvature at a specific train speed. For the WB-EC alignment, archive design information shows that a cant of 85 mm was designed to provide equilibrium for 40 km/h train speed.

[1] London Underground D78 stock, commonly referred to as D stock.

2.2 Methodology

To assess the impact of rail roughness, LUL undertook rail grinding on both rails in the WB-EC tunnel at the end of August 2013. Vibration and roughness measurements were undertaken by Arup (Acoustic Consultants on the ECPL development, advising specifically on structure borne noise and vibration) and LUL, respectively, prior to the grinding work and on three separate future occasions, see Fig. 2 for timescales.

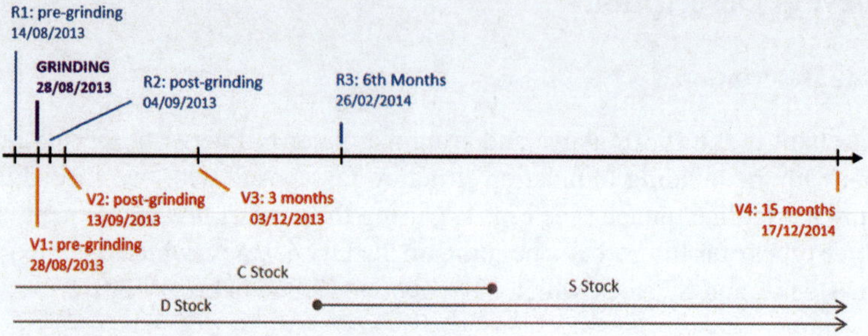

Fig. 2. Timeline describing the roughness (R) and vibration (V) measurements in relation to the grinding work and types of rolling stock operating.

Vibration measurements were carried out inside the former EC1 basement at three locations as shown in Fig. 1. Sacrificial studs were glued to the concrete floor using rapid setting epoxy resin, to which the accelerometers were attached. These studs were left in place from the first survey (i.e. V1) to ensure the exact same locations were used again during the following surveys (i.e. V2, V3 and V4). During each vibration survey, the vertical acceleration time history was continuously recorded. After each survey, train events were identified and extracted from the data record based on train schedule records provided by LUL. These train events were then post-processed to obtain the vibration velocity third octave band spectra $Lv_{max,s}$ and to estimate the train speeds, using Arup's estimation tool [1].

Rail roughness was measured on three separate occasions: before grinding and then one week and six months after grinding. The vertical rail profile was measured with a corrugation analysis trolley (CAT) for approximately 230 m of track and in accordance with the methodology specified by EN15610:2009 [2].

3 Measurement Results

3.1 Train Speed

Using Arup's train speed estimation tool, the speed of the trains travelling in the WB-EC tunnel was accurately derived for each vibration event. As an example, Fig. 3 presents the speed distribution of the trains (C and D stock) recorded during the pre-grinding survey.

Most of the trains travelled at a speed below 40 km/h[2]. The most common speed of the train was around 35 km/h and this speed will be used in Sect. 3 for the vibration predictions.

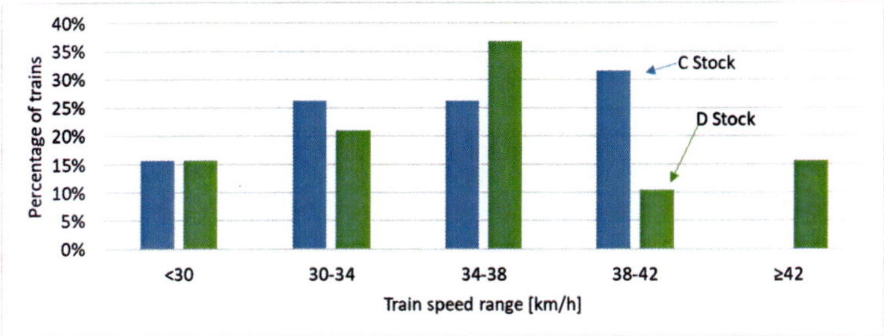

Fig. 3. Speeds of the train in the WB-EC tunnel during the pre-grinding vibration survey V1

3.2 Vibration

Typical vibration spectra were derived for each survey and each location by averaging the measured vibration levels for a number of train passes. These spectra are presented in Fig. 4, 5 and 6. Although for each location and survey the vibration velocity spectra $Lv_{max,s}$ are different, the trains traveling at around the 35 km/h typical speed generated consistent spectra.

Location 1

Just after the grinding work, the vibration levels were reduced by approximately 5 dB at frequency bands below 40 Hz and by 10 dB at higher frequencies. Three months after the grinding work, the vibration levels at 40–63 Hz were similar to pre-grinding work, whereas the vibration levels for the other frequencies were similar to the post-grinding levels. Fifteen months after the grinding work, the vibration levels below 50 Hz were similar to those that occurred post-grinding: the levels at 40–50 Hz were lower than

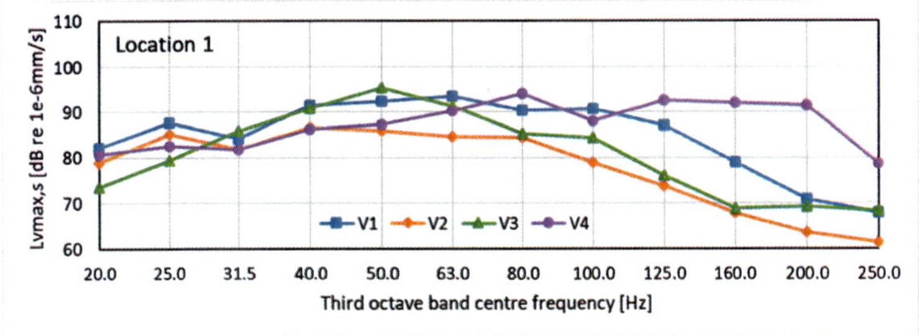

Fig. 4. Vibration spectra measured at location 1 during the four surveys: V1 - pre-grinding, V2 - post grinding, V3 – 3 months after grinding and V4 – 15 months after grinding

[2] 40km/h is the equilibrium cant design speed in the WB-EC tunnel.

those measured during survey V3. The vibration levels above 50 Hz were similar to those before the grinding work. At high frequency the levels remained relatively high. Note that location 1 is located close to an isolated rail joint, which could explain those high levels.

Location 2

Just after the grinding work, the vibration levels at frequencies above 40 Hz were reduced by approximately 10 dB with less change below 40 Hz. The vibration spectral shape remained relatively unchanged during the post grinding surveys. An increase of 3 dB was observed however at 63 Hz and above during surveys V3 and V4. Fifteen months after the grinding work, a peak was observed at 200 Hz, reaching the pre-grinding levels.

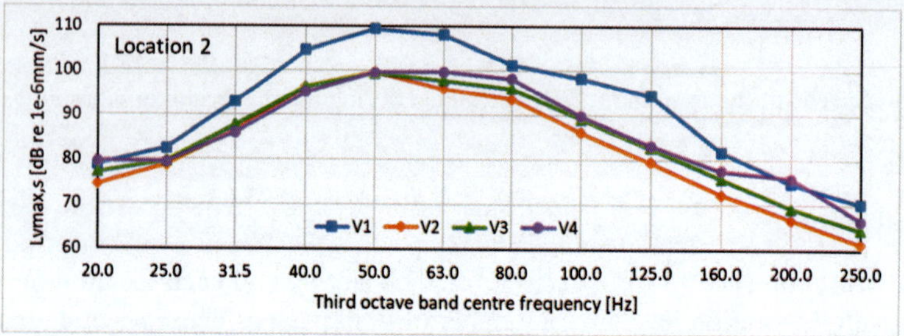

Fig. 5. Vibration spectra measured at location 2 during the four surveys: V1 - pre-grinding, V2 - post grinding, V3 – 3 months after grinding and V4 – 15 months after grinding

Location 3

Just after the grinding work, the vibration levels at frequency bands >25 Hz were reduced by approximately 10 dB. Three months after the grinding work, the vibration levels above 50 Hz returned to the pre-grinding levels. At 125 Hz and 160 Hz, the levels were also 7 dB higher than those before the grinding work. Twelve months later, levels reduced back to those mesaured after the grinding work, with a peak at 50 Hz, which was 6 dB above the post-grinding levels.

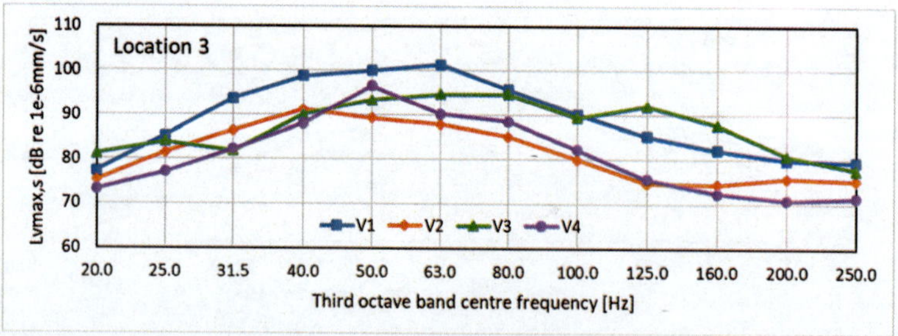

Fig. 6. Vibration spectra measured at location 3 during the four surveys: V1 - pre-grinding, V2 - post grinding, V3 – 3 months after grinding and V4 – 15 months after grinding

3.3 Roughness

The measured vertical rail profiles were processed in accordance to EN15610 standard to derive the outer rail (OR) and inner rail (IR) roughness spectra, as shown in Fig. 7.

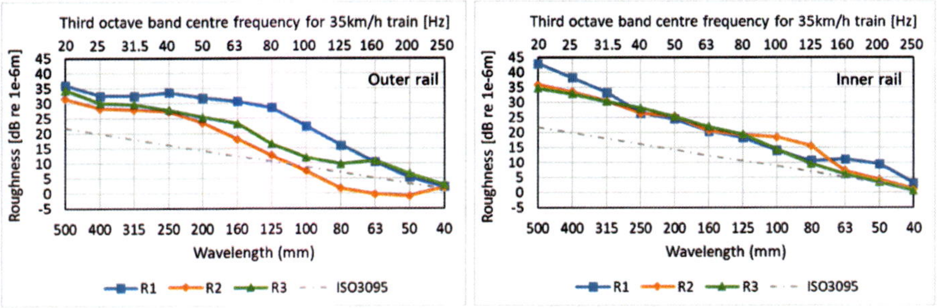

Fig. 7. Rail roughness spectra of the outer rail (left) and inner rail (right)

Prior to the rail grinding, the OR roughness levels were up to 9 dB above the IR for the vibration driving frequency range (40–80 Hz). The rail grinding work reduced significantly the OR roughness levels (up to 16 dB at 125 mm), although it only affected the IR levels at wavelengths smaller than 80 mm and larger than 315 mm. Six months after the grinding work, the OR levels increased for the wavelengths less than 200 mm but remained below the pre-grinding levels. The pre-grinding levels were achieved for wavelengths less than 63 mm. The IR roughness levels remained relatively consistent during the three periods, with the lower levels observed 6 months after the grinding work. The OR appears therefore to be more affected by the train traveling in the WB-EC tunnel (more rapid roughness development and highest reduction due to rail grinding). Figure 3 shows that the majority of the trains travel below the equilibrium cant speed (40 km/h). This is most likely due to the rails being loaded to different levels with the highest load probably occurring on the OR.

4 Limitations

Between the V1 and V4 surveys, rail grinding took place in August 2013. No other major modifications occurred to the track system. However, different rolling stock was operated on the District line during the surveys. For example, C and D stock trains were operated during the V1, V2 and V3 surveys and D and S stock trains were operated during the V4 survey. Analysis of the vibration spectra from each train type indicates no significant difference between the different trains.

5 Predictions

The mechanisms generating structure borne vibration from trains are well understood and documented in a series of international standards [3]. Using this method vibration predictions are obtained from solution of a single rail with the modelled roughness, being commonly the average levels of the OR and IR [4, 5]. This approach is generally accepted because both the OR and IR roughness spectra are usually comparable in straight track. The previous sections suggest that on the WB-EC curved track, the train wheels were not equally loading both rails.

Prior to the rail grinding, the Left-Hand-Rail (LHR) roughness levels were the highest and were approximately 5 dB higher than the two-rail averaged levels. Two weeks after the grinding works, the Right-Hand Rail (RHR) roughness levels were now the highest, especially for wavelength smaller than 160 mm. For these wavelengths, the maximum roughness was up to 8 dB higher than the two-rail averaged levels. Six months after the grinding, both LHR and RHR roughness levels were similar, and therefore there was no significant difference between the maximum and averaged roughness levels.

For the estimated typical train speed (ie 35 km/h), two transfer functions TF_{max} and TF_{avg} can then be derived by considering the maximum or averaged roughness levels:

$$TF_{max}(f) = Lvmax, s(f) - max(R_{RHR}(\lambda, 35\text{km/h}), R_{LHR}(\lambda, 35\text{km/h})) \quad (1)$$

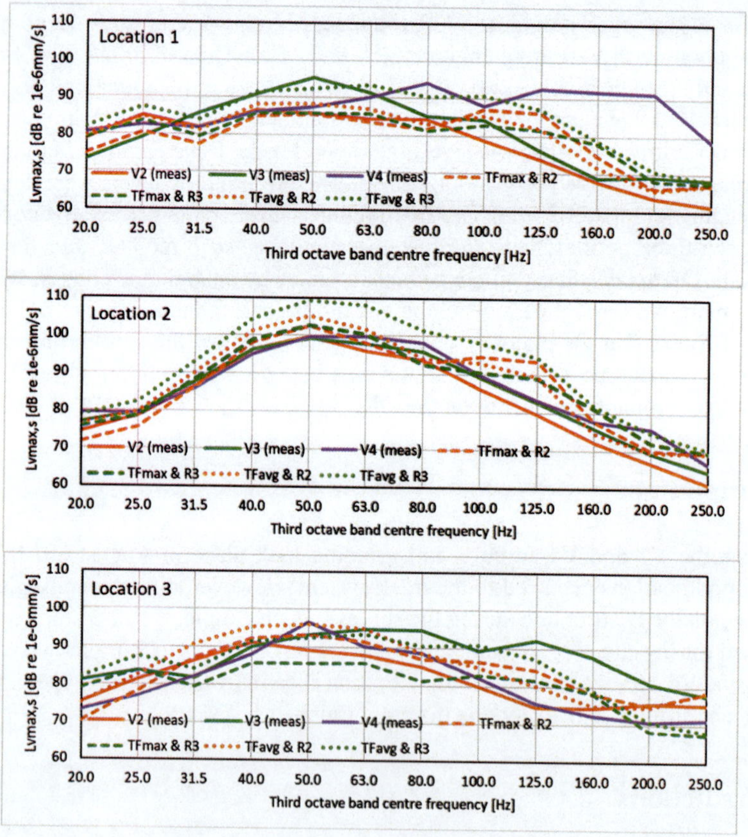

Fig. 8. Predicted (--) and measured (-) vibration spectra for the V2 (-), V3 (-) and V4(-) surveys

$$TF_{avg}(f) = Lvmax, s(f) - average(R_{RHR}(\lambda, 35\text{km/h}), R_{LHR}(\lambda, 35\text{km/h})) \quad (2)$$

Because the OR and IR roughness levels are the highest for the pre-grinding measurement R1, the transfer functions were derived with the corresponding average and maximum roughness spectra.

These transfer functions are then combined with the two post grinding roughness levels (ie R2 and R3) to predict the post grinding vibration levels. These predicted vibration spectra are compared to the spectra measured during the three post-grinding surveys, as shown in Fig. 8.

With the two-rail averaged roughness, the predicted vibration spectra are 6 to 10 dB higher than those measured at locations 2 and 3. The difference between measurement and prediction is smaller, however, when the maximum roughness level is used instead. At location 1, the averaged roughness levels provide better agreement than the maximum roughness spectra, however given that Location 1 is located close to an isolated rail joint, it is likely that is affecting the outcomes given that the isolated rail joint is not included in the predictions.

6 Conclusion

Vibration and rail roughness measurements were undertaken before and after rail grinding took place on the WB-EC track. The following conclusions are drawn:

- In the WB EC tunnel, the majority of the trains are travelling at lower speeds than the design speed for equilibrium cant (ie 40 km/h).
- Because trains are traveling at lower speed (generally around 35 km/h), the wheel load is not equally distributed and the OR wears more rapidly than the IR.
- In prediction models, the common approach is to average the roughness levels of both rails. However for the WB-EC tight curved track, this approach is inconsistent with the measured results.
- Better agreement in the prediction were obtained by using the maximum roughness levels measured for each rail before and shortly after rail grinding, recognizing that roughness levels between the OR and IR could be significantly different.

References

1. Jurdic V., Bewes, O., Burgemeister, K., Thompson, D. J., Train speed estimation from ground vibration measurements using a simple rail deflection model mask. In: Notes on Numerical Fluid Mechanics and Multidisciplinary Design, vol. 139, pp. 181–192. Springer (2018)
2. EN 15610:2009 Railway applications – Noise emission – Rail roughness measurement related to rolling noise generation (2009)
3. EN ISO 3095:2013 Railway applications – Acoustics –Measurement of noise emitted by railbound vehicle (2013)
4. Thompson D.J.: Railway noise and vibration, chap. 13, 1st edn. In: Mechanisms, Modelling and Means of Control. Elsevier (2009)
5. Jurdic, V., Bewes, O., Greer, R., Marshall, T.: Developing prediction model for groundborne noise and vibration from high speed trains running at speed in excess of 300 km/h. In: 21st International Congress on Sound and Vibration on Proceedings, Beijing (2014)

Vibration Emission of Underground Rail Systems: Experimental Assessment, Extrapolation and Parametric Study

Eric Augis(✉), Pierre Ropars, and Alice Ly

SYSTRA, 120 rue Massena, 69006 Lyon, France
eaugis@systra.com

Abstract. Representative vibration emission spectra are key to reliably predict ground borne noise and vibration levels inside dwellings due to the operation of a new metro line. This article first describes an experimental method to reliably assess a 'baseline' vibration excitation spectrum representing the force density level applying on the tunnel invert during the passage of one existing underground metro vehicle. Then, simple rules and modelling techniques are proposed to convert the 'baseline' force density spectrum to other situations (different vehicle, track, rail unevenness, vehicle speed, tunnel or soil parameters). A hybrid model coupling an analytical train-track interaction model to a numerical tunnel-soil interaction model is carried out and validated from comparison to experimental results. The hybrid model is then used to perform a parametric study aiming at analyzing the influence of the tunnel geometry and soil properties on the force density level and on the tunnel response.

Keywords: Vibration emission · Metro · Force density level · Tunnel

1 Introduction

Predicting ground borne noise and vibration levels inside dwellings arising from the operation of a new underground rail system requires the development and calibration of a model able to simulate (a) the vibration emission, (b) vibration transfer mobilities from tunnel to the ground surface near the dwelling, (c) vibration transfer functions from the soil to the floor and resulting re-radiated noise. Representative vibration emission data is not easy to get at the design stage of a project for a new rail line. Excitation data consisting of vibration response levels measured on the tunnel structure (or track slab) for a comparable rail system is highly dependent on site parameters (rail irregularities, track properties, tunnel vibration response, soil properties, etc.) and cannot be reliably considered for other contexts without being corrected. The main objective of this article is to propose 'baseline' force density levels representative of a metro system operating in a common tunnel, and a reliable method and simple tools for extrapolating new sets of excitation data for other vehicle, track, unevenness, speed, tunnel or soil conditions.

G. Degrande et al. (Eds.): *Noise and Vibration Mitigation for Rail Transportation Systems*, NNFM 150, pp. 546–554, 2021.
https://doi.org/10.1007/978-3-030-70289-2_59

2 Experimental Characterization of Force Density Levels

2.1 Method

The methodology for experimentally characterizing vibration emission spectra induced by the operation of an existing underground metro line, uses the same principle as that described in the 'FTA guidance manual' [2]. It is based on the assumption that the vibration response of the tunnel structure during the passage of a metro vehicle is the result of a line of uncorrelated forces applying on the invert during the same time period. Assuming that forces are continuously distributed below the train, it is convenient to characterize a force density (i.e. force per unit train length in $Nm^{-1/2}$). It is also assumed that the force density is somewhat invariant to the invert mobility.

The force density level (FDL) applying on the tunnel invert can be determined from site measurements of (a) vibration velocity levels on the tunnel structure at location i ($L_{Veq,Tp}$ (i)) during train pass-by events, and (b) the line source transfer mobility (LSTM) at location (i) with a force on the invert, using the following equation:

$$\mathrm{FDL} = L_{Veq,Tp}(i) - \mathrm{LSTM}\ (i) \quad (1)$$

2.2 Application to a Commuter Rail Line in Tunnel

The method is applied to an existing commuter rail line in an ovoid single-track tunnel with vehicles operated at 60 km/h. This case, located in central Paris, was previously presented in [1]. The all-concrete tunnel presents a thick concrete invert (see T_a in Tables 2 and 3). The track consists of UIC60 rails fastened on booted concrete sleepers lying on a concrete slab. The track component properties are in Table 1. The train is composed of two units of five cars (3 trailers + 2 powered cars). Each unit has a total length of about 110 m. The unsprung mass is 1.8t/axle (for trailer bogies). The car mass in nominal load conditions is around 65 t (trailer).

First vibration velocity levels ($L_{Veq,Tp}$) measured on the tunnel wall and the invert are averaged over a total of 15 train passages at 60 km/h (see [1]). Then, the LSTM of the tunnel structure at the same locations are calculated in two ways: (a) by integrating point-to-point transfer mobilities (PPTM) measured on site, (b) by simulating the tunnel response using a 2,5D FE-BE numerical model handled with the TRAFFIC software developed by KU Leuven [3]. A view of the FE model of the tunnel (labelled T_a) can be seen in Table 3. The model is calibrated by adjusting the soil parameters. Once calibrated, the model is used to calculate its response to unit forces, PPTM and LSTM. A comparison of measured and calculated PPTM on tunnel invert are given in Fig. 1 for three soils (whose parameters are in Table 4). Soil #1 provides the best agreement between calculation and measurements. Note that a numerical model presents the advantage to provide reliable results at low frequencies (below 20 Hz) where measured PPTM are commonly tainte with errors due to a signal-to-noise ratio that is too low. Figure 2 compares the PPTM and the LSTM for two receptors on the wall and the invert, the force being applied on the invert. The FDL calculated using Eq. (1) is plotted in Fig. 3. It is the proposed 'baseline' FDL which corresponds to a slab track with booted sleepers and a vehicle speed of 60 km/h and whose rail uneveness spectrum (measured on the test site) is shown in Fig. 4 (grey curves). The maximal force amplitude is around 45–46 Hz. Rail irregularities are measured on the test site using the CAT trolley system [1, 6]. Levels are comparable to those of ISO3095 rail roughness limit curve, for wavelength from 10 to 315 mm, meaning that the rail is in a 'good state'. A generic rail uneveness spectrum (gauge) is proposed as well as a generic combined wheel-rail uneveness spectrum which assumes wheels in a 'fairly common wear state'.

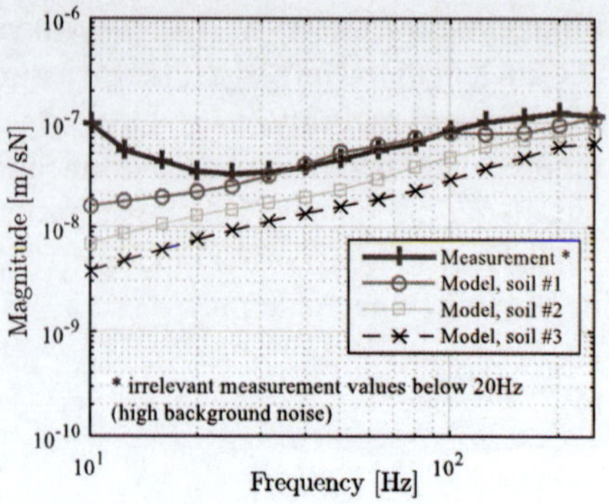

Fig. 1. Comparison of measured and calculated PPTM on tunnel invert

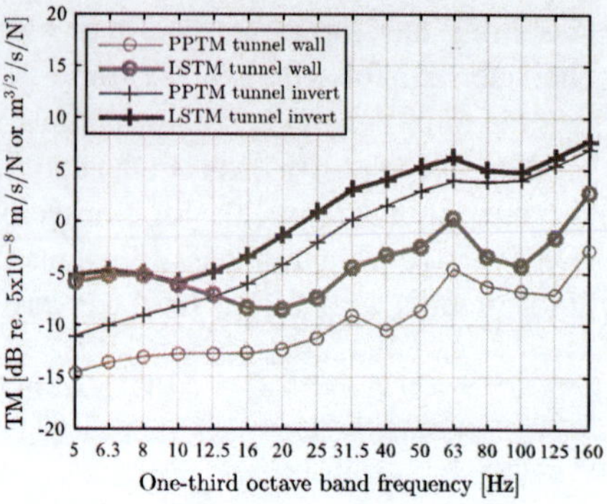

Fig. 2. Comparison of calculated PPTM and LSTM on tunnel invert and wall

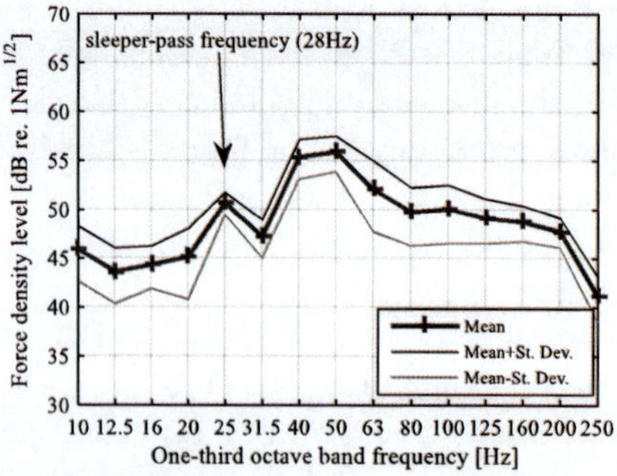

Fig. 3. 'Baseline' FDL at 60 km/h for a booted-sleeper slab track

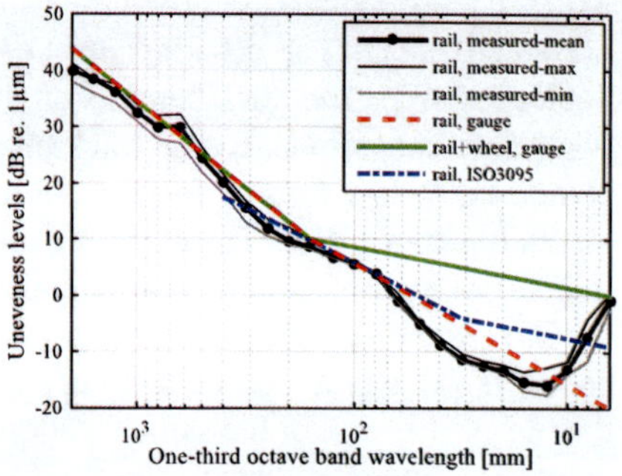

Fig. 4. Rail and combined uneveness levels (site measurement and gauge)

3 Predicting Excitation Spectra for Other Situations

3.1 Method and Simple Rules

The principle of determination of FDL for any new configuration is based on the calculation of a correction function to be applied to the 'baseline' FDL, representing the effect of change of one or several parameters. The effect of change of vehicle speed and/or rail uneveness conditions can easily be determined using simple laws/formula (see below). However, more comprehensive modelling techniques are necessary to assess the effect of change of track, vehicle, tunnel and/or soil conditions. A hybrid model is elaborated and validated in that aim (see description hereafter).

Dynamic forces on a tunnel invert due to the movement of vehicles in tunnel are estimated using S-RIV (Railway Induced Vibration model developed by Systra), which is an analytical model simulating the track-vehicle dynamic interactions due to rail-wheel combined uneveness (see Fig. 5). The track model is based on RODEL developed in the TWINS software [4, 5, 7]: the rail is an infinite Timoshenko beam lying on continuous spring-mass-spring layers respectively representing the railpad, the sleeper, and the under-sleeper pad. The tunnel invert is modelled by its receptance (Y_i) which is calculated with a tunnel-soil interaction model elaborated with TRAFFIC. Y_i is the invert receptance for one half-sleeper area. The train is modelled using a set of springs and rigid masses representing the carbody, the bogie, and the wheelset ('unsprung' mass), and connecting suspensions. The source term is the uneveness spectrum at desired train speed. The output is the force F(x) applying to the slab at any longitudinal position x. The FDL is then obtained by integrating F(x) over a certain distance travelled by the train and then by reducing this force level to a 1m train length.

Effect of Train Speed. Force density levels at a speed V_2 can be derived from those determined at the reference speed (V_{ref} = 60km/h) as follows: (a) convert 1/3-octave rail uneveness spectra from wavelength domain to frequency domain at speeds V_{ref} and V_2, using the fundamental formula $f = V / \lambda$ (f: frequency, V: train speed, λ: wavelength of irregularities); (b) calculate the difference of rail uneveness levels at each 1/3-octave frequency band, due to the change of vehicle speed ($\Delta L_R = L_{R,V2}y$- $L_{R,Vref}$); (c) apply ΔL_R on the FDL at 60 km/h. Using generic curves of rail or combined wheel-rail uneveness instead of the measured rail uneveness spectrum (see Fig. 4) to assess the effect of change of vehicle speed may be found more reliable.

Predicting FDL for Other Uneveness Conditions. As the rail uneveness levels (LR) were measured on the test site, it is easy to derive force density spectra for other rail uneveness spectra (e.g. typical of corrugated rails). Note that the new rail uneveness spectrum must be converted in the frequency domain, at the desired speed.

3.2 Application: Assessment of the Effect of Track and Vehicle Change

The hybrid model used for assessing the effect of change of parameters, is first validated as follows. The track model in S-RIV is calibrated to replicate the rail receptance measured in unloaded track condition. Then the dynamic stiffness of under-sleeper pads is increased so that the 'unsprung mass' frequency (maximal amplitude frequency) appears near 45–46 Hz as observed on measured pass-by vibration velocity spectra and force density spectra (see Fig. 2). The receptance of the tunnel invert Y_i (for one half-sleeper) calculated with TRAFFIC is input in the S-RIV model. The track and vehicle parameters considered in the model are presented in Table 1 (see column 'Booted-sleeper track').

The FDL calculated with the model by considering the measured rail uneveness spectrum is compared to the 'base' FDL determined from measurements in Fig. 6. The agreement is found acceptable.

Table 1. Track and rolling stock main parameters

Component	Property	Booted-sleeper track	Standard slab track
Rail UIC60	Mass per unit length [kg/m]	60	60
	Vertical flexural stiffness EI_V [Nm^2]	6.4 E6	6.4 E6
	Shear coefficient κ [-]	0.4	0.4
	Poisson's ratio ν [-]	0.3	0.3
Railpad	Fastener spacing [m]	0.60	0.60
	Vertical dynamic stiffness K_p [kN/mm]	550	200
	Loss factor η_V [-]	0.25	0.25
Bi-block sleeper	Mass [kg] – half sleeper	100	–
Under-sleeper pad	Vertical dynamic stiffness K_{usp} [kN/mm]	50	–
	Loss factor η_V [-]	0.35	–
Rolling stock	Bogie unsprung mass [kg]	1800	1500

The effect of change of track from booted-sleeper track ('baseline' case) to a standard slab track (SST) with stiff rail pads is calculated with the hybrid model, in terms of track insertion loss (i.e. attenuation level of one track with respect to a SST). The relevant parameters for the SST and the new vehicle are listed in Table 1.

The calculated track insertion loss is shown in Fig. 7 (bold: unchanged bogie unsprung mass, dashed: unsprung mass changed from 1800 kg to 1500 kg per axle). The new FDL for the SST can be seen in Fig. 8.

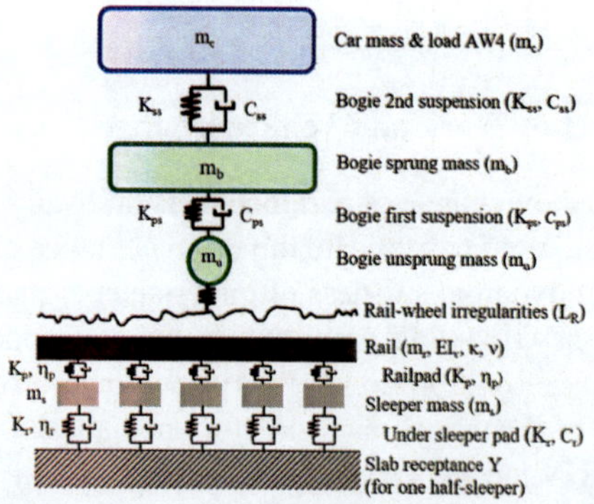

Fig. 5. Model for railway track-vehicle vibration response (S-RIV)

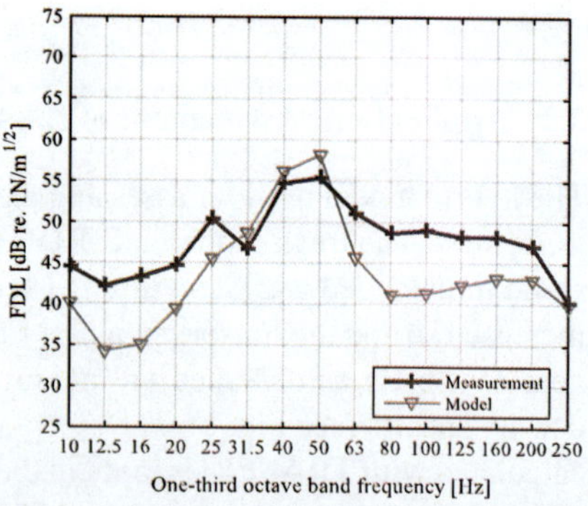

Fig. 6. Comparison of calculated FDL and 'baseline' FDL (test track, 60 km/h)

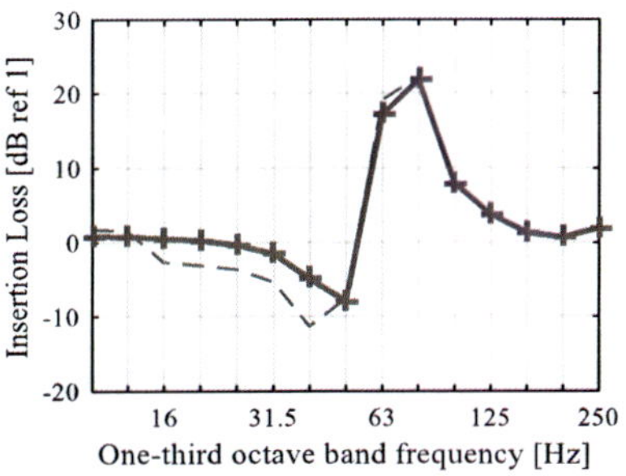

Fig. 7. Track insertion loss calculated for the booted-sleeper track.

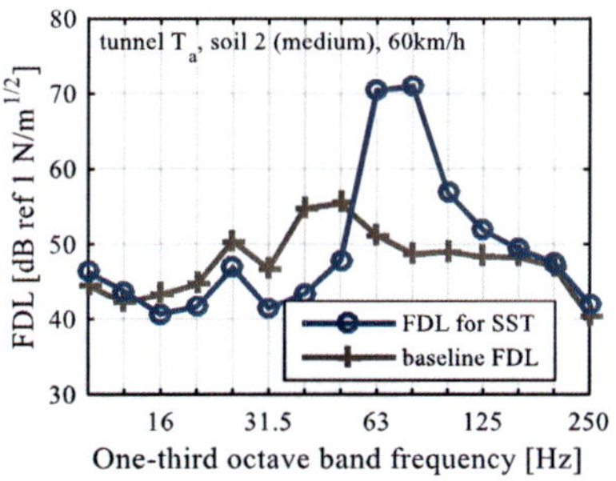

Fig. 8. Comparison of the FDL calculated for the SST and the baseline FDL.

3.3 Effect of Change of Tunnel Features and Soil Properties

The combined effect of changes of the tunnel geometry and soil properties on the FDL is investigated. Six tunnels (labelled T_a to T_f) with various shapes, dimensions and thicknesses are studied (see Tables 2 and 3). All tunnels are made of the same concrete material whose properties are: E = 25 000 MPa, $\nu = 0.3$, $\rho = 2500$ kg/m^3.

Table 2. Main design features of the various tunnel types

Tunnel type		Ta	Tb	T_c	Td	Te	Tf
Interior diameter or width [m]		6.2	11.7	9.3	8.6	8.65	8.5
Wall thickness [m]		0.4	0.6	0.3	0.4	1.5	1
Invert thickness [m]	max	0.65	0.85	0.6	1.75	2	2.25
	min	0.15	0.15	0.15	0.5	1.5	1

The effect of tunnel design features is studied for three soil types, assuming that the tunnel is entirely in one soil layer. Soil parameters are typical for a soft soil (#1 e.g. Plastic clay), a medium soil (#2 e.g. Marl) and a stiff soil (#3 e.g. Beauchamp sand). Soil parameters are in Table 4. The FDL for tunnel T_a and soil #1 is the 'baseline' FDL.

Results, illustrated in Fig. 9 and Fig. 10, show that the FDL calculated is only slightly affected by the tunnel geometry when it is coupled to a medium or stiff soil (#2 and #3). The effect of tunnel geometry is higher when it is coupled to a soft soil (#1). The effects are a shift of the unsprung mass frequency and a variation of the maximal force level of 0.5–3 dB for soils #2 and #3, or 0.5–5 dB for soil #1. The effect of soil on FDL is low for tunnel T_d w/ a thick invert, but somewhat high for tunnel T_c w/ a thinner invert. Figure 11 shows that the effect of tunnel design and soil properties on the tunnel response (LSTM) is high: around 12–15 dB between the worst and the best cases on the tunnel invert.

Table 3. Finite Element models of the various tunnel types

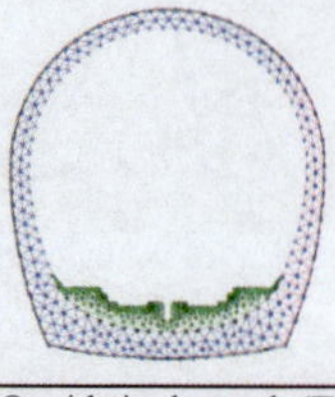	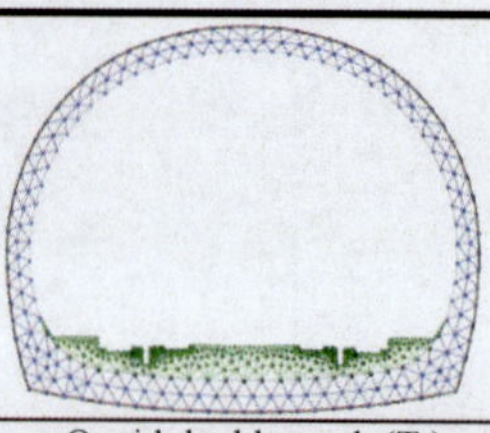	
Ovoid single-track (T_a)	Ovoid double-track (T_b)	NATM double-track (T_c)
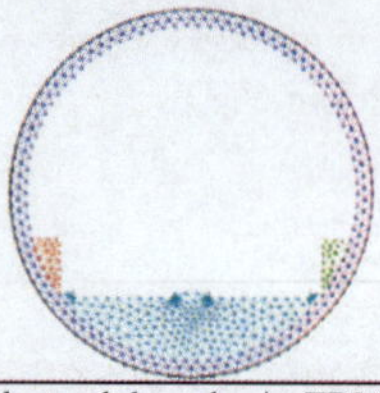	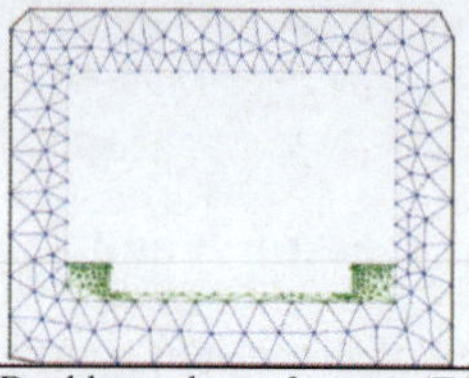	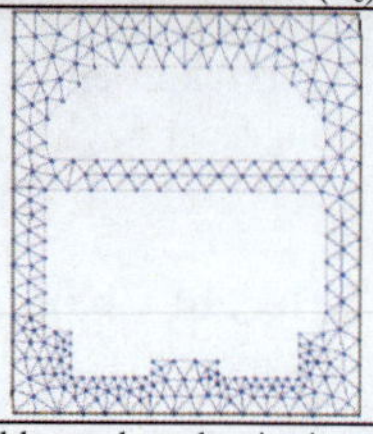
Double-track bored w/ a TBM (T_d)	Double-track cut-&-cover (T_e)	Double-track underpinning (T_f)

Table 4. Soil parameters

Soil	P-wave velocity (m/s)	S-wave velocity (m/s)	Density	Damping
#1 soft (Plastic clay)	1490	310		
#2 medium (Marl)	1720	710	1800 kg/m3	1%
#3 stiff (Beauchamp sand)	2330	1100		

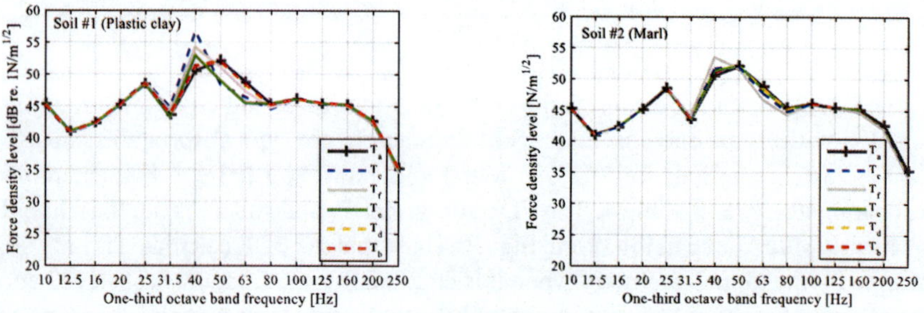

Fig. 9. Force density levels for all tunnels and for soils #1 (left) and #2 (right)

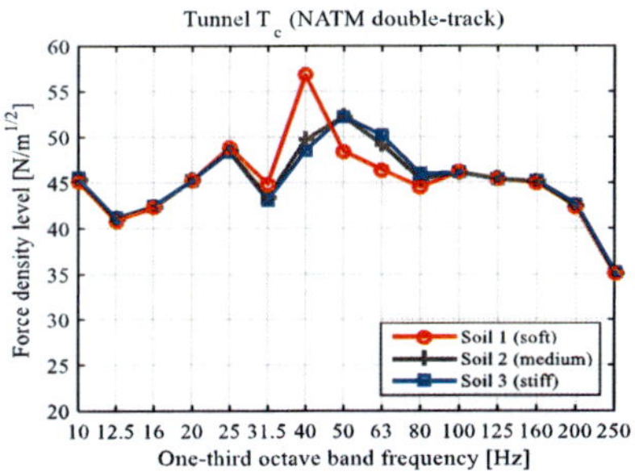

Fig. 10. Force density levels for the three soils and for tunnel T_C (NATM double-track)

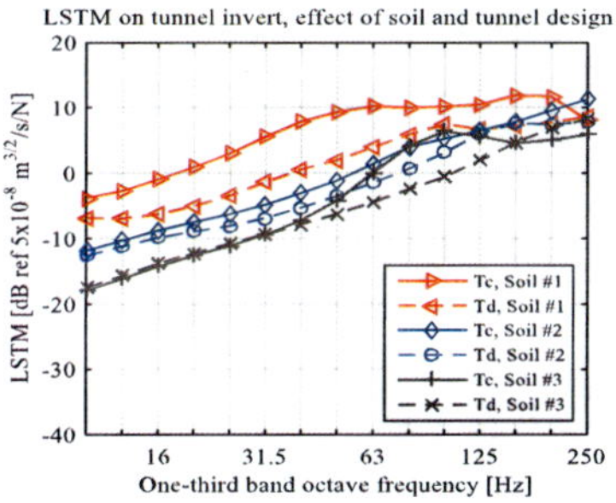

Fig. 11. Effect of soil and tunnel design on LSTM on tunnel wall and invert

4 Conclusion

A ‘baseline’ set of excitation spectra in terms of force density levels (FDL) is provided for a metro in tunnel operating at 60km/h on a booted sleeper track. Rail uneveness spectra associated to the ‘baseline’ FDL are given, facilitating the conversion of the FDL for other train speeds and alternative rail uneveness conditions.

The parametric study confirms that the FDL is a reliable descriptor to characterize the excitation spectrum of a railway system, since it is independent of soil and tunnel geometries (except for specific cases combining a soft soil and a soft tunnel invert), as opposed to the tunnel response which is strongly affected by these parameters.

The study also showed that a train-track interaction model (with or without coupling to a tunnel-soil model) can be used to calculate track insertion loss and then estimate new FDL for other track and/or vehicle configurations. A hybrid model coupling S-RIV (analytical train-track excitation model) and TRAFFIC (2,5D FE-BE model for tunnel-soil interactions) was successfully created and managed to assess the effect of change of track, vehicle, tunnel and soil parameters.

Acknowledgements. The authors wish to thank Franck Leparq, Société du Grand Paris for financially supporting and organizing the tests on a commuter rail line in tunnel, and M. Villot & C. Guigou-Carter, CSTB, for their contributions to the comprehensive analysis of the experimental data, which inspired the work presented in this article.

References

1. Villot, M., Augis, E., et al.: Vibration emission from railway lines in tunnel – characterization and prediction. Int. J. Rail Transp. **4**(4), 208-228 (2016)
2. Transit Noise and Vibration Impact Assessment Manual. FTA Report No. 0123 (2018)
3. Gupta, S., Degrande, G., Lombaert, G.: Experimental validation of a numerical model for subway induced vibrations. J. Sound Vib. **321**(2009), 786–812 (2008)
4. Thompson, D.J.: ‘Wheel/rail noise—theoretical modelling of the generation of vibrations’, Ph.D. thesis, University of Southampton (1990)
5. Thompson, D.J.: Railway Noise and Vibration: Mechanisms. Modelling and Means of Control, Elsevier, Oxford (2009)

6. Manual Corrugation Measurement: CAT. Technical sheet. Rail Measurement Ltd. https://railmeasurement.com/wp-content/uploads/2018/12/CATinfo.2018.pdf
7. TWINS - Track Wheel Interaction Noise Software. TNO technical sheet. https://www.tno.nl/media/2479/twins.pdf

Reduction of Ground Vibration Transmission by Means of Double Wall Barriers

Cédric Van hoorickx[1(✉)], Mattias Schevenels[2], and Geert Lombaert[1]

[1] KU Leuven, Department of Civil Engineering, Structural Mechanics Section, Kasteelpark Arenberg 40, Box 2448, 3001 Leuven, Belgium
cedric.vanhoorickx@kuleuven.be

[2] KU Leuven, Department of Architecture, Architectural Engineering, Kasteelpark Arenberg 1, Box 2431, 3001 Leuven, Belgium

Abstract. Stiff wall barriers can be effective in reducing the transmission of environmental ground vibration. As double walls are used in building acoustics in order to realize a high level of sound insulation, the potential of using double jet-grout walls in reducing ground vibration transmission is investigated in this paper. The three-dimensional free field response due to a point load and a simplified train passage is computed using a two-and-a-half dimensional finite element methodology. In some cases, double jet-grout wall barriers are found to be slightly more effective than single wall barriers, in particular when the thickness of the walls and the intermediate soil matches a quarter Rayleigh wavelength. If there is a large difference between the soil and barrier stiffness, the performance is dominated by the stiffness effect and is similar for single and double wall barriers.

Keywords: Vibration mitigation measures · Double wall barrier · Two-and-a-half dimensional methodology · Quarter wave-stack condition

1 Introduction

Railway induced vibrations in the built environment are a major environmental concern. In order to reduce environmental ground vibration due to railway traffic, mitigation measures at the transmission path, such as stiff wall barriers, can be applied. These mitigation measures aim at impeding propagation of ground vibrations from source to receiver. They are particularly appealing in situations with existing track and buildings, since these can be left unmodified.

In building acoustics, it is well known that double walls can reduce sound levels much better than single walls. For single walls, the sound reduction index increases by 6 dB per octave over the entire frequency range, while for double walls, the sound reduction index increases by 18 dB per octave between the

G. Degrande et al. (Eds.): *Noise and Vibration Mitigation for Rail Transportation Systems*, NNFM 150, pp. 555–562, 2021.
https://doi.org/10.1007/978-3-030-70289-2_60

double wall resonance frequency and the first cavity resonance frequency [1]. In this contribution, the potential of double wall barriers for environmental ground vibration is studied.

The paper is organized as follows. In Sect. 2, the model is described and the 2.5D methodology is explained. Using this model, the three-dimensional free-field response is computed and analyzed in Sect. 3. Next, these results are interpreted in Sect. 4 and the influence of the soil properties is discussed in Sect. 5.

2 2.5D Model of a Double Wall Barrier

The geometry of the longitudinally invariant problem is shown in Fig. 1. This problem consists of a longitudinally invariant (double) wall barrier in a homogeneous halfspace, representing the soil, which is excited at the surface by a point source. In order to define the geometry, a right-handed Cartesian frame of reference is chosen where the origin coincides with the position of the source point. The aim is to study the vibration levels for two receivers located at ($x = 15\,\mathrm{m}, y = 0\,\mathrm{m}, z = 0\,\mathrm{m}$) and ($x = 15\,\mathrm{m}, y = 15\,\mathrm{m}, z = 0\,\mathrm{m}$). These positions will be referred to as receiver 1 and receiver 2, respectively.

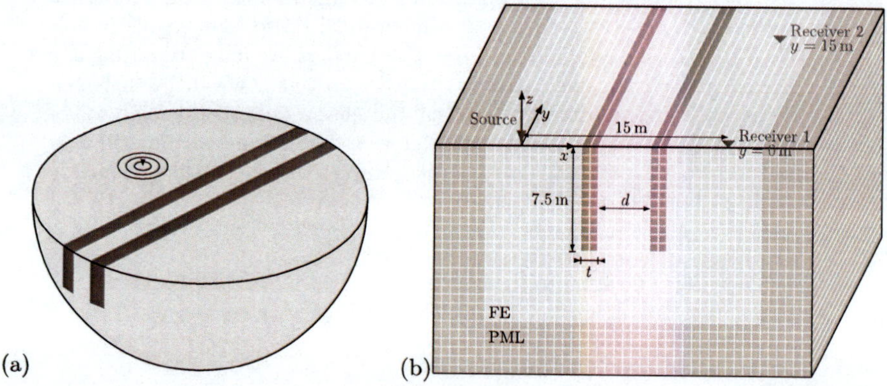

Fig. 1. (a) A semi-infinite halfspace excited by a point load at the surface with double wall barriers introduced to reduce wave transmission. (b) The finite element model with the location of the source, barrier, and receivers indicated. The mesh is added as illustration only and is not the actual finite element mesh.

The soil has a mass density ρ_1 of $1800\,\mathrm{kg/m^3}$, a longitudinal wave velocity C_{p1} of $300\,\mathrm{m/s}$, a shear wave velocity C_{s1} of $150\,\mathrm{m/s}$, and material damping ratios of 2.5% in deviatoric and volumetric deformation. A stiff double wall barrier is introduced between the source and the receivers to reduce the transmission of ground vibration. For the material properties of the barrier, $2000\,\mathrm{kg/m^3}$ is taken for the mass density ρ_2, $950\,\mathrm{m/s}$ for the longitudinal wave velocity C_{p2}, $550\,\mathrm{m/s}$ for the shear wave velocity C_{s2}, and 2.5% for the material damping ratios in deviatoric and volumetric deformation. The double wall barrier is located symmetrically with respect to the center of the source and the first receiver. The double wall barrier is characterized by its depth, in this case a constant value of 7.5 m, the thicknesses of the two walls, denoted by t, and the spacing between the walls, denoted by d. The performance of the double wall barrier is compared with the performance of a single wall barrier with a thickness of $2t$, having the same volume as the double wall barrier.

For longitudinally invariant geometries, a computationally efficient two-and-a-half-dimensional (2.5D) method can be used to calculate the three-dimensional wave field in the soil. In this approach, the longitudinal coordinate y is transformed to the wavenumber k_y by means of a forward Fourier transform: $\mathcal{F}\left[f\left(y\right), k_y\right] = \int_{-\infty}^{\infty} e^{+ik_y y} f\left(y\right) \mathrm{d}y$. The response is then calculated in the frequency-wavenumber domain, where, for every considered frequency ω (rad/s), the dynamic equilibrium equations can be written as:

$$\left(-\omega^2\mathbf{M} + \mathbf{K}_0 - ik_y\mathbf{K}_1 - k_y^2\mathbf{K}_2\right)\underline{\tilde{\mathbf{u}}}\left(k_y, \omega\right) = \underline{\tilde{\mathbf{f}}}\left(k_y, \omega\right) \tag{1}$$

with $\mathbf{M}$ the mass matrix, $\mathbf{K}_0$, $\mathbf{K}_1$, and $\mathbf{K}_2$ the stiffness matrices, $\underline{\tilde{\mathbf{u}}}\left(k_y, \omega\right)$ the response vector, and $\underline{\tilde{\mathbf{f}}}\left(k_y, \omega\right)$ the force vector. The response in the spatial domain is then obtained by performing an inverse Fourier transform from the wavenumber k_y to the longitudinal coordinate y.

At the boundaries of the finite element mesh, perfectly matched layers (PMLs) are added to prevent spurious wave reflections (Fig. 1b). The modified elastodynamic equations include complex-valued coordinate stretching along the direction(s) normal to the interface [2].

3 Analysis of a Double Wall Barrier

In this section, the performance of a single and double wall barrier are compared. The single wall barrier has a depth of 7.5 m and a thickness of 2 m. The center axis of the wall is located at a distance of 7.5 m from the point source. The two walls of the double wall barrier have a depth of 7.5 m and a thickness t of 1 m, resulting in the same volume of material as for the single wall barrier. The distance d between the two walls equals 4 m. The center axis of both walls is located at a distance of respectively 5 m and 10 m from the point source.

The insertion loss is commonly used to quantify the effectiveness of mitigation measures. This is defined as:

$$\hat{\mathrm{IL}}(x, y, z, \omega) = 20 \log \frac{\left|\hat{u}_z^{\mathrm{ref}}(x, y, z, \omega)\right|}{\left|\hat{u}_z(x, y, z, \omega)\right|} \tag{2}$$

where $\hat{u}_z^{\mathrm{ref}}(x, y, z, \omega)$ is the vertical displacement before and $\hat{u}_z(x, y, z, \omega)$ the vertical displacement after introducing the considered mitigation measure in the soil. The insertion loss of the single and double wall barrier for the two receivers in Fig. 1 is shown in Fig. 2. At low frequencies, the depth of the wall barrier is small compared to the penetration depth of the incoming surface waves, resulting in low insertion losses. Above about 20 Hz, corresponding to a shear wavelength equal to 7.5 m, an insertion loss between 6 dB and 20 dB is obtained. For receiver 1, the insertion loss for the double wall barrier is higher than for the single wall barrier for most of the frequency range. However, the improvement disappears at larger distances in the y-direction (receiver 2). This can be mainly attributed to the weaker performance of single wall barriers for points close to the line going through the source point and perpendicular to the barrier, as was reported before [3]. The double wall barrier improves the performance in the important area close to the source while maintaining the effectiveness of the single wall barrier at larger distances.

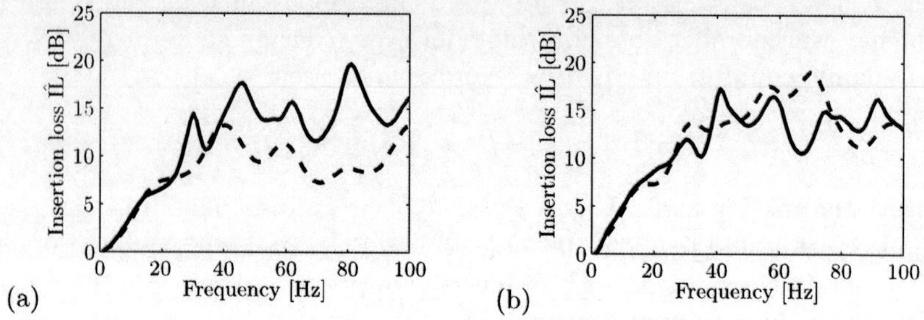

Fig. 2. Insertion loss $\hat{\text{IL}}$ in the case of a point load for the single wall barrier (dashed line) and the double wall barrier (solid line) in the different receiver points: (a) receiver 1 and (b) receiver 2.

As the goal is to mitigate ground vibration caused by passing trains, the displacement field resulting from a passing train is calculated next. This is done in a simplified way, applying the dynamic axle loads at fixed positions as a series of incoherent point loads of the same magnitude, as was done in [4]. Previous studies have indicated that the stationary part of the response can be well approximated in this way [5]. The axle positions are taken from a typical four-car EMU train, with a car length of 26.4 m, and a total length of 105.6 m. The bogie centre distance is 19 m and the distance between two axles of a bogie is 2.7 m. The center of the train is located at the position of the point source indicated in Fig. 1, and the train is oriented parallel to the y-axis.

The resulting insertion loss for the three receivers is shown in Fig. 3 for the considered single and double wall barriers. As a result of geometric attenuation and material damping in the soil, the excitation points closest to the receiver point contribute most to the response at the receiver point.

4 Physical Interpretation

It is known from literature that two phenomena play an important role for the performance of single wall barriers: the stiffness effect and reflection. The role of these phenomena for double wall barriers is investigated in this section.

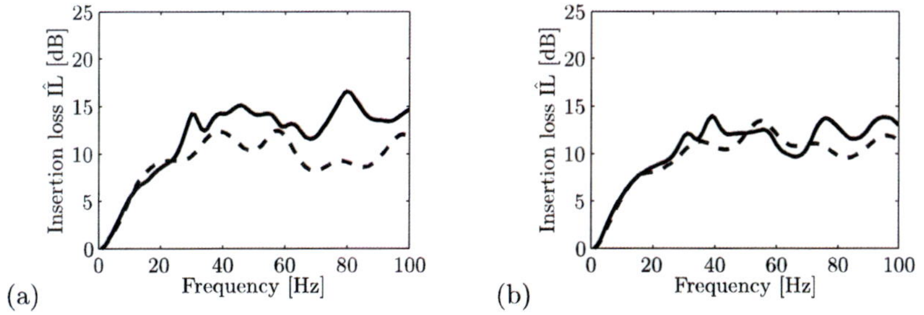

Fig. 3. Insertion loss ÎL in the case of a train load for the single wall barrier (dashed line) and the double wall barrier (solid line) in the different receiver points: (a) receiver 1 and (b) receiver 2.

4.1 The Stiffness Effect

Since the calculations are performed in the frequency-wavenumber domain, the insertion loss at a point can easily be plotted as a function of the frequency f and the longitudinal slowness $p_y = k_y/\omega$. This is done in Fig. 4 for both the single and double wall barriers considered in Sect. 3.

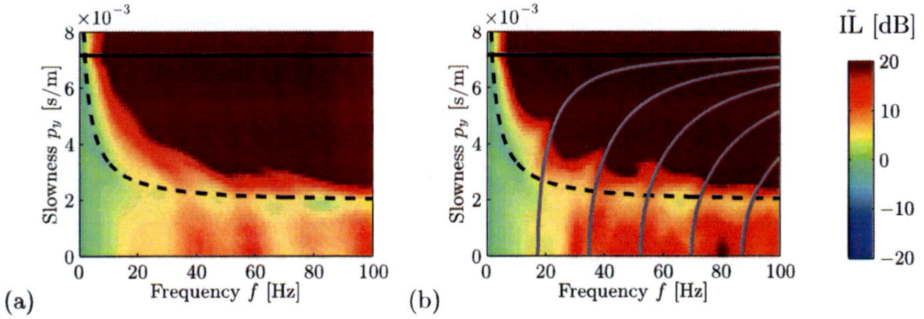

Fig. 4. Insertion loss $\tilde{\mathrm{IL}}(x = 15\,\mathrm{m}, k_y, z = 0\,\mathrm{m}, \omega)$ for (a) the single wall barrier and (b) the double wall barrier. Superimposed are the Rayleigh wave slowness (black solid line) and the analytical dispersion curve for Timoshenko's bending mode around the x-axis (black dashed line). For the double wall barrier, the dispersion curves caused by standing waves between the walls are added (gray solid lines).

When the longitudinal wavenumber k_y exceeds the Rayleigh wavenumber k_R, corresponding to a slowness $p_\mathrm{R} = 7.16 \times 10^{-3}\,\mathrm{s/m}$, the lateral wavenumber $k_x = \sqrt{k_\mathrm{R}^2 - k_y^2}$ is imaginary and the wave in the x-direction becomes evanescent. Consequently, the free field response is very small for these wavenumbers.

For both the single and the double wall barriers, the highest insertion losses are found for slownesses above the dispersion curve for Timoshenko's bending mode around the x-axis, which is given by [6]:

$$-\rho A\omega^2 + EIk_y^4 - \rho I\left(1+\frac{E}{\kappa G}\right)\omega^2 k_y^2 + \frac{\rho^2 I}{\kappa G}\omega^4 = 0 \tag{3}$$

In this region, the bending stiffness of the wall prevents the propagation of waves as the trace wavelength λ_y is smaller than the bending wavelength λ_b of the barrier. The single and the double wall barrier both have the same bending stiffness around the x-axis (as this is proportional to the total barrier width) and therefore the free bending dispersion curves of both barriers coincide.

4.2 Reflection

For the double wall barriers, the insertion loss is affected by resonance caused by standing waves between the two walls. Standing waves appear at the surface when the distance between the two walls is equal to an integer number (m) times half the wavelength in the x-direction:

$$d = \frac{m\lambda_x}{2} = \frac{m\pi}{k_x} = \frac{mC_R}{2f\cos\theta} \tag{4}$$

Here, use is made of the relation $k_x = k_R\cos\theta$. As $\sin\theta = k_y/k_R$, this equation represents a dispersion relation between k_y and ω. At the corresponding frequency-slowness pairs (Fig. 4b), the transmitted wave field is increased, resulting in a lower insertion loss.

Figure 2a shows that for receiver 1, the insertion loss for the double wall barrier is larger than the one for the single wall barrier, especially at some local maxima. At these frequencies, less energy is transmitted, in a way which is similar to stopbands in periodic media. For multiple periodic cells consisting of a layer with thickness d and wavenumber $k_1 = \omega/C_{R1}$ and a layer with thickness t and wavenumber $k_2 = \omega/C_{R2}$, it can be shown that frequencies are located in a stopband if:

$$k_1 d + k_2 t = n\pi \tag{5}$$

for $n = 1\cdots\infty$ [7]. These frequencies coincide with the local maxima in Fig. 2a.

Especially when $k_1 d = k_2 t = n\pi/2$, or $d = \lambda_1/4$ and $t = \lambda_2/4$, called the quarter wave-stack condition, a large reduction of the wave transmission is obtained. For the considered thickness of the walls, the quarter wave-stack condition cannot be satisfied in the frequency range of interest.

Figure 5 shows the performance of the wall barriers for a train load when the thickness of the walls is enlarged to 2 m. A clear improvement is visible as compared to $t = 1$ m (Fig. 3).

5 Influence of the Soil Properties

The effectiveness of the barrier depends, for given barrier properties, on the material properties of the soil. Consider soft clay, with a mass density $\rho =$

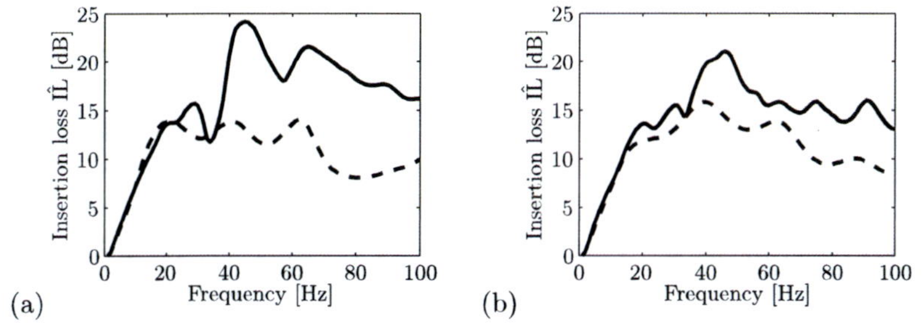

Fig. 5. Insertion loss IL̂ in the case of a train load for the single wall barrier with a thickness of 4 m (dashed line) and the double wall barrier with two walls with a thickness of $t = 2\,\text{m}$ (solid line) in the different receiver points: (a) receiver 1 and (b) receiver 2.

$2500\,\text{kg/m}^3$, a dilatational wave velocity $C_\text{p} = 120\,\text{m/s}$, a shear wave velocity $C_\text{s} = 60\,\text{m/s}$, and material damping ratios $\beta_\text{p} = \beta_\text{s} = 2.5\,\%$. Figure 6a shows the insertion loss as a function of the frequency at receiver 1 for a point load. The insertion loss is rather large for the entire frequency range considered.

In Fig. 6b, the insertion loss is shown as a function of the frequency f and the slowness p_y. As the soil has a lower Rayleigh wave velocity, there are more standing waves in the frequency range considered. Moreover, the Rayleigh wavenumber k_R increases for a softer soil. As a consequence, the response at the receiver positions is increasingly dominated by the effect of the bending stiffness. This results in an increased performance of the barrier. As the stiffness effect is similar for both the single and double wall barrier, the difference in performance is small (Fig. 6a).

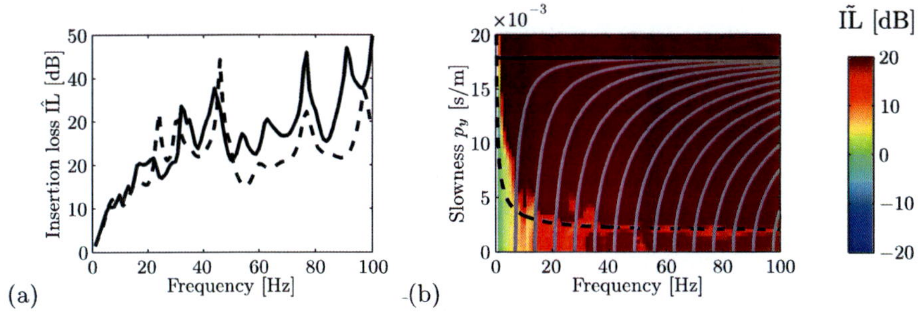

Fig. 6. (a) Insertion loss IL̂ in the case of a point load for the single wall barrier (dashed line) and the double wall barrier (solid line) in soft clay soil at receiver 1. (b) Insertion loss $\tilde{\text{IL}}(x = 15\,\text{m}, k_y, z = 0\,\text{m}, \omega)$ for the double wall barrier. Superimposed are the Rayleigh wave slowness (black solid line), the analytical dispersion curve for Timoshenko's bending mode around the x-axis (black dashed line), and the dispersion curves caused by standing waves between the walls (gray solid lines).

6 Conclusions

In this paper, the performance of double jet-grout wall barriers are compared with single wall barriers with the same material properties and total volume. A two-and-a-half dimensional finite element methodology is used to analyze the performance. For a small difference in stiffness between the soil and the barrier, the double wall barrier is found to perform slightly better than the single wall barrier close to the plane of symmetry of the axle loads. For large thicknesses t and spacings d between the walls, a higher insertion loss is found, especially when the thickness of the walls (t) and the intermediate soil (d) are close to a quarter of the Rayleigh wavelength. For a large difference in stiffness between the soil and the barrier, the stiffness effect dominates and the single and double wall barrier have a similar performance.

Acknowledgements. The research was performed under a doctoral fellowship of the Research Foundation Flanders (FWO). The authors also gratefully acknowledge the support from the Research Council of KU Leuven through the funding of the project C16/17/008 "Efficient methods for large-scale PDE-constrained optimization in the presence of uncertainty and complex technological constraints".

References

1. Fahy, F., Gardonio, P.: Sound and Structural Vibration: Radiation, Transmission and Response, 2nd edn. Academic Press, Oxford (2007)
2. François, S., Schevenels, M., Lombaert, G., Degrande, G.: A 2.5D displacement based PML for elastodynamic wave propagation. Int. J. Numerical Methods Eng. **90**(7), 819–837 (2012)
3. Coulier, P., François, S., Degrande, G., Lombaert, G.: Subgrade stiffening next to the track as a wave impeding barrier for railway induced vibrations. Soil Dyn. Earthquake Eng. **48**, 119–131 (2013)
4. Dijckmans, A., Coulier, P., Jiang, J., Toward, M., Thompson, D., Degrande, G., Lombaert, G.: Mitigation of railway induced ground vibration by heavy masses next to the track. Soil Dyn. Earthquake Eng. **75**, 158–170 (2015)
5. Verbraken, H., Lombaert, G., Degrande, G.: Verification of an empirical prediction method for railway induced vibrations by means of numerical simulations. J. Sound Vib. **330**(8), 1692–1703 (2011)
6. Timoshenko, S.: On the transverse vibrations of bars of uniform cross section. Phil. Mag. **43**, 125–131 (1922)
7. Van hoorickx, C., Schevenels, M., Lombaert, G.: Topology optimization of one-dimensional wave impeding barriers. In: Sas, P., Moens, D., Denayer, H. (eds.) Proceedings of ISMA 2014 International Conference on Noise and Vibration Engineering, Leuven, Belgium, pp. 3557–3571 (2014)

Investigation of Vibration Mitigation by Concrete Trough with Integrated Under Ballast Mats for Surface-Railways

Rüdiger Garburg[1(✉)], Christian Frank[2], and Michael Mistler[3]

[1] Department of Infrastructure Technology, Deutsche Bahn AG, Digitalization and Technology, Europaplatz 1, 10557 Berlin, Germany
ruediger.garburg@deutschebahn.com
[2] DB Netz AG, Gleistechnik, Adam-Riese-Straße 11-13, 60327 Frankfurt am Main, Germany
[3] Baudynamik Heiland & Mistler GmbH, Bergstraße 174, 44807 Bochum, Germany

Abstract. In recent years environmental impact studies including an extensive vibration analysis of the effect of the vibration emissions from the railways on adjacent residents have become more and more important during the planning process of construction of new railway lines as well as for the modification or extension of existing ones. Within such a study, not only the immission have to be predicted, but also appropriate mitigation measures must be proposed in a very early planning phase. For surface-railways, however, there are hardly any measures available to mitigate vibrations. Next to under sleeper pads a solid concrete trough with integrated ballast mats is one of the alternatives. The use of this measure must be provided for more and more vibration-sensitive sections to achieve a certain quality of vibration mitigation. However, the influences of different parameters on the effectiveness of vibration mitigation are not fully understood and therefore an adequate prediction is not yet possible of whether the effectiveness will be as desired. The investigations described are essentially based on measurements on a total of six existing concrete troughs. The procedure and the results of the investigation carried out on six various sites already installed [1] are described. First conclusions are drawn and an outlook with specific recommendations for improved mitigation efficiency is given.

Keywords: Vibrations · Vibration mitigation · Concrete trough · Under ballast mat · Insertion loss

1 Introduction

Besides noise mitigation also ground-borne vibrations induced by railways are another important environmental issue that had to be tackled during the last decade.

Annoyance can occur, particularly for lines in urban areas with small distances to neighboring houses or lines in shallow depth tunnels under buildings. The ground-borne

G. Degrande et al. (Eds.): *Noise and Vibration Mitigation for Rail Transportation Systems*, NNFM 150, pp. 563–570, 2021.
https://doi.org/10.1007/978-3-030-70289-2_61

vibrations can be perceived by the inhabitants via the floor vibrations as well as reradiated noise from structures and ceilings.

In Germany, an environmental impact study including an extensive vibration analysis of the effect of the vibration emissions from the railways on residents must be performed for the construction of new railway lines as well as for the modification or extension of existing ones. Within the study, not only the immission must be predicted, but also suitable mitigation measures must be proposed.

In contrast to vibration-mitigation systems such as the floating-slab system or under ballast mats (UBM) in tunnels which are well documented and predictable (comp. e.g. [3, 4] and many other examples), only a few effective mitigation measures are available for surface-lines. In recent years, therefore, concrete troughs with integrated under ballast mats have been built several times in sensitive areas to mitigate vibrations. Figure 1 shows a typical cross-section. The mass of the concrete trough provides a huge input impedance compared to soft soil. The trough itself ensures that the ballast is put into a frame and therefore not able to flow out as happens when under ballast mats are placed directly onto the soil. Between the concrete and ballast, the UBM is inserted as a ballast protection elastic layer.

Concerning the effectiveness of the mitigation measure, the quality of the prediction is not sufficiently known and the influences of the different parameters on effective vibration mitigation are not fully understood. Therefore, an adequate prediction is not possible yet. For this reason, a systematic investigation of the mitigation effects of different already implemented concrete troughs with under ballast mats was performed.

This investigation aims to provide a measurement database, which can be used as a basis for a dynamic design concept of future concrete troughs with integrated under ballast mats.

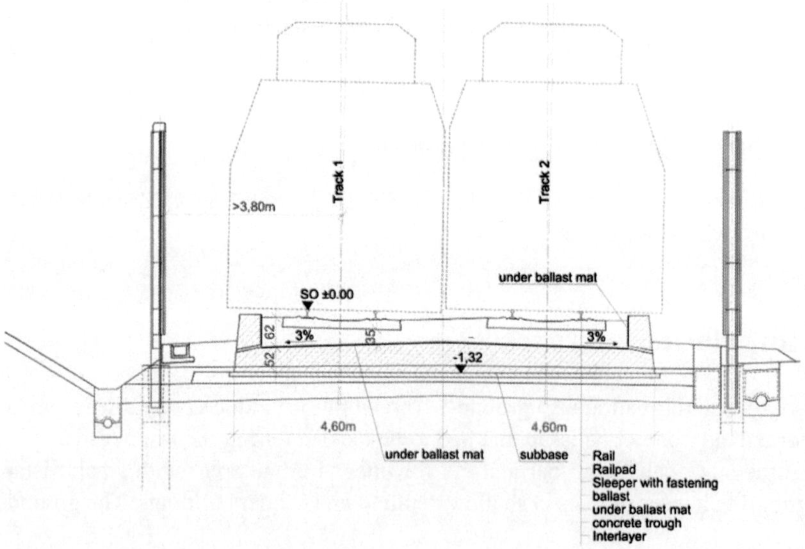

Fig. 1. Typical cross-section of a concrete trough with integrated under ballast mats

According to the current state of knowledge the vibration-mitigation effect of a concrete trough with integrated UBM is mainly influenced by the following mechanisms:

- load distribution effect of the concrete slab in the longitudinal direction due to its bending stiffness,
- load distribution effect in the transverse direction,
- more uniform substructure stiffness under the ballast and thus also under the track as a whole,
- under ballast mat as an elastic intermediate layer.

The actual mitigation effect of the concrete trough depends on both, the interactions with the subsoil as well as its structural behavior itself. Since the subsoil may have a significant impact on the effectiveness of such a system, the dynamic soil properties must also be investigated and considered in the evaluation.

2 Methodology

2.1 Definition of Insertion Loss (IL)

In general, the vibration reduction of mitigation measures is quantified by using the "insertion loss" (IL) which describes the relative abatement effect of a mitigation measure compared to a reference situation. The standard procedure to determine IL is to compare the ground vibration of a train passing at a specific site before and after the installation of a mitigation measure (the so-called "before/afterwards" method) or the same train passing at two adjacent track sections with and without measure (the "left/right" method) as described in [1, 2].

For investigating existing concrete troughs only the left/right method is applicable. But in this case, it must be ensured that the conditions of the subsoil, the ground and within the transmission paths are identical at both sections or these must be normalised by an appropriate correction term. Because this is typically not the case, the guideline [1, 2] proposes a soil correction in the following way:

$$D_e(f) = L_{v,Ref}(f) - L_{v,U}(f) + \left(L_{Y,U}(f) - L_{Y,Ref}(f)\right)[dB] \tag{1}$$

where $D_e(f)$ is the insertion loss (IL) relating to the third-octave spectrum of structure- / ground-borne noise. The term "$L_{v,Ref}(f) - L_{v,U}(f)$" is the difference of the vibration velocity level due to a train passing, measured at the reference section and the section of the concrete trough. "$L_{Y,U}(f) - L_{Y,Ref}(f)$" represents the correction term to consider the different soil conditions. $L_y(f)$ is the frequency-dependent transfer mobility level. The advantage is that it can be easily determined by simple impact test measurements, see Sect. 2.2. Please note that the terms "$L_{v,Ref}(f) - L_{v,U}(f)$" and IL have been often used synonymously in the past and this influence has been often neglected.

2.2 Determination Procedure

The investigations of IL due to the concrete trough with integrated UBM are based on measurements on a total of six already existing troughs at different locations in the south of Germany. The measurement procedure consisted of the following steps and was carried out at all six sites:

- determination of vibrations within the different sections by field measurements during train passages,
- determination of the transfer mobility of the subsoil for both sections by defined impact tests.

Additionally, free field propagation measurement for the investigation of the soil (MASW method) and impact test measurements on the concrete trough itself were carried out.

In principle, the same sensor layout was used for all test sites as shown in Fig. 2.

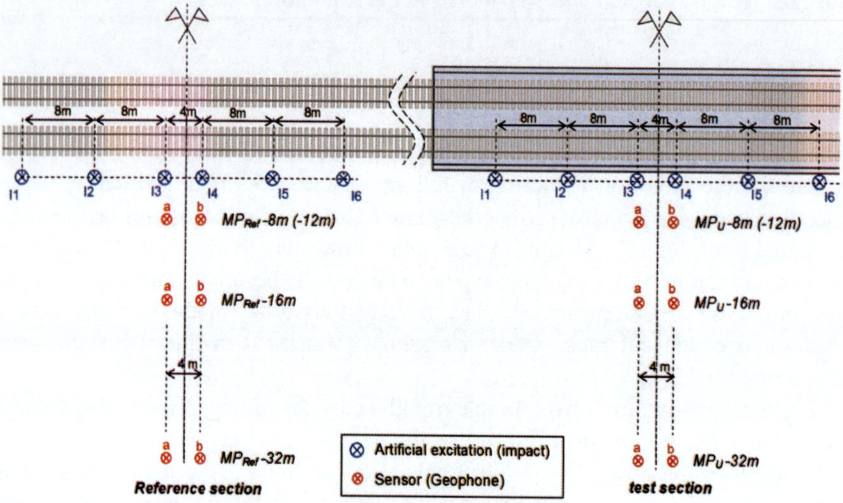

Fig. 2. Measurement layout with measuring positions (MP) at 8 m (or 12 m), 16 m and 32 m from the track centre at the reference section (standard ballast track without concrete trough) and at the test section (with concrete trough). The positions I1 to I6 were used for the impact test measurement.

The transfer mobility was determined by impact test measurements due to six impact locations along the track. For this, DYNPACT® (see Fig. 3) was used which is a drop weight with an implemented load cell. The drop height is 1,0 m and the drop weight is 50 kg. The maximum peak force depends on the impact surface stiffness and lies in the range of 40–150 kN.

The time history of the impact force and the vibration of the measuring points were recorded simultaneously to calculate the mobility function. To be correct it must be considered that the artificial excitation of the drop weight is a point vibration source whereas the passing train is a line vibration source. These types of investigation are described in more detail in the references [5] and [6].

Fig. 3. Application of the drop weight DYNPACT®

3 Results

The evaluation of the measurements of the six existing concrete troughs with integrated UBM shows that even when considering one location at a single site and even within the same type of trains at slightly different speeds, variations of the insertion loss of approximately the same order of magnitude occur. This is shown in Fig. 4 for the example of several high speed (ICE) train passages of the same type of ICE.

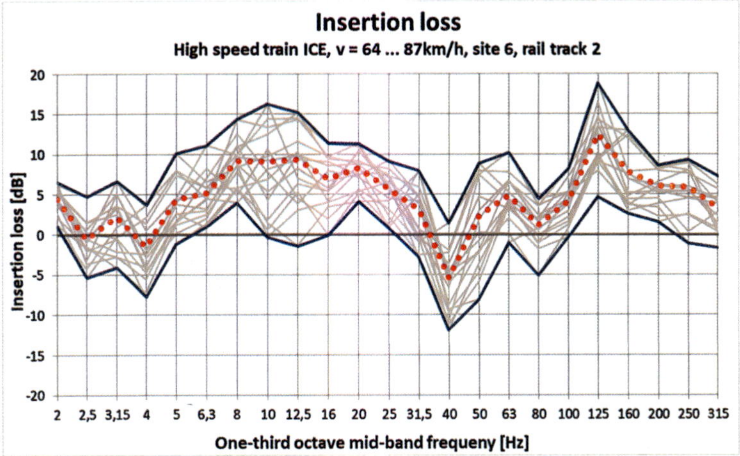

Fig. 4. Measured insertion loss during ICE passing by one of the six locations considered

The evaluations also show that the application of the soil correction is essential for the comparison of the insertion losses of concrete troughs with integrated UBM at different locations. Without the background correction carried out, there would be much greater variability. Figure 5 shows the insertion losses of two identically constructed concrete troughs at two different locations. The results are shown both with and without soil correction. The frequency-dependent soil correction shows different spectral correction values between 0–20 dB depending on the cross-section and its soil properties. From the figures, it can be concluded that the variability without soil correction is much larger than with correction.

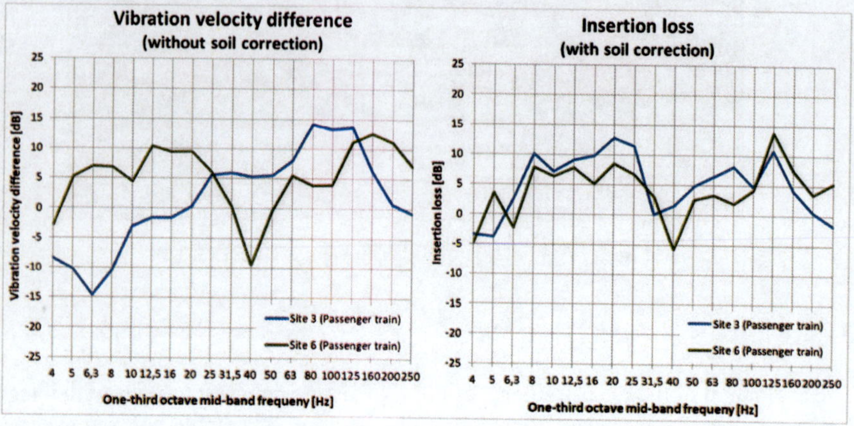

Fig. 5. The insertion loss of two identically constructed concrete troughs at two different locations: left without the influence of soil correction and right with soil correction

The dynamic analysis of the six concrete troughs with integrated UBM in Fig. 6 shows that:

- The insertion loss in the low-frequency range from 2 to 16 Hz leads to averaged values of about 5 dB (except for a minimum in the range of 5 Hz, probably depending on the block length of the concrete trough).
- The insertion loss decreases significantly from 16 Hz and in the range between 25 Hz and 80 Hz no reduction of the vibrations (0 dB) and sometimes even a light vibration amplification due to the concrete trough occurs (values < 0 dB).
- Considering the frequency range above 80 Hz, the averaged insertion loss is higher than 5 dB.
- Although the insertion loss of the individual locations calculated from the measured data is closer to each other due to the soil correction, significant variations are still detectable. The largest scattering occurs in the resonance frequency of the UBM, this affects mainly the frequency range from 25 Hz to 40 Hz.

In the frequency range between 20 Hz and 63 Hz, two outliers (most probably due to an obsolete construction type and an insufficient quantity of trains considered in the measurement) can be seen. The outliers were nevertheless included in the averaging of all measurement results. In [7] a further analysis of the various influences was made.

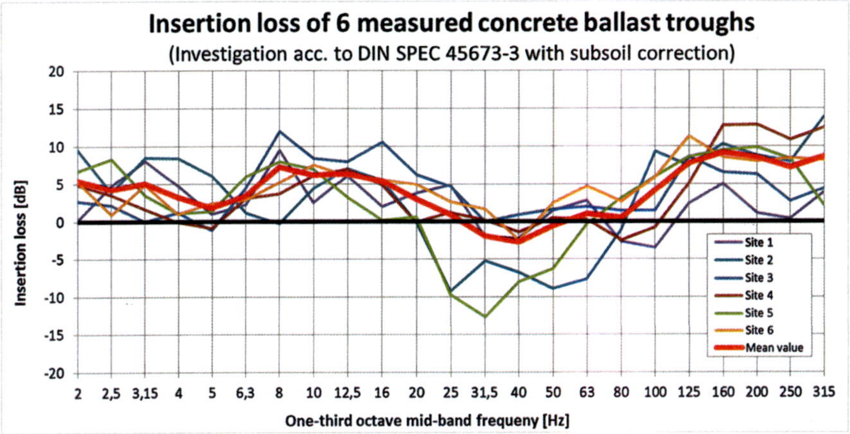

Fig. 6. Measured results for the insertion loss of 6 concrete troughs with under ballast mats including soil correction factors

4 Summary and Outlook

The results show that:

- The insertion loss in the low-frequency range from 2 Hz to 16 Hz is on average approx. 5 dB (except for a dip in the range of 5 Hz, depending on the block length of the concrete trough).
- The insertion loss decreases significantly at frequencies above 16 Hz. In the range between 25 Hz and 80 Hz there is no reduction in vibrations (0 dB) and in some cases even light amplification due to the concrete trough (values <0 dB). Insertion loss increases above 80 Hz and for frequencies above 100 Hz, the average insertion loss is more than 5 dB.
- Although the insertion loss of the individual sites calculated from the measured data shows less variance after subsoil correction, significant scatter can still be determined.

The largest differences between the curves occur in the range of the resonance of a stiff UBM at the frequency range 25 Hz–40 Hz. It is to be assumed that a low natural frequency of the ballast in the trough using soft UBM, at 20 Hz–31 Hz would lead to greater dips than a higher natural frequency at 40 Hz. Accordingly, a preferable stiff UBM would mean a better overall effect of the vibration mitigation.

The following improvement potential is seen for using future concrete troughs:

- The length of the trough has a significant influence on the longitudinal bending stiffness. Block joints should be avoided, and if necessary, only installed at greater distances and generally equipped with transverse force transmission.
- It is important to adapt the stiffness of the UBM to achieve the desired effect of the mitigation. A stiffer tuned UBM is preferable if the reduction of low-frequency vibrations is in focus. It may be different if ground-borne noise is the central issue.

- At the planning stage, it is essential that the transfer admittance is measured on beforehand, so the insertion loss of the mitigation measure can be predicted more accurately together with subsequent measurements.
- These insulation values should be set in relation to the dynamic characteristics (e.g. G-modulus) of the respective subsoil.

Acknowledgements. These investigations were financed from DB Netz AG's funds, for which the authors would like to express their sincere thanks.

References

1. Mistler, M., Heiland, D.: Betontröge mit Unterschottermatte, Messtechnische Untersuchung der erschütterungsmindernden Wirkung von Betontrögen, Baudynamik Heiland & Mistler Bericht-Nr.: 80–10340–01-D3, Bochum 08.12.2017.
2. DIN SPEC 45673–3 (04/2014): Mechanische Schwingungen — Elastische Elemente des Oberbaus von Schienenfahrwegen — Teil 3: Messtechnische Ermittlung der Einfügungsdämmung in eingebautem Zustand (Prüfaufbau und Betriebsgleis) bei Ersatzanregung. Beuth Verlag, Berlin
3. Garburg; R., Jaquet, T.: Measurements and Investigations at the Floating-Track-Bed System in the North-South Tunnel in Berlin; published in Notes on Numerical Fluid Mechanics and Multidisciplinary Design, Volume 99 2008; Noise and Vibration Mitigation for Rail Transportation Systems, Proceedings IWRN 9th (2007)
4. Jaquet, T., Heiland, D., Rutishauser, G., Garburg, R.: Nord-Süd-Verbindung in Berlin (Eisenbahningenieur 57), vol. No. 9 (2006)
5. FTA-VA-90–1003–06: Transit noise and vibration impact assessment. Office of Planning and Environment, Federal Transit Administration U. S. Department of Transportation, Washington DC 20590, USA (2006)
6. Mistler, M., Heiland, D.: Erschütterungsprognose mit Hilfe künstlicher Anregungsmethoden. VDI-Berichte Nr. 2063, 3. VDI-Fachtagung Baudynamik Kassel, 14. und 15. Mai 2009, pp. 555–568, ISBN 978-3-18-092063-4.
7. Breitsamter, N.: Zur erschütterungsmindernden Wirkung von Betontrögen mit Schotteroberbau auf Unterschottermatten, Proceedings DAGA 2018 München

Ground-Borne Vibration from Manchester Metrolink

James Block[✉], Chris Jones, Steve Cawser, Conor Tickner, and Paul Shields

AECOM Infrastructure and Environment, 12 Regan Way, Chetwynd Business Park, Chilwell, Nottingham NG9 6RZ, UK
james.block@aecom.com

Abstract. The Trafford Park Line extension to the Manchester Metrolink light rail system included a planning condition to control the ground-borne noise from operating trams within a building in ITV's studios on Trafford Wharf Road. To determine the expected levels of vibration and noise within the studios, a series of vibration measurements and predictions were carried out during the detailed design stage of the project. These included source vibration measurements on the existing Metrolink network and ground vibration propagation measurements on site. A trailer-mounted sinusoidal vibration source was used in order to quantify the complete propagation from the source to studio, together with the resulting re-radiated noise. This paper describes these measurements and the subsequent analyses and assessment carried out to determine the anticipated ground-borne noise levels within the studio.

Keywords: Light rail · Ground-borne vibration · Ground-borne noise

1 Introduction

The Trafford Park Line (TPL) extension to Manchester Metrolink light rail network includes a section of new track alignment outside ITV's television studios in Trafford Park (Fig. 1). The studios are where the iconic Coronation Street television drama is filmed and include outdoor sets and indoor studios. Approval for the TPL was granted with a planning condition imposed to limit the tram-generated ground-borne noise inside ITV's Studio 4 building at a distance of approximately 32 m from the TPL.

This paper describes the measurements and analysis used to resolve the practical challenges of the project and hence assess the ground-borne noise within the internal studio area.

2 Planning Conditions

Planning Condition 10 [1] for the project is designed to mitigate the ground-borne noise due to trams in Studio 4 using the Noise Rating (NR) curve 25 L_{Fmax} as defined in BS 8233:2014 [2]. A time constant of 1/8 s (0.125 s) is indicated by the letter 'F' for 'Fast', together with a maximum level.

G. Degrande et al. (Eds.): *Noise and Vibration Mitigation for Rail Transportation Systems*, NNFM 150, pp. 571–578, 2021.
https://doi.org/10.1007/978-3-030-70289-2_62

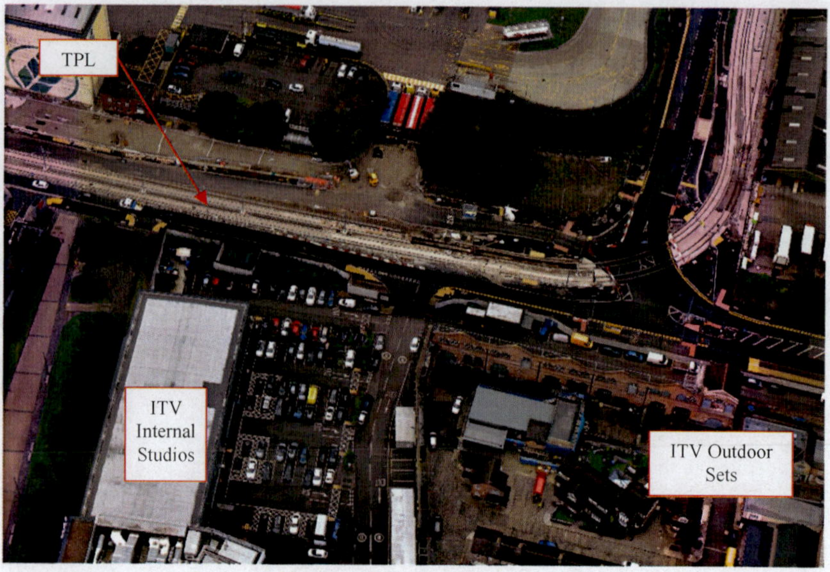

Fig. 1. Aerial view of site

3 Methodology

The methodology to determine the likely ground-borne sound levels was as follows:

- Source vibration: measurements of vibration from the existing Metrolink system;
- Propagation: on site measurements; and
- Response: on site measurements of floor vibration and the resulting noise.

3.1 Source Vibration

Tram vibration source data were obtained from measurements by others on the existing Metrolink alignment at Merrill Street as part of a previous project (2nd City Crossing Project). At both Merrill Street and the future alignment outside Studio 4 the tracks comprise embedded rail with Ri59 section and Vector Boot encapsulation. Measurements were made of 'Single' Metrolink trams (2 vehicles, 3 bogies) travelling at speeds of between 23 and 37 kmh^{-1}.

The vibration acceleration data were re-analysed for a maximum level with a 'Fast' time constant to match the Condition 10 limit (Fig. 2). Individual tram passes are identified as blue lines and the spectrum with the highest level as a dashed pink line. Because of the short time window, the spectral information is poorly defined below the 40 Hz one-third octave band. For similar reasons the NR curve is defined, for the purposes of the limit criterion, down to the 63 Hz octave band.

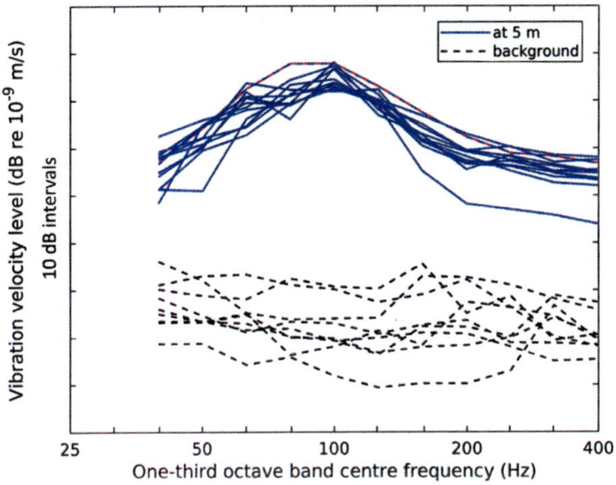

Fig. 2. Measured vibration spectra corresponding to L_{Fmax}

3.2 Propagation

To provide data on the vibration propagation between the prospective alignment on Trafford Wharf Road and Studio 4, measurements were made using a trailer mounted vibration source. Other vibration sources were considered, e.g. an impact hammer, but due to the distance between the alignment and the building, and the propagation characteristics at this location, a larger vibration source was required.

The 'MiniVib' trailer (Fig. 3) was used to create vibration in the ground at three positions along the TPL alignment to represent the approximate length of a 'Single' tram. Vibration excitation was generated in the form of a swept sinusoidal input with frequency linearly increased from 8 to 355 Hz over a period of 40 s, with a force of up to approximately 27 kN. The sweeps were repeated until 10 complete measurements were achieved without extraneous vibration affecting the integrity of data. Background vibration levels without the MiniVib operating were measured at the beginning and end of each set of sweeps.

No valid force signal was available and therefore the use of the Force Density Level/Line Source Transfer Mobility [3] method of vibration prediction was precluded.

During these tests the vibration response was measured outside of the studio in the vertical direction at distances of 5 m, 17 m, 22 m and 32 m from the source. The 32 m distance was adjacent to the external wall of the Studio 4 building. Inside the studio, the vibration response was measured in the vertical direction at three positions, together with measurements of sound at 1.5 m above the floor at a position adjacent to one of the vibration response measurement positions.

Fig. 3. MiniVib trailer

Figure 4 presents a comparison of the vibration levels measured inside the studio at the three positions (labelled M1, M2 and M3). The measurements were repeated for the operation of the MiniVib vehicle at the three positions along the prospective alignment (labelled VP1, VP2 and VP3). For each spectrum, the presented levels are the average of five frequency sweeps. Also included for comparison are the measured background vibration levels without the vehicle operating. The background vibration was thought to be due to the airborne excitation of the building by the noise from the MiniVib. The measured vibration is significantly higher than the background vibration (greater than 10 dB difference) in the frequency bands from 10 Hz to 100 Hz. At frequencies higher than 100 Hz, the measured vibration is not reliably determinable above the background vibration. At these frequencies, the low level of measured vibration during a MiniVib sweep was due to either the magnitude of the force that the MiniVib vehicle can input into the ground or the low propagation of vibration.

Also presented in Fig. 4 are the average of the results from the nine measurements inside the studio (three vehicle positions and three measurement locations) and the average of the same measurements outside the studio (thick black lines). These are similar in level across the frequency range and indicate that there are little or no coupling losses between the ground outside the studio and the floor of the studio.

To quantify the vibration in the studio, the difference in vibration level was required between the studio and the position 5 m from the alignment (Fig. 5). As expected, there is an approximate linear decay trend for most frequency bands. The lower frequencies also decay more slowly than the higher frequencies; this is due to the damping losses in the ground. Also, the higher spectral bands are adversely affected by the background vibration (black line).

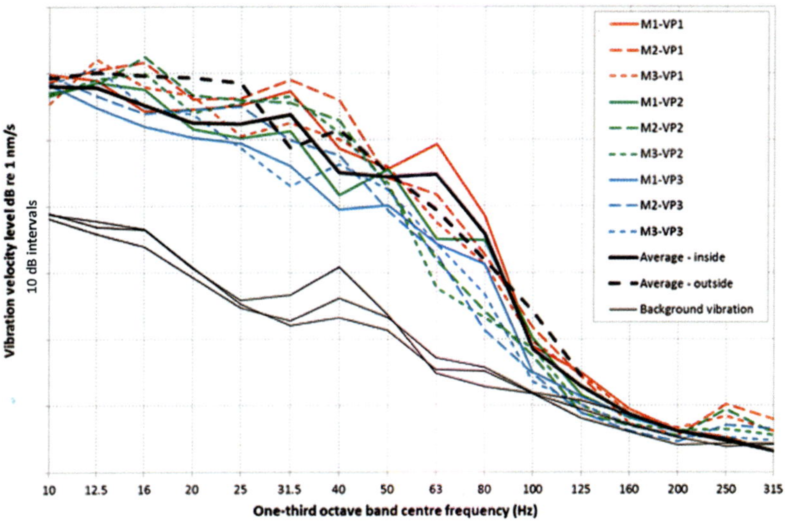

Fig. 4. Measured vibration inside and outside Studio 4

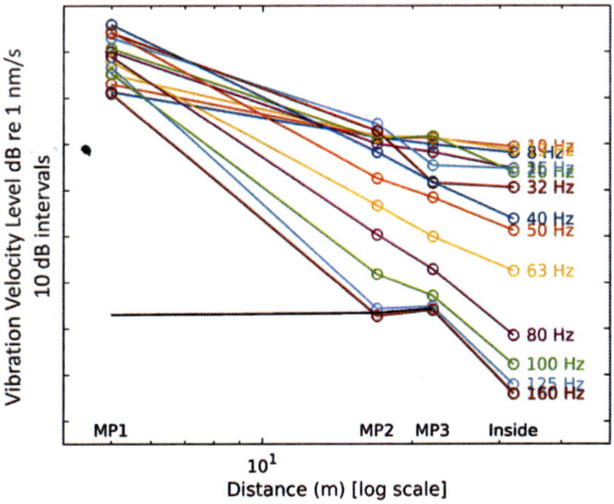

Fig. 5. Measured level difference - source to studio

3.3 Prediction of Sound Level from Floor Vibration

Figure 6 averages the sound level spectra and the studio floor vibration spectra (Fig. 4) in octave bands in order to establish a relationship between the sound level (L_p) and the floor vibration velocity level (L_v). Because of the masking from background levels, only data in the 31 Hz and 63 Hz bands provide a measure of the constant of proportionality between the sound and the vibration velocity. In this case this is approximately −27 dB and this is the relationship used in the subsequent predictions.

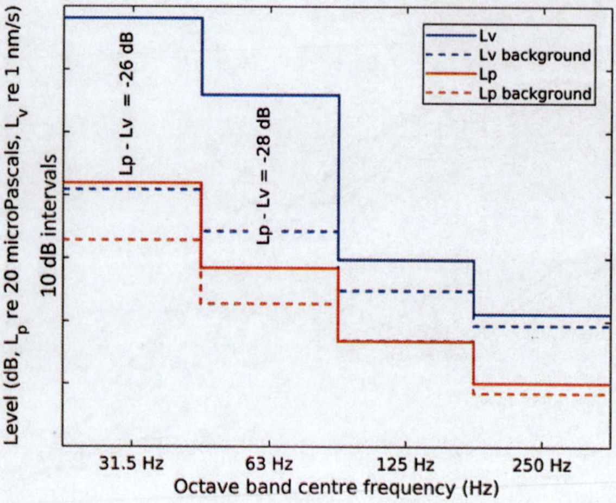

Fig. 6. Sound and floor vibration averaged in octave bands

4 Results

The predicted ground-borne noise (L_{Fmax}) in the studio from the passby of a 'Single' Metrolink tram travelling at the track design speed of 48 kmh^{-1} is presented in Fig. 7, together with the NR 25 curve. The highest level is in the 63 Hz octave frequency band where the tram-generated ground-borne noise is predicted to be 49 dB L_{Fmax} (approximately 6 dB lower than the NR 25 curve). In the other third octave frequency bands, it is more than 6 dB below the curve. Therefore the predicted L_{Fmax} from a 'Single' tram was not expected to exceed the NR 25 curve.

Higher levels of ground-borne noise inside the studio are expected if a 'Double' Metrolink trams passes (comprising four vehicles and six bogies); or two trams pass Studio 4 at the same time, travelling in opposite directions.

The main sources of uncertainty that affect the predicted sound level in the 63 Hz octave band are:

- The variation of low frequency noise inside the studio, due to acoustic modes and the long wavelengths involved, may lead to a variation of sound level despite the generally high degree of acoustic absorption in the studio. This variation was estimated to be between 3 and 7 dB based on measurements carried out by ITV.
- The source vibration used was measured from trams operating on the existing network with nominally good track. If higher rail roughness should develop on the track outside Studio 4 compared to the track on which the source vibration was measured, this may lead to higher levels of sound than have been estimated. This source of variation should however be controllable by rail head grinding should the noise, at any stage, increase to a level approaching the NR 25 limit.

These uncertainties resulted in a risk that the predicted levels could exceed the NR 25 limit in some operational cases.

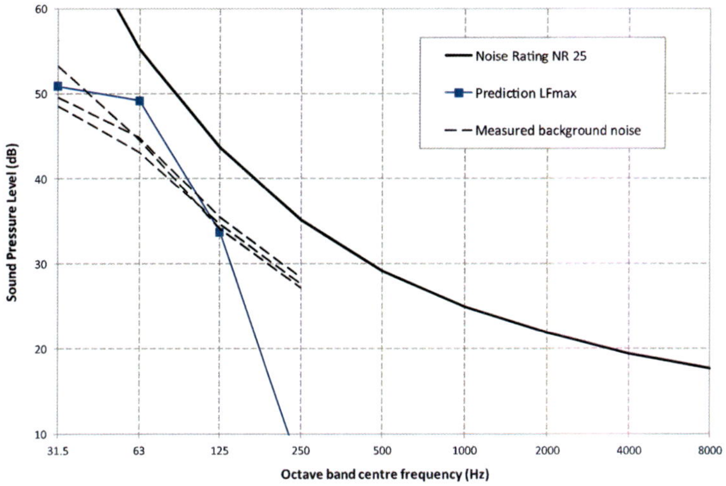

Fig. 7. Predicted L_{Fmax} octave-band sound levels compared with the NR 25 limit curve

5 Mitigation

To address the risk of exceeding the NR 25 limit, mitigation options were investigated, which included:

- Operational procedures (to prevent two trams passing at the same time);
- Speed restrictions; and
- Vibration isolating trackforms.

The project chose to implement a vibration-isolating trackform (a floating slab track) into the length of track outside the ITV studio. The design comprised a resilient mat (30 mm thickness, 0.01 N/mm^2 dynamic bedding modulus) under and to the sides of the TPL standard embedded rail track slab (420 mm depth) to form a system with a natural frequency of 15 Hz. With this design a vibration reduction of 10 dB was predicted at 63 Hz. Transitions of 15 m length were included between the floating and non-floating sections of track, with half mat thickness. Where required, ducting and connections to chambers within the slab were isolated with the same resilient mat.

Due to the additional low frequency noise radiated by the floating slab track compared to standard track, it was also necessary to consider the trade-off between the ground-borne noise in the studio and the airborne sound at the adjacent outdoor sets. As a result, the length of vibration isolating track was optimised to minimise these conflicting requirements.

6 Summary

Ground-borne noise from passing trams on the Trafford Park Line extension to the Manchester Metrolink system has been predicted within ITV's Studio 4 to assess against the relevant limit within the planning conditions.

Due to the distance between the track and the studio and the characteristics of the ground at this location, a vibrating trailer was used to quantify the entire propagation path from source to receiver. These propagation measurements were combined with vibration source data obtained from an existing line of the Metrolink system.

Using worst-case tram source vibration data, it is predicted that the likely ground-borne noise impact of a 'Single' (two vehicle) Metrolink tram passing at design line speed outside of the studio on Trafford Wharf Road would not exceed the required limit of Noise Rating curve NR 25 L_{Fmax}.

It was identified that there was a risk that the limit would be exceeded if:

- The uncertainties in the predictions were allowed for; and
- Adjustments were applied to account for the passby of a 'Double' tram and two trams passing at the same time while travelling in opposite directions.

The uncertainties included the variation of low frequency noise around the studio due to acoustic modes and the long wavelengths involved and the levels of rail head roughness.

Mitigation in the form of operational procedures to prevent two trams passing at the same time, speed restrictions, and vibration-isolating trackforms were investigated and their effectiveness quantified. The project chose to implement a vibration-isolating trackform (a floating slab track) into the length of track outside the ITV studio.

References

1. Town and Country Planning Act 1990: Transport for Greater Manchester (Light Rapid Transit System) (Trafford Park Extension) Order 2016, Department for Transport, reference TWA/14/APP/06 (2016)
2. British Standard BS 8233:2014 Guidance on sound insulation and noise reduction for buildings, British Standards Institution (2014)
3. Hanson, C.E., Towers, D.A., Meister, L.D.: Transit noise and vibration impact assessment. Report FTA-VA-90-1003-06, U.S. Department of Transportation, Federal Transit, Administration, Office of Planning and Environment (2006)

Wheel and Rail Noise

CRoNoS Railway Noise Prediction Tool: Description and Validation Based on Field Tests

Ainara Guiral[1](✉), Maddi Olazagoitia[2], and Egoitz Iturritxa[1]

[1] CAF R&D, N&V Area, Beasain, Spain
ainara.guiral@caf.net
[2] University of Navarra, San Sebastián-Donostia, Spain

Abstract. Railway noise results from the contribution of various sources, where rolling noise is considered to be dominant between 50 and 300 km/h. In the effort to reduce this noise source, accurate prediction tools are needed. CRoNoS rolling noise prediction tool takes advantage of the improvements carried out by the scientific community in further developing the theoretical models to derive the structural vibration and noise radiation of railway tracks and wheels.

A previous publication compared CRoNoS with respect to TWINS results in terms of sound power level. Same results were obtained provided that the models implemented on the tools compared were made equivalent.

This paper presents a comparison of CRoNoS predictions with field test pass-by noise results at two different speeds. The model is found to give reliable results. Spectral results analysis show that the absolute difference in each one-third octave band averaged over bands is less than 0.7 dB and all bands have differences of less than 4.6 dB. Global pass-by values are predicted within a margin of 1.8 dBA.

Keywords: Rolling noise · CRoNoS · Field test validation

1 Introduction

Railway noise results from the contribution of various sources, where rolling noise is considered to be dominant between 50 and 300 km/h. In the effort to reduce this noise source, accurate prediction tools are needed so that improved designs can be proposed and used to support vehicle homologation.

As a result of 10 years of research into rolling noise CRoNoS was born. CRoNoS stands for CAF Rolling Noise Software and is the rolling noise prediction tool developed by CAF S.A. It takes advantage of the improvements carried out by the scientific community in further developing the theoretical models in order to derive the structural vibration and noise radiation of railway tracks and wheels. The first results of the tool have already been given in [1], where CRoNoS and TWINS results in terms of sound power level were presented. In the same reference, a second set of computations was made to analyse the impact of using a more accurate theoretical approach by modelling a whole wheelset instead of a standalone wheel. Significant spectral differences were found.

G. Degrande et al. (Eds.): *Noise and Vibration Mitigation for Rail Transportation Systems*, NNFM 150, pp. 581–588, 2021.
https://doi.org/10.1007/978-3-030-70289-2_63

The aim of this paper is to present a full scale field test representing a validation example of CRoNoS. To this end, pass-by noise results measured at different speeds are compared with predictions from the tool.

Rolling noise predictions require a lot of input information including the wheel and rail roughness spectra, track configuration and its component characteristics, track emission properties, wheel vibration and emission properties and parameters associated with the outdoor sound propagation. The accuracy of the prediction results depends to a large extent on the quality of these input parameters. A methodology is used that allows obtaining the possible input parameters, as accurately as possible, from typical field pass-by noise measurements carried out in the commissioning phase of projects involving the manufacturing of new rolling-stock.

This paper is structured as follows: Sect. 2 describes the theoretical models implemented in CRoNoS as well as its main features. The modelling of the test scenario is described in Sect. 3 with special focus on the treatment of wheel and rail roughness and the wheel and track characterization. In Sect. 4 a comparison between experimental pass-by noise results and predictions is presented. Finally, in Sect. 5, some conclusive remarks are made.

2 Rolling Noise Prediction Model

Figure 1 shows a schematic diagram of the theoretical model implemented in CRoNoS. Each box represents different modules encompassed in one of the following three sub-models: structural vibration, sound radiation and sound propagation.

The **structural vibration sub-model** solves the wheel-rail contact mechanics the main inputs of this being the wheel and rail receptance and roughness data. The outputs obtained are the contact forces and the wheel, rail and sleeper vibrations.

- The wheelset model solves the dynamic response of a flexible wheelset taking into account the inertial and gyroscopic effects due to the main direction of rotation [2]. It enables two modelling options, i.e. wheelset or standalone wheels. In this work the wheelset model is selected.
- The track model enables two additional modelling options for ballasted tracks – continuously or periodically supported – with either bibloc or monobloc sleepers. In this study, the rail vibration has been calculated using a Timoshenko beam model on a continuous double layer support to model pad and ballast. The monobloc sleeper has been modelled using an equivalent uniform Timoshenko beam [3].

The **sound radiation sub-model** calculates the sound radiation in terms of sound power by combining predicted vibration spectra with radiation efficiencies. Wheel radiation efficiencies are taken from reference [4], in which simple formulae are derived from boundary element calculations. Rail and sleeper radiation efficiencies implemented in CRoNoS include the presence of a reflective ground and the effect of the ballast vibration, in contrast to the commonly used free space rail and rigid ground embedded sleepers radiation efficiencies (see [5] for impact on rail radiation efficiency).

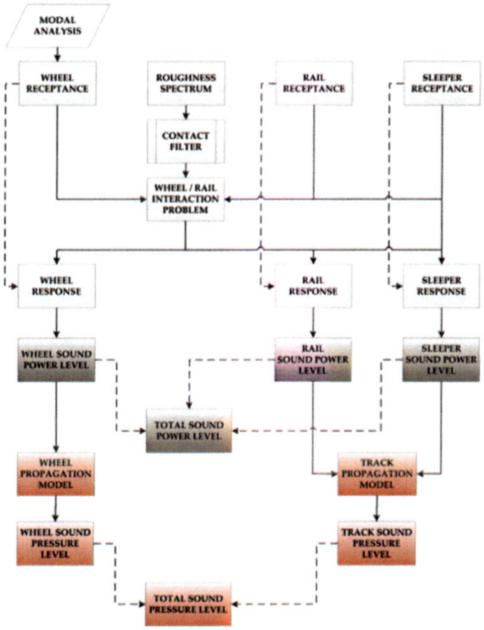

Fig. 1. Schematical diagram of the model implemented in CRoNoS for rolling noise prediction.

Pass-by noise requirements are set in terms of sound pressure level at a certain distance from the track. Therefore CRoNoS integrates a **sound outdoor propagation sub-model**. It accounts for reflections from the partially absorbing ground considering the phase relationship between direct and reflected rays as well as the Doppler effect due to train speed.

3 Modelling the Test Scenario

Exterior noise measurements were performed on an electric regional vehicle with four coaches and Jacobs bogies at two different speeds; 80 and 160 km/h. The measurement position was located at 7.5 m from the centreline of the track 1.2 m above the surface of the rail as required by ISO3095:2013 [6].

3.1 Wheel and Rail Roughness Measurements

Rail roughness was measured using the APT – Rail surface analyser (RSA) trolley [7] shown in Fig. 2 (left). Measurements were made over 3 parallel lines and a total length of 100 m on both rails.

Wheel roughness was measured for all twenty wheels by using the APT – Wheel surface analyser (WSA) [8] shown in Fig. 2 (right). The wheel roughness was measured by means of displacement sensors on 3 parallel lines. For each wheel, a minimum of 8 revolutions of the wheel were measured. The wavelength spectrum is calculated for each revolution and then the energetically averaged spectrum is calculated.

Fig. 2. APT (left) Rail surface analyser (RSA) and (right) Wheel surface analyser (WSA).

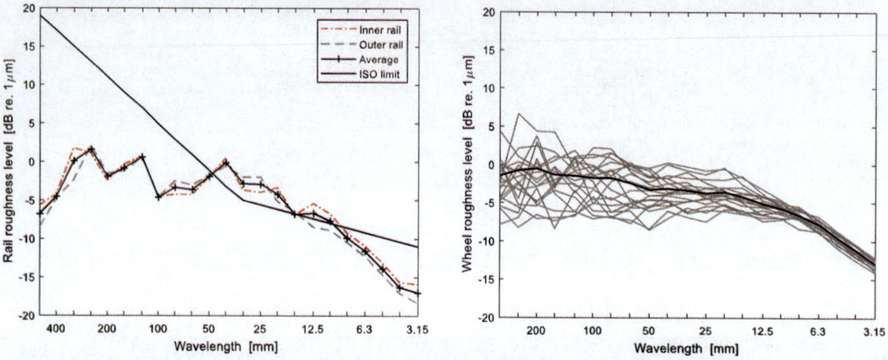

Fig. 3. Rail (left) and wheel (right) measured roughness level.

Figure 3 (left) shows the inner and outer rails roughness spectra compared to the limit proposed by standard ISO 3095:2013 [6]. Figure 3(right) shows the average spectrum of each of the wheels compared to their average spectrum.

3.2 Wheelset Modelling

The mode-shapes and natural frequencies were derived from a modal analysis calculation on a commercial Finite Element Method (FEM) software package. Wheelset modes of vibration are a superposition of stand-alone wheel modes – axial, radial and torsional – combined with axle modes – bending, torsional and axial/compressional. As shown in Fig. 4, one-nodal-diameter wheel modes are usually combined with axle bending modes, while zero-nodal diameter wheel modes combine with axial and torsional axle modes.

Fig. 4. Wheel (left) one-nodal and (right) two-nodal diameter axial mode.

On the contrary, wheel modes with higher number of nodal diameters present negligible axle deformation. For those modes, the wheels vibrate independently; thereby the natural frequency coincides with that of the single wheel.

The wheelset modal description was completed by experimentally deducing the modal damping ratios from a modal analysis carried out in a laboratory.

3.3 Track Modelling

Pad and ballast stiffness and the decay of vibration with distance along the track are determined through measured data. The pad stiffness is adjusted by fitting the predicted and measured Track Decay Rate (TDR) using nonlinear least squares method. Measured vertical and lateral TDR results are shown in Fig. 5 for both rails together with the averaged curve and TSI limits.

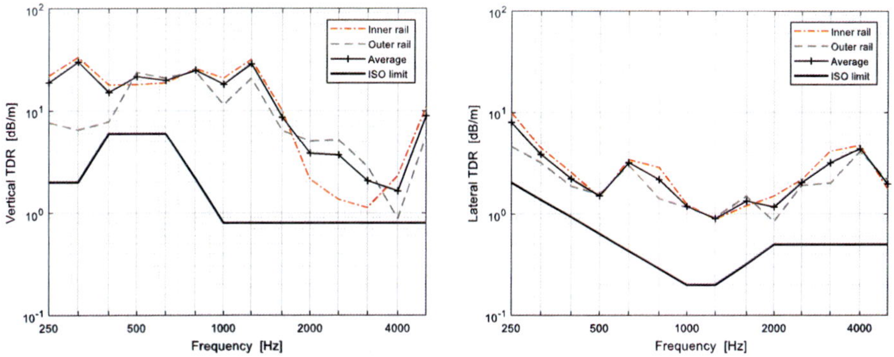

Fig. 5. Measured track decay rate (left) in the vertical and (right) lateral directions.

The noise radiated by the track has been calculated using the measured TDR instead of the predicted one as this procedure has proven to give more reliable results. The ballast stiffness has been defined as a frequency-dependent experimentally measured curve. The rest of the parameters are shown in Table 1.

Table 1. Ballast-track parameters.

Rail type	UIC60	Sleeper type	Monobloc
Rail shear coefficient	0.4	Sleeper spacing	0.6 m
Rail loss factor	0.02	Sleeper mass	300 kg
Vertical pad stiffness	2200 MN/m	Vertical ballast stiffness	Measured
Vertical pad loss factor	0.2	Vertical ballast loss factor	1
Lateral pad stiffness	70 MN/m	Lateral ballast stiffness	100 MN/m
Lateral pad loss factor	0.2	Lateral ballast loss factor	2

3.4 Uncertainties Due to Measurements Error

Figure 6 (left) shows the averaged combined wheel/rail roughness. This is derived by averaging the 40 combined wheel/rail roughness spectra resulting from combining the energetically averaged measurement results of both rails and the measurements of each of the 20 wheels. The shadowed zone indicates the uncertainty due to the roughness measurements. This is calculated by subtracting the averaged combined wheel/rail roughness spectra from each of the 40 combined spectra and calculating the standard deviation. Long wavelengths, above 63 mm, show deviations of 2–3 dB, whereas at short wavelengths the deviations are limited to around 0.5–1 dB.

The TDR is subjected to the same analysis. Figure 6 (right) shows the averaged vertical and lateral decay rates derived from measurements on both inner and outer rails. Results show higher standard deviations at low frequencies and at frequencies above 2000 Hz, were differences of around 10 dB/m are found for the vertical direction.

Additionally, pass-by noise measurement itself has a 2 dB uncertainty. Consequently, predictions can be considered to be validated as long as the difference between measured and predicted levels remains within this range.

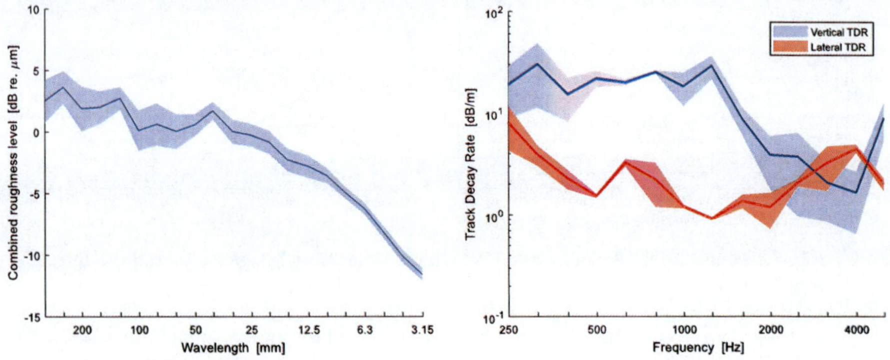

Fig. 6. Averaged (left) combined wheel/rail roughness spectrum and (right) TDR.

4 Application

A total of six pass-bys were recorded; three at 80 km/h and three at 160 km/h. The A-weighted equivalent pass-by time sound pressure level $L_{pAeq,Tp}$ was measured as per [6] and the averaged one-third octave band spectra was computed for each speed.

Figure 7 shows the contribution of wheel, rail and sleeper radiation to the total pass-by noise at 80 km/h. At low frequencies, i.e. up to approximately 250 Hz, sleepers are the dominant source. In this frequency range the sleeper and rail vibrate in phase. The higher the pad stiffness is, the higher the decoupling frequency will be. For the given test scenario, the sleeper maximum contribution is at a 630 Hz. In the mid frequency range the rail becomes the dominant source. The vertical rail component usually predominates over the lateral one. However, given the track stiffness in the vertical direction the lateral rail movement becomes dominant in the largest frequency range. The wheel is

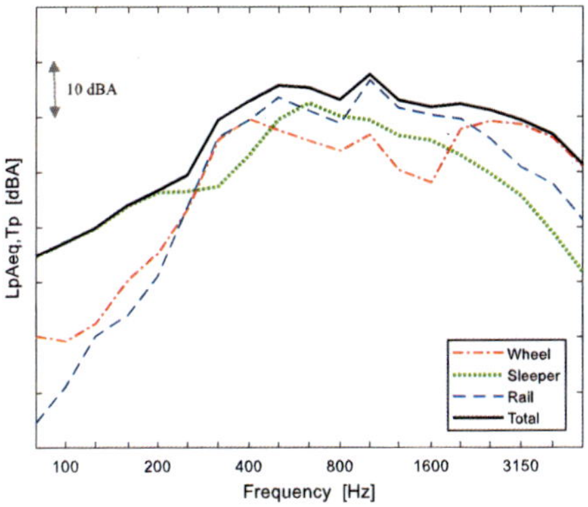

Fig. 7. Predicted pass-by noise at 80 km/h.

the predominant source above 2 kHz. In terms of global value the track is the dominant source.

Predicted and measured pass-by noise spectral results at 80 km/h are shown in Fig. 8 (left). The predicted results match the reference results well. The largest discrepancies are found in the mid-frequency range, where the rail is the main contributor. A way to improve the accuracy would have been to tune the track model through track vibration measurements. In contrast, the goal here is to show the accuracy to be expected when best estimates of the most important track parameters are used.

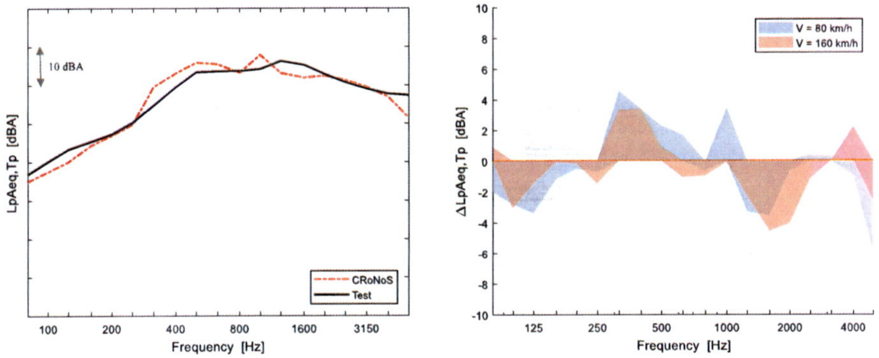

Fig. 8. CRoNoS vs measured spectrum: (left) at 80 km/h, (right) difference at 80 and 160 km/h.

Results are given in Fig. 8 (right) as the difference between predicted and measured spectra for 80 and 160 km/h pass-by speeds. For the 80 km/h pass-by, the absolute difference in each one-third octave band, averaged over bands, is 0.5 dB and all bands have differences of less than 4.5 dB. For the 160 km/h run, the absolute difference in each one-third octave band, averaged over bands, is 0.7 dB and all bands have differences of less than 4.6 dB. In terms of overall pass-by noise levels, the predicted value at 80 km/h agrees within a margin of less than 0.4 dBA while the predicted value at 160 km/h differs by 1.8 dBA.

5 Conclusions

A validation example of the CRoNoS model for a real test scenario at two different speeds has been successfully carried out.

The model is found to give reliable results. Spectral analysis shows that the absolute difference in each one-third octave band averaged over bands is less than 0.7 dB and all bands have differences of less than 4.6 dB. Global pass-by values are predicted within a margin of 1.8 dBA. Accuracy could have been further improved by tuning the track model through track vibration measurements. In contrast, the goal is to show the precision to be expected when best estimates of the available track parameters for certification pass-by tests are used.

References

1. Guiral, A., Iturritxa, E., Alonso, A.: CRoNoS railway rolling noise prediction tool: Wheelset model assessment. In: Euronoise 2018 (2018)
2. Guiral, A., Alonso, A., Giménez, J.G.: Vehicle–track interaction at high frequencies – modelling of a flexible rotating wheelset in non-inertial reference frames. J. Sound Vib. **355**, 284–304 (2015)
3. Grassie, S.L., Cox, S.J.: The dynamic response of railway track with flexible sleepers to high frequency vertical excitation. Proc. Inst. Mech. Eng. **198**(2), 117–124 (1984)
4. Thompson, D.J., Jones, C.J.C.: Sound radiation from a vibrating railway wheel. J. Sound Vib. **253**(2), 401–419 (2002)
5. Zhang, X., Jeong, H., Thompson, D., Squicciarini, G.: The noise radiated by ballasted and slab tracks. Appl. Acoust. **151**, 193–205 (2019)
6. ISO 3095:2013 Acoustics — Railway applications — Measurement of noise emitted by railbound vehicles (2013)
7. APT: APT-RSA Rail Surface Analyzer
8. APT: APT-WSA Wheel Surface Analyzer

Scale and Numerical Modelling of a Metro Rail Viaduct with Sound Absorption to Reduce In-Car and Wayside Noise

Conrad Weber(✉), Terrence Wong, and Benjamin Panarodvong

Renzo Tonin & Associates, PO Box 877, Strawberry Hills, NSW, Australia
conrad.weber@renzotonin.com.au

Abstract. A scale model of a 2-car metro train on a slab track rail viaduct was employed to investigate the reduction of in-car and wayside noise when acoustic treatments were added to the inside face of the edge noise barrier and within the four-foot. The performance of the acoustic treatments was evaluated in full octaves for the full-scale frequency range of 31.5 Hz to 2 kHz. The scale model results were compared with general rail noise modelling prediction formulae and with numerical modelling results using ODEON. The model results confirmed that the interior noise level experienced by passengers is likely to increase with the addition of viaduct edge noise barriers and that wayside noise levels are reduced due to the noise barrier effect. With the addition of acoustic panels on the inside face of the viaduct edge barrier and within the four-foot, in-car and wayside noise levels reduce.

Keywords: Scale model · Sound absorption · ODEON · Viaduct · In-car · Barrier

1 Introduction

In many major cities, the introduction of a new rail system is an efficient way to reduce road traffic congestion and move large numbers of passengers between places of work, recreation and home. However, train operations generate airborne noise, which can cause nuisance to wayside receivers and affect the comfort of train passengers.

This study examines metro train operations on a rail viaduct with ballastless track. Many studies have identified that the noise emission of at-grade and elevated slab track is approximately 2 dB(A) to 5 dB(A) louder than an equivalent ballasted structure [1–3]. The noise increase on elevated structures can be towards the higher end of this range or greater because of the loss of the ground effect [3].

Higher noise levels occur due to additional rail-radiated noise (attributed to the reduced track decay associated with soft rail fasteners) and increased reflections (otherwise absorbed by the ballast).

Many elevated rail systems include noise barriers on the edge of the viaduct to reduce wayside noise at sensitive buildings. However, high noise barriers have negative

G. Degrande et al. (Eds.): *Noise and Vibration Mitigation for Rail Transportation Systems*, NNFM 150, pp. 589–596, 2021.
https://doi.org/10.1007/978-3-030-70289-2_64

visual impacts, can obstruct the view of passengers and can be costly due to increased engineering requirements to withstand wind loads.

The introduction of a reflective noise barrier (without absorption) close to the train results in multiple reflections between the train and barrier (see Fig. 1 from [3]). For a reflective noise barrier close to the train, the performance is reduced at wayside receivers and in-car noise levels are higher. In both cases, this is due to an increase in reverberant noise levels in the area between the train and the noise barrier due to a lack of sound absorption. If acoustic panels are placed between the rails (four-foot) and/or on the inside face of the noise barrier, reflections are minimized, resulting in an improved noise barrier performance and reduced in-car noise levels.

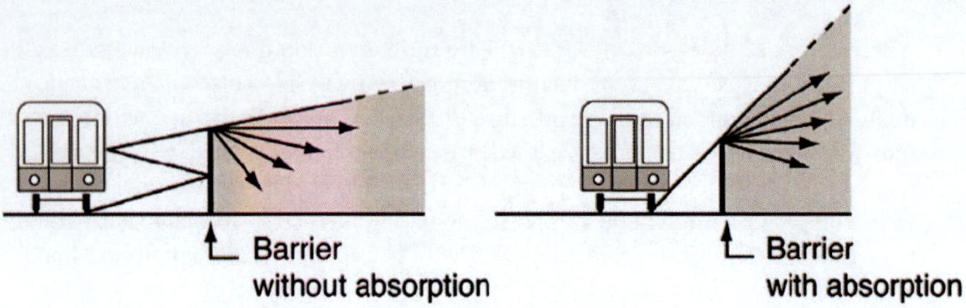

Fig. 1. Noise barrier geometry and effect of close noise barriers without absorption (left) and with absorption (right) from [3]

It follows that the potential noise reduction associated with placing acoustic panels on the inside face of the edge noise barrier and within the four-foot is dependent on variables including the height of the noise barrier, the distance between the train and the barrier and the sound absorption properties of the acoustic materials.

In this paper, a scale model of a 2-car metro train on a slab track rail viaduct is utilized to investigate the reduction of wayside and in-car noise for a specific geometry (see Sect. 3.1). The use of scale modelling allows the 3-dimensional geometry of the train and track system to be examined with the improved accuracy required for detailed design purposes.

Numerical modelling was performed using the ODEON software package to examine the noise reduction with the same geometry and sound absorption treatments. The numerical modelling results are compared with the scale model results and with general rail noise modelling prediction formulae.

The addition of rail dampers to reduce the track decay and associated rail-radiated noise is acknowledged as a cost-effective mitigation measure. The noise benefits of rail dampers are well documented elsewhere and are not discussed further herein.

2 Railway Noise Sources and Spectrum

Most metro systems in major cities operate at speeds up to 100 km/h. At this speed, rolling noise at the wheel-rail interface is the dominant noise source, together with motor and gearbox noise (located in the bogie area of motor cars). Noise from train ancillary equipment and aerodynamic noise is not normally significant at this speed and is not considered further.

To determine the changes in overall dB(A) noise levels, a representative noise spectrum of a train passby is required. The spectrum is based on wayside noise measurements of a metro train operating at 100 km/h on ballast track, with rail roughness and track decay rates (TDR) compliant with [4]. The spectrum in Fig. 2 was derived from at-grade measurements performed at 15 m from the near track centerline.

The study assumes that rail dampers have been applied to the rails on the slab track viaduct to achieve the TDR requirements. The spectrum in Fig. 2 is therefore representative of the rolling and propulsion noise for the base case modelling scenario.

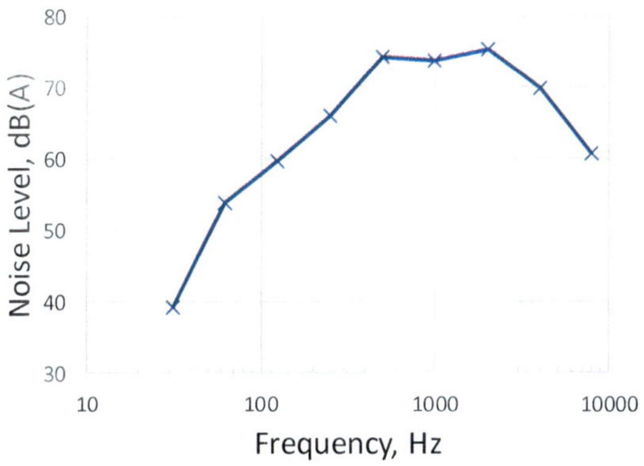

Fig. 2. Reference $L_{pAeq,Tp}$ noise spectrum of representative metro train at 100 km/h and measurement distance of 15 m.

3 Scale Modelling

3.1 Description of Scale Modelling and Test Setup

In acoustic scale modelling, the complex behavior of sound, including reflections, absorption and distance propagation can be reproduced in miniature [5]. For a 1:10 scale model and with the speed of sound in air constant, full-scale frequencies are 10 times higher than the scale model frequencies. The absorption coefficients of the materials used in the model should respond similarly for frequencies 10 times higher than the actual values [6].

The scale model comprised two cars (each 22 m long full-scale) and the equivalent length of viaduct. The train and viaduct were constructed from Melamine MDF board, considered representative of the full-scale materials (i.e. glass and steel for the train body and concrete for the viaduct). Noise barriers on the viaduct edge were positioned 120 mm (1.2 m full-scale) from the near side of the train. The top of the barrier was located 120 mm (1.2 m full-scale) above rail level.

Polyester-based acoustic panels (8 mm thick) were used in the scale model (height of 90 mm for the inside face of the noise barrier and width of 90 mm for the four-foot area) to mimic full-scale panels. The scaled noise reduction coefficient (NRC) of the 8 mm thick panels is similar to typical full-scale acoustic panels with NRC of 0.9. A scale model reverberation room could be used to fine tune the selection of scale model absorption materials [5, 7], but was not available for this study.

Four driver unit speakers (Type TU-631) were used as the sound source. The speakers were placed inside the two cars. Holes were drilled in the bottom of the cars in the virtual position of the bogies, so that the speakers could be directed downwards. The cars were raised 80 mm (800 mm full-scale) above the surface of the viaduct using threaded rods.

Pink noise was used to replicate wheel-rail and motor-gearbox noise at the four bogie positions. Two ½ inch microphones (Class 1) were used for the measurements. One microphone was placed on the outboard side of the train above the bogie area, at a virtual train door position. This location was selected to investigate changes in in-car noise levels. The other microphone was placed at a wayside position 1,500 mm (15 m full-scale distance) from the track centerline, level with the top of the 120 mm high barrier (1.2 m full-scale) and 800 mm (8 m full-scale) above ground level. This position represents an upper floor receiver of a nearby sensitive building.

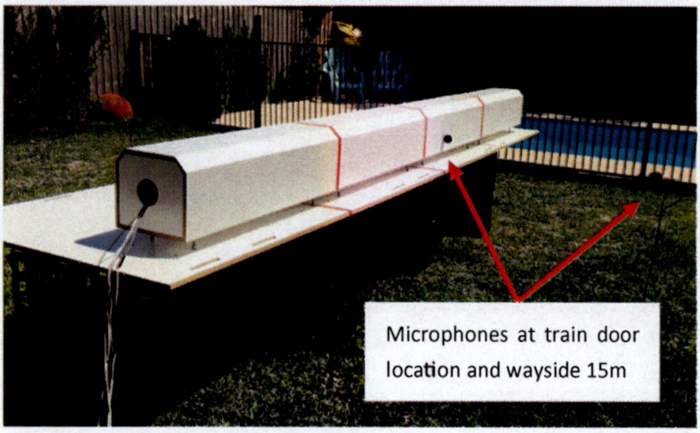

Fig. 3. Picture of scale model test outdoors showing 2-car train on viaduct (without barriers or absorption – base case) and position of wayside and train door microphones

At the train door position, the signal-to-noise ratio was greater than 10 dB over the full frequency range. At the wayside location, there was insufficient signal to obtain valid measurements in the 2 kHz octave frequency band (full-scale). A spark noise source and high-speed data acquisition system are often used in acoustic scale models to generate and measure sound energy above 20 kHz. For this study, the acquisition rate was 48 kHz which limits the upper frequency to 2 kHz (full-scale). Noise levels in the 4 kHz octave band and 2 kHz octave band (for wayside location) were extrapolated from the measurement results to obtain overall A-weighted noise levels.

Testing was performed outdoors in a quiet residential area. In scale modelling, high humidity values increase air absorption values. As the current study is concerned with the relative change in noise levels (i.e. with and without barriers and acoustic panels), air absorption values remain constant and humidity corrections are not required. A picture of the scale model, taken during base case noise testing, is provided in Fig. 3.

3.2 Test Cases and Results

Five test cases were examined as follows:

- Case 1 (base case) – viaduct without noise barrier and no sound absorption
- Case 2 – with noise barrier and no sound absorption
- Case 3 – with noise barrier and sound absorption on the inside face of the barrier
- Case 4 – with noise barrier and sound absorption within the four-foot
- Case 5 – with noise barrier and sound absorption on the inside face of the barrier and within the four-foot.

Data was averaged and extraneous noise events were excluded. The performance of the acoustic treatments was evaluated in full octaves for the scale model frequency range 315 Hz to 20 kHz (full-scale frequency range of 31.5 Hz to 2 kHz). The test results are summarized in Table 1.

Table 1. Scale and numerical model test results

Test case	Noise level change from base case, dB(A)			
	Train door position		15 m from track centerline	
	Scale model	Numerical model	Scale model	Numerical model
Case 1	Base case	Base case	Base case	Base case
Case 2	4.6	6.8	−5.7	−8.3
Case 3	3.7	3.6	−9.0	−14.1
Case 4	1.4	5.0	−8.3	−10.3
Case 5	−1.6	1.8	−12.2	−16.9

3.3 Discussion of Scale Model Results

At the train door measurement position, noise levels increased by 4.6 dB(A) due to the presence of the noise barrier (Case 2 minus base case). It follows that in-car noise levels would be higher in this area of the train. In practice, other noise paths including the gangway, structure-radiated noise and air-conditioning noise influence the overall in-car noise levels. The total increase would be somewhat less than 4.6 dB(A).

With the addition of acoustic panels on the inside face of the barrier and the four-foot (Cases 3 to 5), noise levels at the train door position are reduced. For Case 5 (acoustic panels in the four-foot and inside face of the barrier), the measured noise level at the train door position was lower than for the base case.

At the wayside measurement position, noise levels reduced by 5.7 dB(A) due to the presence of the noise barrier (Case 2 minus base case). With the addition of acoustic panels on the inside face of the barrier and the four-foot (Cases 3 to 5), noise levels at the wayside measurement position were further reduced. For Case 5, the noise barrier insertion loss increased from 5.7 dB(A) to 12.2 dB(A) with the addition of acoustic panels at both locations.

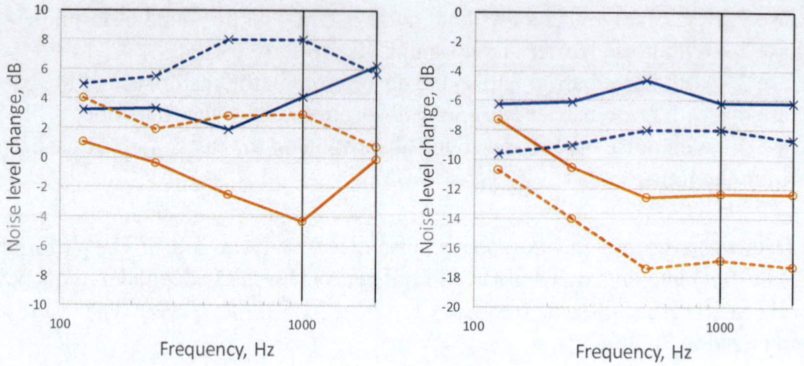

Fig. 4. Change in octave band noise levels (left = train door and right = wayside), x—x scale model Case 2, o—o scale model Case 5, x - - x numerical model Case 2, o - - o numerical model Case 5.

4 Numerical Modelling in ODEON

The room acoustic software ODEON (Version 14.04) was used to perform a numerical analysis of the noise reduction with and without absorption for the test cases described in Sect. 3.2.

ODEON is an energy-based room acoustics modelling package which uses a combination of the image source method, ray tracing and ray radiosity to model sound propagation in large spaces, including the effects of sound absorption, scattering and diffraction. A full-scale 3D model of the test case was created and imported into the software. The model was enclosed with a 100% absorptive box which enables the software to perform an outdoor simulation.

Noise sources were defined in ODEON for two different paths: a line source was used to simulate rail-radiated noise; and point sources were used to model wheel and motor-gearbox noise. Receivers were placed in the model at positions replicating the scale model positions. Absorption coefficients were assigned to each material.

The modelling output provided octave band noise levels for each test case 125 Hz to 4 kHz.

4.1 Discussion of Numerical Model Results

The numerical modelling results in Table 1 identify similar findings as the scale model testing, namely: in-car noise levels increase when a noise barrier is introduced and wayside noise levels decrease; noise levels decrease at both locations with the addition of sound absorption in the four-foot and inside face of the barrier; and noise levels reduce further as more absorption is added.

At the train door position, the numerical model predicts higher noise levels for Case 2 compared with the scale model, and more noise reduction when acoustic panels are added to the inside face of the barrier. This is likely related to the different orientations of the noise sources. In the scale model, the speakers are directed towards the four-foot panels, but in the numerical model, the noise sources are positioned closer to the noise barrier and the orientation is more favorable for generating and absorbing reflections from the barrier.

At the wayside position, the numerical model predicts higher noise barrier reductions compared with the scale model for all test cases. The noise barrier panels are predicted to provide higher noise reductions than for the four-foot panels (the scale model results indicate otherwise for the reason stated above).

Comparisons of the octave band noise differences for the Case 2 and Case 5 reductions (compared with base case) are provided in Fig. 4. Whilst the test results provide similar trends, large differences occur in each octave band. At both measurement locations, the noise difference with and without acoustic panels generally increases with increasing frequency. This trend is in-line with the scale model and real-life absorption materials, which have higher sound absorption coefficients at higher frequencies.

5 Comparison of Noise Barrier Loss with General Rail Noise Formulae

The scale and numerical modelling results are compared with general rail noise barrier formulae for reflective and absorptive barriers in Table 2. The same geometry is used for each scenario. The formulae for reflective and absorptive rail noise barriers for '*Calculation of Rail Traffic Noise (CORTN)*' and '*Nordic 94*' models are based on [8].

The results show that there is a large variation between the calculated noise reductions for reflective and absorptive barriers and that the scale model results are slightly conservative compared with the other three models. The scale and numerical models permit more detailed octave band analysis, whereas the general modelling algorithms are based on overall A-weighed noise levels.

Table 2. Comparison of noise barrier insertion loss values

Test Case	Noise barrier reduction, dB(A)			
	Scale model	Numerical model	CORTN	Nordic 84
Case 2 reflective	−5.7	−8.3	−10.2	−5.9
Case 3 absorptive	−9.0	−14.1	−12.2	−12.4

6 Conclusions

The results indicate that scale modelling is a useful and valid method for evaluating the change in wayside and in-car noise levels for a specific geometry and quantifying the noise reduction from acoustic panels placed in the four-foot and on the inside face of the viaduct edge barrier. A limitation of the 1:10 scale model was the upper frequency limit of 2 kHz. The scale modelling results could be improved by raising the upper frequency limit with an ultrasonic source and higher sampling rates, and by using a scale model reverberation room to fine tune the selection of scale model absorption materials.

Whilst similar trends were observed for numerical modelling with ODEON, large differences exist in each octave band and overall dB(A) reductions differed.

The scale and numerical models confirm that in-car noise levels are likely to increase when a reflective noise barrier is introduced in proximity to the train and that acoustic panels are effective in reducing in-car and wayside noise levels when placed in the four-foot and on the inside face of the noise barrier.

The scale model results were slightly conservative compared with the numerical model and general rail noise modelling algorithms.

Acknowledgements. • University of NSW undergraduate students Hein and Thang for assistance building the scale models and providing preliminary test results.
• Timothy Braunstein from Renzo Tonin & Associates for assistance with field testing.

References

1. Thompson, D., Jones, C.: Noise and vibration from railway vehicles, chap. 10. In: Handbook of Rail Vehicle Dynamics, pp. 279–325. Taylor & Francis Group (2006)
2. Glickman, G. and Bensing, S.: An examination of trackbed sound absorptive panels for minimizing wayside noise from rail transit. In: Proceedings of 25th National Conference on Noise Control Engineering 2011, Noise-Con 11 (2011)
3. Quagliata, A., et al.: For the Federal Transit Administration Office of Planning and Environment, Transit Noise and Vibration Impact Assessment, FTA Report No. 0123 (2018)
4. ISO (the International Organization for Standardization): ISO3095:2013 Acoustics - Railway applications - Measurement of noise emitted by railbound vehicles (2013)
5. Barron, M.: Acoustic scale model testing over 21 years. Acoust. Bull. **22**(3), 5–12 (1997)
6. Eroz, F.: Use of scale modeling for architectural acoustic measurements. Gazi Univ. J. Sci. Part B Art Humanit. Des. Plann. (2013)
7. Rindel, J.: Room acoustic modelling techniques: a comparison of a scale model and a computer model for a new opera theatre. In: Proceedings of the International Symposium on Room Acoustics (2010)
8. SoundPLAN GmbH: SoundPLAN User Manual, chap. 6.

A Comparison of Rolling Noise from Different Tram Tracks

Wenjing Sun[1(✉)], David J. Thompson[2], Martin Toward[2], Marcus Wiseman[2], and Evangelos Ntotsios[2]

[1] Institute of Rail Transit, Tongji University, Shanghai 201804, People's Republic of China
sunwenjing19@gmail.com
[2] Institute of Sound and Vibration Research, University of Southampton, Southampton SO17 1BJ, UK

Abstract. A number of different track forms are studied from one tramway system including tracks in ballast, on slab, and embedded rail in grass. Measurements have been carried out of rail roughness, wheel roughness and track decay rate, which are used as input data for the noise prediction model TWINS. The rail vibration and exterior noise of trams in operation have also been measured for comparison with the prediction model. The model is used to identify the relative contributions of noise radiated by the track and the wheels. The differences between different track forms are determined and the reasons for them are identified.

Keywords: Tramways · Rolling noise · Track decay rate · Rail roughness · Embedded rail

1 Introduction

New tramways are being developed to overcome traffic congestion in urban areas all over the world. They can have both conventional open tracks and paved-in tracks for street-running, see Fig. 1. Even though the speed is relatively low, the rolling noise can be quite significant. The rail roughness is often very high and corrugation appears prevalent [1, 2].

Fig. 1. Tramway track forms: ballasted track, slab track, 'grass track'.

G. Degrande et al. (Eds.): *Noise and Vibration Mitigation for Rail Transportation Systems*, NNFM 150, pp. 597–604, 2021.
https://doi.org/10.1007/978-3-030-70289-2_65

To apply theoretical models of rolling noise to tramways, attention should be given to particular aspects of the tramway system. Resilient wheels are often used, which are generally quieter than mono-block steel wheels, but they can also change the sound radiation from the track. The track decay rate and noise radiation of embedded tracks are different from conventional railways. Due to the low speed, the results are more sensitive to the contact filter [3]; this therefore requires careful attention. In this paper, the rolling noise prediction model TWINS [4] is adapted for application to tramways and predictions are compared with measurements.

2 Track Models

A range of different tramway track forms is studied from one tram system in Europe, including tracks in ballast, on slab and embedded rail in grass.

2.1 Ballast and Slab Track Model

Figure 2 shows the open tram tracks with ballast and slab considered here. The Vignole rail type S49 (49 kg/m) is used for both track forms. Mono-block sleepers are used for the track in ballast and 'booted' sleeper blocks are used for the slab track. In both cases the track is modelled as discretely supported with two layers of support. The acoustic effects of the ballast and rigid slab in proximity to the rail are included in the prediction models [5].

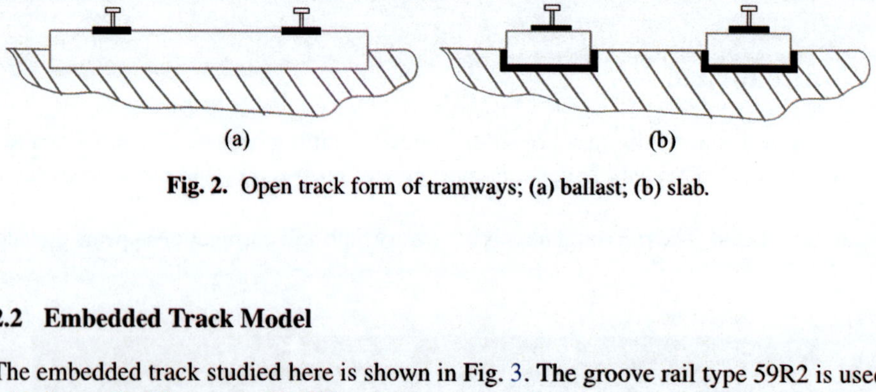

Fig. 2. Open track form of tramways; (a) ballast; (b) slab.

2.2 Embedded Track Model

The embedded track studied here is shown in Fig. 3. The groove rail type 59R2 is used with the embedding material filling between the rail and concrete. The rail is coated with the embedding material and then cast into the concrete. The free surface of the rail head and the filling can both radiate noise. The filling has an influence on both the rail vibration and radiation. Due to the absorptive effect of the grass around the track, this can reduce the sound pressure level at the receiver.

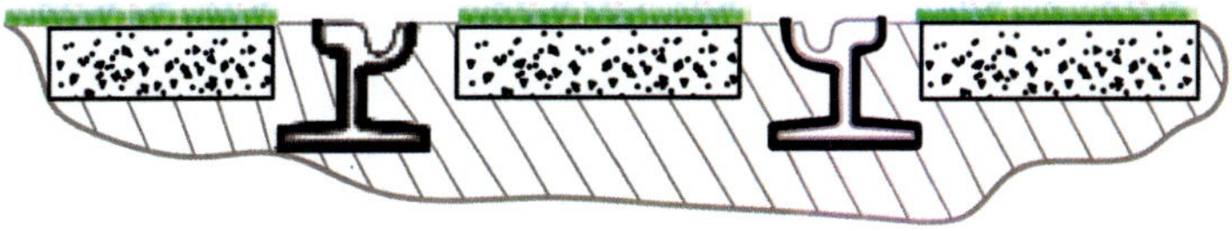

Fig. 3. Embedded track form of tramway: grass.

3 Analysis of Roughness and Track Decay Rate Properties

3.1 Track Decay Rate

Track decay rate measurements have been undertaken on the various kinds of track. The vertical track decay rates of the three different types of tram track are compared in Fig. 4(a). The limit curve from ISO 3095:2013 [6] is also shown for comparison. The ballast track, which is fitted with relatively stiff rail pads, has the highest decay rate over the frequency range between 300 Hz and 1250 Hz. Due to the softer support stiffness,

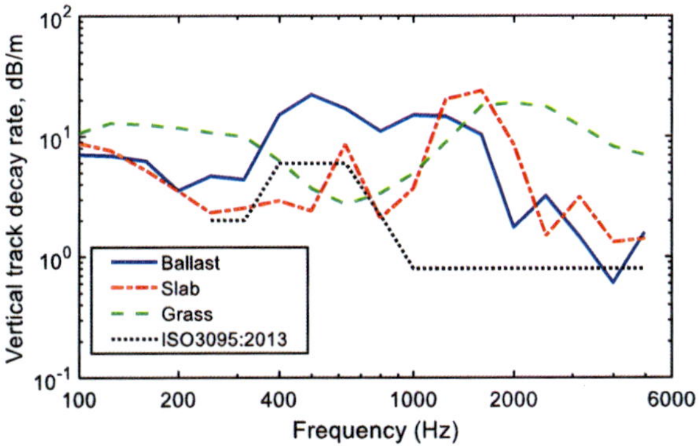

(a)

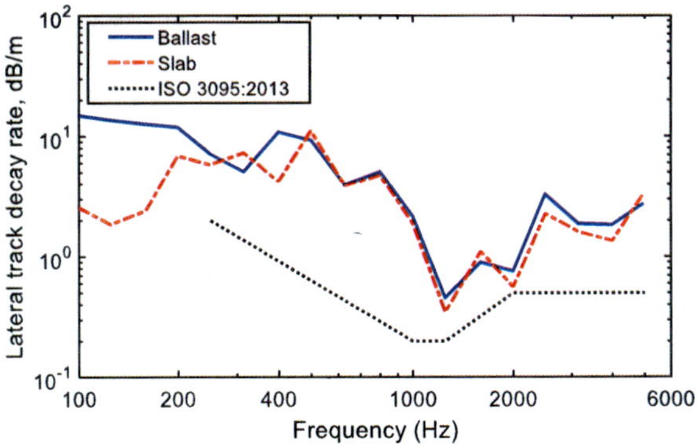

(b)

Fig. 4. Measured track decay rates on ballast, slab and grass track. (a) Vertical; (b) lateral.

the decay rates of the embedded rail are lower for this frequency range, which shows the opposite trend to the ballast track. However, they increase at high frequencies. The vertical track decay rate of the slab track studied here (which has resiliently mounted sleeper blocks) is similar to the ballasted track at frequencies above 1250 Hz, but it is much lower between 250 and 1000 Hz. For the lateral track decay rate, only the ballast and slab tracks were measured. From Fig. 4(b), it can be seen than their results are similar to each other over the whole frequency range above 200 Hz.

3.2 Roughness

The rail roughness was measured using a CAT trolley over lengths of 100 m on each rail. Figure 5 shows a comparison of the roughness for the different track forms (ballast, slab, embedded in grass). The limit curve from ISO 3095:2013 [6] is also shown for comparison. The rail roughness of all these tramway tracks is quite high and exceeds the ISO standard limit for noise type tests by more than 20 dB over much of the wavelength range; this will lead to high noise levels. In these results, there is no distinct peak in the roughness spectra although some short wavelength corrugation could be seen. There is no significant difference between the different tracks; the slab track roughness is slightly lower, possibly due to more recent rail grinding. The wheel roughness was also measured. Compared with the rail roughness, it is much lower.

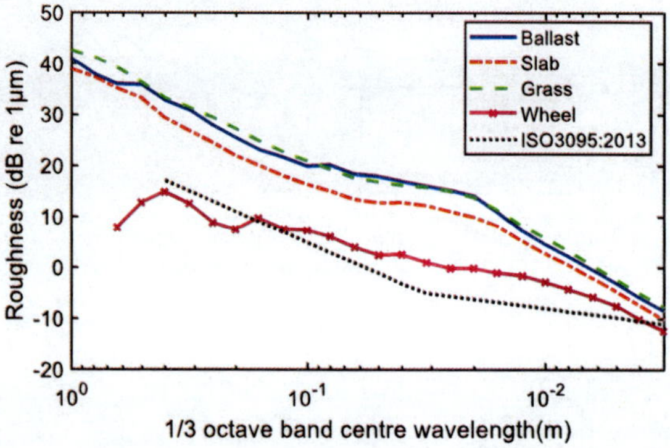

Fig. 5. Rail roughness of different types of tram track and wheel roughness.

4 Rolling Noise Predictions and Measurements

The TWINS model was adapted for predicting the rolling noise for different track structures and different ground surfaces including ballast and rigid slab [5]. It is not possible to predict the noise of the grass track using TWINS, as it requires a more detailed finite element model, e.g. [7], to allow for the dynamic behaviour of the embedded rail. This is beyond the scope of the current paper. Rail pad parameters have been determined by fitting calculated track decay rates to the measured values. The values of the parameters of vehicle and track used for the model are given in Table 1 and Table 2 respectively.

Although similar pads were fitted to both tracks the parameter values obtained by this method differ to some extent. The speed of the tram for the simulations was 55 km/h. The shielding effect is included by calculating the insertion loss of bogie cavity. The tram had resilient wheels and was modelled in terms of its wheel modal basis obtained from a finite element model; this has been validated with simple modal measurements. The main parameters are listed in Table 2. The noise measurement location was 7.5 m from the track centreline and 1.2 m above the rail top [6].

Figure 6(a) shows a comparison of the rolling noise predicted using the TWINS model and field measurements for the ballasted track, which shows a good agreement for most of the frequency range. Figure 6(b) shows the wheel, rail and sleeper contributions to the noise. The rail and wheel contributions are similar in the frequency range 500–1600 Hz.

Table 1. Vehicle parameters

	Parameter	Value
Tramcar	Total length	54 m
	Number of wheelsets	10
	Speed	55 km/h
Wheel	Wheel radius	0.3 m
	Wheel load	40 kN

Figure 7(a) shows a comparison of the rolling noise predicted using the TWINS model and field measurements on the slab track. The prediction is slightly higher than measurement between 500 Hz and 1250 Hz, which is believed to be due to the shielding effect of the tramcar skirt around the bogie area (which will also affect the other tracks). From Fig. 7(b), the sleeper contributes most at low frequencies and the rail contribution is much higher than the wheel at high frequencies.

Figure 8 shows a comparison of the measured rolling noise on all three kinds of tramway track. The speeds of the tramcars at these locations were all very close to 55 km/h. The grass track is found to be the noisiest one. It can be seen that the noise of the grass track is the highest in the frequency range between 500 Hz and 1250 Hz due to its low track decay rate in this frequency range (see Fig. 4). Conversely it has the lowest noise at frequencies above 1250 Hz and below 200 Hz due to its higher track decay rate at these frequencies. The ballast track has the lowest measured sound pressure level. Because the slab track studied here is installed with resiliently mounted sleeper blocks, the track dynamic properties are similar to those of the ballast track. At frequencies above 2000 Hz, the noise from the slab and ballast tracks are almost the same. However, due to the lack of ground absorption, the total noise of the slab track is still higher than the ballast track.

To allow for the differences in rail roughness at the measurement sites, the noise from each track form has been normalized to the rail roughness of the grass track. There is no identified reason for any systematic differences in roughness on different track forms.

Table 2. Track parameters

	Parameter	Ballast	Slab
Rail	Rail type	S49 rail	S49 rail
Subgrade	Subgrade type	Ballast	Slab
Sleeper	Sleeper type	Mono-block	Booted sleeper
	Sleeper mass	280 kg	40 kg per block
	Sleeper spacing	0.755 m	0.755 m
Rail pad	Vertical stiffness	1500 MN/m	1500 MN/m
	Lateral stiffness	300 MN/m	250 MN/m
	Vertical loss factor	0.2	0.4
	Lateral loss factor	0.2	0.1
Ballast/Booted rubber pad (slab track)	Vertical stiffness	Frequency dependent	80 MN/m
	Lateral stiffness	100 MN/m	15 MN/m
	Vertical loss factor	1.0	0.4
	Lateral loss factor	1.0	0.15

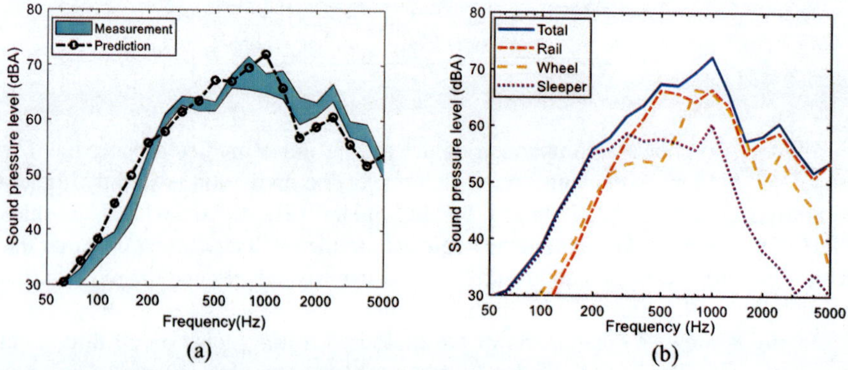

Fig. 6. Rolling noise of ballast tramway; (a) Comparison of TWINS prediction and measured noise; (b) contributions to the predicted noise from different components noise.

This result is shown in Fig. 9. It can be seen that for the same roughness excitation, the slab track is the noisiest one, with much higher levels than the ballast track over most of the frequency range. The noise level of grass track is relatively high at low frequencies, despite having no sleepers, which is believed to be related to the vibration of the concrete around the rail in the embedded track system (see Fig. 3). The vibration level of the embedded rail, its radiation efficiency and the influence of both on the noise should be investigated with a validated model, which should consider the deformation of grooved rail profile in a 3D model [7].

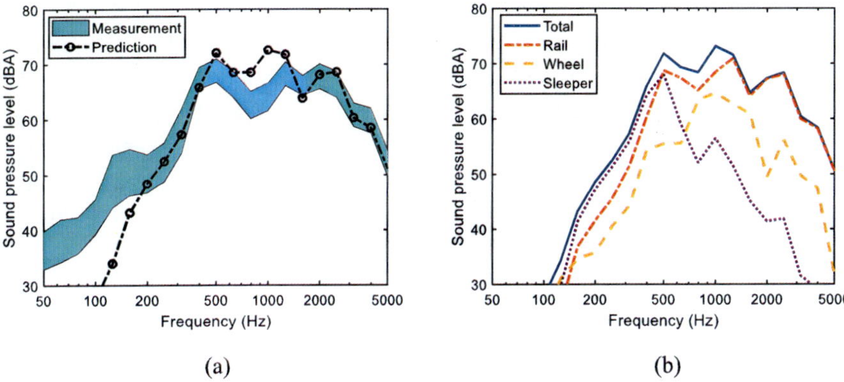

Fig. 7. Rolling noise of slab tramway; (a) Comparison of TWINS prediction and measured noise; (b) contributions to the predicted noise from different components.

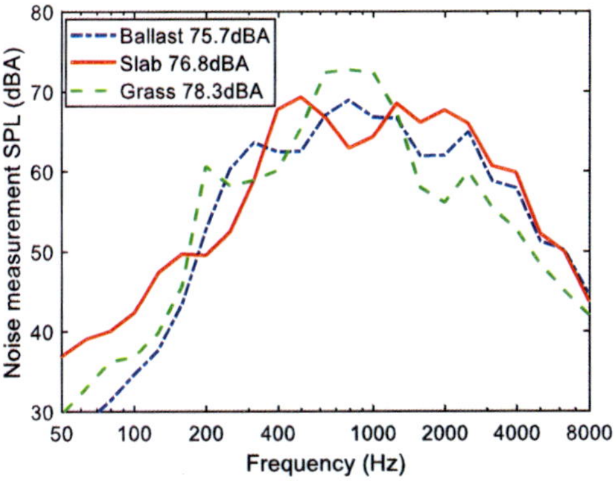

Fig. 8. Average noise of measurement for different tram tracks.

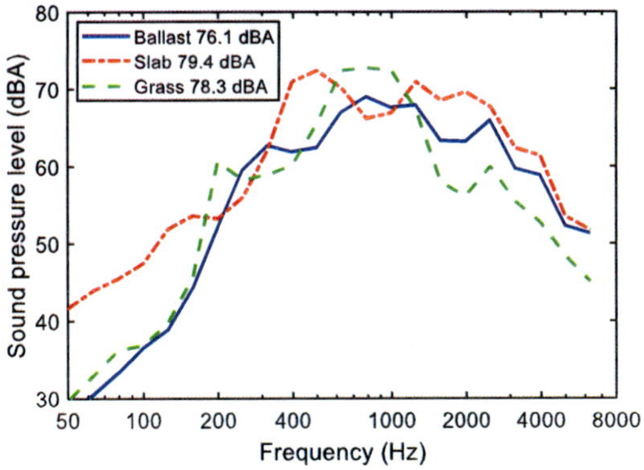

Fig. 9. Noise comparison between different tram tracks after normalizing to the same roughness.

5 Conclusions

The differences in the rolling noise between open and embedded tramway tracks have been investigated with prediction models and measurements. The rolling noise performance of ballasted, slab and grass tram tracks are compared and the relative contributions of wheel and track are identified in each case. The predictions show good agreement with field measurements. In common with other tram lines [1], high levels of rail roughness are found.

The track decay rates of the three kinds of track have been compared. The decay rate of the slab track, with booted sleeper blocks, is similar to that of ballast track and is larger than that of other kinds of slab track at most frequencies. The vertical decay rate of grass track is large at frequencies below 300 Hz and above 1600 Hz; however, it is quite low between 300 Hz and 1600 Hz.

The noise measurement results normalized to the same roughness show the ballast track to be the quietest of the three tram tracks studied here. This is mainly because of its high track decay rate and the absorptive effect of the ballast. The noise level of the slab track is the largest, the main contribution being from the rail. The noise level of the grass track is similar to the slab track due to its low track decay rate in the important frequency range between 300 Hz and 1250 Hz. The contributions of the wheel and track have been identified for the ballast and slab tracks. The rail contribution is dominant for both tracks, especially for the slab track. The sleeper contributes most at low frequencies up to 300 Hz for ballast track and up to 500 Hz for slab track. The noise levels from the resilient wheels of the tramcar are relatively low. For the grass track the noise prediction model of the embedded rail is still required.

Acknowledgements. The work described here has been supported by the EPSRC under the programme grant EP/M025276/1, 'The science and analytical tools to design long life, low noise railway track systems (Track to the Future)'. The first author is an academic visitor at ISVR supported by the China Scholarship Council.

References

1. Chiacchiari, L., Thompson, D.J., Squicciarini, G., Ntotsios, E., Loprencipe, G.: Rail roughness and rolling noise in tramways. J. Phys. Conf. Ser. **744**(1), 012147 (2016)
2. Grassie, S.L.: Rail irregularities, corrugation and acoustic roughness: characteristics, significance and effects of reprofiling. J. Rail Rapid Transit **226**, 542–557 (2012)
3. Remington, P.J.: Wheel/rail noise, part IV: rolling noise. J. Sound Vib. **46**, 419–436 (1975)
4. Thompson, D.: Railway Noise and Vibration: Mechanisms, Modelling and Means of Control. Elsevier, Oxford (2009)
5. Zhang, X., Squicciarini, G., Thompson, D.J.: Sound radiation of a railway rail in close proximity to the ground. J. Sound Vib. **362**, 111–124 (2016)
6. International standard "Railway applications – Acoustics – Measurements of noise emitted by railbound vehicles", ISO 3095:2013
7. Nilsson, C.M., Jones, C.J.C., Thompson, D.J., Ryue, J.: A waveguide finite element and boundary element approach to calculating the sound radiated by railway and tram rails. J. Sound Vib. **321**(3–5), 813–836 (2009)

Numerical Study on Acoustic Characteristics of Embedded Track Used in a Metro

Jian Han[1](✉), Gang Yang[2], Xinbiao Xiao[3], Xiaozhen Sheng[3], and Xuesong Jin[3]

[1] School of Mechanical Engineering, Southwest Jiaotong University, No. 111 North Section 1, 2nd Ring Road, Chengdu 610031, China
super_han@126.com
[2] Chengdu Xinzhu Road and Bridge Machinery Co. Ltd., No. 99 Xingyuan Road 3, Xinjin County, Chengdu 611430, China
[3] State Key Laboratory of Traction Power, Southwest Jiaotong University, No.111 North Section 1, 2nd Ring Road, Chengdu 610031, China

Abstract. With urban rail transit developing fast, there is an increasing demand for ride comfort and reduction of vibration and noise. Metro being the main public transportation facility of urban rail transit in China, faces more severe challenges. Metro tracks use a range of different vibration isolating ballasted track structures. Environmental vibration along metro lines is controlled by using so-called vibration isolating track structures. In doing so, many problems arise in the train track system of metro lines, such as strong vibration and excessive wear of wheel/rail systems, possibly causing vibration fatigue and noise. Embedded track is used as an option to reduce vibration and noise, usually applied for tramways. When applied to metro lines, embedded track faces challenges due to high speed, large axle load and more complex line conditions. When applied to metro lines, the acoustic characteristics of embedded track need to be re-evaluated. These characteristics were investigated through a combined method of simulation and experiment. In this study, the wheel and track acoustic radiation of the embedded track designed for metro was prediction and compared to that of a common track. The study can provide a basis for the application of embedded track in metro systems.

Keywords: Embedded track · Metro · Acoustic characteristics · Wheel-track noise

1 Introduction

Since the first continuously supported embedded track appeared in Deurne of the Netherlands in 1976 [1], its advantages in vibration and noise reduction have been widely focused. Relevant studies have been carried out on the vibration and noise reduction characteristics of embedded track structures.

In 1999, a three-kilometer embedded track test line was established near Best in the Netherlands. The type of the rail is traditional European UIC54. The track was applied for heavy rail. In 1998, this kind of structure was analyzed and optimized further. The

G. Degrande et al. (Eds.): *Noise and Vibration Mitigation for Rail Transportation Systems*, NNFM 150, pp. 605–612, 2021.
https://doi.org/10.1007/978-3-030-70289-2_66

rail is a low rail structure of SA42 (a type of rail), which can reduce noise by 5 dB(A) and save filling materials in the groove. This optimized embedded track was also laid in this line, with a length of 150 m [2–4]. Georgios [5] also analyzed this kind of track, and finally concluded that the track and the related acoustic barrier together, can achieve a noise reduction of about 16 dB(A). Balfour Beatty designed a similar low rail with rectangular cryosection for an embedded track. The rail is symmetrical at the top and the bottom, can be re-used by flipping it in case of severe top wear [6, 7]. The "silent railway traffic" projects using a TWINS model combined with the finite element method, study the acoustic radiation properties of an existing embedded track and the optimized structure [8]. The wheel-track system noise of the optimized embedded track is 4 ~ 6 dB(A) less than ballasted track. In 2009, Nilsson [9] used the wavenumber finite and boundary element methods to calculate the acoustic radiation of an embedded track and a common rail. In the frequency range 500 ~ 2000 Hz where the rail acoustic radiation makes a major contribution, the embedded track has smaller noise radiation. And in the low frequency range, embedded track may emit more noise than common rail. Some tests on the vibration and noise of embedded track in a tramway were carried out. The test results show that the embedded track will provide better vibration and noise reduction effect [10–12].

Until now, basic principle research and analysis of embedded track on vibration and noise reduction have been carried out. Seldom studies focus on the vibration and acoustic radiation characteristics of embedded track applied in a metro system. In this paper, the wheel-track acoustic radiation of embedded track designed for a metro line will be predicted and compared with common track.

2 Simulation of Vibration and Noise of Metro Wheel and Embedded Track

2.1 Introduction of an Embedded Track Structure for Metro

Figure 1 shows the structure of an embedded track which is designed for metro. The embedded track includes the rail, filling material, elastic block, raised pad, rail pad, slab, and concrete base. In order to connect with other common tracks, the same type (CN60) of rail was used. In order to facilitate rail grinding, the rail head is exposed on the outside. The selection of material parameters of each part mainly considers the safety and reliability of the structure, and the characteristics of vibration and noise reduction. The material parameters of each part are shown in Table 1.

2.2 Vibration and Acoustic Model of Embedded Track and Wheel

Figure 2 shows the two-dimensional (2D) section of the finite element (FE) model of embedded track. A fixed constraint is applied at the bottom of the concrete base layer. Since the embedded track developed in 2D has symmetrical characteristics, a symmetrical constraint is adopted at the center of the track. Excitation is loaded at the center of the rail head.

Based on the 2D section of the FE model of the embedded track and material parameters, a three-dimensional (3D) FE model of the embedded track is developed. The

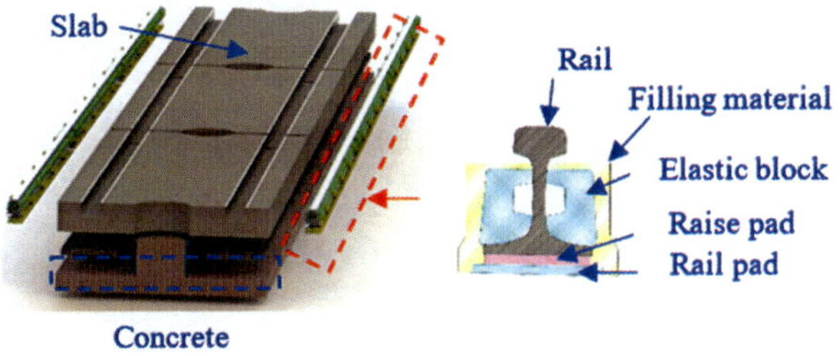

Fig. 1. Structure of embedded track

Table 1. Material parameters of embedded track.

Component	Young's modulus (MPa)	Poisson's ratio	Density (kg/m^3)	Loss factor
Rail	2.1×10^5	0.3	7850	–
Filling material & Raise pad	4	0.38	1000	0.15
Elastic block	20	0.42	1300	0.2
Rail pad	1	0.4	600	0.05
Slab	3.6×10^4	0.2	2400	–
Concrete base	3.25×10^4	0.2	2400	–

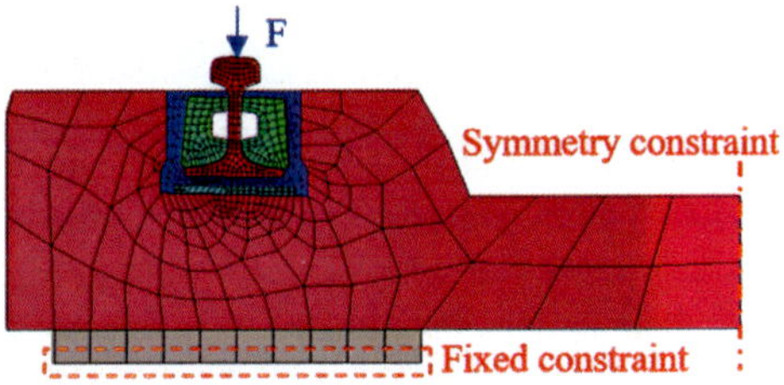

Fig. 2. 2D section of FE model of embedded track

boundary on both ends of rail in longitudinal direction has a symmetry constraint. The rail in the longitudinal direction is continuously supported. The length of the model in the longitudinal direction must be investigated. So, one slab (4.84 m), three slabs (14.52 m), five slabs (24.2 m) and seven slabs (33.88 m) are selected for analyzing the accuracy and efficiency of the 3D model of the embedded track. Figure 3 shows the rail receptance at the excitation point to investigated the influence of the model lengths in longitudinal direction.

It can be seen from Fig. 3 that the rail receptance of the embedded track of different lengths has a significant peak at 140 Hz. While in the middle and high frequency bands above 200 Hz, it is shown that there is a large fluctuation when the track model is short. As

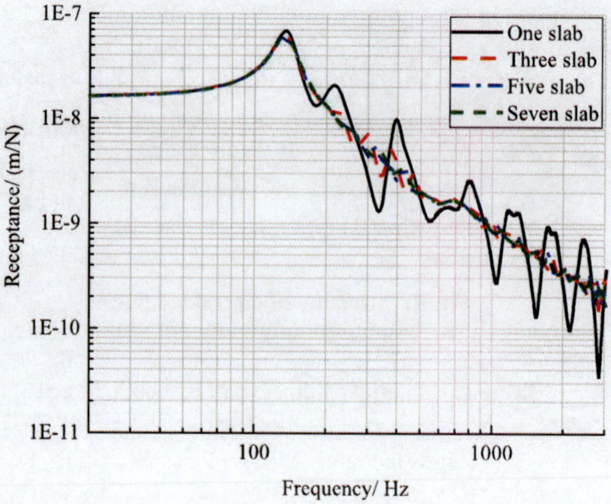

Fig. 3. Rail receptance at the excitation point

the slab length increases gradually, the fluctuation decreases gradually. The fluctuation of medium and high frequency is mainly due to the insufficient length of the track. When the vibration wave propagates to the boundary at both ends, it is reflected. At resonance frequencies a standing wave is formed. However, the actual track is an infinite structure. And rail is supported continuously. Therefore, attention should be paid to avoid the wave reflection phenomenon when a track model with finite length is developed to simulate infinitely long features. When the track length is more than five slabs, the phenomenon of medium and high frequency fluctuations affected by boundary reflection is basically eliminated, and the results are stable. Considering the accuracy of the model and the computational efficiency, the five-slab model is adopted in the subsequent analysis.

The vibration characteristics of the wheel structure can be well simulated by using 3D FEM [13, 14]. The wheel diameter is 840 mm with straight web, the elastic modulus is 210 GPa, the density is 7850 kg/m^3, the poisson ratio is 0.3, and the damping loss factor is 0.04%. The hub hole is fixed and loaded on the nominal rolling circle position of the wheel tread.

The interaction between the wheel and the rail is referenced in [15–17]. The running speed is assumed to be 60 km/h. In calculating the contact stiffness, the wheel diameter is set to 840 mm, the curvature radius of CN60 rail top is assumed to be 300 mm, and the wheel load is 5.5 tonnes. The receptance of wheel, rail, and contact are shown in Fig. 4. The calculation frequency range is 20–5000 Hz. The wheel-rail defined roughness is shown in Fig. 5. Roughness of Class B in Fig. 5 is used in this paper for both wheel and rail.

Once the vibration of the track has been determined using the 3D FE model, the sound radiation from the wheel and track can be evaluated using an appropriate vibro-acoustic prediction method. There are four radiators: the wheel, the rail, the filling material, and the slab. The 3D acoustic integral equation for a sound field of frequency ω can be found in reference [18]. For the BE model of the wheel, the hole in the wheel hub is closed to model the effect of the axle. For the BE model of the track, sound radiation

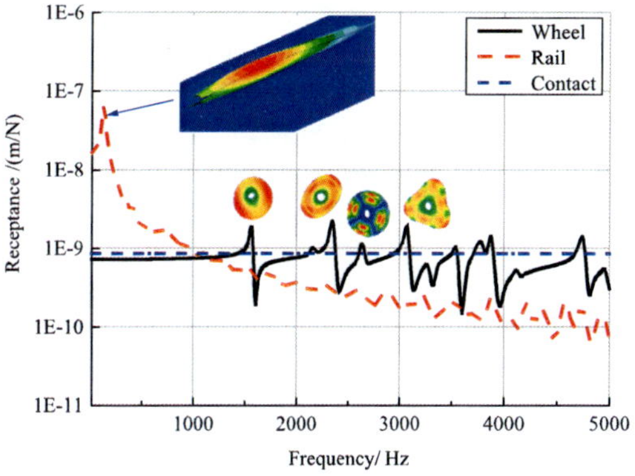

Fig. 4. Receptance of wheel, rail, and contact.

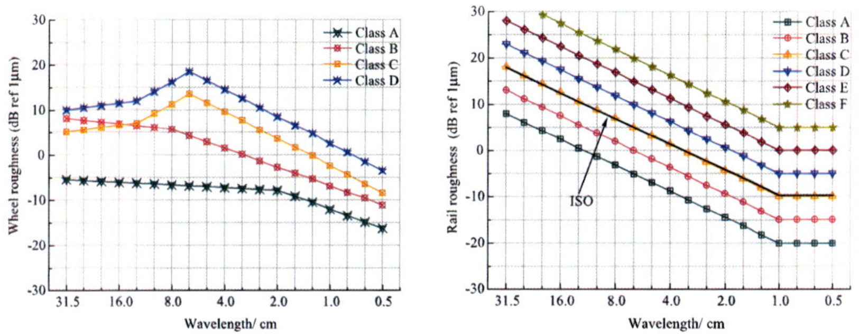

Fig. 5. Wheel/rail roughness for simulation.

only considers the rail, the filling material, and the slab surface. For a railway on the surface of the ground, the acoustic domain may be approximated as a half-space with two acoustically hard boundaries (the sound absorption effect of the ground surface is not considered in this study). These are the horizontal planes beyond the top surface of the slabs. All other boundaries of the acoustic domain are radiating, including the top surface of the slab, the filling material, and the rail.

2.3 Acoustic Characteristics of Embedded Track and Wheel

Figure 6 shows the sound power level spectrum of each part of the embedded track under the excitation of typical wheel/rail roughness at the speed of 60 km/h, and the contribution of each part of the embedded track to the sound power level of the whole embedded track is analyzed. The overall sound radiation of the track is contributed by rail and filling material. And the contribution of the slab is less compared to the overall sound radiation of the track. The contribution is approximately between rail and filling material below 400 Hz. The contribution of filling material is dominating below for 500–630 Hz. The contribution of rail is dominating above 1000 Hz.

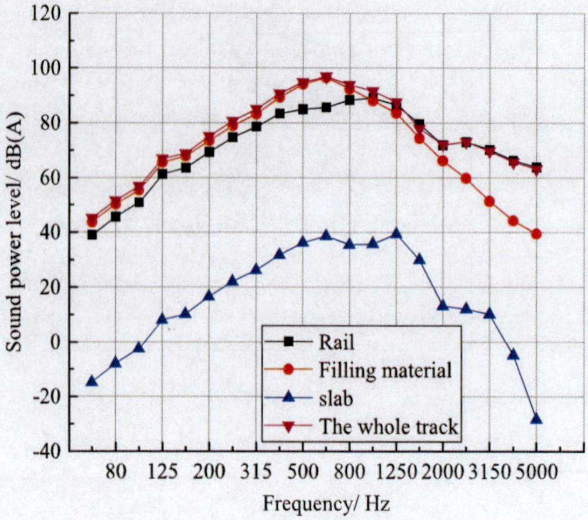

Fig. 6. Sound power level spectrum of each part of the embedded track.

Figure 7(a) includes the contribution of the wheel and embedded track under typical wheel-rail roughness excitation. The contribution of track is dominating below 2000 Hz. The contribution of metro wheel is dominating above 2000 Hz. A local peak appears in the 2500 Hz frequency band, which is mainly due to the significant vibration of the wheel's web at 2640 Hz, leading to a significant peak in the sound power of the wheel. In general, the significant frequency range of the sound radiation of the wheel-track system is 400–1000 Hz. The sound radiation of the wheel-embedded track system is dominated by that of the track. And the reduction of the sound radiation of the track can reduce the total sound radiation of the wheel-track system more effectively than the wheel.

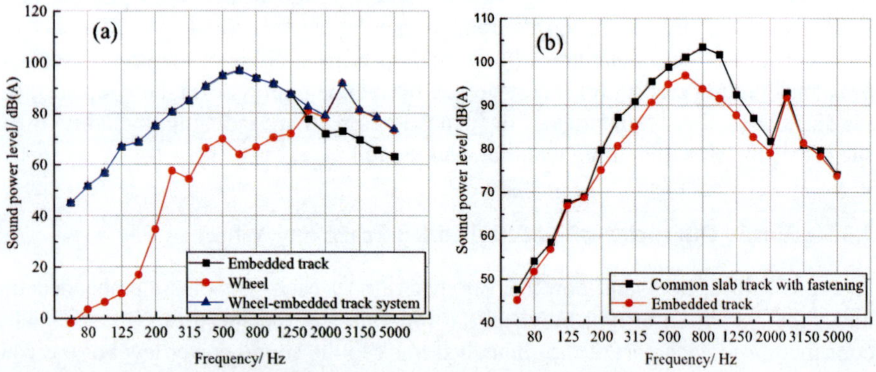

Fig. 7. Sound power level spectrum of the metro wheel-embedded track system: (a) each part of embedded track; (b) comparison with common slab track system.

In order to analyze the noise reduction effect of embedded track on wheel-track system, the simulation results of the total sound power level and spectrum characteristics of wheel-embedded track system and wheel-common slab track system with fastening were compared on the basis of same working condition in 2.3, as is shown in Fig. 7(b). The vertical stiffness of fastener for common slab track is 40 kN/mm. The vertical damping coefficient of fastener for common slab track is 10 N·s/m. The main differences between embedded track and common slab track is embodied in 200–2000 Hz. For embedded track, the sound power level is the most significant in the frequency band of 500–800 Hz. For common slab track with fastening, the sound power level is the most significant in the frequency band of 630–1000 Hz. The total sound power level of the metro wheel-embedded track system is 102 dB(A). The total sound power level of the metro wheel-common slab track system is 107 dB(A). Thus, the sound power of a wheel-track system corresponding to embedded track can be reduced by 5 dB(A).

3 Conclusions

The wheel and track acoustic radiation of the embedded track designed for metro was prediction and compared to that of a common track based on 3D FE and BE method. The relevant conclusions are as follows:

(1) The influence of the length of the model in the longitudinal direction on the accuracy and efficiency of the 3D model of the embedded track was investigated. the five-slab model is adopted in the analysis.
(2) The sound power level spectrum of each part of the embedded track under the excitation of typical wheel/rail roughness, and the contribution of each part to the sound power level of the whole embedded track is analyzed. The contribution is approximately between rail and filling material below 400 Hz. The contribution of filling material is dominating below for 500–630 Hz. The contribution of rail is dominating above 1000 Hz.
(3) The significant frequency range of the sound radiation of the wheel-track system is 400–1000 Hz. The sound radiation of the wheel-embedded track system is dominated by that of the track.

Acknowledgement. This study was supported by the National Key R&D Program of China (2017YFB1201103), Science and Technology Research and Development Program of China Railway (N2019G037), Sichuan Science and Technology Program (2020YJ0076), and the Fundamental Research Funds for the Central Universities (2682020CX35).

References

1. Michas, G.: Slab track systems for high-speed railways. KTH Royal Institute of Technology, Stockholm (2012)
2. Esveld, C.: Recent developments in slab track. Eur. Railway Rev. **9**(2), 81–85 (2003)
3. Markine, V.L., De Man, A.P., Jovanovic, S., et al.: Optimum design of embedded rail structure of high-speed lines. In: Proceedings of the International Conference, Railway Engineering (2000)

4. Markine, V., de Man, A., Jovanovic, S., et al.: Modelling and optimisation of an embedded rail structure. Rail Int. **31**(7), 15–23 (2000)
5. Georgios, M.: Slab Track Systems for High-Speed Railways. Royal Institute of Technology, Sweden (2012)
6. Penny, C.: Balfour Beatty embedded track system. In: Proceedings of Rail-Tech Europe 2003 Conference (2003)
7. Temple, B.P., del Rio Alonso, E.: Design and development of the Balfour Beatty embedded slab track system. In: Railway Engineering, Seventh International Conference, London (2004)
8. Van Lier, S.: The vibro-acoustic modelling of slab track with embedded rails. J. Sound Vib. **231**(3), 805–817 (2000)
9. Nilsson, C.M., Jones, C.J.C., Thompson, D.J., et al.: A waveguide finite element and boundary element approach to calculating the sound radiated by railway and tram rails. J. Sound Vib. **321**(3), 813–836 (2009)
10. Lakusic, S., Bogut, M.: Noise and vibration of tram tracks-differences between discrete and continuous rails instalations. In: 26th Conference on Transportation Systems with International Participation Automation in Transportation 2006, Split (2006)
11. Lakusic, S., Bogut, M., Tkalcevic Lakusic, V.: Noise and vibrations at tram track intersection. J. Acoust. Soc. Am. **123**(5), 3259 (2008)
12. Zhao, Y., Li, X., Lv, Q., Jiao, H., et al.: Measuring, modelling and optimising an embedded rail track. Appl. Acoust. **116**, 70–81 (2017)
13. Han, J., Wang, R., Wang, D., et al.: Effect of wheel load on wheel vibration and sound radiation. Chin. J. Mech. Eng. **28**(1), 46–54 (2015)
14. Han, J., Xiao, X., Wu, Y., et al.: Effect of rail corrugation on metro interior noise and its control. Appl. Acoust. **130**, 63–70 (2018)
15. Remington, P.J.: Wheel-rail noise, Part IV: rolling noise. J. Sound Vib. **46**, 419–436 (1975)
16. Remington, P.J.: Wheel/rail rolling noise, I: theoretical analysis. J. Acoust. Soc. Am. **81**(6), 1805–1823 (1987)
17. Thompson, D.J.: Wheel-rail noise generation, part I: introduction and interaction model. J. Sound Vib. **161**(3), 387–400 (1993)
18. Wu, T.W.: Boundary Element Acoustics. WIT Press, Southampton (2005)

Modelling Wheel/Rail Rolling Noise for a High-Speed Train Running on a Slab Track

Xiaozhen Sheng[1(✉)], Gong Cheng[2], and David J. Thompson[3]

[1] School of Urban Railway Transportation, Shanghai University of Engineering Science, Shanghai 201620, China
shengxiaozhen@hotmail.com

[2] State Key Laboratory of Traction Power, Southwest Jiaotong University, Chengdu 610031, Sichuan, China

[3] Institute of Sound and Vibration, University of Southampton, Southampton SO 17 1BJ, UK

Abstract. About 35,000 km of high-speed railways are in operation in China with a maximum operational speed of 350 km/h. The main track form on the high-speed lines is non-ballasted slab track. Measured data show that at high speeds, rolling noise is still a dominant source for both interior and exterior noise. Although rolling noise modelling has been conducted for more than 30 years, a train running at 350 km/h or higher along a non-ballasted slab track brings many new factors which have not been adequately addressed in the past. The aim of this paper is to describe a new approach that brings together elements that have been developed by the authors to model rolling noise from a high-speed train running on a slab track, and to present some results for a typical high-speed train and track, including rolling noise frequency spectrum, dependence on train speed, and contributions of the wheel, rail and slab.

Keywords: Wheel/rail rolling noise · Slab track · High-speed railway

1 Introduction

China has seen a boom in high-speed railways since 2008. After only 12 years, about 35,000 km of high-speed railways are in operation with a maximum operational speed of 350 km/h. Trains with even higher speeds are also under development. The main track form on the high-speed lines is non-ballasted slab track.

To control the impact of the high-speed railway network on the environment, a detailed understanding of noise generation mechanisms must be achieved. To this end a large number of in-situ noise measurements have been performed. Measured data show that at high speeds, noise from the bogie area is the dominant noise source for both interior and exterior noise [1].

At high speeds, noise from the bogie region mainly consists of two parts. One is rolling noise generated from wheel/rail interaction and the resulting vibration, and the other is generated aerodynamically from interactions between the bogie and air. It is

G. Degrande et al. (Eds.): *Noise and Vibration Mitigation for Rail Transportation Systems*, NNFM 150, pp. 613–620, 2021.
https://doi.org/10.1007/978-3-030-70289-2_67

understood that the former is dominant over the latter at low speeds, whereas aerodynamic noise increases with speed at a higher rate and becomes dominant at a sufficiently high speed [2]. However, the relative importance of the various sources, particularly at speeds of 350 km/h and above, and how they depend on train speed and other design and operational parameters, are still questions to be answered.

It is difficult to answer these questions solely by measurement, since measured noise is a mixture of rolling noise and aerodynamic noise. One way to answer the questions is to develop models which can be used to predict each of the noise sources. This paper focuses on rolling noise modelling for slab track high-speed railways.

Rolling noise modelling has been performed for more than 30 years, mainly for ballasted track using a frequency domain approach [3]. The earliest models were developed by Remington in an analytical manner [4]. These were extended by Thompson combining numerical methods and implemented in the TWINS model which was validated by a number of field tests [5, 6]. Some comparisons were also made with measurements for high speed trains [7]. However, a train running at 350 km/h or higher along a non-ballasted slab track brings new factors which have not been adequately addressed in the past, including, for example, wheel rotation, sound reflection from the slab, and the fast movement of the sources as observed from a fixed receiver.

The aim of this paper is to describe a new approach (Sect. 2) that brings together elements that have been developed in the past few years to model rolling noise from a high-speed train running on a slab track, and present some typical results (Sect. 3).

2 The Modelling Approach

As a train wheelset rotates and moves along a track, relative displacements between the wheels and rails are generated by roughness on the wheel/rail rolling surfaces [8], as well as by parametric excitation due to the periodic variation in track stiffness. These relative displacements induce dynamic forces that excite the wheelset and track to vibrate in a complex manner. Sounds radiated by such vibrations are rolling noise.

Roughness of the right wheel/rail is denoted by $z_R(x)$, and that of the left wheel/rail is denoted by $z_L(x)$. Rolling noise generated by the roughness and the moving axle load can be decomposed into two parts, one due to a symmetric roughness excitation and the moving axle load, and the other due to an anti-symmetric roughness excitation. The symmetric roughness is given by $0.5[z_R(x)+z_L(x)]$, and the anti-symmetric one is given by $0.5[z_R(x) - z_L(x)]$. For rolling noise prediction, the right wheel/rail roughness and the left wheel/rail roughness should be regarded to be incoherent. Thus, rolling noise generated by the symmetric roughness and that by the anti-symmetric one can be added incoherently to give the total rolling noise.

The modelling approach is explained for the symmetric roughness, since it is equally applicable to the anti-symmetric roughness. The modelling approach can be broken into six steps: (1) prediction of wheelset dynamics; (2) prediction of track dynamics; (3) prediction of wheel/rail interaction; (4) prediction of wheelset acoustics; (5) prediction of track acoustics; (6) prediction of wheel/rail rolling noise. They are explained below.

2.1 Prediction of Wheelset Dynamics

In this step, the vibration of a rotating wheelset due to a unit harmonic force applied at each wheel/rail contact point is calculated. The calculation methodology is briefly described here. For more details see Refs. [9, 10]. A 2D mesh-based finite element scheme is combined with the momentum law and the momentum torque law to establish partial differential equations of motion (including rigid body motions and elastic displacements) for the wheelset. The elastic displacement is described in terms of cylindrical coordinates, and is a 2π-periodic function of the circumferential angle about the wheelset axis. By decomposing the elastic displacement, using Fourier series, into components at particular circumferential orders, the partial differential equations become ordinary differential equations.

For a harmonic load rotating (seen from the wheelset) about the wheelset axis, these ordinary differential equations can be solved algebraically for steady-state solutions. It is found that the displacement of the loading point, which is stationary if observed from the train, is also harmonic at the load frequency. Thus receptance can be readily defined and calculated for the wheelset at a wheel/rail contact point. This receptance is used in the calculation of the wheel/rail interaction.

2.2 Prediction of Track Dynamics

In this step, track vibration due to a unit vertical harmonic wheel/rail force on each rail moving along the track is calculated. It is briefly described here. See Refs. [11–13] for details.

A high-speed slab track can be idealised to be a periodic structure, with the period L being the length of a slab which, for rolling noise prediction, can be assumed to be resiliently supported on a rigid foundation. Along the train travelling direction, the slabs are numbered as $-\infty, \cdots, -1, 0, 1, \cdots\infty$. The rail is periodically supported by fasteners and there are the same number of fasteners on each slab. The origin of the x-coordinate is located at the junction between the -1th and 0th slabs. The harmonic load, $\mathbf{p}_0 \mathrm{e}^{\mathrm{i}\Omega t}$, is moving at speed c along the track. At $t = 0$, the load is at $x = x_0$. The response of the track (the rail and the slab) at a given position x is temporally transient, thus having a frequency spectrum. The vibrational velocity spectrum, $\hat{\mathbf{q}}(x, x_0, f, \Omega)$ at spectral frequency f is given by [13]

$$\hat{\mathbf{q}}(x, x_0, f, \Omega) = \frac{\mathrm{i}2\pi f}{c}\left[\sum_{j=-\infty}^{\infty} \hat{\mathbf{Q}}_j(\beta^*, \Omega)\mathrm{e}^{\mathrm{i}\beta_j x}\right]\mathrm{e}^{-\mathrm{i}\beta^* x_0}\mathbf{p}_0 \tag{1}$$

where, $\mathrm{i} = \sqrt{-1}$, $\hat{\mathbf{Q}}_j(\beta^*, \Omega)$ is a matrix whose expression is given in Ref. [13], and.

$$\beta^* = (\Omega - 2\pi f)/c, \quad \beta_j = \beta^* - 2\pi j/L \tag{2}$$

Thus, according to Eq. (1), the vibrational velocity spectrum of the track is the sum of an infinite number of propagating waves at frequency f. The wavenumber of the jth wave is β_j and the amplitude is given by $(\mathrm{i}2\pi f/c)\hat{\mathbf{Q}}_j(\beta^*, \Omega)\mathrm{e}^{-\mathrm{i}\beta^* x_0}\mathbf{p}_0$.

2.3 Prediction of Wheel/Rail Interaction

This step is to calculate wheel/rail forces generated by wheel/rail roughness as wheelsets rotate and move along the track. To do so, the combined wheel/rail roughness is assumed to be periodic in the track direction, with the period being a multiple of the track period (i.e. NL, where N is an integer). Since the track is an infinitely long periodic structure with period L, and the receptance of a wheelset at the wheel/rail contact point can be defined (see Sect. 2.1) even though the wheelset is in rotation, the wheel/rail forces are also temporally periodic with the period being NL/c. In other words, the wheel/rail forces contain components only at the fundamental frequency $c/(NL)$ and its harmonics. Applying the method developed in Ref. [14], these wheel/rail force components can be determined by solving a set of linear algebraic equations. A fine frequency resolution can be achieved by setting a large value for N.

The above method, termed the Fourier-series method in Ref. [14], is not suitable for calculating wheel/rail forces generated by wheel/rail impacts such as when a wheel rolls over a rail indentation, because wheel/rail separation may occur due to the impact. In this case, wheel/rail forces may be predicted using a method based on the so-called time domain moving Green's function of the railway track [15].

2.4 Prediction of Wheelset Acoustics

This step is to evaluate sound radiation of a wheelset due to the vibration predicted in Sect. 2.1. For vibro-acoustics, an acoustic domain must be first defined. For a high-speed railway on the ground surface, the acoustic domain may be approximated to be a half-space. The horizontal surface beyond the slab is assumed to be acoustically hard. Connections between the rails and the slab are removed in acoustic calculation, leaving a uniform gap between the rail and the slab.

To predict sound radiated by the vibration of the wheelset predicted in Sect. 2.1, the rails are removed but the rigid ground is taken into account by including an image wheelset. Thus both the wheelset and the image are in a full space.

From Sect. 2.1, for a harmonic load rotating about the wheelset axis, the response of the wheelset surface seen by air in contact is also harmonic. Thus, by neglecting acoustic interactions between the wheelset and its image, sound radiations from them can be calculated separately using the 2.5D acoustic boundary element method for each circumferential order, as described in Ref. [10]. Summation of sound powers at different circumferential orders gives the total sound power of the wheelset.

For a stationary observer, a wheelset is a moving source; the sound is therefore modified by the Doppler Effect and aero-dynamics. For a high-speed train, these effects can be significant and should be taken into account. However, a complete account of the effects is still difficult and simplifications have to be adopted. To do so, the wheel is simplified to be a point source moving uniformly in the track direction. This point source pulsates at the same frequency as the wheel/rail force and radiates the same sound power as predicted above. The sound pressure frequency spectrum of the moving point source in the half-space can be obtained by introducing an image source using the image principle. This is termed the *Wheel Sound Transfer Function*, denoted by $WSTF(f, \Omega, x_0)$, where, f is the spectral frequency, Ω is the radian frequency of the unit force, and x_0 is the x-coordinate of the wheel centre at $t = 0$.

2.5 Prediction of Track Acoustics

This step is to calculate sound pressure spectrum generated by the track due to the vibration predicted in Sect. 2.2. The sound pressure spectrum is the sum of that generated from slab vibration and that from rail vibration. These two sound pressure components are computed separately.

For sound generated just from slab vibration, the two rails are removed to make the acoustic domain simpler. The sound pressure spectrum at a given location in the acoustic domain may be produced for a slab vibrational velocity wave defined by $\Phi_j(y)e^{i\beta x}e^{i2\pi ft}$ for a range of spectral frequencies (f), wavenumbers (β) and lateral shape $\Phi_j(y)$. Shape functions are introduced to describe variation of slab vibration along the slab width. This will generate a number of so-called *Slab Sound Transfer Functions* (SSTF). The SSTF for the jth shape function $\Phi_j(y)$ is denoted by $SSTF_j(f, \beta)$.

Since the acoustically hard boundaries of the acoustic domain are infinite in the lateral direction, the slab is equivalent to an infinitely long baffled plate. Its sound radiation may be calculated using the classic Rayleigh integral equation Fourier-transformed with respect to the x-coordinate.

For sound radiation from the vibration of a rail the slab is assumed to be acoustically hard. As for the slab, a number of so-called *Rail Sound Transfer Functions* (RSTF) may be produced using the 2.5D BEM for a rail structural wave for a range of spectral frequencies (f) and wavenumbers (β). Another set of shape functions are introduced to express vibration variation along the cross-sectional boundary of the rail. The RSTF is denoted by $RSTF_k(f, \beta)$ for the kth shape function. The image principle is applied to avoid meshing the infinitely wide acoustically hard boundaries.

2.6 Prediction of Wheel/Rail Rolling Noise

In this last step, rolling noise is estimated by a summing process. For brevity, discussion is given to a single wheelset. At $t = 0$, the wheelset is at x_0. As shown above, the wheel/rail force, in addition to the axle load, contains components at fundamental frequency $f_0 = c/(NL)$. The kth component is denoted by $\tilde{P}_k e^{ikf_0t}$. Due to this component, the vibrational velocity spectrum of the slab may be written as (see Eq. (1))

$$v_{Sk}(x, y, f, x_0) = \left[\sum_{j=-\infty}^{\infty}\sum_{m=1}^{\infty}\lambda_{km}\Phi_m(y)e^{i\beta_{kj}x}\right]e^{-i\beta_k^* x_0}\tilde{P}_k \tag{3}$$

where, λ_{km} is a coefficient showing the contribution of 'mode' $\Phi_m(y)$ in the spectrum. Thus, sound pressure spectrum contributed by the slab subject to the kth harmonic component is given by

$$\hat{p}_{Sk}(f, x_0) = \left[\sum_{j=-\infty}^{\infty}\sum_{m=1}^{\infty}\lambda_{km}SSTF_k(f, \beta_{kj})\right]e^{-i\beta_k^* x_0}\tilde{P}_k \tag{4}$$

Similar expressions can be written for the rail (denoted by $\hat{p}_{Rk}(f, x_0)$) and the wheel (denoted by $\hat{p}_{Wk}(f, x_0)$). Now the total sound pressure spectrum due to the kth harmonic component of the wheel/rail force is calculated to be the sum of those three, and the total sound pressure spectrum, $\hat{p}(f, x_0)$, due to all the wheel/rail force harmonic components, is calculated to be the incoherent sum of those due to each harmonic component, i.e.

$$|\hat{p}(f, x_0)|^2 = \sum_k \left(|\hat{p}_{Wk}(f, x_0)|^2 + |\hat{p}_{Rk}(f, x_0) + \hat{p}_{Sk}(f, x_0)|^2 \right) \quad (5)$$

3 Results

Results are produced for a typical Chinese high-speed train and track. The mass of a wheelset is 1105 kg, rolling radius is 0.43 m, and half the axle load is 78.4 kN. For the track, parameters typical of the CRTS II track are used. Slab length is 6.5 m, sleeper spacing is 0.65 m and rail pad stiffness is 5.44 $\times 10^7$ N/m. Other parameters can be found in Ref. [13]. A measured roughness spectrum on a newly opened line is applied. To account for the wheel/rail contact patch filtering effect, the analytic filter, $L_z(\lambda) = 20\log_{10}(1 + 0.25\pi(2\pi a/\lambda)^3)$, presented in Ref. [3] is used, where, λ is roughness wavelength and a the equivalent radius of the contact patch (about 6.7 mm).

A 2D FE mesh of the wheelset is shown in Fig. 1(a). Vertical displacement at the wheel/rail contact point (a unit vertical force is applied at each wheel/rail contact point) is shown in Fig. 1(b) for the wheelset not in rotation by solid line and rotating at 350 km/h by dashed line. It can be seen that rotation splits a peak into two peaks of smaller amplitude. It can be expected that rotation affects wheel/rail interaction.

Sound spectra radiated from four wheelsets in two neighbouring bogies and from the rail and slab are predicted for an observation point at $x = 0$ m, $y = 7.5$ m (from the track central line), and $z = 3.5$ m (above the rail top), as shown in Fig. 1(c). It can be seen that, below about 300 Hz, the slab and wheelsets are nosier and from where up to 2000 Hz, the rail dominates. The wheel is noisier for higher frequencies.

Overall sound pressure levels in dB(A) radiated from the four wheelsets and track components at the above-defined point are calculated for four train speeds, 250, 300, 350 and 400 km/h. The overall sound pressure level is calculated to be an equivalent level based on the passing time of the four wheelsets, which is 9.5 m divided by the train speed. The overall noise level may be approximated to be

$$L = \mu \log_{10}(V/250) + L_0 \quad (6)$$

where V is the train speed in km/h and L_0 is a constant. Values of μ and L_0 are listed in Table 1. For this particular case, noise from the wheel increases with train speed much faster than the rail, although noise from the rail is higher than that from the wheel.

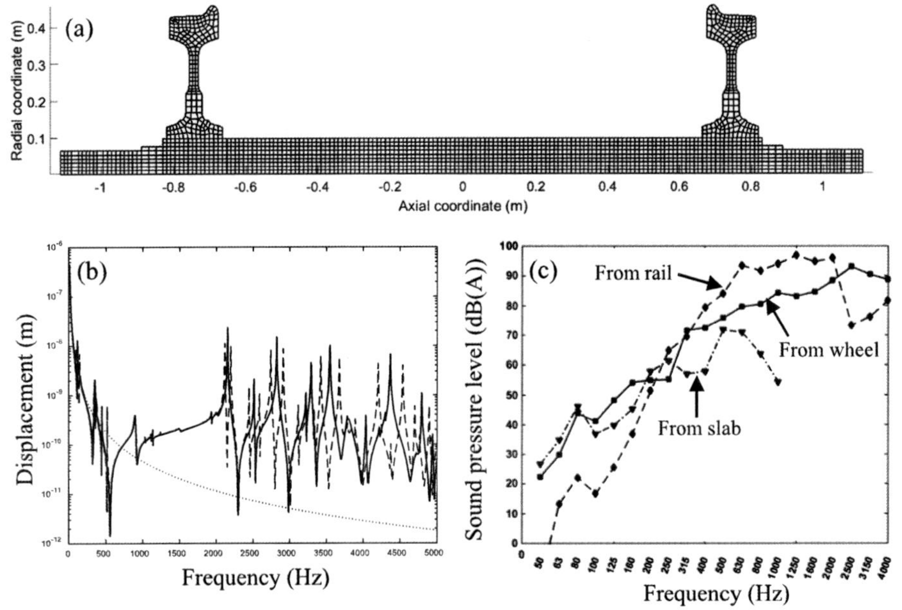

Fig. 1. (a) A 2D FE mesh of the wheelset; (b) receptance of the wheelset (—, nonrotating; – –, rotating at 350 km/h; ……, as a rigid body); (c) A-weighted 1/3-octave band sound pressure spectrum at the observation point (—, due to wheel; – –, due to rail; – · –, due to slab).

Table 1. Values of μ and L_0 for Eq. (6)

	Total	Wheelsets	Rail	Slab
μ	26.2	32.2	25.1	13.6
L_0 (dB(A))	99.0	90.3	98.5	70.2

4 Conclusions

An approach is described for modelling rolling noise for a train running on a slab high-speed railway track. Factors important to rolling noise, such as wheel rotation, periodicity and infinity of the track, reflection of the slab, fast moving of the sources etc. are considered in a sufficiently detailed manner. Characteristics of rolling noise are preliminarily investigated for a typical high-speed train and track, including frequency spectrum, dependence on train speed, contributions of the wheel, rail and slab etc. Validation of, and study with, the model are still underway.

Acknowledgement. This work is funded by the National Key R&D Program of China (2016YFE0205200), the National Natural Science Foundation of China (U1834201), and Scientific Research Foundation of State Key Laboratory of Traction Power (2017TPL_T01).

References

1. Sheng, X., Li, M., Zhu, Z.: Using microphone array data to analyse source contributions of a high-speed train to pass-by noise. In: Proceedings of the 7th All-Russian Research to Practice Conference with International Participation 'Protection against Increased Noise and Vibration', St Petersburg 19–21 March (2019)
2. Thompson, D.J., Lglesias, E.L., Liu, X.W.: Recent developments in the prediction and control of aerodynamic noise from high-speed trains. Int. J. Rail Transp. **3**, 119–150 (2015)
3. Thompson D.J.: Railway Noise and Vibration: Mechanisms, Modelling and Means of Control. Elsevier (2009)
4. Remington, P.J.: Wheel/rail rolling noise, I: Theoretical analysis. J. Acoust. Soc. Am. **81**, 1805–1823 (1987)
5. Thompson, D.J., Hemsworth, B., Vincent, N.: Experimental validation of the TWINS prediction program, part 1: description of the model and method. J. Sound Vib. **193**, 123–135 (1996)
6. Jones, C.J.C., Thompson, D.J.: Extended validation of a theoretical model for railway rolling noise using novel wheel and track designs. J. Sound Vib. **267**, 509–522 (2003)
7. Castel, L., Gautier, P.E., Vincent, N., Goudard, J.P.: 350 kph running tests to assess a new railway noise model. Inter-Noise, pp. 1467–1470 (1993)
8. Thompson, D.J., Jones, C.J.C.: A review of the modelling of wheel/rail noise generation. J. Sound Vib. **231**, 519–536 (2000)
9. Sheng, X., Liu, Y., Zhou, X.: The response of a high-speed train wheel to a harmonic wheel-rail force. In: Journal of Physics: Conference Series, Southampton, vol. 744, p. 012145 (2016)
10. Zhong, T., Cheng, G., Sheng, X., Zhan, X., Zhou, L., Kai, J.: Vibration and sound radiation of a rotating train wheel subject to a vertical harmonic wheel–rail force. J. Mod. Transp. **26**(2), 81–95 (2018)
11. Sheng, X., Jones, C.J.C., Thompson, D.J.: Responses of infinite periodic structures to moving or stationary harmonic loads. J. Sound Vib. **282**, 125–149 (2005)
12. Sheng, X.: Generalization of the Fourier transform-based method for calculating the response of a periodic railway track subject to a moving harmonic load. J. Mod. Transp. **23**(1), 12–29 (2015)
13. Sheng, X., Zhong, T., Li, Y.: Vibration and sound radiation of slab high-speed railway tracks subject to a moving harmonic load. J. Sound Vib. **395**, 160–186 (2017)
14. Sheng, X., Li, M., Jones, C.J.C., Thompson, D.J.: Using the Fourier-series approach to study interactions between moving wheels and a periodically supported rail. J. Sound Vib. **303**, 873–894 (2007)
15. Sheng, X., Xiao, X., Zhang, S.: The time domain moving Green function of a railway track and its application to wheel–rail interactions. J. Sound Vib. **377**, 133–154 (2016)

Estimation of Track Decay Rates and Noise Based on Laboratory Measurements on a Baseplate Fastening System

Boniface S. Hima[1(✉)], David J. Thompson[1], Giacomo Squicciarini[1], Evangelos Ntotsios[1], and David Herron[2]

[1] Institute of Sound and Vibration Research, University of Southampton, Southampton SO17 1BJ, United Kingdom
B.S.Hima@soton.ac.uk

[2] Pandrol, 63 Station Road, Addlestone, Surrey KT15 2AR, UK

Abstract. The dynamic properties of the rail fastening system affect the damping of the rail and the degree of coupling between the rail and the foundation. It is well known that the rate of decay of vibration along the rail is closely linked to the noise performance of the track. For this reason, the track decay rate (TDR) is used as an important measurable input quantity for models that predict railway rolling noise. This paper investigates whether the TDR can be estimated correctly from laboratory measured dynamic properties of rail fastening systems. The system studied in this work is a commercial two-stage baseplate system that is mounted on a slab track fitted with 60E1 rail. Four different types of rail pads were fitted during the track decay rates measurements. The TWINS model was used to predict the rolling noise using the measured and calculated track decay. The comparison has shown a good agreement between the measured and predicted TDR.

Keywords: Track decay rates · Railpad · Baseplate · Fastening systems

1 Introduction

When structural waves propagate along the track they decrease in amplitude with distance from the excitation point. This is quantified by means of the Track Decay Rates (TDR) that express the amplitude decrease in dB/m. They are usually represented in the form of one-third octave band spectra. They can be obtained by means of calculations or they can be measured on track according to standards [1–3]. TDRs are also used in noise prediction models to evaluate the noise emission from the track.

The static and dynamic stiffness of the resilient elements on the railway track play a key role in determining the noise radiated by the track and the vibration isolation [2]. While the lateral stiffness and roll stiffness of the fastening system are of interest, it is usually the vertical stiffness that is of main interest. The vertical stiffness is also important in relation to track deterioration and maintenance requirements. To predict noise, it is the stiffness at higher frequencies and small strains that is required. The

G. Degrande et al. (Eds.): *Noise and Vibration Mitigation for Rail Transportation Systems*, NNFM 150, pp. 621–628, 2021.
https://doi.org/10.1007/978-3-030-70289-2_68

dynamic stiffness of resilient elements can be measured in several ways. In this work, the indirect method is adopted for frequencies up to 1 kHz [4].

1.1 Track Description

Measurements of track decay rate for vertical and lateral excitation were made on a non-operational slab track located at the National College for High-Speed Rail, Doncaster. The slab track has two adjacent sections (20 m long), one fitted with Pandrol rail fastening system as seen in Fig. 1 and the other with another type of rail fastening system (not addressed here). For the Pandrol two-stage system, measurements were made with four different types of railpads fitted. Figure 2 shows a schematic view of the track sections.

Fig. 1. Photographs of the Pandrol two-stage fastening system.

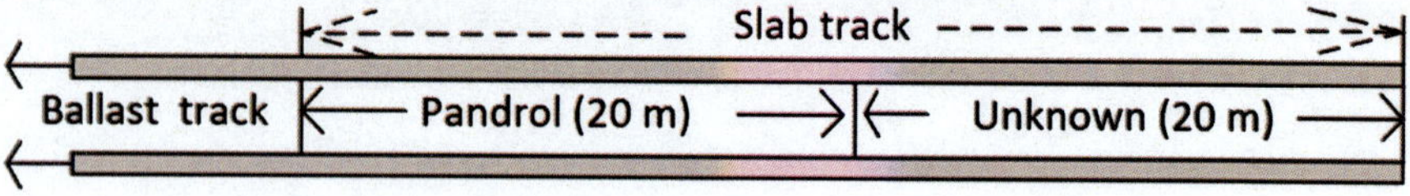

Fig. 2. A schematic view of the sections of the measured track.

2 Methodology

2.1 Dynamic Stiffness

There are several methods for measuring the stiffness of track resilient elements. In [5, 6] an indirect method was established, which will be used here. The indirect method involves a test rig with two blocks of known mass; a resilient element can be inserted between them as shown in Fig. 3. The lower mass is mounted on flexible isolators. A hydraulic actuator applies a static preload and a high frequency excitation is applied by an inertial shaker. This allows the high-frequency transfer stiffness of resilient elements to be measured as a function of frequency in the low amplitude region. The frequency dependent dynamic stiffness and the damping properties of the resilient element can be obtained using the following equation:

$$K = -m_2\omega^2\frac{\ddot{x}_2}{\ddot{x}_1};\ \eta = \tan(\angle K) \quad (1)$$

where K is the transfer stiffness, η is the loss factor, m_2 is the mass of the lower block, $\ddot{x}_1$ is the acceleration of the upper mass, $\ddot{x}_2$ is the acceleration of the lower mass and ω is the circular frequency in rad/s.

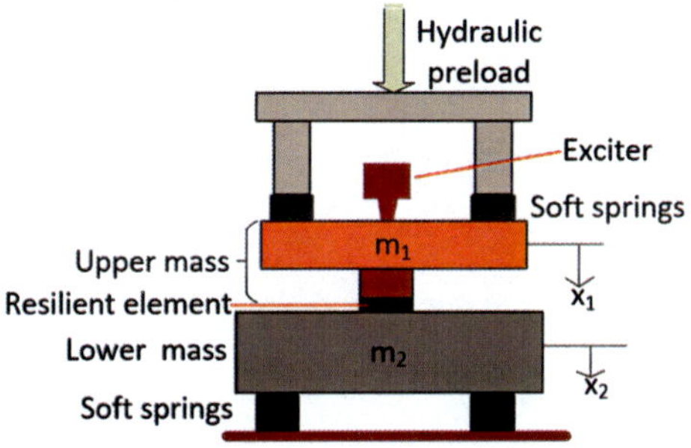

Fig. 3. Test rig for measuring the high-frequency dynamic stiffness of resilient elements [5].

2.2 Track Decay Rates Measurements

TDR measurements were conducted in accordance with EN15461:2008 [1]. The test requires an accelerometer to be attached to the railhead and series of transfer functions to be measured between an input force, delivered with an instrumented impact hammer, and the measured acceleration. The TDR in each one-third octave band is obtained as

$$DR \approx \frac{4.343}{\sum_{n=0}^{n_{max}} \frac{|A(x_n)|^2}{|A(x_0)|^2}\Delta x_n} \quad (2)$$

where DR is the decay rate expressed in dB/m, $A(x_0)$ and $A(x_n)$ are the point and transfer frequency-response functions (FRFs) averaged in each one-third octave band and Δx_n is the spacing between adjacent measurement positions, in m. FRFs in the form of mobility (i.e. ratio between velocity and force) were used throughout the analysis.

2.3 Calculation of Decay Rate from Models

The TWINS model [7–9] predicts the noise from the wheels and track. It can use the measured decay rate as an input parameter to predict the noise from the track or it can calculate the track decay rate. For the calculated decay rates, a model of a Timoshenko beam over a double elastic foundation is used. The rate of attenuation of vibration along the track is obtained from the imaginary part of the wavenumber of the propagating structural wave. The decay rate in dB/m is given by [2]

$$DR = -20\log_{10}\exp(k_i) = -8.686k_i \quad (3)$$

where k_i is the imaginary part of the wavenumber. It can be demonstrated that the TDR is directly related to the noise radiated from the track as follows

$$L_w \approx 10\log_{10}\left(4.343\rho_0 c_0 P\frac{v_{ref}^2}{W_{ref}}\right) + 10\log_{10}\sigma + 10\log_{10}\left(\frac{|v(0)|^2}{2v_{ref}^2}\right) - 10\log_{10} DR \quad (4)$$

where L_w is sound power level in decibels, $\rho_0 c_0$ is the characteristic acoustic impedance of air, P is a perimeter length of the cross-section of which only the part that is projected onto a plane perpendicular to the motion is considered. W_{ref} is the reference value used for the definition of sound power level, v_{ref} is the corresponding reference value for velocity and σ is the radiation ratio which depends on frequency [2]. From this, it can be demonstrated that the sound power is influenced by the decay rate according to

$$L_w = -10\log_{10} DR + \text{const} \quad (5)$$

2.4 Model for Noise Prediction

The TWINS model is used in this study for noise predictions. The TWINS software allows the prediction of vibration levels on the wheels, rails and sleepers as well as the total rolling noise generated in the railway system. The model has been validated with measurements [7–9]. During the passage of a train, the noise can be calculated either in terms of sound power or the average sound pressure at specified positions at the trackside.

3 Results

3.1 Dynamic Stiffness

The dynamic stiffness of several rail-fastening systems, as well as individual railpads, was measured. A single value of dynamic stiffness was estimated from the magnitude of the measured stiffness using a fitted line for each preload (dashed lines in Fig. 4(a)). The single values were extracted from the fitted line at a frequency of 200 Hz (vertical dashed-dotted line, as shown in Fig. 4(a)). The damping loss factor was predicted from the phase angles of the dynamic stiffness following a similar procedure. Example results for dynamic stiffness as a function of static preload for different railpads are shown in Fig. 4(b). A summary of the results for some of the pads tested is presented in Table 1.

3.2 Track Decay Rates

Measured vertical and lateral track decay rates for the different fastening systems and rail pad configurations are shown in Fig. 5(a) and (b).

In comparison with a typical ballasted track with railpad stiffness of 120 MN/m the measured TDRs are lower below 500 Hz and higher above 1 kHz.

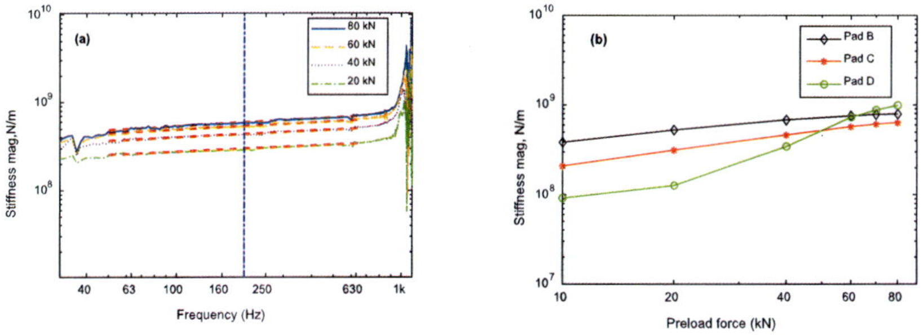

Fig. 4. Dynamic stiffness with frequency for a single railpad Pad C **(a)**, dynamic stiffness at 200 Hz as a function of static preload for different railpads **(b)**.

Table 1. Types of rail pads with the dynamic stiffness and damping loss factor measured in the laboratory. Results correspond to a preload of 20 kN and a frequency of 200 Hz.

Type	Pandrol rail pads	Stiffness (MN/m)	Damping loss factor
A	HDPE EVA plain	1200	0.1
B	21422 EVA studded	528	0.1
C	9970 EVA studded	310	0.1
D	8854 NR Studded	120	0.2
E	Lower pad double studs	50 at 0.5 kN*	0.2

* For the lower pad a nominal preload of 0.5 kN is used to represent the weight of rail above the fastening system.

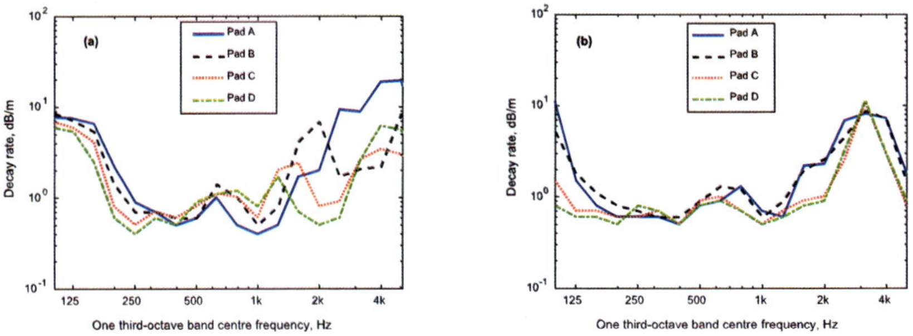

Fig. 5. Measured track decay rate **(a)** vertical, **(b)** lateral.

3.3 Comparison of the Measured and Predicted Decay Rates

The track model implemented in TWINS was used to compare the measured decay rates with analytical predictions. The baseplate was modelled as a rigid mass. In order to predict the decay rates, the stiffness and damping loss factor of the rail pad were taken from the measurements at the preload of 20 kN. The stiffnesses in the lateral direction were not measured in the laboratory. In the prediction, the lateral stiffness was obtained by arbitrarily scaling the vertical stiffness by a factor of 0.2. Figure 6 shows comparisons between the measured and predicted track decay rates obtained for one example rail pad configuration.

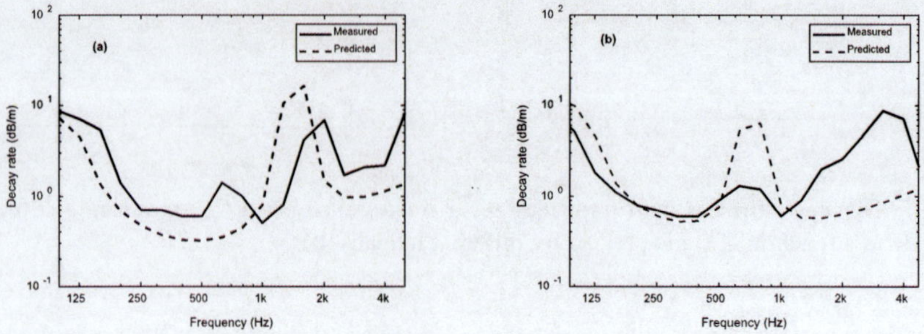

Fig. 6. Comparison between the measured and predicted track decay rate for rail pad B **(a)** vertical, **(b)** lateral.

A reasonably good agreement was found between the measured and predicted TDR using the measured stiffness. However, the measured results suggest that the vertical pad stiffness in the field is higher than the value used in the prediction, which was obtained from the lab measurements. The measured results also show some additional peaks at higher frequencies which could be due to the modal behaviour of the baseplate.

3.4 Noise Level Prediction Using TWINS Model

The sound power levels were predicted using both measured and calculated track decay rates in TWINS. A wheel with diameter 0.84 m and a straight web was used, and a nominal roughness corresponding to cast-iron brake blocks was used. The results are presented for 120 km/h in Fig. 7. In these figures, the results obtained for both the measured decay rates and the analytical ones are shown. Figure 8 shows the total sound power level and the contributions of baseplate, rail, and the wheel for one example case.

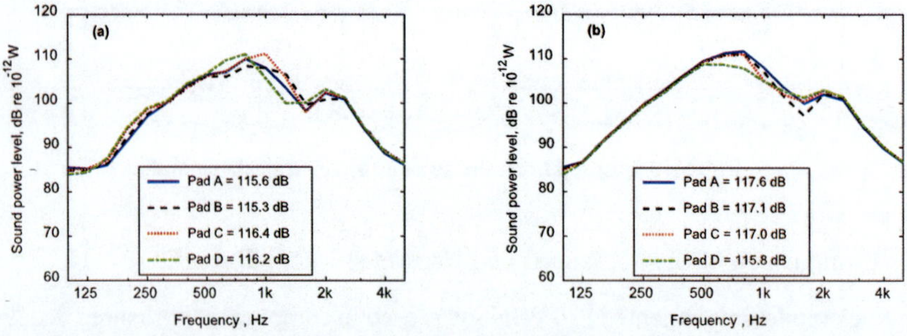

Fig. 7. Total A-weighted sound power level prediction for 120 km/h **(a)** using the measured track decay rate, **(b)** using the calculated track decay rate and the measured dynamic stiffness.

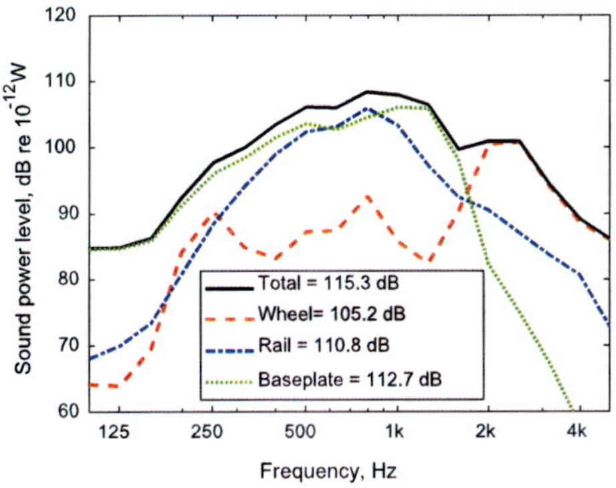

Fig. 8. A-weighted sound power level prediction for rail pad B at 120 km/h using the measured track decay rate, showing the contributions of the baseplate, rail, wheel total.

From these results, there is only a small difference in the noise predicted for the different configurations. The results are summarized in Table 2. Using different rail pad configurations, the predicted noise can differ by up to 2 dB between measured and calculated track decay rates. The configuration with rail pad B gives the lowest noise prediction from measured decay rates whereas pad D is lowest for the calculated ones.

It was discovered that the sound radiation from the baseplate in the model could be quite significant for high values of rail pad stiffness. In order to obtain more reliable results, further analysis of the baseplate vibration will be conducted to allow a better model of the baseplate to be used for noise predictions.

Table 2. The overall sound power levels for 120 km/h.

Sound power level dB(A) Rail pad configuration	Noise from measured decay rates				Noise from the calculated track decay rates		
	Wheel	Rail	Baseplate	Total	Rail	Baseplate	Total
Pad (A)	105.3	112.3	111.9	115.6	114.4	114.2	117.6
Pad (B)	105.2	110.8	112.7	115.3	113.1	114.4	117.1
Pad (C)	105.3	111.1	114.4	116.4	112.8	114.4	117.0
Pad (D)	105.2	111.3	113.9	116.2	112.0	112.7	115.8

4 Conclusions

The results show generally good agreement between the measured and the predicted TDR. The measured stiffness and TDR have been used to predict the track noise. The noise from the baseplate itself is found to be significant, especially for high values of railpad stiffness. Further analysis of the baseplate vibration will be conducted to allow a better model of the baseplate to be developed.

The results from noise predictions using the measured and calculated track decay rates have shown that the radiated noise has quite small differences between the different configurations. The results have shown that using any of the rail pad configurations, the noise prediction will vary by up to 2 dB between measured and calculated track decay rates.

Acknowledgements. The authors gratefully acknowledge Pandrol Track Systems for providing the rail-fastening system, railpads and organising the TDR measurement at the test track located at the National College for High-Speed Rail, Doncaster. The work described here has been supported by the EPSRC and CDT SIS under the programme grant EP/M025276/1, 'The science and analytical tools to design long life, low noise railway track systems (Track to the Future)'.

References

1. European Standard. EN 15461:2008+A1:2010 -Railway applications. Noise emission. Characterisation of the dynamic properties of track selections for pass by noise measurements (2010)
2. Thompson, D.J.: Railway Noise and Vibration mechanisms, modelling and means of Control 2009, Oxford: Elsevier Ltd.
3. Jones, C.J.C., Thompson, D.J., Diehl, R.J.: The use of decay rates to analyse the performance of railway track in rolling noise generation. J. Sound Vibr. **293**(3–5), 485–495 (2006)
4. International Standard ISO 10846–3, Acoustics and vibration-Laboratory measurement of vibro-acoustics transfer properties of resilient elements Part 3: Indirect method for determination of the dynamic stiffness of resilient supports for translatory motion (2002)
5. Thompson, D.J., Verheij, J.W.: The dynamic behaviour of rail fasteners at high frequencies. Appl. Acoust. **52**(1), 1–17 (1997)
6. Thompson, D.J., Van Vliet, W., Verheij, J.: Developments of the indirect method for measuring the high frequency dynamic stiffness of resilient elements. J. Sound Vib. **213**(1), 169–188 (1998)
7. Thompson, D.J.: Wheel-rail noise generation, part II: Wheel vibration. J. Sound Vib. **161**(3), 401–419 (1993)
8. Thompson, D.J., Fodiman, P., Mahé, H.: Experimental validation of the TWINS prediction program for rolling noise, part 2: Results. J. Sound Vib. **193**(1), 137–147 (1996)
9. Thompson, D.J., Hemsworth, B., Vincent, N.: Experimental validation of the TWINS prediction program for rolling noise, part 1: description of the model and method. J. Sound Vib. **193**(1), 123–135 (1996)

Track Decay Rate Analysis and Rail Damper Noise Reduction for Slab Tracks

Briony Croft[1(✉)], Aaron Miller[2], and Christoph Gramowski[3]

[1] SLR Consulting, 200–1620 West 8th Avenue, Vancouver, BC V6J 1V4, Canada
bcroft@slrconsulting.com

[2] SLR Consulting, 202 Submarine School, Sub Base Platypus, North Sydney 2060, Australia

[3] Schrey & Veit GmbH, Graf-von-Sponheim-Straße 2, 55576 Sprendlingen, Germany

Abstract. The STARDAMP software tool enables prediction and assessment of the performance of wheel and rail dampers. This paper supports ongoing validation and testing efforts by examining the use of STARDAMP in relation to assessing the performance of rail dampers on slab track with highly resilient fasteners. This track type is common on transit systems that are increasingly considering rail dampers for noise control. The normal assessment approach of summing a measured undamped TDR with a damped short rail DR may under-predict the rail damper benefit for some slab tracks. Alternatively the damper effect may be estimated by summing the predicted undamped TDRs with a damped DR from a short rail test, supported by investigations of the dominant rolling noise frequencies. The STARDAMP slab track model may limit the accuracy of rolling noise prediction by ignoring baseplate dynamic effects.

Keywords: Stardamp · Track decay rate · Rail · Damper · Noise · Slab · Baseplate

1 Introduction

The STARDAMP project was undertaken with the aim of providing a comprehensive methodology for prediction and assessment of the performance of wheel and rail dampers. The STARDAMP Final Report [1] recommends ongoing validation of the tool by testing it in real situations, with a variety of track and rolling stock types. This paper aims to support this ongoing validation and testing effort by examining the use of the STARDAMP tool for slab track with resilient fasteners and rail dampers.

Jones et al. [2] provided comparisons of predicted and measured track decay rates (TDRs) for three tracks, including a slab track with baseplate fasteners. For this track, they found the predicted vertical TDR was lower than the measured decay rate at frequencies between 200 Hz and 1600 Hz, with a notable peak in the measured TDR at 800 Hz that is not explained by the predicted track behavior. Using a comparison of measured rolling noise at 7.5 m with noise predictions, they observed that the predicted vertical TDRs result in an accurate noise prediction, whereas use of the measured TDRs

G. Degrande et al. (Eds.): *Noise and Vibration Mitigation for Rail Transportation Systems*, NNFM 150, pp. 629–637, 2021.
https://doi.org/10.1007/978-3-030-70289-2_69

resulted in a substantial under-prediction of rolling noise, by around 4 dB in overall A-weighted level [2].

Starnberg [3] investigated different factors associated with undamped TDR measurements on real tracks with the potential to influence the measured results. One particular key finding of [3] is of particular interest in the context of this paper investigating real TDR measured on slab tracks. This is that inhomogeneity along the track could have a considerable influence on the resulting measured TDR with large variations noted from different measurement starting positions along the rail or from the opposite rail. This was the case even when there was no obvious reason for inhomogeneity, and the requirements of the EN15461 standard [4] to avoid loose fasteners, pumping sleepers, welds etc. have been met.

The STARDAMP investigations of factors associated with uncertainty in prediction all involved ballasted track scenarios [1, 3]. This reflects the European context in which rail dampers were developed, where the primary concern is rolling noise from mainline trains (particularly freight trains) on ballasted tracks. In this context rail dampers can often give a relatively small benefit of up to 3 dB, with their effectiveness limited by regional preferences for the stiffness of the rail pads [5]. Elsewhere, the primary interest in rail dampers has been from transit systems, for example in Australia for direct noise control [6, 7], and in North America [8] and Asia [9] for direct noise control and to investigate potential to change corrugation growth rates. In these situations with slab track, resilient fasteners and relatively small metro-style wheels, it is expected that rail dampers will provide a larger relative rolling noise reduction than on European ballasted track systems.

The results in this paper include measurements of undamped TDR on several different slab track systems, comparable theoretical predictions of undamped TDR and a series of STARDAMP investigations of the anticipated rail damper benefit for the various systems. The measured noise benefit achieved by rail dampers in practice is also reported for several cases where results are available from physical implementations. The question posed is whether use of measured undamped TDRs for slab track systems can result in under-prediction of the rolling noise emitted by the track and hence the potential noise reduction from rail dampers, as reported by Jones et al. [2].

2 Methodology

Measured undamped TDRs have been compiled from five different existing slab track systems. These systems have rail types ranging from 54 kg/m up to 60 kg/m, and fastener spacing ranging from 0.65 m up to 1.0 m. The nominal vertical stiffness per support ranges from approximately 5 MN/m up to around 25 MN/m. TDRs have been measured generally following the EN15461 approach [4]. Table 1 describes the parameters of the five different track systems for which TDR data has been compiled.

Predicted baseline undamped TDRs have been calculated using STARDAMP and a track vibration model described in [10], with Timoshenko beam representation of the rail on 2-layer discrete supports with constant fastener spacing. The model parameters used for the discrete support model are shown in Table 2. The STARDAMP model includes a single resilient layer and neglects the baseplate mass.

Table 1. Slab track system parameters with measured undamped TDR results available.

Case	Track description	Fastener spacing	Nominal vertical fastener stiffness
A	54 kg/m rail, Rheda, baseplate fasteners IOARV 300	0.65 m	22 MN/m
B	57 kg/m rail, baseplate fasteners Delkor Alt1 RF191	1.0 m	25 MN/m
C	57 kg/m rail, baseplate fasteners Lord	1.0 m	25 MN/m
D	60 kg/m rail, baseplate fasteners Delkor Alt1 RF192	0.7 m	25 MN/m
E	60 kg/m rail, Pandrol Vanguard fasteners	0.7 m	5 MN/m

Table 2. Parameters for discretely supported beam model to calculate vertical TDR.

Case	Rail pad stiffness	Rail pad loss factor	Baseplate mass (spacing)	Baseplate pad stiffness	Baseplate pad loss factor
A	2500 MN/m	0.2	4 kg (0.65 m)	22 MN/m	0.2
B	2500 MN/m	0.2	4 kg (1.0 m)	25 MN/m	0.2
C	2500 MN/m	0.2	4 kg (1.0 m)	25 MN/m	0.2
D	2500 MN/m	0.2	4 kg (0.7 m)	25 MN/m	0.2
E	n/a	n/a	n/a (0.7 m)	5 MN/m	0.2

Short rail testing of rail dampers uses the approach described in [11]. Results for four configurations of Schrey & Veit (S&V) rail dampers have been used (Table 3).

Table 3. Rail dampers with short rail (6 m) damped TDR results available.

Rail damper	Fastener and damper spacing
S&V VICON AMSA 115 FSV	0.6 m "typical" fastener spacing, one damper per 0.6 m
	1.0 m "extended" fastener spacing, one damper per 1.0 m
	1.0 m "extended" fastener spacing, one damper per 0.5 m
S&V AMSA AS60 VS	0.7 m "typical" spacing, one per crib

3 Results

3.1 Measured Undamped Vertical TDR vs Theory

Figure 1 shows the measured vertical TDR from each track compared with the predicted vertical TDR from STARDAMP and the discretely supported beam model. The measured results for Case A are reproduced from [12]. For this case the measured vertical TDR is high relative to both the predictions, particularly in the 800 Hz and 1000 Hz frequency bands which are often important for rolling noise.

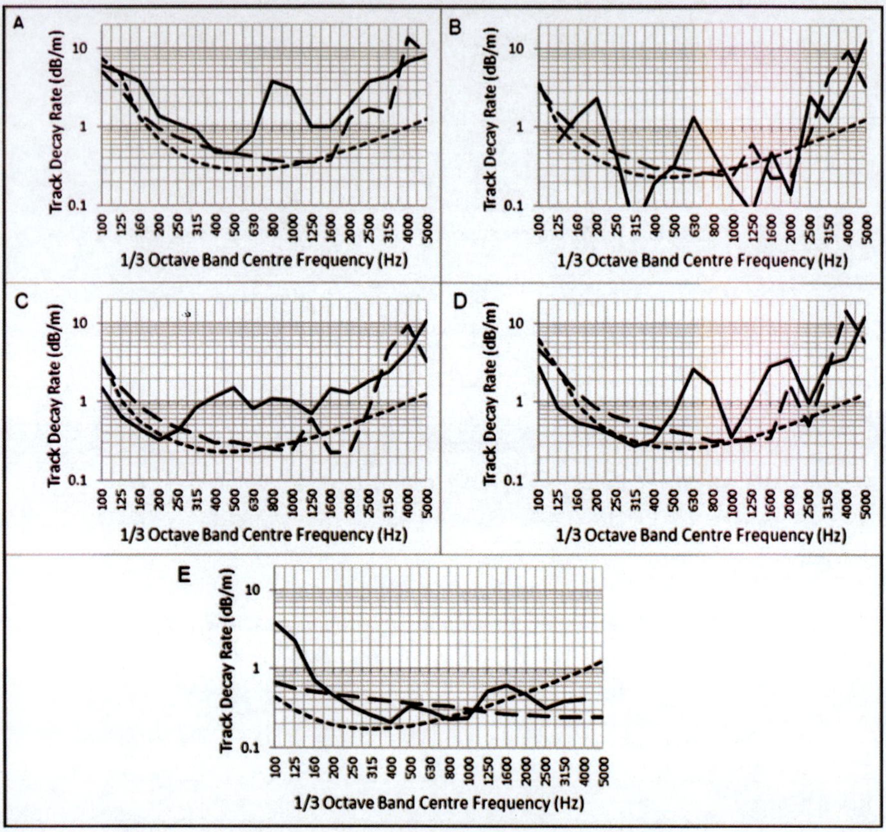

Fig. 1. Measured vs predicted vertical undamped track decay rates, Case A through Case E. ——— Measured; — — — Discrete support model; - - - - - - STARDAMP model.

Case B and C are identical in terms of the nominal vertical fastener stiffness and fastener spacing for prediction of TDR. The measured vertical TDR in Case B shows several distinct peaks, with the peak at 630 Hz likely to be most relevant to predictions of rail damper benefit. The measured vertical TDR for Case C is higher than analytical predictions at frequencies above 315 Hz, and is generally higher than the measured TDR from Case B. Notable differences between these two sites are the age of the track – B is a newly installed line with very consistent fastener spacing, whereas C is a considerably older installation, with fasteners over 20 years old, no discrete pad between rail and baseplate, and some variations from regular fastener spacings.

The results for Case D also show a measured vertical TDR that is high relative to predictions, particularly in the 630 Hz and 800 Hz frequency bands.

Case E shows the best agreement between measured and predicted vertical TDR. This track (Vanguard) incorporates a softer rail support than the other cases examined and has a single resilient layer, with all other cases involving two resilient layer baseplate-type fasteners of fairly typical stiffness for these track forms.

3.2 Measured Short Rail Vertical Damped DR Results

Figure 2 shows the measured vertical damped DR results for four different configurations of rail dampers. It can be seen that DR values below 500 Hz are very low, caused by the modal behavior of the finite rail. As determined during the STARDAMP project execution [13] these frequency bands below 500 Hz from short rail tests are not considered to be reliable for noise prediction.

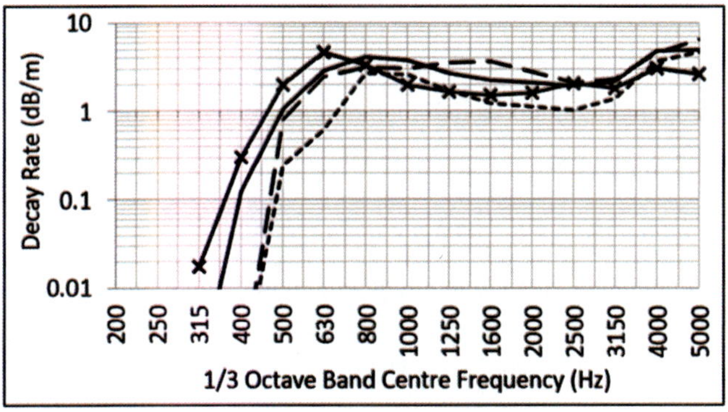

Fig. 2. Measured short rail (6 m) damped decay rates. ——— VICON AMSA 115 FSV 0.6 m spacing; — — — VICON AMSA 115 FSV 0.5 m spacing; ------ VICON AMSA 115 FSV 1.0 m spacing; —x—x AMSA AS60 VS 0.7 m spacing.

3.3 STARDAMP Predictions of Rail Damper Benefit

Table 4 shows predictions of the overall A-weighted rail damper benefit for the five scenarios, with the baseline TDR either measured or calculated analytically.

In Table 4 all cases have 80 km/h train speed, and STARDAMP default disc braked wheel roughness and TSI rail roughness. Results are provided in the form of the calculated reduction in overall sound power level. For each case, three different approaches are used to calculate the benefit from damped short rail DRs. The first uses the measured undamped track baseline TDR. Results have also been generated using the STARDAMP

Table 4. Predicted rail damper noise reduction from measured or modelled TDR baselines.

Track case	Traffic details	Rail damper short rail DR	Sound power level reduction (dBA)		
			Measured TDR	STARDAMP model	Discrete support model
A	STARDAMP regional train default	VICON AMSA 115 FSV 0.6 m	2.1	6.3	5.4
B	Ø 585 mm wheel, 7.7 t axle load, length 17 m	VICON AMSA 115 FSV 1.0 m	5.7	6.0	6.3
C	Ø 585 mm wheel, 7.7 t axle load, length 17 m	VICON AMSA 115 FSV 0.5 m	4.5	8.1	8.3
D	Ø 840 mm wheel, 5.4 t axle load, length 15 m	AMSA AS60 VS 0.7 m	3.1	7.3	6.5
E	Ø 840 mm wheel, 5.1 t axle load, length 24 m	AMSA AS60 VS 0.7 m	7.2	7.7	7.1

calculated baseline TDR, and for another analytical beam model accounting for the discrete nature and 2 resilient layers of the rail fasteners (where applicable). Notably different results are seen when using measured TDR as the baseline rather than modelled TDR.

3.4 Measured Rail Damper Benefit

Measured results for the noise reduction provided from physical trials of S&V rail dampers are available for comparison with cases B, C and E:

- Case B – Measured A-weighted overall noise reduction 4 dB
- Case C – Measured A-weighted overall noise reduction 6 dB
- Case E – Measured A-weighted overall noise reduction 8 dB [7]

In Case B, the noise reduction from the physical rail damper trial was measured indirectly, in the form of in-car noise with the rail damper installation in a tunnel section. The measured damper benefit was less than expected from theoretical predictions using default roughness spectra, with investigations of spectral results indicating rolling noise was dominated by lower frequencies (around 500 Hz), due to residual grinding induced roughness. The rail dampers (installed with one damper per 1.0 m fastener spacing) were more effective at higher frequencies. Similar results were reported in [8] where a 3 dB noise reduction was observed for rail dampers installed on slab track with 36 inch (0.91 m) spacing, limited by dominant noise at 500 Hz.

Case C involved trials of the same dampers as Case B, installed on surface track with two dampers per 1.0 m fastener spacing to improve low frequency performance. The trial location was corrugated with dominant passby noise at 500 Hz. The predicted 4.5 dB noise benefit of the rail dampers when using TSI rail roughness and baseline measured TDR in Case C was 1.2 dB less than for Case B, which is counter-intuitive since the track stiffness is nominally the same, and double the number of rail dampers will increase rail damper performance (as indicated by improved short rail damped DR). The directly measured Case C rail damper noise reduction of 6 dB was greater than predicted based on TSI rail roughness and measured undamped TDR, but remained limited by noise at 500 Hz (corrugation effects). At higher frequencies, the rail dampers had a greater effect. In the 800 Hz frequency band passby noise emissions were reduced by 8–9 dB.

Case E showed good agreement between the measured rail damper benefit of 8 dB and predictions, as expected considering the measured and predicted baseline vertical TDRs are very similar. This result was also measured indirectly in-car for a trial site in a tunnel, and actual rail roughness was comparable to TSI levels.

4 Discussion

For rail dampers, the method developed and implemented in STARDAMP software assumes the damped TDR can be predicted by summing the measured TDR of an undamped real track with a DR measured on a short section of freely suspended rail fitted with the rail dampers. For slab tracks with baseplate type fasteners, the results reported in this paper indicate that this approach may not always be reliable.

Relative to simplified analytical predictions, the measured TDRs for slab track with baseplate fasteners all show relatively high values, either across the spectrum above about 315 Hz or in the form of distinct peaks between 630 Hz and 1000 Hz. These are important frequencies for rolling noise prediction. In 4 of the 5 slab track cases the measured undamped TDR includes areas where the measured TDR is higher than would be expected from theoretical calculations with the baseplate either neglected entirely or represented as a lumped mass (ie no baseplate dynamic effects). Similar measured TDR effects were observed by Jones et al. [2] and Liu et al. [9]. The addition of short rail DRs to these measured vertical TDRs may result in an under-prediction of rail damper benefit in STARDAMP. The use of an analytically calculated TDR seems to give an improved rail damper benefit prediction, however counteracting effects may be involved. Research by Hima et al. [14] indicates that for track with baseplate type fasteners the modal behavior of the baseplate can increase TDR in the frequency range of interest for noise, and it is possible that noise radiated from the baseplate itself contributes to the overall track noise. Both of these effects may limit the effectiveness of rail dampers, but are neglected in STARDAMP slab track models.

In general, the five situations investigated confirm the potential for rail dampers to provide effective noise mitigation for transit systems on slab track, with potential reductions in overall A-weighted noise levels of 4–8 dB confirmed by physical trials. This is the result of the combination of very soft rail fasteners and wheels that are typically smaller and quieter than those used on mainline European systems. Wheel roughness was not investigated during this work, although measurements of multiple train passbys in Case C suggest wheel roughness was not a factor in the results observed. In some circumstances the achievable rail damper benefit can be limited by rail roughness resulting in dominant rolling noise frequencies that are sometimes as low as 500 Hz with speeds around 80 km/h. Existing rail damper designs have typically been optimized to be most effective above 630 Hz, and have limited effect at lower frequencies. This shows the need for tuning rail dampers to specific site condition to maximize the noise benefit for transit systems with slab tracks.

5 Conclusion

The normal STARDAMP assessment approach of summing a real measured undamped TDR with a damped short rail DR has been found to result in uncertainty in predicting the rail damper benefit for some slab tracks with baseplate fasteners. This track type is common on transit systems that are increasingly considering rail dampers for noise control. An alternative estimate of the rail damper benefit is obtained by summing the predicted undamped TDRs with a damped DRs from a short rail test, ie by using default STARDAMP calculation options in place of a measured undamped TDR result. However, this approach requires a simplification omitting baseplate dynamic effects and noise radiated from the baseplate. There are indications these effects may be significant factors in rolling noise from slab tracks. In all cases it is important to supplement predictions with sufficient measurements to understand the rail roughness spectrum and dominant frequency range of rolling noise. On the basis of this investigation, physical measurements of rolling noise spectrum, rail roughness and tuning and trials of rail dampers to confirm the predicted benefit are expected to be more necessary for transit systems with slab tracks than may be the case for mainline ballasted track systems. Further research into the effects of baseplate dynamic effects on noise and rail damper performance is recommended.

References

1. Asmussen, B., Starnberg, M., et.al.: STARDAMP Standardisation of Damping Technologies for the Reduction of Rolling Noise. Final report 27th May 2013.
2. Jones, C.J.C., Thompson, D.J., Diehl, R.: The use of decay rates to analyse the performance of railway track in rolling noise generation. J. Sound Vib. **293**, 485–495 (2006)
3. Starnberg, M.: STARDAMP Project Measuring TDR on undamped real tracks – Best practice rules
4. EN 15461:2008+A1:2010:E. Railway applications. Noise emission. Characterization of the dynamic properties of track selections for pass by noise measurements (2010)
5. de Vos, P.: Railway Noise in Europe State of the art report. International Union of Railways (UIC), March (2016)

6. Weber, C. M., Sburlati, D.: Source noise control to mitigate airborne noise at high rise developments – Epping to chatswood rail link. In: Proceedings of 20th International Congress on Acoustics, Sydney (2010)
7. Zoontjens, L., Welsh, L., Croft, B.: Predicting and managing rolling noise emissions from trains on the Perth metro passenger rail network. Acoustics 2017 Perth, Australia, pp. 19–22 November (2017)
8. Duenas, S., Wolf, S.: Bay Area Rapid Transit (BART) Rail Corrugation Study. American Railway Engineering and Maintenance-of-Way Association Conference, Chicago (2014)
9. Liu, W., Zhang, H., Liu, W., Thompson, D.: Experimental study of the treatment methods for rail corrugation on tracks with Egg fasteners track in the Beijing metro. In: Proceedings of the Institution of Mechanical Engineers Part F, J. Rail Rapid Transit. (2018)
10. Thompson, D.: Railway Noise and Vibration Mechanisms, Modelling and Means of Control, Elsevier, Oxford (2009)
11. Squicciarini, G., Toward, M.G.R., Thompson, D.J.: Experimental procedures for testing the performance of rail dampers. J. Sound Vib. **359**, 21–39 (2015)
12. Gramowski, C.: Gleisseitige Schallemission - Experimentelle Quantifizierung und Bewertung konstruktiver Maßnahmen, Berlin (2012)
13. Kitson, P.: Laboratory Testing of Rail Damper. Stardamp Final Workshop (2012)
14. Hima, B., Thompson, D., Squicciarini, G., Ntotsios, E., Herron, D.: Estimation of track decay rates from laboratory measurements on a baseplate fastening system. IWRN13 Ghent, Belgium, 16–20 September (2019)

Rail Damper Composed of Tuned Mass Damper and Constrained Layer Adopted in Non-symmetric Rail

Sam Lo, Wilson Ho(✉), and Cindy Cheung

Wilson Acoustics Limited, Unit 601, Block A, Shatin Industrial Centre, Yuen Shun Circuit, Shatin, Hong Kong
who@wal.hk

Abstract. A tramway is installed on resilient booted sleepers for controlling groundborne noise at an open cut section connecting to tunnel. However, air-borne noise impact was a concern for the surrounding houses. In order to reduce the air-borne noise emission from the non-symmetric profile groove rail (CEN 53R1), a new type of rail damper, which was consisted of tuned mass dampers and constrained layer, was developed for tackling the dominant noise frequencies from 250 Hz to 1200 Hz. Track decay rate, rail vibration levels, near-field noise and far-field noise levels have been studied for various arrangements of partial installation of rail dampers (four arrangements: baseline without damper, quarter pattern installation, half pattern installation and full installation) and were analyzed in the range of 500 Hz to 2000 Hz in 1/3-octave bands. During passby of two major types of trams in the operational fleet (namely the "old" and "new" trams), near-field noise reduction of 3.7 dB(A) and 4.2 dB(A) and far-field noise reduction of up to 4.3 dB(A) and 3.4 dB(A) were achieved respectively.

Keywords: Groove rail · Rolling noise · Noise and vibration mitigation

1 Introduction

A tramway (groove rail CEN 53R1) on resilient booted synthetic timber sleepers (Fig. 1) was adopted for controlling ground-borne noise at an open cut section connecting to tunnel. However, air-borne noise was marginally exceeding the allowable limit at the surrounding houses. A new type of rail damper has been developed for the non-symmetric groove rail, different from the traditional rail damper adopted in symmetric rail [1, 2]. The damper consists of a constrained layer bar to damp the vibration at the guard of groove rails in addition to nine mass oscillators (Tuned Mass Damper, TMD) tuned to the frequencies of vertical and lateral pinned-pinned resonances, as well as grinding mark induced rail vibration. Track Decay Rate (TDR), noise and rail vibration reduction were measured for quarter pattern damper installation, half pattern damper installation and full damper installation.

G. Degrande et al. (Eds.): *Noise and Vibration Mitigation for Rail Transportation Systems*, NNFM 150, pp. 638–645, 2021.
https://doi.org/10.1007/978-3-030-70289-2_70

Fig. 1. Half damper installation on tramway (groove rail CEN 53R1) on booted sleepers

2 Prototype Damper Design

Before the first overseas site visit, a prototype damper was prepared based on findings from wayside noise and in-saloon noise spectra (sound recording files with spectrograms shown in Fig. 2), drawings and photos of the groove rail. Noise peaks at 630 Hz and 1250 Hz were observed in the wayside noise spectrum.

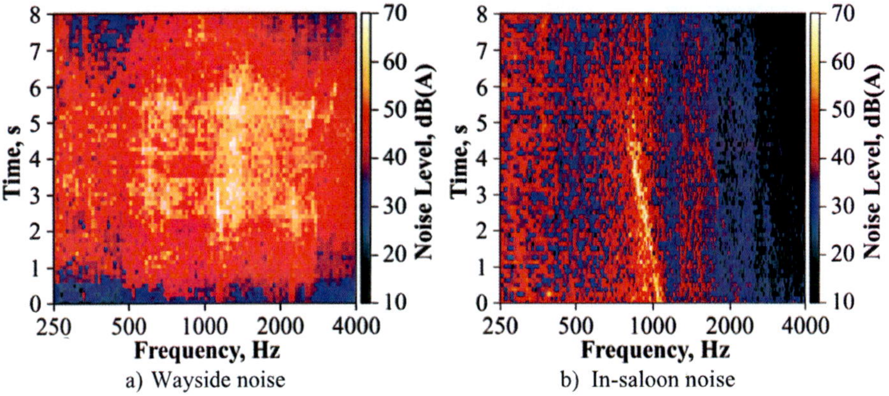

Fig. 2. Spectrogram analysis from sound recording files

The dynamic response of the rail with fasteners at 0.6 m sleeper spacing was simulated by Finite Element Analysis (FEA) and calibrated with a 7 m long rail sample without fasteners. Rail pinned-pinned resonances were found at 657 Hz and 745 Hz respectively for vertical and lateral vibration (Fig. 3). The 1250 Hz peak is suspected to be generated from the grinding marks induced vibration of the rail.

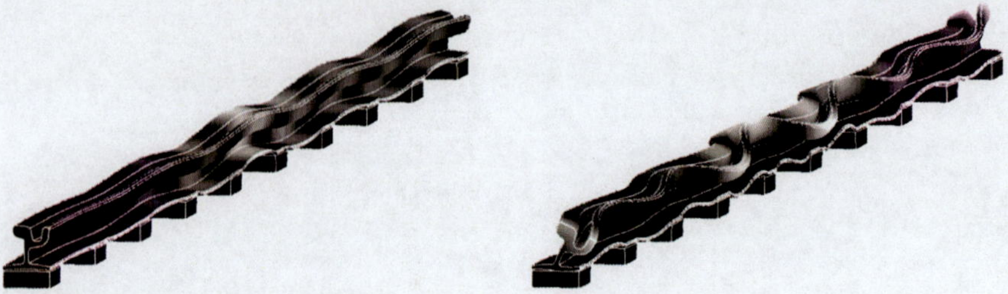

a) Vertical (657 Hz) b) Lateral (745 Hz)

Fig. 3. Simulated pinned-pinned resonance modes in vertical and lateral directions

Based on the FEA simulation results, a prototype damper consisting of nine TMD oscillators (tuned to the frequencies of vertical and lateral pinned-pinned resonances and grinding mark frequencies) and a constrained layer bar suppressing vibration of the guard of the groove rails, was tested (Fig. 4). A wax layer of about 0.5 mm thick was applied along the constrained layer interface (Fig. 4a).

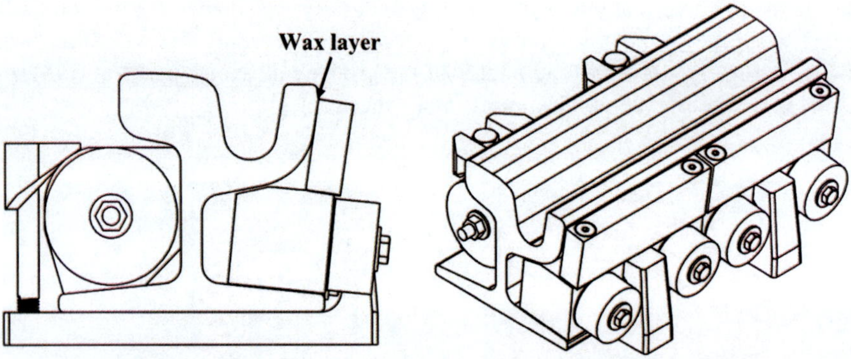

a) Cross-sectional view b) Isometric view

Fig. 4. Drawings of the damper prototype

The prototype damper was installed during the site visit for trial. The effect of a single damper on the operating rail was then measured at the trial site in order to verify the damper design and anticipated performance of the TMD oscillators.

3 Single Damper Effect on the Operating Rail

A single prototype rail damper with tuned frequencies (Fig. 5) was installed on one track (far track) to test on the operating tramway. Rail surface roughness and rail vibration during tram passby were measured.

Rail surface roughness is of interest as it has a role in wheel-rail interaction [3] and was measured to be about 3 μm rms at 50 mm wavelength and about 1.2 μm rms at 8 mm wavelength. Since rolling noise could be excited by wheel/rail roughness, the noise frequency would be the train speed divided by wavelength [3]. The speed of passby trams varied between 22 km/h and 45 km/h. Figure 6 shows the root-mean-square rail roughness in 1/3-octave bands and the corresponding 1/3-octave frequencies to the wavelengths.

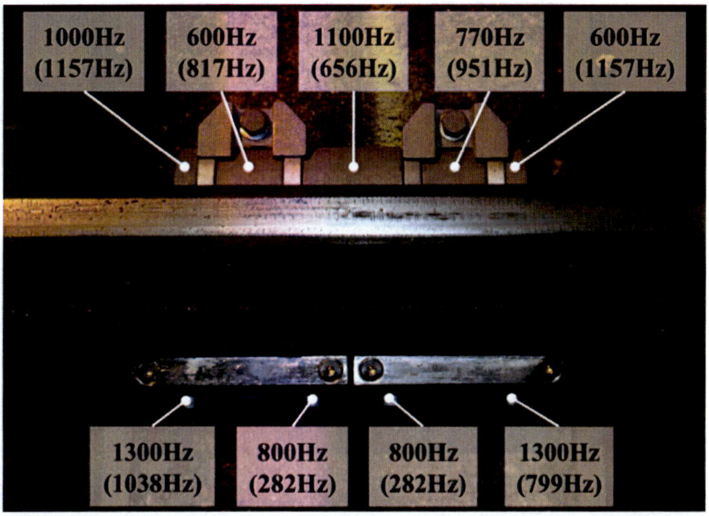

Fig. 5. Prototype for single damper trial installation top view on the groove rail showing the TMD frequencies for each oscillator. Values in brackets are the tuned frequencies for final installation.

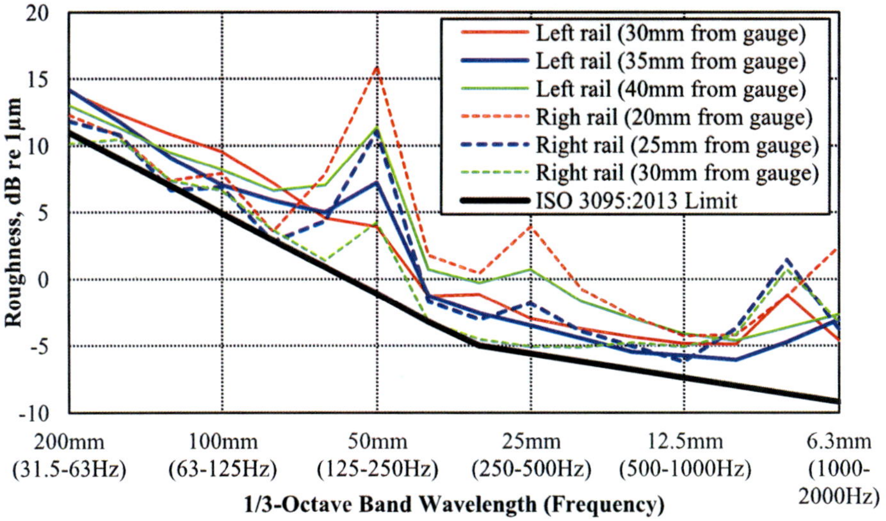

Fig. 6. Rail roughness against wavelength with the frequencies in brackets corresponding to the tram speeds of 22 km/h and 45 km/h.

There were two major types of trams in the operational fleet, namely the "old" and "new" trams. Due to differences in noise characteristics, the data analysis for these two major types was conducted separately.

During the trial of single damper installation, rail vibration levels at the tramway were reduced by more than 10 dB in the vertical direction and around 5 dB in the lateral direction at critical frequencies (Fig. 7). The peak of vibration at 1250 Hz is consistent with the peak of rail roughness at 8 mm, which is likely to correspond to the grinding mark.

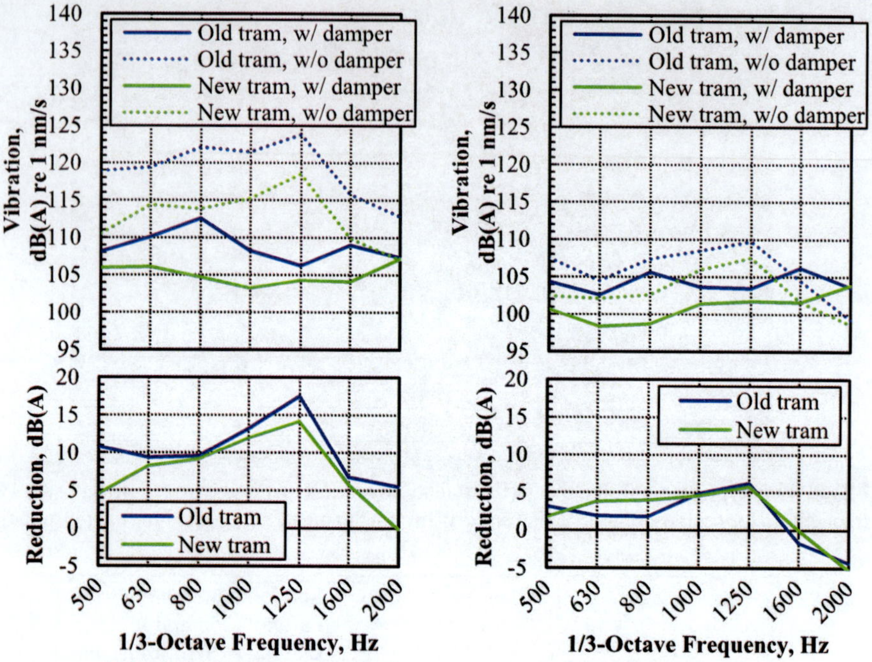

a) Vertical vibration at foot tip b) Lateral vibration at rail web

Fig. 7. Tram passby rail vibration velocity levels (average of 10 events) with and without installation of the single set of rail damper

4 Performance of Partial and Full Installation of Rail Dampers

Rail dampers were installed at tracks in both directions over a total track length of 200 m progressively from baseline, quarter pattern (damper installed at every 4 sleeper spacings), half pattern (damper installed at alternate sleeper spacings) and full installation (damper installed at every sleeper spacing).

Track Decay Rates (TDRs) were measured according to EN 15461 [4] and are shown in Fig. 8. The TDR at 1000 Hz in vertical direction was increased from 0.46 dB/m to 6.9 dB/m, 7.8 dB/m and 13 dB/m with quarter, half and full damper installation respectively. TDRs in both vertical and lateral directions were raised beyond the TSI Limit with half and full installation of rail dampers. The denser the rail damper installation along the rail, the higher the TDRs and the greater the reduction of noise and vibration were obtained.

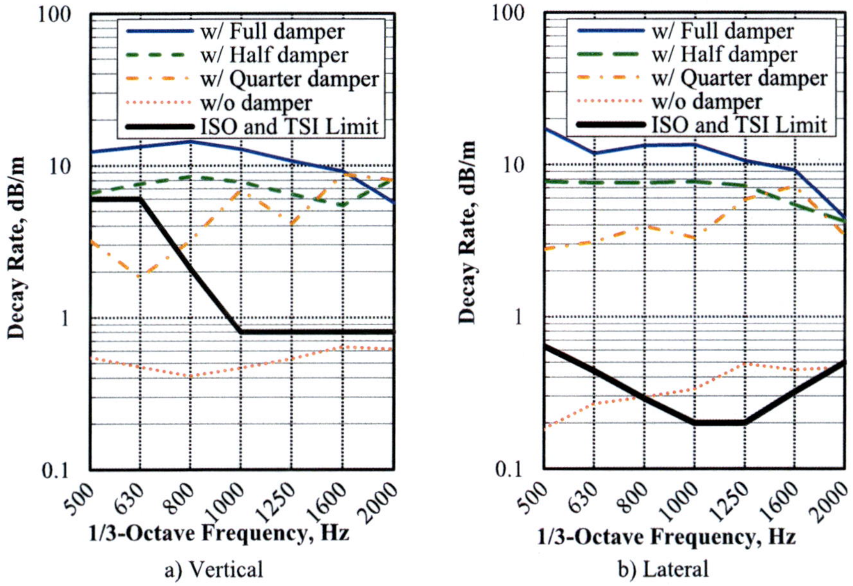

Fig. 8. Track decay rate of the rail in vertical and lateral directions

Figure 9 shows a sketch of the vibration and noise measurement setup at the site. An accelerometer was set under the rail and microphones were set at 0.2 m from rail at near track and at high level (4.7 m from the ground, which is approximately 6 m from the near track and 8 m from the far track).

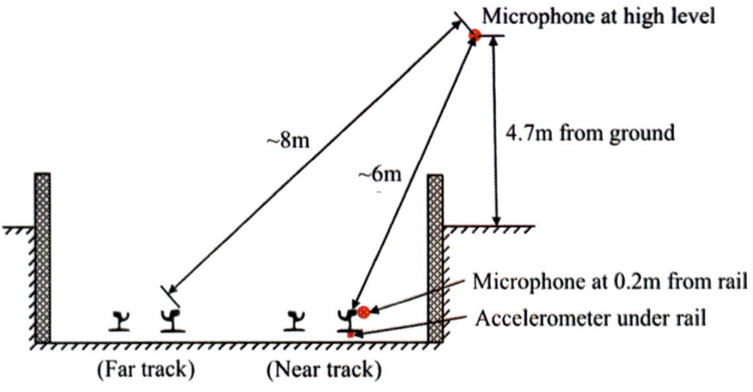

Fig. 9. Sketch of microphone positions

Figure 10 shows the rail vibration, near-field and far-field near track passby noise levels for quarter, half and full damper installation. The vibration reduction over 500 Hz to 2000 Hz during near track old tram and new tram passby was respectively 8.1 dB(A) and 8.7 dB(A) at quarter damper installation. In the situation of old tram passby with full installation, vibration reduction of 11.5 dB(A) and near-field noise reduction of 3.7 dB(A), and far-field noise reduction of 2.7 dB(A) were achieved in the range of 500 Hz to 2000 Hz, and in the situation of new tram passby, the vibration reduction, near-field noise and far-field noise reduction were 12.8 dB(A), 4.2 dB(A) and 3.4 dB(A) respectively. The peak of vibration at 630 Hz in Fig. 10b does not correspond with any peaks of rail roughness (Fig. 6), which suggests that it is not caused by rail roughness. Also, the 800 Hz peak in near-field noise in Fig. 10b does not correspond with any rail vibration peaks, suggesting this is caused by other noise sources (not the rail).

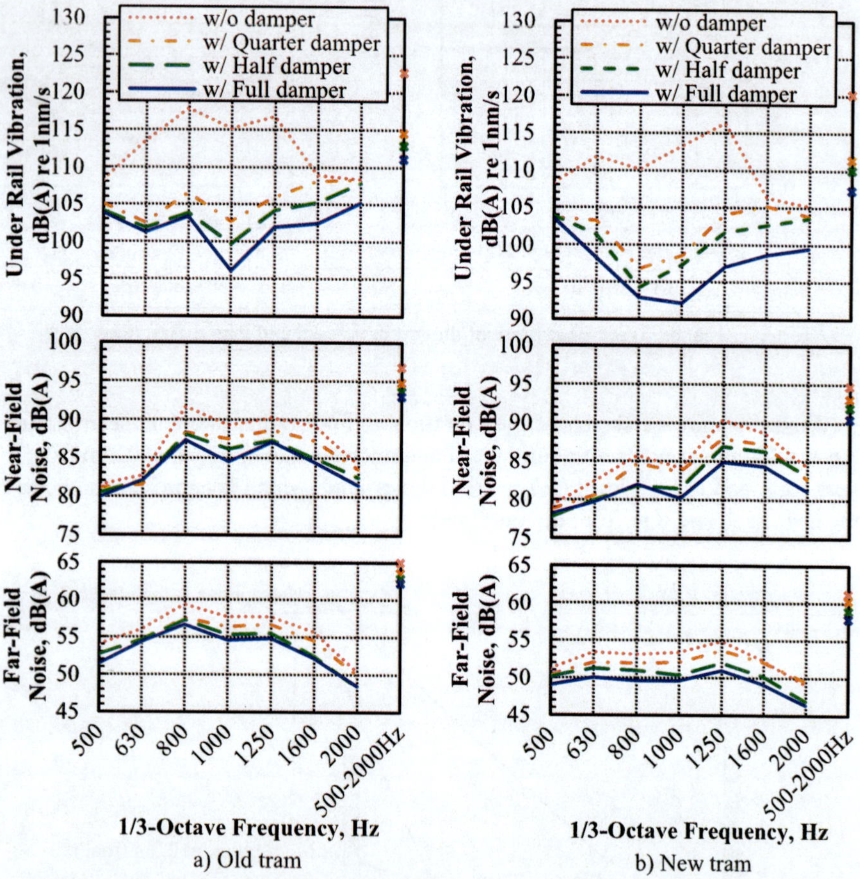

Fig. 10. Near track passby noise and vibration spectra of old and new trams

The far-field noise levels for the far track passby are shown in Fig. 11. Due to the smaller barrier effect of the shielding by the trackside wall, the noise levels of the far track passby were generally higher and the noise reductions by the rail dampers were more significant. Far-field noise reduction of 4.3 dB(A) and 3.4 dB(A) were achieved in the range of 500 Hz to 2000 Hz for the old tram and new tram passby at the far track.

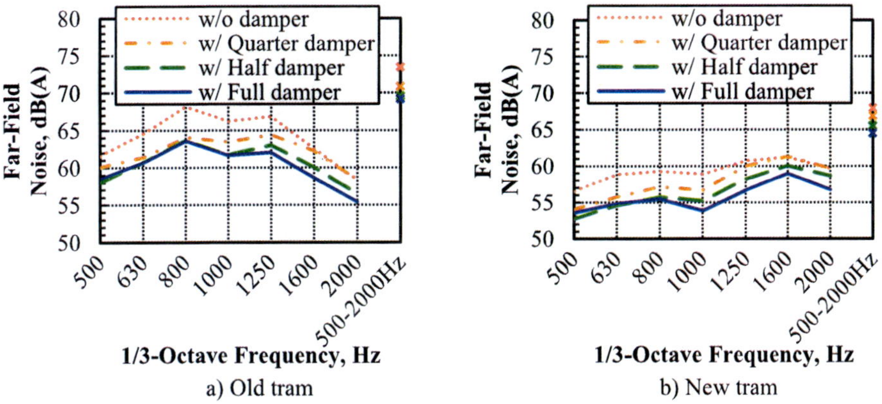

Fig. 11. Far track far-field passby noise spectra of old and new trams

5 Conclusions

New rail dampers, which each consists of 9 TMD oscillators and a constrained layer bar, have been applied in groove rail for tramway rail noise control. Near-field noise reduction of 3.7 dB(A) and 4.2 dB(A) and far-field noise reduction of up to 4.3 dB(A) and 3.4 dB(A) were achieved for old tram and new tram passby in the range of 500 Hz to 2000 Hz in 1/3-octave bands. Track decay rates in both vertical and lateral directions were raised beyond the TSI Limit. Vibration at the tramway is demonstrated to be controlled by rail dampers.

References

1. Ho, W., Wong, B., Tam, P.L., Cai, T.: Tuned mass damper design optimization using FEA. In: ICSV22 on Proceedings, Florence, Italy (2014)
2. Ho, W., Wong, B., England, D.: Tuned mass damper for rail noise control. In: The 10th International Workshop on Railway Noise on Proceedings, Nagahama, Japan (2010)
3. Thompson, D.: Railway Noise and Vibration - Mechanisms, Modeling and Means of Control, 1st edn. Elsevier, Oxford (2009)
4. EN 15461:2008+A1:2010: Railway applications—Noise emission—Characterization of the dynamic properties of track selections for pass by noise measurements (2009)

Check for updates

Entering the Real Operation Phase: Design, Construction and Benefit Verification of Freight Wheel Noise Absorber

Christoph Gramowski[1](✉) and Thomas Gerlach[2]

[1] Schrey & Veit GmbH, Graf-von-Sponheim-Straße 2, 55576 Sprendlingen, Germany
christoph.gramowski@sundv.de

[2] Gutehoffnungshütte Radsatz GmbH, Gartenstraße 40, 46145 Oberhausen, Germany

Abstract. Until today, freight train noise is the main annoying railway noise in many places around the world. Numerous approaches were started in the past to establish mitigation measures. An integrated system of a silent wheel design with additional vibration absorbers has been established. Different tests have proven the noise level $L_{Aeq,Tp}$ (according ISO 3095) is reduced by about 4 dB. Also, the homologation according TSI Wagon was confirmed, enabling a European wide operation. Practical barriers for the implementation are now the missing motivation for freight wagon owners to improve their fleets' noise below TSI Noise limits. This paper introduces technical data and background as well as thoughts on the implementation and current system constraints.

Keywords: Railway noise · Freight train · Acoustically optimized wheel · Wheel noise absorber

1 Introduction

Until today, freight train noise is the main annoying railway noise in many places around the world. Numerous approaches were started in the past to establish mitigation measures. The launch of composite brake blocks, enabling high noise reductions, has shifted the need for further measures by about ten years into the present days.

However, freight train noise reduction is determined by very different constraints compared with passenger trains. A key issue is the use of block brakes which triggers additional requirements on the materials. Also, the return on investments via increased income for freight operators is not realistic. Therefore, the implementation of any measure must be triggered and supported by an external support, usually a public authority. All mitigation technology suppliers need to keep this in mind because the implementation process requires much more time. Also, it is possible that product specifications will require minimum noise reduction to confirm the benefit. Similar developments are visible for railway track installed noise mitigation measures, e. g. from Deutsche Bahn [1].

G. Degrande et al. (Eds.): *Noise and Vibration Mitigation for Rail Transportation Systems*, NNFM 150, pp. 646–653, 2021.
https://doi.org/10.1007/978-3-030-70289-2_71

To reach lower noise levels (after brake block replacement from cast iron to composite blocks), “Wheel Dampers” can also be considered. This technology is already listed by the UIC as a rolling stock measure [2]. The main advantage compared to disc brakes (leading to slightly higher noise reductions) is that wheel dampers allow the ongoing use of the bogie. Disc brakes require another braking device where block brake parts cannot remain in use. Usually, this causes different installation points at the bogie frame which lead to a replacement of the whole bogie.

One has to keep in mind that the Switzerland national legislation is in fact requiring composite brake blocks from 2020 on. The German legislative approach is already started, but other EU member states are still able to block it. Therefore, the freight vehicle owners are forced to change the brake blocks up to 2020. No other measure is required from the legislative point of view.

Since a couple of years, a special adjusted wheel noise absorber is being designed for block braked freight vehicles. From a R&D project, funded by the Swiss Federal Office for Environment (BAFU), an important step towards the real operation phase was taken. This includes design improvements and also field measurements for the standardized benefit quantification. The development methodology and results will be presented. Additionally, current barriers to widespread adoption are summarized, leading to ideas on improving next steps.

This paper gives an overview of the current end user’s view together with political/co-funding options, on the technical/acoustical requirements and presents an example of a ready-to-market product. It provides a deeper understanding of this railway acoustic issue for technical researchers as well as for market stakeholders. All research work described here was integrated in different R&D cooperation projects.

2 Methodology

2.1 Technical Requirements on Freight Train Wheel Absorber

Several design requirements are applicable on freight vehicle wheelsets with vibration dampers. The operational conditions lead to highly increased requirements on the mechanical absorber strength compared to light rail vehicles, tram or metro applications. These are mainly caused by heavy axle loads and impacts from switch frogs, rail gaps and road crossings. The resistance to chemical impacts is important when the related loads are carried. Also, railway typical oils, fuels as well as ultraviolet and ozone impacts have to be considered. The requirements of fire protection according EN 45545-2 [3] have to be taken into account when using damping material such as elastomers.

A very special requirement for vibration absorbers is the resistance to high temperature loads up to 500 °C, caused by the block braking system. From this constraint, the useable material types are strongly limited. Also, it has to be considered that wheelset mounted components have to fulfill the requirements of EN 61373 [4]. Further requirements (mechanical, thermo-mechanical and acoustical) regarding the technical approval for wheels are defined in UIC 510-5 [5] and EN 13979-1 [6].

2.2 Wheel Vibration Reduction Principles

At first, general absorbing principles applicable to train wheels, together with technology constraints are shown, followed by an assessment of their operational usability.

From previous R&D projects and also theoretical analysis, it became clear that wheels are an important source for freight train rolling noise and also curve squealing. Because the reduction (e. g. shielding) of airborne noise from freight wagons is only possible at limited outer wagon areas [7], the obvious way for reducing the noise is the decrease of wheel disc vibration.

In general, two common principles for wheel vibration reduction are applicable:

- damping (by internal viscosity or friction)
- tuned mass damper

For both principles, different commercial products are available on the market. Some of these products use both principles. However, none of these are in a widespread use in freight cars today.

The friction damping was assessed to fail the long-term stability requirement under freight train conditions. Different impacts, like the movement of corrosive freight, lead to undefined surface conditions. Also, global corrosion can prohibit the function at the connection of both friction parts. Beside wheel mounted vibrational damping devices, radial steering wheelsets can also decrease the squealing noise in curves.

2.3 Wheel Damper Type VICON RASA RSI

The setup of this freight wheel absorber type was started in 2000. The initial designs were derived from passenger train applications. Subsequent work led to the conclusion that a much more robust design is necessary. Here, the key feature was kept for the next design phase: a tunable mass-spring system which can be adjusted to the specific wheel.

For a wheel design which conforms to TSI Wagon [8] a new wheel type had to be developed, labelled "BA 308". There, different wheel mounting positions for the absorbers were investigated. The absorbers are based on sandwich stacked steel and elastomer plates. Their proportions are determined by a laboratory development process.

Fig. 1. Production of BA 308 wheelsets and absorbers VICON RASA RSI

These 'blocks' are placed on 'ring elements' which are inserted in the inner side of the wheel, see Fig. 1. No glue or other contact material is used.

Due to special selected materials the absorbers withstand high temperatures up to 500 °C. From the results of a constant impact on the wheel vibration, it was confirmed that neither the temperature impact nor the repeated load limits the absorber performance. However, the standardized temperature impact is linked to the lengthy braking scenario at the Swiss Gotthard slope, which is now avoided due to the base tunnel opened in 2016.

3 R&D Work on Wheels and Absorbers

Most of the development of wheels and vibration absorbers was contained in two R&D cooperation projects.

3.1 German R&D Project Low Noise Train on Real Track

In subproject B2 of the R&D project "Leiser Zug auf realem Gleis" ('Low Noise Train on Real Track' LZarG) funded by the German Federal Ministry for Economic Affairs and Technology BMWi [9], extensive R&D investigations had been executed, aiming at a noise emitting reduced freight train wheelset with vibration absorbers. Different tasks had been worked on by GHH-Radsatz, Technische Universität Dresden and also Schrey & Veit (development of temperature resisting absorber).

The wheels are designed by innovative approaches in terms of weight, mechanical and thermal strength and also acoustic parameters. The optimal combination of all parameters like shape, material and coating leads to maximum availability, reliability and lifetime – with positive impact on the LCC. Of course, all relevant standards are met and therefore the homologation according TSI Wagon is confirmed. The interfaces of wheel design BA 308 are compatible with standard freight wagon wheelsets.

Inside the LZarG project, a set of 12 BA 308 wheelsets (including wheel absorbers) had been produced and were tested in Y25 bogies. After these measurements, a trial operation of more than three years at DB Cargo was started. The wagon type was a bulk freight vehicle of type Tanoos 896, shifting potash salt from Hamburg port to Zielitz. At running distances of 170 000 km for some wagons, no defects etc. are visible on the wheels. Thus, the general ability for operation was demonstrated.

3.2 Swiss R&D Project Railway Noise Investigation

The call for tender "Bahnlärmforschung 2016" ('Railway Noise Investigation 2016') of the BAFU offered the chance to realize further improvements on the whole system. The main focus was laid on the corrosion protection at both wheels and absorbers. Also, the installation and removal process were improved. Therefore, repeated acoustic measurement runs became necessary.

The project funding enables the application of these wheelsets on two freight wagons. Hence, eight wheelsets had been produced and were installed in two container wagons of type "CTW light" Sgnss (bogie type Y25) of wagon rental company Wascosa. Two container wagons of the same type and wheel type BA004 (widely used in Europe) had been used for reference comparison. To allow a fast loading, both vehicles were loaded with so called "E-type swap bodies" of the Wascosa flex freight system®. These open-roof swap boxes are designed for bulk loads. The test train is sketched in Fig. 2.

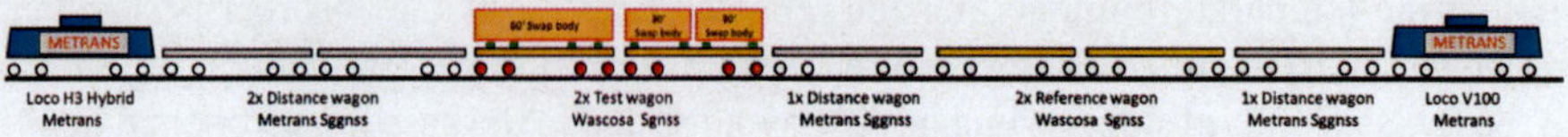

Fig. 2. Test train configuration, BA 308 wheelsets in red

Beside standard requirements (train speeds, acoustic free field conditions, measurement procedure etc.), two track related parameters are relevant: roughness of the rail head and vibrational behavior of the whole track, quantified in the "track decay rate". Both parameters have to be assessed according EN 15610 [10] and EN 15461 [11].

A further advantage of the standardized measurement conditions is that the results can be used for a technology qualification in public funding schemes (for details see also Sect. 4.1). Test sites which meet these demands are difficult to find throughout Europe and mainly located on state railway networks with dense traffic. To enable a quick and flexible test operation, a 'private' network was preferred, considering the requirements above. An appropriate site was found at the EVB network in northern Germany, see Fig. 3 (left).

Additionally, the impact of this wheel/absorber combination on the probability of curve squealing became relevant in the assessment process. This extended the test site requirements by the accessibility of a 'sharp' curve with similar acoustic conditions to free field. For this, a section with a 200 m radius double bended curve was found inside the nearby train station, see Fig. 3.

Fig. 3. Test runs (left: straight track, right: curve section)

4 Results

The tests of the combination of wheel BA 308 and absorber VICON RASA RSI under different conditions have led to a couple of results.

To enable an objective assessment by airborne noise results, the wheel roughness has to be checked to meet similar levels at test and reference wagons. The 16 wheels of the two test and reference bogies, which are inside the analysis timeframe, had been analysed in this way. Almost all wheels are showing low roughness levels (common range for composite block braked wheels) except single ones. These are located at the test and also reference bogies, therefore no relevant impact is expected on the overall acoustic results.

In the 'Low Noise Train on Real Track' project, acoustic qualification measurements were executed twice: initial measurements with the original absorber design and a second run with an optimized absorber design. A reduction of 3.5 dB in $L_{Aeq,Tp}$ was found according ISO 3095 [12] and TSI Noise [13]. A third measurement run with innovative freight bogies DRRS 25L (instead of Y25 bogies) resulted in an increased noise reduction of 4.5 up to 5 dB.

Here, the reduction in $L_{Aeq,Tp}$ is 3.9 and 4.3 dB at 80 and 100 km/h for the wagon without any box/load (compared to the reference wagon), see Fig. 4 (left). If the wagon is loaded with the (empty) swap body, the reduction in $L_{Aeq,Tp}$ is 2.5 dB at 80 km/h. The reason can be seen in the additional noise emission of the swap body, dominating the overall rolling noise.

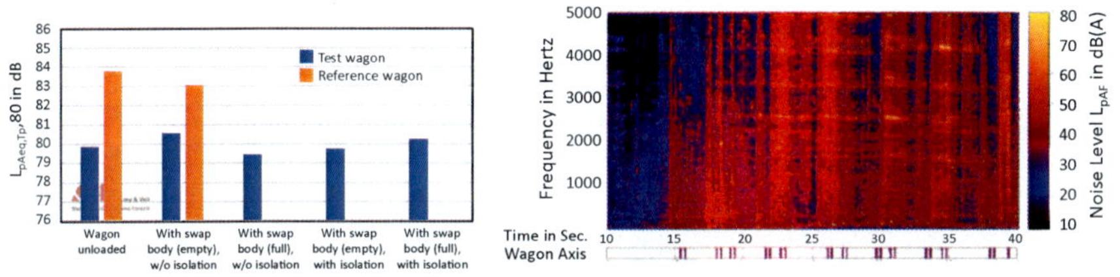

Fig. 4. Measurement results, left: straight track, right: curve section

Curve squealing test results also show advantages for the BA 308 wheel together with the absorbers: Compared to the reference wagons, which start to squeal at a speed of 20 km/h, the test wagons did not squeal (test speeds up to 40 km/h). Figure 4 (right) clearly shows the increased noise levels at the single time points and 2.6, 3.4 and 4.2 kHz, when the reference wagon wheels are passing the microphone. During the early part of the train passby, no sound levels emerge significantly compared to other frequency ranges. This confirms the expectation of squeal avoidance for this vehicle type, including wheel absorbers, at this track type. By a high possibility, similar results would be found at similar curved tracks.

4.1 Ongoing Actions

As of October 2019, two container vehicles are equipped with wheel absorbers of type VICON RASA RSI. These vehicles are in commercial service under different operational conditions throughout Europe with mileages now of up to 20 000 km/month. The operation also includes severe winter conditions in Scandinavian regions. To monitor the absorber performance, a concentrated inspection regime is executed. Results are integrated in the operation manual and will be considered at upcoming construction improvements. Up to now, no indications for such an update had been found.

The practical dissemination of wheel-based noise mitigation measures is strongly linked to developments of the legal framework. From today's experience, only the advanced decrease of noise limits, which requires additional measures, would pave the way. Due to the required investments (see also ongoing action on brake block replacement, e. g. by CEF funding [14]), a public funding of implementing these measures should be discussed. Other economic parameters also have to be considered as well as the experience from the current retrofitting process. This includes also noise related track access charges with an appropriate noise reduction stepping.

Apart from this, two funding guidelines are already established to realize additional freight wagon noise reduction measures:

- "Förderrichtlinie TSI Lärm +" ('Funding Guideline TSI Noise +') of German Federal Ministry of Transport and Digital Infrastructure BMVI [15]
- "Investitionsbeihilfe an besonders lärmarmen Güterwagen" ('Investment Grant at Particular Silent Freight Wagons') of Swiss Federal Office for Environment BAFU [16].

5 Discussion

To implement a further decrease of freight train noise, additional steps beside the brake block replacement are necessary. A wheelset/absorber system was introduced and assessed to add a further reduction of $L_{Aeq,Tp}$ according ISO 3095 by about 4 dB. Repeated testing at slightly different test conditions (which meet ISO 3095/TSI Noise specifications) confirm these results also in real train operation.

In general, the end user of each vehicle-based mitigation measures is the wagon owner. From his view, the most effective approach would be chosen, leading to a competition with alternative technologies. In this process, he has also to decide if it would make sense to replace the whole bogie (e. g. due to its limited remaining lifetime) – resulting in the choice to integrate disc brake systems with new bogies. However, to keep the existing bogie in service, he is looking to replace the minimum number of parts. Here, the wheelset/absorber system is a competitive approach. Standard interfaces to the bogie ease the retrofit work and the required workshop time.

In the current market environment, no wagon owner would start this process from an intrinsic motivation: Compared to other technology advancements, noise measures do not lead to operational improvements like increased freight loads or train speeds. Therefore, the triggering has to be done by an entity who represents the beneficiaries of this process. Because the railway line residents would profit from these measures but are in no contractual relation to the wagon owner, their representative is in charge of actions, here their (federal) governance.

6 Conclusion

Freight vehicle wheelsets with vibration dampers are subjected to several design requirements. From R&D projects an important step towards the real operation phase was taken.

This includes design improvements and also field measurements for a standardized quantification of the acoustic benefit.

It is shown that this combination of wheel type BA 308 and vibration absorber VICON RASA RSI reduces the rolling noise level $L_{Aeq,Tp}$ by about 4 dB compared to a BA 004 wheel type (also with composite brake blocks). Furthermore, the curve squealing is prohibited at different train speeds and it is likely that this effect will also occur at other track/speed conditions.

Review the barriers for widespread usage of this system (and also related technologies) shows that the wagon owners' views are important, focusing on the motivation for implementation. Similar to other noise reduction measures, no additional benefit on the vehicle's revenue can be expected. This opens the field to support from a third party. Because railway line residents are the starting point for noise reduction, it is obvious that their representatives have to cut down these barriers. The decision is at their table to choose the most effective way, while considering legal constraints.

References

1. Deutsche Bahn-Standard 918 290: Technische Lieferbedingungen Schienenstegdämpfer (SSD) (2017)
2. UIC: Railway Noise Technical Measures Catalogue (2013)
3. EN 45545-2 Railway applications - Fire protection on railway vehicles - Part 2: Requirements for fire behavior of materials and components
4. EN 61373 Railway applications - Rolling stock equipment - Shock and vibration tests
5. UIC 510-5 Trailing stock: wheels and wheelsets. Conditions concerning the use of wheels of various diameters
6. EN 13979-1 Railway applications – Wheelsets and bogies - Monobloc wheels - Technical approval procedure – Part 1: Forged and rolled wheels
7. Czolbe, C.; Born, K.: BMVI Forschungsprojekt Innovative Güterwagen. In: 5. Fachtagung Bahnakustik, Planegg (2018)
8. Commission Regulation (EU) No 321/2013 concerning the technical specification for interoperability relating to the subsystem rolling stock—freight wagons of the rail system in the European Union (TSI Wagon)
9. R&D project Leiser Zug auf realem Gleis, term 2008–2010, funded by the German Bundesministerium für Wirtschaft und Technologie, grant number 19U7020E
10. EN 15610:2009 Railway applications—Noise emission—Rail roughness measurement related to rolling noise generation
11. EN 15461:2008 Railway applications – Noise emissions, Characterization of the dynamic properties of track sections for pass by noise measurements
12. ISO 3095:2013 Acoustics—Railway applications—Measurement of noise emitted by railbound vehicles
13. Commission Regulation (EU) No 1304/2014 on the technical specification for interoperability relating to the subsystem rolling stock—noise (TSI Noise)
14. 2019 CEF call for proposals. https://ec.europa.eu/inea/en/connecting-europe-facility/cef-transport/apply-funding/2019-cef-transport-call-proposals-reduction-rail-freight-noise. Accessed 31 May 2019
15. Funding guideline "Förderrichtlinie des Bundesministeriums für Verkehr und digitale Infrastruktur zur Förderung der Beschaffung neuer Güterwagen oder des Umbaus von Bestandsgüterwagen, die den Grenzwert für das Vorbeifahrgeräusch der TSI Fahrzeuge – Lärm unterschreiten (Förderrichtlinie 'TSI Lärm+')", BMVI (2017)
16. Funding guideline: Der besonders lärmarme Güterwagen Pflichtenheft, BAFU/BAV (2017)

Half Installation of Rail Dampers

Wilson Ho(✉), Cindy Cheung, Morgan Cheng, and Lyn Lin

Acoustics Innovation Limited, Unit 106, Block A, Shatin Industrial Centre, Yuen Shun Circuit, Shatin, N.T., Hong Kong
who@aihk.hk

Abstract. Rail dampers are generally installed at every sleeper spacing with the aim of reducing noise by 3 to 5 dB(A). However, some damper installation projects with limited budget may not be able to support full installation of rail dampers for the entire track section of concern. Damper installation at alternate sleeper spacing, namely half installation of rail dampers, is proposed as a compromise between cost and performance. Half installation with high performance is made feasible using strong dampers with tuned mass damping at multiple frequencies with multiple oscillators. Measured track decay rate, train pass by rail vibration and trackside noise results from six sites are presented in this study following progressive installation of rail dampers. Rail vibration data at damper positions and at adjacent mid-span positions without damper installed are also compared. Half installation successfully raised track decay rates in both vertical and lateral directions to meet the TSI and ISO standard, and achieved trackside noise reduction of1 to 6 dB(A). Half Installation can serve as a good alternative to Full Installation under limited project budget or tight timeframe.

Keywords: Rail damper · Tuned mass damping · Track decay rate · Vibration reduction · Noise reduction

1 Introduction

Over the years, rail dampers by the tuned-mass damping mechanism have been proven effective in rail noise and vibration control especially at rail pinned-pinned resonance and to slow down rail corrugation growth [1, 2].

Generally, when a track section is identified to have excessive rail noise, it is desired to have rail dampers installed at every sleeper spacing (typically 0.6 to 0.7 m) for the entire track section to achieve the best damper performance in noise reduction. Given limited budget of some damper retrofitting projects, the length of track section for damper installation and thus the overall noise reduction effect are often compromised by selecting only part of the concerned track section for damper installation. In some cases, tight timeframes for damper installation may also be one of the site constraints. Apart from the straight-forward option of full damper installation over a shorter track section, it is worthwhile investigating the noise and vibration reduction effect of half installation of rail dampers, i.e. damper installation at alternate sleeper spacings. By means of half installation, it is aimed to lengthen the extent of damper installation under project cost and time constraints to achieve a balanced compromise in the overall noise performance.

G. Degrande et al. (Eds.): *Noise and Vibration Mitigation for Rail Transportation Systems*, NNFM 150, pp. 654–661, 2021.
https://doi.org/10.1007/978-3-030-70289-2_72

2 Damper Sites and Progressive Installation Scenarios

In the current study, noise and vibration measurement results are presented from six sites during progressive installation of rail dampers. Background information for the sites is listed in Table 1. Among the sites, three are located at tunnel sections (T1, T2 and T3), two are located at open sections (OS1 and OS2) and the remaining one is located at a viaduct section (VD1). Five out of the six sites are located along operational railway lines with mixed fleets of two types of trains, namely the "New" and "Old" types. Progressive installation scenarios in the study include: Quarter, Half and Full Installation (with damper installation arrangement as illustrated in Fig. 1).

Table 1. Background information of the six damper sites

Site	T1	T2	T3	OS1	OS2	VD1
Environment	Tunnel	Tunnel	Tunnel	Open section	Open section	Viaduct
Trackform	Booted concrete block	Booted concrete block	Booted concrete block	Ballast	Booted sleeper	Floating slab track
Rail type	UIC60	UIC60	UIC60	UIC54	53R1	UIC60
Radius of curvature	330 m	425 m	330 m	1100 m	Straight	2000 m
Number of train Types	2	2	2	2	2 (Tram)	1
Operational speed	70 km/h	70 km/h	60 km/h	60 km/h	48 km/h	70 km/h
Extent of damper installation	50 m	100 m	100 m	200 m	100 m	80 m

The sites were installed with the shear-type tuned-mass rail dampers which effectively absorbs and dissipates rail vibration energy through the tuned-mass damping mechanism in both vertical and lateral directions. A typical design of the rail dampers is shown in Fig. 2.

3 Findings from Noise and Vibration Measurement Results

3.1 Track Decay Rate

The vertical and lateral track decay rates measured based on EN 15461 [3] before and after Half Installation of rail dampers are shown in Fig. 3 and Fig. 4 and compared with the TSI and ISO limit for a standard test track [4, 5]. Before installation of rail dampers, the vertical track decay rate at all sites did not meet TSI and ISO standard especially at

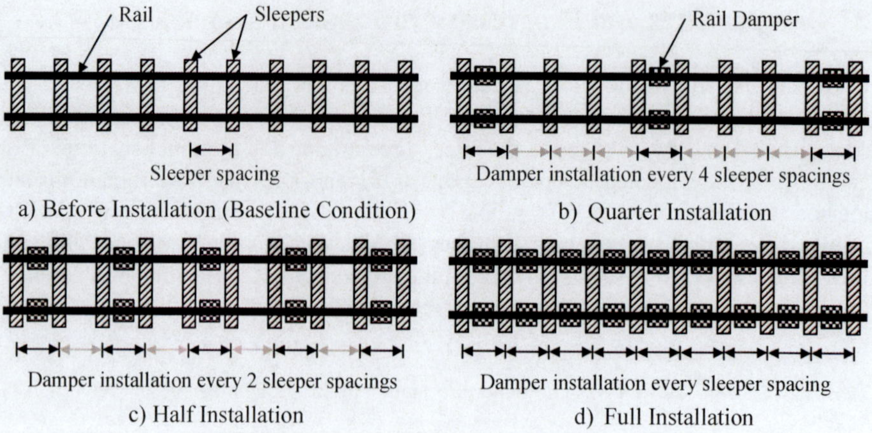

Fig. 1. Illustration of damper installation scenarios

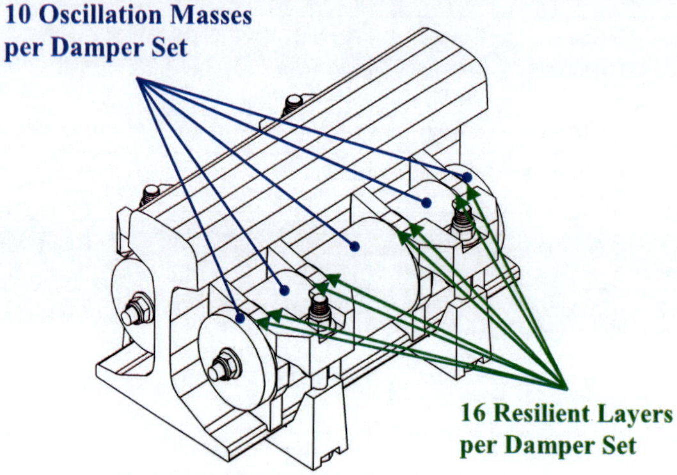

Fig. 2. Shear-type tuned-mass rail damper design

frequency bands from 400 Hz to 800 Hz. After Half Installation of dampers, the vertical track decay rate at frequency bands from 250 Hz to 2500 Hz increased significantly by up to 10 dB/m and horizontal track decay rate by up to 8 dB/m. The track decay rates after half installation are found to meet the TSI and ISO standard.

Track decay rates across progressive damper installation scenarios at Sites OS1 and OS2 are shown in Fig. 5 and Fig. 6. Quarter Installation at these two sites improved track decay rates at frequency bands from 250 Hz to 2500 Hz in both vertical and lateral directions but the effects were not sufficient to raise the track decay rates above the TSI and ISO limit. Full Installation further improved track decay rates in general when compared with Half Installation.

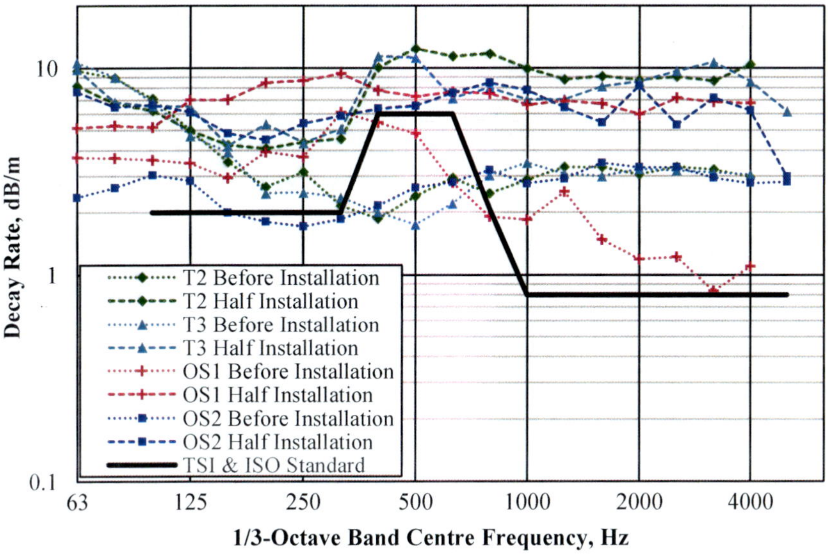

Fig. 3. Vertical track decay rate before and after half installation of rail dampers

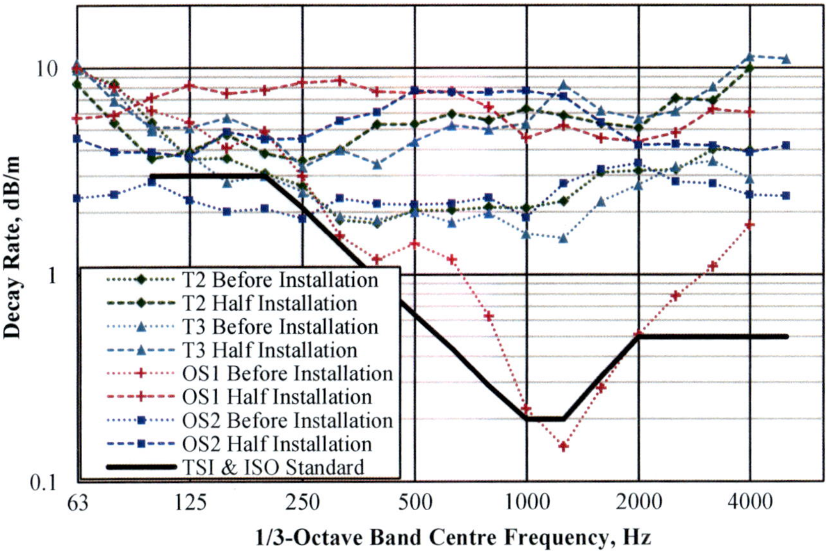

Fig. 4. Lateral track decay rate before and after half installation of rail damper

3.2 Train Pass by Noise and Vibration

The reduction in train pass by rail vibration velocity and noise (from 20 Hz to 20,000 Hz with A-weighting) across the progressive damper installation scenarios are presented in Table 2 and Table 3. Noise reductions by Half Installation at these sites are around

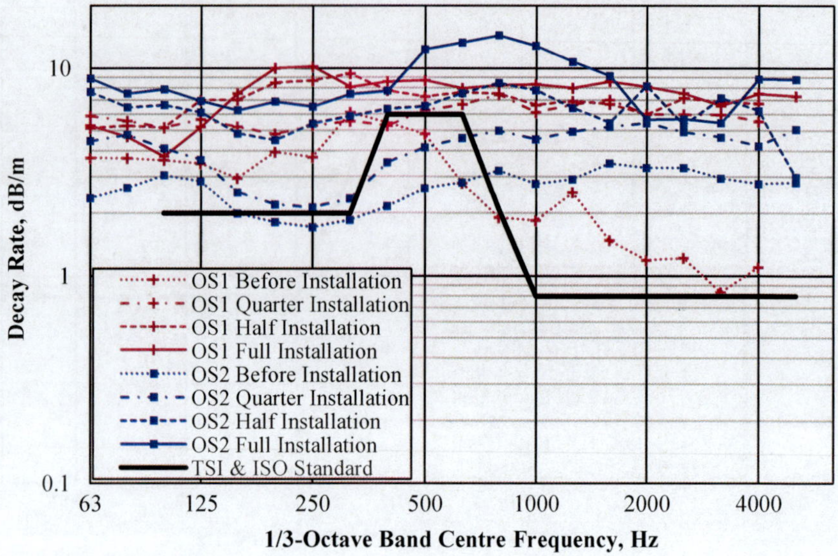

Fig. 5. Vertical track decay rate across progressive damper installation scenarios

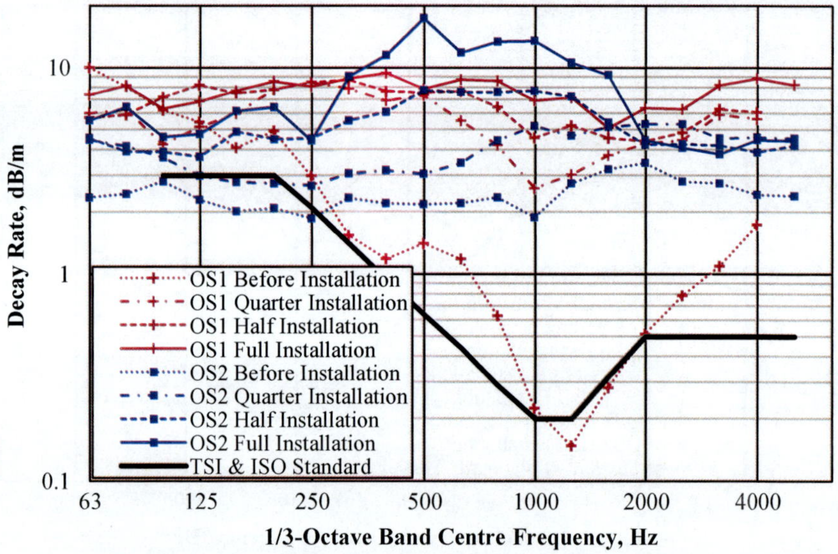

Fig. 6. Lateral track decay rate across progressive damper installation scenarios

1 to 5 dB(A). It can be noted that vibration reductions of damper for Old trains are basically similar to those for New trains, but noise reductions are less due to relatively noisier undercar equipment (e.g. from ventilation equipment of older model and/or at deteriorated condition) which are not mitigated by rail dampers.

Table 2. Reduction in train pass by rail vibration velocity at damper positions, in dB(A)

Site			T1		T2		T3		OS1		OS2		VD1
Train type			New	Old	New	Old	New	Old	New	Old	New	Old	Old
Under Rail (Vertical)	(High) rail	Quarter	–	–	2.3	2.4	–	–	–	–	5.0	7.0	–
		Half	–	–	3.8	3.2	4.8	5.4	–	–	5.4	7.4	7.2
		Full	4.3	5.0	–	–	–	–	–	–	5.7	7.9	–
	Low rail	Quarter	–	–	4.1	2.2	–	–	–	2.7	–	–	–
		Half	2.3	2.0	4.6	2.8	4.9	4.5	–	5.2	–	–	6.4
		Full	2.6	2.3	–	–	–	–	–	6.1	–	–	–
Foot Tip (Vertical)	(High) Rail	Quarter	–	–	4.1	3.4	–	–	–	–	7.5	9.7	–
		Half	–	–	5.4	5.0	10.4	10.6	–	–	8.7	9.8	–
		Full	–	–	–	–	–	–	–	–	9.0	10.5	–
	Low Rail	Quarter	–	–	4.1	2.4	–	–	–	–	–	–	–
		Half	2.8	2.3	5.2	3.3	5.9	2.8	–	7.0	–	–	–
		Full	3.3	2.8	–	–	–	–	–	7.9	–	–	–
Rail Web (Lateral)	(High) Rail	Quarter	–	–	3.0	3.1	–	–	–	–	4.5	6.0	–
		Half	11.2	12.3	5.3	5.7	11.8	12.5	–	–	5.4	6.7	5.4
		Full	13.9	14.2	–	–	–	–	–	–	6.5	6.8	–
	Low Rail	Quarter	–	–	2.0	2.2	–	–	–	–	–	–	–
		Half	2.9	2.0	3.8	3.6	–	–	–	9.2	–	–	3.8
		Full	5.7	4.6	–	–	–	–	–	11.2	–	–	–

Table 3. Reduction in train pass by noise, in dB(A)

Site		T1		T2		T3		OS1		OS2		VD1
Train type		New	Old	New	Old	New	Old	New	Old	New	Old	Old
0.2 m from (High) Rail	Quarter	–	–	2.0	1.2	–	–	–	–	1.4	1.7	–
	Half		–	2.3	2.0	6.3	5.0	–	–	3.5	2.9	6.6
	Full	–	–	–	–	–	–	–	–	3.9	4.1	–
2 m (T1–3)/ 6 m (OS2)/ 10 m (OS1)/ 34 m (VD1)	Quarter	–	–	1.1	0.3	–	–	–	–	0.6	1.2	–
	Half	1.0	0.7	1.9	1.6	4.8	3.6	–	1.5	1.8	1.6	3.6
	Full	1.6	1.4	–	–	–	–	3.0	2.4	–	–	–

In addition to the vibration data at the damper position presented in Table 2, vibration data at 1 and 2 sleeper distances from the reference damper position were collected at Sites OS1 and OS2. Reassembled vibration pattern under rail at mid-span between sleepers at OS1 is shown in Fig. 7 as an example. The Quarter Installation pattern shows that the highest rail vibration levels, which are at 2 sleeper distances from the damper position, were less than the baseline vibration level. This improvement could be the result of both the vibration reduction at source through effective energy dissipation by the tuned-mass damping mechanism and the improved track decay rates.

Energy-average values of mid-span vibration levels within each damper installation pattern was also evaluated and plotted in dotted lines in Fig. 7. A significant improvement in vibration reduction from 1.4 dB(A) to 5.1 dB(A) was found across Quarter Installation and Half Installation scenarios.

The trends of vibration reduction at damper positions, the mid-span average, and noise reduction at rail-side towards noise sensitive receivers shown in Fig. 8 and Fig. 9 provide further support for Half Installation as a budget alternative to Full Installation.

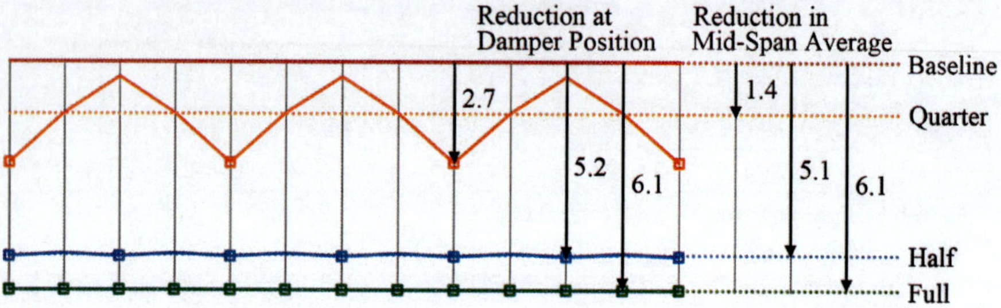

Fig. 7. Under rail vibration reduction at site OS1 (old train), in dB(A) (—, reassembled vibration pattern at mid-span between sleepers; □, damper position; ···, mid-span average vibration velocity level)

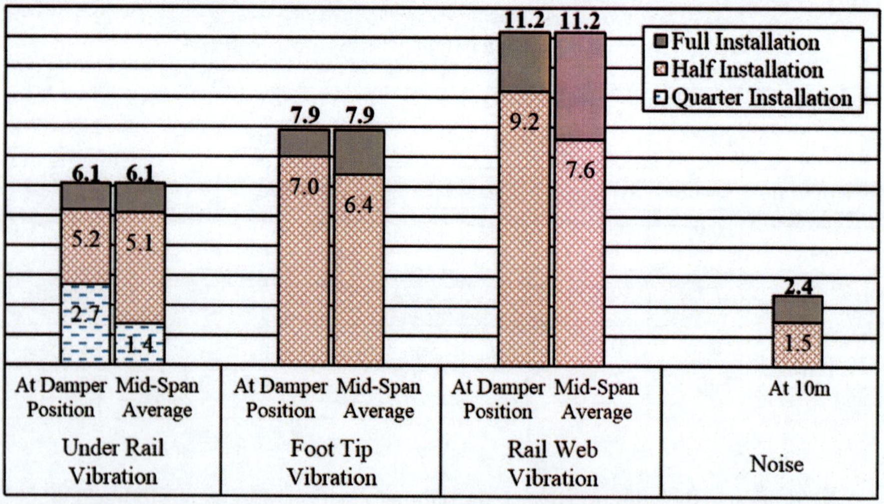

Fig. 8. Progressive noise and vibration reduction at Site OS1 (Old Train), in dB(A)

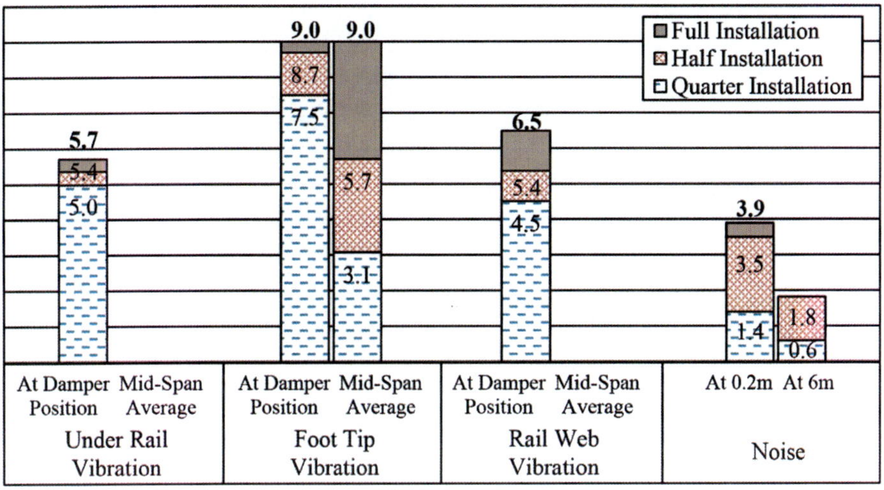

Fig. 9. Progressive noise and vibration reduction at site OS2 (New Tram), in dB(A)

4 Conclusion

Track decay rate, train pass by noise and vibration data were collected from six sites during progressive installation of tuned-mass rail dampers. Damper performance in noise and vibration reduction for three progressive damper installation scenarios were investigated. The results showed Full Installation of rail dampers at every sleeper spacing provided the best performance undoubtedly, but Half Installation (damper installation at alternate sleeper spacings) also provided promising performance between 1 to 6 dB(A) trackside noise reduction. Half Installation can be a good alternative to Full Installation under limited project budget or tight timeframe.

References

1. Thompson, D.J., Jones, C.J.C., Waters, T.P., Farrington, D.: A tuned damping device for reducing noise from railway track. Appl. Acoust. **68**(1), 43–57 (2007)
2. Ho, W., Wong, B., England, D., Pang, A.: Tuned mass damper for rail noise control. In: Noise and Vibration Mitigation for Rail Transportation Systems – Proceedings of the 10th International Workshop on Railway Noise, Nagahama, Japan, 18–22 October 2010, pp. 89–96. Springer (2011). https://doi.org/10.1007/978-4-431-53927-8_11
3. EN 15461:2008+A1:2010: Railway applications—Noise emission—Characterization of the dynamic properties of track selections for pass by noise measurements (2009)
4. ISO 3095:2013: Acoustics—Railway applications—Measurement of noise emitted by rail-bound vehicles (2013)
5. Commission Regulation (EU) No 1304/2014 of 26 November 2014 on the technical specification for interoperability relating to the subsystem 'rolling stock — noise' amending Decision 2008/232/EC and repealing Decision 2011/229/EU Text with EEA relevance (2014)

Vehicle and Track Noise Separation Methodology Based on Advanced Transfer Path Analysis Technique

A. Malkoun[1](✉), Egoitz Iturritxa[2], E. Cierco[3], Ainara Guiral[2], Joan Sapena[4], and F. X. Magrans[3]

[1] ALSTOM, 6 route de Strasbourg, 67110 Reichshoffen, France
antoine.malkoun@alstomgroup.com
[2] CAF, José Miguel Iturrioz, 26, 20200 Beasain, Spain
[3] ICR, C/ Berruguete, 52, 08035 Barcelona, Spain
[4] ALSTOM, 48 rue Albert Dhalenne, 93400 Saint-Ouen, France

Abstract. The radiation mechanism for rolling noise is shared between the vehicle wheels and the track (rails and sleepers), both excited by the combined roughness of wheels and rail. Depending on the track characteristics and vehicle speed, track noise can often be the dominant source during a pass-by. In the context of EU-funded project ROLL2RAIL WP7, a methodology was developed for the separation of track and vehicle rolling noise based on the Advanced Transfer Path Analysis (ATPA). Field tests were carried out under controlled conditions and results showed separation of track and vehicle noise could be obtained for specific rolling stock on a specific track. This paper describes the work performed and presents the detailed results obtained by the ATPA method. The outcome of this work shows a possible application to a wider range of cases, in order to achieve track and vehicle noise separation for different types of rolling stock on different types of track.

Keywords: Rolling noise · Source separation · Advanced Transfer Path Analysis

1 Introduction

During the pass-by of a train, the overall noise emitted is the combination of track and vehicle noise. Equipment and aerodynamic noise sources are inherent to the vehicle, while rolling noise is due to both, with the wheels and the track excited by the combined roughness of wheels and rail [1].

Constant speed pass-by type test results are influenced by the acoustic track characteristics (ATC) and include both track and vehicle contributions. Nevertheless, the noise contribution from the vehicle is often masked by the track contribution and as a consequence, improved vehicle designs are not properly assessed. The reference track in EN ISO 3095 [2] is defined with an ATC that consists of an upper limit of the rail roughness and a lower limit of the track decay rate which are intended to minimize the influence of

G. Degrande et al. (Eds.): *Noise and Vibration Mitigation for Rail Transportation Systems*, NNFM 150, pp. 662–670, 2021.
https://doi.org/10.1007/978-3-030-70289-2_73

the track on the measurement. However, even within these limits, track noise can often be the dominant source over vehicle noise during a pass-by.

Several methods for performing vehicle/track noise separation have been proposed and tested within previous EU projects STAIRRS (2000–2002) and ACOUTRAIN (2011–2014). In the context of the EU-funded project ROLL2RAIL WP7 (2015–2017), rolling stock manufacturers, national train operators and research partners developed different new methodologies [3], mainly experimental, in order to separate track noise and vehicle noise contributions during a train pass-by. Field tests were carried out under controlled conditions and the results showed that separation of track and vehicle noise could be obtained for specific rolling stock on a specific track. Among these methods, the methodology described in this paper that uses principles based on Advanced Transfer Path Analysis (ATPA) has been tested; the method was originally proposed by Malkoun et al. [4].

This paper aims at completing the work summarized in ROLL2RAIL WP7 deliverable [3] and previous publication by Thompson et al. [5] by presenting in more detail the results obtained by the ATPA method, and the possible application to a wider range of cases, in order to achieve track and vehicle noise separation for different types of rolling stock on different types of track.

2 ATPA Methodology

Advanced Transfer Path Analysis (ATPA) takes its origins in the need to find the contributions of the displacements or accelerations of each part of a mechanical system to the pressure at a microphone target location. This general noise contribution problem was initially posed in the 1980's by Magrans [6], with the objective to separate the contributions of different parts.

The objective of the method is to obtain the decomposition of the sound pressure P_M at a target location as the sum of N noise contributions. In a system discretized with N subsystems, the ATPA method performs the following decomposition of the total noise P_M [6]:

$$P_M = \sum_{k=1}^{N} a_k \cdot T^D_{k \to M} \tag{1}$$

where $a_k \cdot T^D_{k \to M}$ is the contribution of each subsystem k to the total noise. This contribution is obtained as the product of the operating vibration signal a_k with a given function $T^D{}_{k\text{-}M}$. This function T^D is the so-called Direct Transfer Function from subsystem k to target noise M, which describes the noise produced at M by the vibration of the subsystem k when all other subsystems are blocked (forced to have zero acceleration). The concept of 'Direct Transfer Function' (DTF) refers to the signal going from a subsystem to the target location without passing through any other subsystem. The concept of 'Global Transfer Function' (GTF) refers to the signal going from one subsystem to another, passing through all other subsystems.

All these functions are characteristics of the physical system and do not depend on operating conditions. GTFs can be measured experimentally, when performing for example impact testing with forced response on a subsystem; but DTFs cannot be measured, since they represent the subsystem response when all other subsystems are blocked. How GTFs and DTFs are related is explained in [3] and [5].

In the application of ATPA for vehicle and track noise separation, the track system is decomposed in separate subsystems considered independently. The measurement process is divided into static tests, from which measured GTF are used to calculate the DTF; and dynamic tests, where the contribution of each subsystem is obtained based on train pass-bys. This technique enables the direct evaluation of total noise radiated by the track, by summing the individual contribution of all the instrumented subsystems.

3 Measurement Campaign

3.1 Complete Instrumentation

The ATPA method was applied in real conditions during the ROLL2RAIL WP7 project [3]. The instrumentation used for the application of the ATPA technique is shown in Fig. 1. A target microphone (M3) was located in front of the instrumented track section at a distance of 3.55 m from the track centerline and at a height of 0.5 m above the rail head. The track section instrumented was separated into six contiguous regions with a spacing of twice the distance between sleepers (1.2 m), leading to a total instrumented section length of 7.2 m. For each region, five accelerometers were installed: one vertical and one lateral accelerometer on each rail and one vertical accelerometer at the center of one sleeper. A total of 30 accelerometers were positioned on the track (24 on the left and right rails, and 6 on sleepers). 8 accelerometers were placed on the 4 wheels of the bogie placed on the track section used during the static tests (W1 and W2 for left rail). These accelerometers on the wheels were installed in order to obtain DTFs from the track subsystems without the contribution from the wheel radiation.

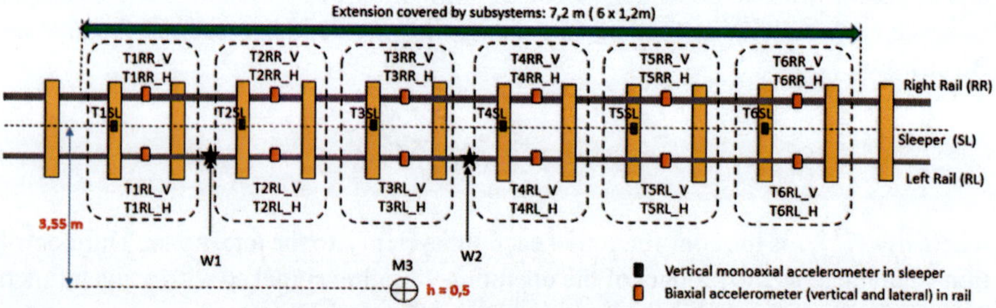

Fig. 1. Scheme of the complete ATPA instrumentation on the track, for static tests

The distance of 3.55 m between track centerline and target microphone was chosen to ensure that the instrumented length would be sufficient to capture at the target location most of the track contribution on the 7.2 m-length instrumented section; a longer instrumented length would be required if the standard pass-by test position of 7.5 m would have been used as target position. A specific model was built to simulate the passage of a train coach consisting of two wheels on one single rail, simulating rail vibration and noise for each consecutive position of the train. This simplified model enabled computation of the noise received at a specific distance from the rail as a function of the track extension length considered. These computations help to find the compromise between instrumented length on the track and distance of the target microphone with the aim of reducing overall error to less than 0.5 dB.

Some simplifications of the ATPA instrumentation setup have been investigated in which a more limited set of transducers is used in both static and dynamic tests. The aim of these simplifications is to reduce the number of sensors required for application of the method, and the time required for testing, especially for the static tests.

3.2 Static Tests

The first step of the ATPA method consists in the experimental acquisition of GTFs, that relate subsystems with each other. Then the DTFs are calculated using GTFs through a matrix inversion technique, in order to find the unitary response of each subsystem when all other subsystems are blocked.

The GTFs were obtained by exciting each track section and measuring the transfer functions between the acceleration of all subsystems and the sound pressure at the target microphone (see Fig. 2). Two different hammers were used, allowing combination of GTFs and obtain a good coherence on the whole frequency range from 100 Hz to 5 kHz. The hammer impacts were successively applied to all subsystems and the response at all sensors was recorded synchronously.

In order to allow for the effects of the presence of the train (load influence) on the vibration paths and on the acoustic propagation paths (screening influence), a train coach was placed over the track section during the static measurement. Another set of static measurements was performed without the presence of the train. The possibility to use multiple microphones as targets in order to study symmetry effects was also considered in order to reduce instrumentation setup and testing time in future applications of the method [4].

Different sets of DTFs were computed, using all measured GTF information, or partial information in view of the use of a more simplified instrumentation setup. The results from the simplified setup were comparable to the results from the complete setup (see Sect. 4.2).

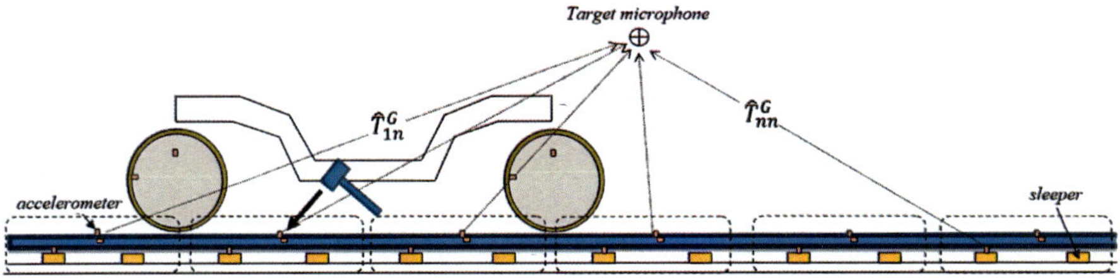

Fig. 2. Scheme representing static test GTF measurements with presence of a train

3.3 Dynamic Tests

In the application of the ATPA method, the second step consists in measuring the accelerations of each subsystem during operational train passage. The instrumentation was the same as that used for the static tests. The field tests were carried out at 40 km/h, 80 km/h and 160 km/h with the test train "Schallmesswagen" from DB consisting of 1 locomotive and 3 coaches, see Fig. 3. The instrumented wheels of the bogies on which the rolling noise separation was performed and compared between the different methodologies developed in ROLL2RAIL WP7 is indicated by the red rectangle in Fig. 3. This section was chosen to avoid any other noise source apart from pure rolling noise, for validation purposes. However, track noise can be obtained with ATPA even in the presence of other noise sources like traction equipment.

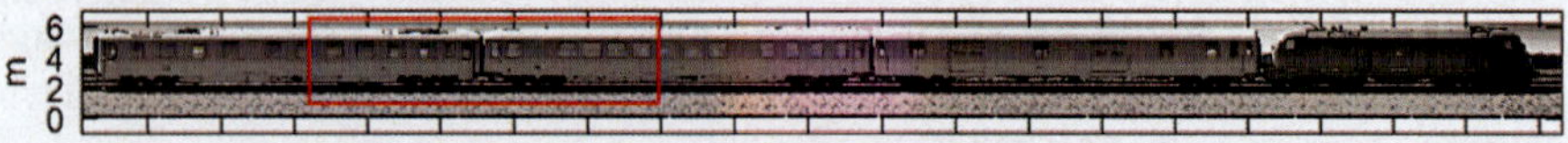

Fig. 3. Test train with three coaches used for the study [3]

In addition to the full instrumentation, a simplified setup was also used, see Fig. 4. Indeed, assuming that the track transfer functions are already characterized for a specific test site, this would allow employing a much simpler instrumentation setup than the one for static tests. By shifting in time the appropriate portion of signals measured on the simplified setup, depending on vehicle speed, operational signals could be reconstructed at non-instrumented subsystem locations. This exercise has been performed in the study and shows very similar results to the complete setup (see Sect. 4.2).

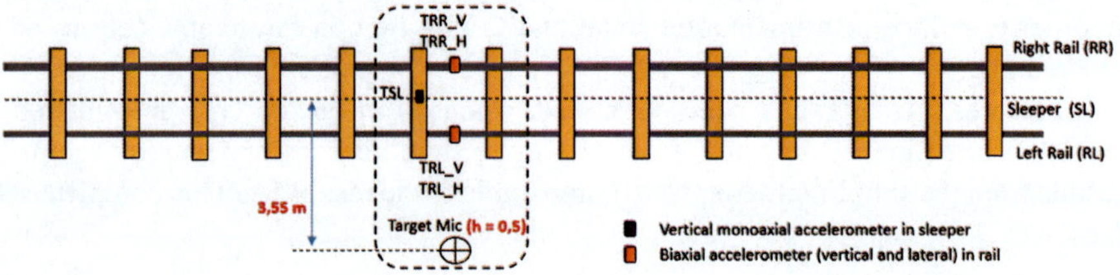

Fig. 4. Scheme of simplified ATPA instrumentation on the track, for dynamic tests

4 ATPA Separation Results

4.1 Full ATPA Method

The results obtained by the ATPA method give a direct evaluation of total noise radiated by each instrumented subsystem. Summing up contributions of subsystems related to the same macro system provides left/right (outer/inner) rail and sleeper contributions, see Fig. 5. Finally, summing up all subsystem contributions represents the total track noise, referred as "synthesized" in Fig. 5. By comparing the track noise to the total noise measured at the target location, an estimation of the noise emitted by the vehicle alone is obtained.

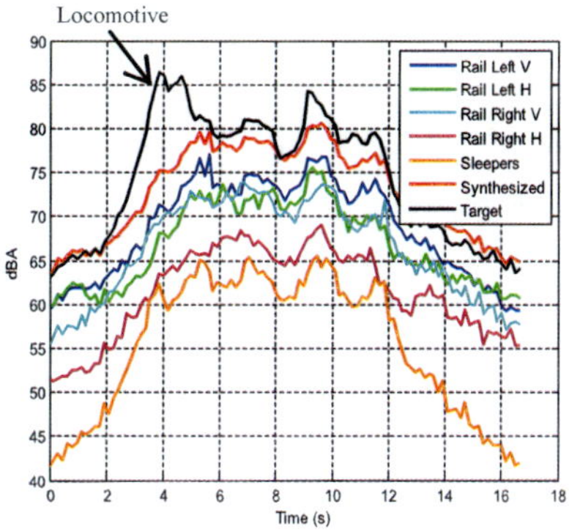

Fig. 5. Time evolution during a pass-by at 40 km/h: measured noise (in black) and synthesized track contribution presented per subsystem groups

Frequency analysis was carried out for the specific section representing the instrumented coach passage: the red rectangle in Fig. 6 on the left. Figure 6 on the right shows the 1/3rd octave analysis of the reconstructed contributions in that period of time, from 100 Hz to 5 kHz. The sleepers contribute the least (71 dBA) to the overall level (86.6 dBA). Then the horizontal motion of the right (inner) rail is also seen as a low contributor (73 dBA), probably shielded by the train during the pass-by. The largest contributors are the horizontal motion of left (outer) rail (80.3 dBA) and vertical motion of both rails (82.4 dBA for left rail and 81.4 dBA for right rail).

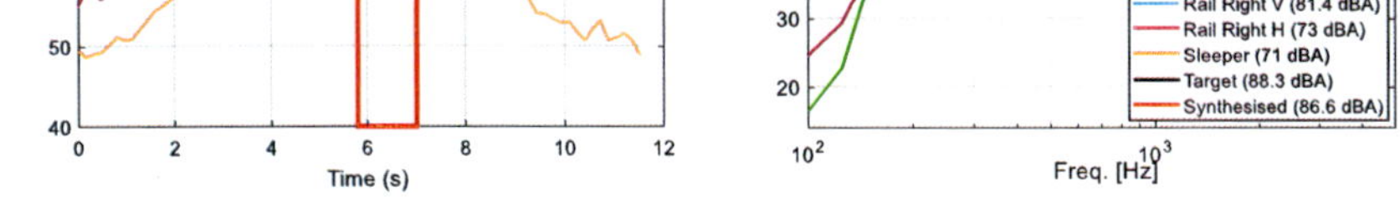

Fig. 6. Left: Time evolution during a pass-by at 80 km/h; Right: 1/3rd octave analysis during the passage of the instrumented bogies (red rectangle)

An in-depth comparison of the ATPA separation results with a reference TWINS simulation tuned-model using wheel and track vibration measured inputs has been done

[3]. The comparison between reference TWINS calculated spectra with ATPA method is shown in Fig. 7 for vertical rail contributions (average of both rails), horizontal rail contributions and sleeper contributions. This comparison shows a very good agreement, within 1 dB in terms of overall levels. At high frequency above 2.5 kHz, ATPA results are a little higher than the reference results, which suggests that the track noise could be underestimated by the simulation model in that frequency range due to the neglect of rail cross-section deformation in the TWINS model [5].

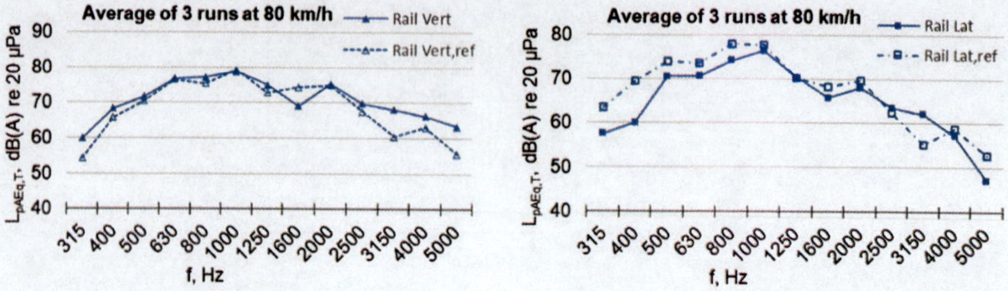

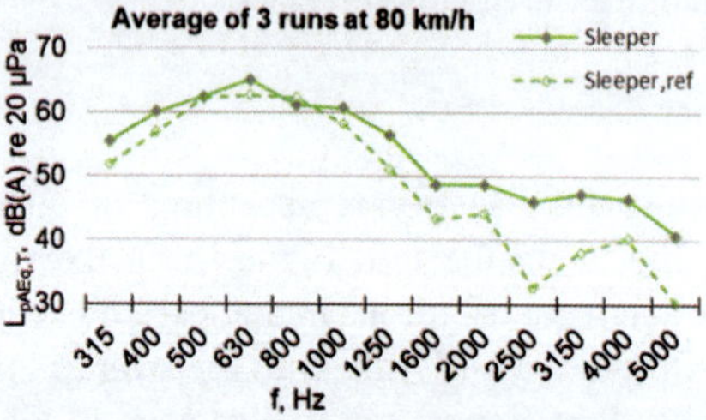

Overall levels 80km/h (dBA)	ATPA method	TWINS reference
Rail vertical	84.8	84.3
Rail horizontal	80,.9	83.3
Sleeper	69.9	68.4
TOTAL Track	86.4	86.9

Fig. 7. ATPA noise separation results compared with reference TWINS simulation results for pass-by runs at 80 km/h

4.2 Simplified ATPA Method

A simplification of the instrumentation setup used for the static tests has been considered which consists of not including the horizontal subsystems on both rails. With this simplification, on average for all runs and speeds, the induced error is below 0.2 dB. The further simplification of excluding the subsystems on the right (inner) rail and assuming that both rails vibrate similarly was studied. This induces an error of 0.4 dB. However not taking into account the presence of the train during static tests was quantified to lead to an error of 2.4 dB. Concerning dynamic tests, the impact of having only a reduced instrumentation setup and using a time shift reconstruction for the other signals as explained in Sect. 3.3 induced a negligible error (0.0 dB).

5 Conclusion and Method Extension

This paper presents the methodology developed in ROLL2RAIL WP7 based on the ATPA technique that discretizes a track in multiple subsystems. This technique enables an experimental evaluation of track noise during a train pass-by. The method requires static tests to identify track transfer functions and dynamic tests. Measurements were carried out with the full method and with simplified alternatives that aim at reducing the number of sensors (see instrumentation in Fig. 1). The full method was compared to reference TWINS calculations tuned with measured vibration inputs and gives very similar results, within 1 dB accuracy. The static test simplified alternatives show errors below 0.2 dB when considering only vertical rail and sleeper radiation and 0.4 dB with sleeper and left (outer) rail radiation only, showing good potential for applicability of a simplified method. Nevertheless, using transfer functions measured with no train on the track leads to a significant error of 2.4 dB. Furthermore, in the case a track is already characterized for static test transfer functions, it has been seen that realizing the dynamic tests with a reduced instrumentation setup (5 accelerometers and 1 microphone, see Fig. 4) induces a negligible error.

A proposal to extend the methodology to a wider range of cases could be considered, with the goal to achieve track noise evaluation with minimum testing efforts, and then be able to deduce noise from vehicle only during a pass-by. Currently, compared to the current EN ISO 3095 standard [2], the ATPA method can allow for measurements in different types of track without the need to comply with the ATC of the standard. If the static tests of the ATPA method are performed on several types of track to evaluate typical track transfer functions, the method could then be applied easily through simple dynamic tests. This approach could be used as an alternative to current pass-by TSI certification method, with the benefit to identify noise of the vehicle alone, and not limited to TSI tracks. Such a method would allow for a clear identification of quiet rolling stock versus noisy designs. A continuation of this work and further development of the method is foreseen in the context of the JU SHIFT2RAIL initiative where several types of track and vehicle will be evaluated.

Acknowledgements. The work described in this paper has been carried out as part of the ROLL2RAIL project, which has been financially supported by the European Union's Horizon 2020 grant, agreement No. 636032. The authors are grateful to the staff of Deutsche Bahn for the field tests organization and execution, as well as for the partners for the constructive collaboration and good leadership of the project.

References

1. Thompson, D.: Railway Noise and Vibration: Mechanisms, Modelling and Means of Control, 1st ed. Elsevier (2009)
2. EN ISO 3095:2013 - Measurement of noise emitted by railbound vehicles
3. Thompson, D., Squicciarini, G., Wändell, J., Arteaga, I.L., Marcano, E.Z., Malkoun, A., Iturritxa, E., Guiral, A., Gutiérrez, I., Schleinzer, G., Stangl, M., Lopez, B.M., Chaufour, C.: Roll2Rail Deliverable D7.4 – Assessment and validation of source separation methods, April 2017
4. Malkoun, A., Sapena, J., Arcas, K., Magrans, F.X.: Vehicle and rail noise separation method proposal based on transfer path analysis techniques. In: 21st International Congress on Sound and Vibration, Beijing (2014)

5. Thompson, D., Squicciarini, G., Zhang, J., Arteaga, I.L., Zea, E., Dittrich, M., Jansen, E., Arcas, K., Cierco, E., Magrans, F.X., Malkoun, A., Iturritxa, E., Guiral, A., Stangl, M., Schleinzer, G., Lopez, B.M., Chaufour, C., Wändell, J.: Assessment of measurement-based methods for separating wheel and track contributions to railway rolling noise. Appl. Acoust. **140**, 48–62 (2018)
6. Magrans, F.X.: Method of measuring transmission path. J. Sound Vib. **74**(3), 321–330 (1981)

Rail Vibration and Rolling Noise Reduction Using Tuned Rail Damper for Vulcanized Bonded Baseplate

Ziyan Ma[1], Anbin Wang[1,2](✉), Xiaogang Gao[1], Xiaohan Gu[1], Longhua Ju[1], and Zhiqiang Wang[2]

[1] School of Urban Railway Transportation, Shanghai University of Engineering Science, 333 Longteng Road, Shanghai 201620, People's Republic of China
wangab725@163.com

[2] Luoyang Ship Material Research Institute, Luoyang 471003, People's Republic of China

Abstract. Work was conducted on a curved track section of 350 m radius on an underground line where excessive in-car noise was experienced by passengers. Vulcanized bonded baseplates were installed on the section. Investigation was performed comprising initial in-car noise and carriage floor vibration level assessments, testing of rail surface roughness, rail vertical and lateral vibration responses during train normal operation, and measurements of track decay rate. A rail web damper designed for in-car noise control was evaluated on a trial section of track. All experimental data were analysed, showing that the highest L_{Amax} was about 95 dB and the highest L_{Aeq} 91 dB for the Egg Fasteners Section. 1/3 octave band peak frequencies of dominant in-car noise is at 250 Hz and 315 Hz, consistent with those of rail roughness and rail vibration spectrum. Reduction of rail vibration velocity is not obvious in these frequency bands on the track with rail web damper installed.

Keywords: Rail vibration · Rolling noise · Rail corrugation

1 Introduction

To relieve the pressure of road traffic congestion and associated air pollution in large modern cities, the capacity of underground train network is continuing to grow in China. With rapid increase of total mileage, environmental and line operating conditions become more complicated, which result in serious structural wearing and damage to components of both rolling stock and track systems. A typical phenomenon is the wearing and fatigue of rail and wheel, i.e., the corrugation, which increases the wheel-rail contact force and rolling noise, one of its consequences is excessive in-car noise of underground trains.

Research on the initiation and development mechanism of corrugation have found many influencing factors [1, 2], such as dynamic characteristics of train and track systems, interaction between vehicle and track, track layout, fastening system etc. Based

G. Degrande et al. (Eds.): *Noise and Vibration Mitigation for Rail Transportation Systems*, NNFM 150, pp. 671–679, 2021.
https://doi.org/10.1007/978-3-030-70289-2_74

on the study on support stiffness changes of rail system and wheel-rail dynamic receptance relationship, the mechanism of "variation of difference between wheel and rail receptance" was proposed, which explains the reason of generation of short wavelength corrugation in frequency range between 200 Hz and 1200 Hz [3, 4].

This paper presents an investigation carried out on an underground line, due to excessive noise level experienced by passengers. It is found through initial assessment of in-car noise level that high noise levels are most evident along a curved segment with small radius of 350 m. At this curve, Steel Spring Floating Slabs Track (SSFST) and Vulcanized bonded baseplates (also called Egg Fasteners) were used. Noise level, rail surface roughness, track decay rates, and rail vibration response to wheel-rail force excitation were all measured and analysed. The evaluation of the suitability of a type of rail web damper intended for in-car noise control was then performed. Dampers were attached against rail webs at Egg Fasteners section. The same set of measurements without proposed treatment were repeated, then compared against measurement results before treatment, for the effectiveness of proposed treatment.

2 Details of Track Form Under Investigation

The metro line under investigation has been operating since 2010. It has a total track length of 36 km. Each train consists of 6 carriages with a total length of around 140 m and a maximum axle load of 16 ton. The maximum design speed of the line is 80 km/h.

The line is underneath some highly populated commercial and residential areas, where more stringent environmental standards of permissible vibration and noise levels were imposed. Therefore several types of track vibration isolation system were installed at different track sections, including SSFST and Egg Fasteners bonded resilient baseplate for vibration and noise control purpose; and common standard fastenings were installed on the rest of the track.

However, during normal operation, high in-car noise levels were experienced by passengers when trains running pass some of the track sections installed with noise and vibration control fastening systems.

3 In-car Noise Reduction Method

To fully understand the contributing factors of high in-car noise levels, experimental investigation was carried out. Firstly, an initial assessment was performed, i.e., in-car noise and vibration measurements, rail surface condition inspection and rail roughness evaluation. Based on the initial investigation, a target section of the problem line was selected. The chosen section was between two stations, station A & station B, and was located at one of the popular areas in the old part of the city. More importantly, there was a small radius curve segment with the radius of 350 m, and this is where in-car noise level was highest compared to other parts of the line. At this curved segment, steel spring floating slab and Egg Fasteners baseplate fastening system were installed for track lengths of 430 m and 120 m respectively, instead of common fasteners. Further experiments were conducted at the track section with Egg Fasteners baseplates, including rail surface roughness, rail vibration response to wheel-rail excitation under normal train operation condition, and track decay rates.

A rail web damper designed for rolling noise reduction was recommended as the remedial solution for in-car noise reduction. The tuned system is installed onto rail tracks by means of a specifically designed metal clippers, as well as the application of polyurethane adhesives between rail web and dampers.

After tuned dampers were installed at the Egg Fasteners section and put into service for a while, all experiments were repeated; results for before and after web dampers installation were compared, and are discussed in detail in the following sections. An evaluation of the suitability of this type of damper was also made based on experimental outputs and evidences from comparisons (Fig. 1).

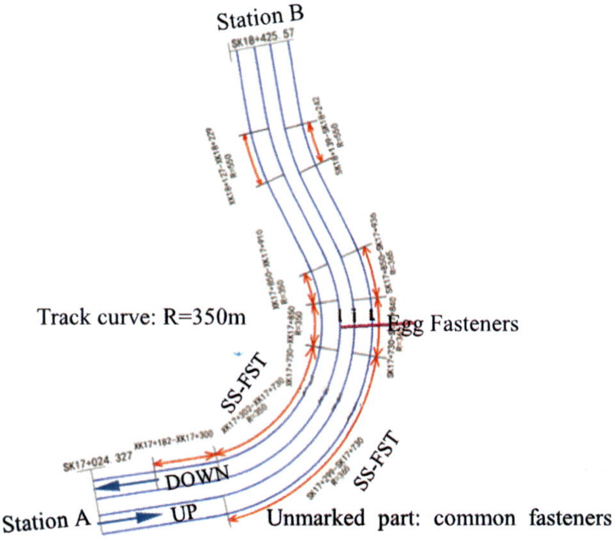

Fig. 1. Selected curved section for investigation

4 Initial In-car Noise and Vibration Assessments

Noise and vibration measurements were carried out inside train carriages for two 3-h periods during normal operating hours, i.e. 14:00–17:00 and 19:00–22:00. Typical measurement interval is 5 min between 2 trains. To make sure data collected truly represents the relevant sections of track, measurements covered several running trains. Above measurement duration and time apply to both in-car noise measurement and vibration measurement.

In-car noise and vibration levels were measured by sound level meters and accelerometers, which were arranged at the central cross section and longitudinal central line of the middle carriage of the train[1]. Microphones were located at 1.2 m above the carriage floor, while accelerometers were fixed onto carriage floor.

In-car noise and vibration levels were measured for several train trips for both up and down lines, with an average train running speed of around 60 km/h. Figure 2a shows the time history of measured in-car noise levels of both up and down trips. Results demonstrate that in-car noise levels at the track sections with SSFST and Egg Fasteners are much higher than that at sections with common fastenings. The highest L_{Amax} was about 95 dB(A) and highest L_{Aeq} was 91dB(A) at the Egg Fasteners sections.

In-car noise 1/3 octave band spectra are shown in Fig. 2b, including results from both before and after the installation of rail web dampers at the Egg Fasteners section. The maximum peaks are at 250 Hz and 400 Hz for the Egg Fasteners and the SSFST section respectively. For the Egg Fasteners section, in-car noise peaks are at 63 Hz, 250 Hz, 315 Hz and 500 Hz; major noise reduction occurred at frequency regions above 800 Hz.

Vibration measurements show that higher acceleration responses were also at the SSFST and Egg Fasteners sections. Table 1 below summarises overall noise and vibration levels. It shows that the installation of rail web dampers at Egg Fasteners section has limited benefit to in-car noise reduction at this particular section, in that this type of damper is more effective for the frequency range above 800 Hz as demonstrated in Fig. 2b, but higher noise levels for this particular line were at 250–315 Hz. Furthermore, noise and vibration levels for SSFST section, where dampers were not installed, were higher after the installation of dampers installed on Eggs section, this is most likely due to external factors during measurements, rather than the impact from installation of dampers.

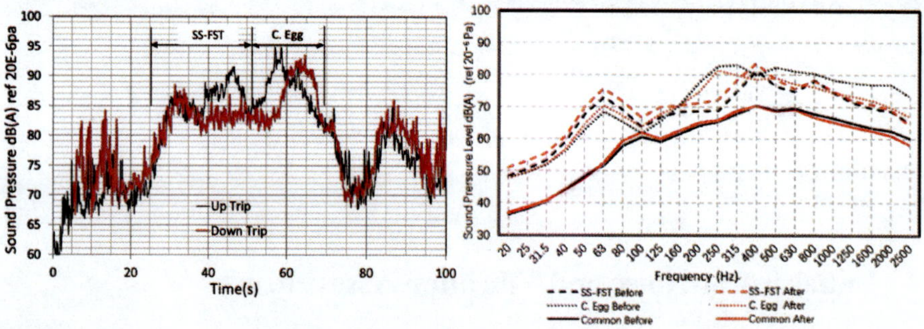

a) b)

Fig. 2. Time history and frequency contents of measured in-car noise levels (up and down trips). a) In-car noise time history[2] b) In-car noise spectra (up trip)

[1] Although there were some passengers in the train carriage, the difference between background noise measured at the stations when trains were stationary and rail noise at heavily corrugated sections are more than 20 dB(A), therefore background noise can be considered negligible in the assessment.

[2] The time is reversed for the down-line trip.

Table 1. Overall in-car noise and floor vibration levels with and without rail web dampers

Track Form	SSFST		Egg Fasteners	
Trip direction	Up	Down	Up	Down
Rail web damper installation	Before/After	Before/After	Before/After	Before/After
Sound pressure level dB(A) ref 20^{-6} Pa	87/88	84/85	91/88	91/90
Vertical Acceleration (dB) ref 10^{-6} m/s^2	122/124	124/125	124/124	124/124
Lateral Acceleration (dB) ref 10^{-6} m/s^2	108/110	109/111	111/111	111/111

5 Rail Roughness, Decay Rate, and Vibration Response

Corrugation marks were found through the inspection at the selected small radius track section; confirming that the increase of in-car noise levels when trains passing through this section was due to corrugation induced wheel-rail contact force; leading to the increase of rolling noise.

Corrugation is a series of wavelike wearing marks on the surface of wheel-rail contact surface. Its roughness and wavelength λ can be measured and quantified. Therefore, the excitation frequency f_c due to corrugation can be calculated when the train speed v is known, per Eq. (1) below.

$$f_c = \frac{v}{\lambda} \tag{1}$$

In addition, train structures and rail components respond to corrugation excitation, consequently lead to different dynamic behaviour.

It is found, via the initial assessment of in-car noise and vibration monitoring descried in above sections, that the worst case of noise and vibration levels is at the Egg Fasteners section when train is on its up trip, so further investigation was focused on the upline track at the Egg Fasteners section.

5.1 Rail Corrugation on the Track with Egg Fasteners Baseplates

Rail corrugation is evaluated based on rail surface roughness, which was measured using Bi-CAT. At the section with Egg Fasteners baseplates, wavelength spectra of measured rail roughness, before and after the installation of rail web dampers, are plotted and compared in Fig. 3. Overall roughness level on the low rail is higher than that on the high rail. The low rail roughness level is above the ISO3095 criteria for almost the entire wavelength range and was peaked at the wavelength of 80 mm and 400 mm. As for the high rail, rail roughness level is peaked at 63 mm wavelength.

To directly compare frequency contents of rail roughness and those of in-car noise level, wavelength spectrum of measured roughness was converted to A-weighted frequency spectrum using Eq. (1). The average train speed during in-car noise measurement along selected track section was approximately 60 km/h.

It shows that rail roughness has major peaks at frequency 250 Hz, equivalent to corrugation wavelength of 63 mm. This is the dominant excitation frequency to the rail system due to rail corrugation; and agrees with in-car noise peak frequency of 250 Hz as given in Fig. 2b for the Egg Fasteners section.

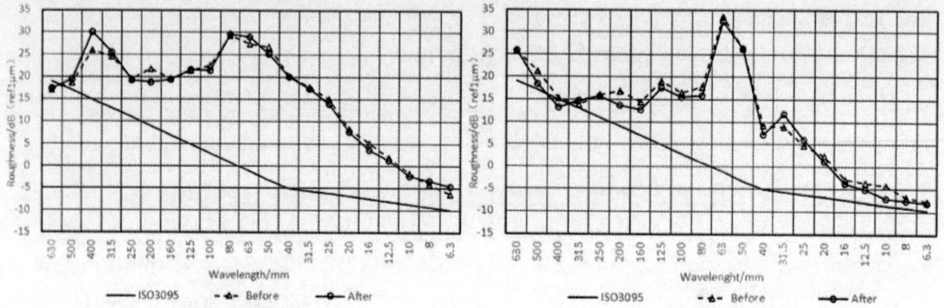

a) Low Rail b) High rail

Fig. 3. Rail roughness wavelength spectra – Egg Fasteners Section Upline

5.2 Measurement of Track Wave Decay Rate

The track decay rate characterizes the acoustic performance of acoustically optimised track components and fastening system. A track system with high wave decay rate can reduce the propagation distance of vibration and rolling noise waves along rail tracks, implying a reduction of dynamic energy transferring to surrounding structures.

According to EN 15461:2008 + A1 2010 [7], Track decay rate was measured for the Egg Fasteners section upline. Measured data are shown in Fig. 4. For vertical vibration wave, the maximum decay rates at 250 Hz were indifferent with and without rail web dampers.On the other hand, at 800 Hz, the decay rate with rail web dampers is doubled comparing to that without dampers. Furthermore, the decay rate dropped quickly between 63–500 Hz with a minimum value of 0.1 dB/m at 250 Hz. For the lateral velocity wave, the overall decay rate was also low. Even with rail dampers, maximum decay rate was just over 4 dB/m at 630 Hz.

5.3 Rail Vibration Response to Wheel-Rail Excitation

Dynamic impact forces from wheels to rail cause vibration of the train and track system and associated rolling noise radiation. The appearance of rail surface corrugation increases wheel-rail impact force and results in dynamic problem for rolling stock.

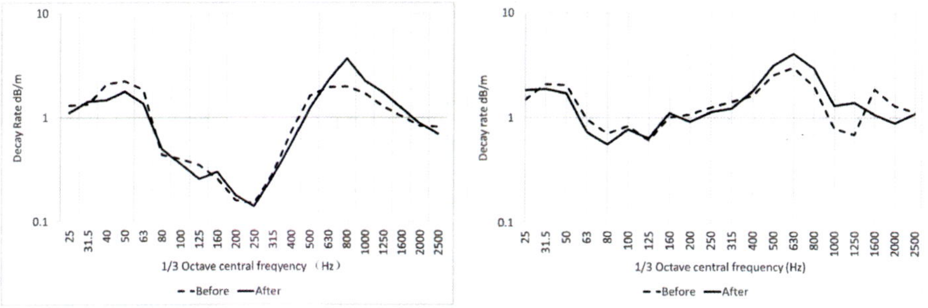

a) Vertical b) Lateral

Fig. 4. Track decay rates - Egg Fasteners Section Upline

To this end, rail vibration responses of the upline track at the Egg Fasteners section were measured at a mid-span cross section between sleepers. Both low and high rails were tested. Accelerometers were arranged as illustrated in Fig. 6. Experiments were conducted during train's normal operation hours, rail vertical and lateral accelerations were measured.

For a direct comparison with in-car noise spectra, measured accelerations were converted to A-weighted velocity levels. Resultant velocity spectra at 1/3 octave bands are plotted in Fig. 6. Before rail web dampers were installed, the rail vertical velocity response peaks were at 200–315 Hz and 800 Hz, overall velocity level is 120 dB(A); lateral velocity peaked at 250–315 Hz and 500–1250 Hz with overall level of 116 dB(A). After rail dampers installed on upline tracks at the Egg Fasteners section, both vertical and lateral velocity responses reduced significantly at frequency regions above 800 Hz, but not at 250 Hz, which is the dominant frequency of measured in-car noise level, rail roughness and decay rate. Overall velocity reductions were 3.9 dB(A) and 2.4 dB(A) for vertical and lateral respectively. The major peaks of rail vibration levels in both vertical and lateral directions were still at 250–315 Hz (Fig. 5).

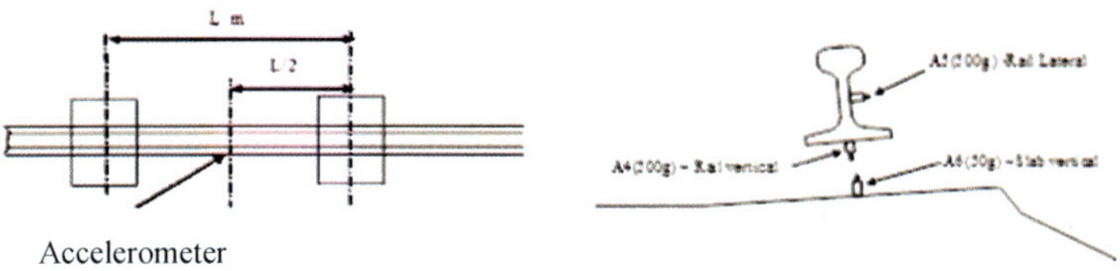

Fig. 5. Rail vibration measurement setup – Egg Fasteners Section Upline

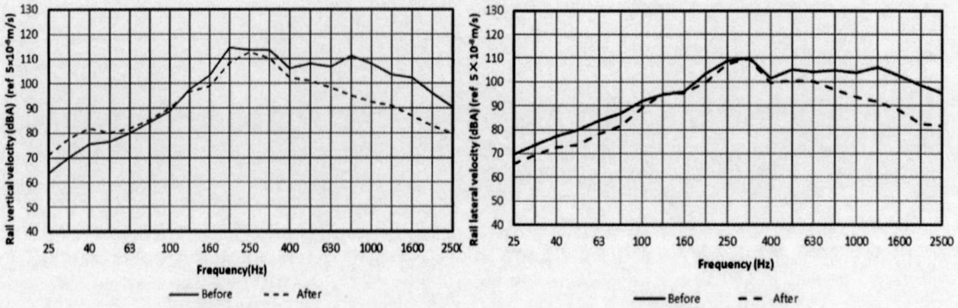

a) Vertical b) Lateral

Fig. 6. Rail vibration velocity spectrum - Egg Fasteners Section Upline

6 Conclusions

This paper presents the investigation carried out at a curved track section with small radius of 350 m of a city underground line, where steel spring floating slab and vulcanized bonded baseplates were installed. The investigation included initial in-car noise and carriage floor vibration level assessments, measurements of rail surface corrugation, decay rates of vibrational waves along the track, and rail vertical and lateral vibration responses. Suitability of using a type of rail web damper for in-car noise control was also evaluated. According to analysis of experimental data, the following conclusions were made for the Egg Fasteners section:

a) Through in-car noise assessment, it is found that the maximum peak Sound Pressure Level (SPL) was about 95 dB(A) and overall SPL 91 dB(A). The dominant 1/3 octave band frequency of in-car noise level was at 250 Hz.
b) Rail surface roughness, rail vertical and lateral velocity were all peaked at 250 Hz 1/3 octave frequency band.
c) there is no reduction in overall vibration level but there is a clear reduction in the spectrum of vibration at frequencies above 500 Hz. This is also seen in the noise spectrum.
d) Rail vibration wave decay rate were low even after rail web dampers were installed, the maximum values were around 4 dB/m. Especially for the vertical direction, the decay rate was merely 0.1 dB/m at 250 Hz.
e) There is no obvious reduction in rail vibration velocity after rail web dampers were installed. The overall noise level was reduced by 3 dB(A) for up trip from 91 dB(A) to 88 dB(A).

References

1. Collette, C., Horodinca, M., Preumont, A.: Rotational vibration absorber for the mitigation of rail rutting corrugation. Veh. Syst. Dyn. **47**(6), 641–659 (2009)

2. Grassie, S.L.L., Cox, S.J.: The dynamic response of railway track with flexible sleepers to high frequency vertical excitation. Proc. Inst. Mech. Eng. **198D**(7), 117–124 (1984)
3. Wang, A., Wang, Z., Zhang, P., Xu, N., Zhang, Z., He, K.: Study on the mechanism of discontinuous support stiffness on the development of rail corrugation. In: 21st International Congress on Sound and Vibration, vol. 1, pp. 1926–1933, Beijing, China, July 2014
4. Wang, A., Wang, Z., Zhang, Z., Xu, N., Qiao, X., Du, M.: Effects of rail lateral dynamic deflection and vibration level on rail corrugation development. In: Anderson, D., et al. (eds.) Noise and Vibration Mitigation for Rail Transportation Systems. NNFMMD, vol. 139, pp. 355–366. Springer, Cham (2018). https://doi.org/10.1007/978-3-319-73411-8_26
5. Railway Applications - Noise Emission - Characterisation of the Dynamic Properties of Track Sections for Pass by Noise Measurements, EN 15461:2008 + A1 2010 (2008)

Effect of Ground Conditions and Microphone Position on Railway Noise Measurement Results

David J. Thompson(✉), Xianying Zhang, and Giacomo Squicciarini

Institute of Sound and Vibration Research,
University of Southampton, Southampton SO17 1BJ, UK
djt@isvr.soton.ac.uk

Abstract. According to ISO 3095, railway pass-by noise should be measured at 7.5 m from the track and at 1.2 m or 3.5 m above the rail head or at 25 m from the track and 3.5 m height. The level of the ground surface should be between 0 m and –2 m relative to the top of the rail. The level of the ground and nature of the ground surface can affect the spectral content of the measured sound. In this paper the TWINS model is used for a range of situations to illustrate the effect of ground height and equivalent flow resistivity on the sound pressure spectrum. The location of the ground dip is shown to depend on the ground properties. Measurement results support these results. For the particular cases illustrated it is shown that the differences between the overall A-weighted noise levels at 7.5 m and 25 m can vary between 4.6 and 7.4 dB for the cases considered. Moreover, the noise level differences between different track types can also vary by several dB. In particular, the level differences between two ballasted tracks and a slab track are found to vary in a range of about ± 2 dB.

Keywords: Rolling noise · Pass-by noise · Ground reflection · Variability

1 Introduction

According to ISO 3095 [1], railway pass-by noise should be measured at 7.5 m from the track centre and at a height of 1.2 m or 3.5 m above the rail head or at 25 m from the track and 3.5 m height. Whereas the 25 m position was previously favoured for high speed train measurements, later versions of the TSI Noise [2] require measurements at 7.5 m, although for speeds at and above 250 km/h both microphone heights should be used. The level of the ground surface should be between 0 m and 2 m below the top of the rail. As noted by the standard, the level of the ground and nature of the ground surface can affect the spectral content of the measured sound. Of particular importance is the so-called ground dip that occurs when direct and reflected sound interfere due to their path-length difference. The purpose of this paper is to use a theoretical model to investigate the extent of this effect as well as variations between microphone positions. Only rolling noise sources are considered but similar effects can be expected for other sources.

G. Degrande et al. (Eds.): *Noise and Vibration Mitigation for Rail Transportation Systems*, NNFM 150, pp. 680–687, 2021.
https://doi.org/10.1007/978-3-030-70289-2_75

2 Methodology

The TWINS model [3] allows predictions of rolling noise to be made from the properties of the wheel and track, including their surface roughness. The model produces the sound power spectra in one-third octave bands from the radiation of the wheel and track. Moreover, a simple sound propagation model is included to allow the average sound pressure spectra over the pass-by to be determined in the vicinity of the track. The wheel is considered as a moving point source (using both monopole and dipole directivities) and the rail as a coherent line source. The ground is represented as a flat plane at a specified height below the rail. It has an impedance defined by the Delany-Bazley model [4], which relies on a single parameter, the equivalent flow resistivity. Spherical wave reflection is taken into account [5]. When comparing measured results with TWINS predictions it is often possible to adjust the ground height and equivalent flow resistivity used in the model to ensure satisfactory agreement. Here the model is used for a range of situations in order to illustrate the effect of ground height and equivalent flow resistivity on the sound pressure level. In addition the predicted sound pressure levels at the different microphone positions are compared. Measurement results at a range of microphone heights are also presented for comparison.

3 Effects of Ground Properties on the Track Sound Radiation

The parameters used for two ballasted tracks and one slab track are listed in Table 1. These have been obtained by matching the decay rate and point mobility predicted by the track model with the corresponding measured results. The two ballasted tracks have been measured previously [6]. Track 1 had stiff rail pads whereas Track 2 had soft pads. Both tracks had monobloc concrete sleepers. For the slab track, the fastening system has two resilient layers with an intermediate mass; this is represented by the same model but the 'sleeper' now represents the intermediate baseplate; the slab is assumed to be rigid. More details on the validation of these three tracks can be found in [7]. Measured track decay rates are used in the predictions of the sound radiation from these three tracks. A common roughness spectrum is assumed, based on the limit curve in ISO 3095:2013 [1]. For the slab track the noise radiated by the slab itself is not included; it has been found only to be significant below 100 Hz [7].

The properties of the ground, including its height below the rail and its flow resistivity, affect the sound propagation to receiver positions at the trackside, especially below 1000 Hz. Some results were presented by Nielsen et al. [8] which showed differences of several dB in overall level depending on the ground height and flow resistivity.

The effects of the ground height and the flow resistivity on the noise propagation at the trackside are shown in Fig. 1 for the standard receiver position at 7.5 m from the centre of the track and 1.2 m above the rail. In Fig. 1(a) three values are used for the ground height: 0.5 m, 1.1 m and 1.5 m; a fourth case is also considered in which there are no ground reflections which is used as a reference. Using the track parameters from Table 1, the total rolling noise for ballasted track 1 is predicted with the different ground heights, for a pass-by speed of 120 km/h. A flow resistivity of 3×10^5 Pa.s.m^{-2} is used initially for the ground.

Figure 1(a) presents these results as level differences in decibels relative to the case with no ground. The ground dip moves to a lower frequency as the ground height is increased. In Fig. 1(b) corresponding results are shown for three values of ground flow resistivity: 3×10^4 Pa.s.m^{-2}, 3×10^5 Pa.s.m^{-2} and 3×10^6 Pa.s.m^{-2}, in each case with an assumed ground height of 1.1 m. It can be seen that the three results are similar above 1 kHz but the flow resistivity also affects the position of the ground dip. Figure 2 presents the corresponding results for the position at 25 m away from the track and a height of 3.5 m. Similar trends are found but the details are different.

The influence of the ground properties on the differences in the sound pressure level between the two microphone positions, 7.5 m and 25 m from the track, is explored next. The sound pressure level of the three tracks presented in Table 1 are predicted using TWINS at the two receiver positions, when the train passes at 80 km/h, 120 km/h, 200 km/h and 300 km/h. Five values, 0.25 m, 0.5 m, 1.1 m, 1.5 m and 2 m, are used as the ground reflection height in the calculations when the ground flow resistivity is 3×10^5 Pa.s.m^{-2}. Flow resistivities of 3×10^4 Pa.s.m^{-2} and 3×10^6 Pa.s.m^{-2} are additionally considered when the reflection height is set as 1.1 m.

Table 1. Parameters used for the three tracks

Parameters	Track 1	Track 2	Slab track
Rail cross section	UIC60	UIC60	UIC60
Rail vertical bending stiffness	6.42 MNm^2	6.42 MNm^2	6.42 MNm^2
Rail lateral bending stiffness	1.07 MNm^2	1.07 MNm^2	1.07 MNm^2
Rail vertical loss factor	0.02	0.02	0.02
Rail lateral loss factor	0.025	0.02	0.02
Sleeper model	Beam	Beam	Mass
Sleeper mass	300 kg	300 kg	6.0 kg
Sleeper spacing	0.65 m	0.65 m	0.65 m
Pad vertical stiffness	800 MN/m	120 MN/m	1730 MN/m
Pad lateral stiffness	30 MN/m	15 MN/m	150 MN/m
Pad vertical loss factor	0.25	0.25	0.2
Pad lateral loss factor	0.25	0.25	0.15
Ballast vertical stiffness per half sleeper	125 MN/m	75 MN/m	75 MN/m*
Ballast lateral stiffness per half sleeper	100 MN/m	75 MN/m	35 MN/m*
Ballast vertical damping per half sleeper	62.2 kNs/m	37.3 kNs/m	0.2
Ballast lateral damping per half sleeper	199 kNs/m	149 kNs/m	0.15

*Baseplate pad vertical/lateral stiffness

In Fig. 3, the predictions of the sound pressure level at 7.5 m are compared with those at 25 m. Figure 3(a) plots the overall A-weighted levels against each other for all the cases. The results at 7.5 m are higher than those at 25 m, with a mean difference of

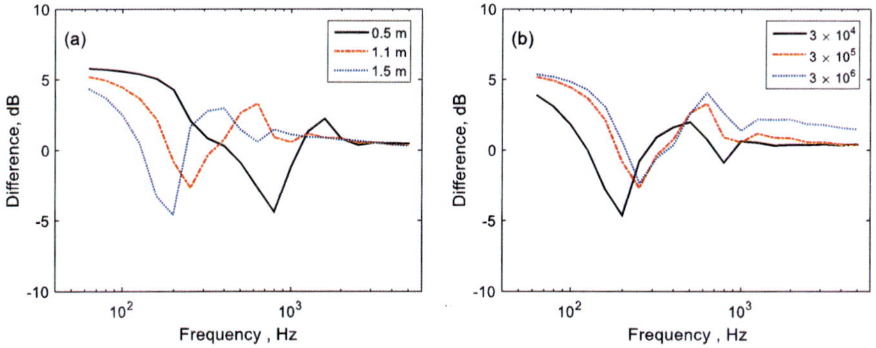

Fig. 1. Level difference at the receiver position at 7.5 m distance and 1.2 m above top of rail between cases with and without ground reflections. (a) Different heights of ground below the rail head for flow resistivity of 3×10^5 Pa.s.m^{-2}; (b) different ground flow resistivities for ground height 1.1 m below rail head.

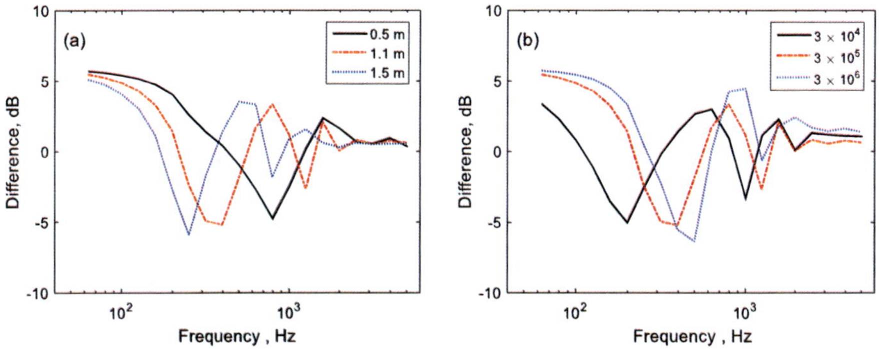

Fig. 2. Level difference at the receiver position at 25 m distance and 3.5 m above top of rail between cases with and without ground reflections. (a) Different heights of ground below the rail head for flow resistivity of 3×10^5 Pa.s.m^{-2}; (b) different ground flow resistivities for ground height 1.1 m below rail head.

+6.0 dB. This can be compared with the result expected for geometric spreading from a line source of 5.2 dB. The spread of these results has a range, expressed as ± two times the standard deviation, of ±1.4 dB. This range is also shown in the figure. Thus the difference varies between 4.6 and 7.4 dB. These results are independent of speed. Figure 3(b) shows the results as a function of frequency. Especially below 1 kHz, where the main ground dip occurs, the differences can be as large as ±7 dB.

The differences in the sound radiation between ballasted track 1 and the slab track are presented in Fig. 4 for the same range of speeds, ground heights and flow resistivities. This figure contains the results for both 7.5 and 25 m. The sound radiation from the slab track is on average 6.3 dB greater than that from ballasted track 1, which has stiff rail pads. The spread of these results, given by two times the standard deviation, is ±2.1 dB. The difference reduces slightly with increasing speed (the results with higher levels generally correspond to the higher speed). The mean value for the results at 80 km/h is

6.8 dB, whereas it is 5.5 dB for 300 km/h. In terms of spectra, the slab track is up to 10 dB noisier around 500 Hz, but is quieter at low frequency where the sleeper component dominates the noise from the ballasted track.

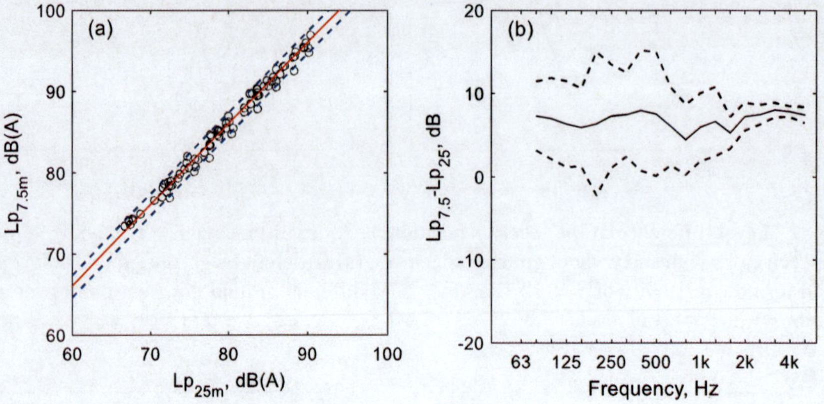

Fig. 3. (a) Predicted total sound pressure level at the receiver position at 7.5 m plotted against the predicted noise at 25 m (o, noise predictions). (b) Level difference between 7.5 m and 25 m. —, mean difference; - - -, mean ± two standard deviations.

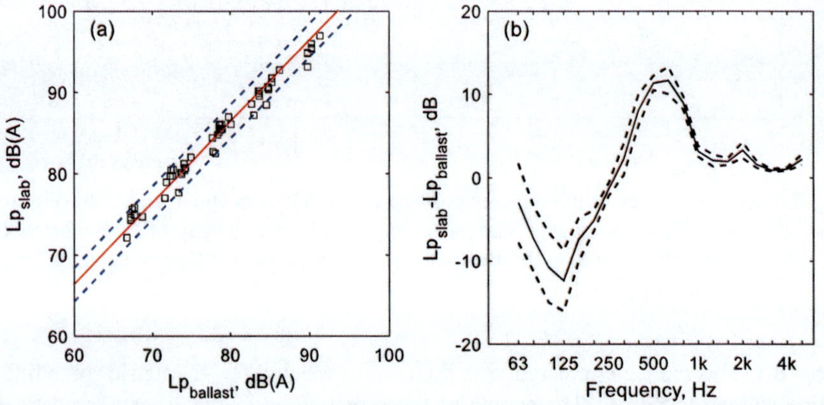

Fig. 4. (a) Predicted total sound pressure level for ballasted track 1 plotted against the sound radiation from the slab track (□, noise predictions). (b) Level difference between slab and ballasted track 1. —, mean difference; - - -, mean ± two standard deviations.

In Fig. 5 the results for the slab track are compared with the corresponding results for ballasted track 2. In this case the mean value of the level difference is 2.0 dB, with a range (two times the standard deviation) of ±1.6 dB; i.e. the range is from 0.4 to 3.6 dB. The difference reduces with increasing speed in this case as well. The mean value is 2.4 dB for 80 km/h, and only 1.5 dB for 300 km/h. Similar trends are seen in the spectra but the differences are smaller than for ballasted track 1.

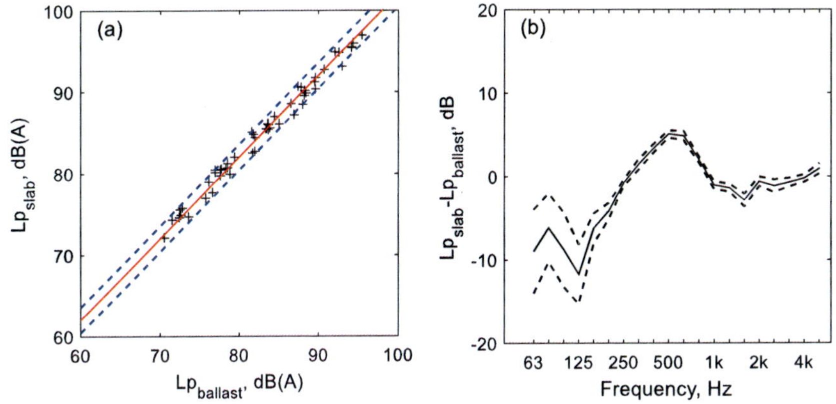

Fig. 5. (a) Predicted total sound pressure level for ballasted track 2 plotted against the sound radiation from the slab track (+, noise predictions). (b) Level difference between slab and ballasted track 2. —, mean difference; - - -, mean ± two standard deviations

4 Measurement Validation

In the measurement campaign for the Roll2Rail project [9], extensive measurements were made of the rolling noise from a test train at a site near Munich. This included the sound pressure level measured by using an array of microphones at 7.5 m from the nearest rail (8.25 m from the track centre) for train passages at various speeds. Four microphones, which have heights of 1.2 m, 1.9 m, 3.3 m and 5.1 m above the ground, are chosen to study the effect of the ground height on the measured data. The ground height is assumed to be 1.5 m below the rail head, while the flow resistivity used is 3×10^5 Pa.s.m^{-2}.

The properties of this ballasted track are similar to those of track 2 in Table 1. The vertical rail pad stiffness is 105 MN/m and the lateral pad stiffness is 13 MN/m; in both cases the damping loss factor is taken as 0.2. The ballast stiffnesses per sleeper are 105 MN/m and 35 MN/m for the vertical direction and lateral direction, respectively with damping loss factors of 1.0 and 2.0. In the noise predictions the measured track decay rates are used.

The level differences between the sound pressure level at four receiver positions and the corresponding case without ground reflections as predicted in TWINS are shown in Fig. 6(a). The location of the ground dip moves towards lower frequencies as the height of the receiver position is increased.

In Fig. 6(b) results are shown that are based on the measured noise for a train speed of 80 km/h. This shows the level differences between the *measured* sound pressure level at the same four receiver positions and the corresponding *predicted* noise without ground reflections. There are some systematic differences at high frequency, which are consistent between the receiver locations. Below 100 Hz the measured noise is consistently higher than the prediction, possibly due to the influence of traction noise from the locomotive. However, despite these features, the location of the ground dip can be clearly seen in these results between 160 and 500 Hz, in each case at a similar frequency as in Fig. 6(a).

Such comparisons can be used to verify the effective ground height and/or flow resistivity for use in the model.

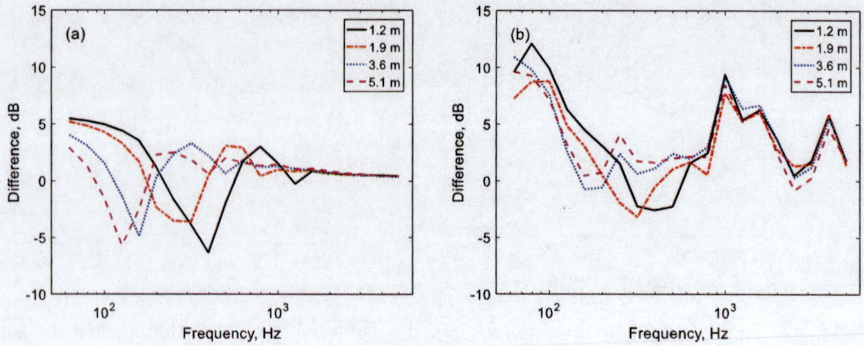

Fig. 6. Level difference at four receiver positions at 8.25 m from the track centre between cases with and without ground reflections. Heights shown are above ground level. (a) Predicted; (b) measurement at 80 km/h minus the predicted levels without ground reflections.

Results for another example are shown in Fig. 7. These measurements were used in the Acoutrain project to test transposition techniques [10]. In Fig. 7(a) the level difference is shown between measured noise spectra and predictions with no ground. A clear ground dip can be seen at 400 Hz for all speeds. Figure 7(b) shows predictions for different ground heights, which are compared with the measurement result for 80 km/h. From this, a height of 1 m below the rail is seen to be most appropriate.

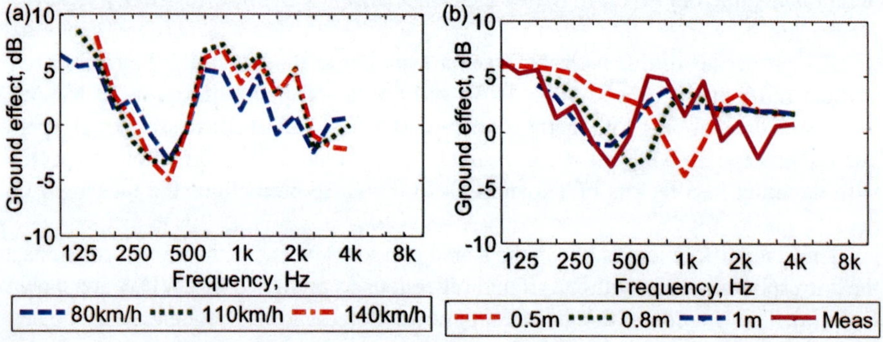

Fig. 7. (a) Level difference at 7.5 m from the track centre between measured noise spectra and predictions with no ground. (b) Level difference between cases with and without ground reflections from TWINS for different heights of the ground below the rail.

5 Conclusions

Using simulations with the TWINS model, the results presented show the sensitivity of measured pass-by noise to ground height, ground absorption and microphone position.

Measurement results are also shown which support these findings and which can be used to determine the effective ground height and/or flow resistivity for use in the model. For the particular cases illustrated it is shown that the differences between noise levels at 7.5 m and 25 m can vary between 4.6 and 7.4 dB. Moreover, the noise level differences between different track types can also vary by several dB. As an example the level differences between each of two ballasted tracks and a slab track are found to vary in a range of about ± 2 dB.

Acknowledgements. The work described here has been supported by the EPSRC under the programme grant EP/M025276/1, 'The science and analytical tools to design long life, low noise railway track systems (Track to the Future)'.

References

1. ISO 3095–2013 Acoustics – Railway applications – Measurement of noise emitted by railbound vehicles, International Organization for Standardization
2. Commission Regulation No 1304/2014 of 26 November 2014 on the technical specification for interoperability relating to the subsystem 'rolling stock — noise'. Official Journal of the European Communities 12.12.2014 L356/421–437.
3. Thompson, D.J., Hemsworth, B., Vincent, N.: Experimental validation of the TWINS prediction program for rolling noise, part 1: description of the model and method. J. Sound Vib. **193**, 123–135 (1996)
4. Delany, M.E., Bazley, E.N.: Acoustical properties of fibrous absorbent materials. Appl. Acoust. **3**(2), 105–116 (1970)
5. Thompson, D.: Railway noise and vibration mechanisms, modelling and means of control. Elsevier, Oxford (2008)
6. Zhang, X., Thompson, D.J., Li, Q., Kostovasilis, D., Toward, M.G.R., Squicciarini, G., Ryue, J.: A model of a discretely supported railway track based on a 2.5D finite element approach. J. Sound Vibration **438**, 153–174 (2019)
7. Zhang, X., Jeong, H., Thompson, D.J., Squicciarini, G.: The noise radiated by ballasted and slab tracks. Appl. Acoust. **151**, 193–205 (2019)
8. Nielsen, J.C.O., Augis, E., Biebl, F.: Reproducibility of railway noise measurements – influence of weather and test site conditions. In: Anderson, D., et al. (eds) Notes on Numerical Fluid Mechanics and Multidisciplinary Design, vol. 139, pp. 167–179 (2018)
9. Thompson, D.J., Squicciarini, G., Zhang, J., Lopez Arteaga, I., Zea, E., et al.: Assessment of measurement-based methods for separating wheel and track contributions to railway rolling noise. Appl. Acoustics **140**, 48–62 (2018).
10. Jansen, H.W., Dittrich, M.G.: TNO Technical Sciences. Acoutrain Deliverable D2.5 Source separation and transposition techniques. EC Contract No. FP7 – 284877 (2014)

Bridge Noise and Vibration

Comparison of Vibration and Noise Characteristics of Urban Rail Transit Bridges with Box-Girder and U-Shaped Sections

Qi Li[1](✉), Hong-yao Huang[1], and David J. Thompson[2]

[1] Department of Bridge Engineering, Tongji University, 1239 Siping Road, Shanghai, China
liqi_bridge@tongji.edu.cn

[2] Institute of Sound and Vibration Research, University of Southampton, Southampton SO17 1BJ, UK

Abstract. Rolling noise and structure-borne noise from the viaducts usually lead to complaints from nearby residents. Concrete box section bridges are traditionally adopted for the viaduct but those with U-shaped sections have become popular recently due to their more attractive form and reduced visual impact. One question that often arises in relation to the choice of the section types is their relative noise performance. This study aims to compare the vibration and noise characteristics of concrete bridges with different sections by using the same noise prediction method. A coupled track-bridge model is introduced to obtain the rail vibration and the power input to the bridge through the rail fasteners. A three-dimensional acoustic finite element method is applied to obtain the noise radiated from the bridge and the rail subjected to given multiple forces acting on them. The averaged mean squared vibration velocity of the coupled wheel-track-bridge model subjected to roughness excitation is used to scale the noise from the acoustic model. This noise prediction method is then validated by comparison with field measured noise from a U-shaped bridge in Shanghai. Comparative investigations are conducted among the U-shaped bridge, a box girder bridge with single cell and a twin-box girder bridge with two cells. It is found that the rail noise is about 10 dB larger than the bridge noise for the three bridges at positions to the side of the bridge. The U-shaped girder generally leads to higher total noise levels than the box girders by about 1 dB(A) when they have the same height of noise barriers. In terms of the bridge noise, the single-box and twin box girders produce an average of 2.4 and 4.2 dB(A) less noise than the U-shaped girder.

Keywords: Urban rail transit · Concrete bridges · Vibration and noise · Wheel-rail interaction · Finite element method

1 Introduction

For rail transit traffic in China, viaducts are often adopted in suburban areas instead of underground tunnels for economic reasons. However, high levels of rolling noise and structure-borne noise from the viaducts may lead to complaints from nearby residents.

G. Degrande et al. (Eds.): *Noise and Vibration Mitigation for Rail Transportation Systems*, NNFM 150, pp. 691–698, 2021.
https://doi.org/10.1007/978-3-030-70289-2_76

Concrete box section bridges are traditionally adopted for the viaduct but those with U-shaped sections have become popular recently due to their more attractive form and reduced visual impact.

Many prediction methods have been proposed for noise from concrete bridges. Zhang et al. [1] developed a hybrid method to predict bridge noise by combining the finite element method and statistic energy analysis. Li et al. [2] proposed a two-and-a-half dimensional (2.5D) method for the prediction of bridge noise to improve computational efficiency compared with a fully-three-dimensional approach. Song et al. [3] extended this method to allow the prediction of noise from multi-span bridges. Later Song et al. [4] adopted the same method to investigate both bridge and rail noise. Li et al. [5] presented a simplified acoustic model based on a two-dimensional (2D) approach for the prediction of bridge and rail noise above 200 Hz. Song et al. [6] combined the 2.5D and 2D methods and compared the bridge and rail noise between a U-shaped girder and a box girder. The noise barriers or crash barriers on the box girder were ignored in this study. Wu and Liu [7] have made comparisons between a U-shaped girder and a box girder using 3D finite element and boundary element models and found that the structure-borne sound power radiated from the box girder is lower than that from the U-shaped girder by about 2.5 dB, but they did not discuss the rail noise.

This study aims to provide guidelines on the selection of viaduct types in terms of their noise characteristics including the noise from both the bridge and the rail. A procedure is proposed to simulate the vibration of the coupled rail-bridge system as well as the noise radiation. A three dimensional acoustical model is used for the bridges and the rails. The effect of noise from multi-span bridges is taken into account by a simple reciprocity method. This procedure is validated by comparison with field test results. It is then applied to a comparative study of the vibration and noise from a U-shaped bridge, a single box-girder bridge and a twin box-girder bridge.

2 Methodology

To calculate the vibration of the rail and the bridge, an infinite Timoshenko beam is used to represent each rail, and a 3D finite element model is developed for the bridges. An example for the U-shaped bridge of length 30 m is shown in Fig. 1(a). The maximum element size in the model was 250 mm, which is sufficiently small compared with the structural wavelength (about 2.0 m) of vibration modes of the bridge at around 1000 Hz. These models are used to obtain the mobilities at the positions of the rail fasteners. The rail fasteners are assumed to be discrete springs with constant stiffness and damping loss factor. The compatibility equation of the track-bridge system is expressed as [5]:

$$(\mathbf{Y}_{\mathrm{B}} + \mathbf{Y}_{\mathrm{R}} + \mathbf{Y}_{\mathrm{s}})\mathbf{F} + \mathbf{Y}_{\mathrm{p}}P = 0 \quad (1)$$

where P is the external concentrated force applied on the rail, which is the function of frequency; $\mathbf{F}$ is the unknown vector of forces applied by the fastener springs; $\mathbf{Y}_{\mathrm{s}}$ is a diagonal matrix containing the mobilities of the springs; $\mathbf{Y}_{\mathrm{R}}$ and $\mathbf{Y}_{\mathrm{B}}$ are respectively matrices containing the mobilities of the free rail and the bridge disconnected from the fasteners, when subjected to unit fastener forces at the connection points; and $\mathbf{Y}_{\mathrm{p}}$ is the vector of transfer mobilities of the free rail from the wheel/rail contact point to the fastener connection points. The mobilities of the rail coupled to the bridge can be further obtained by considering the effects of all the fastener forces on the rail.

The wheel/rail contact positions change with the motion of the train. In this study two specific loading positions are considered: one where the connection between two vehicles is at the middle of the bridge span and the other where the centre of a vehicle is at the middle of the span. The vibration energy is then averaged over these two loading positions. Four wheels are generally acting on each of the two rails within the span length. For a series of wheel/rail contacts, the compatibility equation of the wheel-track-bridge system can be written as:

$$(\mathbf{Y}_{\mathrm{r}} + \mathbf{Y}_{\mathrm{c}} + \mathbf{Y}_{\mathrm{W}})\mathbf{F}_{\mathrm{c}} = \mathrm{j}\omega\mathbf{r} \tag{2}$$

where $\mathrm{j} = \sqrt{-1}$; ω is the circular frequency; $\mathbf{r}$ is the vector of wheel-rail combined roughness amplitudes at the wheel/rail contacts, they have the same spectrum at each wheel and are regarded as incoherent; $\mathbf{F}_{\mathrm{c}}$ represents the vector of vertical wheel-rail contact forces, it is calculated based on the separate roughness excitation at each wheel/rail contact position and then summed up according to the energy superposition principle; $\mathbf{Y}_{\mathrm{r}}$ contains the driving point and transfer mobilities of the rail under all wheels, which is computed from the infinite Timoshenko beam model of the rail subjected to a unit force and the fastener forces obtained from Eq. (1); $\mathbf{Y}_{\mathrm{c}}$ and $\mathbf{Y}_{\mathrm{W}}$ are respectively the mobilities of the wheel-rail contact spring and wheel.

The same finite element model of the bridge is also utilized to obtain the spatially averaged mean squared vibration velocity of the bridge when it is subjected to several point forces of unit amplitude but different phases at the fastener locations. The phase of these forces is found to have insignificant effect on the vibration of the bridge above 50 Hz. The vibrating boundary is thereafter used in an acoustic model with finite element mesh for the air and an automatic matched layer enclosing the air mesh. Figure 1(b) shows a cross-section of the acoustic model for the U-shaped bridge developed in the software Virtual Lab. The size of the acoustic finite elements are determined by the software to obtain results up to 1 kHz. The spatially averaged mean squared vibration velocity of the bridge obtained from the coupled wheel-track-bridge model subjected to roughness excitation is then used to scale the sound pressures obtained from the acoustic model.

The sound pressures caused by the vibrating rails are computed by using an acoustic model including the vibrating boundary of the rails and the reflections from both the bridge and the car-body. Figure 2 shows the acoustic model for the prediction of rail noise. This model uses the harmonic vibration responses of the rails obtained from the Timoshenko beam sitting on the rail fasteners and subjected to multiple unit vertical forces of different phases as the vibrating boundary; it also includes the bridge and car-body as rigid boundaries. The rails are located above the concrete bridge with a small gap so the whole rail surface can radiate sound. The effect of the ground below the bridge is not included in the study.

The three-dimensional acoustic model for the noise prediction just includes the bridge or rails in a single span. However, the noise in the far field is influenced by multiple bridge spans that are excited into vibration during the passage of the train. To account for the effect of multi-span bridges on noise radiation, it is assumed the bridges and rails of each span are incoherent sources. The sound pressure produced by a multi-span

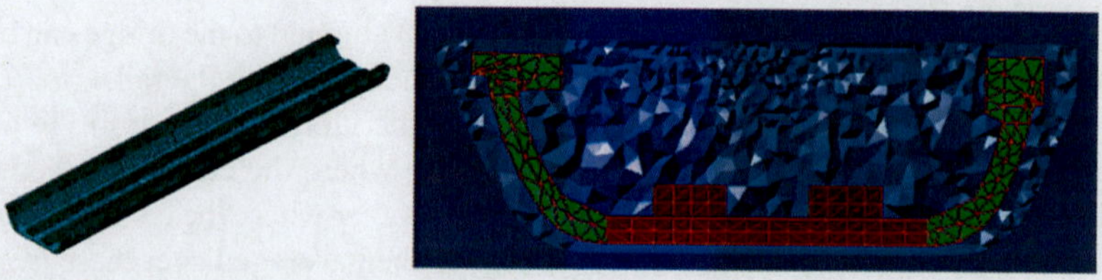

Fig. 1. Three dimensional finite element and acoustic model of the U-shaped bridge: (a) finite element model; (b) acoustic model mesh

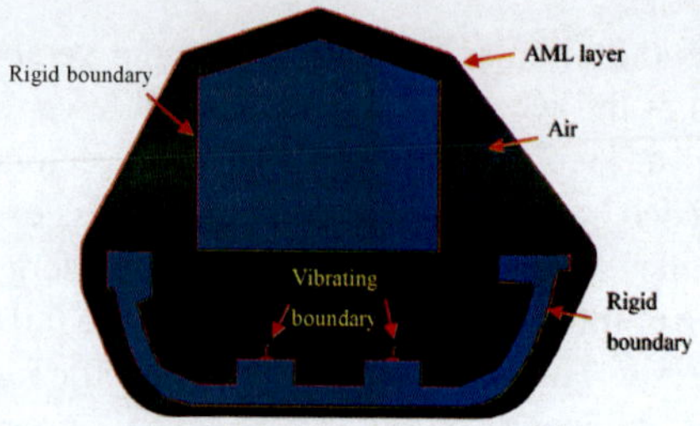

Fig. 2. Acoustic model for the prediction of rail noise

bridge can be calculated with the weighted summation of sound energy from each span. In this study, the vibration and sound radiation of the bridge and the rail are calculated at each frequency from 4 to 1024 Hz with a frequency interval of 4 Hz. The results are then converted to and displayed in one-third octave bands.

3 Results

The method developed for noise prediction is firstly validated by the comparison with noise measured in the field for the U-shaped concrete bridge shown in Fig. 3(a). The elastic modulus, damping loss factor and density of the concrete are respectively 40 GPa, 0.04 and 2600 kg/m^3. The wheel-rail combined roughness was estimated from the measured rail vibration similarly with the method used in Refs. [5, 8]. The roughness level (reference to 1 μm) in one-third octave band is about 12 dB at a wavelength of 100 m and −11 dB at 20 mm. Figure 4 gives the noise from the U-shaped girder bridge (bridge A) at 7.5 m away and 1.5 m above the centre and top of the two rails. It can be seen from Fig. 4 that the predicted noise matches well with the measured one. The rail noise exceeds the bridge noise above 250 Hz.

Comparative results will also be shown for a box girder bridge with single cell (bridge B) and a twin-box girder bridge with two cells (bridge C) as depicted in Figs. 3(b) and 3(c). The spans of all three bridges are the same. The same roughness is adopted for all three bridges for comparative study. The webs of the U-shaped girder can work as both noise barriers and crash barriers. Additional crash barriers are often installed on the sides of the box girders to prevent trains from falling off in case of a derailment. The

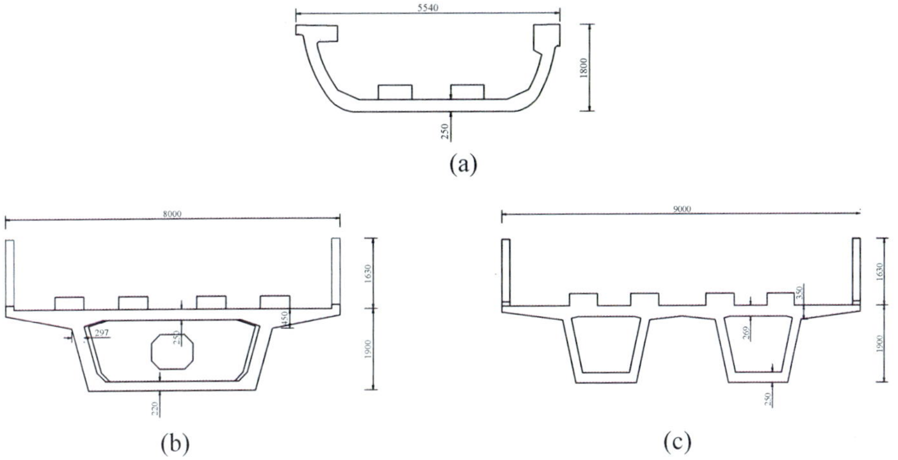

Fig. 3. Sectional views of three bridges at the middle span (unit: mm): (a) A; (b) B; (c) C

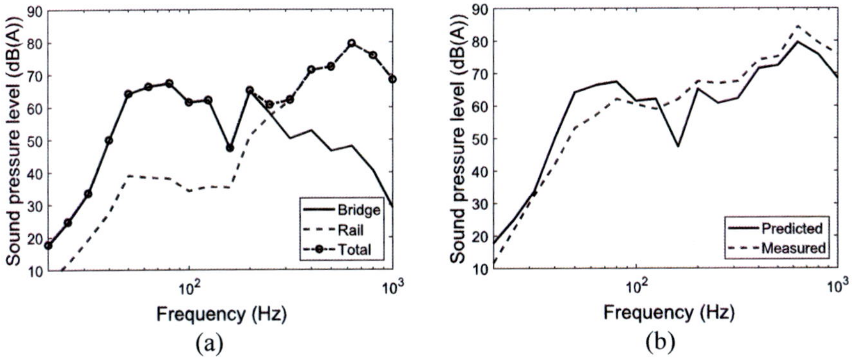

Fig. 4. Noise at 7.5 m away and 1.5 m above the rail centre: (a) noise from rail and U-shaped bridge; (b) comparison between predicted and measured total noise

height of the crash barriers on the box girders are set to be 1.0 m above the top of the rail as the top of the U-shaped girder is also 1.0 m above the top of rail. The crash barriers on the box girder can also work as noise barriers. This effect is taken into account but the vibration of the crash barriers is neglected.

Figure 5(a) shows the track decay rates of the track in each case. It can be seen that the bridges have little influence on the track decay rate above 200 Hz. At lower frequencies below 100 Hz, the track decay rate for bridge A is about 2 dB/m less than those for the other two bridges owing to the low stiffness of the deck of bridge A. Figure 5(b) gives total power input to the three bridges through 8 wheels of a vehicle subjected to the same estimated roughness spectrum. It can be observed that the power input to bridge A is about 3 dB larger than that for bridge B below 300 Hz. The power input to bridge C is similar to that for bridge B below 100 Hz but is about 3 dB lower than that for bridge B between 200 Hz and 500 Hz.

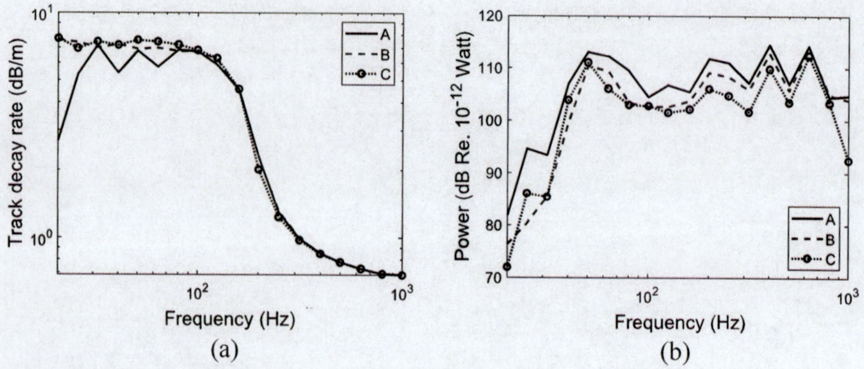

Fig. 5. Track decay rates and power input to bridges: (a) track decay rates; (b) power input to bridges due to 8 wheel/rail contacts

Figure 6 depicts the radiation efficiency and radiated power of the three bridges under the given wheel-rail combined roughness. It can be seen from Fig. 6 that the U-shaped bridge has a lower radiation efficiency than the box girders below 630 Hz. As a result, the radiated power of the U-shaped bridge is only slightly larger (about 1.4 dB) than that of the box girders at 50 Hz.

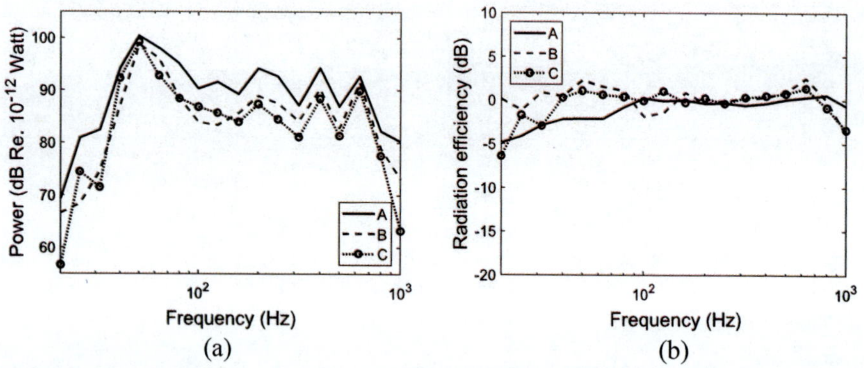

Fig. 6. Radiation efficiency and power of various bridges: (a) radiation efficiency; (b) power

To investigate the difference in sound radiated to the far field, 15 field points are defined as listed in Table 1. The horizontal distance of these points from the track centre ranges from 7.5 m to 200 m, and the vertical distance of them measured from the top of rail ranges from −10 m to 5 m.

Table 1 also lists the difference in overall noise level between bridges B and A, as well as between C and A. It can be seen that, for the field points more than 25 m away from the track centre, the A-weighted total noise levels from bridges B and C are about 1–2 dB lower than that from bridge A. The average difference over all positions is 1 dB in both cases. The rail noise level difference has a similar tendency to the total noise, but with a smaller level difference (0.7 dB) averaged over all positions. This difference is

Table 1. Difference of noise level from bridge A, B and C at various field points

Number of field point	Distance from rail centre and top (m)		A-weighted total noise (dB)		A-weighted rail noise (dB)		A-weighted bridge noise (dB)		Non-weighted bridge noise (dB)	
	Horizontal	Vertical	B-A	C-A	B-A	C-A	B-A	C-A	B-A	C-A
S1	7.5	−10	−2.6	−2.6	−1.2	−1.2	−4.9	−4.9	−6.8	−5.1
S2	7.5	1.5	0.9	0.7	1.0	1.0	−0.7	−3.9	0.4	−3.6
S3	7.5	5	−0.8	−0.9	−0.8	−0.8	0.3	−3.8	2.5	−1.1
S4	25	−10	−1.1	−1.2	−0.1	−0.1	−5.8	−6.6	−6.8	−8.5
S5	25	1.5	−0.4	−0.5	−0.3	−0.3	−2.6	−3.6	−1.6	−4.5
S6	25	5	0.7	0.6	0.8	0.8	−1.9	−3.0	−0.3	−3.3
S7	50	−10	−1.1	−1.1	−0.5	−0.5	−5.2	−5.4	−5.0	−4.5
S8	50	1.5	−1.3	−1.3	−1.2	−1.2	−2.9	−5.3	−1.9	−5.3
S9	50	5	−0.7	−0.8	−0.6	−0.6	−2.0	−4.8	−0.9	−4.8
S10	100	−10	−1.5	−1.7	−1.1	−1.1	−3.7	−5.2	−2.5	−4.4
S11	100	1.5	−1.3	−1.6	−1.3	−1.3	−1.6	−4.4	0.0	−3.5
S12	100	5	−1.4	−1.7	−1.4	−1.4	−1.0	−4.1	0.8	−3.1
S13	200	−10	−1.2	−1.3	−1.1	−1.1	−2.3	−3.1	−0.3	−1.6
S14	200	1.5	−1.2	−1.4	−1.2	−1.2	−1.2	−2.7	1.0	−1.1
S15	200	5	−1.2	−1.4	−1.3	−1.3	−0.9	−2.6	1.4	−0.9

mainly caused by the shape and location of the barriers although the heights of the barriers above the rail are the same in each case. The differences in A-weighted bridge noise are larger, with differences of up to 6 dB. Generally, the box girders radiate significantly less bridge noise than the U-shaped girder. The average difference over all positions relative to bridge A is 2.4 dB for bridge B and 4.2 dB for bridge C. Although the bridge noise is significantly different among the three bridges, the A-weighted overall noise levels do not have obvious differences. This is because the rail noise is about 9–12 dB larger than the bridge noise for the three bridges in this study at the positions considered.

4 Conclusion

The U-shaped girder generally leads to higher total noise levels than the box girders by about 1 dB(A) when they have the same height of noise barriers. The rail noise is about 10 dB larger than the bridge noise for the three bridges when averaged over the field positions considered in this study. The total noise is therefore dominated by the rail noise which is mainly influenced by the barrier effect but not affected by the stiffness of the bridge structure. In terms of the bridge noise, the single-box and twin box girders produce an average of 2.4 and 4.2 dB(A) less noise than the U-shaped girder because the mobilities of the U-shaped bridge beneath the rails are greater than those of the box girders. As a result, less power is input to the box girders than to the U-shaped girder. Moreover, the mass of the box girders is larger than the U-shaped girder. However, this effect is partly compensated by their larger radiating areas and radiation efficiency.

Acknowledgements. The study has been supported by the National Natural Science Foundation of China (grant number 51878501).

References

1. Zhang, X., Li, X.Z., Zhang, J.Q., Song, L.Z., Li, Y.D.: A hybrid model for the prediction of low–frequency noise emanating from a concrete box-girder railway bridge. Proc. Inst. Mech. Eng. Part F J. Rail Rapid Transit **230**(4), 1242–1256 (2016)
2. Li, Q., Song, X.D., Wu, D.J.: A 2.5-dimensional method for the prediction of structure-borne low-frequency noise from concrete rail transit bridges. J. Acoustical Soc. Am. **135**(5), 2718–2726 (2014)
3. Song, X.D., Wu, D.J., Li, Q., Botteldooren, D.: Structure-borne low-frequency noise from multi-span bridges: a prediction method and spatial distribution. J. Sound Vib. **367**, 114–128 (2016)
4. Song, X.D., Li, Q., Wu, D.J.: Investigation of rail noise and bridge noise using a combined 3D dynamic model and 2.5 D acoustic model. Appl. Acoustics **109**, 5–17 (2016)
5. Li, Q., Li, W.Q., Wu, D.J., Song, X.D.: A combined power flow and infinite element approach to the simulation of medium-frequency noise radiated from bridges and rails. J. Sound Vib. **365**, 134–156 (2016)
6. Song, X.D., Li, Q., Wu, D.J.. Prediction of rail and bridge noise in near- and far-field: a combined 2.5-dimensional and two-dimensional method. ASME. J. Vibr. Acoust. **139**(1), 011007–011007 (2017)
7. Wu, T., Liu, J.: Sound emission comparisons between the box-section and U-section concrete viaducts for elevated railway. Noise Control Eng. J. **60**(4), 450–457 (2012)
8. Li, Q., Thompson, D.J.: Prediction of rail and bridge noise arising from concrete railway viaducts by using a multilayer rail fastener model and a wavenumber domain method. Proceedings of the Institution of Mechanical Engineers, Part F: Journal of Rail and Rapid Transit **232**(5), 1326–1346 (2018)

A New Guideline to Reduce Noise Radiated by Railway Bridges in Germany

D. Stiebel[1(✉)], C. Gerbig[1], B. Schewe[2], M. Neudeck[3], Christian Frank[4], P. Tecklenburg[4], and B. Asmussen[5]

[1] DB Systemtechnik GmbH, Völckerstraße 5, 80939 Munich, Germany
dorothee.stiebel@deutschebahn.com
[2] DB Netz AG, Caroline-Michaelis-Straße 5-11, 10115 Berlin, Germany
[3] DB Netz AG, Mainzer Landstraße 181, 60327 Frankfurt, Germany
[4] DB Netz AG, Richelstr. 3, 80634 Munich, Germany
[5] DB Netz AG, Theodor-Heuss-Allee 5-7, 60486 Frankfurt, Germany

Abstract. When trains pass over railway bridges, noise can be radiated by the bridge structure. Since this so-called "bridge noise" can lead to significant annoyance of residents, mitigation measures were intensively investigated in recent years. Within the project "ZIP Brückenleitfaden" guidelines for the application of mitigation measures at new, renewed or refurbished bridges were developed. This paper summarizes results of a recent investigation.

Keywords: Noise radiation · Railway bridge · Guideline

1 Introduction

When trains pass over railway bridges, vibrations are generated at the wheel-rail contact leading to rolling noise that may be different from rolling noise at the adjacent plain track. In addition, vibrations caused by train passages are transmitted into the bridge structure (see Fig. 1, left). The vibrating bridge parts radiate sound that is particularly distinct in the range of low frequencies. Both components, the specific rolling noise of the train on the bridge and the noise radiated by the bridge structure, contribute to the so-called bridge noise. Since bridge noise can lead to significant annoyance of residents, mitigation measures were intensively investigated in Germany within the last twenty years [1, 2].

To assess bridge noise, the sound pressure level during the train pass-by can be measured next to the bridge and next to the adjacent open track (see Fig. 1, right). In the high frequency range, the sound pressure level is mainly determined by the rolling noise. In the considered situation, the rolling noise is comparable at the bridge and the open track. In the low frequency region, the sound pressure level measured next to the bridge is significantly increased due to the sound radiation of the bridge. To describe the bridge noise by a single-number value, the so-called bridge correction factor can be

G. Degrande et al. (Eds.): *Noise and Vibration Mitigation for Rail Transportation Systems*, NNFM 150, pp. 699–706, 2021.
https://doi.org/10.1007/978-3-030-70289-2_77

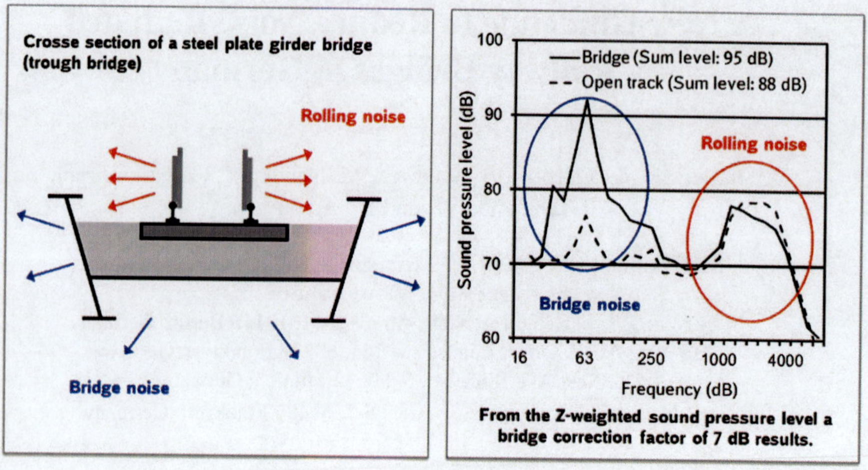

Fig. 1. Sound radiation of bridges, left: schematic illustration of a bridge during the pass-by of a train, right: measured Z-weighted sound pressure levels next to the bridge and next to the adjacent open track during the pass-by of the same train and assessment of bridge noise

calculated as the difference of the Z-weighted sound pressure level at the bridge and at the open track. In the example, the bridge correction factor has a value of 7 dB.

Based on measured bridge correction factors, four different categories of bridges are considered in the German calculation scheme "Schall 03" [3] for railway noise. Steel bridges without ballast bed are described by a level addition K_{BR} of 12 dB, steel bridges with ballast bed by 6 dB, concrete bridges or special low-noise steel bridges with ballast bed by 3 dB and bridges with slab tracks by 4 dB. In addition, mitigation measures are considered. Resilient rail fasteners reduce the noise radiated by steel bridges without ballast by 6 dB and under-ballast mats reduce noise radiated by steel and concrete bridges with ballast by 3 dB.

In Germany, noise calculations based on the calculation scheme Schall 03 must be performed in the planning phase for new lines or upgrading existing lines. Based on the calculation, mitigation measures must be planned if the calculated noise levels exceed the limit values set by the national noise legislation. But for refurbishment of a bridge or replacement of a bridge by a similar bridge construction, calculations are not mandatory and therefore mitigation measures were not necessarily considered. This was one of the reasons for the development of the guideline within the project "Brückenleitfaden" [4].

2 Study on the Need for Mitigation Measures

The project started with a state-of-the-art review including existing results on the sound radiation of bridges. Based on an identification of typical railway bridges within the German network, further investigations were planned. To predict critical areas around steel bridges where annoyance of residents cannot be excluded, an extensive measurement campaign concerning the sound radiation of a typical steel girder bridge without ballast bed during train pass-by was performed (see Fig. 2). In the next step, the results of the measurement campaigns were used to calibrate a simulation model based on the CadnaA software [5] to predict critical areas. For the calculation of the train pass-by, the sound radiation of the wheel and the rail was represented by line sources and the sound radiation of the bridge construction by area sources. The comparison of calculated and measured bridge correction factors as a function of frequency is shown in Fig. 2. Here, measured and simulated bridge correction factors were compared for four different positions which are also shown in Fig. 2. The simulated directivity of the radiated bridge noise was in a good agreement with the measured data.

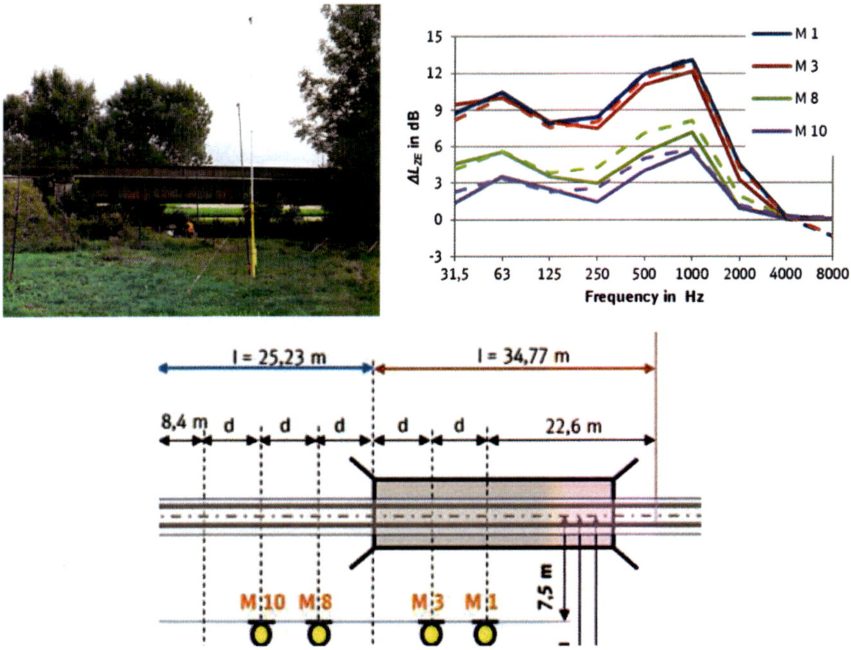

Fig. 2. Investigation of critical areas, the photo (top left) shows the measurement of the sound radiation of the bridge near Langenau, the diagram (top right) shows the comparison of measured (solid line) and calculated (dashed line) bridge correction factor as a function of frequency for four different measurement positions and the figure (bottom) shows the measurement positions M1, M3, M8 and M10 situated next to the bridge and next to the open track [6]

The validated model considered mainly the radiation of vibrating plates and the rolling noise of the trains. It was used to simulate e.g. results for 25 m and 50 m long steel bridges with and without ballast bed (see Fig. 3). The isophones of the bridge correction factors are shown as colored lines. Here, it is assumed that annoyance of residents cannot be excluded in areas where the bridge correction factor exceeds 3 dB. Various simulations were performed resulting in different sizes of the critical areas depending on bridge length. To simplify the application, the critical areas derived from the simulations were described by geometric objects. For steel bridges with ballast bed, the critical areas were estimated by rectangular areas (black dashed lines in Fig. 3, left

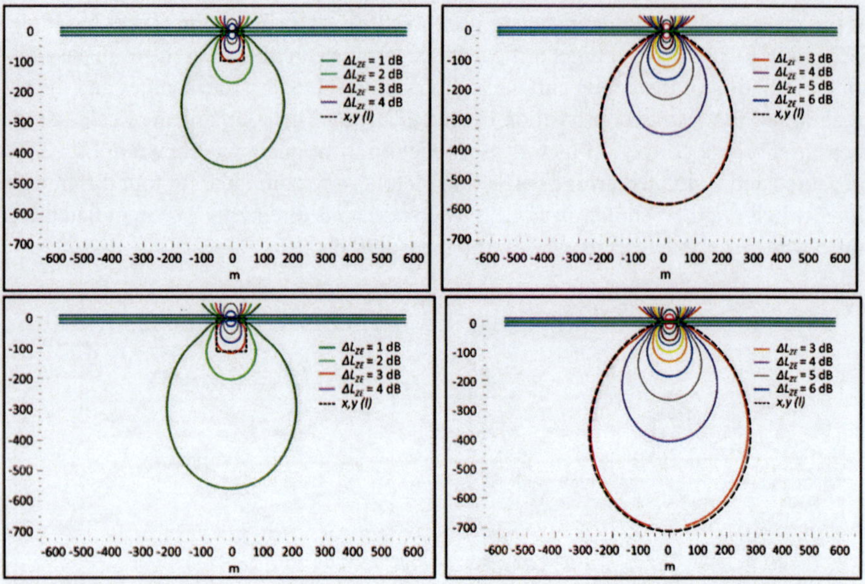

Fig. 3. Calculated bridge correction factors of $l = 25$ m long steel bridges with ballast bed (top left) and without ballast bed (top right) and of $l = 50$ m long steel bridges with ballast bed (bottom left) and without ballast bed (bottom right) [6]. The middle of the bridge is situated at (0,0). The critical areas where annoyance of residents cannot be excluded are fitted by the black dashed lines.

diagrams). For steel bridges without ballast, the critical areas were described by fitted curves (black dashed lines in Fig. 3, right diagrams).

From the simulations, the decision tree as shown in Fig. 4 was derived. Here, the critical areas as illustrated by the dashed lines in Fig. 3 are described. If a building with entitlement to noise reduction is situated within the critical area, mitigation measures should be checked in terms of feasibility and costs, otherwise no measures are required.

3 Track Measure to Reduce Bridge Noise

To decide which mitigation measures should be chosen, the decision trees from Fig. 5 can be used for steel bridges with and without ballast bed. As shown, the optimum choice depends on bridge type and the planned work (building a new bridge, renewing or refurbishing an existing bridge). In the decision trees, mitigation measures are divided into measures already approved by the calculation scheme Schall 03 [3] and measures which are not yet approved. The latter requires confirmation measurements for each individual application.

In Germany, resilient track elements are widely used to reduce bridge noise. Therefore, simulations of their effectiveness were performed using the calculation model RIM-B [7] which is mainly based on modeling the train and the track by masses, dampers and springs. For the application of under-ballast mats at different bridge types, the requirements for the static and the dynamic stiffness were derived from simulations. The aim

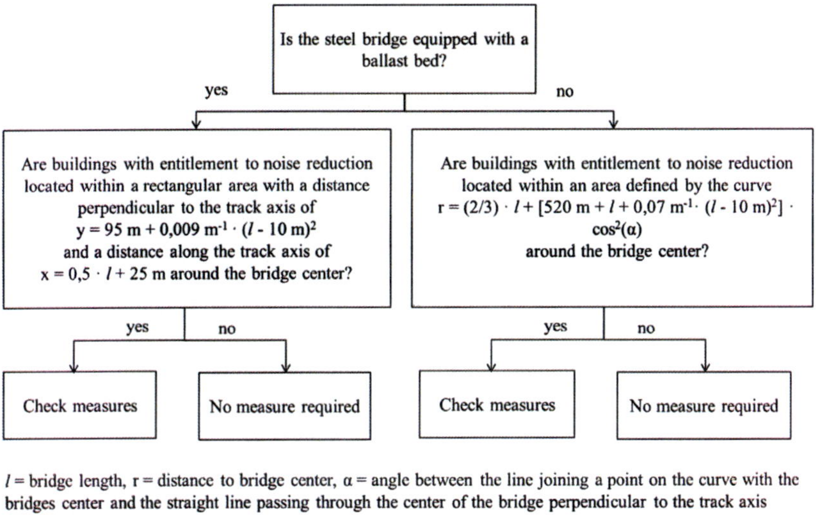

Fig. 4. Decision tree to determine whether mitigation measures should be considered [4].

of the investigation was to decide which material category can be used to obtain the effectiveness of 3 dB required by the calculation scheme Schall 03 [3].

For the simulations, the under-ballast mats were divided into three different categories defined by the dynamic stiffening at 40 Hz according to DIN 45673-5 [8]. Materials of category 1 are defined up to a dynamic stiffening of 1.1 ± 0.3, materials of category 2 are defined up to a dynamic stiffening of 1.9 ± 0.3 and materials of category 3 up to a dynamic stiffening of 2.6 ± 0.3. Table 1 shows the results of the simulations.

Simulations of the effectiveness of resilient rail fasteners with a static stiffness of 10 kN/mm could verify the possible bridge noise reduction of 6 dB predetermined by by the calculation scheme Schall 03 [3]. In the past, an increase in rolling noise due to the installation of resilient rail fasteners was not observed at steel railway bridges.

4 Constructive Measure to Reduce Bridge Noise

The influence of construction details on the sound radiation of different steel bridge types was also investigated. It was shown that especially steel plate girder bridges with a low construction height and stiffened web plates have a low noise radiation during train pass-by. It was concluded that the so-called “thick plate bridge” – a steel plate girder bridge with trough cross-section, 100 mm thick deck plate and ballast bed – should be regarded as low-noise construction. An example of this bridge type can be found in Fig. 6.

To prove this hypothesis, extensive measurements of the sound radiation of two thick plate bridges were performed during the pass-by of regular trains. Besides noise measurements, vibration measurements on the bridge structure and also measurements of the rail roughness were carried out. It could be shown that this bridge type can be allocated to the category of the low-noise steel bridges with ballast bed and a level

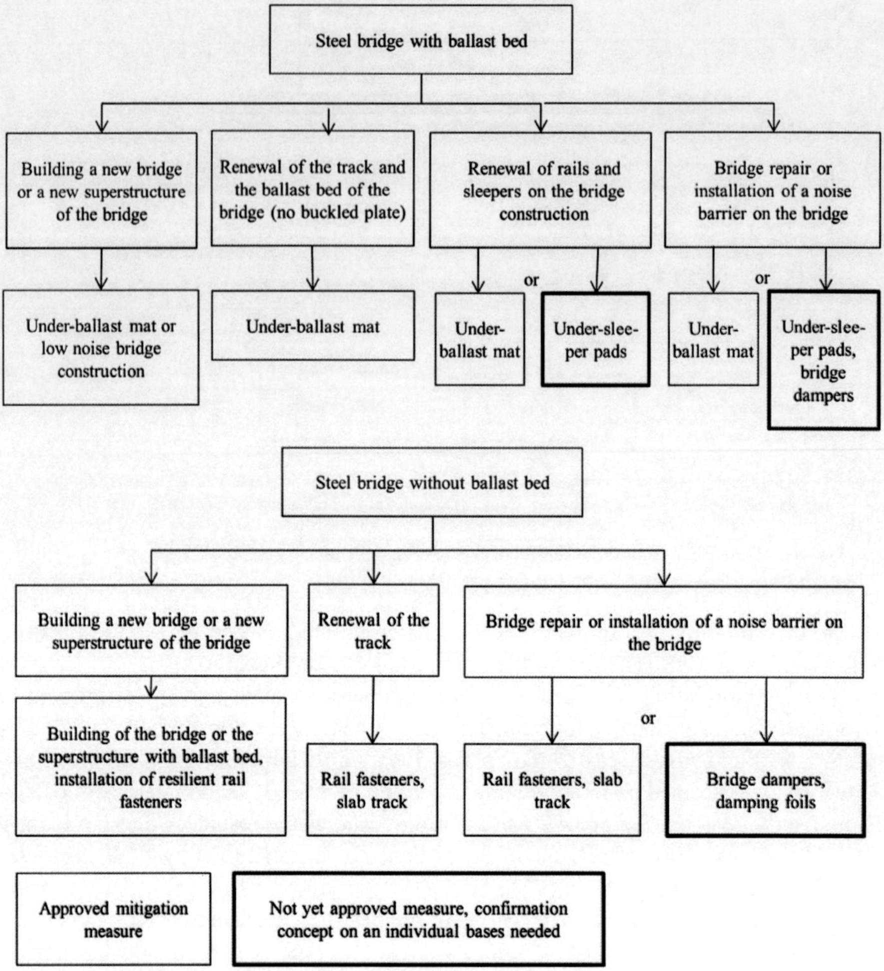

Fig. 5. Decision trees for the choice of measures to reduce bridge noise depending on the bridge type and the work planned [4].

addition K_{BR} of 3 dB according to Schall 03 [3]. By also using an acoustic under-ballast mat, the increase in noise due to the bridge disappears.

Table 1. Requirements for the use of under-ballast mat materials derived from simulations.

Bridge type	Train speed	Under-ballast mats to be used for noise reduction			
		Static bedding modulus	Category 1	Category 2	Category 3
Bridge with massive deck plate (concrete or 100 mm thick steel plate) and ballast bed	≤ 120 km/h	0.03 N/mm^3	Appropriate	Appropriate	Appropriate
	120 km/h < v < 200 km/h	0.06 N/mm^3	Appropriate	Appropriate	Appropriate*
	≥ 200 km/h	0.10 N/mm^3	Appropriate*		
Bridge with steel deck plate and ballast bed	≤ 120 km/h	0.03 N/mm^3	Appropriate	Appropriate	Appropriate*
	120 km/h < v < 200 km/h	0.06 N/mm^3	Appropriate		
	≥ 200 km/h	0.10 N/mm^3	Appropriate to a limited extent**		

*Requirements are not reached in all third-octave bands, but a general reduction of 3 dB is reached.
**Requirements are not reached in several third-octave bands, but a reduction of 2.6 dB is reached.

Fig. 6. Thick plate bridge (steel plate girder bridge with trough cross-section, 100 mm thick deck plate and ballast bed, Photo: D. Stiebel)

5 Conclusion

Within the project "Brückenleitfaden" measures to reduce the sound radiation of steel bridges were compiled and evaluated. In addition, critical areas around steel bridges where annoyance of residents cannot be excluded were determined dependent on the bridge length. With these results, guidance is given on the decision whether a mitigation measure is needed to reduce bridge noise for a specific bridge and which mitigation measure fits best for the construction type. Furthermore, simulations were performed considering the effectiveness of mitigation measures leading to the conclusion how to adapt the mitigation measures to the present conditions. In addition, it was shown that special steel trough bridges can be allocated to the category of the low-noise steel bridges with ballast bed. At the end of the project, the results were included into a public guideline [4] to support engineers with planning and construction of noise reduction measures on bridges.

Acknowledgements. The authors would like to thank the Federal Ministry of Transport and Digital Infrastructure (BMVI) for funding the project.

References

1. Stiebel, D., Lölgen, T., Gerbig, C.: Innovative measures for reducing noise radiation from steel railway bridges. In: Proceedings of the 11th IWRN, Udevalla, Sweden, Published in Notes on Numerical Fluid Mechanics and Multidisciplinary Design (NNFM), vol. 126, pp. 409–416 (2015)
2. Lutzenberger, S., Stiebel, D., Gerbig, C., Wettschureck, R.G.: Luftschall aus dem Schienenverkehr. chapter 17. In: Müller, G., Möser, M. Taschenbuch der Technischen Akustik. Springer-Verlag, Berlin Heidelberg (2017)
3. Berechnung des Beurteilungspegels für Schienenwege (Schall 03): Annex 2 of 16. Bundesimmissionsschutzverordnung (BImSchV) in Bundesgesetzblatt, Part I, no. 61 (2014)
4. DB Netz AG: Einsatz von Schallminderungsmaßnahmen an bestehenden und neuen Eisenbahnbrücken im Netz der DB AG – Brückenleitfaden (Version 2018) (2018)
5. Program CadnaA, Datakustik GmbH, Version (2017)
6. Stiebel, D., Gerbig, C., Asmussen, B.: Untersuchung der Betroffenheit von Anwohnern im Hinblick auf die Schallabstrahlung von Eisenbahnbrücken. Z. Lärmbekämpfung **3**, 79–84 (2018)
7. Müller-BBM Gmbh: Report No. 37332/6, user manual RIM-B. Version 2.08 (2000)
8. DIN 45673-5: Mechanical vibration – resilient elements used in railway tracks – Part 5: Laboratory test procedures for under-ballast mats (2013)

Noise Radiation from Concrete Box Girders Generated by a High-Speed Train Running Along a Track-Bridge System as an Infinitely Long Periodic Structure

Yuanpeng He[1], Jian Han[1], Gong Cheng[1], Xiaozhen Sheng[2(✉)], and Qingsong Feng[3]

[1] State Key Laboratory of Traction Power, Southwest Jiaotong University, Chengdu, Sichuan, China
674954626@qq.com

[2] School of Urban Railway Transportation, Shanghai University of Engineering Science, Shanghai, China
shengxiaozhen@hotmail.com

[3] Engineering Research Center of Railway Environment Vibration and Noise, East China Jiaotong University, Nanchang, Jiangxi, China

Abstract. Bridges with concrete box girders are widely used on China's high-speed railways. Although the bridge structure has a high global static stiffness, its natural frequencies are still low due to the wall-shaped structure, raising concerns about low frequency noise radiation. An effective approach to addressing these concerns is to predict bridge vibration and noise radiation by establishing a dynamic-acoustic vehicle-track-viaduct model. This is the topic of this paper. The dynamics of the vehicle-track-viaduct system is modelled in the frequency domain based on the periodic structure theory and the Fourier-series approach. Predicted vibrations of all the box girders are decomposed into propagating waves, so that sound radiation from the box girders can be predicted using the 2.5D acoustic boundary element method. Comparison between prediction and measurement is performed, confirming the usefulness of the model. The model is then used to study noise radiation from the concrete box girders as a high-speed train passes.

Keywords: Vehicle-bridge interaction · Bridge noise · Railway noise

1 Introduction

With the development of high-speed railways in China, railway noise problems have attracted more and more attentions. Bridges with concrete box girders are widely used on Chinese high-speed railways. Although the bridge structure has a high global static stiffness, its natural frequencies are still low due to the wall-shaped structure, raising concerns about low frequency noise radiation [1]. According to previous studies [2], such bridge noise can be 10 dB higher than other railway noise at low frequencies, and

G. Degrande et al. (Eds.): *Noise and Vibration Mitigation for Rail Transportation Systems*, NNFM 150, pp. 707–715, 2021.
https://doi.org/10.1007/978-3-030-70289-2_78

its attenuation is low and propagation distance is large. Therefore, bridge noise is an important part of high-speed railway noise which should be addressed adequately.

Noise from high-speed railway bridges can be investigated through in-situ measurement or model prediction, or a combination of them. Pure field measurement is difficult, or impossible, to quantitatively determine factors important to bridge noise, making model prediction more and more widely employed. From previous research on bridge noise, it can be seen that predictions of bridge vibration and noise can be divided into two independent parts: one is to analyse vehicle-track-bridge system interaction, and the other is to calculate bridge sound radiation.

Vehicle-track-bridge system interaction models developed in the past can be grouped into moving roughness models and moving vehicle models [3]. Differential equations of these models can be solved either in the time-domain [4, 5] or in the frequency-domain [6]. The moving roughness model [1], normally working in the frequency domain, can largely simplify analysis involved, and therefore has a high computational efficiency. However, wheel-rail interactions caused by axle loads moving along the periodic track structure cannot be considered in such a model.

The moving vehicle model is most consistent with the actual train operation. Since differential equations derived from such a model are of time-varying coefficients, a time-domain method is normally employed to solve the equations, often requiring a huge computational cost and possibly leading to excessive accumulation of errors [7, 9]. To overcome the shortcomings of the moving vehicle model, a circular track model is proposed in Ref. [10] to simulate an infinitely long track. In that model, rails are connected at the beginning and the end. However, the radius of the circular track may become too large when an entire train is included.

Once bridge vibration has been worked out, sound radiation from the bridge can be predicted. Given the complexity of the box girders, numerical methods, such as FEM and BEM, are currently the primarily applied ones. The statistical energy analysis (SEA) has also been applied, especially for steel bridges [8].

Zhang et al. [9, 11, 12] calculated noise from a box girder by using the FEM-BEM method in three steps: (1) obtaining bridge vibration by solving the train-track-bridge interaction problem in the time-domain; (2) transforming bridge vibration from the time domain to the frequency domain using fast Fourier transform. The resulted vibrational velocity on the bridge surface provides the acoustic boundary condition for sound radiation prediction in the next step; (3) predicting sound radiation using frequency domain acoustic BEM. Since step 3 is a 3D acoustic problem, the computation time is often too long, especially when the vibration and sound radiation of multi-span bridges are considered, as in the case of a high-speed train.

Li et al. [13] also used the FEM to model transient train–track–bridge interactions, while the frequency-dependent modal acoustic transfer vector (MATV) method, which is based on the BEM, was applied for acoustic analysis. According to a follow-up study by Li et al. [14], the computation of MATVs using the conventional three-dimensional BEM required a prohibitively long time; thus, they presented a 2.5D BEM-based procedure. In their study, two-dimensional MATVs were calculated for many wavenumbers and then transformed into three-dimensional MATVs using space-wave numerical transforms. However, sound radiation from adjacent spans was neglected, causing predicted sound levels lower than measured ones at far-field points. On the basis of Ref [14], Song et al. [15] established a finite element model of three-span bridges and used the FEM to model transient train-track-bridge interactions. Bridge noise was calculated according to the method in Ref [14]. The results show that the sound pressures calculated from the multi-span bridge model match the measured results in both the time and frequency domains for both near-field and far-field points.

This paper is also concerned with the prediction of noise radiation from concrete box girders generated by a high-speed train. The model is presented in Sect. 2. Different from previous work, the vehicle-track-viaduct dynamics is modelled in the frequency-domain based on the periodic structure theory and the Fourier-series approach. Vibrations of all the box girders are decomposed into propagating waves, so that sound radiation from all the box girders can be predicted using the 2.5D acoustic boundary element method. Section 3 shows in-situ measurement of bridge vibration and noise, and a comparison between prediction and measurement is made in Sect. 4. An investigation is also performed in Sect. 4 using the model on the noise radiation characteristics of the concrete box girders as a high-speed train passes.

2 Development of the Model

In this section, the Fourier-series approach, originally developed in Ref [6] is extended and used to study wheel-rail interaction for a train running along a track on viaduct. The track-bridge system is represented by an infinitely long periodic structure with the period equal to the bridge span and rail irregularities are assumed to be also periodic but with the period equal to a number of bridge spans. Vehicles are modelled as a rigid-body system. Three steps are involved to predict wheel/rail forces: (1) prediction of vehicle dynamics, i.e. to calculate the vibration of vehicles due to a unit harmonic wheel/rail force, generating a receptance matrix for the vehicles; (2) prediction of track-bridge dynamics, i.e. to calculate track-bridge vibration due to a unit harmonic wheel/rail force moving along the track as an infinitely long periodic structure. To perform this step, receptance seen by the rail at each fastener is required and this is calculated using the conventional finite element method; (3) prediction of wheel/rail forces by solving a set of linear algebraic equations.

Vehicle Dynamics. Vehicles are modelled as multi-rigid body systems with linear stiffness and damping, as in Ref. [16]. Only vertical dynamics is considered. Mass, stiffness and damping matrices can be derived for the vehicles. Based on these matrices, a receptance matrix associated with vertical degrees of freedom at wheel/rail contact points can be worked out.

Track-Bridge Dynamics. The track-bridge system is idealized to be an infinitely long periodic structure with the period being the bridge span. Following a procedure similar to that described in Ref. [17], responses of the track/bridge system due to a unit harmonic force moving along the rail can be calculated, generating a 'receptance matrix' associated with displacements of the rail at the wheel/rail contact points. However, there are some differences from Ref. [17]: when considering the track-bridge system, the required receptance seen by the rail at each fastener within a bridge span has to be calculated using the conventional finite element method, since the bridge is now taken into account.

Wheel/Rail Interaction. By assuming rail irregularity to be periodic with the period being a multiple of bridge spans, wheel/rail forces can be calculated based on the receptance matrices mentioned above by solving a set of linear algebraic equations, as described in Ref. [6]. Thus, bridge vibration due to the irregularity can be worked out.

Bridge Sound Radiation. Bridge vibrations worked out in the last step satisfy a periodicity condition and, therefore, Fourier transform can be performed on vibrations of all the bridges as a whole facility to generate the vibration frequency spectrum. As described in Ref. [17], the frequency spectrum of bridge vibration can be expressed to be the sum of a number of propagating waves at particular wavenumbers, and sound radiation from these propagating structural waves can be predicted using the 2.5D acoustic boundary element method.

3 In-Situ Measurement of Bridge Vibration and Noise

3.1 Description of the Measurement

The box girder for the measurement is shown in Fig. 1. The railway is a double-track line supported by box girders at a height of 7.2 m. Box girders are resiliently supported by abutments which are assumed to be rigid on the ground. The ground is rather absorbing and therefore the bridges are assumed to be in a free space acoustically. The high-speed train runs at 350 km/h. Vibration and noise sensors are mounted in the mid-cross section at the top (A1), the wing (A2, S2) and the bottom (A3, S3), where A stands for acceleration and S for sound pressure.

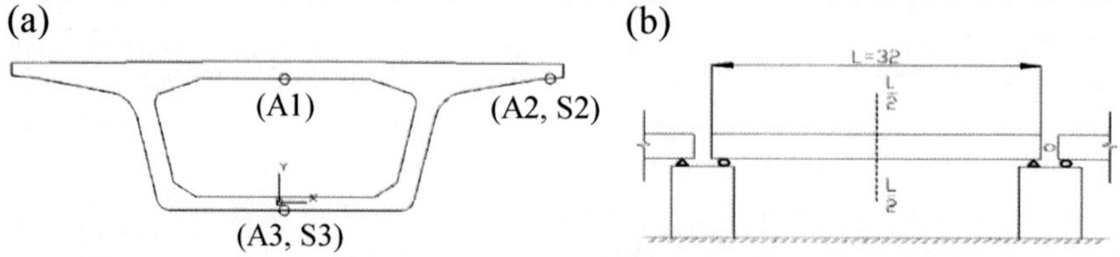

Fig. 1. Schematic diagram of bridge vibration and noise measurement points.

3.2 Measured Bridge Vibration

Accelerations at A1, A2 and A3 in one-third octave bands are shown in Fig. 3 which can be seen that the maximum vibration acceleration spectral levels at measurement points A1, A2 and A3 are 87.6 dB, 84.6 dB and 86.4 dB, respectively, occurring at different frequencies. A broad dip is seen at all the three points around 16 Hz. Below this frequency several peaks can be identified which correspond to modal frequencies of the bridge.

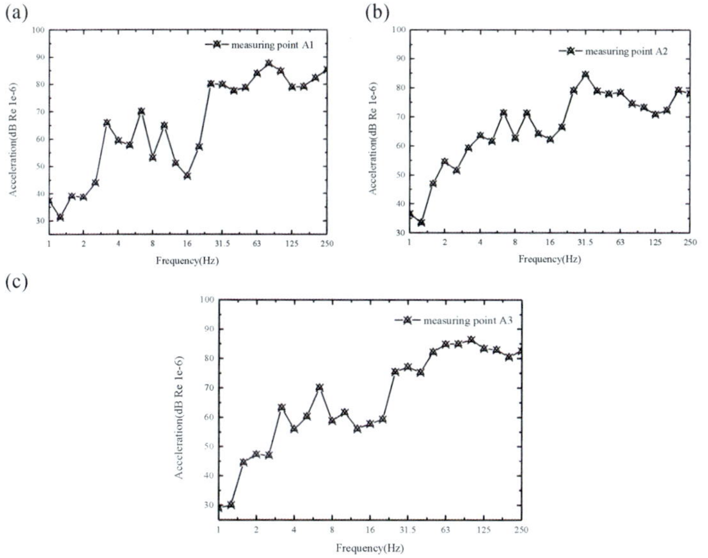

Fig. 2. Measured acceleration (Speed = 350 km/h): (a) at A1; (b) at A2; (c) at A3.

3.3 Measured Bridge Noise

According to Ref. [2], the structural noise of concrete bridges is mainly distributed between 20 Hz and 200 Hz. In order to more accurately analyse measured bridge structure noise, sound pressure data are pre-processed by using a 10–1000 Hz bandpass filter. The sound pressure spectra at S2 and S3 are shown in Fig. 4. It can be seen that at S2 the highest sound pressure level is 84.6 dB at 31.5 Hz, and at S3, the maximum is 87 dB at 50 Hz. The spectrum at S2 is much flatter than that at S3.

4 Model Prediction and Comparison with Measurement

Table 1 lists parameters of the vehicles. The train speed is 350 km/h, the track irregularity is measured during the test described in Sect. 3. For wavelengths greater than 2 m, rail irregularity data measured on a similar line is used [18]. The period of the irregularity has been assumed to be 32 m, the length of one bridge span.

Vibration and noise are predicted using the model for the measurement points shown in Fig. 2 and compared with measurement. Comparison in noise can be made for these points because it may be assumed that the noise is mainly from bridge vibration. The comparisons are shown in Fig. 5 and Fig. 6. According to (a), (b) and (c) of Fig. 5, the predicted and measured acceleration at A1, A2 and A3 agree relatively well. The acceleration level at the centre of the deck (A1) is larger than that at A2 and A3, which is consistent with the measured results. According to (a) and (b) of Fig. 6, the predicted sound pressure results are basically consistent with the measured values in the range of 20–250 Hz, indicating that the prediction model can be used to study the vibro-acoustic behaviours of the box girders subject to a moving high-speed train.

Table 1. Parameters used in the prediction.

Item	Value	Item	Value	Item	Value
Car body mass (kg)	40000	Primary suspension stiffness (kN/m)	1280	Secondary suspension stiffness (kN/m)	165
Bogie mass (kg)	2624	Primary suspension damping (kNs/m)	5	Secondary suspension damping (kNs/m)	12
Wheelset mass (kg)	1560	Wheel base (m)	2.5	Distance between bogie centres (m)	17.8

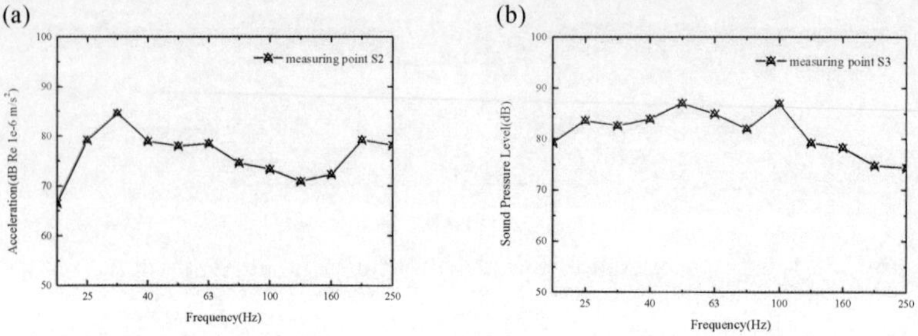

Fig. 3. Noise test results (Speed=350 km/h): (a) at S2; (b) at S3.

Bridge noise is also predicted for a point at which train pass-by noise is normally measured. The point is within the mid-span plane of a bridge, 25 m away from the track central line and 3.5 m above the rail top surface. The result is shown in Fig. 7 in comparison with the measured sound pressure level. As can be seen, the contribution of bridge noise below 100 Hz to the total sound pressure level is not insignificant. However, for frequencies above 100 Hz, other noise sources such as rolling noise and/or aerodynamic noise are dominating.

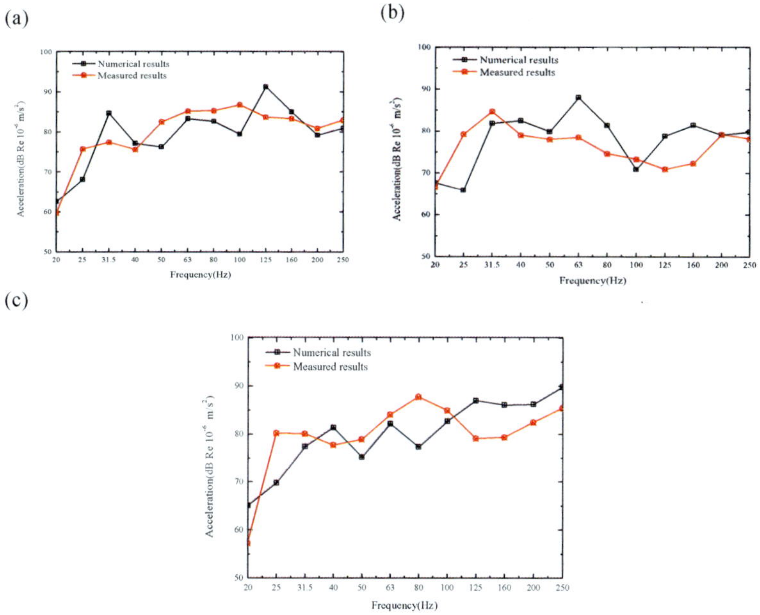

Fig. 4. Comparison in bridge vibration between measurement and prediction. (a) at A1; (b) at A2; (c) at A3.

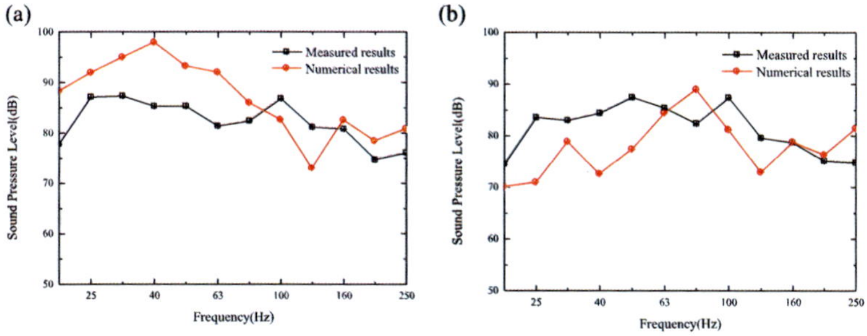

Fig. 5. Comparison in bridge noise between measurement and prediction. (a) at S2; (b) at S3.

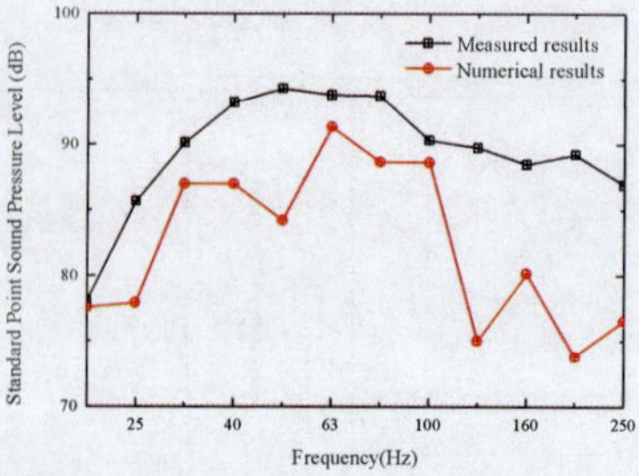

Fig. 6. The contribution of bridge noise to sound pressure at standard point.

5 Conclusion

An approach is described for studying wheel/rail force and noise radiation from concrete box girders when a high-speed train runs along a track-bridge system. Factors important to noise radiation from concrete box girders, such as periodicity, infinite extent of a track-bridge system etc. are considered in a sufficiently detailed manner. Comparison between prediction and measurement indicates that the method can be used to predict bridge noise. It is shown that contribution of bridge noise to the total train pass-by noise is not insignificant below 100 Hz.

Acknowledgement. This work is funded by the National Key R&D Program of China (2016YFE0205200), the National Natural Science Foundation of China (U1834201), and Sichuan Science and Technology Program (20YYJC3285).

References

1. Li, X., Liang, L., Wang, D.: Vibration and noise characteristics of an elevated box girder paved with different track structures. J. Sound Vibr. **425**, 21–40 (2018)
2. Thompson, D.J.: Railway Noise and Vibration: Mechanisms. Elsevier, Modelling and Means of Control (2009)
3. Knothe, K., Grassie, S.L.: Modelling of railway track and vehicle/track interaction at high frequencies. Veh. Syst. Dyn. **22**, 209–262 (1993)
4. Zhai, W., Wang, K., Cai, C.: Fundamentals of vehicle–track coupled dynamics. Veh. Syst. Dyn. **47**, 1349–1376 (2009)
5. Chen, Z., Zhai, W., Cai, C., et al.: Safety threshold of high-speed railway pier settlement based on train-track-bridge dynamic interaction. Sci. China **58**, 202–210 (2015)
6. Sheng, X., Li, M., Jones, C.J.C., et al.: Using the Fourier-series approach to study interactions between moving wheels and a periodically supported rail. J. Sound Vibr. **303**, 873–894 (2007)
7. Li, X., Yang, D., Chen, G., et al.: Review of recent progress in studies on noise emanating from rail transit bridges. J. Mod. Transp. **24**, 237–250 (2016)
8. Li, X., Liu, Q., Pei, S., et al.: Structure-borne noise of railway composite bridge: Numerical simulation and experimental validation. J. Sound Vibr. **353**, 378–394 (2015)
9. Zhang, X., Li, X.Z., Liu, Q.M., et al.: Theoretical and experimental investigation on bridge-borne noise under moving high-speed train. Sci. China: Tech. Sci. **56**, 917–924 (2013)

10. Baeza, L., Ouyang, H.: A railway track dynamics model based on modal substructuring and a cyclic boundary condition. J. Sound Vibr. **330**, 75–86 (2011)
11. Zhang, X., Li, X., Hao, H., et al.: A case study of interior low-frequency noise from box-shaped bridge girders induced by running trains: its mechanism, prediction and countermeasures. J. Sound Vibr. **367**, 129–144 (2016)
12. Zhang, X., Zhai, W., Chen, Z., et al.: Characteristic and mechanism of structural acoustic radiation for box girder bridge in urban rail transit. Sci. Total Environ. **627**, 1303–1314 (2018)
13. Li, Q., Xu, Y.L., Wu, D.J.: Concrete bridge-borne low-frequency noise simulation based on train–track–bridge dynamic interaction. J. Sound Vibr. **331**, 2457–2470 (2012)
14. Li, Q., Song, X., Wu, D.: A 2.5-dimensional method for the prediction of structure-borne low-frequency noise from concrete rail transit bridges. J. Acoust. Soc. Am. **135**, 2718–2726 (2014)
15. Song, X., Wu, D., Li, Q., et al.: Structure-borne low-frequency noise from multi-span bridges: A prediction method and spatial distribution. J. Sound Vibr. **367**, 114–128 (2016)
16. Zhai, W.: Dynamics of Coupled Vehicle-Track Systems. Science Press, in Chinese) (2015)
17. Sheng, X., Zhong, T., Li, Y.: Vibration and sound radiation of slab high-speed railway tracks subject to a moving harmonic load. J. Sound Vibr. **395**, 160–186 (2017)
18. Li, J., Liu, L., Kou, D.: Wu Guang high-speed rail track irregularity power spectrum analysis. Appl. Mech. Mater. **638–640**, 1224–1228 (2014)

Author Index

A
Abe, Kazuhisa, 360
Augusztinovicz, Fülöp, 267, 368
Anderson, Dave, 496
Andringa, Tjeerd, 150
Arcos, Robert, 445, 454, 504
Arteaga, Ines Lopez, 186
Asmussen, B., 277, 699
Auersch, Lutz, 403, 411
Augis, Eric, 546

B
Baeza, Luis, 186
Bewes, Oliver, 385
Bezin, Yann, 420
Bistagnino, A., 118, 225
Block, James, 571
Bouvet, Pascal, 41, 186
Brechbühl, Yves, 403
Brick, Haike, 158, 167
Brunel, Jean-François, 339

C
Carels, Patrick, 267, 368
Carrigan, Tobias D., 286
Cawser, Steve, 571
Cettour-Janet, Raphaël, 478
Ceulemans, Jonathan, 259
Chai, Xiaodong, 214
Chaufour, Claire, 295
Chen, Yu, 57
Cheng, Gong, 613, 707
Cheng, Morgan, 654
Cheung, Cindy, 638, 654
Chiello, Olivier, 295, 339
Cierco, E., 662
Clot, Arnau, 445, 454, 504
Collina, Andrea, 251, 303
Coquel, Guillaume, 351
Corradi, Roberto, 251, 303
Cottereau, Régis, 521
Croft, Briony, 629

D
Dahl, Birgitte M., 429
de Vos, Paul H., 150
Dec, Patryk, 521
Decroos, Kris, 259
Degrande, Geert, 512, 529
Della Neve Longo, Roberto, 470
Denia, Francisco D., 109, 186
Di Gialleonardo, Egidio, 251, 303
Dinhobl, Günter, 142, 234
Dufrénoy, Philippe, 339

E
Ekblad, Alf, 487

F
Fast, Dennis, 158
Faure, Baldrik, 295, 521
Feng, Qingsong, 707
Fiala, P., 368
Frank, Christian, 563, 699
Furuta, Masaru, 360

G. Degrande et al. (Eds.): *Noise and Vibration Mitigation for Rail Transportation Systems*, NNFM 150, pp. 717–720, 2021.
https://doi.org/10.1007/978-3-030-70289-2

G
Gao, Xiaogang, 214, 671
Gao, Yang, 57
Garburg, Rüdiger, 563
Gerbig, C., 699
Gerlach, Thomas, 646
Germonpré, Matthias, 529
Ghangale, Dhananjay, 445, 504
Giner-Navarro, Juan, 109, 186
Gomes Neves, A. L., 225
Gramowski, Christoph, 629, 646
Grassie, Stuart L., 320
Gu, Xiaohan, 671
Gu, Xiaohan Phrain, 214
Guerrero, Anne, 206
Guigou-Carter, Catherine, 351, 394, 403
Guiral, Ainara, 158, 581, 662

H
Han, Jian, 126, 605, 707
Hanson, David, 496
He, Yuanpeng, 707
Herron, David, 621
Hima, Boniface S., 621
Hirasawa, Toru, 81
Ho, Wilson, 638, 654
Hu, Zhiwei, 66
Huang, Hong-yao, 691
Huguenet, Pierre, 98
Hutterer, Helmut, 234

I
Iida, Akiyoshi, 73
Inoue, Tatsuya, 331
Isert, Nathan, 176
Iturritxa, Egoitz, 581, 662

J
Jarillo, Julián Martín, 186
Jean, Philippe, 351, 394
Jin, Xuesong, 605
Johansson, Peter, 487
Jones, Chris, 571
Ju, Longhua, 214, 671
Jurdic, Vincent, 538

K
Kanemoto, Hiroyuki, 331
Karantonis, Peter, 377
Kasess, Christian H., 142
Kawaguchi, Tsugutoshi, 331
Kaynia, Amir M., 429
Kempinger, T., 277
Kendl, F., 277
Kim, Hyungjun, 134
Kirchner, Karl-Richard, 158
Kitagawa, Toshiki, 73, 331
Kohrs, Torsten, 158
Koopman, Arnold, 487
Koro, Kazuhiro, 360
Kostovasilis, Dimitrios, 420
Kropp, Wolfgang, 90
Kümmritz, Sebastian, 234
Kuo, Kirsty, 512, 529
Kurita, Takeshi, 81
Kuzuta, Masahito, 331

L
Lai, Van Vuong, 339
Lange, S., 277
Latorre Iglesias, Eduardo, 225
Lever, Rob, 538
Li, Chaowei, 57
Li, Hui, 109
Li, Qi, 691
Li, Qianqian, 251, 303
Li, Xiaofeng, 57
Lin, Lyn, 654
Liravi, Hassan, 504
Liu, Xiaowan, 66, 109, 186
Lo, Sam, 638
Lombaert, Geert, 512, 529, 555
Lurcock, Daniel, 66
Lütke, B., 277
Lutzenberger, Stefan, 403
Ly, Alice, 546

M
Ma, Ziyan, 214, 671
Maes, Kristof, 529
Magrans, F. X., 662
Malkoun, A., 662
Maly, Thomas, 142
Martner, Otto, 176
Masoumi, Hamid, 267, 368
Miller, Aaron, 629
Mistler, Michael, 563
Mizushima, Fumio, 81
Moreno, Juan, 186
Müller, Angela D., 134
Müller, Roger, 394, 403
Müller, Roland, 403

N
Nagy, A. B., 368
Nakajima, Shinji, 81
Nelain, Brice, 437
Nelson, James Tuman, 312, 462, 470
Neudeck, M., 699
Nielsen, Jens C. O., 3
Noori, Behshad, 454, 504
Norén-Cosgriff, Karin, 429
Ntotsios, Evangelos, 420, 597, 621

O
Oertli, Jakob, 197
Ögren, Mikael, 487
Olazagoitia, Maddi, 581
Ostermann, Michael, 142

P
Pallas, Marie-Agnès, 295
Panarodvong, Benjamin, 589
Papadopoulos, Manthos, 267, 529
Park, Joonsang, 429
Passmore, Adrian, 538
Persson Waye, Kerstin, 487
Petz, Martin, 234
Pieringer, Astrid, 3, 90, 242
Poisson, Franck, 206
Puckeridge, Hayden, 377

R
Rajaram, Shankar, 312, 462
Reumers, Pieter, 512
Reynaud, Emanuel, 295, 437
Rissmann, Martin, 41, 109, 186
Romeu, Jordi, 445, 454, 504
Ropars, Pierre, 546
Rothhämel, Jörg, 242, 277
Rüst, Philipp, 167
Ryue, Jungsoo, 134

S
Said, Samir, 403
Saito, Mitsuru, 81
Sapena, Joan, 118, 225, 662
Saurenman, Hugh, 312, 462, 470
Sawamura, Yoichi, 73
Schevenels, Mattias, 555
Schewe, B., 699
Scotter, Mark, 538
Sheng, Xiaozhen, 605, 613, 707
Shields, Paul, 571
Shimada, Sav, 496
Sica, Gennaro, 98, 385
Sievi, Alex, 167
Solé, Jaume, 98
Squicciarini, Giacomo, 66, 109, 420, 621, 680
Stallaert, Bert, 259
Stangl, Matthias, 242
Stiebel, D., 699
Suehara, Michiko, 360
Sueki, Takeshi, 331
Sun, Wenjing, 597

T
Talbot, James P., 286
Tecklenburg, P., 699
Theyssen, Jannik S., 90
Thompson, David J., 3, 66, 109, 134, 186, 420, 597, 613, 621, 680, 691
Thornely-Taylor, Rupert, 385
Tickner, Conor, 571
Tielkes, Thorsten, 242
Torstensson, Peter T., 3
Toward, Martin, 66, 597
Trévisan, Benjamin, 478
Tufano, Anna Rita, 295

U
Uda, Toki, 73

V
van Hagen, Mark, 150
Van hoorickx, Cédric, 555
Vanhonacker, Tom, 259
Veelhaver, Bram, 267
Verachtert, Ramses, 529
Villot, Michel, 394, 403, 478
Vincent, Nicolas, 295, 437

W
Wakabayashi, Yusuke, 81
Wang, Anbin, 214, 671
Wang, Qi, 126
Wang, Yigang, 57
Wang, Zhiqiang, 671
Waubke, Holger, 142
Weber, Conrad, 377, 589
Wiseman, Marcus, 597
Wong, Terrence, 589
Wu, Yue, 126

X
Xiao, Xinbiao, 126, 605
Xie, Gang, 186
Xie, Suming, 57

Y
Yamada, Koya, 360
Yamada, Sota, 360
Yang, Gang, 605
Yokoyama, Hiroshi, 73
Yoshitake, Sho, 360

Z
Zhang, Jie, 529
Zhang, Jin, 66
Zhang, Xianying, 680